Ivo D. Dinov, Milen Velchev Velev

Data Science

Time Complexity, Inferential Uncertainty, and Spacekime Analytics

DE GRUYTER

Authors
Prof. Dr. Ivo D. Dinov
University of Michigan
426 North Ingalls Str.
Ann Arbor 48109-2003
USA
dinov@umich.edu

Assoc. Prof. Dr. Milen Velchev Velev
"Prof. Dr. A. Zlatarov" University
1 Prof. Yakimov bul.
8010 Burgas
Bulgaria
Milen.Velev@gmail.com

ISBN 978-3-11-069780-3
e-ISBN (PDF) 978-3-11-069782-7
e-ISBN (EPUB) 978-3-11-069797-1

Library of Congress Control Number: 2021933301

Bibliographic information published by the Deutsche Nationalbibliothek
The Deutsche Nationalbibliothek lists this publication in the Deutsche Nationalbibliografie; detailed bibliographic data are available on the Internet at http://dnb.dnb.de.

Typesetting: Integra Software Services Pvt. Ltd.
Printing and binding: CPI books GmbH, Leck

www.degruyter.com

. . . dedicated to all known polymaths, forgotten intellectuals, and unknown scholars that came before us, whose giant shoulders support this scientific work and all other contemporary research inquiries, as well as, all prospective, yet to be unearthed discoveries . . .

Preface

Acquiring information, processing data, assessing risk, and projecting forecasts in space and time have always paralleled human evolution and the ever-changing social dynamics that accompany it. The need for reliable accounting, robust record keeping, common time, space and celestial referencing, and building of useful tools, weapons, and structures emerged with the dawn of mankind's scientific quest to understand nature. Object coding, canonical representations, and mathematical operations arise in many of man's early earthly activities, e.g., tracking numerical data (Mesopotamia, c. 3500 BC), fractional arithmetic (Old Egypt Kingdom, c. 2500 BC), medical texts (ancient Egypt, c. 1600 BC), empirical astronomy (Babylonia, c. 700 BC), exploring geometric properties (ancient Greece, c. 550 BC), natural philosophy (Plato's Academy, c. 380 BC), etc. [1, 2]. Although all of these ingenious advances were based on real, tangible, and physical observations, a very basic level of information gathering, calculation, and communication was also required to establish an early scientific epistemology. Historical records indicate that many discoveries were lost, rediscovered, or reformulated over the centuries. However, during the Renaissance (c. 1400 AD) and the Enlightenment (c. 1600 AD) periods, a rapid acceleration of knowledge accumulation led to developing a more solid foundation for all contemporary scientific advances. As a result of this chain-reaction and trackable discovery provenance, most contemporary scholars can reliably trace their academic genealogy to Sharaf al-Dīn al-Ṭūsī (Mesopotamia, c. 1160 AD, https://mathgenealogy.org/extrema.php, accessed January 29, 2021) [3, 4].

Over the past 5,000 years, the scientific rise of humanity reflected the transformation of human civilization from a hunter-gatherer (c. 10000 BC), to pastoral (c. 7500 BC), agricultural (c. 3000 BC), feudal (c. 900 AD), industrial (c. 1700 AD), and digital information (c. 1980) societies. The 21st century represents a new era where information generation, processing, transmission, and rapid interpretation become fundamental sources of productivity, power, wealth, and affluence. This shift from industrialism to data-informationalism directly ties to current and prospective global, cultural, environmental, and socioeconomic dynamics.

Since 2010, there has been compelling evidence illustrating extremely rapid, ubiquitous, and disruptive quantization of all human experiences. The rate of increase of the amount of data we collect doubles every 12–14 months (Kryder's law) [5–7], the rate of expansion of computational power also grows exponentially (Moore's law) [8], and the universal proliferation of information digitalization (digital transformation) [9] now covers all facets of life. This digital transformation represents the natural progression from continuous analogue communication to exchanges of discrete information bits. All life forms, including humans, have perfectly evolved to interpret intrinsically discrete processes as continuous patterns. Now, human civilization is swinging the pendulum in the opposite direction, toward quantizing information detection, aggregation, communication, and interpretation. This leads to interesting and disruptive

https://doi.org/10.1515/9783110697827-202

human-machine interfaces that promise to pool the exceptional human capabilities of reason, emotion, intelligence, and creativity, with the unique computer ability for rapid *en masse* processing of data, algorithmic precision, and impassive treatment of complex information. The result of this is an explosion of new machine learning techniques, artificial intelligence tools, augmented reality experiences, and human-machine interfaces. Most of these advances require novel data science methods, computational algorithms, and inferential strategies.

Today, contemporary digital transformation is now seamlessly blending technological advances within all human experiences; impacting how we perceive the environment, transforming our value system, and disrupting traditional daily interactions. Beyond that, a cultural change is also taking place requiring individual, social, governmental, public, and private organizations to continually challenge the status quo, explore alternatives, and get comfortable with some degree of failure in the relentless march toward digitalization.

The momentum of volume and complexity growth of digital information appears to be exceeding the increase in our ability to interpret it holistically. This imbalance between the natural progression of digital transformation and mankind's capacity to derive value from this information stresses resources and inhibits knowledge-driven actions. As an example, it's now common for scientists to gather heterogeneous data on the order of a petabyte, $1PB = 10^{15}$ bytes. To grasp the enormity and the complexity of handling such voluminous information, consider it in relation to the Milky Way galaxy, which has approximately 2×10^{11} stars. If each star represents a byte, then one petabyte of data corresponds to 5,000 Milky Way galaxies. Just as the spatial location, luminosity, chemical composition, and momentum of each star in the galaxy changes with time, and in relation to other galactic bodies, the scientific information collected by researchers has longitudinal patterns and spatial associations that further complicate the interpretation of such complex data archives.

There is a scientific revolution underway to develop adaptive, semi-supervised, self-reinforced, and reliable computational and data science methods capable of transfer-learning supporting effective data-driven decision support systems. All of the efforts to design, implement, and validate such new techniques require an in-depth basic scientific knowledge across many applied, experimental, and theoretical disciplines, as well as transdisciplinary team-based open-science collaborations with amalgamation of exploratory and confirmatory scientific discoveries.

This book provides motivation along with substantial mathematical background, examples of promising approaches, and a new methodological foundation for data science that extends the concept of event order (time) to the complex plane. It also includes hands-on demonstrations of several biomedical and econometric case studies, implementation protocols, and applications. Supplementary online materials include datasets; an R package with a suite of libraries for managing and interrogating raw, derived, observed, experimental, and simulated big healthcare datasets; and other web-based services for data visualization and processing. The supporting websites provide additional interactive content, learning modules, case studies, and demonstrations (https://spacekime.org, accessed January 29, 2021 and https://tciu.predictive.space, accessed January 29, 2021).

The content of this book may be appropriate for a variety of upper-division undergraduate or graduate science, technology, engineering, and mathematics (STEM) courses. To address the specific needs of their students, instructors can reshuffle and present the materials in alternative ways. The book's chapters are organized sequentially illustrating one intuitive way of covering the topics; however, the content can be restructured and tailored to fit specific curricular objectives or audience requirements. This is not a general textbook for learning the foundations of computational and data sciences. As of 2021, there are many other textbooks that cover the breadth and depth of data science and predictive analytics. Capitalizing on the existing wealth of knowledge, this book develops the spacekime analytics technique and includes illustrative examples of data science applications; inferential uncertainty; complex-time representation of large, high-dimensional, and longitudinal datasets.

This book utilizes the constructive definition of "Big Data" provided in Data Science and Predictive Analytics (https://DSPA.predictive.space, accessed January 29, 2021) [10], which is based on examining the common characteristics of many biomedical, social-science, and healthcare case studies. The seven key "Big Data" characteristics identified from such studies include large size, heterogeneity, incongruency, incompleteness, multiscale format, time-varying nature, and multisource origins. These properties play a critical role in many scientific inference processes representing compound systematic investigations. Specifically, many scientific workflows include critical steps to hand each of the Big Data characteristics from the initial study-design, to subsequent data collections, wrangling, and management; model-based statistical analyses, model-free machine learning methods, outcome predictions, trend forecasting, derived computed phenotyping, algorithm fine-tuning, assessments, comparisons; and scientific validations.

The authors are profoundly indebted to all of their students, mentors, advisors, and collaborators for inspiring this study, guiding the courses of their careers, nurturing their curiosity, and providing constructive and critical research feedback. Among these scholars are Guentcho Skordev (Sofia University); Kenneth Kuttler and Anant Godbole (Michigan Tech University); De Witt L. Sumners and Fred Huffer (Florida State University); Jan de Leeuw, Nicolas Christou, and Michael Mega (UCLA); Arthur Toga (USC); Brian Athey, H.V. Jagadish, Kathleen Potempa, Janet Larson, Patricia Hurn, Gilbert Omenn, and Eric Michielssen (University of Michigan).

Particularly useful feedback and constructive recommendations were provided by Yueyang Shen, Yuxin Wang, Zijing Li, Daxuan Deng, Yufei Yang, Christopher Hale, and Yupeng Zhang. Many other colleagues, students, researchers, and fellows have shared their expertise, creativity, valuable time, and critical assessment for generating, validating, and enhancing these open-science resources. Among these are Yongkai

Qiu, Zhe Yin, Rongqian Zhang, Yuyao Liu, Yunjie Guo, Jinwen Cao, Reza Soroushmehr, Yuming Sun, Lingcong Xu, Simeone Marino, Alexandr Kalinin, Kalyani Desikan, Christoph Köhn, Vimal Rathee, Daniel Rowe, Manthan Mehta, Hristo Pavlov Pavlov, Vesselin Gueorguiev, and many others.

In addition, students and colleagues from the Statistics Online Computational Resource (SOCR), the Michigan Institute for Data Science (MIDAS), and "Prof. Dr. Asen Zlatarov" University in Burgas provided encouragement, support, and valuable suggestions. Insightful comments, suggestions, corrections, and constructive critiques from many anonymous reviewers significantly improved and clarified the material. We welcome broader contributions, comments, feedback, and input from the entire scientific community via an online webform (https://tciu.predictive.space, accessed January 29, 2021) and through an open-source software version control platform (https://github.com/SOCR/TCIU, accessed January 29, 2021).

The research methods, computational developments, and scientific applications reported in this book were partially supported by the US National Science Foundation (grants 1916425, 1734853, 1636840, 1416953, 0716055 and 1023115), US National Institutes of Health (grants P20 NR015331, U54 EB020406, UL1TR002240, P30 DK089503, R01 CA233487, R01 MH121079, R01 MH126137, T32 GM141746), "Prof. Dr. Asen Zlatarov" University in Burgas, and the University of Michigan.

Ivo D. Dinov
(Ann Arbor, Michigan, USA)
Milen V. Velev
(Burgas, Bulgaria, EU)

Foreword

Before diving into this book, all readers, scholars, instructors, formal and informal learners, and working professionals are encouraged to review the basics of mathematical modeling, statistical inference, computational data science, and scientific visualization. This book assumes no deep expertise in these disciplines, but some basic scientific exposure will certainly be helpful to all audiences. The chapters of this book were organized in a way that was intuitive for the authors; however, the material may be read and/or covered in an alternative order that fits the audience's needs.

The introductory **Chapter 1 (Motivation)** presents the mission and objectives of this book and provides some basic definitions, driving motivational problems, and issues with the classical definition of *time* as a non-negative univariate measure of event order. Here, we also define the seven characteristics of Big Datasets and explain the synergies between data science, predictive analytics, and scientific inference. This chapter provides a simple yet illustrative example of the core idea of utilizing complex time for data analytics. We demonstrate prospective forecasting of the Michigan Consumer Sentiment Index (MCSI) and contrast the complex-time analytics to traditional model-based multivariate longitudinal statistical analyses.

Chapter 2 (Mathematics and Physics Foundations) provides some basic theoretical formulations necessary to illustrate the concepts of data representation, mathematical operators, and statistical inference. This chapter starts with some fundamental quantum physics definitions such as wavefunctions, Dirac bra-ket notation, and commutator operators. It covers some well-known properties of position, momentum, and energy operators. Some of these mathematical formalisms relate to partial differential equations (PDE), functional analysis, and quantum mechanics operators.

In **Chapter 3 (Time Complexity)**, we will build on the earlier mathematical foundation to extend various quantum physics concepts to data science. For instance, *inference-functions* in data science correspond to the quantum mechanics notion of *wavefunctions*. This translation will facilitate lifting the 4D Minkowski spacetime to 5D spacekime by extending time to the complex plane. Specifically, we will define the notions of complex time (kime) and complex events (kevents). We review the Kaluza-Klein theory, formulate kime-velocity, define the spacekime *metric tensor*, and show its invariance under Lorentz transformations of spacekime inertial frames. The notions of amplitudes and phases of the forward and reversed Fourier transformation will play an important role in this spacekime data analytic process. This chapter also presents the Heisenberg's uncertainty principle in spacekime and the dichotomy between the Copenhagen and spacekime interpretations of the observed collapse of the wave, or inference, functions in real experiments. Finally, we discuss the philosophy of data science analytics using complex time and the causal structure of the spacekime manifold.

https://doi.org/10.1515/9783110697827-203

Chapter 4 (Kime-series Modeling and Spacekime Analytics) illustrates strategies to transform classical time-series into complex-time-indexed kime-series (or kime-surfaces). It also derives the generalized likelihood ratio test for complex-valued kime-indexed univariate processes. The intensities of these complex-valued processes over kime are called kintensities, or complex-intensities over complex-time. In this chapter, complex-valued functional magnetic resonance imaging (fMRI) data provide a key motivational challenge driving the theoretical developments and the practical demonstrations. We also discuss the Laplace transform of longitudinal data, which supports the analytic duality between longitudinal spacetime processes (time-series) and their spacekime counterparts (kime-surfaces).

Chapter 5 (Inferential Uncertainty) defines the data-to-inference duality principle where inference functions map pairs of observables and analytical strategies into probabilistic decision spaces. Rather than yielding a specific analytical decision or practical action based on the data, inference functions effectively encode and represent problem systems in a probabilistic sense. This chapter explicates the parallels between various core quantum mechanics concepts and their data science counterparts. We provide several alternative formulations of uncertainty in data science and derive an embedding of classical 4D spacetime into the 5D spacekime manifold, which yields an extra force that is parallel to the 4-velocity. We explore synergies between random sampling in spacetime and spacekime and present a Bayesian formulation of spacekime analytics. The chapter appendix includes bra-ket formulations of multivariate random vectors, time-varying processes, conditional probability, linear modeling, and the derivation of the cosmological constant (Λ) in the 5D Ricci-flat spacekime.

The final **Chapter 6 (Applications)** illustrates several examples of applying the spacekime transformation to analyze complex multivariate data. The differences between spacekime data analytics and spacetime data modeling and inference are driven by the choice of sampling strategy. In spacetime, independent and identically distributed (IID) samples intend to cover the underlying probability distribution, whereas in spacekime inference, the sampling of the kime-phases requires an effective kime-phase representation to approximate the population characteristics of interest. When the exact kime-phase distribution is known, or can be accurately estimated, reliable spacekime inference may be obtained based on only a few spacetime samples. We demonstrate the process of spacekime data analytics based on case studies involving functional magnetic resonance imaging (fMRI) data, multi-source brain data, and financial market and economics indices.

Throughout the book, there are statements of open problems and suggested challenges, which are indicated by this puzzle-piece icon. Some of these problems may be easy to prove or disprove and some may be hard. Different types of readers may consider these problems conjectures that may or may not be true. Readers may wish to explore, reformulate, validate, or search for counter examples of these challenges, some of which may be only partially known or not yet fully understood.

The book's online appendices (https://SpaceKime.org, accessed January 29, 2021) contain continuously updated and expanded additional content, datasets, case study references, source-code, and scripts used to generate the demonstrated analytics, graphs, and example applications. Throughout the book, there are cross-references to appropriate chapters, sections, datasets, web services and live demonstrations, and other peer-reviewed scholarly work. The sequential arrangement of the chapters provides a suggested reading order. However, readers and instructors are encouraged to explore alternative material presentations and customized coverage pathways to fit their specific intellectual interests or particular curricular needs.

Contents

Use and Disclaimer

The methods, techniques, software, and other resources presented in this book are designed to help scientists, trainees, students, and professionals learn, experiment with, extend existing, and build novel computational and data science instruments. They also provide some pedagogically relevant practical applications and protocols for dealing with complex datasets. Neither the authors nor the publisher has control over, or can make any representation or warranties, expressed or implied, regarding the use of these resources by educators, researchers, users, patients, or their representatives or service provider(s), or the use or interpretation of any information stored on, derived from, computed with, suggested by, or received through any of the materials, code, scripts, or applications demonstrated in this book. All users are solely responsible for the results of deriving, interpreting, and communicating any information using these techniques and the supporting resources.

Users, their proxies or representatives (e.g., clinicians) are solely responsible for reviewing and evaluating the accuracy, relevance, and meaning of any information stored on, derived by, generated by, or received through the application of any of the software, protocols, or techniques. The authors and the publisher cannot and do not guarantee said accuracy. These resources, their applications, and any information stored on, generated by, or received through them, are not intended to be a substitute for professional or expert advice, diagnosis, or treatment. Always seek the advice of an appropriate service provider (e.g., physician) or other qualified professional with any questions regarding any real case study (e.g., medical diagnosis, conditions, prediction, and prognostication). Never disregard professional advice or delay seeking it because of something read in this book or learned through the use of the material or any information stored on, generated by, or received through SOCR or other referenced resources.

All readers and users acknowledge that the copyright owners or licensors, in their sole discretion, may from time to time make modifications to these materials and resources. Such modifications may require corresponding changes to be made in the mathematical models, algorithmic code, computational protocols, learning modules, interactive activities, case studies, etc. Neither the authors and publisher, nor licensors shall have any obligation to furnish any maintenance, support, or expansion services with respect to these resources. These resources are intended for education and scholarly research purposes only. They are neither intended to offer or replace any professional advice nor to provide expert opinion. Please contact qualified professional service providers if you have any specific concerns, case studies, or questions. Persons using any of these resources (e.g., data, models, algorithms, tools, or services) for any medical, social, healthcare, or environmental purposes should not rely on accuracy, precision, or significance of the reported results. While the materials and resources may be updated periodically, users

https://doi.org/10.1515/9783110697827-205

should independently check against other sources, latest advances, and most accurate peer-reviewed scientific information.

Please consult appropriate professional providers prior to making any lifestyle changes or any actions that may impact you, those around you, your community, or various real, social, and virtual environments. Qualified and appropriate professionals represent the single best source of information regarding any biomedical, biosocial, environmental, and health decisions. None of these resources have either explicit or implicit indication of approval by the US Food and Drug Administration (FDA)! Any and all liability arising directly or indirectly from the use of these resources is hereby disclaimed. These resources are provided "as is" and without any warranty expressed or implied. All direct, indirect, special, incidental, consequential, or punitive damages arising from any use of these resources or materials contained herein are disclaimed and excluded.

Glossary, Common Notations, and Abbreviations

The table below includes some of the common notations used throughout this book.

Notation	Description
Common terms and abbreviations	
1D, 2D, 3D, 4D, . . .	One, two, three, four, and higher-dimensions. Typically used to denote the dimension of a manifold or the complexity of a dataset
df (DOF)	Degrees of freedom
[1], [2], [3], . . .	In-text citations to bibliographical references provided in the end
1 °C, 34° F	Temperature in Celsius or Fahrenheit
a.k.a.	Also known as
w.r.t.	With respect to
AD, ADNI	Alzheimer's disease, Alzheimer's disease neuroimaging initiative
AR, ARMA, ARIMA	Auto-regressive integrated moving average models of longitudinal data with and without exogenous variables (invariant with time)
ARIMAX	Auto-regressive integrated moving average with eXogenous variables, referring to specific model-based statistical method for longitudinal data analysis
CBDA	Compressive Big Data Analytic, a meta-learning algorithm
CO_2, CH_4, . . .	Various chemical formulas, e.g., carbon dioxide and methane
CPU/GPU	Central or Graphics Processing Unit (referring to computer chipsets)
CSI and MCSI	(Michigan) consumer sentiment index
DSPA	Data science and predictive analytics
Fed	US Central Bank, the Federal Reserve
FT, IFT	The forward Fourier transform and its counterpart, the inverse Fourier transform, used for analysis and synthesis of analytical functions, discrete signal, or multivariate datasets
GDP	Gross domestic product, a measure of the total economic output of countries, typically annualized
GLM	Generalized linear model
gLRT, LRT	(Generalized) likelihood ratio test
GTR	General theory of relativity
ICS	Index of consumer sentiment

https://doi.org/10.1515/9783110697827-206

(continued)

Notation	Description
IID (iid)	Independent and identically distributed, typically referring to random variables, observations, or samples
IoT	Internet of Things
Kevent [*keivent*]	Complex-event, an extension of the concept of linearly ordered events
Kime [*kaim*]	2D complex-plane extension of time in terms of kime-order (time) and kime-direction (phase). It is used to provide a framework for advanced predictive analytics and scientific inference
Kimesurface	A 2D complex-time surface parameterized by 2D kime, extends 1D longitudinally (time-indexed) time-series
Kintensity	Complex-valued and kime-indexed intensity, e.g., of an image or a volume
LT, ILT	The forward and inverse Laplace transforms
Manifold	A space with a topology that locally resembles flat Euclidean spaces, but globally may be significantly curved, e.g., circles, spheres, tori, etc.
MAP	Maximum a posteriori, an estimation technique for approximating a quantity by the mode of the posterior distribution
MCMC	Markov Chain Monte Carlo, an empirical approach for sampling from a process or a probability distribution
MRI	Magnetic resonance imaging, including structural (sMRI), functional (fMRI), spectral MRI, and diffusion (dMRI)
Neural Networks	Deep learning (DL), deep neural networks (DNN), artificial neural networks (ANN), multilayer perceptrons (MLP), unsupervised auto encoding-decoding, convolutional (CNN), recurrent (RNN), long short-term memory (LSTM), and so on
NIH	National Institutes of Health
ODE and PDE	Ordinary and partial differential equations
OLS	Ordinary least squares, a method for parameter estimation
PCA, ICA	Principal component analysis, independent component analysis, linear dimensionality reduction techniques
PDF, CDF	Probability and cumulative distribution functions
QM	Quantum mechanics

(continued)

Notation	Description
QR code	Quick response code which extends the common product barcode to a 2D matrix code that is a machine-readable optical label that can store complex product meta-data
RFID	Radio-frequency identification tagging, which utilizes electromagnetic fields to automatically codify and identify object characteristics
SOCR	Statistic online computational resource, a multi-institutional research laboratory based at the University of California, Los Angeles, and the University of Michigan, Ann Arbor
Spacekime	The 5D extension of 4D Minkowski spacetime to complex time
STR, GTR	Special (and general) theory of relativity
SVM	Support vector machine(s), a class of supervised machine learning algorithms for classification and regression analyses
TCIU	Time complexity and inferential uncertainty, referring to the subtitle of this textbook and the corresponding R package
t-SNE	t-distributed stochastic neighbor embedding, a non-linear dimensionality reduction technique
TSV	Tab separated value, a data file format structure
UK, BG, US, . . .	Standard 2-character references to countries (country codes), e.g., United Kingdom, Bulgaria, United States of America, etc.
Mathematical notations	
$\sum_{k=1}^{K} x_k$	The sum of K elements, x_k, $1 \le k \le K$, which may be numeric, vectors, matrices, tensors, or other objects for which the addition operation is well defined
$\prod_{k=1}^{K} x_k$	The product of K elements, x_k, $1 \le k \le K$, which may be numeric, vectors, matrices, tensors, or other objects for which the multiplicative operation is well defined
$\lceil\ \rceil$, $\lfloor\ \rfloor$	The ceiling and floor functions
$\mathbb{R}^d$, $\mathbb{C}^d$	The flat Euclidean real and complex spaces of dimension d
$\langle\vert$ and $\vert\rangle$	Dirac bra-ket notation
f', f'', f'''	Function (f) derivatives of orders 1 (prime), 2 (double-prime), 3 (triple prime), etc.
$\frac{\partial f}{\partial x}, \frac{\partial^2 f}{\partial x^2}, \frac{\partial^k f}{\partial x^k}$	First, second, and higher order partial derivative of a function

(continued)

Notation	Description
$\nabla f = \left(\frac{\partial f}{\partial x_1}, \ldots, \frac{\partial f}{\partial x_k}\right)$	The gradient of a function (f), which evaluated at a point $p = (x_1, x_2, \ldots, x_k)$ represents the direction of the greatest rate of increase of the function ($\nabla f(p)$). The magnitude of the gradient represents the maximum rate of change of the function at a given point
$\Delta f = \nabla^2 f = \sum_{j=1}^{k} \frac{\partial^2 f}{\partial x_j^2}$	The Laplacian of a function. Evaluated at a point p, the Laplacian represents the rate at which the average value of f, over spheres centered at p, deviates from $f(p)$ as the radius of the sphere grows
i	The imaginary unit ($i^2 = -1$)
$\forall, \exists, \in$	Mathematical symbols – for all, exists, and belongs to (element of)
$\bar{\cdot}, \cdot^*$	Complex conjugate, e.g., $\overline{a+ib} = a - ib;\ a, b \in \mathbb{R}$
$\int_a^b \psi(x)dx \equiv \int_a^b dx\ \psi(x)$	Integral of a function over a specified interval $[a, b]$
$\oint_{z_1}^{z_2} f(z)dz$	Path integral
$a.b = \langle a \vert b \rangle$	Vector inner product
$\cdot^T, \cdot^t, \cdot'$	Transpose operator
$\cdot^\dagger$	The adjoint operator, which is the transposed conjugate
$\hat{\cdot}$	Operators associated with observables, or estimated quantities, or the forward or the inverse Fourier transforms
$\Box * \Box$	The convolution operator, or product, depends on context
$\breve{\cdot}, \tilde{\cdot}, \check{\cdot}, \widehat{\cdot}$	Parameter estimates (for scalars, vectors, matrices, or tensors)
$\Box^{-1}$	Multiplicative inverse operator, associated with object products
H^*	The dual space, $H^* = \left\{ f{:}H \xrightarrow{continuous} \mathbb{C} \right\}$
$\delta(x)$	Point source, Dirac delta function
$\Psi, \Phi, \varphi, \psi$	Wavefunctions or inference functions, depending on context
$\varphi, \phi, \nu, \theta$	Kime-phases or complex-intensity value phases, context dependent
$h, \hbar$	Planck constant $(h = 6.62607015 \times 10^{-34}\ J{\cdot}s)$, and the reduced Planck constant $\left(\hbar = \frac{h}{2\pi}\right)$, measured in Joule by second
$\lVert \cdot \rVert, \lvert \cdot \rvert$	(Various) norms, functions satisfying scalability, additivity, and inequality properties, that assign positive real numbers to each argument that is an element in a vector space over a field

(continued)

Notation	Description
$\langle A \rangle$	Expectation of an operator, typically a linear self-adjoint (Hermitian) operator corresponding to an observable, like position, momentum, energy, spin, etc.
$[A, B] = AB - BA$	The commutator operator
$\Delta A = A - \langle A \rangle$	Uncertainty, or deviance, of an operator A
σ_A^2	Mean square uncertainty of an operator A
$Re(\cdot)$, $Im(\cdot)$	Functions returning the real and imaginary part of the argument
$\text{sgn}(\cdot)$	The sign function
$\mathbb{V}/\mathbb{Q}$	Quotient space, $\mathbb{V}$ *modulo* $\mathbb{Q}$
$HankelH1(\cdot \quad \cdot)$ and $HankelH2(\cdot \quad \cdot)$	The Hankel function of the first and second kinds
$J_n(\cdot)$, $Y_n(\cdot)$	Order n Bessel functions of the first and second kind
$I_n(\cdot)$, $K_n(\cdot)$	Order n *modified* Bessel functions of the first and second kind
$AiryAi()$	The Airy function, represents a solution an ODE $y'' - xy = 0$
$\square \times \square$	Context-specific multiplication (product operator)
$\square \circ \square$	Operator composition, or outer product operator
$\square . \square = \langle \square, \square \rangle$	Inner product for appropriate objects like vectors, matrices, and tensors
$\square \otimes \square$	Kronecker product
$\square \odot \square$	Khatri-Rao tensor dot product
$\equiv ,:=$	Definition
$\Lambda_{-}^{\downarrow \rightarrow}$	Orthochronous transformation notations
$O(\cdot \quad \cdot)$, $SO(\cdot \quad \cdot)$	The orthogonal and special orthogonal groups of distance-preserving transformations of a Euclidean space that preserve a fixed point
$\square \perp \square$	Orthogonal, independent components
$\text{erf}(\cdot)$, $\text{erfi}(\cdot)$	Real and imaginary error functions
$l = \Lambda$ and $ll = \log \Lambda$	Likelihood and log-likelihood functions. Λ may also refer to the cosmological constant representing the energy density of space (vacuum energy)
H_o and H_1	The null and alternative research hypotheses, associated with statistical tests and inference problems
$\mathbb{V}(\cdot)$	The tensor vectorization operator

(continued)

Notation	Description
$\mathbb{E}(\cdot)$ or $\langle\cdot\rangle$	Expectation (of a random variable, vector, matrix, or operator)
$\arg\min_C F$, $\arg\max_C F$	The argument optimizing (minimizing or maximizing) the objective function F over the constrained space C
$g^{\mu\alpha}$, $g_{\alpha\beta}$	The *contravariant* and *covariant* metric tensors, which are inverse to each other, i.e., $\delta^{\mu}{}_{\beta} = g^{\mu\alpha} g_{\alpha\beta} = g_{\beta\alpha} g^{\alpha\mu} = \delta_{\beta}{}^{\mu}$
$\propto$	Proportional, e.g., $a \propto b$ suggests that $a = k\ b$, for some constant k
$k : l$	A sequence of consecutive numbers (typically integers indexing arguments) from k to l, i.e,. $k, k+1, k+2, \cdots, l \in Z$.

Chapter 1
Motivation

This book is about data science, inferential uncertainty, and the enigmatic concept of complex time (*kime*), an extension of the common notion of event order (time) to the complex plane. It includes a mathematical–physics treatise of *data science*, a recently established scientific discipline that complements the three other pillars of scientific discovery – experimental, theoretical, and computational sciences [11]. Specifically, we present the concepts of time complexity and inference uncertainty in the context of data-driven scientific inquiry.

In broad terms, most scientific investigations that are based on large-scale datasets utilize advanced analytic strategies, which can be classified into inferential (retrodictive) or forecasting (predictive) subtypes. In both cases, uncertainty and time considerations play vital roles in the derivation of the final results, the scientific discovery, and the phenomenological interpretation of the data analytics [12]. For instance, uncertainty is always present in the processes of prediction, resolving power of instruments, or assessing reliability of modeling techniques. Intrinsically, all quantitative measures always include uncertainties and many data analytic methods either rely significantly on stochasticity or introduce uncertainties during various data interrogation steps. As we will present later in this book, the uncertainty of the time direction (phase angle of time) also contributes to the uncertainty of classical inferential models and impacts the reliability of forecasting approaches. Around 500 BCE, the Greek philosopher Heraclitus noted [13] that "*No man ever steps in the same river twice, for it is not the same river and he is not the same man.*" Similarly, contemporary data scientists can't expect to always get perfectly identical outcomes by repeating a modern large-scale study, "*for it is not the same data and the analytic processes are not necessarily deterministic.*"

We will provide a constructive definition of Big Datasets and show biomedical, health and economic examples of the challenges, algorithms, processes, and tools necessary to manage, aggregate, harmonize, process, and interpret such data. In data science, time complexity frequently manifests as sampling incongruence, heterogeneous scales, or intricate interdependencies. We will present the concept of 2D *complex time (kime)* and illustrate how the kime-order (time) and kime-direction (phase) affect advanced predictive analytics and the corresponding derived scientific inference. The kime-representation provides a mechanism to develop novel spacekime analytics that increase the power of data-driven inference. It also solves some of the unidirectional arrow-of-time problems, e.g., psychological arrow of time (which reflects the irrevocable past-to-future flow) and thermodynamic arrow of time (which reflects the closed systems' relentless growth of entropy). While kime-phase angles may not always be directly observable, we will illustrate how they can be estimated and used to improve the resulting space-kime modeling, boost trend forecasting, and perform enhanced predictive data analytics. We will use simulated data, clinical observations

https://doi.org/10.1515/9783110697827-001

(e.g., neurodegenerative disorders), multisource census-like datasets (e.g., UK Biobank), and European economic market data to demonstrate time complexity, inferential uncertainty, and spacekime analytics.

1.1 Mission and Objectives

The mathematical foundations of data science are not yet fully developed. There are a number of efforts underway to propose a canonical mathematical formulation of data science that will allow reliable, reproducible, verifiable, consistent, and complete treatment of a wide array of data science problems. Despite the complexities associated with developing uniform data representation, homogeneous modeling, and inferential concordancy, there are some recent fruitful directions. Examples of recent advances include topological data analyses [14], compressive Big data analytics [15], tensor representations [16], neural networks (NNs) [17], and deep learning (DL)[18, 19], to name but a few.

One-dimensional time is in the core for all of these methods and other classical statistics including model-based and data-driven inductive inference techniques. This type of scientific reasoning is extremely well understood for random samples from specific families of distributions and performs quite well on empirical (traditional) observations. One of the key benefits of utilizing a classical statistical model is the ability to reliably assess uncertainties, for either Bayesian or frequentist statistics, and provide theoretically consistent and unbiased estimates along with replicable practical conclusions. However, their applications are somewhat limited for massive, multisource, multi-scale, and heterogeneous datasets. Extending the notion of positive real time to complex time provides a mechanism to effectively embed prior knowledge into the analytic process, compress information, and obtain reliable inference. Mathematically, classical time represents a positive cone over the field of the real numbers, i.e., time forms a subgroup of the multiplicative group of the reals, whereas complex time (kime) describes an algebraic prime field that naturally extends time. Although time is ordered and kime is not, the intrinsic time ordering is preserved by the kime magnitude.

In data science, algorithmic result *reproducibility* implies that the same numerical output is generated for the same input data. For analytical function definitions, the functional values $y_o = f(x_o)$ are always the same for the same input x_o. However, in practice, many model-based and mode-free techniques utilize stochastic algorithms and their outputs depend on extrinsic (environmental) or intrinsic (methodological) conditions. Hence, the corresponding algorithmic outputs and numerical results may vary for repeated invocations of the same process using the same input data. Of course, reproducibility suggests stability and robustness of the algorithm or method.

On the other hand, result *replicability* represents a stronger declaration where the essence of the conclusion of the inference remains the same when a new (random) sample drawn from a specific (joint) distribution, i.e., input data, is fed into the algorithm, under the same conditions, to generate the output result. Replicability of findings does not demand reproducibility of the numerical outputs. It suggests that in an independent experiment, using the same conditions but different sample observations from the same phenomena or process, the study findings are expected to be successfully replicated, up to some quantifiable statistical error. Statistical inference, analysis, and forecasting assume the availability of appropriate models and their efficient algorithmic implementations that provide closed-form parameter estimates as well as quantify the variability of these estimates, i.e., provide upper bounds on the statistical errors.

Method replicability depends on some assumptions such as sample homogeneity (identical distribution) and independence. Thus, classical inference requires that subsets and supersets follow the same distributions and include independent observations. Often, in practical big data analytics, we deal with samples including heterogeneous, associated, or correlated observations. This violates the parametric (homogeneity and independence) assumptions. In such situations, statistical error estimates may be biased (incorrect) and imprecise (widely varying), which affects the finding replicability and the stability of the ultimate scientific conclusions.

A common example illustrating the importance of replicability is the stability of the final inference obtained from sub-samples of a large dataset or super-samples augmenting or appending the original dataset. When the research findings using subsets or supersets of the data agree with the inference based on the original dataset, this successful replication suggests the method's stability and validates the technique's power to study the underlying phenomenon. However, statistical robustness and result replicability often require either very large datasets or rely on tenuous assumptions about smaller datasets.

In this book, we attempt to address some of these big data challenges by transforming the notion of multiple samples acquired in the 4D Minkowski spacetime into a 5D spacekime extension manifold. This embedding of the 4D space into a 5D spacekime manifold will facilitate a new kind of data analytics, which naturally reduce to their classical 4D spacetime analogues associated with unobserved kime-phases (time-directions).

A simple example may provide motivation, explain the basic idea, and contextualize the process of 5D spacekime inference. While there is no common definition of intelligence, nor is there a direct way to observe it, attempts to measure it and explain its origin still captivate human imagination. In general, *intelligence* is a cognitive capacity for rational, logical, and pragmatic understanding of the environment including self-awareness, ability to learn, adapt, and interpret emotional cues, reflect on the past, reason for the future, organize or create structure, and solve challenges. Intelligence is a latent process that cannot be natively and holistically observed. It's tracked mostly qualitatively as abilities to perceive events, memorize or retain knowledge, infer conclusions that can be applied to adapt behaviors, plan for expected events, or predict future conditions in a specific environment. **Figure 1.1** includes the generic process of scientific inference with a specific example of studying intelligence across species.

Suppose the evidence we collect about cross-species intelligence is in the form of a complex dataset D:

$$D = \{d_{s,k,f,t} | s = \text{species},\ k = \text{case},\ f = \text{feature},\ t = \text{time}\},$$

where $d_{s,k,f,t}$ is a univariate or multivariate, real or complex, discrete or continuous, binary or categorical observation indexed across time by the species, case, and feature identifiers.

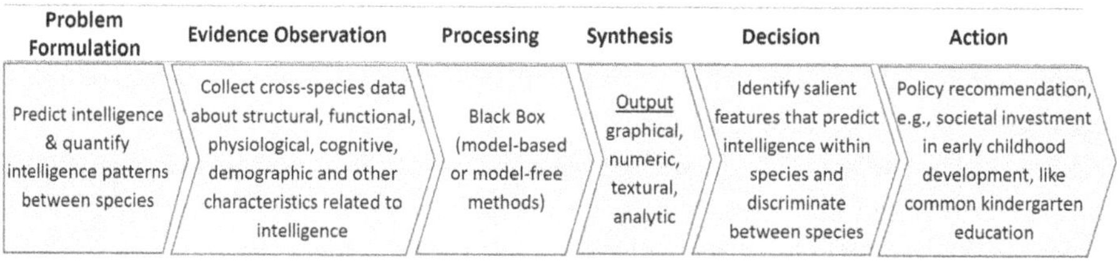

Figure 1.1: The general process of scientific inference with a heuristic example examining intelligence.

If we have a mechanism to understand the process within one species, the between-species effects can be interpreted by contrasting the quantitative characterizations across species. Let us focus on one species, humans, and assume we apply model-free clustering to obtain automated computed phenotypes that can be interpreted in relation to human intelligence.

Let $D^{(0)} = \{d_{\text{human},k,f,t} | k = \text{case},\ f = \text{feature},\ t = \text{time}\}$ represent the raw human data and $I(D^{(0)})$ be the resulting inference, e.g., derived cluster labels that may be associated with human intelligence. Let $FT(D^{(0)}) = \hat{D}^{(0)}$ be the Fourier transformation (FT) of the data into k-space [20]. Clearly, we are assuming the data is quantitative. If not, we can preprocess and quantize the data in advance. Preprocessing steps transforming unstructured data into quantitative elements may be accomplished in many different ways, including mapping elements to structured tensors (e.g., using text mining, natural language processing (NLP), image processing), introducing dummy variable coding, and by other methods.

As we will show later, the Fourier representation of the data $(\hat{D}^{(0)} = \hat{D}^{(0)}(\delta, \omega))$ is complex-valued in terms of the wavenumbers (δ, ω) representing the spatial and temporal frequencies. We can consider the time-frequency feature transformation as a complex-time (kime) variable $\omega = re^{i\theta} \in \mathbb{C}$, where $\boldsymbol{\omega} = (\omega_1, \omega_2, \omega_3, \ldots, \omega_K)$, $\boldsymbol{r} = (r_1, r_2, r_3, \ldots, r_K)$, and $\boldsymbol{\theta} = (\theta_1, \theta_2, \theta_3, \ldots, \theta_K)$. By pooling the kime-phases (θ) across all cases, we can obtain an estimate $\hat{\theta}$, representing an aggregate univariate kime-phase for $D^{(0)}$. This *aggregation* can be accomplished in many different ways, e.g., via the arithmetic-mean aggregator $\hat{\theta}_1 = \frac{1}{K}\sum_{k=1}^{K}\theta_k$, the geometric-mean aggregator $\hat{\theta}_2 = \left(\prod_{k=1}^{K}\theta_k\right)^{1/K}$, the median aggregator[1] $\hat{\theta}_3 = \frac{\theta_{\left(\left\lceil\frac{(K+1)}{2}\right\rceil\right)} + \theta_{\left(\left\lfloor\frac{K+1}{2}\right\rfloor\right)}}{2}$, the kurtosis aggregator $\hat{\theta}_4 = \frac{\frac{1}{K}\sum_{k=1}^{K}(\theta_k - \hat{\theta}_1)^4}{\left(\frac{1}{K}\sum_{k=1}^{K}(\theta_k - \hat{\theta}_1)^2\right)^2} - 3$, the scrambling phase aggregator $\hat{\theta}_5 = \theta_{\pi(k)}$, where $\pi(\cdot)$ is a random permutation of the phase indices, etc.

We can substitute the phase aggregator estimate $\hat{\theta}$ for the phases of the individual cases (θ_k) in $\omega_k' = r_k e^{i\hat{\theta}}$, and denote the pooled-phase data in k-space by $\hat{D}'(0) = \hat{D}^{(0)}(\omega')$. Next, we can invert the FT synthesizing the morphed data back into the native space. Let's denote the reconstructed data by $\hat{\hat{D}}^{(0)} = IFT(\hat{D}'(0))$. In addition to using statistical phase aggregators, we can use model-based phase estimation using specific probability density or distribution functions. We will also show an analytical strategy to derive the enigmatic phases. For instance, we will show that the Laplace transform can be applied to any longitudinal signal, i.e., time series, to obtain its Laplace analytic dual kimesurface.

Now, note that in the native space, the original data ($D^{(0)}$) and the corresponding morphed reconstruction ($\hat{\hat{D}}^{(0)}$) may be similar, but generally not identical, $D^{(0)} \neq \hat{\hat{D}}^{(0)}$. We now can apply the same (model-based or model-free) data analytic strategy to the original and the reconstructed datasets and compare the results. There are obviously many alternative analytical strategies that may be appropriate to use for the specific case study. As we indicated earlier, in this illustration, we can apply model-free clustering to obtain automated computed phenotypes that can be interpreted in relation to human intelligence. In general, $D^{(0)} \neq \hat{\hat{D}}^{(0)}$ and we can expect the resulting clusters derived from the original and morphed data to be somewhat distinct. Later, in **Chapter 6**, we will demonstrate that applying and validating this spacekime analytics protocol to real datasets yields different results, which in some cases may be better than the corresponding results based on the initial spacetime data.

In other words, different inference methods, such as identifying salient features associated with intelligence, forecasting human intelligence based on other observational traits, or quantifying the cross-species intelligence characteristics, may all be improved by transforming the data into spacekime. This process approximates the unobserved time-directions by aggregating the estimated kime-phases and models the morphed reconstructed data in the native (spacetime) domain.

In this book, we will provide the technical details and the mathematical foundation for this data-analytic approach which utilizes a generalization of the concept of

---

1 Median aggregator is defined in terms of the ceiling and the floor integer functions, $\lceil\square\rceil$ and $\lfloor\square\rfloor$, and the order statistics $\theta_{(k)}$.

time to obtain parameter estimates that yield more accurate and replicable scientific findings.

1.2 Internet of Things

Since 2010, there has been a relentless push to embed digital tracking (e.g., chips, bar or QR codes, wearables) in all objects (including animals and humans) and central processing units (CPUs) in all devices, no matter how small, insignificant, or impactful these entities may be. Even a basic refrigerator model sold in 2013 has a motherboard and a pair of CPUs. This trend is spurred by accelerated automation, rapid information and communication technology advances, and manufacturing cost reductions. The internet of things (IoT) is a virtual network of connected devices (e.g., phones, vehicles, and home appliances) having core electronics components and supporting common communication protocols. IoT connectivity enables instantaneous, efficient, and reliable data connection that can be used for effective decision-making, resource optimization, information exchange, and predictive forecasting [21, 22].

IoT systems connect billions of devices (some standard like phones, laptops, and high performance computing systems, and some less common like radio frequency identification microchips) that are in constant flux in space, time, and state. Aside from security, privacy, and policy issues, this avalanche of information presents massive challenges related to the seven characteristics of Big Data and their handling, management, processing, interpretation, and utilization. The Kryder law (doubling of storage in 12–14 months) overtakes Moore's law (doubling computing power each 18 months) reflecting the faster rate of data collection compared to the rate of increase of our computational capabilities [6]. Going forward, all of the generated digital information cannot be physically processed or analyzed with current technologies. The breadth of coverage and the depth of penetration of IoT will certainly change all aspects of human life. However, to extract meaningful information, derive new knowledge and supply effective decision support systems based on the wealth of IoT information will demand radically different data science methods and truly innovative predictive analytic techniques.

1.3 Defining Characteristics of Big Datasets

Many large and complex datasets share a set of seven common properties that are referred to as the seven dimensions characterizing big data [10]. **Table 1.1** summarizes each of these aspects and provides hints to their importance.

These characteristics of big data can be observed in virtually all application domains, e.g., all contemporary environmental, physical, medical, social, economic, and biological studies collect and interpret multiplex, voluminous, inharmonious,

Table 1.1: Common characteristics of complex big data.

Big Data Dimensions	Specific Challenges
Size	Harvesting and management of vast amounts of information reflecting the complete digitalization of all human experiences
Complexity	Need for data wranglers for processing of heterogeneous data and dealing with varying file formats, modalities, and representations
Incongruency	Data harmonization and aggregation are paramount; these require new generic protocols for consistent data homology and metadata matching
Multisource	Ability to transfer and jointly model disparate data elements is critical
Multiscale	A process may be observed via different lenses, which generates multiple proxy characteristics of the phenomenon. It's important to have a holistic interpretation of multiresolution views of the process of interest (from macro to meso, micro, and nano scale observations)
Time	Data analytical methods need to appropriately account for longitudinal patterns in the data (e.g., seasonality, autocorrelation)
Incomplete	Big Data are never complete, exploring missing patterns and effectively imputing incomplete cases are vital for the subsequent data analytics

and impure observations. Of course, in practice, for each specific study, the data characterization loadings on each of these dimensions may vary substantially.

1.4 High-Dimensional Data

There is no generic, canonical, and consistent representation theory of Big Data that facilitates the management, modeling, analysis, and visualization of all observable datasets. In the light of Gödel's incompleteness theorem [23], it is not even clear if such a theory exists. At the same time, there are algebraic, topological, probabilistic, and functional analytic strategies that are being developed and tested to interrogate specific types of big data challenges. Some examples are summarized below, but there are many others and the scientific community is constantly introducing and validating new techniques.

Topological data analysis (TDA) employs geometric, topological and computational techniques to analyze high-dimensional datasets [14]. It is based on information extraction from incomplete and noisy multivariate observations. The general TDA framework captures the intrinsic structural information in the data and enables dimensionality reduction and analysis irrespective of the particular metric chosen to quantify distances between data points. The foundation of TDA exploits the topological nature of the complete dataset via effective algebraic and computational manipulations that preserve the topological and shape characteristics of the original data structure. In practice, TDA connects geometric and topological techniques to estimate the persistent homology of the dataset considered as a point cloud.

In **Chapter 4**, we will see that tensors represent a very useful mathematical framework for formulating and solving many types of problems via algebraic manipulations of geometric operators [24, 25]. In practice, a tensor of order n is an n-way array $X \in \mathbb{R}^{I_1 \times I_2 \times I_3 \times \ldots \times I_n}$ with elements $x_{i_1, i_2, i_3, \ldots i_n}$, indexed by $i_j \in \{1, 2, 3, \ldots, I_j\}$, $1 \leq j \leq n$. Tensors extend the notions of scalars (zeroth-order tensors), vectors (first-order tensors), matrices (second-order tensors), and so on to higher orders. Later on, we will define mathematical operators on tensors, like inner, outer and Kronecker products, Frobenius norm, and others. They allow tensor algebraic manipulations that lead to functional and analytic representations used for mathematical modeling, statistical inference, and machine learning-based forecasting based on observed data.

Manifold embedding methods, e.g., locally linear embedding or non-linear t-distributed stochastic neighbor embedding (t-SNE) [26, 27] and uniform manifold approximation and projection (UMAP) [28], assist with analyzing large amounts of multivariate data by reducing the problem dimensionality. These techniques effectively allow discovery-driven exploration of compact proxy representations of the original high-dimensional data. Such reductions of complexity can be paired with both supervised and unsupervised learning algorithms; however, they are distinct from other clustering methods that also facilitate local dimensionality reduction, but some embedding methods do not rely on local minima optimizations, which may be suboptimal. Many embedding methods encode, learn, and exploit the local and global symmetries of linear reconstructions of the geodesic metric distances between data points in the manifold.

A Statistic Online Computational Resource (SOCR) t-SNE/UMAP interactive webapp, based on TensorBoardJS, provides a dynamic mechanism to ingest complex data (e.g., thousands of cases and thousands of features) directly in the browser, see **Figure 1.2**. The app defaults to embedding 10,000 cases of the UKBB archive (https://socr.umich.edu/HTML5/SOCR_TensorBoard_UKBB, accessed January 29, 2021). Each case has 5,000 clinical, demographic, and phenotypic structural data elements as well as structural magnetic resonance imaging 3D brain volumes, which yield over 3,000 derived neuroimaging morphometric measures. At the start, the webapp spherizes the 3D linear embedding using principal component analysis (PCA), but also allows 2D and 3D linear and non-linear (e.g., t-SNE) embeddings of the 8,000-dimensional UKBB dataset. This specific web-demonstration allows users to import their own high-dimensional data (TSV format) and dynamically interrogate it directly into the client browser.

Ensemble methods for modeling high-dimensional data with a large number of features tend to be more robust with respect to noise as they aggregate families of independent, possibly weak, learning algorithms to reduce the dimension complexity, select salient features, and provide reliable predictions. Integrating many

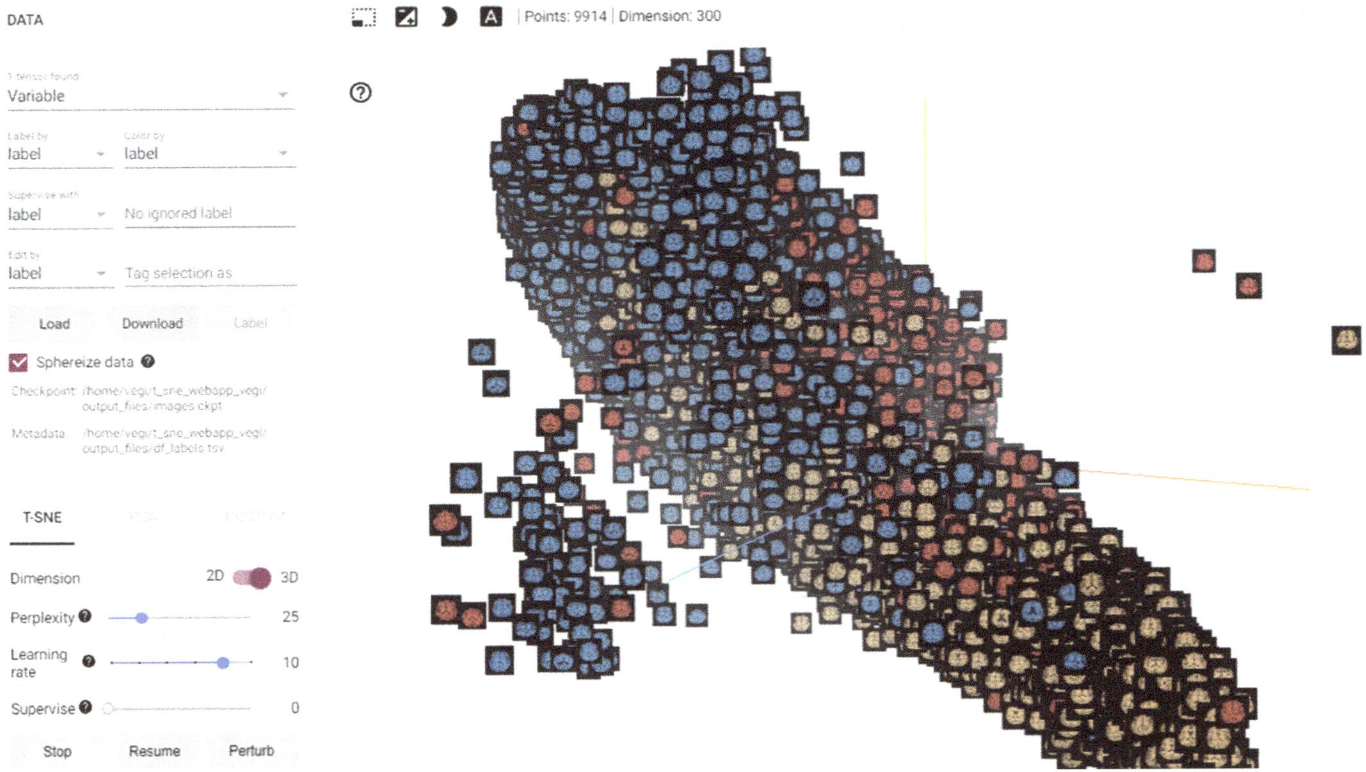

Figure 1.2: SOCR webapp (https://socr.umich.edu/HTML5/SOCR_TensorBoard_UKBB, accessed January 29, 2021) illustrating dynamic visualization of non-linear manifold embedding as dimensionality reduction method using t-SNE.

different feature selection algorithms yields stable feature subsets that represent local or global optima in the feature spaces [29]. Ensemble methods select salient independent features approximating the optimal subset of variables that may be highly informative to forecast specific outcomes. Many regression, clustering, and classification algorithms address big data challenges due to either a very large sample size or very high number of dimensions, but are impractical when both conditions are present [30, 31].

There are many alternative approaches to employ graph representation of high-dimensional data. Most of these techniques rely on graph formulation of supervised or semi-supervised learning methods [32]. For example, probit classification [33] generalizes level-set methods for Bayesian inverse problems with graph-theoretic approaches and general optimization-based classification into a Bayesian framework. This strategy capitalizes on efficient numerical methods tuned to handle large datasets, for stochastic Markov chain Monte Carlo-based sampling as well as gradient-based Maximum a posteriori (MAP) estimation. Graph-based semi-supervised learning methods tend to be computationally intensive, but yield high classification accuracy and provide a means to quantize uncertainty of the forecasting models.

Partial differential equations and various operators, like linear Laplacian or quadratic Hessian eigenmaps [34], provide low-dimensional representations of complex high-dimensional data that preserve local geometric characteristics and distances between data elements [35]. Such methods have a solid mathematical manifold foundation, but may not be practical for some real-world data that exhibit uneven data sampling, out-of-sample problems, or extreme heterogeneity.

1.5 Scientific Inference and Forecasting

Scientists often aim to predict what might happen in the future state of a process, guide the knowledge discovery, and recommend appropriate actions by examining what happened in the past and bring that knowledge to the present. The two distinct tasks of data-driven *retrodiction* and *prediction* require using information about the present to infer possibilities about different periods of time. However, these tasks differ in the direction of time they try to explain. Retrodiction aims to describe the past, whereas prediction attempts to forecast the future.

There is some evidence suggesting that for humans, the task of prediction is constantly and automatically in play. We always try to estimate prospective outcomes based on present knowledge (e.g., balance risk and benefits in planning for intrinsically stochastic events, determine when to buy and sell products, securities, or real estate). Predicting the near future allows us to plan our subsequent actions.

Both human mental models as well as artificial machine intelligence algorithms base these predictions on specific prior experience, artificial intelligence (AI), machine learning and training, and specific model states (e.g., complexity, parameters).

These a priori assumptions allow us to extrapolate, simulate and foresee possible future states of the phenomenon. Forward-looking prospective models are based on how we, or algorithms, expect the process to evolve, unfold or change across time, space, and conditions. Prior research on human prospection (elaborative prediction) suggests that the task of predicting future trends (extrapolating the world forward) relies on various cognitive mechanisms including remembering the past, understanding the present, and dreaming about the impending unknown [36].

1.6 Data Science

Data science is truly a twenty-first century scientific discipline that grew organically in response to the rapid, precipitous, and ubiquitous immersion of digital information in all aspects of human experience [10, 37–41]. Data science is an extremely transdisciplinary area bridging between the theoretical, computational, experimental, and biosocial areas. It provides techniques for interrogating enormous amounts of complex, incongruent, and dynamically changing data from multiple sources. Most data science efforts are focused on generating semi-automated decision support systems based on novel algorithms, methods, tools, and services for ingesting and manipulating complex datasets. Data science techniques are used for mining data patterns or motifs, predicting expected outcomes, deriving clusters, or computed phenotypes using retrospective data or labeling of prospective observations, computing data signatures or fingerprints, extracting valuable information, and providing evidence-based actionable knowledge. Data science protocols integrate techniques for data manipulation (wrangling), data harmonization and aggregation strategies, exploratory or confirmatory data analyses, predictive analytics, fine-tuning, and validation of results.

Predictive analytics is the process of utilizing data science along with advanced mathematical formulations, statistical computing, and software tools to represent, interrogate, and interpret complex data. For instance, predictive analytics processes aim to forecast future trends, predict prospective behavior, or prognosticate the forward process characteristics by examining large and heterogeneous data as a proxy of the underlying phenomenon.

By identifying intervariable relationships, associations, arrangements, or motifs in the dataset, predictive analytics can determine the entanglement of space, time, and other features, exploit the predicament duality of information compression and reduction of the dimensionality of the data, as well as generate derived phenotypic clusters. Data science and predictive analytics help uncover unknown effects, provide parameter, and error estimates, generate classification labels, or contribute other aggregate or individualized forecasts. The varying assumptions of each specific predictive analytics technique determine its usability, affect its expected accuracy, and guide the actions resulting from the derived forecasts. To facilitate the predictive analytics process, data science relies on supervised and unsupervised, model-based and model-free, classification and regression, deterministic, stochastic, classical inference, and machine learning-based techniques. Often, the expected inferential outcome type (e.g., binary, polytomous, probability, scalar, vector, tensor, etc.) determines which techniques may be most appropriate and controls the resulting prediction, forecasting, labeling, likelihoods categorization, or grouping conclusion.

1.7 Artificial Intelligence

Human intelligence is commonly understood as the individual or collective intellectual ability of people or mankind that is uniquely enabling us to construct, interpret, express, communicate, and react to complex stimuli [42]. Across species, (natural) intelligent behavior depends on multiple intrinsic and extrinsic factors, such as genetics, environment, and evolutional serendipity. The degree of intelligence may be associated with cognitive acuity, learning style, concept formation and understanding, capacity for logical abstraction, rational reasoning, pattern recognition and perception, ideation, planning, effective decision-making, formation and retrieval of short and long-term memory, information synthesis, and effective communication.

AI, also known as synthetic or machine intelligence, goes farther than simple automation, high-throughput computation (e.g., brute-force solutions), and deterministic algorithmization. At the same time, the line between automation and AI appears to be fluid. As the relentless embedding of AI in all aspects of human experiences continues to grow, prior AI successes tend to be discounted as basic automations. We tend to refer to contemporary AI as proposed solutions to challenging tasks that are difficult or impossible to solve at the present time via state-of-the-art technologies. For instance, recall the exceptional AI solutions such as the Apollo Guidance Computer that led to the Apollo's successful landing on the Moon in 1969 [43, 44], the AI clinical decision support system that advised physicians in antimicrobial therapy in 1973 [45], the Deep Blue's defeat of Garry Kasparov in 1997 [46], and Google DeepMind AI network (*AlphaFold*), which leaped forward in solving grand structural biology problem to determine a protein's 3D shape from its amino-acid sequence [47]. At their times, each of these success stories represented enormous AI breakthroughs. Present computational, machine learning, and AI advances dwarf each of these prior feats of ingenuity. Many consider the moment of successfully conquering a previously difficult challenge as the dynamic boundary where prior AI innovations lose their *intelligent* status to become more routine human-driven inventions [48].

Broadly speaking, AI deals with non-trivial reasoning problems, information extraction, representation of knowledge, effective planning, forecasting, structure understanding, NLP, perception, temporal dynamics, etc. AI relies on numerical methods, optimization theory, logic, probabilistic reasoning, classification and regression techniques, statistical and DL methods, artificial NNs, and engineering tools. Its applications are virtually ubiquitous.

Some AI methods are model-based and some are model-free [49–52]. Some are supervised and some are unsupervised. Some are applicable for structured or unstructured data elements, low or high dimensional data, **Figure 1.3**. AI algorithms may demand substantial backend computational support infrastructure (storage, computing and network) that make them uniquely tailored to specific challenges and dependent on high-end services. For instance, DL extends classical feedforward artificial NN and multilayer perceptrons to more elaborate networks such as unsupervised auto encoding–decoding, convolutional, recurrent, and long short-term memory [53–55]. The pragmatics of these complex graph network topologies and their model-free structure typically demand exceptional computational foundation, substantial memory capacity, and high bandwidth network resources to complete the optimization problem and generate the final classification, prediction, recommendation, decision, or forecast. The extremely large number of unknowns in such deep neural networks (DNNs) leads to difficult computational challenges. For instance, a substantial stumbling block in DNN prediction is the process of hyperparameter optimization. It typically involves initializing, normalizing, learning (estimating), and tuning a vast number of free parameters that are possibly interdependent and collectively have non-linear effects on the final DL results. Recent advances in graphics processing unit computing, scalable Cloud services, and increases in channel communication bandwidth facilitate the proliferation of complex DNN modeling, analytics, and applications.

1.8 Examples of Driving Motivational Challenges

All sectors of the economy are beginning to be affected by the IoT information wave and the data science disruptive revolution. Even traditionally labor-intensive areas such as agriculture, construction, the humanities, and the arts are deeply impacted by digital sensors, IoT instruments, and data-driven forecasts to improve crop yield, reduce transportation costs, source commodities, improve material properties, and optimize scheduling. **Figure 1.4** shows estimates of (1) big data potential index, (2) expected annual growth, and (3) the relative GDP contributions for a number of industry sectors, based on 2016 data from the US Bureau of Labor Statistics and the McKinsey Global Institute.

Below we will summarize several specific examples illustrating the role and impact of big data in biomedical, environmental, and socioeconomic applications.

1.8.1 Neuroimaging-Genetics

A recent Alzheimer's disease (AD) study [15] utilized multiplex data from the Alzheimer's Disease Neuroimaging Initiative [56] to formulate a generic big data representation that facilitates the mathematical modeling, statistical inference, and

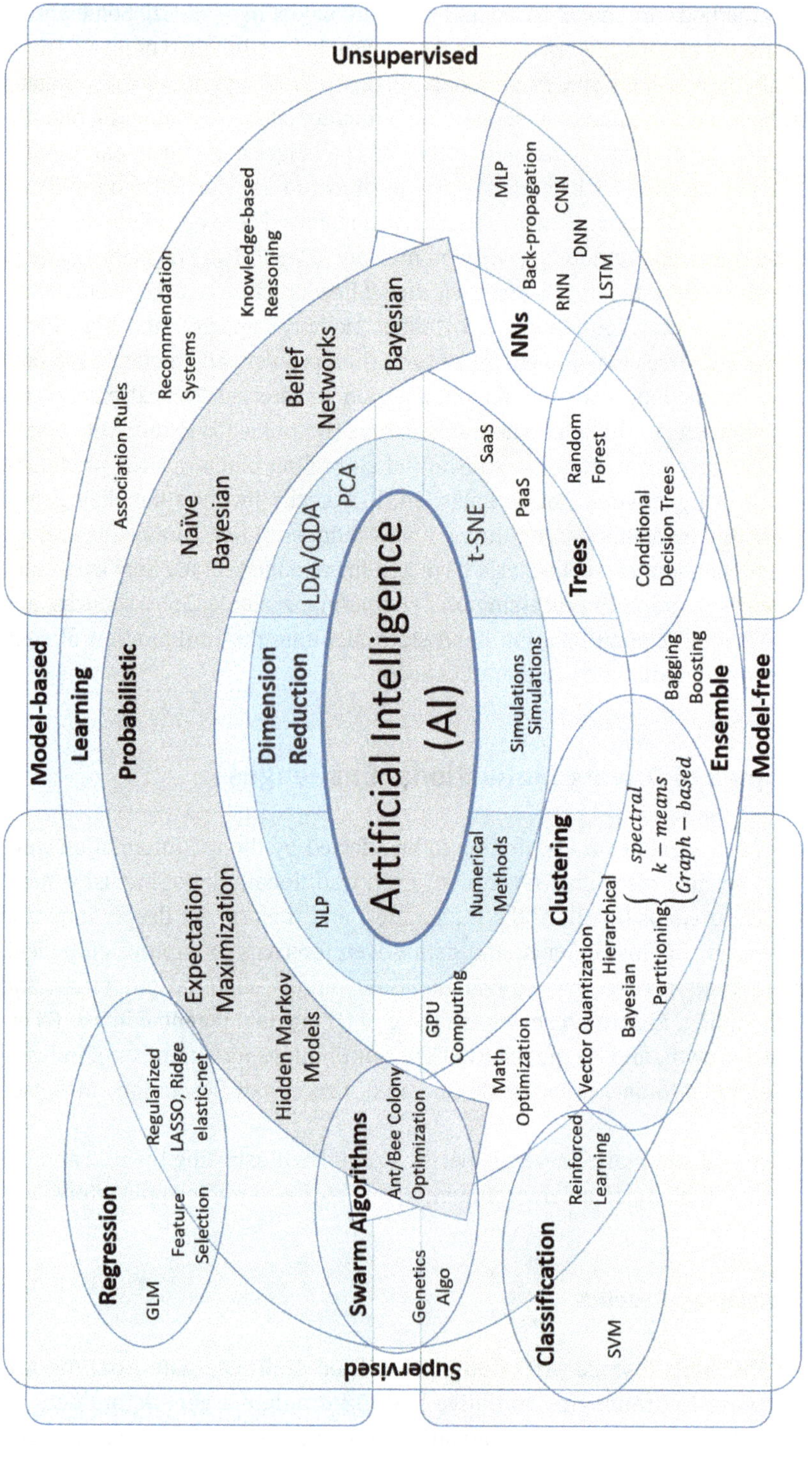

Figure 1.3: Schematic representation of different types of popular AI methods, techniques, services, and algorithms.

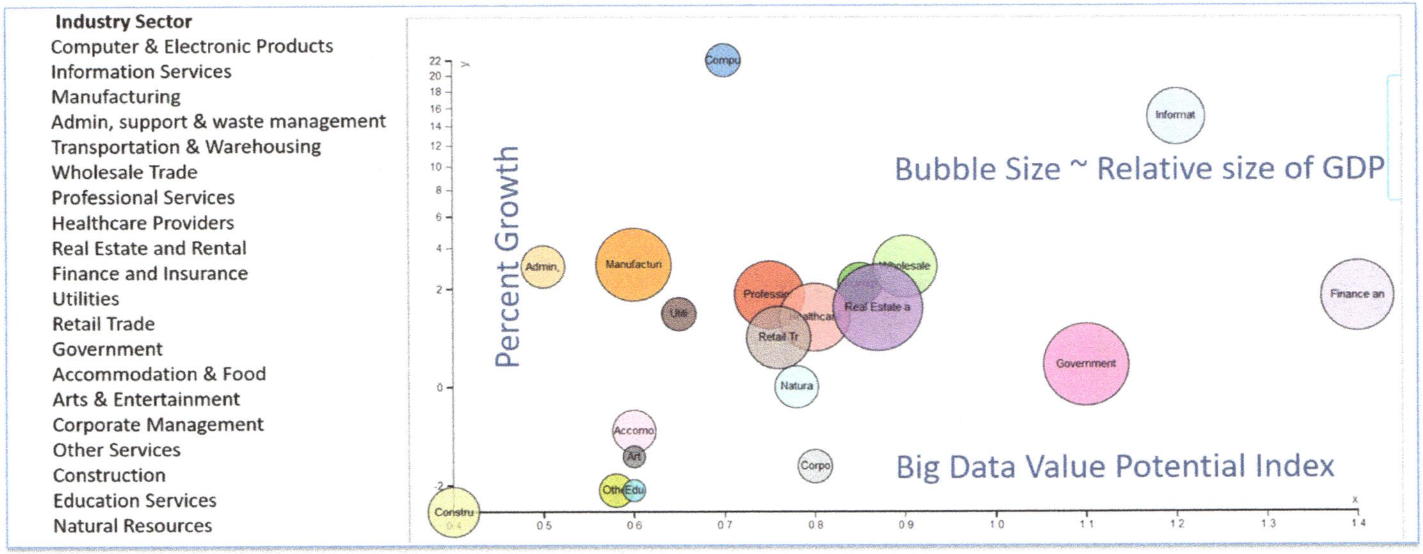

Figure 1.4: Big Data impact on various sectors of the US economy in terms of value-potential and expected growth. Each blob represents an economic sector whose size is proportional to its annual GDP contribution. The horizontal (x) and vertical (y) axes represent the potential of Big Data to disrupt the sector operations and expected data-driven sector annual growth, respectively.

machine-learning forecasting of clinical phenotypes. The data consisted of clinical measurements, brain images, and genetic markers for a large sample of elderly volunteers representing three cohorts of patients: early onset Alzheimer disease (EO-AD, of size 406), normal controls (of size 747), and early onset mild cognitive impairment (EO-MCI of size 1, 347).

Employing global and local shape analyses of the neuroimaging data, the authors derived imaging-biomarkers based on 56 regions of interest. Then, they used a compressive big data analytic technique to forecast the participant clinical diagnosis and validate the prediction accuracy. This problem was particularly challenging because of normal variations in cognitive aging and the heterogeneity of dementia. In addition, the complexity of the data with most data elements being represented in different incongruent spaces required special data science wrangling methods and advanced machine learning classifiers to jointly model and holistically interpret the wealth of information.

The reported results identified the 10 most salient features including a blend of clinical, demographic and derived imaging covariates (clinical dementia rating global score, weight, sex, age, right cingulate gyrus, functional assessment questionnaire total score, left gyrus rectus, right putamen, cerebellum and left middle orbitofrontal gyrus) that affected the diagnostic prediction of the clinical phenotypes of the participants. An internal statistical cross-validation confirmed a high accuracy of predicting the correct diagnosis, $accuracy = 90\%$ with confidence interval CI: $[87\%, \ 93\%]$.

1.8.2 Census-Like Population Studies

From an individual or family perspective, human health and well-being are very personal issues. However, as the 2018 measles outbreak among communities with low child vaccination rates within the US State of Washington shows, assuring the health of the public goes beyond focusing on the health status of individuals and demands population-wide epidemiological approaches. Strategies to holistically understand the temporal patterns of health and disease across location, demographics, and socioeconomic status are vital to understanding the factors that influence population well-being.

The United Kingdom Biobank (UKBB) is a unique and powerful open-access health resource enabling international researchers and scholars to examine, model, and analyze census-like multisource healthcare data. UKBB presents a number of Big Data challenges related to aggregation and harmonization of complex information, feature heterogeneity and salience, and predictive health analytics. Investigators at the University of Michigan SOCR recently used UKBB to examine mental health in a general population [57]. By using 7,614 imaging, clinical, and phenotypic features of 9,914 participants the researchers performed deep computed phenotyping using unsupervised clustering and derived two distinct sub-cohorts. Parametric and nonparametric tests identified the top 20 most salient features contributing to the AI-based clustering of participants. This approach generated decision rules predicting the presence and progression of depression, anxiety, and other mental illnesses by jointly representing and modeling the significant clinical and demographic variables along with the derived salient neuroimaging features. **Table 1.2** shows the results of an internal statistical cross-validation for a random forest classifier for a set of four clinical outcomes: *sensitivity/hurt feelings, ever depressed for a whole week, worrier/anxious feelings,* and *miserableness.*

Table 1.2: Results of a population-wide exploratory analytics study using random forest classifier applied to the UKBB data.

Clinical outcomes	Accuracy	95% CI of accuracy	Sensitivity	Specificity
Sensitivity/hurt feelings	0.699	(0.675, 0.723)	0.653	0.742
Ever depressed (≥ 1 week)	0.782	(0.760, 0.803)	0.937	0.618
Worrier/anxious feelings	0.730	(0.706, 0.753)	0.720	0.739
Miserableness	0.740	(0.716, 0.762)	0.863	0.549

While not perfect, these results illustrate how completely automated methods can derive complex phenotypic traits in large populations and identify the salient data elements that provide important contextual information about clinically relevant idiosyncrasies.

The study reported high consistency and reliability of the automatically-derived computed phenotypes and the top salient imaging biomarkers that contributed to the unsupervised cohort separation, **Figure 1.5**. The outcome of the study is a clinical decision support system that utilizes all available information and identifies the most critical biomarkers for predicting mental health, e.g., anxiety. Applications of this technique on different populations may lead to reducing healthcare expenses and improving the processes of diagnosis, forecasting, and tracking of normal and pathological aging.

1.8.3 4D Nucleome

In 2015, the National Institutes of Health launched the 4D Nucleome initiative [58], which aims to understand the fundamental principles of 3D special organization and 1D temporal morphogenesis of the cell nucleus. To untangle this complex problem, investigators examine gene expression, imaging, and cellular recordings data, and track the observed changes of shape, size, form, function, and physiology of multiple cellular and subcellular structures.

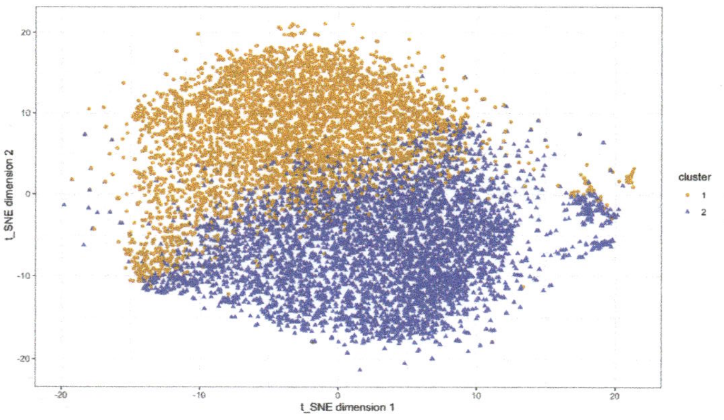

A lower-dimensional plot of the high-dimensional data, with orange and blue k-means generated clustering labels, obtained by non-linear projection of the data onto a 2D t-SNE manifold. The clearly visible correspondence between label colors and blobular geometric structure suggests that there is energyin the datasetthat can be exploited to epidemiologically examine mental health.

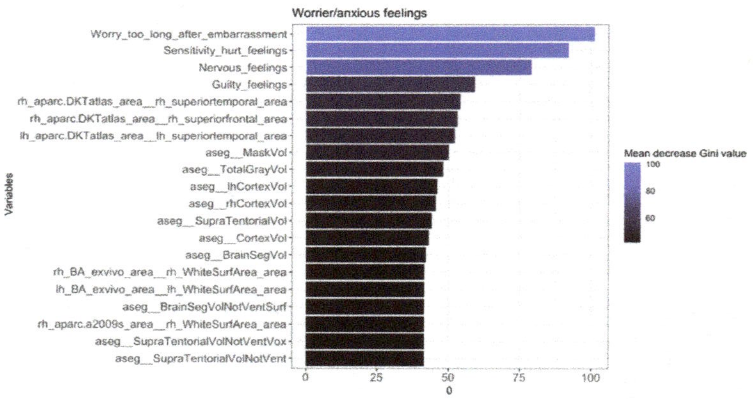

Variable (y-axis) importance plot for predicting clinical anxiety, based on mean decrease Gini values (x-axis) of a random forest classifier.

Figure 1.5: Example of an automated clinical decision support system applied to the UKBB data to generate computed phenotypes (left) and identify the most salient data elements associated with anxiety (right).

Two recent studies illustrate some of the data-complexity intricacies and analytical opportunities to enhance our basic knowledge about cellular growth, maturation and aging in both health and disease (e.g., cancer). The first study performed experiments, collected multimodal heterogeneous data, and applied data and computational science techniques to measure genome conformation and nuclear organization [59]. The authors combined biophysical approaches to generate quantitative models of spatial genome organization in different biological states for individual cells and cell populations, and investigated the relations between gene regulation and other genome functions. This research demonstrated that advanced mathematical modelling and computational statistics applied to complex genomics data provide mechanisms for precision health and biomedical forecasting.

Another study [60], proposed a new approach to integrate data modeling, analysis, and interpretation with the intrinsic 3D morphometric characteristics of cell nuclei and nucleoli. This study required a substantial preprocessing of high-volume of 3D tissue imaging data to generate (1) volumetric masks of cells and sub-cellular organelles, and (2) robust surface reconstruction that allows accurate parametric approximation of the intrinsic 3D object boundaries. A number of computed geometric morphological measures were utilized to characterize the shape, form and size of cell nuclei and nucleoli. Augmenting the phenotypic and experimental conditions data with these imaging-derived features allowed the modeling, comparison, and analysis of nuclei and nucleoli of epithelial and mesenchymal prostate cancer cells as well as serum-starved and proliferating fibroblast cells.

The study developed an end-to-end computational protocol that starts with the multisource raw data and images, goes through the data harmonization and modeling, learns the affinities of the relations among imaging, structural organization and experimental phenotypes on training data, and applies the automated decision support system to predict cell types on independent (testing) data. The excellent classification results report accuracy of 95.4% and 98% for discriminating between normal and cancer cells using sets of 9 and 15 cells, respectively, **Figure 1.6.** This was one of the first attempts to completely automatically process multisource cellular data, combine 3D nuclear shape modeling methods, apply morphometry measures, and build a highly parallel pipeline workflow that forecasts the cell phenotypes.

1.8.4 Climate Change

It is common to confuse shifting *weather patterns* and *climate change* in casual conversations. In short time periods, the former does cause a lot of anxiety in local geographic regions under specific conditions. Dynamic and extreme weather patterns typically have counterbalancing effects at different spacetime locations. However, alterations in the Earth's climate represent global, lasting, compounding, and inertial effects with prolonged periodicity that typically result in drastic environmental perturbations.

Mastering the forecasting of weather patterns represents perhaps the greatest accomplishment of human ingenuity. Annually, across the globe, extreme weather kills approximately 500,000 people and injures tens of millions. Despite that, millions of lives are saved each year by accurate forecasts of extreme weather events. As of 2019, the costliest weather disaster in the US was the 2017 Hurricane Harvey, which reduced the GDP by over $190B. Reliable forecasting of the trajectory, severity, and impact of such weather systems is critical. It enables advance warnings, accurate projections of their paths, and appropriate road condition advisories. Such models also facilitate governmental planning, emergency response coordination, and organizational management to mitigate many types of natural disasters.

Climate change is much more enigmatic. Most climate-change models agree that the Earth is warming rapidly, most likely due to human causes (e.g., extreme fossil fuel consumption and increase of CO_2 or CH_4 emissions in the atmosphere). Predicting a category 3 hurricane when a category 5 storm makes landfall will have a devastating, yet very localized effect. Forecasting a global average temperature increase of 2 °C by 2100, but observing an actual 3 °C change in 2100 will be catastrophic to all life on Earth. The effect of such "moderate" change of 1 °C may result in a similar climate to the last epoch of the Tertiary period, the Pliocene, 3 million years ago. Permafrost will melt, trees will grow in the Arctic, all continental glaciers will disappear, and sea levels will rise by 25 meters displacing billions of people.

Global climate change and its impact on life may be the greatest challenge in the twenty-first century. It requires an urgent response, massive amounts of historical, contemporary and prospective data, innovative model-based statistical and model-free machine learning techniques to understand the interactions of the many terrestrial (e.g., energy use, agriculture), and extraterrestrial (e.g., Sun radiation) factors. There is a stark contrast between the enormous success of weather prediction and detailed forecasting of climate change using big data. The root cause of this discrepancy lies in the complex nature of climate data (e.g., extremely long periodicity and extremely complex interdependencies) and the existential nature of the climate science challenges.

The Howard Hughes Medical Institute BioInteractive resource includes an Earth Viewer that dynamically illustrates the longitudinal climate change relative to temperature, atmospheric greenhouse particles (e.g., CO_2 and CH_4), and other factors, **Figure 1.7**. It demonstrates unmistakably the effects of climate change, as well as represents an oversimplified version of a big data-driven model for estimating retrospective relations and forward prediction of the complex longitudinal interrelations between multiple factors affecting and characterizing the Earth's climate.

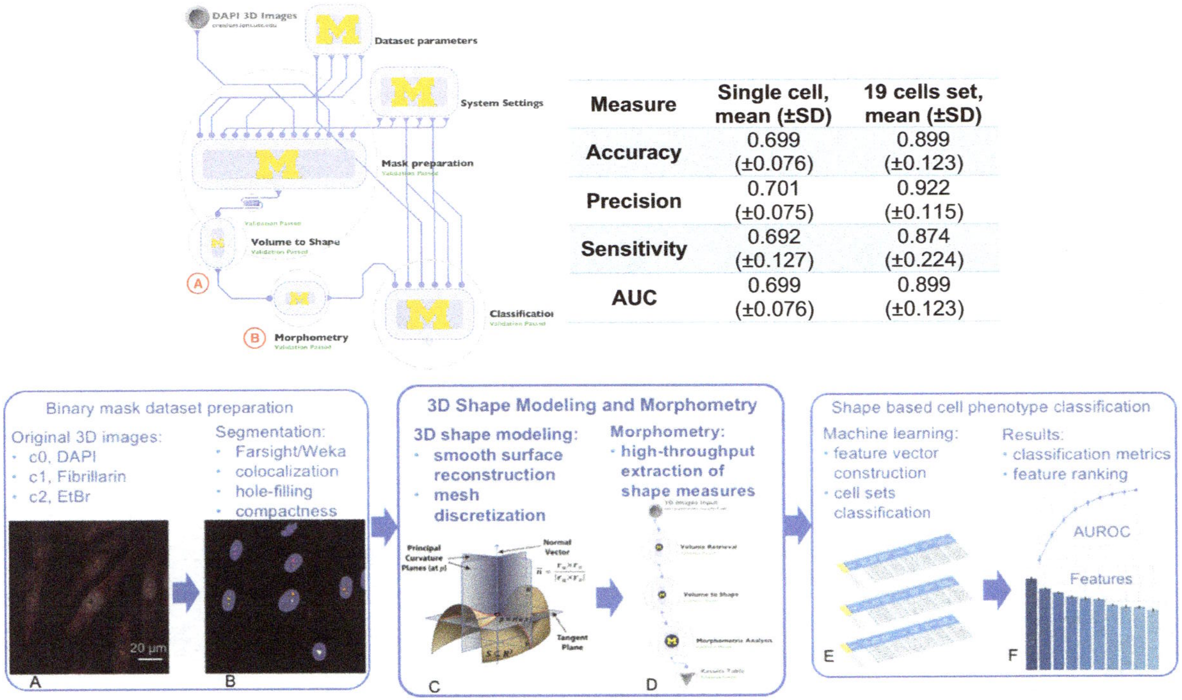

Measure	Single cell, mean (±SD)	19 cells set, mean (±SD)
Accuracy	0.699 (±0.076)	0.899 (±0.123)
Precision	0.701 (±0.075)	0.922 (±0.115)
Sensitivity	0.692 (±0.127)	0.874 (±0.224)
AUC	0.699 (±0.076)	0.899 (±0.123)

Figure 1.6: Automated decision support system for phenotypic cell classification. The top row shows a graphical representation of the complete pipeline workflow protocol (left) and some of the validation metrics (right). The bottom row illustrates the complex transdisciplinary data processing and analytics steps: **A**: raw microscopy input data and phenotypic meta data; **B**: image processing and region of interest masking in 3D; **C**: mathematical representation and modeling of biologic shape, form and size; **D**: calculation of derived intrinsic and extrinsic, geometric and topological measures; **E**: machine learning-based trait quantification; **F**: result visualization and interpretation.

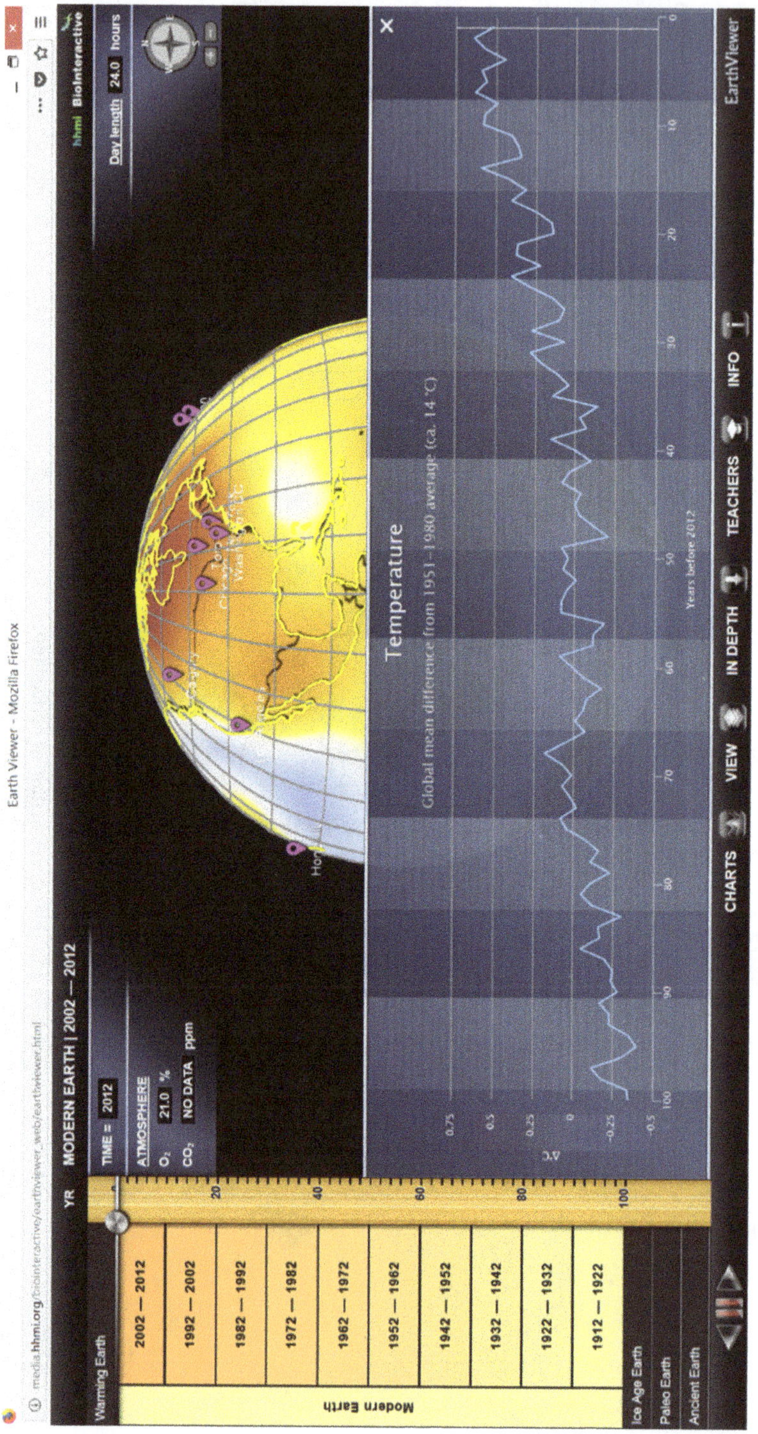

Figure 1.7: HHMI BioInteractive activity (https://www.hhmi.org/biointeractive/earthviewer, accessed January 29, 2021) includes climate data (https://www.hhmi.org/biointeractive/earthviewer-data-files, accessed January 29, 2021) and a dynamic Earth Viewer webapp (https://media.hhmi.org/biointeractive/earthviewer_web/earthviewer.html, accessed January 29, 2021) for exploring the relations between space, time, temperature, CO_2, and other factors related to global climate.

1.9 Problems of Time

The concept of time is a bit mysterious. In general, time refers to the continuous progress of existence where observable events occur in an *irreversible* succession from the past, to the present, and further into the future [61]. In essence, time is the positive continuous univariate measure we associate with the ordering of all observable phenomena. Time-derivative measures allow us to compare events, quantify various phenomenological measurements like time duration (distance in time, interval) and rates of change of other measurable physical quantities, e.g., speed and acceleration, and interpret conscious experiences. As the fourth dimension of the Minkowski spacetime manifold, time plays a key role in the special and general theories of relativity. At present, precise measurement of time is critical in all aspects of human life as it underpins virtually all human communications, spatial–temporal localizations, and all computing processes including the basic read and write of a single bit of information.

The contemporary interpretation of time is based on the early 1900s idea of Einstein to axiomatically define the speed of light as a constant value for all observers, which effectively allows intrinsic measurements of distances in 4D spacetime. This notion extended the Newtonian interpretation of classical mechanics. According to the special theory of relativity, two events in separate inertial frames can be simultaneous yet the spatial-distances may appear compressed and time-intervals may appear lengthened to observers in these frames.

Distances in space can be measured by the length of time light travels that distance, which leads to an intrinsic definition of a spatial yardstick, a meter, as the distance traveled by light (in vacuum without distortions) in a small fraction of time (*the constant speed of light = 299,792,458 meters/second*). In Minkowski 4D spacetime, a pair of events can be separated by a spacelike, lightlike, or timelike invariant interval. The timelike separated events cannot occur simultaneously in any frame of reference, as their order of occurrence has a non-trivial temporal component. Spacelike separated events may occur simultaneously in time in some frame of reference, but the events will be observed at different spatial locations. In different inertial frames of reference, local observers may estimate different distances and different time intervals between two events, however, invariant intervals between events do not depend on the observer frame of reference or the inertia velocity.

In its most general form, the *problem of time* conceptualizes a conflict between general relativity, which considers the flow of time as malleable and relative, and quantum mechanics, which regards the flow of time as universal and absolute. In simple terms, the problem of time relates to the paradox of time as being a real, measurable, distinctly interpretable, and reversible phenomenon. In fact, general relativity hints to the illusive nature of time as a uniform and absolute quantity. According to the general theory of relativity, the dynamic nature of spacetime breaks the uniformity and immutability of time by suggesting that gravity morphs the 4D spacetime geometry.

The limited scope of the definition of time yields characterization of event localization purely in terms of their "order of occurrence," which leads to some paradoxical results. To circumvent some of these intrinsic limitations of a univariate positive time, we will extend the notions of events and their order (time) to complex events (*kevents*) and their corresponding 2D complex time (*kime*). This generalization will capture both order and direction of kevent appearances and resolve many of the problems of time. This 2D kime representation naturally reduces to the standard 1D concept of time expressing the traditional event longitudinal order (time). However, kime also reflects more accurately the general states of kevents in terms of their sequential arrangement (kime-time), as well as their orientation (kime-phase). Of course, the key part of this generalization will be the extension from time-dependence to kime-dependence of various fundamental concepts, mathematical functions, linear operators, and physical properties. Examples of such spacekime generalizations that we will derive include the concepts of rate of change, intervals, velocity, derivatives, integrals, Newton's equations of motion, Lorentz transformations, etc., which will be defined in the spacekime manifold. Finally, we will demonstrate the impact of the spacekime generalization to data science problems.

1.10 Definition of *Kime*

At a given spatial location, complex time (*kime*) $\kappa = re^{i\varphi} \in \mathbb{C}$, where the magnitude represents the dual nature of the *event order* ($r > 0$, kime order) and characterizes the longitudinal displacement in time, and the *event phase* ($-\pi \le \varphi < \pi$, kime phase) that can be interpreted as an angular displacement or event direction. Later we will show that there are multiple alternative parameterizations of the complex plane that can be used to represent kime. **Figure 1.8** schematically illustrates the **spacekime** universe ($\mathbb{R}^3 \times \mathbb{C}$). Readers should examine the similarities and differences between various pairs of spacekime points. For instance, $(\boldsymbol{x}, k_1)$ and $(\boldsymbol{x}, k_4)$ have the same spacetime representation, but different spacekime coordinates, $(\boldsymbol{x}, k_1)$ and $(\boldsymbol{y}, k_1)$ share the same kime, but represent different spatial locations, and $(\boldsymbol{x}, k_2)$ and $(\boldsymbol{x}, k_3)$ have the same spatial-locations and kime-directions, but appear sequentially in order.

For a pair of events separated by a negative spacetime interval (i.e., the invariant Minkowski metric tensor is less than zero) the spatial distance between the events is less than the distance that can be traveled by a photon in the time interval separating the two events. Such events are called *timelike separated*, as it is possible to send a *causal signal* (e.g., physical particle carrying information) from the early to the later events. Conversely, when the spacetime metric is positive, the spatial distance separating the two events is larger than the distance that can be traveled by a causal signal (whose speed is less than or equal to the speed of light) during the corresponding time interval. Such a pair of events is said to be *spacelike separated*, as no conventional causal signal can travel from one event to the other, and thus, it is impossible for the

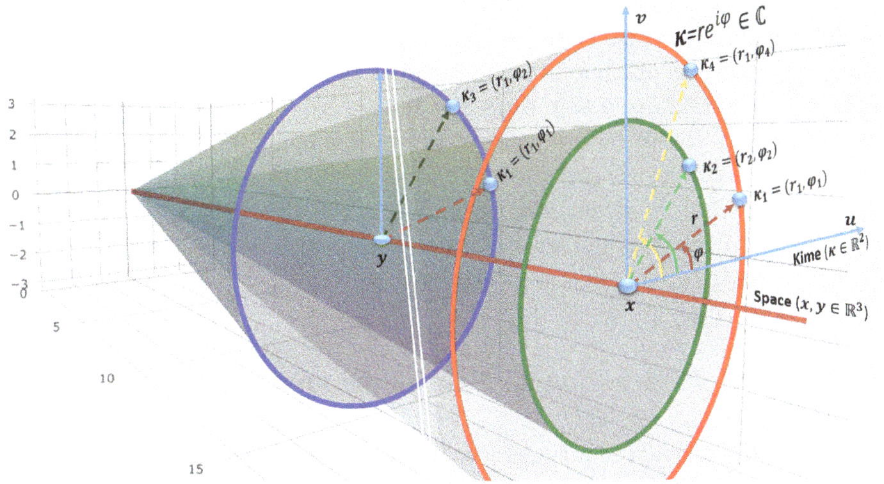

Figure 1.8: A schematic of the 5D spacekime manifold. The graphic pictorially illustrates kevent separation by a *kimelike* interval, $ds^2 < 0$ (e.g., (x, k_1) and (x, k_2)), a *spacelike* interval, $ds^2 > 0$ (e.g.,$(\boldsymbol{x}, k_1)$ and $(\boldsymbol{y}, k_1)$), or a *lightlike* interval, $ds^2 = 0$ (e.g., $(\boldsymbol{x}, k_2)$ and some $(\boldsymbol{y}, k_3)$, where the time difference $c\|r_1 - r_2\|$ is chosen to be equal to the spatial interval $\|\boldsymbol{x} - \boldsymbol{y}\|$).

earlier event to be the "cause" of the latter one. A trivial metric value suggests that the two events are *lightlike separated*, and the two events are a spatial distance apart that equals the exact distance light can travel in the given time period.

Multiple time dimensions have been previously proposed and applied in physics [62–64], music [65], performing arts [66], engineering [67], and economics [68]. There is a difference between "complex-valued" processes, e.g., complex time series, purely imaginary time, e.g., Wick rotation [69], and complex time (kime) indexed processes. It is important to draw early this distinction between *imaginary time*, complexifying the domain of a process (*kime*), and *complex* representation of the *range* of a process. Most prior work has been focused on complex-valued (range) processes. However, Bars [62], Wesson [70], Overduin [63], and Köhn [64] have each formulated alternative approaches to account for 2D time in physics.

In **Chapter 3**, we will see how this kime-manifold definition leads to the extension of the concept of an event in time to a complex event (kevents) in kime, as well as the remediation of some of the problems of time by the extension of Minkowski 4D spacetime to the 5D spacekime manifold. Since there are infinitely many distinct differential structures on the Minkowski 4D spacetime [71, 72], differential topology provides another justification for the 5D spacekime manifold extension. Topologically, the uncountably many incompatible calculus-of-differentiation possibilities on 4D (i.e., infinitely many ways to measure rate of change) are reduced to a finite number of distinct differentiable structures on 5D+ (multiple time and space) dimensions [73]. Under certain conditions, e.g., the 5D+ manifold is a flat Euclidian space, there is actually a unique rate-of-change structure on the manifold [71, 74]. In other words, there are nice 4D manifolds with distinct smooth structures, e.g., there are uncountably many manifolds homeomorphic to $\mathbb{R}^4$ (i.e., each neighborhood in the space is locally flat and Euclidean) that are not diffeomorphic (i.e., the local neighborhood isomorphism maps are not necessarily smooth) [75].

1.11 Economic Forecasting via Spacekime Analytics

Deep market analytics and continuous economic forecasting involve a blend of theoretical models of how the economy works (e.g., cause and effect relations), practical observations of market conditions and the state of the economy (e.g., surveys of experts, balance of supply and demand), and data-driven modeling of various economic metrics (e.g., fiscal and monetary policies in relation to the rate of GDP growth and general business activity). In the developed world, where consumer spending accounts for a large share of economic activity, consumer purchasing, investment, and saving decisions determine the principal trends of future economic conditions. Data from experts, observations of the state of markets and economic conditions (e.g., employment, jobs growth, interest and inflation rates), and numerous private and government indicators all play critical roles in prognosticating the future state of the economy. Central banks (e.g., US Federal Reserve and the European Central Bank) use a wealth of indicators to set interest rates, and government financial departments (like the US Treasury) determine the optimal fiscal policies (e.g., taxes, budget spending, money in circulation) to influence short- and long-term unemployment, inflation rates, and the liquidity balance between monetary supply and demand.

The University of Michigan provides several economic indices, like the Michigan Consumer Sentiment Index (MCSI), normalized to 1966, which capture the current sentiment and forecast the future consumer confidence. These monthly economic forecasts are derived by conducting hundreds of telephone interviews with expert panelists covering 50 core questions [76]. MCSI plays a key role in the market reviews and policy decisions of the Federal Reserve Eighth District Bank of St. Louis, which is part of the US Federal Reserve System (Fed). The Fed open market committee makes quarterly monetary decisions by tracking indices like MCSI, the circulation of money, the state of banking, the trends of various financial indicators, and the broader economic activity in the US.

Data collected for computing the MCSI is openly available via the Michigan Surveys of Consumers website (https://data.sca.isr.umich.edu, accessed January 29, 2021). It contains longitudinal data of consumer sentiment since 1966, comprising over $n = 280,000$ cases and more than $p = 110$ features. The archive includes demographics (e.g., age, gender, marital status, family size, level of education, home ownership, and employment status), personal micro-economic indicators (e.g., family income, personal finances, investments), macro-economic measures (e.g., sentiment of economic conditions, expected unemployment, home ownership sentiment), and consumer trends (e.g., consumer price index, purchases of durable goods, and home price index).

Figure 1.9 depicts the trajectories of several MCSI demographic, economic, and sentiment indicators over the past 41 years. In this graph, the data is smoothed over time and across participants to show the global macroeconomic trends. The MCSI archive includes index of consumer sentiment (ICS), index of current economic conditions (ICC), GOVT (government economic policy), durables buying attitudes (DUR), home buying attitudes (HOM), vehicle buying attitudes (CAR), AGE (age of respondent), and EDUC (education of respondent). Much like other large, complex, and longitudinal economic sentiment and market index datasets, this archive is continuously expanded, and provides interesting computational and data-analytic challenges. Solutions to such forecasting problems require novel approaches for extraction of useful information that can be translated to actionable knowledge driving effective monetary policies and responsible fiscal regulations.

Let us use this case study to illustrate the basic idea behind complex time and kime-phase estimation and draw the parallels between traditional statistical longitudinal modeling and data analytics in the spacekime manifold. In the interest of providing a very specific driving motivational example that illustrates the end-to-end spacekime data analytic protocol, we are suppressing the technical details of this experiment and only showing the idea and results. The mathematical foundations of spacekime analytics will be provided in the coming chapters, along with additional examples elucidating the power of time-complexity modeling to explicate inferential uncertainty.

Fluctuations in the US Index of Consumer Sentiment are closely monitored by investors, policy makers, economists, private corporations, mutual and hedge funds managers, and public organizations. For example, between January and February 2019, University of Michigan scholars reported that the US consumer sentiment rose to 95.5 in February of 2019 compared to 91.2 in January of 2019. A deeper inspection of this sudden monthly ICS increase in February 2019 reflected the end of the partial US government shutdown coupled with a shift in consumer expectations about the US Central Bank (Federal Reserve) halting short-term interest rate hikes.

Accurate, timely, and reliable ICS estimates have significant impact on investment and broader market activity. ICS is a derived measure; it is entangled with many other macro socioeconomic conditions and macroeconomic financial benchmarks. This interdependence suggests that ICS can be (1) modeled and predicted using available economic data and information on other consumer trends, (2) used along with other covariates to predict higher-order derivative measures, and (3) used to validate

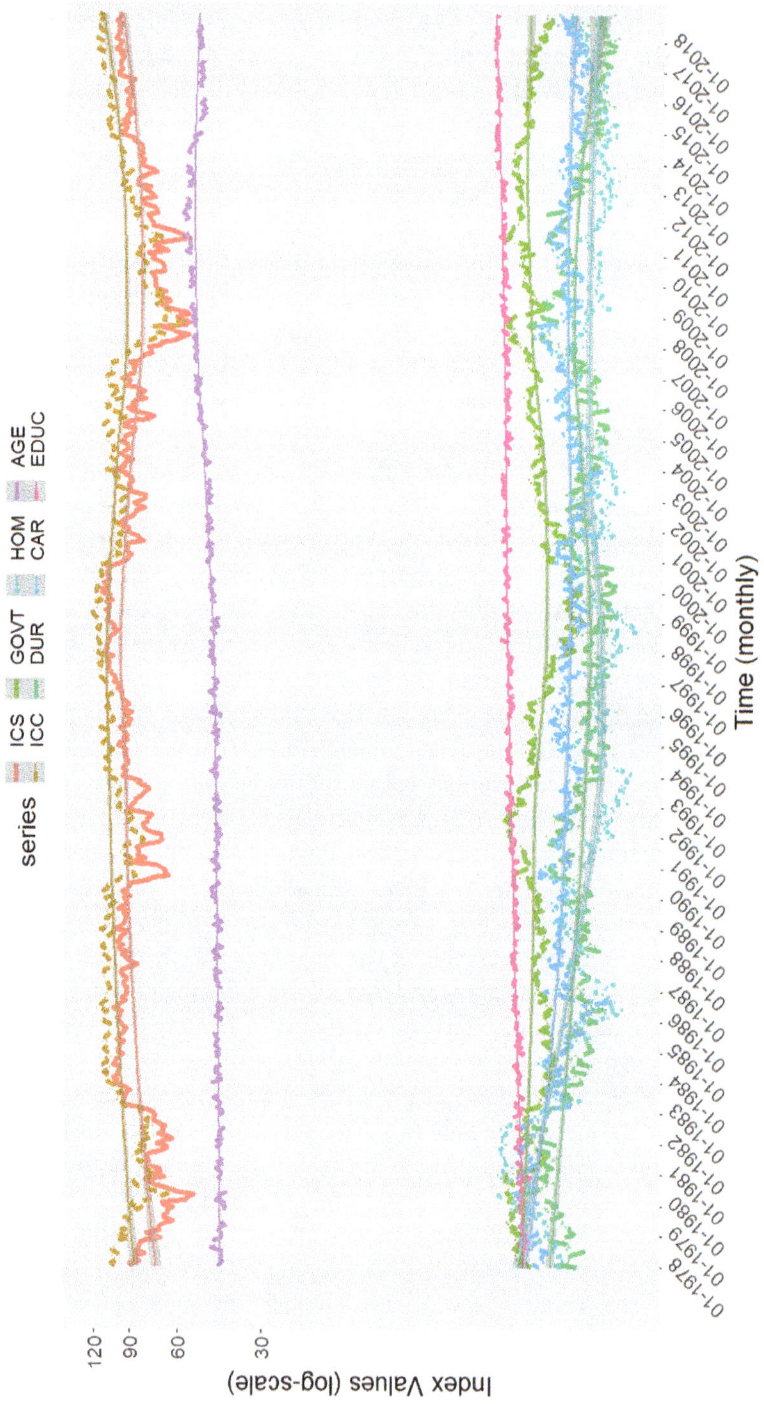

Figure 1.9: Time courses of several MCSI demographic, economic and sentiment indicators over the past 41 years (1978–2018), smoothed over time and across participants. For each indicator, the graph shows data-point scatter and gray confidence bands indicating 95% confidence limits of the indicator moving average (solid lines). The *y*-axis is log-transformed to better display the complete range of all indicators. The data is acquired and plotted monthly, but for readability the x-axis labels are only printed yearly.

other fiscal measures, calibrate monetary conditions, or support or hinder specific market expectations.

With this in mind, let us use the MCSI dataset to forecast ICS using exogenous variables autoregressive integrated moving average (ARIMAX) modeling [77]. We will derive ICS forecasts based on traditional spacetime prediction methods and compare these against new spacekime analytical strategies. Specifically, we will estimate the best ARIMAX longitudinal model of the consumer sentiment indicator using six observed covariates: GOVT, DUR, HOM, CAR, current total household income (income), and education level (Educ). For validation purposes, we will split the multivariate longitudinal data into two sets: (1) a *training* set (covering the period from 1978 to 2016) that will be the base for estimating the ARIMA and the regression parameters of the ARIMAX model, and (2) an independent *testing* dataset that will be used for validation of the resulting forecasts (covering the last two years, 2017–2018). The optimal ARIMAX model of ICS using the above covariates is shown in **Figure 1.10**.

The ARIMAX model is trained on a span of 39 years (1978–2016), and then its effect to predict the prospective ICS is assessed on the last 2 years (2017–2018). **Figure 1.10** graphically depicts the process of traditional statistical model-based inference – from training data (green), to parameter estimation (top panel), statistical significance of effect sizes, model-based prediction (red), and the prospective MICS-reported consumer sentiment index (gray). The 24-month period on the right also shows the 80% and 90% confidence bands around the mean ARIMAX ICS-model prediction. Clearly, this type of statistical inference is appealing, as it has nice theoretical properties, provides an efficient algorithmic strategy for computing the forecast, and is adaptive to variations in the exogenous covariates. The computed ARIMAX model slightly underestimates the ICS that is reported prospectively by MCSI. However, this observed versus predicted difference is not statistically significant as the 2017–2018 ICS trajectory is enclosed completely within the ARIMAX model confidence bands (see top-right in **Figure 1.10**).

We will now demonstrate the process and results of applying a spacekime analytics strategy to the same challenge – prospective prognostication of the consumer sentiment index. To ensure a fair comparison of the alternative ICS forecasts and direct compatibility of the corresponding inference or decision-making, we will still rely on ARIMAX modeling. For simplicity, we will denote the training and testing outcomes by $Y_{train} = ICS$, over 1978–2016, and $Y_{test} = ICS$, over 2017–2018, respectively. Similarly, we will denote the covariate features used to predict the corresponding training and testing outcomes by:

$$X_{train} = \{GOVT, DUR, HOM, CAR, INCOME, EDUC\} \text{ over } 1978-2016, \text{ and}$$

$$X_{test} = \{GOVT, DUR, HOM, CAR, INCOME, EDUC\} \text{ over } 2017-2018.$$

```
Arima_Train <- auto.arima(MCSI_ Y_train, xreg = X_train)
Coefficients of regression with ARIMAX(2,0,1)(1,0,1)[12]
```

Parameter	ar1	ar2	ma1	sar1	sma1	intercept	GOVT	DUR	HOM	CAR	INCOME	EDUC
mean	1.4	−0.5	−0.7	0.6	−0.5	152.5	−11	−11	−2.8	−0.8	0.0	1.6079
SE	0.2	0.2	0.2	0.4	0.4	8.7	1.2	1	0.9	1	0.0	1.8702

Model diagnostics: $\hat{\sigma}^2 = 8.564$, log-likelihood $= -1161$, AIC $= 2348.6$, AICc $=$ 2349.4, BIC $= 2402.5$.

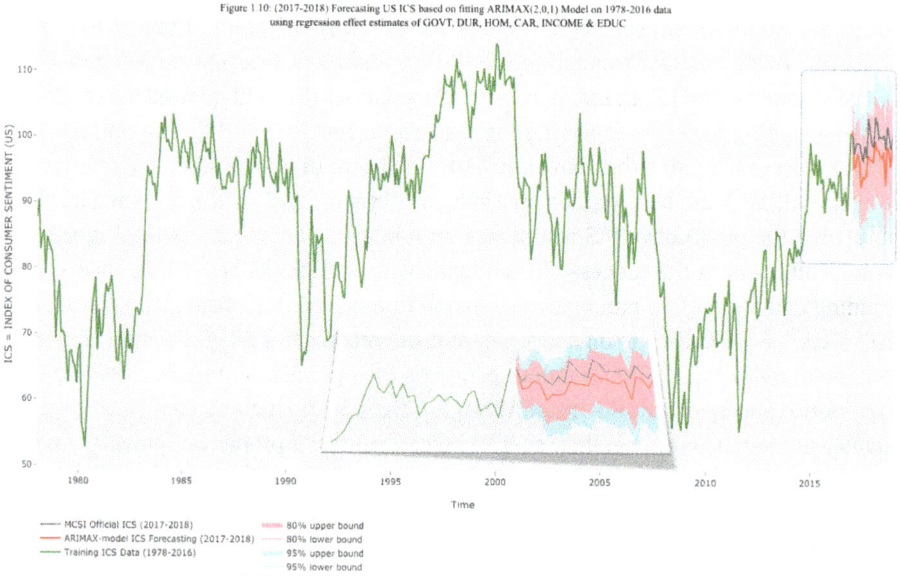

Figure 1.10: (*Traditional statistical modeling and prediction*) Estimation of the optimal ARIMAX model of ICS. The top panel shows the analytical ARIMAX model description (results are rounded to 10^{-1}). The bottom graph visually illustrates the performance of the ARIMAX model trained on the 39-year span (1978–2016) and tested on 2-year prospective longitudinal data (2017–2018). The training data (green) covers 39 years, and the 24-month period on the right side shows the official MCSI prospective ICS index (gray), the model prediction (red), and the 80% and 90% confidence bands around the ARIMAX model mean prediction. Although the ARIMAX model appears to somewhat underestimate the actual reported ICS, the model confidence interval does contain the prospective ICS value for all 24 months (2017–2018).

The FT allows us to analyze the training data by mapping it to the frequency k-space, where we can employ alternative kime-aggregators ($\mathscr{K}$) to estimate the missing kime-phases. Then, we will use the inverse Fourier transform (IFT) to synthesize new reconstructions of the training data back in spacetime. Based on alternative kime-aggregators, the IFT will generate different training data reconstructions that will be used to refit the ARIMAX model.

These spacekime-derived data analytics can be compared against the results of the traditional spacetime statistical inference we saw in **Figure 1.10**. Symbolically, we will be comparing ICS predictions based on the classical $ARIMAX(\{X_{train}, Y_{train}\})$ model and the results of various $ARIMAX(\{X_{train}^{kime}, Y_{train}^{kime}\})$ models, using alternative kime-aggregators. Symbolically, the spacekime-reconstructed data, $\{X_{train}^{kime}, Y_{train}^{kime}\}$, represents the following composite transformation of the native training data, $\{X_{train}, Y_{train}\}$:

$$\left\{X_{train}^{kime}, Y_{train}^{kime}\right\} = \underbrace{IFT}_{\substack{Fourier \\ synthesis}} \left(\underbrace{\mathcal{K}}_{\substack{spacekime \\ phase \\ aggregation}} \left(\underbrace{FT}_{\substack{Fourier \\ analysis}} \left(\{X_{train}, Y_{train}\} \right) \right) \right).$$

Figure 1.11 shows schematically the design of the spacetime and spacekime analytics for this ICS predictive forecasting demonstration. The *observed*, *computed*, and *derived* labels reference raw datasets, kime-transformed datasets, and analytically forecasted outcomes, respectively. While this example illustrates the use of a model-based inferential strategy (ARIMAX modeling), this framework allows any supervised or unsupervised, model-based or model-free, parametric or non-parametric approach to obtain the desired inference in terms of specific predictions, forecasts, classification, regression, or clustering.

Data Forecasting	Domain/Region	
Analytics Manifold	Training/Learning (1978-2016)	Testing/Validation (2017-2018)
4D Spacetime	$\underbrace{\{X_{train}, Y_{train}\}}_{observed}$	$\underbrace{\{X_{test}, Y_{test},}_{observed} \underbrace{Y_{Predicted}\}}_{derived}$
5D Spacekime	$\underbrace{\{X_{train}^{kime}, Y_{train}^{kime}\}}_{computed}$	$\underbrace{\{X_{test}^{kime}, Y_{predicted}^{kime},}_{computed} \underbrace{Y_{test}\}}_{observed}$

ARIMAX Model-based/Model-free Approaches

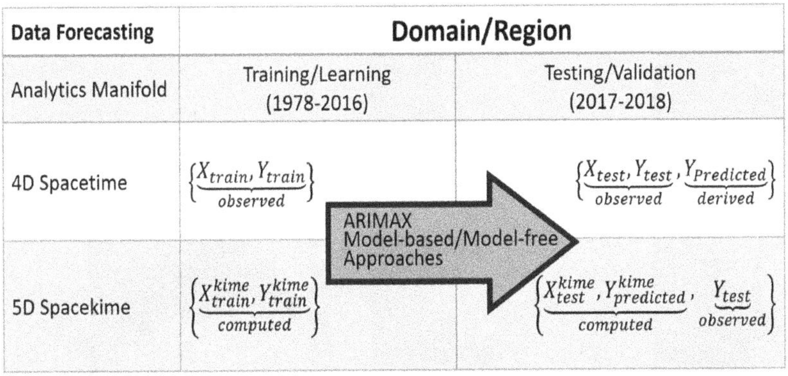

Figure 1.11: Schematic illustration of the observed data, computed information, and derived predictions for the consumer sentiment indicator study using traditional spacetime forecasting and spacekime analytics.

Figure 1.12 shows the distributions of the kime-phase angles for each of the seven covariates (6 predictors and one outcome, ICS). Note the following three properties of the distributions of different features: (1) all features have different phase distributions,

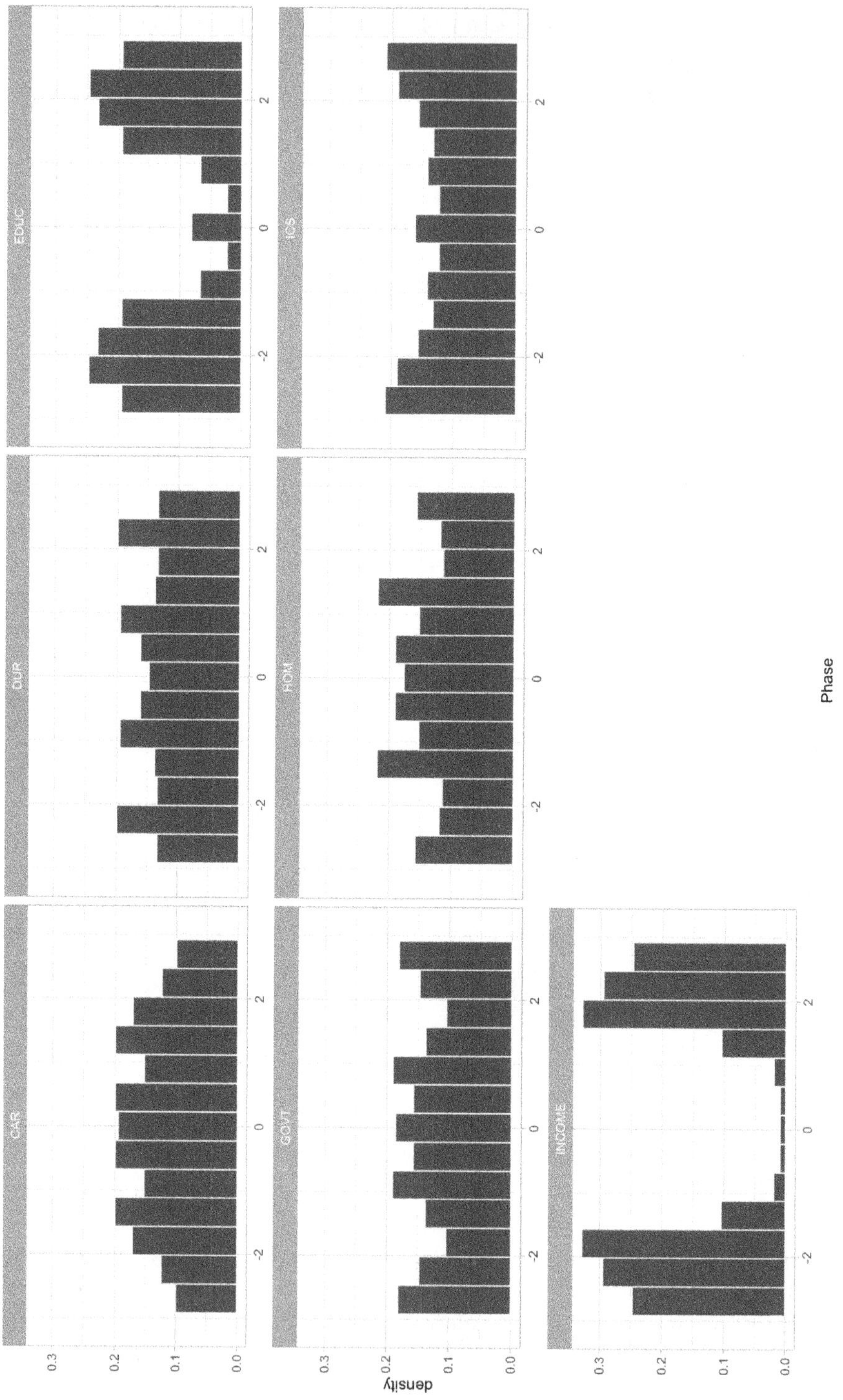

Figure 1.12: Kime-phase angle distributions of the MCSI features.

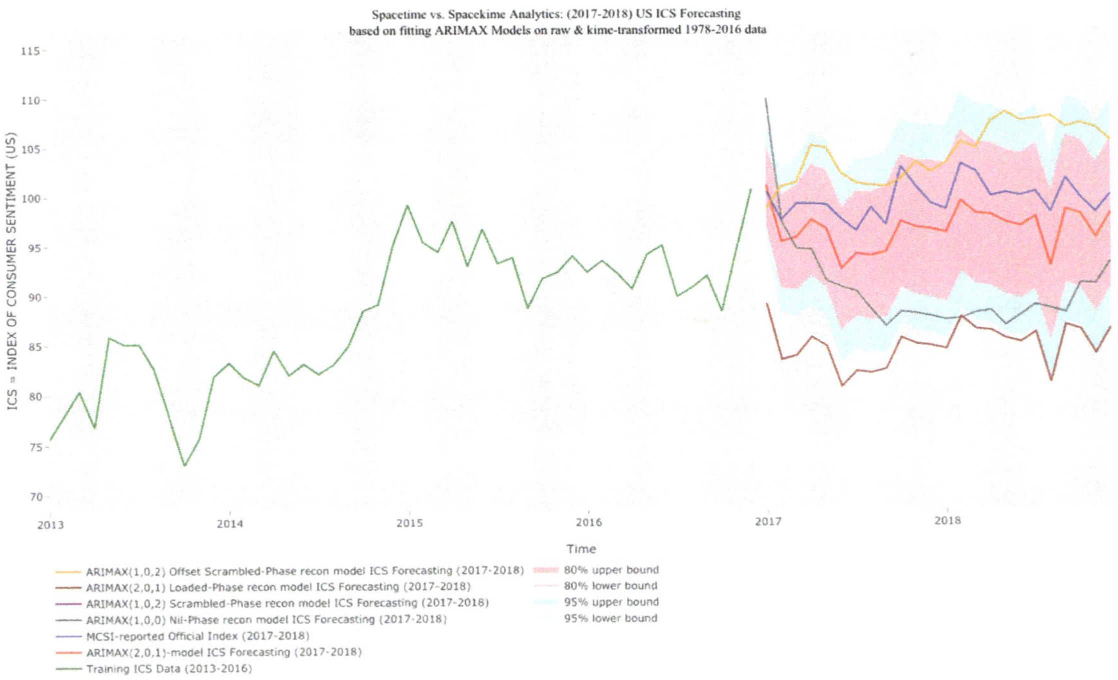

Figure 1.13: Results of the traditional (spacetime) statistical modeling and spacekime prospective longitudinal ICS forecasting using exogenous variables autoregressive integrated moving average (ARIMAX) models (see text for details). The results illustrate that statistical modeling using spacekime-transformed data provides reasonable forecasting that can be tuned by selecting appropriate phase-aggregator operators. The insert image shows the ICS time course over the entire 41-year timespan (1978–2018).

(2) as expected, all phases are in the range $[-\pi, +\pi)$, and (3) the phase distributions appear to be zero-mean and symmetric. This is not really surprising, as kime-phases are not observed and the FT is synthetically generating unbiased directional phase estimates.

For each covariate feature, we can investigate the results on the subsequent data analytics following the spacekime phase manipulations. Examples of such phase manipulations include (1) scrambling (for each feature, we randomly sample from the corresponding phase distribution), (2) aggregating the phases (e.g., replacing the phases by a constant value like zero or the phase sample mean), and (3) loading the phases (e.g., by introducing a random positive offset).

Figure 1.13 illustrates the effects of several alternative kime-phase estimation strategies on the reconstruction of the training data and, subsequently, their direct effect on the ultimate ICS forecasts. The figure shows the official (prospectively reported) MCSI index (red curve) along with four ICS forecasting models including the classical spacetime ARIMAX (2,0,1) statistical model (blue, with 80% and 90% confidence limits), and three spacekime manifold models based on different kime-phase estimators. The latter include: (1) ARIMAX(1,0,0) model estimated from spacetime covariate reconstructions using nil-phase estimates (green); (2) ARIMAX(1,0,2) model derived by using scrambled kime-phases, randomly sampled from the distributions of the corresponding 7 covariates (purple); (3) ARIMAX(2,0,1) model fit using loaded phases $\left(loading\ factor = \frac{\pi}{8}\right)$ offsetting the phase centrality (brown), which we earlier observed empirically, see the phase distributions in **Figure 1.12**; and (4) ARIMAX(1,0,2) offset scrambled kime-phase reconstructed data (orange).

More details about strategies to construct kime-aggregators are shown in later chapters. This example is included early on to provide a flavor of the spacekime analytical protocol and motivate the subsequent mathematical formulations, computational inference, and experimental applications that will be presented later.

The first motivational examples, including neuroimaging-genetics, 4D nucleome, and climate change studies illustrate the need for novel predictive Big Data analytical strategies to interrogate large, multiplex, and heterogeneous datasets. The last example of economic forecasting of consumer sentiment showed a concrete demonstration of interrogating Big Data using spacekime analytics. The latter approach relies on a commonly used mathematical trick of lifting the space of a problem to identify solution paths that are not clearly visible in lower dimensional spaces. Specifically, we are lifting the time dimension to the complex plane where the use of complex time allows us to derive spacekime-based forecasts for the original analytical problem.

Chapter 2
Mathematics and Physics Foundations

In this chapter, we introduce all fundamental concepts that will play important roles in the subsequent mathematical formulations extending Minkowski 4D spacetime to 5D spacekime. These represent classical mathematical abstractions, important analytical expressions, and well-known physics constructs like wavefunctions, function inner products, Dirac bra-ket notation, linear operators, commutators, etc. Their relevance to spacekime analytics and the uncertainty of scientific inference will become apparent in **Chapters 3** and **4** when we present their analogous data science counterparts.

2.1 Wavefunctions

Physicists have very cleverly defined the fundamentals of quantum mechanics using two complementary approaches, each with its advantages and disadvantages [78]. The first one is an analytic strategy (a.k.a. Schrödinger picture) that relies on differential equations. The other one is based on linear algebra on vector spaces (a.k.a. Heisenberg picture). The agreement of both of these formulations and their cross-sectional reinforcement of each other into a valid, coherent, compelling, and highly applicable theory is an incredible feat of human ingenuity.

The mathematical description of wave motion can be derived in multiple ways. For a particle of mass m, one approach is based on the relation between the particle's energy (E) and its momentum (p): $E = \frac{p^2}{2m}$. The Planck constant and the reduced Planck constant are defined by $h = 6.6 \times 10^{-34}$ $(m^2\ kg\ s^{-1})$ and $\hbar = \frac{h}{2\pi}$. A dual representation of the same wave motion can be expressed as a frequency-based dispersion relation $\omega = \frac{\hbar}{2m} k^2$, where $E = \hbar\omega$, $p = \hbar k$, and the wave number $k = \frac{2\pi}{\lambda}$ and the angular frequency $\omega = 2\pi\nu$ are related to the wavelength λ and the frequency ν, respectively. The dispersion relation expresses the dependence of the wave propagation velocity as a function of its wavelength or frequency where a wave packet consists of different wavelengths diverging (dispersing) over time.

If $\psi(\mathbf{x}, t)$ denotes the complex-valued amplitude and phase of a wave, the classical partial differential equation (PDE) approach for describing light, surface, acoustic, elastic, and spring waves is expressed by these wave equations (in 1D and 3D, respectively):

$$\left| \begin{array}{c} 1D: \frac{\partial^2 \psi(x,t)}{\partial t^2} = c^2 \frac{\partial^2}{\partial x^2} \psi(x,t), x \in \mathbb{R}, t \in \mathbb{R}^+ \\ 3D: \frac{\partial^2 \psi(\mathbf{x},t)}{\partial^2 t} = c^2 \nabla^2 \psi(\mathbf{x},t) = c^2 \left(\frac{\partial^2}{\partial x^2} \psi(\mathbf{x},t) + \frac{\partial^2}{\partial y^2} \psi(\mathbf{x},t) + \frac{\partial^2}{\partial z^2} \psi(\mathbf{x},t) \right), \mathbf{x} = (x,y,z) \in \mathbb{R}^3, t \in \mathbb{R}^+ \end{array} \right.$$

https://doi.org/10.1515/9783110697827-002

The linearity of the wave equations yields the superposition principle that guarantees that a linear combination of potential functions (wave PDE solutions) is also a solution. That is, if ψ_1 and ψ_2 are two wavefunctions satisfying the PDE, then any linear combination $\psi = \alpha\psi_1 + \beta\psi_2$ is another wave solution of the same PDE.

Note also that if the wavefunction is separable, $\psi(x,t) = f(t)g(x)$, then the derivation of solutions to linear PDEs is much easier. For instance, $\psi(x,t) = \psi_o e^{i(kx-\omega t)}$ is a separable function, which represents a potential function satisfying the wave equation, for all constants ψ_o, k, ω. This is because plugging in a separable solution into the wave equation yields $f''(t)g(x) = c^2 f(t)g''(x)$, i.e., $\frac{f''(t)}{f(t)} = c^2\frac{g''(x)}{g(x)}$. Since both sides of the equation must be constants (each depends only on one variable, but not the other), $g(x) = e^{\pm\frac{1}{c}\alpha x}$ and $f(t) = e^{\pm\alpha t}$, for any constant $\alpha \in \mathbb{C}$. As all solutions must be bounded for each $x, t, \alpha = i\omega$, for $\omega \in \mathbb{R}$. Thus, defining $k = \frac{\omega}{c}$ yields $g(x) \cong e^{\pm ikx}$ and $f(t) \cong e^{\pm i\omega t}$, and the superposition of constant multiples of these functions, $\psi(x,t) = \psi_o e^{\pm ikx} e^{\mp i\omega t}$, are (traveling wave) solutions to the wave equation.

A *wavefunction* is a mathematical representation describing the quantum state of an isolated quantum system. The wavefunction is a complex-valued probability amplitude, which allows the derivation of the probabilities of all possible outcomes or measurements made on the system.

The simple case of harmonic traveling waves that propagate linearly may be represented as $\psi(x,t) = \psi_o e^{i(kx-\omega t)}$, where ψ is the wave position, ψ_o is the constant wave amplitude, x is the spatial location, λ is the wavelength of the wave, ω is the frequency of the wave, t is time, $k = \frac{2\pi}{\lambda}$, and $\omega = 2\pi\nu$ is the angular frequency (in Hertz). The harmonic periodicity of such waves is the result of the oscillatory properties of the trigonometric functions. In other words, $\psi(x,t)$ is a periodic function in both, space and time; periodicity in time is $\Delta t = \frac{2\pi}{\omega}$, as $\omega\Delta t = 2\pi$, and periodicity in space is driven by the wavelength, $\Delta x = \frac{2\pi}{k}$ (**Figure 2.1**).

In quantum mechanics, the particle's position (x) at time order (t) may again be represented via square integrable, twice differentiable, and complex-valued position-space wavefunctions $\psi(x,t) \in L^2$, which rapidly decay to zero as x tends to infinity. Let us look at an example of a particle with energy E, which is spatially moving along the x-axis subject to being constrained in the region $0 \le x \le u$. We can estimate the normalization constant of a wavefunction that describes the spatial location of the particle at a given time:

$$\psi(x,t) = \begin{cases} Ae^{-\frac{iEt}{\hbar}} \sin\frac{\pi x}{u}, & 0 \le x \le u \\ 0, & x \in \mathbb{R}\backslash[0,u] \end{cases}.$$

To ensure the probabilistic (measure-theoretic) interpretation of the wavefunction, we can set:

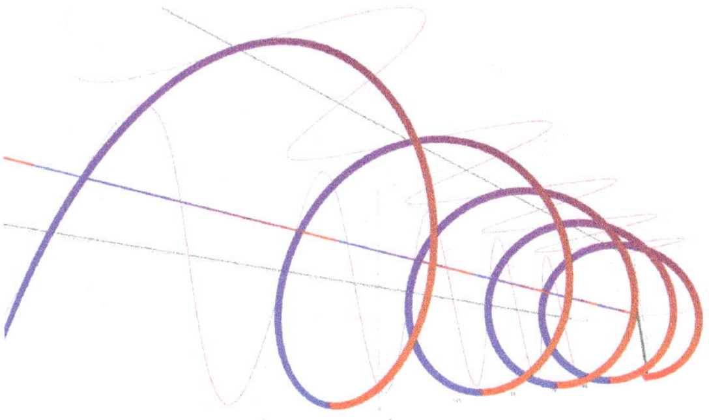

Figure 2.1: Schematic illustration of the propagation of a 1D wave as a complex amplitude, in terms of its real and imaginary parts. The wavefunction magnitude represents the probability of finding the particle at a given point *x* is spread out like a waveform. There is no unique definite position of the particle along space (*x*). As the amplitude oscillates (increases and decreases), the result is a wave with an alternating amplitude.

$$1 \equiv \int_{-\infty}^{\infty} \|\psi(x,t)\|^2 dx = \int_0^u \overline{\psi(x,t)}\psi(x,t)dx = \int_0^u \underbrace{Ae^{+\frac{iEt}{\hbar}}\sin\frac{\pi x}{u}}_{\overline{\psi(x,t)}} Ae^{-\frac{iEt}{\hbar}}\sin\frac{\pi x}{u}dx =$$

$$A^2\int_0^U \sin\frac{\pi x}{u}\sin\frac{\pi x}{u}dx = A^2\int_0^U \sin^2\frac{\pi x}{u}dx = A^2\frac{u}{2}.$$

and determine the wavefunction normalization constant, $A = \sqrt{\frac{2}{u}}$.

This intuitive probabilistic interpretation of the wavefunction for a particle in 1D space is based on its square modulus, i.e., the probability amplitude, which is always a positive real number:

$$\|\psi(x,t)\|^2 dx = \overline{\psi(x,t)}\psi(x,t)dx = \rho(x,t)dx.$$

The amplitude represents the probability density at time t that the particle is at spatial position x, where the complex conjugate is denoted by $\bar{\square}$. In quantum mechanics, when measuring the particle's position at time t, the wavefunction does not provide one unique spatial location. Rather, it describes the particle's location as a probability distribution, $\rho(x,t)dx$. For instance, the likelihood that the particle position x, measured at time t, will be in the interval $a \le x \le b$ can be computed using the magnitude of the wavefunction, i.e., by integrating the density over the interval:

$$P_{\{a \le x \le b\}}(t) = \int_a^b \|\psi(x,t)\|^2 dx.$$

As at any point in time when we measure the particle, the probability that it will be somewhere in space must be 1, therefore, we have to have a wavefunction normalization condition:

$$\int_{-\infty}^{\infty} \|\psi(x,t)\|^2 dx = 1.$$

2.2 Dirac bra-ket Notation

Dirac **bra-ket** notation [78] expresses the inner product $\langle\varphi|\psi\rangle$ as a pair of components. The right component, <u>*ket*</u>, $|\psi\rangle$, is a vector, e.g., a column vector, and the left part <u>*bra*</u>, $\langle\varphi|$, is the Hermitian conjugate of the *ket*, e.g., a row vector. When the ket part $|\psi\rangle$ is an element of a vector space V, the bra part $\langle\varphi|$ is an element of its dual space V^* (see Riesz representation theorem [79]).

More details will be provided later, but for now, we will just point out that wavefunctions are defined in Hilbert spaces, where the (finite dimensional) notion of *vector dot product* is generalized to (infinite dimensional) *function inner product*, as will be explained below. We will use Dirac *bra-ket* notation to define a canonical representation of inner-product for either finite dimensional vectors or infinite dimensional function spaces. Let us see an example of the dot product as an inner product of finite dimensional vectors. The inner product of a pair of arbitrary vectors $(a, b \in \mathbb{R}^n)$, $\underbrace{\langle a|}_{\text{bra}} = (a_1^*, a_2^*, \ldots, a_n^*)$ and $\underbrace{|b\rangle}_{\text{ket}} = (b_1, b_2, \ldots, b_n)^T$ can be expressed as:

$$a^*.b = \langle a|b\rangle = \sum_{i=1}^{n} a_i^* b_i,$$

where $\square^T$ denotes the transpose and $\square^*$ denotes the complex conjugate of a vector. For instance, in $\mathbb{R}^3 \ni a, b$, if $a = (3, -i, \sqrt{5}-i)^T$ and $b = (2, -i, \sqrt{5}-i)^T$, then:

$$\underbrace{a^*.b}_{\text{dot product}} = \langle a|b\rangle = \sum_{i=1}^{3} a_i^* b_i = \underbrace{\left(3, +i, \sqrt{5}+i\right)}_{a_i^*} * \underbrace{\begin{pmatrix} 2 \\ -i \\ \sqrt{5}-i \end{pmatrix}}_{\text{b}} =$$

$$(3*2) + (i*(-i)) + (5 - i^2) = 6 + 1 + 5 + 1 = 13.$$

In quantum mechanics, the possible atomic states of a particle are represented by unitary state vectors in a complex separable Hilbert state space. For particle position or momentum state types, the Hilbert state space represents the space of all square-integrable functions ($f \in L^2$). Each observable position or momentum measurement represents a self-adjoint linear operator acting on the corresponding position or momentum Hilbert state space. The corresponding eigenstates of observables are eigenvectors of the respective linear operators associated with eigenvalues representing the observed measurement in that eigenstate.

For infinite-dimensional Hilbert function spaces, we use $\square^\dagger$ and $\widehat{\square}$ to represent the adjoint of a state (i.e., the complex conjugate of the corresponding function) and the adjoint of an operator (i.e., its conjugate transpose). The inner product of two functions f and g, over a given interval $[c, d]$, is defined by:

$$f^\dagger.g \equiv \langle f|g\rangle = \int_c^d f^*(x)g(x)dx.$$

Note the morphing of the summation operator in the finite dimensional space into integration in the infinite dimensional functional space. For instance, if $f = \sin(kx)$ and $g = \cos(mx)$, their inner product over $[0, 2\pi]$ is:

$$\langle f|g\rangle = \int_o^{2\pi} f^*(x)g(x)dx = \int_o^{2\pi} \sin^*(kx)\cos(mx)dx =$$

$$\frac{1}{2}\int_o^{2\pi} (\sin((k+m)x) + \sin((k-m)x))dx = 0, \forall k, m \in \mathbb{Z}.$$

The last equality confirms that the (trigonometric functions) *sine* and *cosine* are orthogonal. They actually represent the *base functions* of the Fourier representation.

The bra-ket notation is the same for both finite and infinite dimensional spaces. For a *finite-dimensional* vector space V with a fixed orthonormal basis, the bra-ket is just the inner product of a row vector with a column vector:

$$\langle\varphi|\psi\rangle \equiv \varphi_1^*\,\psi_1 + \varphi_2^*\,\psi_2 + \cdots + \varphi_n^*\,\psi_n = (\varphi_1^*,\ \varphi_2^*,\ \cdots,\ \varphi_n^*)\begin{pmatrix}\psi_1\\ \psi_2\\ \cdots\\ \psi_n\end{pmatrix} = \langle\varphi|\psi\rangle\ ,$$

where concatenating bra $(\varphi_1^*,\ \varphi_2^*,\ \cdots,\ \varphi_n^*)$ and ket $|\,\psi\rangle = \begin{pmatrix}\psi_1\\ \psi_2\\ \cdots\\ \psi_n\end{pmatrix}$, i.e., bra next to a ket notation, signifies matrix multiplication. We also define the Hermitian conjugate (complex-conjugate transpose) connecting the bra to the ket by: $\langle\varphi\,|^\dagger = |\,\varphi\rangle$ and $|\,\psi\,\rangle^\dagger = \langle\psi|$. This is because $\langle\varphi\,|^\dagger \equiv \left(\overline{\langle\varphi|}\right)^t = \overline{(\langle\varphi\,|^t)} = |\,\varphi\rangle$, which can be explicated in the simplest situation of any matrices $A_{m\times n}$ and $B_{n\times m}$, and any complex vectors $x \in \mathbb{C}^n$ and $y \in \mathbb{C}^m$: Hermitian conjugation means $\langle Ax|y\rangle = \langle x\,|\,A^\dagger y\rangle$, and similarly $\langle x|By\rangle = \langle B^\dagger x|y\rangle$, i.e.,

$$\langle A| = |A^\dagger\rangle \text{ and } |B\rangle = \langle B^\dagger|.$$

For *infinite dimensional* spaces, the bra term represents linear functionals on the space of kets – the quantum state vector. A *Hilbert space* is a generic vector space where the distance measure (norm) is derived from an inner product. Many infinite dimensional function spaces are Hilbert spaces, e.g., flat manifolds (Euclidean spaces), and the space of square-integrable functions.

If H is a Hilbert space of quantum states and its dual space is $H^* = \{f : H \xrightarrow{\text{continuous}} \mathbb{C}\}$, then $\forall\, x, y \in H$, the function $\varphi_x(y) \equiv \langle x|y\rangle$, $\varphi_x(\cdot) \in H^*$, where the operator $\langle \cdot|\cdot\rangle$ represents the inner product in H. By the Riesz representation theorem, all $f \in H^*$ can be uniquely represented as inner products for some specific bra $\langle x_f|$ terms.

The bra-ket notation represents linear transformations (*bra*) acting on quantum-state-vector inputs (*ket*) and outputting complex numbers. If the linear functional $\langle B|$ is the bra Riesz representation of the ket $|B\rangle$, then $\langle A|B\rangle = \langle B|(|A\rangle)$. Thus, the bra-ket yields the same complex number output as the inner product. The Riesz representation theorem states that the vector space of the bras is the dual space (H^*) to the vector space of the kets (H), establishing explicit correspondence between the bra and ket terms.

Position-space **wavefunctions** describe particle states by $\Psi : \mathbb{R}^3 \times \mathbb{R} \to \mathbb{C}$, where $\boldsymbol{x} \in \mathbb{R}^3$ and $t \in \mathbb{R}$ represent the particle position $\boldsymbol{x}$ at time t. The wavefunction may be interpreted as a probability amplitude and its magnitude, i.e., square modulus, is:

$$||\Psi(\boldsymbol{x}, t)||^2 = \Psi^\dagger(\boldsymbol{x}, t)\Psi(\boldsymbol{x}, t) = \langle \Psi(\boldsymbol{x}, t)|\Psi(\boldsymbol{x}, t)\rangle = \rho(\boldsymbol{x}, t) \geq 0.$$

In other words, the magnitude is the probability density of the particle being at position $\boldsymbol{x}$. *The wavefunction does not precisely determine the particle's position, rather it describes its spatial probability distribution.* For instance, the probability that the particle is in a region $\boldsymbol{\Omega} \subseteq \mathbb{R}^3$ at time t, is computed by integrating the density over the region:

$$P_{\boldsymbol{x}\in\boldsymbol{\Omega}}(t) = \int_{\boldsymbol{\Omega}} ||\Psi(\boldsymbol{x}, t)||^2 d\boldsymbol{x} = \int_{\boldsymbol{\Omega}} \langle \Psi|\Psi\rangle d\boldsymbol{x} \equiv \int_{\boldsymbol{\Omega}} \Psi^\dagger \Psi d\boldsymbol{x}.$$

Wavefunctions form an infinite-dimensional space H, as there is no finite set of base functions whose linear combinations would yield every possible wavefunction. The norm in this Hilbert space is derived from the inner product $\langle \cdot|\cdot\rangle$. In other words, for a pair of wavefunctions Ψ_1 and Ψ_2 the bra-ket operator at time t represents the inner product, which leads to the norm, or modulus, i.e., the probability amplitude:

$$\langle \Psi_1(\boldsymbol{x}, t)|\Psi_2(\boldsymbol{x}, t)\rangle = \int_{\mathbb{R}^3} \Psi_1^\dagger(\boldsymbol{x}, t)\Psi_2(\boldsymbol{x}, t) d\boldsymbol{x}.$$

The inner product of a wavefunction Ψ with itself is always a positive real number, $\Psi^{\dagger}(\boldsymbol{x},t)\ \Psi(\boldsymbol{x},t) = \langle \Psi(\boldsymbol{x},t)\ |\ \Psi(\boldsymbol{x},t)\rangle = \rho(\boldsymbol{x},t) \geq 0$. However, in general, the bra-ket of a pair of wavefunctions is a complex number, $\langle \Psi_1(\boldsymbol{x},t)|\Psi_2(\boldsymbol{x},t)\rangle \in \mathbb{C}$.

When a wavefunction Ψ is represented as a linear combination of a finite number of orthonormal (base) wavefunctions $\{\Psi_i\}_{i=1}^n$, then $\Psi(\boldsymbol{x},t) = \sum_{i=1}^{n} a_i \Psi_i(\boldsymbol{x},t)$. In this expression, the linear term coefficients (weights) are defined by:

$$a_i = \underbrace{\frac{1}{\langle \Psi_i(\cdot,t)|\Psi_i(\cdot,t)\rangle}}_{\text{normalization constant}} \times \underbrace{\langle \Psi_i(\cdot,t)|\Psi(\cdot,t)\rangle}_{\substack{\text{inner product} \\ (\Psi \text{ projection into the base function } \Psi_i)}}$$

At a fixed moment in time, t_o, the values of the wavefunction $\Psi(\boldsymbol{x},t_o)$ represent uncountably many components of a vector in the infinite dimensional Hilbert state-space, which can be expressed in bra-ket notation as:

$$|\Psi(t)\rangle = \int_{\mathbb{R}^3} \Psi(\boldsymbol{x},t)|\boldsymbol{x}\rangle d\boldsymbol{x}.$$

This vector of components represents the particle *quantum-state* vector. Notice that to simplify the notation, we can drop the time parameter t. Then, the ket $|\boldsymbol{x}\rangle$ represents an orthonormal basis of H :

$$\langle \boldsymbol{x}'|\ \boldsymbol{x}\rangle = \delta(\boldsymbol{x}'-\boldsymbol{x}) = \begin{cases} 0, & \boldsymbol{x}' \neq \boldsymbol{x} \\ 1, & \boldsymbol{x}' = \boldsymbol{x} \end{cases},$$

$$\langle \boldsymbol{x}'|\Psi\rangle = \int_{\mathbb{R}^3} \Psi(\boldsymbol{x})\langle \boldsymbol{x}'|\boldsymbol{x}\rangle d\boldsymbol{x} = \Psi(\boldsymbol{x}'),$$

$$|\Psi\rangle = \int_{\mathbb{R}^3} |\boldsymbol{x}\rangle\langle \boldsymbol{x}|\Psi\rangle d\boldsymbol{x} = \underbrace{\left(\int_{\mathbb{R}^3} |\boldsymbol{x}\rangle\langle \boldsymbol{x}| d\boldsymbol{x}\right)}_{\text{identity operator}} |\Psi\rangle.$$

The identity operator $I = \int_{\mathbb{R}^3} |\ \boldsymbol{x}\rangle\langle \boldsymbol{x}|\ d\boldsymbol{x}$ in the space H expresses the abstract state explicitly as the inner product between two quantum-state vectors.

This interpretation of the wavefunction description of the particle state starts with any ket $|\Psi\rangle$ in the Hilbert space of square integrable functions and the definition of a complex-valued scalar function of the state (e.g., position), $\boldsymbol{x}$, $\Psi(\boldsymbol{x}) \equiv \langle \boldsymbol{x}|\Psi\rangle : \mathbb{R}^3 \to \mathbb{C}$, where the ket $|\Psi\rangle$ represents a superposition of kets, $|\boldsymbol{x}\rangle$, with relative coefficients specified by $\langle \boldsymbol{x}|\Psi\rangle$.

The representation of wavefunctions as quantum state vectors in the Hilbert space allows capturing the particle states as $|\Psi\rangle$, i.e., as quantum superpositions or vector sums of the constituent states. For instance, an electron superposition of states $|a\rangle$, and $|b\rangle$ may be expressed as the quantum state $|a\rangle + i|b\rangle$.

Wave equation system measurements and dynamics are directly associated with linear operators (also known as *observables*) on the Hilbert space of quantum states. For example, in the Schrödinger picture representation, there is a linear *time evolution* operator U specifying the future state of an electron that is currently in state $|\psi\rangle$, as $U|\psi\rangle$, for each possible current state $|\psi\rangle$. The time evolution of a closed quantum system is unitary and reversible. This implies that the state of the system at a later point in time, t, is given by $|\psi(t)\rangle = U(t)|\psi(t_o)\rangle$, where $U(t)$ is a unitary operator, i.e., its adjoint $U^\dagger \equiv (U^*)^T$ operator is the inverse: $U^\dagger = U^{-1}$. The integral equation $|\psi(t)\rangle = U(t)|\psi(t_o)\rangle$ relates the state of the particle at the initial time t_o with its state at time t. Locally, we can express the position of a inertia particle at time t is $x(t) = x(t_o) + v \times (t - t_o)$, where v is the constant speed and $x(t_o)$ is the initial position, i.e., $\frac{dx}{dt} = v$. The (time-dependent) Schrödinger equation, $i\hbar \frac{\partial}{\partial t}\psi(x,t) = H\psi(x,t)$, represents a generalization of this (ordinary) differential equation, where the particle system Hamiltonian is H and the PDE solution is the particle wavefunction $\psi(x,t)$, which describes the particle state (e.g., position) at time $t \geq t_o$, given its initial position $\psi(x,t_o)$.

The wavefunction definition requires that it vanishes at infinity sufficiently fast to ensure square integrability. It may naturally be extended on the *spacekime* manifold, $\Psi: \mathbb{R}^3 \times \mathbb{R}^2 \to \mathbb{C}$. It still expresses the system states, i.e., particle position $\boldsymbol{x}$ at kime $k = (t, \phi)$. Again, the spacekime wavefunction is interpreted as a probability amplitude and its square modulus, $||\Psi(\boldsymbol{x},k)||^2 = \Psi^\dagger(\boldsymbol{x},k)\Psi(\boldsymbol{x},k) = \langle \Psi^\dagger(\boldsymbol{x},k)|\Psi(\boldsymbol{x},k)\rangle = \rho(\boldsymbol{x},k) \geq 0$, representing the probability density that the particle is at position x. The spacekime wavefunction describes the spatial probability distribution of the particle's position and specific kime order and phase. As an example, the probability that the particle is at position $\boldsymbol{x} \in \boldsymbol{\Omega} \subseteq \mathbb{R}^3$ at kime k, is computed by the integral of the density over the spatial region $\boldsymbol{\Omega}$:

$$P_{\boldsymbol{x}\in\boldsymbol{\Omega}}(k) = \int_{\boldsymbol{\Omega}} \langle \Psi|\Psi\rangle d\boldsymbol{x} = \int_{\boldsymbol{\Omega}} \Psi^\dagger \Psi d\boldsymbol{x} = \int_{\boldsymbol{\Omega}} |\Psi(\boldsymbol{x},k)|^2 d\boldsymbol{x}.$$

2.3 Operators

Next, we will clarify the synergies between classical and quantum physics. There is a one-to-one correspondence between physical *observables* and *linear operators* in the dual of the Hilbert space of functions. This will be explained in more detail below. As a result, the measured value of each physical observable can be obtained from the corresponding wavefunction by taking the expectation value of the operator associated to the specific observable and acting on the wavefunction.

Let us start with a simple illustration of the observable-to-operator correspondence, using a simple wave equation that can be expressed as

$$\psi = Ae^{-i(Et - px)/\hbar},$$

where the observables E and p are the energy and momentum of the particle. If we differentiate the wavefunction ψ with respect to space (x) and time (t), respectively, we will obtain

$$(observables)\left|\begin{array}{l} \frac{\partial\psi}{\partial x} = \frac{i}{\hbar}p\psi \\ \frac{\partial\psi}{\partial t} = -\frac{i}{\hbar}E\psi \end{array}\right. \Rightarrow \left|\begin{array}{l} \frac{\hbar}{i}\frac{\partial\psi}{\partial x} = p\psi \\ i\hbar\frac{\partial\psi}{\partial t} = E\psi \end{array}\right. \Rightarrow (operators)\left|\begin{array}{l} \hat{p} = \frac{\hbar}{i}\frac{\partial}{\partial x} \\ \hat{E} = i\hbar\frac{\partial}{\partial t} \end{array}\right. .$$

Observables are the results of specific measurements. Their connection (one-to-one mapping) with their operator counterparts is expressed in linear operator form. Specifically, the observable quantities are the eigenvalues, and the wavefunctions are the eigenfunctions, of the corresponding self-adjoint operators. A measurable observed quantity is *not* equal to the operator itself; rather, it is *equivalent* to it in terms of the measurable $\leftrightarrow$ operator correspondence. The act of measuring an observable for some state is characterized mathematically as the action of the corresponding operator on the state vector. The recorded value associated with this operator action is one of the eigenvalues of the operator, i.e., the actual measured experimental value is an eigenvalue of the operator.

Occasionally, this equivalence relation may be slightly abused by short-hand notations equating an operator $\hat{\square}$ to an observable quantity $\square$, e.g., $\hat{p} = p$ and $\hat{E} = E$. However, such notations are most of the time clear from the context as an observed number (measurement) does not actually equal, but it is equivalent, to its corresponding operator. These operators can be thought of as functors mapping objects from one space of physical states, e.g., wavefunction states, onto another.

In **Chapter 5**, the concepts of observables, states, and wavefunctions will be translated to their corresponding data science counterparts – features, data, and inference functions, respectively. However, it may be useful to work out a very simple, yet, illustrative data analytic example previewing this translation of the concepts of states, inner products, linear operators, and wavefunctions, into functional inference spaces.

Suppose the observed data is $O = \{X_1, Y\}$:

$$O = \begin{cases} X_1 = \{-2, -1, 0, 1, 2\} \\ Y = \{-3, -2, 1, 2, 4\} \end{cases}.$$

In general, we are looking for an analytical description of an inferential state representing the observed data, i.e., the evidence (O), e.g., a simple linear model $Y = Y(X_1) = \beta_o + \beta_1 X_1 + \varepsilon$. In the simplest case, we may have closed-form analytical solutions (via *least squares*) to the problem of predicting the outcome (Y) based on a specific analytical strategy using the independent feature X_1. In general, there may be several covariate features, $(X_1, X_2, \cdots, X_k)$, that jointly explain the outcome Y.

Specifically, the inference function, $\psi = \psi\left(X, Y | \underbrace{\text{Linear Model}}_{\text{analytical strategy}}\right)$, quantifies the effects of all independent features (X) on the outcome (Y) via the ordinary least squares (OLS) estimate:

$$\hat{\beta} = \hat{\beta}^{OLS} = \langle X|X\rangle^{-1}\langle X|Y\rangle \equiv \left(X^T X\right)^{-1} X^T Y.$$

In our simple example, we can translate the modeling problem $(Y = \beta_o + \beta_1 X_1 + \varepsilon)$ into a set of linear equations reflecting the observations, O:

$$\left| \begin{array}{l} \beta_o - 2\beta_1 + \varepsilon_1 = -3 \\ \beta_o - \beta_1 + \varepsilon_2 = -2 \\ \beta_o + 0 \times \beta_1 + \varepsilon_3 = 1 \Rightarrow \underbrace{X\beta + \varepsilon = Y}_{\text{linear model}}, \text{ where } \quad \underbrace{\beta = \begin{pmatrix} \beta_o \\ \beta_1 \end{pmatrix}}_{\text{model parameter vector}}, \\ \beta_o + \beta_1 + \varepsilon_4 = 2 \\ \beta_o + 2\beta_1 + \varepsilon_5 = 4 \end{array} \right.$$

$$\underbrace{X = (\mathbf{1}, X_1) = \begin{pmatrix} 1 & -2 \\ 1 & -1 \\ 1 & 0 \\ 1 & 1 \\ 1 & 2 \end{pmatrix}, \quad Y = \begin{pmatrix} -3 \\ -2 \\ 1 \\ 2 \\ 4 \end{pmatrix}}_{\text{evidence}}, \quad \text{and } \varepsilon = \underbrace{\begin{pmatrix} \varepsilon_1 \\ \varepsilon_2 \\ \varepsilon_3 \\ \varepsilon_4 \\ \varepsilon_5 \end{pmatrix}}_{\text{error}}.$$

From the given observed data, the OLS solution of $X\beta + \varepsilon = y$ for β is

$$\left| \begin{array}{l} \hat{\beta} = \hat{\beta}^{OLS} = \left(X^T X\right)^{-1} X^T Y \\ X^T X \hat{\beta} = X^T Y \end{array} \right. ,$$

$$\underbrace{\begin{pmatrix} 1 & 1 & 1 & 1 & 1 \\ -2 & -1 & 0 & 1 & 2 \end{pmatrix}}_{X^T} \underbrace{\begin{pmatrix} 1 & -2 \\ 1 & -1 \\ 1 & 0 \\ 1 & 1 \\ 1 & 2 \end{pmatrix}}_{X} \hat{\beta} = \underbrace{\left(\begin{pmatrix} 1 & 1 & 1 & 1 & 1 \\ -2 & -1 & 0 & 1 & 2 \end{pmatrix}\right)}_{X^T} \underbrace{\begin{pmatrix} -3 \\ -2 \\ 1 \\ 2 \\ 4 \end{pmatrix}}_{Y},$$

$$\begin{pmatrix} 5 & 0 \\ 0 & 10 \end{pmatrix} \begin{pmatrix} \beta_o \\ \beta_1 \end{pmatrix} = \begin{pmatrix} 2 \\ 18 \end{pmatrix} \Rightarrow \hat{\beta} = \begin{pmatrix} \beta_o \\ \beta_1 \end{pmatrix} = \begin{pmatrix} 0.4 \\ 1.8 \end{pmatrix}.$$

Hence, (part of) the inference is captured by the derived linear relation $Y = 0.4 + 1.8 X_1 = (1, X_1)\hat{\beta} = X\hat{\beta}$. Therefore, this linear operator represents the model correspondence between the observable data $O = \{X_1, Y\}$ and the state space of the inference function.

Before we generalize the action of the observable-related operators in infinite dimensional spaces, let us consider their finite-dimensional counterparts – matrices, i.e., second-order tensors. Square matrices are special cases of linear transformations acting on vectors by "matrix-multiplication." Again, let us assume $|b\rangle = (b_1, b_2, \ldots, b_n)^T \in H$ and $A = A_{n\times n}$. Then, the action of the linear operator $A : H \to H$, i.e., $A \in H^*$ (the dual space), is defined by:

$$A|b\rangle = A_{n\times n} b_{n\times 1} = \begin{pmatrix} a_{1,1} & \cdots & a_{1,n} \\ \cdots & \cdots & \cdots \\ a_{n,1} & \cdots & a_{n,n} \end{pmatrix} \begin{pmatrix} b_1 \\ \cdots \\ b_n \end{pmatrix} = \left(b_1', b_2', \ldots, b_n'\right)^T,$$

where $b_i' = \sum_{j=1}^n a_{i,j} b_j,\ \forall 1 \le i \le n$.

If the Hilbert space is infinite dimensional, all linear transformations are part of the dual space and act on Hilbert space *functions*. Thus, each operator acts on, or maps, functions (inputs) assigning new functions (outputs). An example of an infinite dimensional operator is the *derivative operator*, $\frac{d}{dx}$, which is a linear operator on differentiable functions:

$$\frac{d}{dx}\left(\underbrace{f}_{\text{input}}\right) = \frac{df}{dx} = \underbrace{f'(x)}_{\text{output}}.$$

Recall that multiplication by a constant (e.g., the weights $\langle\psi_i|\phi\rangle$) is commutative. Thus, if $\{\psi_i\}_i$ is a complete orthonormal basis of H, i.e., $\langle\psi_i|\psi_j\rangle = \delta_{i,j}$, then for each $\phi \in H$:

$$|\phi\rangle = \sum_i \underbrace{\langle\psi_i|\phi\rangle}_{\substack{\text{projection of } \phi \\ \text{onto } \psi_i}} \underbrace{|\psi_i\rangle}_{\text{base}} = \sum_i |\psi_i\rangle\langle\psi_i|\phi\rangle = \underbrace{\left(\sum_i |\psi_i\rangle\langle\psi_i|\right)}_{\hat{1}} |\phi\rangle$$

The *expectation of the position*, in either finite or infinite dimensional state spaces, is defined by:

$$\bar{x} = \langle x\rangle = \textstyle\sum_{i=1}^l \tilde{n}_i x_i, \text{ (finite dimensional spaces)}$$
$$\langle x\rangle = \textstyle\int x\, \rho(x)\, dx \text{ (infinite dimensional spaces).}$$

In the discrete case, n_i represents the observed (raw) frequencies of outcome x_i in l repeated experiments measuring the system state, and $\tilde{n}_i = \frac{n_i}{N}$ are the relative frequencies, which, according to the law of large numbers, in the limit $(l \to \infty)$ become the (theoretical) probabilities of the chances of observing the outcome states x_i.

In the continuous case, the probability density function is $\rho(x)$. In both cases, the expectation represents the weighted average of all possible outcome states according to either observed frequencies or theoretical probability distribution weights. For instance, if $\psi(\mathbf{x}, t)$ represents the wave motion of a particle, or a solution to the Schrödinger's time-dependent equation, then the probability density representing the likelihoods of all spacetime states $\rho(\mathbf{x}, t) = \|\psi(\mathbf{x}, t)\|^2$. Furthermore, the *mean position* of the particle in the y-direction is defined by:

$$\langle y \rangle = \int \rho(\mathbf{x}, t) y dV = \int \|\psi(\mathbf{x}, t)\|^2 y dV = \int \psi^*(\mathbf{x}, t)\psi(\mathbf{x}, t) y dV, \text{ where } \int \rho(\mathbf{x}, t) dV = 1.$$

Similarly, the expectation of a linear operator $\hat{A}$ is defined in terms of the bra-ket notation:

$$\langle \hat{A} \rangle = \langle \psi | \hat{A} | \psi \rangle = \int_{-\infty}^{\infty} \psi^*(x)\ \hat{A}\psi(x) dx.$$

Analogously to matrix complex conjugate transposing, when $\hat{A}$ is self-adjoint (Hermitian), its expectation can also be expressed as:

$$\langle \hat{A} \rangle = \langle \psi | \hat{A} | \psi \rangle = \int_{-\infty}^{\infty} \psi^*(x)\hat{A}\ \psi(x) dx = \int_{-\infty}^{\infty} \left(\hat{A}^\dagger \psi\right)^*(x)\psi(x) dx = \langle \hat{A}^\dagger \psi | \psi \rangle = \langle \hat{A}^\dagger \rangle.$$

We noted earlier that all observables, like position, momentum, energy, etc., are in one-to-one correspondence with Hermitian (self-adjoint) operators, i.e., operators that are equal to their conjugate transpose operators. Since the expected values of Hermitian operators are always real, these Hermitian operators conveniently describe the expectation values of all observable quantities. The spectral theorem yields that the eigenvectors of Hermitian operators, which are called *eigenfunctions* in general Hilbert spaces, form a complete set [81].

Given a (finite or infinite dimensional) vector space, V, any linear operator $A: V \to V$ is also a linear operator on the algebraic dual space, $V^* = \left\{\varphi : V \xrightarrow{\text{linear}} \mathbb{C}\right\}$, i.e., $A: V^* \to V^*$. This is because:

$$\forall \underbrace{\langle c|}_{\text{ket vector}} \in V, A(|c\rangle) \equiv A|c\rangle \equiv |Ac\rangle \in V,$$

$$\forall \underbrace{\langle b|}_{\substack{\text{bra} \\ \text{linear operator}}} \in V^*, A(\langle b|) \equiv \langle b|A \in V^*,$$

$$\forall |c\rangle \in V,\ \underbrace{\langle b|A}_{\text{operator}}\ \underbrace{(|c\rangle)}_{\text{argument}} \equiv \langle b|A|c\rangle \equiv \langle b|Ac\rangle \equiv \langle A^\dagger b|c\rangle \in \mathbb{C}.$$

For a given pair of a ket $|c\rangle \in V$ and a bra $\langle b| \in V^*$, the linear operator $A \equiv |c\rangle\langle b|$ acts on vectors in V and on other operators in V^*. Clearly, $A: V \to V$, as $\forall |v\rangle \in V$,

$$A(|v\rangle) \equiv |c\rangle\langle b||v\rangle \equiv |c\rangle \underbrace{\langle b|v\rangle}_{\mathbb{C}} = \langle b|v\rangle|c\rangle \in V.$$

And similarly, $A : V^* \to V^*$, as $\forall\ \langle w| \in V^*$,

$$A(\langle w|) \equiv \langle w|A = \langle w||c\rangle\ \langle b| = \underbrace{\langle w|c\rangle}_{\mathbb{C}}\langle b| \in V^*.$$

This extends to all linear operators on the vectors space, as each linear operator may be expressed as a combination of scalar multiples of such base kets in V and bras in V^*. For a finite orthonormal basis of kets $\{|i\rangle\}_{i=1}^{n} \in V$, linear operators are square matrices $A = (A_{i,j}) \in V^*$:

$$A = \sum_{i,j} |i\rangle A_{i,j}\langle j|,\ A_{i,j} = \langle i|A|j\rangle,$$

$$\forall 1 \le i', j' \le n,\ \langle i'|A|j'\rangle = \sum_{i,j} \langle i'||i\rangle A_{i,j}\ \langle j||j'\rangle = \sum_{i,j} A_{i,j}\langle i'|i\rangle\langle j|j'\rangle$$

$$= \sum_{i,j} A_{i,j}\delta_{i',i}\delta_{j',j} = A_{i',j'}.$$

The unitary operator, $\hat{1}$, mapping each element in the Hilbert space to itself is called *resolution of the identity*:

$$\underbrace{\hat{1}}_{I} = \sum_{i} |i\rangle\langle i|,$$

where

$$\langle k|l\rangle = \langle k|I|l\rangle = \sum_{i,j} \langle k|i\rangle\delta_{i,j}\langle j|l\rangle = \sum_{i}\langle k|i\rangle\langle i|l\rangle = \sum_{i} \delta_{k,i}\delta_{i,l} = \delta_{k,l}.$$

For uncountably many eigenfunctions, consider an orthonormal basis of eigenfunctions, $\{\varphi\}$. The resolution of identity is naturally transformed from a sum into an integral:

$$\hat{1} = \int |\varphi\rangle\langle\varphi| dx.$$

In an infinite dimensional Hilbert space with a basis, $\{|\varphi\rangle\}$, applying the identity operator to any state ψ yields:

$$|\psi\rangle = \hat{1}|\psi\rangle = \int \underbrace{|\varphi\rangle}_{\substack{\text{base}\\ \text{functions}}} \quad \underbrace{\langle\varphi|\psi\rangle}_{\substack{\psi\ \text{projections}\\ \text{onto the bases}}} \quad dx.$$

For instance, we can express the eigenfunctions of the position operator in the momentum basis and vice versa, the momentum operator in terms of the position basis.

This dichotomy reflects state representation in terms of either sharp positions (space-time) or sharp momenta (Fourier frequency). The position representation of $|\psi\rangle$ is $\psi(x) = \langle x|\psi\rangle$, in particular, $\varphi(x) = \langle x|\varphi\rangle$, and $\psi^*(x) = \langle x|\psi\rangle^* = \langle x|\psi\rangle^\dagger = \langle\psi|x\rangle$. Thus, $|x'\rangle$ has the following position representation $\langle x|x'\rangle = \delta(x - x')$, sharp in space representation of each precisely localized state, i.e., point-mass distribution of the spatial measurement of a quantum object.

In the other extreme, the momentum representation $|k\rangle$, characterized by sharp localization of the momentum ($p = k\hbar$) is expressed by $\langle x|k\rangle = \frac{1}{\sqrt{2\pi}} e^{ikx}$, and its adjoint yields the momentum representation of a state defined by a sharp position $\langle k|x\rangle = \langle x|k\rangle^\dagger = \frac{1}{\sqrt{2\pi}} \left(e^{ikx}\right)^* = \frac{1}{\sqrt{2\pi}} e^{-ikx}$.

Before we explore various uncertainty relations, e.g., time-energy, momentum-position, we need to review different interpretations of time in quantum theory [82]. First, (*external*) time may simply be considered as a free parameter in the Schrödinger equation that can be measured by some external experimental clock. Second, (*intrinsic*) time can be considered as a dynamic duration lapse defined by the behavior of the quantum particles themselves. Third, (*observable*) time may reflect a standalone measurable characteristic of event ordering. Depending on the specific meaning of time, Hamiltonian mechanics may or may not be extended to include a *time operator*, which would lead to interpretation of a time-energy uncertainty relation.

If time is assumed to be an observed measurable quantity, we can define time to be the eigenvalue of the *time operator*, $\hat{t}$. In this case, both the classical *position* x and *time* t remain unaffected when they are transformed to operators:

$$\left| \begin{array}{l} x \to \hat{x} = x \\ t \to \hat{t} = t \end{array} \right. . \tag{2.1}$$

This is because for each specific realization of a position (x) and a time (t), the corresponding operators (denoted by $\hat{}$) on the Hilbert space determined by the wavefunctions, act as:

$$\hat{x}(\psi(x,t)) = x\psi(x,t)$$

and

$$\hat{t}(\psi(x,t)) = t\psi(x,t).$$

In other words, the position operator, $\hat{x}$, and the time operator, $\hat{t}$, act on a wavefunction by multiplying it by x and by t, respectively.

However, classical *energy* and *momentum* are transformed in quantum terms to non-identical differential operators:

$$\left|\begin{array}{l} E\to\hat{E}\equiv\underbrace{\hat{H}}_{\text{Hamiltonian}}=i\hbar\frac{\partial}{\partial t} \\ p_x\to\hat{p}=-i\hbar\frac{\partial}{\partial x}\end{array}\right. . \tag{2.2}$$

The *derivation of the momentum operator* in a quantum setting is based on a translation operator $\hat{T}$, shifting the position $\psi(x,t)$ to the right by a small amount Δx :

$$\psi(x-\Delta x,t)=\hat{T}(x+\Delta x,x)\psi(x,t).$$

Around $\Delta x=\mathbf{0}$, $\hat{T}=\hat{1}$, the *identity operator*, and we can expand the position function ψ, using a first-order Taylor expansion in the neighborhood of x :

$$\psi(x-\Delta x,t)\cong\psi(x,t)-\underbrace{\hat{A}}_{\substack{\text{operator}\\(\textit{unknown})}}\psi(x,t)\Delta x\ \Rightarrow\ \psi(x,t)-\psi(x-\Delta x,t)\cong\hat{A}\psi(x,t)\Delta x.$$

Let us see why the unknown operator $(\hat{A})$ is a multiple of the momentum operator, $\hat{p}$, $\left(\hat{A}\equiv\frac{i}{\hbar}\hat{p}\right)$ where the factor $\frac{i}{\hbar}$ ensures that the $\hat{p}$ operator is Hermitian. Denoting $\hat{p}=-i\hbar\frac{\partial}{\partial x}$, i.e., $\left(\frac{i}{\hbar}\hat{p}\right)\psi=-\frac{i^2\hbar}{\hbar}\frac{\partial}{\partial x}\psi=\frac{\partial\psi}{\partial x}$, then:

$$\begin{array}{l}\left(\frac{i}{\hbar}\hat{p}\right)\psi(x,t):=\frac{\partial\psi}{\partial x}=\lim\limits_{\Delta x\to o}\frac{\psi(x,t)-\psi(x-\Delta x,t)}{\Delta x}\\ \left(\frac{i}{\hbar}\hat{p}\right)\psi(x,t)\Delta x\cong\psi(x,t)-\psi(x-\Delta x,t)\end{array}. \tag{2.3}$$

Thus, as indicated in (2.3), the momentum operator is $\hat{p}=-i\hbar\frac{\partial}{dx}$ and its action is to move the wavefunction, ψ, by an infinitesimal amount in x space, just like its classical counterpart does.

Similarly, we can explicitly identify the mapping of the three types of angular momenta observables into their corresponding quantum mechanical operators – *orbital* $\left(\hat{L}\right)$, *spin* $\left(\hat{S}\right)$ and *total* $\left(\hat{J}=\hat{L}+\hat{S}\right)$ angular momenta:

$$\left|\begin{array}{c}(\text{spatial quantization})\,L\equiv r\times \mathrm{p}\ \to\ \hat{L}=-i\hbar(r\times\nabla)\\ (\text{spin quantization})\,S\equiv\frac{n}{2}\ \to\ \hat{S}=-\hbar\sigma,\ \sigma\in\{-\mathrm{s},\ -(\mathrm{s}-1),\ \ldots,0,\ \ldots,+(\mathrm{s}-1),\ +\mathrm{s}\}.\\ (\text{total quantization})\,J\equiv L+S\ \to\ \hat{J}=\hat{L}+\hat{S}=\hbar\sqrt{|\sigma|(|\sigma|+1)}\end{array}\right.$$

Note that the angular momenta operators also have nice compact representations in spherical coordinates. For instance, the orbital angular momentum $\hat{L}=-i\hbar$ $\left(\hat{\phi}\frac{\partial}{\partial\theta}-\frac{\hat{\theta}}{\sin(\theta)}\frac{\partial}{\partial\phi}\right)$, where $0\le\phi<\pi$ and $0\le\theta<2\pi$ are the angular parameters of the spherical coordinate system, and $\hat{\phi}=(-\sin\phi,\ \cos\phi,0)$ and $\hat{\theta}=(\cos\theta\ \cos\phi,\cos\theta\ \sin\phi,$ $-\sin\theta)$. Also, for an arbitrary unit vector, $\vec{u}=(u_x,u_y,u_z)$, the spin angular momentum operator, which measures the spin along $\vec{u}$, relies on the Pauli spin matrices $\hat{S}_u=\frac{\hbar}{2}(u_x\sigma_x+u_y\sigma_y+u_z\sigma_z)$. The parity between measurable values of observables and their corresponding linear quantization operators suggests that the eigenvalues of $\hat{S}_u$

are $\pm\frac{\hbar}{2}$, which correspond to the usual spin matrices. Of course, computing the spin operator in an arbitrary direction generalizes to higher spin states by using the dot product of the spin direction and a 3D vector for each of the coordinate axis directions x, y, z.

The quantum mechanical operators in equation (2.2) act on wavefunctions $\psi(x,t)$, which represent complex-valued probability amplitudes describing the likelihoods for all possible results of measurements made on the system. A 1D classical physics spring-mass system represents a simplified demonstration of this dichotomy between measurable quantities and operators. Given a fixed spring constant k, the *total energy* of an object attached to a spring, with mass m, oscillating up-and-down is:

$$\underbrace{E}_{\text{total energy}} = \underbrace{\frac{1}{2}mv^2}_{\text{kinetic energy}} + \underbrace{\frac{1}{2}kx^2}_{\text{potential energy}} = \frac{p_x^2}{2m} + \frac{1}{2}kx^2. \quad (2.4)$$

In quantum mechanical terms, this equation is converted to the *time-dependent Schrödinger equation*, $i\hbar\frac{\partial}{\partial t}|\psi(x,t)\rangle = \hat{H}|\psi(x,t)\rangle$, describing a quantum mechanical oscillator. This equation represents a PDE operator modeling an oscillating spring:

$$\underbrace{i\hbar\frac{\partial}{\partial t}}_{\hat{E}}\psi(x,t) = \left[\frac{1}{2m}\left(\underbrace{-i\hbar\frac{\partial}{\partial x}}_{\hat{p}}\right)^2 + \frac{1}{2}k\underbrace{x^2}_{\hat{x}}\right]\psi(x,t) = -\frac{\hbar^2}{2m}\frac{\partial^2}{\partial x^2}\psi(x,t) + \frac{1}{2}kx^2\psi(x,t). \quad (2.5)$$

More generally, the Schrödinger equation whose solution (a wavefunction, $\psi(x,t)$) describes the motion of a single particle in 3D, e.g., modeling the vibrating motion of an electron around an atomic nucleus, may be expressed in the position basis as:

$$\overbrace{i\hbar\frac{\partial}{\partial t}\psi(x,t)}^{\text{Total Energy}} = \overbrace{-\frac{\hbar^2}{2m}\left(\underbrace{\frac{\partial^2}{\partial x^2} + \frac{\partial^2}{\partial y^2} + \frac{\partial^2}{\partial z^2}}_{\text{Laplacian, } \Delta = \nabla^2}\right)\psi(x,t)}^{\text{Kinetic Energy}} + \overbrace{\frac{1}{2}kx^2\psi(x,t)}^{\text{Potential Energy}}.$$

This vector operator equation permits a valid representation in any other complete basis of kets in the Hilbert space. Using the notations $\langle p|\psi\rangle = \Psi(p,t) = \int \psi(x,t)e^{-ipx}dx$, $v(x,t) = \frac{1}{2}kx^2$, and $V(p,t) = \int v(x,t)e^{-ipx}dx$, and the convolution $(V{*}\Psi)(p,t) = \int_{-\infty}^{\infty} V(y,t)\Psi(p-y,t)dy$, the Schrödinger equation can also be represented in the *momentum space* basis:

$$i\hbar\frac{\partial}{\partial t}\Psi(p,t) = \frac{p^2}{2m}\Psi(p,t) + (V{*}\Psi)(p,t).$$

This momentum representation of the equation can be derived from first principles by multiplying both hand sides of the position formulation of Schrödinger equation by e^{-ipx}, integrating over space, applying integration by parts twice, and interchanging the order of integration.

2.4 Commutator

In this section, for simplicity we are suppressing the operator hat $\widehat{\square}$ notation and using capital letters to denote operators. The *commutator* of two operators (A and B,), e.g., the PDE operators in (2.2), is defined by:

$$[A,B]=AB-BA. \tag{2.6}$$

Trivial commutators, $[A,B]=0$, correspond to commutative (Abelian) operators and tell us what pairs of physical observables are potentially simultaneously measurable with infinite precision, see examples below. Conversely, non-trivial commutators, $[A,B]\neq 0$, correspond to pairs of physical observables that can't be simultaneously measured with infinite precision. This latter type of commutator exemplifies Heisenberg's uncertainty principle.

The mathematical abstraction of (operator) commutator is directly related to the statistical concept of independence and the quantum theory notion of kinematic independence of observables [83]. These concepts reflect the properties of coexistence, compatibility, and inter-relations between two, or more, observables. Statistical and kinematic independence are logically distinct [84]. The former notion suggests that a pair of quantum systems are independent when each of them can be prepared in any state irrespective of how the other system is prepared. Kinematic independence of two observables (or two quantum systems) requires that as members of the corresponding C^*-algebras their elements commute [83]. Bell's inequalities provide an upper bound on the strength of expected correlations between systems that are not presently interacting, although they may have interacted in the past [85]. These relations quantify the degree of independence of observables, or the commutation of their corresponding operators [86].

The *dispersion* measures the uncertainty in the state of the particle. Let us assume we have two non-commuting operators A and B. The *expectation* of an operator, $\langle A\rangle$, is defined by:

$$\langle A\rangle \equiv \langle\psi|A|\psi\rangle = \int_{\mathbb{R}} \psi^*(x)A\psi(x)dx = \int_{\mathbb{R}} \langle\psi|A\psi\rangle\, dx. \tag{2.7}$$

The *uncertainty operators*, or deviances, $\Delta A = A-\langle A\rangle$ and $\Delta B = B-\langle B\rangle$, and the *mean square uncertainty values* $(\sigma_A)^2$ and $(\sigma_B)^2$ are generally defined for a given state, ψ, by:

$$\left|\begin{aligned} \Delta A &\equiv A - \langle A\rangle = A - \langle\psi|A|\psi\rangle \\ \Delta B &\equiv B - \langle B\rangle = B - \langle\psi|B|\psi\rangle \\ (\sigma_A)^2 &\equiv \langle\psi|(\Delta A)^2|\psi\rangle = \langle\psi|(A-\langle A\rangle)^2|\psi\rangle \\ (\sigma_B)^2 &\equiv \langle\psi|(\Delta B)^2|\psi\rangle = \langle\psi|(B-\langle B\rangle)^2|\psi\rangle \end{aligned}\right. . \tag{2.8}$$

As the expectation is always a number, not a random variable, commutators of operators with constants are trivial. Thus, the commutator of the operators A and B and their uncertainties (dispersions or deviances) ΔA and ΔB are identical. That is,

$$[\Delta A, \Delta B] = [A - \langle A\rangle, B - \langle B\rangle] = [A,B] - \underbrace{[\langle A\rangle, B] - [A, \langle B\rangle] + [\langle A\rangle, \langle B\rangle]}_{\text{trivial commutators}} = [A,B]. \tag{2.9}$$

Let us look at the inner product of $\Delta A\psi + i\lambda\Delta B\psi$ with itself:

$$\begin{gathered} 0 \le \|\Delta A\psi + i\lambda\Delta B\psi\|^2 = \langle\Delta A\psi - i\lambda\Delta B\psi|\Delta A\psi + i\lambda\Delta B\psi\rangle = \\ \langle\psi|(\Delta A)^2\psi\rangle + \lambda^2\langle\psi|(\Delta \mathrm{B})^2\psi\rangle + i\lambda\langle\Delta A\psi|\Delta B\psi\rangle - i\lambda\langle\Delta B\psi|\Delta A\psi\rangle = \\ \langle\psi|(\Delta A)^2\psi\rangle + \lambda^2\langle\psi|(\Delta \mathrm{B})^2\psi\rangle + i\lambda(\langle\Delta A\psi|\Delta B\psi\rangle - \langle\Delta B\psi|\Delta A\psi\rangle) = \\ (\sigma_A)^2 + \lambda^2(\sigma_B)^2 + i\lambda\langle\psi|[\Delta A, \Delta B]\ \psi\rangle. \end{gathered} \tag{2.10}$$

The minimal value of (2.10) with respect to λ can be obtained by setting the partial derivative to zero:

$$0 = \frac{\partial}{\partial\lambda}\left((\sigma_A)^2 + \lambda^2(\sigma_B)^2 + i\lambda\langle\psi\ |[\Delta A, \Delta B]\ \psi\rangle\right) = 2\lambda(\sigma_B)^2 + i\langle\psi\ |[\Delta A, \Delta B]\ \psi\rangle.$$

This minimum is attained at $\lambda = -\frac{i\langle\psi\ |[\Delta A, \Delta B]\ \psi\rangle}{2(\sigma_B)^2}$, and we can plug it into (2.10) to get:

$$\begin{gathered} 0 \le (\sigma_A)^2 - \frac{1}{4}\frac{(\langle\psi|[\Delta A, \Delta B]\psi\rangle)^2}{(\sigma_B)^4}(\sigma_B)^2 + \frac{\langle\psi\ |[\Delta A, \Delta B]\psi\rangle}{2(\sigma_B)^2}\langle\psi\ |[\Delta A, \Delta B]\psi\rangle = \\ (\sigma_A)^2 - \frac{1}{4}\frac{(\langle\psi|[\Delta A, \Delta B]\psi\rangle)^2}{(\sigma_B)^2} + \frac{(\langle\psi|[\Delta A, \Delta B]\psi\rangle)^2}{2(\sigma_B)^2}. \end{gathered}$$

Rearranging the terms, we obtain the uncertainty principle in terms of the commutator of the pair of uncertainties, $\Delta A, \Delta B$:

$$(\sigma_A)^2(\sigma_B)^2 \ge -\frac{1}{4}(\langle\psi|[A,B]\ \psi\rangle)^2 = -\frac{1}{4}\langle[A,B]\rangle^2 = \left(\frac{1}{2i}\langle[A,B]\rangle\right)^2. \tag{2.11}$$

In other words, $\sigma_A\sigma_B \equiv \sqrt{(\sigma_A)^2(\sigma_B)^2} \ge \frac{i}{2}\langle[A,B]\rangle$. The next section presents uncertainty in terms of the special case of the position and momentum operators, where

$$\sigma_A\sigma_B \equiv \sqrt{(\sigma_A)^2(\sigma_B)^2} = \sqrt{-\frac{1}{4}\langle[\hat{x},\hat{p}]\rangle^2} = \sqrt{-\frac{1}{4}\left[x, -i\hbar\frac{\partial}{\partial x}\right]^2} =$$
$$\sqrt{-\frac{1}{4}(i\hbar)^2} = \sqrt{\frac{1}{4}\hbar^2} = \frac{1}{2}\hbar > 0.$$

2.4.1 Example 1: Non-Trivial Commutator (Position/Momentum)

Let us revisit the specific example of a non-trivial commutator that corresponds to the quantum transformations of the position and momentum operators. We will show that $[A,B] = \left[x, -i\hbar\frac{\partial}{\partial x}\right]$, and thus, using equation (2.11), $\sigma_A\sigma_B \geq \frac{i}{2}\langle[A,B]\rangle$. Of course, when the commutator is constant, like in the position-momentum case, $\langle[A,B]\rangle = \langle[p,x]\rangle = [p,x]$:

$$\sigma_A\sigma_B \geq \frac{i}{2}\langle[A,B]\rangle = \frac{i}{2}[x,p] = \frac{i}{2}\left[\underbrace{x}_{A}, -\underbrace{i\hbar\frac{\partial}{\partial x}}_{B}\right] = \frac{\hbar}{2} > 0.$$

Using the chain rule for differentiation we can exactly compute the commutator $[A = position,\ B = momentum] = AB - BA$, and derive Heisenberg's principle, for position and momentum of a particle whose motion is described by the wavefunction $\psi(x,t)$:

$$[A,B]\psi(x,t) = \left[x, -i\hbar\frac{\partial}{\partial x}\right]\psi(x,t) =$$
$$-i\,x\,\hbar\frac{\partial}{\partial x}(\psi(x,t)) - \left(-i\hbar\frac{\partial}{\partial x}\right)(x\psi(x,t)) = \tag{2.12}$$
$$-i\hbar x\frac{\partial\psi(x,t)}{\partial x} - \left(-i\hbar\psi(x,t) - i\hbar x\frac{\partial\psi(x,t)}{\partial x}\right) = i\hbar\psi(x,t).$$

In other words, $\left[x, -i\hbar\frac{\partial}{\partial x}\right] = i\hbar$, where $\hbar$ is the reduced Planck constant, suggesting that position and momentum operators do not commute. Therefore, the uncertainty principle applies to position and momentum. Specifically, the uncertainty in position, $\Delta x = \sigma_x$, and in momentum, $\Delta p = \sigma_p$, satisfies this lower bound:

$$\underbrace{\Delta x}_{\sigma_x}\ \underbrace{\Delta p}_{\sigma_p} \geq \frac{\hbar}{2}. \tag{2.13}$$

The more precisely we measure one of these observables, the higher the uncertainty about the other will be.

2.4.2 Example 2: Trivial Commutator (Energy/Momentum)

On the flip side, by the commutative property of partial differentiation, the commutator between two other observables, *energy* and *momentum*, is trivial. Thus, in principle, we could simultaneously measure the energy and the momentum with infinite precision:

$$(AB - BA)\psi(x,t) \equiv [A,B]\psi(x,t) = \left[i\hbar\frac{\partial}{\partial t}, -i\hbar\frac{\partial}{\partial x}\right]\psi(x,t) =$$
$$i\hbar\frac{\partial}{\partial t}\left(-i\hbar\frac{\partial}{\partial x}\right)\psi(x,t) - \left(-i\hbar\frac{\partial}{\partial x}\right)\left(i\hbar\frac{\partial}{\partial t}\right)\psi(x,t) = \tag{2.14}$$
$$\hbar^2\frac{\partial}{\partial t}\frac{\partial}{\partial x}(\psi(x,t)) - \hbar^2\frac{\partial}{\partial x}\frac{\partial}{\partial t}(\psi(x,t)) = 0.$$

Later on, in **Chapter 5**, we will explore the "data science" inferential analogues to the quantum mechanics concepts of wavefunctions, transformations, operators, commutators, etc.

Chapter 3
Time Complexity

This chapter illustrates the fundamentals of the complex-time extension of time, explains how some of the problems of time may be resolved in the 5D spacekime manifold, and sets the stage for spacekime data-driven analytics. We begin this chapter by reviewing the basics of Fourier space-frequency function transformations, Minkowski spacetime, types of variance, and Kaluza-Klein theory. Then we define the concepts of complex time (kime), complex events (kevents), and the spacekime metric tensor.

We will also generalize the kime extensions of rate of change, velocity, equations of forward and backward motion, Lorentz transformations, the spacekime velocity addition law, and the Doppler effect. The kime-specific derivations of the Wirtinger derivative and kime-calculus are presented along with several kime parameterizations and the causal structure of spacekime. The chapter concludes with a formulation of a spacekime analogue of the Copenhagen interpretation and some spacekime Data Science applications.

3.1 Introduction

Classical definitions of time involve measurements of oscillatory, regular-frequency occurrences, or events of a periodic nature, e.g., hourly, daily, monthly, seasonal, or annual recurrences driven by stellar motions, geophysical processes, chemical reactions, or physical observations such as radioactive decay. There are interesting synergies between time and gravity. For instance, just like gravity, time is entangled with space, it's mostly detected, effective, and interpreted locally, and the *up* and *down* directions of gravity parallel the forward/future and back/past orientations in time (both are isometric to $\mathbb{R}^+$).

Even Richard Feynman was perplexed about various time-related quantum mechanics paradoxes, "*I cannot define the real problem, therefore I suspect there's no real problem, but I'm not sure there's no real problem.*" [87] To date, there is no physical explanation of a separate "time-direction," or kime-phase, although there is some indirect evidence of alternative and complementary spacetime foci, which may be altered for a fixed point in Minkowski spacetime, e.g., consciousness states, illusions, imaginations, creativity, and virtualization. Time-direction, a separate dimension in the 5D spacekime, can be experienced differently by separate observers. For instance, pairs of novel writers and readers, performing musicians and listeners, artists and dilettantes, idealists and materialists, dreamers and psychologists, and many others reflecting on the same 4D spacetime phenomena always internalize their experiences differently, subjected to their unique idiosyncratic kime-phase

https://doi.org/10.1515/9783110697827-003

interpretations. Groups of individuals immersed in a virtual environment, simulation, or augmented reality may also differentially experience shared stimuli, which by design can be deterministic or stochastic in the intrinsic nature of spacetime.

In a closed system, most observers jointly experiencing a common spacetime phenomenon may perceive the encounter through similar time-directions (i.e., kime-phases). However, in reality, while the kime-phases may be drawn from one common distribution, they are not identical; each individual in the group will discern the experience in a slightly different and unique spacekime. Paraphrasing the two-and-a-half-millennia old Heraclitus observation, which we saw in **Chapter 1**, "*No two people can ever perceive the exact same experiences, for they can't share the same spacekime location and they do not represent the same observer.*" Often, the slight variations of different kime-phase experiences become clear in post-hoc discussions of joint group experiences, reflections, of shared encounters. Results of common shared experiences reflect the perceptions of the stimuli. The information content interpreted by different people in most evidence-based decision making situations will vary, sometimes widely.

3.1.1 Wavefunctions and the Fourier Transformation

Figure 3.1 shows the extension of the representation of a 1D constant-wavelength varying-amplitude wave to the 2D space of constant-amplitude and complex phase. This provides a striking parallel to the extension of the longitudinal 1D event time order into 2D kime space representing the complex structure of events in terms of their *order* and *phase*. **Figure 3.1.B** shows the 2D projections of the 3D fixed-amplitude complex-phase representation of the wavefunction onto the 2D planes $y+3=0$ and $x+3=0$. These projections resemble exactly the oversimplified representation of the wavefunction as a varying amplitude over space (vertical z-axis), **Figure 3.1.A**. The 2D wavefunction representation gives an illusion of the oscillatory wave repeatedly losing its amplitude at regular points in space, which are associated with the wave period and its frequency.

The more realistic complex amplitude wave representation in 3D enables multiplication by a constant, which represents the time evolution of waves. For instance, multiplying the wave curve by the imaginary unit i pushes the amplitudes forward and suggests direct wave propagation in space, which characterizes the basic principle of time evolution of quantum theory. This multiplication by i has the effect of helical rotation of the wave amplitude by one quarter turn $\left(\frac{\pi}{2}\text{ radians}\right)$ around the space axis. Therefore, mathematical multiplication by i is equivalent to the physical motion of turning a quarter of the wavelength in the direction of the wave propagation. Each complex-number multiplication affects only the amplitude at one fixed point. Cumulatively aggregating these effects across 3D space results in the entire wave advancing forward along the depth axis.

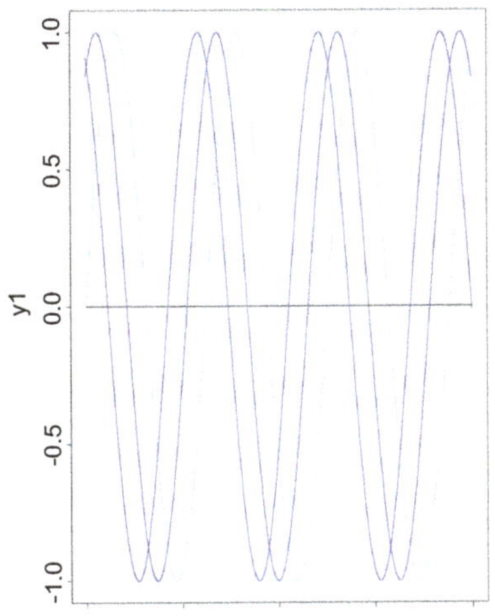

A. 2D (oscillatory amplitude)

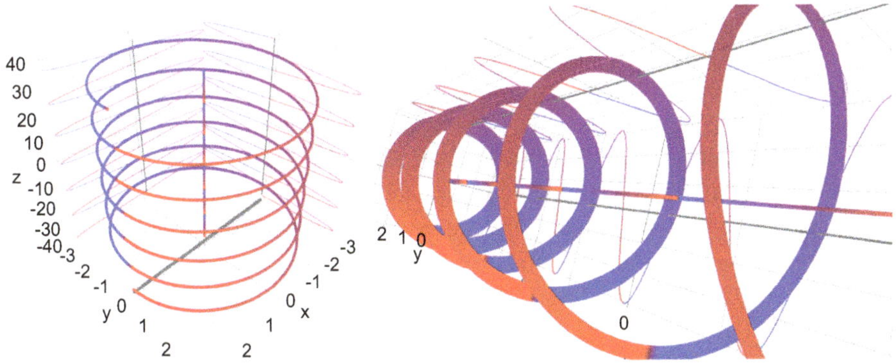

B. 3D (fixed amplitude, complex phase)

Figure 3.1: Wavefunction representations. Panel A: 2D representation of wavefunctions by varying amplitudes over space (horizontal x-axis). Panel B: 3D fixed-amplitude complex-phase representation of the wavefunction, whose 2D cardinal planar projections coincide with the oscillatory patterns on Panel A. The position of the particle is along the depth axis, the phase of the wave is color coded, $[-\pi$ (red), π(blue)], and the colored wave projections on the 2D planes and the space-axis illustrate the corresponding phases.

For waves with fixed wavelengths, this dimension-lifting approach resolves the problems with interpreting addition, constant-multiplication, and interference of wave amplitudes in 2D projections. A similar approach for lifting the longitudinal order of events (time dimension) into complex-time (kime manifold) aims to resolve some of the problems of time (as uni-directional positive real event order).

To explicate the directional kime-phase component of spacekime as an observable quantity, we can borrow ideas from Röntgen (X-ray) crystallography. Just like magnetic resonance imaging (MRI), X-ray crystallography tries to indirectly resolve substrate atomic structure by capturing spectral frequency motifs. In MRI, these motifs represent resonance frequencies (typically of hydrogen atoms) resulting from radio frequency pulses disrupting high-strength homogeneous magnetic fields. In Röntgen X-ray crystallography, the structure of crystals is observed indirectly using diffraction patterns resulting from substrate excitation by high-energy X-ray beams. The forward and inverse Fourier transforms (FT/IFT) are used for spacetime reconstruction of both MRI signals and X-ray crystal structures. These transformations are also referred to as *Fourier analysis* (decomposing a spacetime signal into its corresponding harmonics in the frequency space, k-space) and *Fourier synthesis* (using the frequency magnitudes and phases to reconstruct the spacetime representation of the signal), respectively.

For instance, the FT of a 2D image (a.k.a. picture), which represents explicitly a sample of a waveform signal (e.g., brain MRI scan), decomposes or breaks down the waveform signal into a superposition of harmonics (pure trigonometric sine waves) of different frequencies. Each sine wave is characterized by (1) its *magnitude* measuring how much that particular frequency participates in the image, and (2) its *phase*, recording the starting point of each sine wave. **Figure 3.2** shows a screenshot of a SOCR Java Applet that provides an interactive example of the space-frequency correspondence for 1D oscillatory functions. This Fourier game Java applet requires a Java-enabled browser (http://www.socr.ucla.edu/htmls/game, accessed January 29, 2021).

Before we formally define the Fourier transform (FT), let's review some of the important FT characteristics: (1) the highest meaningful sine wave frequency of the signal analysis is half the data acquisition frequency, i.e., FT yields a list of frequencies up to the acquisition frequency, with only the first half of the sequence being useful, and (2) the FT frequency resolution depends on the sampling time, i.e., the longer the sampling (analysis) interval, the finer the resolution, where the frequency resolution is $\frac{1}{\text{sampling rate } (sec)}$.

The (hat) notation, $\widehat{\Box}$, indicates applying the Fourier transform, or its inverse, on the argument, depending on the context. So, if $FT(f) = \hat{f}$, then $f = IFT(\hat{f}) = \hat{\hat{f}}$. As an example, let's consider the states of a pair of 2D images representing a *square* and a *circle*, **Figure 3.3**. The information content (energy) of each image is identical to that of its FT, i.e., $\|f^2\| = \|\hat{f}\|^2$. Hence, we can consider having the observed images in the k-space (a.k.a. Fourier domain). Just as in the case of X-ray crystallography, assume only the *magnitude* of the FT is observable $\sqrt{\sum_{i,j}\left(Re(\hat{f})^2 + Im(\hat{f})^2\right)}$, where the 2D image coordinates are indexed by i, j, but the phases $\varphi = \arctan\left(\frac{Im(\hat{f})}{Re(\hat{f})}\right)$ are not observable. In the 2D kime manifold, the observed 2D FT magnitude corresponds to kime *order* (r), i.e., the usual time in Minkowski spacetime, and the phase-angle of the FT corresponds to the kime *direction* (φ).

It helps to keep in mind that the Fourier transform maps spacetime functions, $f(\mathbf{x}, t)$, into k-frequency functions, i.e., functions of the wavenumber (spatial frequency) and the angular frequency. Of course, the inverse Fourier transform maps in exactly the opposite direction, i.e., it recasts k-frequency functions into spacetime functions.

Symbolically, the spatial frequency k is a vector, also called the wavenumber, whose components are related to wavelengths in the different spatial directions, $\mathbf{x} = (x, y, z)$, the same way the function period is related to the corresponding angular frequency.

R-code illustrating some of the above experiments is provided in the Supplementary Materials. In **Chapter 6** (Applications), we will present supporting evidence, and several examples, of the impact of knowing the kime direction to obtain data-driven scientific inference.

By the separability property, the 4D spacetime Fourier transform may be represented as four separate 1D Fourier transforms, one for each of the four spacetime dimensions, (x, y, z, t). Although the FT sign may be chosen either way, traditionally the sign convention represents a wave with angular frequency ω that propagates in the wavenumber direction $\mathbf{k}$. i.e., the frequency argument $k = (\mathbf{k})$. Let's use the Minkowski spacetime metric signature $(\underbrace{+,+,+}_{\text{space}},\underbrace{-}_{\text{time}})$ and the corresponding Minkowski metric tensor $\eta = \text{diag}(+1, +1, +1, -1)$. Then, $k \cdot x \equiv k'\eta x = \mathbf{k} \cdot \mathbf{x}$. Below we show three equivalent definitions of the forward and inverse Fourier transforms for a 4D $(n = 4)$ spacetime function $f: \{x = (\mathbf{x}, t) \in \mathbb{R}^3 \times \mathbb{R}^+\} \to \mathbb{C}$.

- Classical representation (engineering):

$$FT(f) = \hat{f}(k) = \int f(x) e^{-ik\cdot x} dx = \int f(\mathbf{x}, t) \underbrace{e^{i(\omega t - \mathbf{k}\cdot\mathbf{x})}}_{e^{-ik\cdot x}} \underbrace{dt d^3\mathbf{x}}_{dx},$$

$$IFT(\hat{f}) = \hat{\hat{f}}(x) = \hat{\hat{f}}(\mathbf{x}, t) = \frac{1}{(2\pi)^n} \int \hat{f}(x) e^{i\,k\cdot x} dk = \frac{1}{(2\pi)^n} \int \hat{f}(\mathbf{k}, \omega) e^{-i(\omega t - \mathbf{k}\cdot\mathbf{x})} d\omega d^3\mathbf{k}.$$

- Radian frequency unitary representation (mathematics):

$$FT(f) = \hat{f}(\mathbf{k}, \omega) = \frac{1}{(2\pi)^{\frac{n}{2}}} \int f(\mathbf{x}, t) \underbrace{e^{i(\omega t - \mathbf{k}\cdot\mathbf{x})}}_{e^{-ik\cdot x}} \underbrace{dt d^3\mathbf{x}}_{dx},$$

$$IFT(\hat{f}) = \hat{\hat{f}}(\mathbf{x}, t) = \frac{1}{(2\pi)^{\frac{n}{2}}} \int \hat{f}(\mathbf{k}, \omega) e^{-i(\omega t - \mathbf{k}\cdot\mathbf{x})} d\omega d^3\mathbf{k}.$$

- Symmetric Hertz frequency unitary representation (signal processing):

$$FT(f) = \hat{f}(k) = \int f(x) e^{-i2\pi \mathbf{k}\cdot x} dx = \int f(\mathbf{x}, t) \underbrace{e^{i2\pi(\omega t - \mathbf{k}\cdot\mathbf{x})}}_{e^{-i2\pi k\cdot x}} \underbrace{dt d^3\mathbf{x}}_{dx},$$

$$IFT(\hat{f}) = \hat{\hat{f}}(x) = \hat{\hat{f}}(\mathbf{x}, t) = \int \hat{f}(x) e^{i2\pi\,k\cdot x} dk = \int \hat{f}(\mathbf{k}, \omega) e^{-i2\pi(\omega t - \mathbf{k}\cdot\mathbf{x})} d\omega d^3\mathbf{k}.$$

Figure 3.3 illustrates the relevance of the phase-angle directions in correctly interpreting the signal energy. Clearly, using only the FT magnitudes captures some of the image energy; however, no phase information, or incorrect phase angles, would distort the representation of the imaging data in spacetime. This observation applies

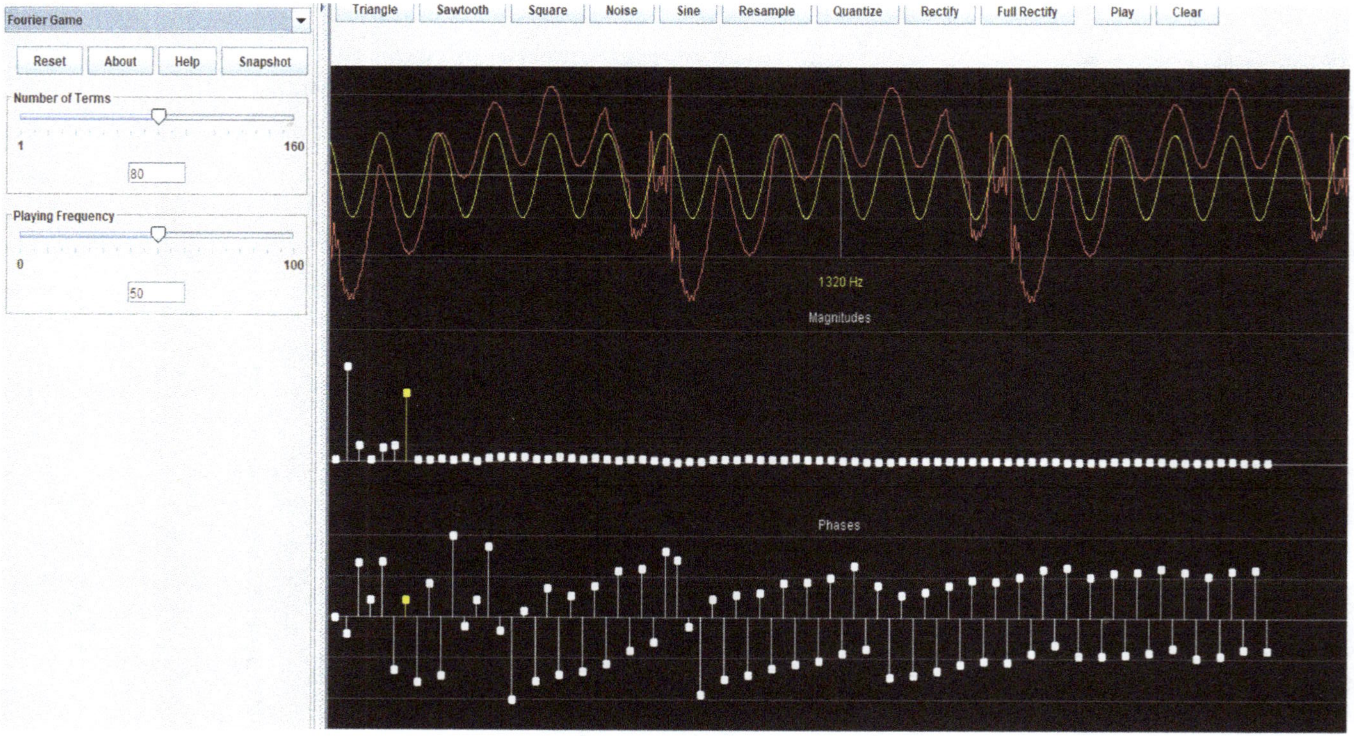

Figure 3.2: SOCR example of a 1D Fourier signal decomposition into *magnitudes* and *phases*. Top-panel represents the original signal (image), white-color curve, which is drawn manually by the user, and the reconstructed synthesized (IFT) signal, red-color curve, which is computed using the user-modified magnitudes and phases. Note that the red-reconstruction curve may overlay the original white-curve drawn by the user. The two bottom panels represent the Fourier analyzed signal (FT) with its magnitudes and phases. The user may alter or manipulate individual magnitudes as well as phases and observe their direct effects on the reconstructed synthesized signal (red-color) on the top panel.

directly to spacekime, where knowledge of the event order (kime-magnitude, time, t) only provides partial information about the kevent state. At the same time, when the kime direction (φ) is not directly observable, we need to indirectly estimate it, just like we do in X-ray crystallography for solving viral crystal structures on the Angstrom scale, by only using the magnitudes of the observed diffraction patterns [88].

In its most general form, the Fourier transformation of an integrable function $f:\mathbb{R}\to\mathbb{C}$, and its inverse, is analytically represented by:

$$FT(f)=\hat{f}(\omega)=\int_{-\infty}^{\infty} f(x)e^{-2\pi i \omega x}dx,$$

$$IFT\left(\hat{f}\right)=\hat{\hat{f}}(x)=\int_{-\infty}^{\infty} \hat{f}(\omega)e^{2\pi i \omega x}d\omega.$$

This directly generalizes to multivariate functions $g:\mathbb{C}^n\to\mathbb{C}^n$:

$$FT(g)=\hat{g}(\boldsymbol{\omega})=\int_{\mathbb{C}^n} g(\mathbf{x})e^{-2\pi i \langle \mathbf{x},\mathbf{w}\rangle}d\mathbf{x}.$$

Some of the basic properties of the Fourier transformation are included below:

- *Linearity*: The Fourier transformation is a linear functional:

$$FT(\alpha f+\beta g) = \alpha FT(f)+ \beta FT(g).$$

- *Translation/offset* property: if a is a constant and $h(x)=f(x-a)$, then

$$FT(h)=\hat{h}(\omega)=e^{-2\pi i a\omega}\hat{f}(\omega)=e^{-2\pi i a\omega}FT(f).$$

- *Scaling* property: if a is a non-zero constant and $h(x)=f(ax)$, then

$$FT(h)=\hat{h}(\omega)=\frac{1}{|a|}\hat{f}\left(\frac{\omega}{a}\right).$$

- *Differentiation* property:

$$FT\left(\frac{df(t)}{dt}\right)(\omega) = 2\pi\ i\ \omega \times FT(f)(\omega).$$

  And more generally, the Fourier transform of the kth derivative is

$$\widehat{\left(f^{(k)}\right)}(\omega)=(2\pi\ i\ \omega)^k\times \hat{f}(\omega).$$

- *Convolution* $\left((f*g)(x)=\int f(y)g(x-y)dy\right)$ *and multiplication* $\left((f\times g)(x)=f(x)\ g(x)\right)$ are dual operations:

$$\widehat{(f \times g)}(\omega) = \hat{f}(\omega) * \hat{g}(\omega),$$

$$\widehat{(f^* g)}(\omega) = \hat{f}(\omega) \times \hat{g}(\omega).$$

- *Energy preservation*: The Fourier transform contains all information of the original function:

$$\int_{\mathbb{R}^n} \|f(\boldsymbol{x})\|^2 d\boldsymbol{x} = \int_{\mathbb{R}^n} \|\hat{f}(\boldsymbol{\omega})^2\| d\boldsymbol{\omega}.$$

- *Duality*: $FT(FT(f))(\omega) = f(-\omega)$.
- *Impulse functions*: The Fourier transform of a shifted impulse function $\delta(x+\lambda)$ is a complex exponential function $d(\omega) = \int_{-\infty}^{+\infty} \delta(x+\lambda) e^{-ix\omega} dx = e^{i\lambda\omega}$. And conversely, the inverse Fourier transformation of a complex exponential function $d(\omega) = e^{i\lambda\omega}$ is an impulse function $\hat{d}(x) = \frac{1}{2\pi} \int_{-\infty}^{+\infty} e^{i\lambda\omega} e^{ix\omega} d\omega = \delta(x+\lambda)$.

3.1.2 Fourier Amplitudes and Phases

Let's try to examine the importance of the Fourier spectra, i.e., the amplitudes (magnitudes) and the phases. Suppose we have a 2D image with intensities stored as a second-order tensor (array) of dimensions $n \times k$, $x = \{x_{a,b}\}$, $1 \le a \le n$, $1 \le b \le k$. Then, the 2D discrete Fourier transform of the image will be

$$FT(x) = X = \{X_{p,q}\}_{p=1,q=1}^{n,k} = \left\{ \sum_{a=1}^{n} \sum_{b=1}^{k} x_{a,b}\, e^{-2\pi i \left(\frac{ap}{n} + \frac{bq}{k} \right)} \right\}_{p=1,q=1}^{n,k}$$

$$= \left\{ \underbrace{A_{p,q}}_{\text{magnitudes}} \times e^{\left(i\, \underbrace{\varphi_{p,q}}_{\text{phases}} \right)} \right\}_{p=1,q=1}^{n,k}.$$

The phase tensor $\{\varphi_{p,q}\}$ contains significant information about the image that can be used to explore the distribution of the underlying process that generated the observed image instance, $\{x_{i,j}\}$. Let's try to explicate the properties of two images, $x = \{x_{i,j}\}$ and $y = \{y_{i,j}\}$, that have identical phase tensors $\varphi^x = \{\varphi^x_{p,q} = \varphi^y_{p,q}\} = \varphi^y$.

Any phase differences won't explain scaling of the images since the phases are invariant to positive rescaling of the images, as scaling only affects the magnitudes. For instance, for $\lambda > 0$,

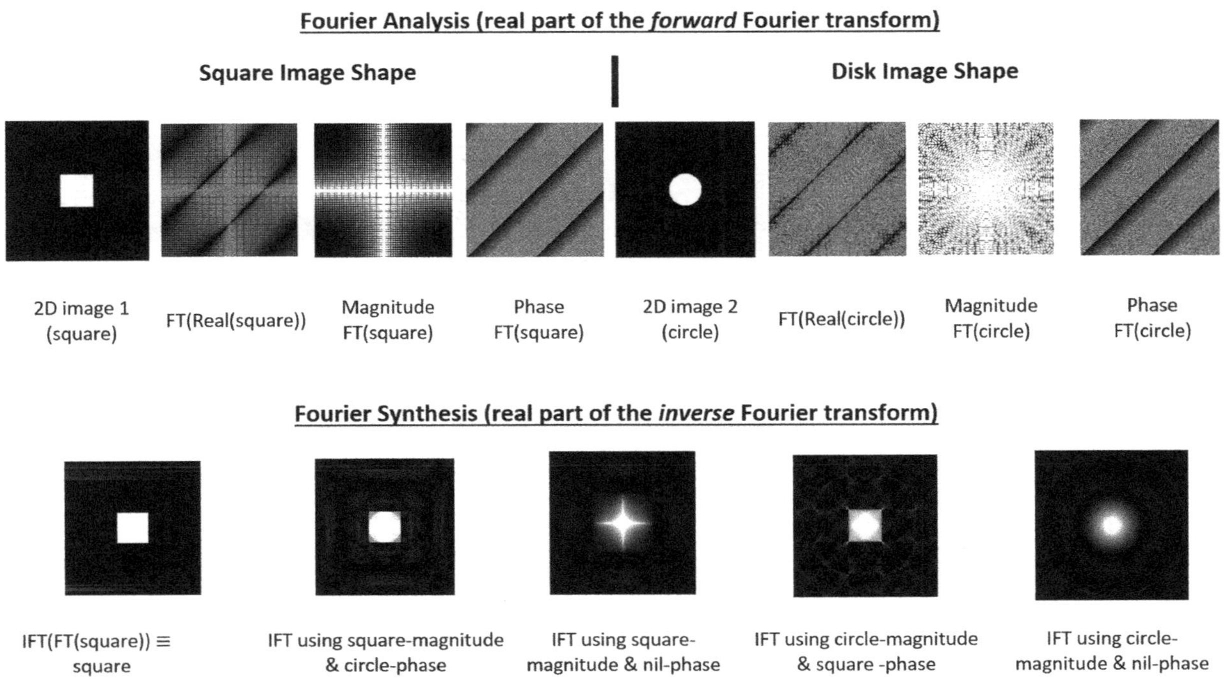

Figure 3.3: Understanding the structure of kime in terms of 2D Fourier magnitudes and phases. This composite figure shows the Fourier analysis (forward transform) and synthesis (inverse transform) for a pair of simple 2D images – square (left) and circle (right). The *Top* row illustrates the original images followed by the real parts, the magnitudes, and the phases of their Fourier transforms. The *Bottom* row depicts the inverse Fourier transforms using three alternative phases – the correct phases, trivial (nil) phases, or swapping the square and circle phases. This simple illustration demonstrates the effects of phase estimation on the spacetime reconstruction of the 2D images. The concept of kime-phase estimation plays a critically important role in data science applications that will be shown later, which utilize complex time (kime) to extend spacetime to the 5D spacekime manifold.

$$FT(\lambda \times x) = \left\{ \underbrace{\lambda \times A_{p,q}}_{\text{magnitudes}} \times e^{\left(i \underbrace{\varphi_{p,q}}_{\text{phases}} \right)} \right\}_{p=1,q=1}^{n,k}.$$

However, the phase moments can be used as signatures quantifying the similarity, or the level of difference, between two images based on their observed, or estimated, phase tensors.

3.1.3 Phase Equivalence

Let's assume the *equivalence of the phases* of a pair of continuous real-valued signals, $f_1, f_2 : \mathbb{R} \to \mathbb{R}$, $\varphi_{f_1}(\omega) = \varphi_{f_2}(\omega), \forall \omega$. This implies that

$$FT(f_1)(\omega) = \hat{f}_1(\omega) = a_{f_1}(\omega) e^{\left(i\varphi_{f_1}\right)} = a_{f_1}(\omega) e^{\left(i\varphi_{f_2}\right)} =$$

$$= \underbrace{\frac{a_{f_1}(\omega)}{a_{f_2}(\omega)}}_{s(\omega)} \underbrace{a_{f_2}(\omega) e^{\left(i\varphi_{f_2}\right)}}_{FT(f_2)} = s(\omega)\hat{f}_2(\omega) = s(\omega) FT(f_2)(\omega),$$

where $s(\omega)$ is a real-valued magnitude-scaling function. As $\hat{s}(x)$ is the inverse Fourier transform of a real-valued scaling function, $s(\omega)$, it must be Hermitian, i.e., its complex-conjugate $\overline{\hat{s}(x)} = \hat{s}(-x)$ and it's real and imaginary parts are respectively *even* (the graph of the real part is symmetric with respect to the vertical axis) and *odd* (the graph of the imaginary part is symmetric with respect to the origin) functions:

$$\overline{\hat{s}(x)} = \frac{1}{2\pi} \int \overline{s(\omega) e^{i(x\omega)}} d\omega = \frac{1}{2\pi} \int \underbrace{\overline{s(\omega)}}_{\text{real}} e^{-i(x\omega)} d\omega =$$

$$= \frac{1}{2\pi} \int s(\omega) e^{i(-x\omega)} d\omega = \hat{s}(-x).$$

Applying the inverse Fourier transform and using the convolution-multiplication property we obtain:

$$f_1(x) = IFT\left(\hat{f}_1\right)(x) = \left(IFT(s) * \underbrace{IFT(FT(f_2))}_{f_2} \right)(x) = \left(\hat{s} * f_2\right)(x).$$

Therefore, $\hat{s}(x) = \underbrace{e(x)}_{\text{even}} + i \underbrace{o(x)}_{\text{odd}}$ and we can plug in the relation $f_1(x) = \left(\hat{s} * f_2\right)(x)$ to obtain:

$$f_1(x) = (e * f_2)(x) + i \underbrace{(o * f_2)(x)}_{\text{trivial}}.$$

As the spacetime functions $f_1(x)$ and $f_2(x)$ are real-valued, the imaginary part above is trivial, and the pair of original (signals) functions are related by convolution ($*$) with an even real function $e(x)$:

$$f_1(x) = (e * f_2)(x), \forall x.$$

Therefore, the real function $\hat{s}(x) = e(x)$ being even and positive is a necessary and sufficient condition for the equivalence of the Fourier phases of the corresponding signals. An additional constraint for the even function to be positive is required to avoid problems with counter examples like the real even function, $s(\omega) = \cos(\omega)$, $e(x) = \hat{s}(\omega) = IFT(\cos(\omega)) = \frac{1}{2}(\delta(x-1) + \delta(x+1))$, which is not always positive, where $\hat{e}(\omega) = FT(e(x)) = \cos(\omega) \equiv \cos(-\omega)$, $\widehat{f_1}(\omega) = \hat{e}(\omega) \times \widehat{f_2}(\omega) = \cos(\omega)\widehat{f_2}(\omega)$, however, the phase of f_1 may not always be equal to the phase of f_2.

As a corollary, in certain cases, we can infer signal properties based purely on knowing the Fourier phases but not the corresponding magnitudes, and vice-versa, knowing the magnitudes, but not the phases. Specifically, when we know the phases, we can discern signal characteristics that are preserved when convolving with an even function. For instance:

- If one of f_1 and f_2 is even, so is the other,
- If the distribution of f_1 is heavy-tail, so will be the distribution of f_2, and
- If f_1 is positively skewed, as f_2 is convolved with an even function, f_2 will also be positively skewed.

Prior reports have documented that distribution reconstructions using only the phases and ignoring the magnitudes (amplitudes) are highly correlated with the original distributions [89].

3.1.4 Amplitude Equivalence

Similarly, we can investigate the effect of *equivalence of the amplitudes (magnitudes)*. This time, we'll assume the amplitudes of a pair of continuous real-valued signals are equal, $f_1, f_2 \colon \mathbb{R} \to \mathbb{R}$, $a_{f_1}(\omega) = a_{f_2}(\omega)$, $\forall \omega$. This is important, as most of the time, the amplitudes are observed, yet the phases are not known. Then,

$$FT(f_1)(\omega) = \hat{f_1}(\omega) = a_{f_1}(\omega)e^{\left(i\varphi_{f_1}\right)} = a_{f_2}(\omega)e^{\left(i\varphi_{f_1}\right)} =$$

$$= a_{f_2}(\omega)e^{\left(i\varphi_{f_1} + i\varphi_{f_2} - i\varphi_{f_2}\right)} = \underbrace{a_{f_2}(\omega)e^{\left(i\varphi_{f_2}\right)}}_{FT(f_2)} \times \underbrace{e^{i\left(\varphi_{f_1} - \varphi_{f_2}\right)}}_{d} = \hat{f_2}(\omega)d(\omega) = d(\omega)FT(f_2)(\omega),$$

where the phase difference is $\Delta\varphi(\omega) = \varphi_{f_1}(\omega) - \varphi_{f_2}(\omega)$ and

$$d(\omega) = e^{i\left(\varphi_{f_1}(\omega) - \varphi_{f_2}(\omega)\right)} = e^{i\Delta\varphi(\omega)} = \cos(\Delta\varphi) + i\sin(\Delta\varphi)$$

is a complex-exponential function of the phase differential.

The convolution theorem connecting function multiplication and convolution through the Fourier transform yields:

$$FT(f_1)(\omega) = d(\omega)FT(f_2)(\omega), \forall\omega \underset{\text{if and only if}}{\Longleftrightarrow} f_1(x) = \left(\hat{d} * f_2\right)(x), \forall x.$$

Given that f_1 and f_2 have the same Fourier amplitudes, we can infer some information about the relation between their Fourier phases. Each function $f\colon \mathbb{C} \to \mathbb{C}$ can be expressed as a sum of real (Re) and imaginary (Im) parts of even (f_e) and odd (f_o) components:

$$f(x) = f_e(x) + f_o(x) = Re[f_e(x)] + i \times Im[f_e(x)] + Re[f_o(t)] + i \times Im[f_o(t)].$$

The Fourier transformation of each part is:

$$FT\{Re[f_e(x)]\} = \int_{-\infty}^{\infty} Re[f_e(x)]e^{-i\omega x}dx = 2\int_{0}^{\infty} Re[f_e(x)]\cos(\omega x)dx\ .$$

$$\text{real and even} \xrightarrow{FT} \text{real and even},$$

$$FT\{i\ Im[f_e(x)]\} = i\int_{-\infty}^{\infty} Im[f_e(x)]e^{-i\omega x}dx = 2i\int_{0}^{\infty} Im[f_e(x)]\cos(\omega x)dx.$$

$$\text{imaginary and even} \xrightarrow{FT} \text{imaginary and even},$$

$$FT\{Re[f_o(t)]\} = \int_{-\infty}^{\infty} Re[f_o(t)]e^{-i\omega x}dx = -2\,i \times \int_{0}^{\infty} Re[f_o(t)]\sin(\omega x)dx.$$

$$\text{real and odd} \xrightarrow{FT} \text{imaginary and odd},$$

$$FT\{i\ Im[f_o(t)]\} = i\int_{-\infty}^{\infty} Im[f_o(t)]e^{-i\omega x}dx = 2\int_{0}^{\infty} Im[f_o(t)]\sin(\omega x)dx.$$

$$\text{imaginary and odd} \xrightarrow{FT} \text{real and odd}.$$

3.1.5 Fourier Transform Effects on Phases and Magnitudes

If we denote the Fourier transformation of $f(x)$ by $\hat{f}(\omega) = R(\omega) + i\,X(\omega) = A(\omega)e^{i\varphi(\omega)}$, then **Table 3.1** shows some of the properties of the Fourier transform ($\hat{f}$), based on the type of the original function (f).

Table 3.1: Properties of real and imaginary, even and odd functions.

Function type (f)	Even	Odd	In general
Purely Real-Valued	$\hat{f}(\omega)$ is real & even $\hat{f}(\omega)=\hat{f}(-\omega)$ $R(\omega)=R(-\omega)$ $X(\omega)=0$ $A(\omega)=A(-\omega)$ $\varphi(\omega)=\varphi(-\omega)=0$ or $\pm\pi$	$\hat{f}(\omega)$ is imaginary & odd $\hat{f}(\omega)=-\hat{f}(-\omega)$ $R(\omega)=0$ $X(\omega)=-X(-\omega)$ $A(\omega)=A(-\omega)$ $\varphi(\omega)=-\varphi(-\omega)=\pm\frac{\pi}{2}$	When $f(x)$ is **real**, $\hat{f}(\omega)$ is Hermitian $\hat{f}(\omega)=\hat{f}^*(-\omega)$ $R(\omega)=R(-\omega)$ $X(\omega)=-X(-\omega)$ $A(\omega)=A(-\omega)$ $\varphi(\omega)=-\varphi(-\omega)$ $A(\omega)$ is even $\varphi(\omega)$ is odd
Purely Imaginary	$\hat{f}(\omega)$ is imaginary & even $\hat{f}(\omega)=\hat{f}(-\omega)$ $R(\omega)=0$ $X(\omega)=X(-\omega)$ $A(\omega)=A(-\omega)$ $\varphi(\omega)=\varphi(-\omega)=\pm\frac{\pi}{2}$	$\hat{f}(\omega)$ is a real & odd $\hat{f}(\omega)=-\hat{f}(-\omega)$ $R(\omega)=-R(-\omega)$ $X(\omega)=0$ $A(\omega)=A(-\omega)$ $\varphi(\omega)=-\pi\cup 0$ and $\varphi(\omega)+\varphi(-\omega)=-\pi$	When $f(x)$ is **imaginary**, $\hat{f}(\omega)$ is anti-Hermitian $\hat{f}(\omega)=-\hat{f}^*(-\omega)$ $R(\omega)=-R(-\omega)$ $X(\omega)=X(-\omega)$ $A(\omega)=A(-\omega)$ $\varphi(\omega)+\varphi(-\omega)=\pi$ $A(\omega)$ is even

In most cases, when $f_1(x)$ and $f_2(x)$ are both real-valued functions, we know that $\varphi_1(\omega)$ and $\varphi_2(\omega)$ are odd functions, which imply that the corresponding phase distributions Φ_1 and Φ_2 are symmetric and zero-mean. Thus, the inverse Fourier transformation of $d(\omega)=e^{i\left(\varphi_{f_1}(\omega)-\varphi_{f_2}(\omega)\right)}$, $h(x)=\hat{d}(x)$ is a real function. In summary, given that the amplitudes are identical, the two real-valued functions f_1 and f_2 are related by:

$$f_1(x)=(h*f_2)(x),\ \forall x.$$

$$\hat{f}_1(\omega)=\hat{h}(\omega)\hat{f}_2(\omega)=d(\omega)\hat{f}_2(\omega),\ \forall\omega.$$

Let's look at some examples of phase difference functions, under the assumption of equivalent amplitudes.

- When the phase differential is a *linear function*, $\Delta\varphi=\varphi_{f_1}(\omega)-\varphi_{f_2}(\omega)=\lambda\omega$, $\lambda\in\mathbb{R}$, then the inverse Fourier transformation of $d(\omega)=e^{i\Delta\varphi}=e^{i\lambda\omega}$ is an impulse function

$$h(x)=\hat{d}(x)=\frac{1}{2\pi}\int_{-\infty}^{+\infty}e^{i\lambda\omega}e^{ix\omega}d\omega=\delta(x+\lambda).$$

In spacetime, $f_1(x)=(h*f_2)(x)=f_2(x+\lambda)$. Thus, $f_1(x)$ is derived by shifting $f_2(x)$ by λ units along the x-axis.

In frequency space, $FT(f_1)(\omega) = e^{i\lambda\omega} FT(f_2)(\omega)$. In other words, $FT(f_1)(\omega)$ is derived by counterclockwise rotation of $FT(f_2)(\omega)$ by $\lambda\omega$. Of course, this confirms the Fourier transformation property that a shift by λ in spacetime corresponds with multiplication by $e^{i\lambda\omega}$ in frequency space, i.e., $FT(f(x-\lambda)) = e^{-i\lambda\omega} FT(f(x))$.

- When the phase differential is *constant*, up to a sign, $\varphi_{f_1}(\omega) - \varphi_{f_2}(\omega) = \lambda\, \text{sgn}(\omega)$, $\lambda \in \mathbb{R}/[-\pi, \pi]$. Recall that $h = \hat{d} = \widehat{e^{i\Delta\varphi}}$ is the inverse Fourier transform of the complex exponential differential. Then,

$$h(x) = \left(\sqrt{2\pi}\cos(\lambda)\delta(x)\right) - \left(\frac{\sqrt{\frac{2}{\pi}}\sin(\lambda)}{x}\right).$$

In spacetime, $f_1(x) = (h * f_2)(x) = \sqrt{2\pi}\cos(\lambda) f_2(x) - \left(\frac{\sqrt{\frac{2}{\pi}}\sin(\lambda)}{x} * f_2(x)\right)$. Therefore, $f_1(x)$ is derived from $f_2(x)$ by subtracting from $f_2(x)$ a multiple of a reciprocal function convolved with $f_2(x)$.

In frequency space, $FT(f_1)(\omega) = e^{i\,\lambda\,\text{sgn}(\omega)} FT(f_2)(\omega)$. When $\omega > 0$, $FT(f_1)(\omega)$ is derived by counterclockwise rotation of $FT(f_2)(\omega)$ by λ. In the other case, when $\omega < 0$, $FT(f_1)(\omega)$ is derived by clockwise rotation of $FT(f_2)(\omega)$ by λ.

- When the phase differential is a *cubic* (an example of an odd degree polynomial), $\varphi_{f_1}(\omega) - \varphi_{f_2}(\omega) = \lambda\,\omega^3$. Then, the inverse Fourier transformation of $d(\omega) = e^{\lambda\omega^3}$ is

$$h(x) = \frac{(-x + |x|)AiryAi\left(-\frac{|x|}{3\lambda^{1/3}}\right) + (x + |x|)AiryAi\left(\frac{|x|}{3\lambda^{1/3}}\right)}{2 \times 3\lambda^{1/3}|x|} = \frac{AiryAi\left(\frac{x}{3\lambda^{1/3}}\right)}{3\lambda^{1/3}},$$

where the *Airy function* of the first kind, $AiryAi(x)$, is a solution of the Stokes ordinary differential equation $y'' - xy = 0$ and $AiryAi(x) \underset{x\to\infty}{\longrightarrow} 0$. An interesting relation to quantum mechanics is that the Airy function is also a solution to the Schrödinger‘s equation (**Chapter 2**) describing the dynamics of particles confined within a triangular potential well and particles restricted to a 1D constant force field. The Airy function can also be represented as a Riemann or a path integral:

$$AiryAi(x) = \frac{1}{\pi}\lim_{u\to\infty}\int_0^u \cos\left(\frac{t^3}{3} + xt\right)dt,\ \forall x \in \mathbb{R}$$

$$AiryAi(z) = \frac{1}{2\pi i}\int_P e^{\left(\frac{t^3}{3} - zt\right)}dt,\ \forall z \in \mathbb{C},$$

where the latter path integral is over a curve P with starting and ending points at infinity corresponding to initial and terminal path-parameter arguments of $-\frac{\pi}{3}$ and $+\frac{\pi}{3}$.

- When the phase differential is the *multiplicative inverse* (reciprocal) function, $\varphi_{f_1}(\omega) - \varphi_{f_2}(\omega) = \frac{\lambda}{\omega}$. Then, the inverse Fourier transform $d(\omega) = e^{\frac{i\lambda}{\omega}}$ is

$$h(x) = 2\pi\delta(x) - \frac{\sqrt{\lambda|x|}(x+|x|)(HankelH1(1, 2\sqrt{\lambda|x|}) + HankelH2(1, 2\sqrt{\lambda|x|}))}{4x^2}$$

$$= 2\pi\delta(x) - \frac{\sqrt{\lambda|x|}(x+|x|)Y_1(2\sqrt{\lambda|x|})}{2x^2}$$

$$= \begin{cases} 2\pi\delta(x) - \dfrac{\sqrt{\lambda}Y_1\left(2\sqrt{\lambda x}\right)}{\sqrt{x}}, & x > 0 \\ 0, & x \le 0 \end{cases},$$

where the Hankel function of the first kind is defined by $H_n^{(1)}(z) = J_n(z) + i\ Y_n(z)$, and the Hankel function of the second kind is defined by $H_n^{(2)}(z) = J_n(z) - i\ Y_n(z)$, where $J_n(z)$ is a Bessel function of the first kind and $Y_n(z)$ is a Bessel function of the second kind.
- When the phase differential is a *square-integrable odd* function $\Delta\varphi(\omega) \in L_2$, it can be expressed as a Taylor series of odd-power terms

$$\Delta\varphi(\omega) = \sum_{k=1}^{\infty} a_k\omega^{(2k-1)} = \sum_{k=1}^{n} a_k\omega^{(2k-1)} + \varepsilon_n \approx \sum_{k=1}^{n} a_k\omega^{(2k-1)},$$

Hence,

$$d(\omega) = e^{i\sum_{k=1}^{n} a_k\omega^{(2k-1)}} = \prod_{k=1}^{n} e^{ia_k\omega^{(2k-1)}}.$$

Therefore,

$$FT(f_1)(\omega) = d(\omega)FT(f_2)(\omega) = \prod_{k=1}^{n} e^{ia_k\omega^{(2k-1)}} \cdot FT(f_2)(\omega)$$
$$= e^{ia_1\omega} \cdot e^{ia_3\omega^3} \cdot \ldots \cdot e^{ia_{2n-1}\omega^{2n-1}} \cdot FT(f_2)(\omega).$$

This allows us to explicate the relation between f_1 and f_2 in spacetime as

$$f_1(x) = (h_1 * h_3 * \cdots * h_{2n-1} * f_2)(x),$$

where $h_k(x)$ is the inverse Fourier transform of $e^{ia_k\omega^k}$, i.e., $h_k(x) = IFT\left(e^{ia_k\omega^k}\right)$.

This implies that in the frequency domain, $FT(f_1)(\omega)$ is derived by counterclockwise rotating $FT(f_2)(\omega)$ sequentially by $a_{2n-1}\omega^{2n-1}, a_{2n-3}\omega^{2n-3}, \cdots, a_3\omega^3, a_1\omega$.

Similarly, in spacetime, $f_1(x)$ is derived by iterative nested convolution by the kernels $h_{2n-1}(x), h_{2n-3}(x), \cdots, h_3(x), h_1(x)$. For instance, for a simple odd cubic,

$$\varphi_{f_1}(\omega) - \varphi_{f_2}(\omega) = \omega + \omega^3,\ d(\omega) = e^{i(\omega+\omega^3)},$$

$$FT(f_1)(\omega) = e^{i(\omega+\omega^3)} FT(f_2)(\omega) = e^{i\omega} \cdot e^{i\omega^3} \cdot FT(f_2)(\omega)$$

$$f_1(x) = (h_1 * h_3 * f_2)(x)$$

$$= \delta(x+1) * \frac{AiryAi\left(\frac{x}{3\lambda^{\frac{1}{3}}}\right)}{3\lambda^{\frac{1}{3}}} * f_2(x)$$

$$= \frac{AiryAi\left(\frac{x+1}{3\lambda^{1/3}}\right)}{3\lambda^{1/3}} * f_2(x).$$

Figure 3.4 shows the graphs of $d(\omega) = e^{i(\omega+\omega^3)}$ and $h(x) = (h_1 * h_3)(x)$ and we can see that $h(x)$ is derived by shifting $\frac{AiryAi\left(\frac{x}{3\lambda^{1/3}}\right)}{3\lambda^{1/3}}$ by $\lambda = 1$ along the x-axis.

In the most general case, given an explicit form of the phase difference $d(\omega)$ function, we may be able to compute $h(x)$, the inverse Fourier transformation of $d(\omega) = e^{i\Delta\varphi}$, and then derive $f_1(x)$ or $FT(f_1)(\omega)$ based on $f_2(x)$ or $FT(f_2)(\omega)$, respectively, according to the following formulas:

$$f_1(x) = (h * f_2)(x),\ \forall x \text{ (spacetime)}$$

$$FT(f_1)(\omega) = d(\omega)FT(f_2)(\omega) = e^{i\left(\varphi_{f_1}(\omega) - \varphi_{f_2}(\omega)\right)} FT(f_2)(\omega),\ \forall \omega \text{ (frequency space)}.$$

For any pair of real-valued signals (functions), the above expressions provide the *necessary and sufficient conditions* for the equivalence of their Fourier amplitudes.

Finally, we can examine the effects of linear transformations of continuous real-valued signals $f_1, f_2 : \mathbb{R} \to \mathbb{R}$. Suppose their phases are linearly related, $\varphi_{f_1}(\omega) = \lambda\varphi_{f_2}(\omega) + b,\ \forall\omega$.

This implies that:

$$FT(f_1)(\omega) = \hat{f}_1(\omega) = a_{f_1}(\omega)e^{\left(i\varphi_{f_1}\right)} = a_{f_1}(\omega)e^{\left(i\left(\lambda\varphi_{f_2}+b\right)\right)} =$$

$$= \underbrace{\frac{a_{f_1}(\omega)}{a_{f_2}(\omega)^\lambda}}_{s(\omega)} \left(\underbrace{a_{f_2}(\omega)e^{\left(i\varphi_{f_2}\right)}}_{FT(f_2)}\right)^\lambda e^{ib} = s(\omega)\left(\hat{f}_2(\omega)\right)^\lambda e^{ib} = s(\omega)(FT(f_2)(\omega))^\lambda e^{ib},$$

where $s(\omega) = \frac{a_{f_1}(\omega)}{a_{f_2}(\omega)^\lambda}$ is a real-valued magnitude scaling function.

Applying the inverse Fourier transform and using the convolution-multiplication property we obtain:

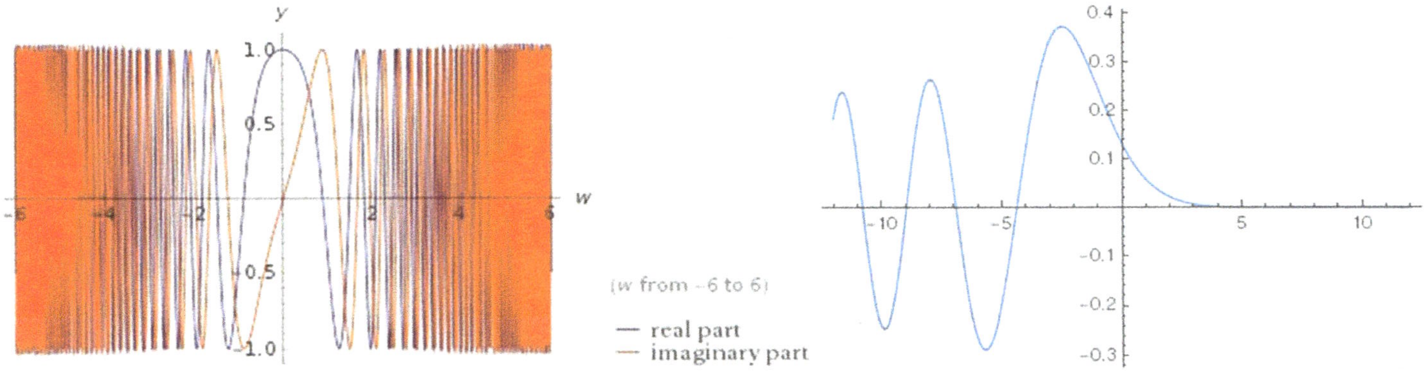

$$y \equiv d(\omega) = e^{i(\omega+\omega^3)}$$

$$h(x) = (h_1 * h_3)(x)$$

Figure 3.4: Graphs of complex-exponential of an odd-cubic phase differential, $\boldsymbol{d}(\omega)$, (left) and it's real-valued inverse Fourier transform, $\boldsymbol{h}(\boldsymbol{x})$, (right).

$$f_1(x) = IFT\left(\hat{f}_1\right)(x) = e^{ib}\left(IFT(s) * IFT\left((FT(f_2))^{\lambda}\right)\right)(x) = e^{ib}(\hat{s} * h_2)(x).$$

where $h_2(x) = IFT\left((FT(f_2))^{\lambda}\right)$. As the conjugation and exponentiation-by-λ operations can be swapped, if $f_2(x)$ is a real function, **Table 3.1** suggests that $FT(f_2)$ is Hermitian, $FT^*(f_2)(\omega) = FT(f_2)(-\omega)$, and $(FT(f_2))^{\lambda}$ is Hermitian:

$$(FT^*(f_2))^{\lambda}(\omega) = \left((FT(f_2))^{\lambda}\right)^*(-\omega).$$

Therefore, $h_2(x) = IFT\left((FT(f_2))^{\lambda}\right)$ is real valued.

As $g(x) = \hat{s}(x)$ is the inverse transform of a real-valued scaling function, $s(\omega)$, $g(x)$ must be Hermitian, i.e., its complex-conjugate $\overline{g(x)} = g(-x)$, and it's real and imaginary parts are *even* and *odd* functions, respectively.

Therefore, $g(x) = \underbrace{e(x)}_{\text{even}} + i\,\underbrace{o(x)}_{\text{odd}}$ and we can plug in the relation $f_1(x) = e^{ib}(\hat{s} * h_2)(x)$ to get:

$$f_1(x) = e^{ib}(e * h_2)(x) + e^{ib} \cdot i(o * h_2)(x).$$

Given that $h_2(x)$ is real we have:

$$f_1(x) = [\cos(b)(e * h_2)(x) - \sin(b)(o * h_2)(x)] + i\,\underbrace{[\sin(b)(e * h_2)(x) + \cos(b)(o * h_2)(x)]}_{\text{trivial}}.$$

As the spacetime functions $f_1(x)$ and $f_2(x)$ are real valued, the imaginary part above is trivial, and the pair of functions are related by:

$$f_1(x) = \cos(b)(e * h_2)(x) - \sin(b)(o * h_2)(x), \forall x.$$

Similarly, assuming that $h_2(x)$ is purely imaginary, i.e. $h_2(x) = il_2(x)$ where $l_2(x)$ is a real function, we obtain:

$$f_1(x) = ie^{ib}(e * l_2)(x) - e^{ib}(o * l_2)(x) =$$

$$[-\sin(b)(e * l_2)(x) - \cos(b)(o * l_2)(x)] + i\,\underbrace{[\cos(b)(e * l_2)(x) - \sin(b)(o * l_2)(x)]}_{\text{trivial}} =$$

$$-\sin(b)(e * l_2)(x) - \cos(b)(o * l_2)(x).$$

However, the value of the power exponent λ, which is the scaling factor of the original linear phase transformation, may also cause $(FT(f_2))^{\lambda}$ to be an arbitrary function, e.g., non-odd and non-even function, which makes the general case more complex.

Similarly we can explore continuous real-valued signals $f_1, f_2: \mathbb{R} \to \mathbb{R}$ that have *linearly related amplitudes*, i.e., $a_{f_1}(\omega) = \lambda a_{f_2}(\omega) + b, \forall \omega$. Then,

$$FT(f_1)(\omega) = \hat{f}_1(\omega) = a_{f_1}(\omega) e^{\left(i\varphi_{f_1}\right)} = \left[\lambda a_{f_2}(\omega) + b\right] e^{\left(i\varphi_{f_1}\right)} =$$

$$\lambda a_{f_2}(\omega) e^{\left(i\varphi_{f_1}\right)} + b e^{\left(i\varphi_{f_1}\right)} = \lambda a_{f_2}(\omega) e^{\left(i\varphi_{f_1} + i\varphi_{f_2} - i\varphi_{f_2}\right)} + b e^{\left(i\varphi_{f_1}\right)} =$$

$$\lambda \underbrace{a_{f_2}(\omega) e^{\left(i\varphi_{f_2}\right)}}_{FT(f_2)} e^{i\left(\varphi_{f_1} - \varphi_{f_2}\right)} + b e^{\left(i\varphi_{f_1}\right)} =$$

$$\lambda \hat{f}_2(\omega) d(\omega) + b e^{\left(i\varphi_{f_1}\right)} = \lambda d(\omega) FT(f_2)(\omega) + b e^{\left(i\varphi_{f_1}\right)},$$

where $d(\omega) = e^{i\left(\varphi_{f_1}(\omega) - \varphi_{f_2}(\omega)\right)} = \cos(\Delta\varphi) + i \sin(\Delta\varphi)$. Inverting the Fourier transform yields:

$$f_1(x) = IFT(\hat{f}_1)(x) = \lambda\left(IFT(d) * \underbrace{IFT(FT(f_2))}_{f_2} \right)(x) + b\, IFT\left(e^{\left(i\varphi_{f_1}\right)} \right) =$$

$$\lambda(\hat{d} * f_2)(x) + b\, IFT\left(e^{\left(i\varphi_{f_1}\right)} \right).$$

Since $f_1, f_2: \mathbb{R} \to \mathbb{R}$ are both real functions, $\varphi_{f_1}(-\omega) = -\varphi_{f_1}(\omega)$ and $\varphi_{f_2}(-\omega) = -\varphi_{f_2}(\omega)$ are odd functions, then $\Delta\varphi = \varphi_{f_1}(\omega) - \varphi_{f_2}(\omega)$ is an odd function.

Therefore, both $\hat{d} = IFT(d) = IFT\left(e^{i\left(\varphi_{f_1}(\omega) - \varphi_{f_2}(\omega)\right)}\right)$ and $IFT\left(e^{\left(i\varphi_{f_1}(\omega)\right)}\right)$ are the inverse Fourier transformation of the form $e^{(i \times \text{odd function})}$, the result of which was discussed above.

In summary, an amplitude, $A(\omega)$, is always an even function, but we can't infer all properties of the corresponding phase function $\varphi(\omega)$ simply based on knowing the amplitude. The characterizations above show that if we have information about the type of the function $f_1(x)$ (spacetime observation) along with some additional evidence, then we can make reliable spacekime inference, prediction, classification, and clustering. For instance, we can use phase-aggregation to model the expected discrepancy between the unobserved kime phases ($\varphi_1(\omega)$) and their model-based kime-phase estimates ($\varphi_2(\omega)$). This will allow subsequent spacekime inference about f_1, based on a spacekime reconstruction (f_2) obtained using the inverse Fourier transform of the observed amplitude ($A_1(\omega)$) and the approximate phases ($\varphi_2(\omega)$).

Our results, and prior reports [89], also empirically document the symmetry of the phase distributions and the fact that reconstructions using only the phases, i.e., ignoring the magnitudes (amplitudes), are highly correlated with the original signals.

3.1.6 Minkowski Spacetime

At the turn of the twentieth century, the mathematician Henri Poincaré suggested that space and time can be integrated into a contiguous 4D spacetime by augmenting the 3D space metric with time represented as an imaginary fourth coordinate $i \cdot c \cdot t$ [90, 91]. Poincaré used the speed of light (c) and the imaginary unit i to apply a Lorentz transformation rotating the 4D $(\boldsymbol{x}, t)$ coordinates. This process obviously requires arithmetic operations jointly on space and time values, which become unitless (actually adopts the spatial distance unit) under the time transform:

$$\tau = i \cdot c \cdot t.$$

This idea led to the special relativity definition of a *metric tensor d* that is invariant under Lorentz transformations of inertial frames:

$$d = x^2 + y^2 + z^2 + \tau^2 = x^2 + y^2 + z^2 - c^2 t^2.$$

In this 4D spacetime, Lorentz transformations appear as ordinary rotations preserving the quadratic form:

$$x^2 + y^2 + z^2 + \tau^2,$$

where the time t in an inertial system is measured by a clock stationary in that system, and $\tau = ict$. Suppose $X = (x, \ y, \ z, ct)^T$ and $X' = (x', \ y', \ z', ct')^T$ represent two inertial frames of reference in a flat Minkowski 4D spacetime and the Minkowski's square metric tensor is:

$$\eta = \begin{pmatrix} 1 & 0 & 0 & 0 \\ 0 & 1 & 0 & 0 \\ 0 & 0 & 1 & 0 \\ 0 & 0 & 0 & -1 \end{pmatrix}.$$

Then, the spacetime interval is represented by a (3,1)-signature quadratic form

$$\langle X | X \rangle = X^T \eta X \underbrace{\qquad = \qquad}_{\substack{\text{Lorentz Transform} \\ \text{invariance}}} X'^T \eta X'.$$

A few years later, Hermann Minkowski used Henri Poincaré's idea to illustrate the invariance of the Maxwell equations in 4D spacetime under the Lorentz transformation and represented Einstein‘s special theory of relativity in this new 4D Minkowski spacetime [92]. Poincaré's and Minkowski's ideas of converting between spatial and temporal measuring units unified the 4D spacetime continuum and ultimately led to Einstein's special and general theories of relativity.

3.1.7 Events

Events are points (i.e., states) in the 4D Euclidean Minkowski spacetime universe. According to particle physics, *events* are quantum-mechanically the instantaneous results immediately following a fundamental interaction of subatomic particles. These results occur in a very short time interval at a specific spatial location. Various quantum mechanical calculations utilize commonly employed quantities as event-proxies to analyze the events. Such proxy measures include differential cross section, flux of beams, rates, particle numbers, characteristics, distributions, and luminosity or intensity of the experiment that generated the observed event.

According to the theory of relativity, each event is *relativistically* defined by where and when it occurs, i.e., its precise spatial location position, $x = (x, y, z)$, and its exact time, t. Both the spatial location and the longitudinal order are relative to a certain reference frame. The theory of relativity aims to explain the likelihood and intensity of one event influencing another. Metric tensor calculations facilitate the exploration of such event interactions and the determination of the causal structure of spacetime. Specifically, the metric tensor quantifies the interval between two events, i.e., the *event difference*, and classifies these intervals into spacelike, lightlike, or timelike event-separations. Two events may be causally related, or influence one another, if and only if they are separated by a lightlike or a timelike interval.

3.1.8 Kaluza-Klein Theory

In 1921, Theodor Kaluza developed an extension of the classical general relativity theory to 5D [93]. This included the metric, the field equations, the equations of motion, the stress-energy tensor, and the cylinder condition. In 1926, after the reports of the work of Heisenberg and Schrödinger, Oskar Klein interpreted Kaluza's five-dimensional theory in quantum mechanical space and proposed that the fifth dimension was curled up and microscopic, i.e., the cylinder condition [94]. Klein suggested a circular geometry for the extra fifth dimension with a radius of 10^{-30}cm and calculated a scale for the fifth dimension based on the quantum of charge. The small size of the fifth dimension ensures that it is below the Planck resolution and therefore cannot be traversed, albeit, the angular direction of S^1 (unit circle) still provides a compass indication of time orientation.

Thus, the topology of the 5D **Kaluza-Klein spacetime** is $K_2 \cong M^4 \times S^1$, where M^4 is a 4D Minkowski spacetime and S^1 is a circle. **Figure 3.5** geometrically shows a way to imagine the 5D K_2 space that has one extra dimension representing small circles at each 4D spacetime coordinate. Note that the S^1 dimension is directional only, it's physically so small as to prohibit spatial traversal. More recently, complex time has been employed in electromagnetism for modeling, representation, and calculus of complex-valued functions. For instance, McLaughlin used complex time to represent contour-independent path integrals and calculate ray barrier penetration [95]. However, there is a distinct difference between a 1D complex time $(0 \pm it)$ [96] and 2D kime (r, φ).

Czajko [97, 98] suggested another geometric interpretation of the existence of a second dimension of time as a corollary of the inverse Lorentz transformation of time rate, which may imply the presence of a tangential gravitational potential. Another interesting analogy of generalizing time to 2D is presented by Haddad and colleagues [99]. It is used for calculation of the spatial Green's functions for a horizontal electric dipole over a multilayer medium. This method relies on exponential series to estimate the spectral Green's function. Applying the inverse Hankel's transform along with Sommerfeld's identity to the resulting exponential expansion yields a series consisting of a number of complex images that can be used to approximate the spatial Green's function.

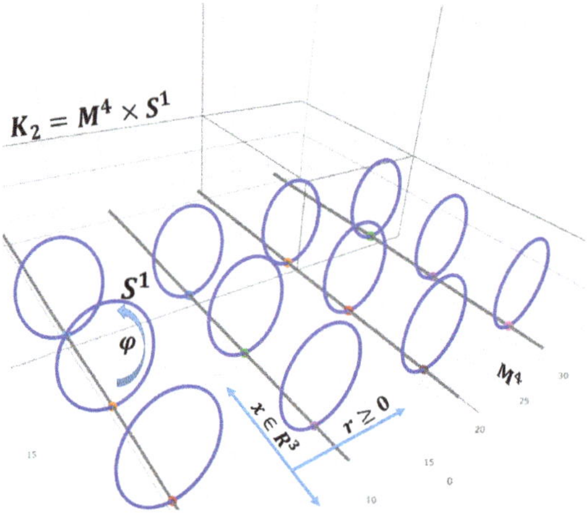

Figure 3.5: Graphical representation of 5D Kaluza-Klein spacetime, $K_2 \cong M^4 \times S^1$, where M^4 is a 4D Minkowski spacetime and S^1 is a non–traversable circle (below Planck resolution scale), representing a curled orientations dimension.

3.1.9 Coordinate Transformations, Covariance, Contravariance, and Invariance

Before we define complex time (kime), complex events (kevents), and the spacekime manifold, we will briefly summarize the basic ideas underlying coordinate transformations, covariance, contravariance, invariance, and tensor arithmetic.

Coordinate transformations allow us to describe quantitatively various changes of the bases used to localize objects in a topological manifold or a linear vector space. For instance, a *basis* for a 5D real vector space is a set of 5 (basis) vectors $(\vec{\alpha}_1, \vec{\alpha}_2, \ldots, \vec{\alpha}_5)$ that have the property that

$$\left\{\forall \alpha \in \mathbb{R}^5,\ \ \exists \text{unique}\,5-\text{tuple}\{\lambda_i\}_{i=1}^5 \in \mathbb{R},\ \ \text{such that}\ \ \ \alpha = \sum_{i=1}^{5} \lambda_i \vec{\alpha}_i\right\}.$$

In other words, every location in the manifold has a unique representation as a linear combination of the *basis vectors*, $\{\vec{\alpha}_i\}_{i=1}^5$. More generally, a set of B elements of a B-dimensional vector space $\{\beta_i\}_{i=1}^B \in V$ is called a *basis*, if all elements $v \in V$ have unique representations as a finite linear combination of the B elements, $v = \sum_{i=1}^B \lambda_i \beta_i$. The coefficients λ_i of this linear combination represent the *coordinates* of v in the basis $\{\beta_i\}_{i=1}^B$ and the elements of the basis are called *basis vectors*. A vector space V may have multiple alternative bases, each representing linearly independent spanning sets; however all the bases of V have an equal number of elements that define the dimension of the vector space. The more general manifolds are topological spaces that are *locally* homeomorphic to a topological vector space over the reals.

Different analytical, modeling, and computational operations on vector spaces and manifolds may be simplified by utilizing specific alternative bases. Hence, it is very important to formalize the algebraic manipulations necessary to transform different coordinate-wise representations of vectors, locations, and operators from one set of basis vectors to another – this transformation of the basis is called *change of basis*.

Suppose we have a linear transformation $T: V_n \to W_m$ between a pair of vector spaces, V_n, W_m, and we introduce changes of bases in both spaces from the initial bases $\{v_1, \ldots, v_n\}$ of V_n and $\{\omega_1, \ldots, \omega_m\}$ of W_m to some new bases $\{v_1', \ldots, v_n'\}$ and $\{\omega_1', \ldots, \omega_m'\}$. Let γ_v and γ_v' be the coordinate isomorphisms transforming the usual $\mathbb{R}^n$ basis to the first $\{v_i\}_{i=1}^n$ and second $\{v_i'\}_{i=1}^n$ bases for V_n. Similarly, we will denote γ_ω and γ_ω' to be the coordinate isomorphisms transforming the usual $\mathbb{R}^m$ basis to the first $\{\omega_j\}_{j=1}^m$ and second $\{\omega_j'\}_{j=1}^m$ bases for W_m. Then, we can define a pair of mappings $T_1 = \gamma_\omega^{-1} \circ T \circ \gamma_v: \mathbb{R}^n \to \mathbb{R}^m$ and $T_2 = (\gamma_\omega')^{-1} \circ T \circ \gamma_v': \mathbb{R}^n \to \mathbb{R}^m$, both have matrix representations, that express the change-of-coordinates automorphisms $(\gamma_v')^{-1} \circ \gamma_v: \mathbb{R}^n \to \mathbb{R}^n$ and $(\gamma_\omega')^{-1} \circ \gamma_\omega: \mathbb{R}^m \to \mathbb{R}^m$, see **Figure 3.6.**

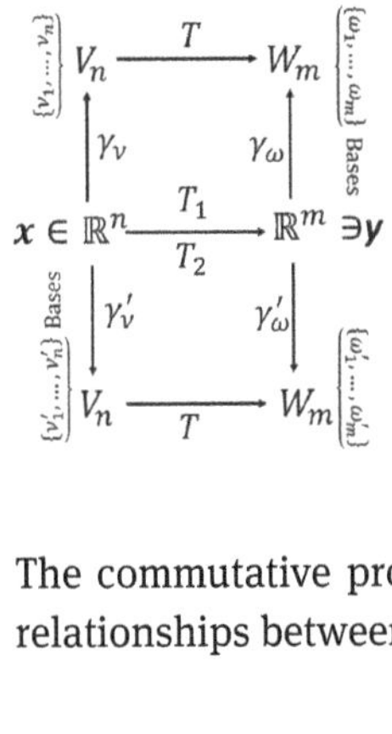

Figure 3.6: Commutative diagram of change of coordinates.

The commutative properties of this mapping diagram illustrate the corresponding relationships between the transformations and we have the following relation:

$$T_2 = (\gamma'_\omega)^{-1} \circ T \circ \gamma'_v = (\gamma'_\omega)^{-1} \circ \underbrace{\gamma_\omega \circ T_1 \circ (\gamma_v)^{-1}}_{T} \circ \gamma'_v.$$

A composition of linear coordinate transformation maps corresponds to matrix multiplication and we have:

$$T_2 = \underbrace{Q}_{(\gamma'_\omega)^{-1} \circ \gamma_\omega} \times T_1 \times \underbrace{P^{-1}}_{(\gamma_v)^{-1} \circ \gamma'_v}.$$

When $V_n \equiv W_m = V$ and the same change of basis is applied, $\gamma_v = \gamma_\omega$ and $\gamma'_v = \gamma'_\omega$, then there is one change-of-basis matrix for the vector space V, $Q \equiv P$, $T_2 = P \times T_1 \times P^{-1}$, and matrices T_1 and T_2 are similar.

For each point on a differentiable manifold, $p \in M$, there are two derived vector spaces:

(1) the *tangent space*, T_pM, representing the space of *contravariant vectors* (kets) $|x\rangle \in T_pM$, derived by taking directional derivatives at the point $p \in M$, and
(2) the *cotangent space*, T_p^*M, this is the dual to the tangent space consisting of the covariant vectors (bras) $\langle x| \in T_p^*M$, which are linear functions $\langle x|:T_pM \to \mathbb{R}$.

More specifically, for a real vector space V, the set of all linear functions $L:V \to \mathbb{R}$ is the dual space to V, which is a vector space denoted by V^*. When V is finite dimensional, then V^* has the same dimension as V. This pairing of all vector spaces with their duals requires disambiguation of elements or *vectors* in V and those in V^*. Vectors in V are called *contravariant vectors* or *kets*, whereas vectors in V^* are called dual vectors, covectors, *covariant vectors*, or *bras*.

This terminology conveys that covariant components transform synergistically ("co") in the same way as the basis vectors, but contravariant components transform anti-synergistically ("contra") in the opposite way to basis vectors. *Transformation of basis vectors* refers to the conversion of the basis vectors in the original coordinate system into new basis vectors, whereas *transformation of vector components* refers to the change in the vector components (coefficients) relative to two different sets of coordinate axes. More succinctly:

$$\underbrace{\begin{pmatrix} \text{New} \\ \text{basis vectors, } v' \end{pmatrix} = \begin{pmatrix} \text{Transform} \\ \text{matrix, } T \end{pmatrix} \times \begin{pmatrix} \text{Original} \\ \text{basis vectors, } v \end{pmatrix}}_{\text{components covary}},$$

$$\underbrace{\begin{pmatrix} \text{Vector components} \\ \text{in the new coordinate system, } u_{v'} \end{pmatrix} = \begin{pmatrix} \text{Inverse Transpose} \\ \text{Transform matrix, } (T^t)^{-1} \end{pmatrix} \times \begin{pmatrix} \text{vector components} \\ \text{in the original coordinate system, } u_v \end{pmatrix}}_{\text{components contravary}}.$$

In 2D, suppose the initial basis $v = \{v_1, v_2\}$ and the new basis $v' = \{v_1', v_2'\}$ are related by the change of basis matrix:

$$T = \begin{bmatrix} T_1^1 & T_1^2 \\ T_2^1 & T_2^2 \end{bmatrix}.$$

Then,

$$v' = Tv,$$

$$\left| \begin{array}{l} v_1' = T_1^1 v_1 + T_1^2 v_2 \\ v_2' = T_2^1 v_1 + T_2^2 v_2 \end{array} \right. .$$

Using Einstein summation notation we see that the pair of bases *covary*, $v_i' = T_i^s v_s$.

Transposition of the transform matrix (T) is necessary when we move between mapping of the *bases* and mapping of the *components* (coordinates) since:

$$\Bigg\langle \underbrace{\begin{pmatrix} u^1 \\ u^2 \end{pmatrix}}_{\text{coordinates}} \Bigg| T \underbrace{\begin{pmatrix} v_1 \\ v_2 \end{pmatrix}}_{\text{basis}} \Bigg\rangle = \Bigg\langle T^t \begin{pmatrix} u^1 \\ u^2 \end{pmatrix} \Bigg| \begin{pmatrix} v_1 \\ v_2 \end{pmatrix} \Bigg\rangle = \Bigg\langle \left((u^1, u^2) T \right)^t \Bigg| \begin{pmatrix} v_1 \\ v_2 \end{pmatrix} \Bigg\rangle .$$

The transposed basis transformation matrix

$$T^t = \begin{bmatrix} T_1^1 & T_2^1 \\ T_1^2 & T_2^2 \end{bmatrix}$$

can be used to transform the coefficient *components* of a vector $u_v = u^1 v_1 + u^2 v_2$ from the old coordinate system, v, to components in the new basis, v':

$$u_{v'} = \left(T^t\right)^{-1} u_v \Leftrightarrow u_v = T^t u_{v'}.$$

Again, this shows that the coefficient components of vectors in the old and new coordinate systems *contravary* as they are related by the inverse coordinate transformation matrix, $(T^t)^{-1}$. The last matrix equation expresses the vector components in the new coordinate system in terms of a product of the inverse transpose base transformation and the vector components in the original (old) coordinate system. That is, a vector u may be represented by the column vectors u_v and $u_{v'}$ in the two bases, v and v':

$$u_{v'} = \left(T^t\right)^{-1} u_v \Leftrightarrow \begin{bmatrix} u_{v'}^1 \\ u_{v'}^2 \end{bmatrix} = \begin{bmatrix} T_1^1 & T_2^1 \\ T_1^2 & T_2^2 \end{bmatrix}^{-1} \begin{bmatrix} u_v^1 \\ u_v^2 \end{bmatrix},$$

$$u_v = T^t u_{v'} \Leftrightarrow \begin{bmatrix} u_v^1 \\ u_v^2 \end{bmatrix} = \begin{bmatrix} T_1^1 & T_2^1 \\ T_1^2 & T_2^2 \end{bmatrix} \begin{bmatrix} u_{v'}^1 \\ u_{v'}^2 \end{bmatrix}.$$

Let's look at one specific example where the change of basis transformation is:

$$\left| \begin{aligned} v_1' &= v_1 + 2v_2 \\ v_2' &= v_1 + 3v_2 \end{aligned} \right.,$$

$$T = \begin{bmatrix} T_1^1 & T_1^2 \\ T_2^1 & T_2^2 \end{bmatrix} = \begin{bmatrix} 1 & 2 \\ 1 & 3 \end{bmatrix}, \ T^{-1} = \begin{bmatrix} 3 & -2 \\ -1 & 1 \end{bmatrix}, \ T^t = \begin{bmatrix} T_1^1 & T_2^1 \\ T_1^2 & T_2^2 \end{bmatrix} = \begin{bmatrix} 1 & 1 \\ 2 & 3 \end{bmatrix}, \ \left(T^t\right)^{-1} = \begin{bmatrix} 3 & -1 \\ -2 & 1 \end{bmatrix},$$

$$v' = Tv, \ \text{ i.e.,} \ \left| \begin{aligned} v_1' &= v_1 + 2v_2 \\ v_2' &= v_1 + 3v_2 \end{aligned} \right. \Leftrightarrow \ v = T^{-1}v', \ \text{ i.e.,} \ \left| \begin{aligned} v_1 &= 3v_1' - 2v_2' \\ v_2 &= -v_1' + v_2' \end{aligned} \right. .$$

Substituting v_1 and v_2 in the old components of the vectors $u_v = u^1 v_1 + u^2 v_2$ yields the following expression for the components of a vector u in the new coordinate system, $u_{v'} = u_{v'}^1 v_1' + u_{v'}^2 v_2'$:

$$u_v = \underbrace{u^1}_{u_v^1} v_1 + \underbrace{u^2}_{u_v^2} v_2 = \left\langle \begin{pmatrix} u^1 \\ u^2 \end{pmatrix} \middle| \begin{pmatrix} v_1 \\ v_2 \end{pmatrix} \right\rangle = u^1 v_1 + u^2 v_2 = u^1\left(3v_1' - 2v_2'\right) + u^2\left(-v_1' + v_2'\right) =$$

$$\left\langle \begin{pmatrix} u^1 \\ u^2 \end{pmatrix} \middle| \underbrace{\begin{pmatrix} 3 & -2 \\ -1 & 1 \end{pmatrix}}_{T^{-1}} \begin{pmatrix} v_1' \\ v_2' \end{pmatrix} \right\rangle = \left\langle \underbrace{\begin{pmatrix} 3 & -1 \\ -2 & 1 \end{pmatrix}}_{(T^{-1})^t} \begin{pmatrix} u^1 \\ u^2 \end{pmatrix} \middle| \begin{pmatrix} v_1' \\ v_2' \end{pmatrix} \right\rangle =$$

$$\left\langle \left[\left(u^1, u^2\right) \underbrace{\begin{pmatrix} 3 & -2 \\ -1 & 1 \end{pmatrix}}_{T^{-1}} \right]^t \middle| \begin{pmatrix} v_1' \\ v_2' \end{pmatrix} \right\rangle =$$

$$\left\langle \begin{pmatrix} 3u^1 - u^2 \\ -2u^1 + u^2 \end{pmatrix} \middle| \begin{pmatrix} v_1' \\ v_2' \end{pmatrix} \right\rangle = \underbrace{\left(3u^1 - u^2\right)}_{u_{v'}^1} v_1' + \underbrace{\left(-2u^1 + u^2\right)}_{u_{v'}^2} v_2' = u_{v'}.$$

Recall from **Chapter 2** that for any matrix $A = A_{m \times n}$ and any complex vectors $x, y \in \mathbb{C}^n$ the bra-ket notation $\langle \cdot | \cdot \rangle$ and the use of Hermitian conjugation $\dagger$, which in the real value case is just the matrix transpose, to move the matrix between the bra and the ket: $\langle A^\dagger x | y \rangle = \langle x | Ay \rangle$. Thus, it can be expressed in matrix form representing the components of the vector u from the new basis, v', into components in the old basis, v, and vise-versa.

Invariance is another property indicating that a quantity does not change with a linear transformation of the reference axes. For a particle of mass m moving at a constant speed v, the four-velocity is $u = (u^1, u^2, u^3, u^4) = \gamma(v_x, v_y, v_z, c)$, where $\gamma = \frac{1}{\sqrt{1-\frac{v^2}{c^2}}}$ is the Lorentz factor associated with v, c is the speed of light, and the contravariant four-momentum $p^\mu = mu^\mu = \gamma m(v_x, v_y, v_z, c)$. Thus, the particle rest mass is the ratio of 4-momentum to 4-velocity $\left(m = \frac{p^\mu}{v_\mu}\right)$ and is measured in mass units that do not depend on distance. Hence, mass is a distance measuring unit invariant. Direction or velocity vectors are basis-independent when their components contravary with a change of basis to compensate, i.e., the transform matrix that maps the vector components must be the inverse of the matrix that transforms the basis vectors (contravariance). Position of an object relative to an observer and derivatives of position with respect to time represent vectors with contravariant components and are denoted in Einstein's notation by *upper indices*. Covectors are basis-independent when their components covary with a change of basis to remain representing the same convector, i.e., the components are transformed by the same matrix as the change of basis matrix (covariance) and the covariant components are denoted with *lower indices*.

During change of basis, a basis-independent vector has components that contravary with a change of basis, i.e., the matrix transforming the vector components is the inverse of the matrix transforming the basis vectors. Thus, the component coordinates of basis-invariant vectors are contravariant and denoted with upper indices. For instance, the position of an object relative to a basis $v = \{e_\mu\}_\mu$ is $\boldsymbol{x} = x^\mu e_\mu$ in Einstein's summation notation, or more explicitly $x_v = \sum_\mu x^\mu e_\mu$. Similarly, component coordinates of basis-independent covectors covary with a change of the basis and are transformed by the same matrix as the change of basis matrix. In Einstein notation, $\boldsymbol{x} = x_\mu e^\mu = \sum_\mu x_\mu e^\mu$, where the gradient of a function represents an example of such a covector whose covariant components x_μ are denoted with lower indices relative to the partial derivative basis $\left\{e^\mu = \frac{\partial}{\partial x^\mu}\right\}_\mu$. To summarize, the components of the position $\boldsymbol{x}$ and the gradient $\nabla = \frac{\partial}{\partial x^\mu}$ represent a contravariant 1-tensor (vector) and a covariant 1-tensor (covector), respectively.

Raising and lowering indices tensors are just a form of index manipulation in tensor arithmetic. These operations change the type of the tensor and are performed by multiplying by the covariant or contravariant metric tensor and then contracting indices by setting a pair of indices to be equal and then summing over the repeated indices (Einstein notation). For instance, if we multiply a (nonsingular) *contravariant* metric tensor (upper indices) g^{ij} with a vector A and then contracting indices yields a new upper index covariant (covector), $g^{ij}A_j = A^i$. Analogously, multiplying the covariant metric tensor and a covector with contracting lowers an index, $g_{ij}A^j = A_i$. As raising and lowering the same index are inverse operations corresponding to the covariant and contravariant metric tensors, $\delta_i^k = g^{kj}g_{ji} = g_{kj}g^{ji} = \delta_k^i$, $\forall i, k$, which represents the identity metric tensor.

In 4D Minkowski spacetime, we can show a concrete example using the 4-position vector:

$$X_\mu = \left(\underbrace{x}_{X_1}, \underbrace{y}_{X_2}, \underbrace{z}_{X_3}, \underbrace{-ct}_{X_4} \right).$$

For this specific tensor signature: $\left(\overbrace{\underbrace{+,+,+}_{i},-}^{\mu} \right)$, the flat Minkowski metric tensor is $g_{\mu\nu} = (g^{\mu\nu})^{-1} \equiv g^{\mu\nu} = \begin{pmatrix} 1 & 0 & 0 & 0 \\ 0 & 1 & 0 & 0 \\ 0 & 0 & 1 & 0 \\ 0 & 0 & 0 & -1 \end{pmatrix}$.

Then to raise the index of the position vector (first-order tensor) we multiply the vector by the tensor and contract indices $X^\mu = g^{\mu\nu}X_\nu = g^{\mu i}X_i + g^{\mu 4}X_4$, for $\mu = 4$ we have the time component $X^4 = g^{4i}X_i + g^{44}X_4 = -X_4$ and for the corresponding space components $X^j = g^{ji}X_i + g^{j4}X_4 = \delta^i_j X_i = X_j$, $\forall 1 \le j \le 3$. Therefore, $X^\mu = g^{\mu\nu}X_\nu = \left(\underbrace{x}_{X_1}, \underbrace{y}_{X_2}, \underbrace{z}_{X_3}, \underbrace{ct}_{X_4} \right)$.

Raising and lowering the indices works similarly for second-order tensors (matrices) by multiplying by the covariant metric tensor $g^{\mu\nu}$ and contracting appropriately the indices:

$$\underbrace{A^{\mu\nu} = g^{\mu\gamma}g^{\nu\delta}A_{\gamma\delta}}_{\text{index raising}} \text{ and } \underbrace{A_{\mu\nu} = g_{\mu\gamma}g_{\nu\delta}A^{\gamma\delta}}_{\text{index lowering}}.$$

3.2 *Kime*, *Kevents* and the Spacekime Metric

In **Chapter 1**, we defined *kime* for a given spatial location as complex time, $k = re^{i\varphi}$, representing the dual nature of *kevents order* ($r > 0$), characterizing the kevent longitudinal displacement in time, and *kevent momentum*, reflecting the direction of change ($-\pi \le \varphi < \pi$), see **Figure 1.7**. This leads naturally to the 5D *spacekime* manifold ($\mathbb{R}^3 \times \mathbb{C}$). There are several complementary representations of the kime manifold that use different coordinate descriptions of the same two kime degrees of freedom, **Figure 3.7**. Having alternative isomorphic ways to represent complex time is useful for symbolic computing and analytical representations. For example, the ability to easily transition between (Descartes) Cartesian and polar coordinates describing the same particle state or kevent helps to simplify notations, utilize known multivariate properties, or employ various physical laws.

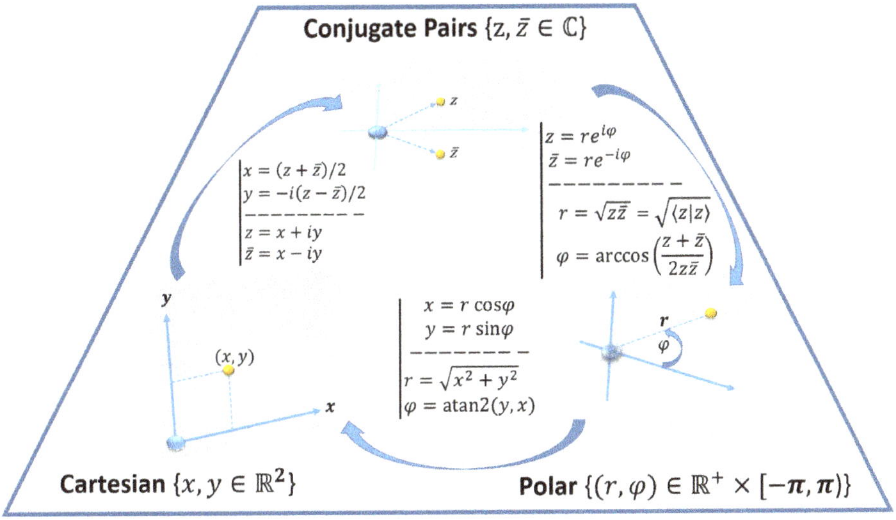

Figure 3.7: Examples of isomorphic kime manifold parameterizations.

Let's extend the relativistic and particle physics definitions of events to complex time. *Kevents* (complex-time events) are points (or states) in the spacekime manifold X. Relativistically, each kevent is defined by where it occurs in space ($\boldsymbol{x} = (x, y, z)$), what is its causal longitudinal order (r), and in what kime-direction it takes place in (φ).

The scaled polar coordinate transformation maps kime order (r) and phase direction (φ) into meters:

$$\left| \begin{array}{l} r = t \\ k_1 = r\cos(\varphi) \\ k_2 = r\sin(\varphi) \end{array} \right. .$$

Note that the trigonometric functions transform the kime phase (φ) into scalars and the speed of light (c) transforms kevent longitudinal order (in seconds) into meters. Thus, we can do 5D spacekime arithmetic (in meters).

In relativistic terms, kevents are characterized by their spacekime (SK) coordinates:

$$(\boldsymbol{x}, c\boldsymbol{k}) = \left(\underbrace{x^1, x^2, x^3}_{\text{space}}, \underbrace{ck_1 = x^4, ck_2 = x^5}_{\text{kime}} \right) \in SK.$$

To extend the particle-physics notion of *event*, referring to fundamental instantaneous particle interactions, to *kevent*, we need to generalize the Minkowski 4D tensor metric to the 5D spacekime manifold.

An *invariant interval* between two kevents, $a,\ b \in SK$, is defined by:

$$ds^2 = \left(x_a^1 - x_b^1\right)^2 + \left(x_a^2 - x_b^2\right)^2 + \left(x_a^3 - x_b^3\right)^2 - \left(x_a^4 - x_b^4\right)^2 - \left(x_a^5 - x_b^5\right)^2,$$

$$ds^2 = \left(dx^1\right)^2 + \left(dx^2\right)^2 + \left(dx^3\right)^2 - \left(dx^4\right)^2 - \left(dx^5\right)^2,$$

where each term represents a classical 1D interval in $\mathbb{R}^1$. These intervals are invariant because other observers using the same metric but different coordinate systems, e.g., $\left(x^{1'}, x^{2'},\ x^{3'}, x^{4'}, x^{5'}\right)$, would measure the same interval independent of the coordinate system:

$$(ds')^2 = \left(dx^{1'}\right)^2 + \left(dx^{2'}\right)^2 + \left(dx^{3'}\right)^2 - \left(dx^{4'}\right)^2 - \left(dx^{5'}\right)^2 = ds^2.$$

The general spacekime metric tensor may also be expressed more concisely as:

$$ds^2 = \sum_{i=1}^{5} \sum_{j=1}^{5} \Lambda_{ij} dx^i dx^j = \Lambda_{ij} dx^i dx^j,$$

where the $3+2$ indices $(1, 2, 3, 4, 5)$ correspond to the space + kime basis, and the implicit summation notation on the right-hand side implies that if an index is repeated as a subscript and superscript, then it represents a summation over all its possible values. The general Minkowski 5×5 metric tensor $\left(\Lambda_{ij}\right)_{i=1,\ j=1}^{5,5}$ characterizes the geometry of the *curved* spacekime manifold. However, in Euclidean (flat) spacekime, the spacekime metric permits a simpler tensor representation:

$$\left(\Lambda_{ij}\right) = \begin{bmatrix} 1 & 0 & 0 & 0 & 0 \\ 0 & 1 & 0 & 0 & 0 \\ 0 & 0 & 1 & 0 & 0 \\ 0 & 0 & 0 & -1 & 0 \\ 0 & 0 & 0 & 0 & -1 \end{bmatrix} = \underbrace{\operatorname{diag}\left(\underbrace{1, 1, 1}_{\text{space}}, \underbrace{-1, -1}_{\text{kime}}\right)}_{\text{tensor signature } (+\,+\,+\,-\,-)}.$$

The invariant intervals are analogous to distances in flat 3D spaces. As the delta change in kime can exceed the aggregate delta change in space, the interval distances can be positive, trivial, or even negative:

- *Spacelike* intervals correspond to $ds^2 > 0$, where an inertial frame can be found such that two kevents $a,\ b \in SK$ occur simultaneously in kime. An object cannot be present at two kevents which are separated by a spacelike interval.
- *Lightlike* intervals correspond to $ds^2 = 0$. If two kevents are on the line of a photon, then they are separated by a lightlike interval and a ray of light could travel between the two kevents.
- *Kimelike* intervals correspond to $ds^2 < 0$. An object can be present at two different kevents, which are separated by a kimelike interval.

The concept of multidimensional time is not new [62, 100, 101]. The real, imaginary, and psychological aspects of multiple time dimensions were debated by Dobb, Chari, Bunge, and others [102–104]. In the early 1980's, Hartle and Hawking investigated the effect of an isometric transformation of time into purely imaginary time ($t \to i\tau$). This approach solves some problems related to singularities of Einstein's equations [105].

From a topological point of view, Chari contrasted the idea of extending unidimensional to multi-dimensional time to the extension of finite dimensional Euclidean spaces to infinite dimensional Hilbert spaces [104]. The major drawback of lifting time to higher dimensions comes from the fact that such extensions may break classical probability axioms, e.g., certain events may have negative likelihoods as higher-dimensional temporal spaces may permit multidirectional "time" travel. However, imposing certain symmetry conditions on the position and momentum of particles may resolve the negative likelihoods problem [62, 101].

Bars connected 2D-time field theory directly to the physical world by extending the standard 4D spacetime model to 6D and resolving the strong charge conjugation parity (CP) symmetry violation problem [106]. In this work, the underlying 2D time reflects the local symmetry of the symplectic group $Sp(2, R)$, a non-compact, simply-connected Lie group, where position and momentum are indistinguishable at any instant. As we are not equipped to directly perceive higher-dimensional spacetime, we can interpret the ambient 4D spacetime environment as a mere silhouette of an actual higher-dimensional world with multidimensional time.

3.3 Some Problems of Time

In **Chapter 1** (Motivation), we already discussed the mysterious nature of time and pointed out some of the paradoxes associated with time. Some of the key problems of time are summarized below.

- Speed of light phenomenon; relativistically time is a local characterization of phenomena, there is no global universal time.
- The unidirectional arrows of time (psychological arrow of time representing inexorable past to future flow; thermodynamic arrow of time reflecting the growth of entropy; and cosmological arrow of time representing the expansion of the universe) [107].
- "The *problem of time*" concept has different meanings in quantum mechanics (flow of time as universal and absolute) and general relativity (flow of time as malleable and relative). In 1960, the physicist Bryce DeWitt described the quantum concept of time by "different times are special cases of different universes" [108]. The Wheeler-DeWitt equation predicts that time is an emergent phenomenon for internal observers, which is absent for external observers of the universe. Evidence in support of this was provided in 2013 by researchers at the

Istituto Nazionale di Ricerca Metrologica (INRIM) in Turin, Italy, via the Page and Wootters experimental test [109, 110].
- Time dilation, i.e., the action of moving objects affects (slows) passage of time [111].

3.4 Kime-Solutions to Time-Problems

If we pick (any) point $k_o \in \mathbb{C}$ as a kime-origin serving as a reference, then the kime, k, is a universal characteristic (relative to k_o) just like space ($\mathbb{R}^3$) is universal, relative to a specific spatial origin ($\mathbf{0} \in \mathbb{R}^3$). However, the spacekime manifold is still only locally Euclidean (flat) and globally curved. Just like time is always a local observable, there is no global and canonical universal kime.

Kime resolves the problem with the unidirectionality of time, as all kimelike intervals connecting a pair of kevents in the complete 2D kime space can be traversed by uncountably many trajectories (e.g., parametric curves). Furthermore, using appropriate gauge symmetries, the 2D kime extension resolves all problems with *unitarity* (time-evolution of a physical system is equivalent to a unitary operator, i.e., relative to the present, a past point and a future point are unique) and *causality* (in the light cone of an event, an effect always occurs after its cause) [112]. Using the 2D time-lifting framework, the classical 4D Minkowski spacetime is naturally and consistently embedded in the 5D spacekime manifold.

The concept of multidimensional time is beginning to gather support and play an increasingly important role in modern theories of physical systems [62, 113–115]. According to the Big Bang inflation theory, the visible universe is only a small part of a larger multiverse, possibly an uncountable collection of many loosely connected individual universes governed by distinct physical laws, different ambient-conditions, and weak interactions between them [116]. So far, no physical principle has been identified that determines the possible number of spatial or temporal dimensions. Therefore, there is no specific law that dictates a precise number of dimensions or limits the possibilities to only what is perceived in the human-observable universe. The number of spatial and temporal dimensions in our universe is more likely to be a result of chance, coupled with our intrinsic understanding, which is shaped by our perceptions, observations, operations, and immersive experiences into the known universe.

As shown in some studies, it is possible to formulate physically significant theories using two, three, and higher temporal dimensions [62, 101, 113]. As the physicist Itzhak Bars noted, "*the physics of two-dimensional time can be seen as a means of better understanding the physics of one-dimensional time; moreover, the physics of two-dimensional time offers new perspectives in the quest for unified theory, raising deep questions about the meaning of space-time*" [106]. For theoretical systems that are not yet fully constructed or understood, e.g., membrane (a.k.a. brane) theory, the physics of two-dimensional time suggests a possible approach to a more symmetric and more meaningful formulation. For instance, constructing universal representations using 11 spatial and 2 temporal dimensions may lead to deeper insights into the nature of space and time and the implications for long-distance communication and travel. In addition, such lifting of dimensionality may facilitate one possible way of constructing the most symmetrical version of a fundamental theory [62, 106].

Intuitively, time is the fourth dimension of the unified Minkowski spacetime. There are common characteristics of space and time, but time, as a positive real event-ordering characteristic, has some specific properties that distinguish it from space. First, time is constantly "running," i.e., it is advancing and diffusing independently of being observed. We perceive the world as naturally and dynamically changing over time. Second, the time is unidirectional and, relative to the illusive *present*, there is a clear asymmetry between the past and the future. This "*arrow of time*" contrasts the observed space isotropicity of various geometric and topological characteristics. Third, the $3 + 1$ spacetime dimensions are somewhat imbalanced as space covers a field of three dimensions without boundary, whereas time occupies a one-dimensional space with a clearly delineated boundary.

If the concept of two-dimensional complex time (kime) does represent a more general and realistic universal description, these differences between space and kime mostly disappear, restoring a global 5D spacekime symmetry. Two-dimensional time may explain the possibility of smooth traversal of the spacekime manifold and thus resolve the time-travel paradoxes. Instead of simply marching relentlessly forward, kime traversal explains the natural trajectories of spacekime motion, including kime-travel as simple closed 2D curves. By varying kime-phases (directions of time), kime arcs can describe perceptions of traveling back or forth in time.

Complex time can also solve the "grandfather paradox." If one goes back in time and kills their grandfather, before the time the latter met the traveler's grandmother, then one of the traveler's parents would not be born and by consequence, the traveler themselves would have never been born. That logical impossibility is driven by the fact that all events in spacetime are assumed to take place in the same kime direction.

Complex time allows for multiple event trajectories, e.g., multiple story lines, alternative narratives, complementary experiences at the same spacetime locations. In a measure-theoretic sense, traveling through the 2D kime manifold has trivial likelihood of revisiting an exact same point in the past. This is clear as passing through one fixed kime point, $\kappa \in \mathbb{C}$, is as likely as randomly choosing one specific real number (cf. Axiom of Choice [117]), an event of trivial measure.

In the 5D Minkowski spacekime dimensional lift, the classical four spacetime dimensions are familiar and are associated with common and intuitive interpretations. However, the spacekime theory suggests that these four dimensions represent a universal silhouette, a "shadow" of reality. Clearly, all fundamental physical laws and concepts may need to be reviewed and extended as lower dimensional projections of complex manifolds do exhibit similar characteristics, but also have unnatural constraints that make the simplified lower-dimensional representations of reality different from their native counterparts. For instance, the Heisenberg principle states that it may be possible to measure precisely either the momentum or the position of a particle, but not both at the same time.

What is the mysterious theoretical limitation that restricts this possibility? Would lifting the concept of time to 2D complex time allow us to jointly and perfectly record both the particle position and momentum? This clearly requires redefining the momentum (p) as $p = m\ v$, where m is the particle mass and the velocity (v) is defined in terms of kime, not time. Itzhak Bars suggests that the position and momentum of a particle are indistinguishable at any given kime moment [118]. When swapping the momentum and position, physics remains the same. But if the momentum and the position are interchangeable, it implies that everyone (universal observer) can get their own unique direction of time. Perhaps a hidden treasure of the universe is wrapped in the extra time-direction (kime-phase) dimension. Understanding this additional time dimension may help us answer the question "*Why, after a thorough search, have we still not been able to find the composites of dark matter?*" and, perhaps, enhance our data science and analytical inference abilities based on spacetime-observed evidence.

Open Problems

1) Does kime have the same interpretation in quantum mechanics and in general relativity (relative to a specified origin), just like the spatial coordinate references? In other words, is kime universal and absolute?
2) We know time, by itself, is excluded from the Wheeler-DeWitt equation. Is this true for kime as well? That is, does the Wheeler-DeWitt equation depend on kime the same way it depends on the particle location?
3) Is there kime-dilation, reminiscent of time-dilation? In other words, does the action of moving objects affect (e.g., slow) kime? How? Some suggestions are made below.
4) Explore the relations between various spacekime principles (e.g., space-kime motion and PDEs with respect to kime) and Painlevé equations in the complex plane [119, 120].
5) Explore the relation of kime to the four arrows of time (epistemological, mutability, psychological, and explanation-causation-counterfactual). How does the arrow of time relate to increasing the kime magnitude (time) and the corresponding growth

of the kime arc-length associated with a fixed kime-phase range? As the kime magnitude increases, there is a perception of a natural spacekime inflation (linear in terms of time) of the kime arc-length and the corresponding spacekime volume. This is related to the second law of thermodynamics and the observed time-dependent increase of the entropy of closed systems in spacetime. Is there a relation between the spacekime volume rate of change, $\frac{d}{dt}V(t) = \Delta x\, \Delta\varphi$, and the accelerating expansion of the universe? The accelerating expansion [121] is defined in terms of the second *time* derivative of the cosmic scale factor, $a''(t)$, and the increasing with time Hubble parameter defined along with its derivative by:

$$H(t) = \frac{a'(t)}{a(t)},$$

$$\frac{dH}{dt} = \frac{a''(t)}{a(t)} - \left(\frac{a'(t)}{a(t)}\right)^2 = -H^2(t)(1+q),$$

$$q \equiv -\frac{a''(t)}{a(t)H^2(t)} = -\frac{a(t)\, a''(t)}{(a'(t))^2}.$$

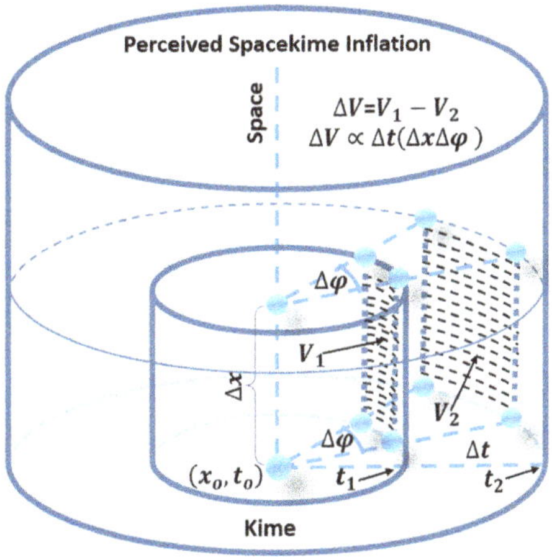

Some empirical observations suggest $q \cong -0.55$, implying the acceleration is $a''(t) > 0$ and $\frac{dH}{dt} < 0$.

3.5 The Kime-Phase Problem

The kime-phase problem in data science is analogous to the crystallographic phase problem. X-ray crystallography uses a detector to measure the diffraction intensities, which are the scattering amplitudes losing all information about the phases of the scattered wavefield. Similarly, in data science we use observed spacetime information without recording the crucial kime-directions needed to successfully reconstruct the data object in spacekime. This "data science kime-phase problem," which is due to a lack of explicit mechanisms to directly measure the kime-direction, may be tackled using the same phase-estimation strategies applied in many other crystallographic and non-crystallographic, imaging and signal processing techniques. The simple strategy, which reflects the current state-of-the-art approaches in all spacetime data analytics, ignores the kime-phases. In other words, spacetime analytics assume $\varphi = 0$ (or a constant) and correspond to utilizing only the Fourier magnitudes, which as we saw in **Chapter 1** generate oversimplifications of the real data objects. In fact, discarding the Fourier magnitudes and utilizing solely the kime-phases lead to more recognizable representations of the original data objects. Thus, recovery or estimation of the unknown kime-phase is expected to boost the information content in the data and hence improve the resulting data analytics. The quality of the data-analytic results (e.g., prediction strength, reproducibility, replicability, and forecasting variability) depends on the spacekime reconstruction of the data object, i.e., robust kime-phase retrieval. There are many kime phase-aggregator methods for phase-estimation that can be constructed, implemented, and validated.

In practice, these data science techniques for kime-phase estimation may resemble some of the iterative X-ray diffraction imaging approaches used for phase retrieval. The simplest of all algorithms would reconstruct data objects simply by synthesizing the data in spacekime using the observed Fourier intensity measurements (k-space transformed data) along with some fixed-phase value, $-\pi \leq \varphi_o < \pi$. Alternatively, we can solve this inverse problem using the Gerchberg-Saxton algorithm applied to the intensity measurements in the Fourier domain and the real spacetime [122]. More complex schemas may be based on iteratively enforcing constraints in spacetime and Fourier domain [123–125]. Alternative types of constraints may be necessary to reflect the characteristics of the observed data object and the specific analytical strategy.

Using the same notation we employed earlier, suppose the observed data object is an $n \times k$ tensor of cases and features. Then, there will be a set of $n \times k$ equations that can be used to solve the inverse constrained problem of recovering the spacekime representation of the data object. However, due to time complexity, the spacekime data object representation consists of n^2k unknowns, accounting for k features for each of the 2 independent time dimensions. One can include additional constraints to reduce the number of unknowns to the number of equations, which may eventually lead to solving for the missing kime-phases. These constraints can be designed so that some are specified on the Fourier-transformed (analyzed) data, e.g., a Fourier modulus constraint in inverse domain, and some are formulated on the spacekime (synthesized) data, e.g., support or overlap constraints. Other types of constraints, like the X-ray crystallographic atomicity constraint and the object histogram constraint [126], may also be useful.

3.5.1 The Effects of Kime-Magnitudes and Kime-Phases

Jointly, the *amplitude spectrum* (magnitudes) and the *phase spectrum* (phases) uniquely describe the spacetime representation of a signal. However, the importance of each of these two spectra is not equivalent. In general, the effect of the phase spectrum is more important compared to the corresponding effects of the amplitude spectrum. In other words, the magnitudes are less susceptible to noise or the accuracy of their estimations. The effects of magnitude perturbations are less critical relative to proportional changes in the phase spectrum. For instance, the zero-level-set of spacetime locations where the signal is zero, the information can be reconstructed (by the IFT) relatively accurately using incorrect magnitudes solely by using the correct phases [127]. For a real-valued signal f, suppose the amplitude of its Fourier transform, $FT(f)=\hat{f}$, is $A(\omega)>0, \forall\omega$, then:

$$f(x)=IFT\left(\hat{f}\right)=Re\left(\frac{1}{2\pi}\int_{\mathbb{R}}\underbrace{A(\omega)e^{i\phi(\omega)}}_{\hat{f}(\omega)}e^{i\omega x}d\omega\right)=Re\left(\frac{1}{2\pi}\int_{\mathbb{R}}A(\omega)e^{i(\phi(\omega)+\omega x)}d\omega\right)=$$

$$=\frac{1}{2\pi}\int_{\mathbb{R}}A(\omega)\cos(\phi(\omega)+\omega x)d\omega.$$

Thus, the roots of $f, f(x)=0$, occur for $\omega x+\phi(\omega)=\pm(2k-1)\frac{\pi}{2}, k=1,2,3,\cdots$.

A solely amplitude-driven reconstruction would preserve the total signal energy but yield worse results:

$$f_A(x)=IFT\left(\hat{f}\right)=\frac{1}{2\pi}\int_{\mathbb{R}}\underbrace{A(\omega)}_{\text{no phase}}e^{i\omega x}d\omega.$$

Whereas a solely phase-based reconstruction would change the total signal energy however it may generate better reconstruction results:

$$f_\phi(x)=IFT\left(\hat{f}\right)=\frac{1}{2\pi}\int_{\mathbb{R}}\underbrace{e^{i\phi(\omega)}}_{\substack{\text{unitary}\\\text{amplitude}}}e^{i\omega x}d\omega.$$

The latter would include some signal-recognizable features, e.g., contours, as the zero-level-curves of the original f preserved by the phase-only reconstruction f_ϕ. This suggests that the *Fourier phase* of a signal is more informative than the *Fourier amplitude*, i.e., the magnitudes are robust and the phases are more susceptible to errors or perturbations. Figure 3.8 shows example reconstructions of the 2D square image using different amplitude and phase estimates.

To resolve the 3D structure of small proteins using X-ray crystallography, crystal structures are bombarded by particles/waves, which are diffracted by the crystal to yield the observed diffraction spots or patterns. Each diffraction spot corresponds to a point in the reciprocal lattice and represents a particle wave with some specific amplitude and a relative phase. Probabilistically, as the particles (e.g., gamma-rays or photons) are reflected from the crystal, their scatter directions are proportional to the square of the wave amplitude, i.e., the square of the wave Fourier magnitude.

X-rays capture these amplitudes as counts of particle directions, but miss all information about the relative phases of different diffraction patterns.

Spacekime analytics are analogous to X-ray crystallography, DNA helix modeling, and other applications, where only the Fourier magnitudes (power spectra) are observed, but not the phases (kime-directions). Thus, the phases need to be estimated to correctly reconstruct the intrinsic 3D object structure, in our case, the correct spacekime analytical inference. Clearly, signal reconstruction based solely on either the amplitudes or the phases is an ill-posed problem, i.e., there will be many *alternative solutions*. In practice, such *signal* or *inference* reconstructions are always application-specific, rely on some a priori knowledge of the process (or objective function), or depend on information-theoretic criteria to derive conditional solutions. Frequently, such solutions are obtained via least squares, maximum entropy criteria, maximum a posteriori distributions, Bayesian estimations, or simply by approximating the unknown amplitudes or phases using prior observations, similar processes, or theoretical models.

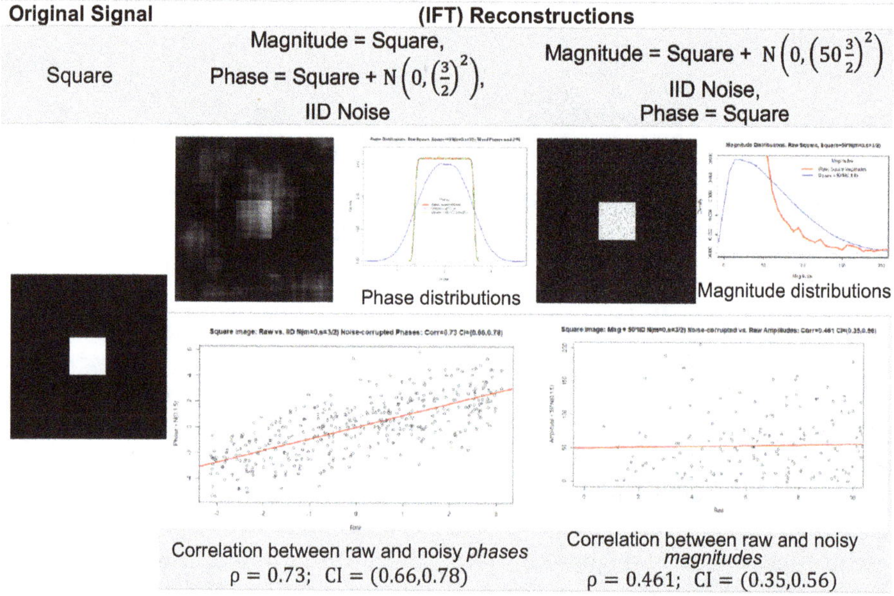

Figure 3.8: Examples of square image reconstructions illustrating the noise robustness and susceptibility of the Fourier magnitudes and phases. Note that relative to the ranges of the phases, $(-\pi, \pi)$, and magnitudes, $(0, 200)$, the synthesized signals are subjected to substantial noise levels, $\sigma_1 = \left(\frac{3}{2}\right)$ and $\sigma_2 = 50\frac{3}{2}$, respectively. Therefore, the phase may encode more of the energy in a signal than the magnitude of the signal.

3.5.2 Strategies for Solving the Missing Kime-Phase Problem

There are many alternative solutions to the problem of estimating the unobserved kime-phases. All solutions depend on the quality of the data (e.g., noise), the signal energy (e.g., strength of association between covariates and outcomes), and the general experimental design. There can be rather large errors in the phase reconstructions, which will in turn affect the final spacekime analytic results. Most phase-problem solutions rely on using some *prior knowledge* about the characteristics of the experimental design (case-study phenomenon) and the desired inference (spacekime analytics). For instance, if we artificially *load the energy* of the case-study (e.g., by lowering the noise, increasing the signal-to-noise ratio, or increasing the strength of the relation between explanatory and outcome variables), the phases computed from such stronger-signal dataset will be more accurate representations than the original phase estimates. Examples of phase-problem solutions include *energy modification* and *fitting and refinement* methods.

3.5.2.1 Energy Modification Strategies

In general, *energy modification* techniques rely on prior knowledge, testable hypotheses, or intuition to modify the dataset by strengthening the *expected* relation we are trying to uncover using spacekime analytics.

Kime-Phase Noise Distribution Flattening: In many practical applications, only a portion of the dataset (including both cases and features) may include valuable information. The remaining part of the data may include irrelevant, noisy, or disruptive information. Clearly, we can't explicitly untangle these two components, however, we do expect that the irrelevant data portion would yield uninformative/unimportant kime-phases, which may be used to estimate the kime-phase noise-level and noise-distribution. Intuitively, if we modify the dataset to flatten the irrelevant kime-phases, the estimates of the corresponding true-signal kime-phases may be more accurate or more representative. We can think of this process as using kime-phase information from some known strong features to improve the kime-phase information of other particular features. Kime-phase noise distribution flattening requires that the kime-phases be good enough to detect the boundaries between the strong and the weakly-informative features.

Multi-Sample Kime-Phase Averaging: It's natural to assume that multiple instances of the same process would yield similar analytics and inference results. For large datasets, we can use ensemble methods (e.g., SuperLearner [128], and CBDA [15, 129]) to iteratively generate independent samples, which would be expected to lead to analogous kime-phase estimated and analytical results. Hence, we expect that when salient features are extracted by spacekime analytics based on independent samples, their kime-phase estimates should be highly associated, albeit perhaps not identical. However, weak features would exhibit exactly the opposite effect – their kime-phases may be highly variable (noisy). By averaging the kime-phases, noisy-areas in the dataset may cancel out, whereas, patches of strong-signal may preserve important kime-phase details, which would lead to increased kime-forecasting accuracy and improve the reproducibility of kime-analytics.

Histogram Equalization: As common experimental designs and similar datasets exhibit analogous characteristics, the corresponding spacekime analytics are also expected to be synergistic. Spacekime inference that does not yield results in some controlled or expected range may be indicative of incorrect kime-phase estimation. We can use histogram equalization methods [130] to improve the kime-phase estimates. This may be accomplished by altering the distribution of kime-phases either to match the phase distribution of other similar experimental designs or to generate more expected spacekime analytical results.

3.5.2.2 Fitting and Refinement

Related to *energy modification* strategies, the *fitting and refinement* technique capitalizes on the fact that strong energy datasets tend to have a smaller set of salient features. So, if we construct case studies with some strong features, the corresponding kime-phases will be more accurate, and the resulting inference/analytics will be more powerful and highly reproducible.

Various classification, regression, supervised and unsupervised methods, and other model-based techniques allow us to both fit a model (estimate coefficients and structure) and apply the model for outcome predictions and forecasting. Such models permit control over the characteristics of individual features and multivariate interrelations that can be exploited to gather valuable kime-phase information. Starting with a reasonable initial guess (kime-phase prior), the *fitting and refinement* technique is an iterative process involving:

1) Reconstructing the data into spacetime using the current iteration kime-phase estimates,
2) (Re)fitting or (re)estimating the spacekime analytical model,
3) Comparing the analytical results or evaluating the inference to expected outcomes, and
4) Refining the kime-phase estimator aiming to gain better outcomes (#3).

Indeed, at each new iteration, alternative *energy modification* strategies (e.g., averaging or flattening) can be applied before building a new model (#1 and #2).

This is a very active research area and for brevity and simplicity of the presentation, our examples in **Chapter 6** show applications on several alternative kime-phase estimators – phase-aggregators. These include *constant phases*, like the *nil-phase* (assuming trivial phases, $\varphi_o = 0$), *swap-phases* which use the observed phases from other analogous datasets, and *random-phases*, where we *randomly sample* the phases from the synthetically generated Fourier-transformation (analyzed) phase distribution.

3.6 Common Use of Time

In this section, we will describe some of the common roles of time, which we will try to extend later to complex-time.

3.6.1 Rate of Change

The average rate of change around a value $x = a$ is a generic difference quotient for a function $y = f(x)$ that represents the average rate of change at a for that function relative to the change of the argument:

$$Avg\ Rate\ Change(a) = \frac{\Delta y}{\Delta x} = \frac{f(a+\Delta x) - f(a)}{\Delta x}.$$

$$Instantaneous\ Rate\ Change(a) = \lim_{\Delta x \to 0}\left(\frac{\Delta y}{\Delta x}\right) = \lim_{\Delta x \to 0}\frac{f(a+\Delta x) - f(a)}{\Delta x} = f'(a).$$

3.6.2 Velocity

The velocity is the rate of spatial position change over time,

$$\boldsymbol{v} = \frac{dx}{dt},\ \boldsymbol{x} \in \mathbb{R}^3 \text{ (spatial position } \boldsymbol{x} = (x, y, z)),\ t = time.$$

Motion is a change in position of an object over time. Suppose $\boldsymbol{x}_o$ and $\boldsymbol{x}$ are the particle's initial ($\boldsymbol{x}_o = \boldsymbol{x}(0)$) and final ($\boldsymbol{x} = \boldsymbol{x}(t)$) positions, $\boldsymbol{v}_o$ and $\boldsymbol{v}$ are the particle's initial $\boldsymbol{v}_o = \frac{dx}{dt}|_{t=0}$ and final $\boldsymbol{v} = \frac{dx}{dt}$ velocities, respectively, $\boldsymbol{a} = \frac{dv}{dt}$ is the particle's acceleration, and t is the time interval. Then, for a 3D particle moving linearly along a straight line with a constant acceleration, the *equations of motion* are provided below.

3.6.3 Newton's Equations of Motion

Newton's Equations of Motion	Derivation
$v = at + v_o$ $x = x_o + v_o t + \frac{1}{2}at^2$ $v^2 = 2a(x - x_o) + v_o^2$	The equations on the left are derived from the definition of velocity and acceleration: – Equ 1: As $a = \frac{dv}{dt}$, integrating both sides yields $\int a dt = \int dv$. Since the acceleration is constant in time, $v = a \int dt = at + v_o$, where v_o is a constant representing the initial velocity. – Equ 2: $v = \frac{dx}{dt} = at + v_o$ (from equ 1), integrating both sides of the equation we get $\int dx = \int at dt + v_o \int dt$. As a is constant, this yields: $x = a \int d\frac{t^2}{2} + v_o t = a\frac{t^2}{2} + v_o t + C$, where the constant $C = x_0$ by setting $t = 0$. – Equ 3: $a = \frac{dv}{dt} = \frac{dv}{dx} \times \frac{dx}{dt} = \frac{dv}{dx} \times v = v\frac{dv}{dx}$. Again integrating over $[0, t]$, we get $\int_0^t a dx = \int_0^t v dv$ and under the initial condition ($v_o = v(0)$) this becomes $2a(x - x_0) + v_o^2 = v^2$.

3.6.4 Position (x) and Momentum (p)

Position and momentum represent finite dimensional vector spaces. The n-dimensional *position space* includes position vectors, spatial locations, or points, $\boldsymbol{x} \in S \equiv \mathbb{R}^n$. When the position vector of a point particle changes with time, its trace is a path (trajectory) in S. *Momentum space* represents all possible momenta vectors $\boldsymbol{p} = m\boldsymbol{v}$ in the system, describing the particle's momentum, measured in units of $\frac{mass(kg) \times length(m)}{time(s)}$, corresponding to its motion over time.

The mathematical duality between *position* and *momentum* is explained by the Fourier transform. For an integrable function $f(x): S \to \mathbb{C}^n$, its Fourier transform $\hat{f}$ represents the wavefunction in momentum space, $\hat{f}(\boldsymbol{p})$. Conversely, the inverse transform of a momentum space function is a position space function, $f(x)$. The position (x) and momentum ($\boldsymbol{p}$) are complementary (or conjugate) variables, connected by the Heisenberg uncertainty principle, $\sigma_x \times \sigma_{\boldsymbol{p}} \geq \frac{h}{4\pi}$, where σ_x and $\sigma_{\boldsymbol{p}}$ represent the standard deviations of the position and momentum, respectively, and h is the Planck constant. A non-zero function and its Fourier transform cannot both be sharply localized in both space and frequency. For example, compare trigonometric functions and polynomial functions; the trigonometric functions are perfectly localized in the frequency domain, but infinitely supported in the spatial domain. Their counterparts, the polynomial functions, exhibit exactly the opposite property – they are perfectly localized in space (polynomial basis) and permit infinite series expansion using the trigonometric basis in the frequency domain.

Thus, the physical state of the particle can either be described by a position (wave) function of x or by the momentum function of $\boldsymbol{p}$, but not by a function of both variables. In the special 1D case, the spatial displacement is a function of *time* (in seconds), $f = f(t)$, the *velocity* $v = \frac{df}{dt}$ is measured in $\left(\frac{length(m)}{time(s)}\right)$, the displacement vector is $f = \int v dt = \int \frac{df}{dt} dt$, and the Fourier transform $\hat{f} = \hat{f}(\xi)$ is a function of the variable ξ, *frequency* in Hertz, $1Hz = 1\ \xi\ s^{-1}$.

3.6.5 Wavefunctions

All wavefunctions (ψ or $\Psi: S \to \mathbb{C}$) are integrals over the entire universe $S = \mathbb{R}^3 \times \mathbb{R}^+$ or $S = \mathbb{R}^5 \cong \mathbb{R}^3 \times \mathbb{C}$, depending if they are defined over spacetime or spacekime. Quantum theory implies that every particle in the universe exists probabilistically, i.e., simultaneously in all points (locations) in S. It may be extremely unlikely to find it in most locations as the probability could be extremely small, albeit rarely absolutely zero in a mathematical sense. The wavefunction defines the state of the wave at each spatial position and each time point.

The degree of freedom (DoF), i.e., the number of independent parameters that define the wavefunction, ψ, and determine the state of the particle, corresponds to a set of commuting observables, e.g., $(\boldsymbol{x}, \boldsymbol{k}) = \left(\underbrace{x, y, z}_{space}, \underbrace{r, \varphi}_{kime}\right)$. Once such a representation is chosen, the wavefunction can be derived from the quantum state.

For a given system, the choice of which commuting degrees of freedom to use is not unique, and correspondingly the domain of the wavefunction cannot be uniquely specified. For instance, it may be taken to be a function of all the position coordinates of the particles over position space, or the momenta of all the particles over momentum space; the two are related by a Fourier transform.

The inner product between two wavefunctions measures the overlap between the corresponding physical states. The relation between state-transition probabilities and inner product represents the probabilistic interpretation of quantum mechanics, the *Born rule* [131]. In essence, at a fixed time the squared modulus of a wavefunction, $\|\psi\|^2 = \langle\psi|\psi\rangle$, is a real number representing the likelihood (probability density) of observing the particle at a specific place, either in space or in frequency. Probability axioms dictate the normalization condition on the integral of the wave magnitude over all the system's degrees of freedom, $\int_\Omega \|\psi\|^2 = 1$, e.g., in 4D, $\Omega \subseteq \mathbb{R}^3 \times \mathbb{R}^+$.

3.6.6 Schrödinger Equation

Self-adjoint "Hermitian" linear operators (A) on Hilbert spaces: A is self adjoint if its conjugate transpose invariant (i.e., $A^* = A$):

$$\langle A\ x|y\rangle \equiv \langle x\ |A^* y\rangle = \langle x|A\ y\rangle, \text{ for all } x, y \in H.$$

The *Schrödinger equation* describes the evolution of wavefunctions over time [132]. Specifically, the Schrödinger equation explains the changes of a quantum mechanical system over time, as a first-order PDE in time (t) and a second-order in space $(\boldsymbol{x})$. Its solution, the wave-function ψ, is a function of all the particle coordinates and time, and delineates the time evolution of a quantum state. Thus, the solution describes the future wave amplitude $(\psi(t))$ from the present value $(\psi(t_o))$. In 3D, the position-basis form of the Schrödinger equation for one particle is:

$$i\hbar \frac{\partial \psi(\boldsymbol{x}, t)}{\partial t} = -\frac{\hbar^2}{2m} \nabla^2 \psi(\boldsymbol{x}, t) + V(\boldsymbol{x}, t)\psi(\boldsymbol{x}, t)$$

or

$$i\hbar \frac{\partial \psi}{\partial t} = -\frac{\hbar^2}{2m} \left(\underbrace{\frac{\partial^2}{\partial x^2} + \frac{\partial^2}{\partial y^2} + \frac{\partial^2}{\partial z^2}}_{\nabla^2 = \Delta} \right) \psi + \underbrace{V(x, y, z, t)}_{\text{potential}}\, \psi.$$

In this equation, h and $\hbar = \frac{1}{2\pi} h$ represent the Planck constant and the reduced Planck constant, m is the particle's mass, $V(\boldsymbol{x}, t)$ is the potential energy, and ∇^2 is the Laplacian, a second-order differential operator.

In 1D, the Schrödinger's equation for a time-varying wavefunction, $\psi(x, t)$, which is not subject to external forces $(V(x, t) = 0)$, is given by:

$$i\hbar \frac{\partial \psi(x, t)}{\partial t} = \frac{-\hbar^2}{2m} \frac{\partial^2 \psi(x, t)}{\partial x^2}.$$

More generally,

$$i\hbar \frac{\partial \psi}{\partial t} = \widehat{H}\psi,$$

$$i\hbar \frac{\partial}{\partial t} |\psi(\boldsymbol{x}, t)\rangle = \widehat{H}\ |\psi(\boldsymbol{x}, t)\rangle,$$

where $\widehat{H} = \underbrace{\frac{\hat{p}.\hat{p}}{2m}}_{\text{kinetic}} + \underbrace{V(\boldsymbol{x}, t)}_{\text{potential}}$ is the Hamiltonian operator characterizing the total energy of the system, and $\hat{p} = -i\hbar\nabla$ is the momentum operator. Also note that the Schrödinger equation is vector-based and has a valid representation in any complete basis of *kets* in a Hilbert state space. For instance, in the momentum-space basis, the Schrödinger's equation is:

$$i\hbar\frac{\partial f(\mathbf{p})}{\partial t}=\frac{p^2}{2m}f(\mathbf{p})+\left(\hat{V}*f\right)(\mathbf{p}),$$

where $|\mathbf{p}\rangle$ represents the plane wave of momentum $\mathbf{p}$, $\langle \mathbf{p}|\psi\rangle\equiv f(\mathbf{p})=\int_{\mathbb{R}^3}\psi(\mathbf{x})e^{-i\mathbf{p}\mathbf{x}}d\mathbf{x}$, $\hat{V}$ is the Fourier transform of V, and $*$ denotes the convolution operator.

3.6.6.1 Derivation of Schrödinger Equation

Schrödinger's equation cannot be exactly *derived*, as it was postulated using logical arguments with its core principle supported by many experimental observations. The rationale of the equations is related to the law of conservation of energy: $E=K+P$, where K and P are the *kinetic* and *potential* energies. An intuitive explanation of the equation may be obtained by linear algebra.

Let $\psi(t)$ represent a wavefunction at time t. By the linearity of quantum mechanics, relative to time t, the wavefunction at time t' must be $\psi(t')=U(t',t)\psi(t)$, where $\hat{U}(t',t)$ is a linear operator. By the law of preservation of energy, time-evolution must preserve the energy (i.e., the norm) of the wavefunction. Thus, the time-evolution operator $\hat{U}(t',t)$ must belong to the *unitary group of operators* acting on wavefunctions:

$$\|\psi(t')\|=\sqrt{\left\langle \hat{U}(t',t)\psi(t)\ \middle|\hat{U}(t',t)\psi(t)\right\rangle}=\sqrt{\left\langle\psi\ \middle|\hat{U}^*\hat{U}\psi\right\rangle}=\left\|\hat{U}\right\|\sqrt{\left\langle\psi\ \middle|\psi\right\rangle}=\left\|\hat{U}\right\|\|\psi(t)\|.$$

Note that if $t'=t$, we must have $\hat{U}(t,t)=\hat{1}$, and $\hat{U}$ is a unitary operator. Since the Lie algebra corresponding to the *unitary group of operators* is generated by skew-Hermitian operators, if $\widehat{H}^*=\widehat{H}$ is Hermitian, then $-i\widehat{H}$ is skew-Hermitian, $\left(-i\widehat{H}\right)^*=i\widehat{H}^*=i\widehat{H}$. Hence, we can express it as an exponential map, $\hat{U}(t)=e^{-\frac{i}{\hbar}\widehat{H}(t)}$ for some Hermitian operator $\widehat{H}=\widehat{H}^*$ [133].

Thus, for small time increments near t, $(\Delta t=t'-t)$, we can use the first-order Taylor expansion of the linear operator at time t:

$$\hat{U}(t',t)\cong\hat{U}(t,t)-\frac{i}{\hbar}\times\widehat{H}\times(t'-t)=\hat{1}-\frac{i}{\hbar}\times\widehat{H}\times(t'-t),$$

where $\widehat{H}$ is a self-adjoint Hermitian operator (whose expectation represents the Hamiltonian total energy of the system) [134].

Then, $\psi(t')=U(t',t)\psi(t)=\left\langle 1-\frac{i}{\hbar}\times\widehat{H}\times(t'-t)\middle|\psi(t)\right\rangle=\psi(t)-\frac{i}{\hbar}\times(t'-t)\times\left\langle\widehat{H}|\psi(t)\right\rangle$. Rearranging the terms, the previous equation simplifies to $i\hbar\ \frac{\psi(t')-\psi(t)}{t'-t}=\left\langle\widehat{H}|\psi(t)\right\rangle$.

Letting $t' \to t$ (i.e., $\Delta t \to 0$) yields the Schrödinger equation:

$$i\hbar \frac{\partial}{\partial t}\psi(t) = i\hbar \lim_{t' \to t} \frac{\psi(t') - \psi(t)}{t' - t} = \widehat{H}\ \psi(t).$$

3.6.7 Wave Equation

The 1D wave equation is a second-order PDE: $\frac{\partial^2 f(x,t)}{\partial^2 x} = \frac{\partial^2 f(x,t)}{\partial^2 t}$. There are infinitely many solutions, e.g., $\cos(2\pi\xi(x \pm t))$, $\sin(2\pi\xi(x \pm t))$ and the more general $Ae^{i(\xi \times x - w \times t)}$. Thus, the main challenge in solving the wave equation is "*solving the fixed boundary conditions equation.*" In other words, given some known functions $g(x)$ and $h(x)$, we need to find a solution, which satisfies specific boundary conditions like:

$$\left| \begin{array}{l} f(x,0) = g(x), \text{ initial location} \\ \frac{\partial f(x,0)}{dt} = h(x), \text{ initial velocity} \end{array} \right. .$$

In these situations, rather than finding the solution f directly, it is often easier to find the Fourier transform $\hat{f}$ of the solution [135]. The second-order partial differential wave equation in f is mapped (by the Fourier transform) into an algebraic equation in $\hat{f}$: $\xi^2 \hat{f}(\xi,\theta) = \theta^2 \hat{f}(\xi,\theta)$. This *spectral method* approach effectively employs the Fourier transform to map *differentiation* in x to *multiplication* by $2\pi i\xi$, and *differentiation* with respect to t to *multiplication* by $2\pi i\theta$, where θ is the frequency.

3.6.8 Lorentz Transformation

Hyperbolic spatial frames rotation may be expressed via the *Lorentz transformation* [136]. The Lorentz transformation illustrates the relationship between two reference frames, $(\boldsymbol{x}, t)$ and $(\boldsymbol{x}', t') \in \mathbb{R}^4$, in relative motion, whose x axis points in the direction of the relative velocity [137]. The transformation mixes *space* ($\boldsymbol{x} = (x, y, z) \in \mathbb{R}^3$) and *time* ($t \in \mathbb{R}^+$) the same way a planar Euclidean rotation around the z axis blends the x and y coordinates in $\mathbb{R}^2$. In the Lorentz transformation, *the speed of light* $c = 299{,}792{,}458\ m/s$ *represents a measuring-unit conversion factor that enables blended arithmetic of space and time variables.* That is, since length in space is measured in meters (m) and length in time is measured in seconds (s), the speed of light converts between length in space and length in time. Denote $v = v(t)$ to be the velocity between the two frames and the Lorenz factor $\gamma = \frac{1}{\sqrt{1 - v^2/c^2}}$, where v/c is the relative velocity of the two reference frames normalized to the speed of light. Then, the Lorentz transformation is defined by:

$$\left| \begin{array}{c} x' = \gamma(x - v \times t) \\ y' = y \\ z' = z \\ t' = \gamma(t - (v \times x)/c^2) \end{array} \right. .$$

The Lorenz transformation suggests that the (local) time in the *moving reference frame* $(\mathbf{x}', t')$ runs slower compared with the time in the *stationary reference frame* $(\mathbf{x}, t)$. This time-dilation phenomenon can be quantified by substituting $\Delta x' = 0$ and determining the relation between $(\mathbf{x}, t)$ *frame* time change, Δt, and the corresponding $(\mathbf{x}', t')$ *frame* time change, $\Delta t'$, according to the Lorentz transformation. If Δt is the measure of time between the events A and B in the stationary reference frame $(\mathbf{x}, t)$, v is the speed of the moving *frame* $(\mathbf{x}', t')$ relative to $(\mathbf{x}, t)$, and c is the speed of light, then, $\frac{\Delta t}{\Delta t'} = \gamma$, i.e.,

$$\Delta t = \left(\frac{1}{\sqrt{1 - v^2/c^2}} \right) \Delta t'.$$

The Lorentz transformation expands and generalizes the Newtonian science Galilean transformation, which effectively assumes an infinite speed of light $(c = \infty)$, i.e., $\Delta t = \Delta t'$ [138].

3.6.9 Euler–Lagrange Equation

Newton's second law of motion may be symbolically expressed as $F = ma$, where F is the force, m is the mass, and a is the acceleration. In the simplest example of modeling the motion of a mass attached to the end of a spring where $f(t)$ is the position of the mass at time t, the kinetic energy (due to the motion of the mass), $K = \frac{m(f'(t))^2}{2}$, and the potential (stored) energy, $P = \frac{kf^2(t)}{2}$, where k is the spring constant accounts for the spring stiffness. The law of total energy preservation suggests that $T = K + P = const$, i.e., $\frac{d}{dt}(K + P) = 0$. The Lagrangian is the difference of the kinetic and potential energies, $L = K - P = \frac{m(f'(t))^2}{2} - \frac{kf^2(t)}{2}$. This leads to the simplest (spring) form of the second-order Euler–Lagrange differential equation:

$$\underbrace{\frac{d}{dt}\Bigg(\underbrace{\frac{\partial}{\partial f'} L}_{mf'(t)} \Bigg)}_{mf''(t)} = \underbrace{\frac{\partial}{\partial f} L}_{-kf(t)} .$$

The more general Euler–Lagrange (EL) equation [139, 140] represents a second-order PDE with solutions that are the functions ($f \in C^1$) optimizing (minimizing or maximizing) a differentiable operator (L). As differentiable operators are stationary at their local extrema (maxima or minima), the EL equation provides a mechanism for solving optimization problems in which the operator (L) is known and we are looking for smooth functions (f) that minimize or maximize L. This directly parallels the fundamental theorem of calculus, which guarantees that points (x) where differentiable functions (f) attain a local extremum correspond to trivial derivatives ($f'(x) = 0$).

Let $f: \Omega \subseteq \mathbb{R}^1 \to \mathbb{R}^1$ be a real-valued and differentiable function, and its derivative is $f'(t)$. Then f is a stationary point of (optimizing) an operator called *action functional* $S(f) = \int_\Omega L(t, f(t), f'(t))\, dt$, provided that f is a solution to the EL equation:

$$\underbrace{L_2(t,f(t),f'(t))}_{\frac{\partial}{\partial f}L(t,f(t),f'(t))} - \frac{d}{dt}\underbrace{L_3(t,f(t),f'(t))}_{\frac{\partial}{\partial\left(\frac{df}{dt}\right)}L(t,f(t),f'(t))} = 0,$$

where $L_2 = \frac{\partial}{\partial y}L(t,y,z)$ and $L_3 = \frac{\partial}{\partial z}L(t,y,z)$ represent the second- and third-parameter partial derivatives of $L(t, f(t), f'(t))$.

This real-valued function formulation of the EL equations may be generalized to multivariate functions $f: \Omega \subseteq \mathbb{R}^n \to \mathbb{R}^1$, which represent high-dimensional surfaces. Then, f solves the EL-equations below when it is a stationary (optimizing) point of this operator:

$$S(f) = \int_\Omega L\left(x_1, x_2, \ldots, x_n; f(\mathbf{x}); \frac{\partial f}{\partial x_1}, \frac{\partial f}{\partial x_2}, \ldots, \frac{\partial f}{\partial x_n}\right) d\mathbf{x}:$$

$$\frac{\partial L}{\partial f} - \sum_{i=1}^{n} \frac{\partial}{\partial x_i}\left(\frac{\partial L}{\partial z_i}\right) = 0,$$

where $\mathbf{x} = (x_1, x_2, \ldots, x_n)$, $z_k = \frac{\partial f}{\partial x_k}$, and $\frac{\partial L}{\partial z_i} = \frac{\partial}{\partial z_i}(L(x_1, x_2, \ldots, x_n; f(\mathbf{x}); z_1, z_2, \ldots, z_n))$. The EL optimization strategy may be further generalized to more complex functions $f: \Omega \subseteq \mathbb{R}^n \to \mathbb{R}^m$.

3.6.10 Wheeler-DeWitt Equation

Wheeler-DeWitt equation [141] is a functional differential equation providing a unifying framework for relativity and quantum mechanics. It includes no time reference, suggesting time may not be a universal intrinsic characteristic. The Wheeler-DeWitt equation only depends on the particle's position and a universal wavefunctional:

$$\widehat{H}\,(\mathbf{x})|\psi\rangle = 0,$$

where $|\psi\rangle$ is a wavefunctional and $\widehat{H}(\boldsymbol{x})$ is the Hamiltonian constraint. There is a difference in this notation relative to the Schrödinger equation, presented earlier. The Wheeler-DeWitt equation expresses *timelessness*, where $|\psi\rangle$ is not a spatial complex-valued wavefunction defined on a 3D space, but a functional of field configurations on the entire spacetime, which contains all of the information about the geometry and matter content of the universe. While the Hamiltonian $\widehat{H}$ is still an operator acting on the Hilbert space of wavefunctions, it is not the same Hilbert space as in the Schrödinger's equation, and the Hamiltonian no longer determines the evolution of the system.

The potential function in the Wheeler-DeWitt equation, i.e., the wavefunction that solves this functional differential equation, doesn't evolve in time! In Einstein's special theory of relativity, spacetime is one 4D whole with no "universal time". This reflects that something that is happening to one observer now that can be simultaneously observed by all others.

The connection between spacekime and the Wheeler-DeWitt equation is linked to the quantum gravity interpretation of this timeless equation, which suggests that the wavefunction of the entire universe doesn't change. As illustrated by DeWitt [142], the strong operational foundation of quantum theory and general relativity represents an "extraordinarily economical" theoretical framework that offers very specific answers to exact questions and nothing more. For instance, quantum physics will not provide answers about time, unless a specific clock tracking longitudinal progression mechanism is provided, nor will it answer geometric questions about the universe unless a specific device (e.g., a material object, gravitational waves, or some other form of radiation) is specified to inform when and where the geometry is to be measured. Considering the kime-phase as an independent quantum variable leads to a non-local property of quantum physics, where a single particle can occupy two separate spatial locations at the same time and the de Broglie wavelength is directly linked to the existence of a second time dimension [64, 143].

3.7 Analogous Kime Extensions

Next, we will generalize to complex time some of the common principles, time-varying equations, and physical concepts that we discussed above.

3.7.1 Rate of Change

The domain of the kime variable (k) is the complex plane parameterized by pairs of Descartes Cartesian coordinates, conjugate-pairs coordinates, or polar coordinates:

$$\mathbb{C} \equiv \mathbb{R}^2 = \{z = (x,y) | x,y \in \mathbb{R}\} \equiv \{(z,\bar{z}) | z \in \mathbb{C}, z = x + iy,\ \bar{z} = x - iy,\ x,y \in \mathbb{R}\}$$
$$\equiv \{\ k = r\ e^{i\varphi} = r(\cos\varphi + i\sin\varphi)\ |\ r \geq 0,\ -\pi \leq \varphi < \pi\}.$$

The *Wirtinger derivative* of a continuously differentiable function (f) of kime (k), $f(k)$, and its conjugate are defined as *first*-order linear partial differential operators:

- In Cartesian coordinates:

$$f'(z) = \frac{\partial f(z)}{\partial z} = \frac{1}{2}\left(\frac{\partial f}{\partial x} - i\frac{\partial f}{\partial y}\right) \text{ and } f'(\bar{z}) = \frac{\partial f(\bar{z})}{\partial \bar{z}} = \frac{1}{2}\left(\frac{\partial f}{\partial x} + i\frac{\partial f}{\partial y}\right).$$

- In conjugate-pair basis: $df = \partial f + \bar{\partial} f = \frac{\partial f}{\partial z} dz + \frac{\partial f}{\partial \bar{z}} d\bar{z}$.
- In polar kime coordinates:

$$f'(k) = \frac{\partial f(k)}{\partial k} = \frac{1}{2}\left(\cos\varphi\frac{\partial f}{\partial r} - \frac{1}{r}\sin\varphi\frac{\partial f}{\partial \varphi} - i\left(\sin\varphi\frac{\partial f}{\partial r} + \frac{1}{r}\cos\varphi\frac{\partial f}{\partial \varphi}\right)\right)$$
$$= \frac{e^{-i\varphi}}{2}\left(\frac{\partial f}{\partial r} - \frac{i}{r}\frac{\partial f}{\partial \varphi}\right)$$

and

$$f'(\bar{k}) = \frac{\partial f(\bar{k})}{\partial \bar{k}} = \frac{1}{2}\left(\cos\varphi\frac{\partial f}{\partial r} - \frac{1}{r}\sin\varphi\frac{\partial f}{\partial \varphi} + i\left(\sin\varphi\frac{\partial f}{\partial r} + \frac{1}{r}\cos\varphi\frac{\partial f}{\partial \varphi}\right)\right)$$
$$= \frac{e^{i\varphi}}{2}\left(\frac{\partial f}{\partial r} + \frac{i}{r}\frac{\partial f}{\partial \varphi}\right).$$

Notes:

- The derivatives in terms of the polar coordinates are obtained by transforming the Cartesian complex variable $z = (x, y)$ into the complex-time (kime) variable $k = (r, \varphi)$ using polar transformations:

$$\left| \begin{array}{l} x = r\ \cos\varphi \\ y = r\ \sin\varphi \end{array} \right., \left| \begin{array}{l} r = \sqrt{x^2 + y^2} \\ \varphi = \arctan\left(\frac{y}{x}\right) = \arctan(y, x) \end{array} \right., \left| \begin{array}{l} \frac{\partial}{\partial x} = \cos\varphi\frac{\partial}{\partial r} - \frac{1}{r}\ \sin\varphi\frac{\partial}{\partial \varphi} \\ \frac{\partial}{\partial y} = \sin\varphi\frac{\partial}{\partial r} + \frac{1}{r}\ \cos\varphi\frac{\partial}{\partial \varphi} \end{array} \right., \quad \text{see[143]}.$$

- Using the chain-rule of differentiation, we can derive the Cartesian coordinate derivatives by transforming the conjugate-pairs basis

$$(x,y) \rightarrow \left(\frac{1}{2}(z + \bar{z}),\ \frac{1}{2i}(z - \bar{z})\right),$$
$$\frac{\partial}{\partial z} = \frac{\partial}{\partial x}\frac{\partial x}{\partial z} + \frac{\partial}{\partial y}\frac{\partial y}{\partial z}.$$

Therefore, $\frac{\partial x}{\partial z} = \frac{1}{2}\frac{\partial(z+\bar{z})}{\partial z} = \frac{1}{2}$ and $\frac{\partial y}{\partial z} = \frac{1}{2i}\frac{\partial(z-\bar{z})}{\partial z} = \frac{1}{2i} = -\frac{i}{2}$.

Similarly,

$$\frac{\partial x}{\partial \bar{z}} = \frac{1}{2}\frac{\partial(z+\bar{z})}{\partial \bar{z}} = \frac{1}{2} \text{ and } \frac{\partial y}{\partial \bar{z}} = \frac{1}{2i}\frac{\partial(z-\bar{z})}{\partial \bar{z}} = -\frac{1}{2i} = \frac{i}{2}.$$

This explains the Cartesian coordinate derivatives:

$$f'(z) = \frac{\partial f(z)}{\partial z} = \frac{1}{2}\left(\frac{\partial f}{\partial x} - i\frac{\partial f}{\partial y}\right) \text{ and } f'(\bar{z}) = \frac{\partial f(\bar{z})}{\partial \bar{z}} = \frac{1}{2}\left(\frac{\partial f}{\partial x} + i\frac{\partial f}{\partial y}\right).$$

Below, we present the core principles of *Wirtinger differentiation* and *integration*:

- Complex conjugation ($\bar{z} \in \mathbb{C}$) for $z = (x+iy) \in \mathbb{C}$ is defined by $\bar{z} = x - iy$, so that the square norm of z is: $z\bar{z} = (x+iy)(x-iy) = x^2 + y^2 - ixy + ixy = x^2 + y^2 = \|z\|^2$. Solving for x and y, in terms of z and $\bar{z}$ we get:

$$\left| \begin{array}{l} x = \frac{1}{2}(z+\bar{z}) \\ y = \frac{1}{2i}(z-\bar{z}) \end{array} \right. .$$

- We can effectively change the variables: $(x,y) \to (z,\bar{z})$. Thus, all complex functions $f{:}\mathbb{C} \to \mathbb{C}$ can be thought of as $f = f(x,y)$ or as $f = f(z,\bar{z})$.
- *Wirtinger differentiation*: The Wirtinger derivative of f, df_z is an $\mathbb{R}$-linear operator on the tangent space $T_z\mathbb{C} \cong \mathbb{C}$, i.e., df_z is a differential 1-form on $\mathbb{C}$. However, any such $\mathbb{R}$-linear operator (A) on $\mathbb{C}$ can be uniquely *decomposed* as $A = B + C$, where B is its *complex-linear part* (i.e., $B(iz) = iBz$), and C is its *complex-antilinear part* (i.e., $C(iz) = -iCz$). The reverse (*composition*) mapping is $Bz = \frac{1}{2}(Az - iA(iz))$ and $Cz = \frac{1}{2}(Az + iA(iz))$.
- For the Wirtinger derivative, this duality of the decomposition of $\mathbb{R}$-linear operators characterizes the conjugate partial differential operators ∂ and $\bar{\partial}$. That is, for all differentiable complex functions $f{:}\mathbb{C} \to \mathbb{C}$, the derivative can be uniquely decomposed as $df_z = \partial f + \bar{\partial} f$, where ∂ is its *complex-linear part* ($\partial iz = i\partial z$), and $\bar{\partial}$ is its *complex-antilinear part* ($\bar{\partial} iz = -i\bar{\partial} z$).
- Applying the operators $\frac{\partial}{\partial z}$ and $\frac{\partial}{\partial \bar{z}}$ to the identify function ($z \to z = x + iy$) and its complex-conjugate ($z \to \bar{z} = x - iy$) yields the natural derivatives: $dz = dx + idy$ and $d\bar{z} = dx - idy$. For each point in $\mathbb{C}$, $\{dz, d\bar{z}\}$ represents a conjugate-pair basis for the C cotangent space, with a dual basis of the partial differential operators:

$$\left\{\frac{\partial}{\partial z}, \frac{\partial}{\partial \bar{z}}\right\}.$$

- Thus, for any smooth complex functions $f{:}\,\mathbb{C} \to \mathbb{C}$,

$$df = \partial f + \bar{\partial} f = \frac{\partial f}{\partial z} dz + \frac{\partial f}{\partial \bar{z}} d\bar{z}.$$

- *Wirtinger calculus*: The *path-integral* is the simplest way to integrate a complex function $f: \mathbb{C} \to \mathbb{C}$ on a specific path connecting $z_a \in \mathbb{C}$ to $z_b \in \mathbb{C}$. Generalizing Riemann sums:

$$\lim_{|z_{i+1} - z_i| \to 0} \sum_{i=1}^{n-1} (f(z_i)(z_{i+1} - z_i)) \cong \oint_{z_a}^{z_b} f(z_i) dz.$$

This assumes the path is a polygonal arc joining z_a to z_b, via $z_1 = z_a, \ z_2, \ z_3, \ \ldots, \ z_n = z_b$, and we integrate the piecewise constant function $f(z_i)$ on the arc joining $z_i \to z_{i+1}$. Clearly the path $z_a \to z_b$ needs to be defined and the limit of the generalized Riemann sums, as $n \to \infty$, will yield a complex number representing the Wirtinger integral of the function over the path. Similarly, we can extend the classical area integrals, indefinite integral, and Laplacian:

- Definite area integral: for $\Omega \subseteq \mathbb{C}$, $\int_\Omega f(z) dz d\bar{z}$.
- Indefinite integral: $\int f(z) dz d\bar{z}$, $df = \frac{\partial f}{\partial z} dz + \frac{\partial f}{\partial \bar{z}} d\bar{z}$,
- The Laplacian in terms of conjugate pair coordinates is $\nabla^2 f \equiv \Delta f = 4 \frac{\partial}{dz} \frac{\partial f}{d\bar{z}} = 4 \frac{\partial}{d\bar{z}} \frac{\partial f}{dz}$.

More details about Wirtinger calculus of differentiation and integration are provided later.

3.7.2 Kime Motion Equations

Prior work by Wesson, Ponce De Leon, Overduin, and others described the equations of motion in 5D space-time-matter, which is an unrestricted 5D manifold theory that uses the extra (induced matter) dimension to geometrically explain the origin of 4D spacetime [63, 70, 100, 145]. The kime equations of motion describe the behavior of a physical system in terms of its motion as a function of kime. In this section, we will start by defining the kime velocity and acceleration, the Newtonian equations of motion, the most general spacekime motion equations, and connect the Lagrangian and Eulerian frames of reference. In **Chapter 5, Section 5.4**, we will consider the 5D Space-Time-Matter theory, which extends Einstein's theory of General Relativity to five dimensions. As part of their 5D space-time-energy Riemannian space formulation (Deformed Relativity in Five Dimensions, DR5), Cardone and Mignani also defined the geodesic equation of motion in a general Kaluza-Klein model [146].

3.7.2.1 K-velocities

In the case of a two-dimensional kime, the velocities of a particle according to different kime dimensions cannot be defined as a set of partial derivatives of the independent variables $k_1,\ k_2$. Indeed, the movement of a point-like particle in the general case can be presented as a one-dimensional (time-like) world line in the $(3+2)$-dimensional spacekime. Let us set $x = x(k_1,\ k_2)$, where $x = (x_1, x_2, x_3)$, define the spatial position of the particle $(x_1 = x, x_2 = y, x_3 = z)$ and, accordingly, $\boldsymbol{dx} = \frac{\partial x}{\partial k_1} dk_1 + \frac{\partial x}{\partial k_2} dk_2$. Then, using the special indexing $\eta = 1, 2, 3$, we have $x_\eta = x_\eta(k_1,\ k_2)$ and $dx_\eta = \frac{\partial x_\eta}{\partial k_1} dk_1 + \frac{\partial x_\eta}{\partial k_2} dk_2$.

In the general case, each one of the spatial coordinate functions $x_\eta = x_\eta(k_1,\ k_2)$ must be represented by a two-dimensional *surface* in the $(3+2)$-dimensional spacekime $x_\eta, k_1,\ k_2$, i.e., it will not present a simple one-dimensional world line. Therefore, the functions $x_\eta = x_\eta(k_1,\ k_2)$ could not describe the movement of a point-like particle in the spacekime, but rather a radial wave ripple through the 2D kime manifold. Each one-dimensional world line in the $(3+2)$-dimensional spacekime x_η, k_1, k_2 is described as a path on the kime-surface via a system of 6 implicit equations:

$$\begin{aligned} F_{1\eta}(x_\eta, k_1, k_2) &= 0 \\ F_{2\eta}(x_\eta, k_1, k_2) &= 0 \end{aligned}.$$

Uniform and rectilinear movement of the particle through a straight world line can be described by the implicit linear functions $F_{1\eta}, F_{2\eta}$. From here, we can derive the explicit equalities: $x_\eta = f_{1\eta}(k_1) = f_{2\eta}(k_2)$, where $f_{1\eta}, f_{2\eta}$ are different explicit (linear) functions of the variables $k_1,\ k_2$ accordingly. Thus, we have:

$$dx_\eta = f'_{1\eta}(k_1) dk_1$$

$$dx_\eta = f'_{2\eta}(k_2) dk_2.$$

When the particle moves along a straight world line, the derivatives $f'_{1\eta}$ and $f'_{2\eta}$ are constants defining the velocities of the particle in relation to k_1 and k_2, respectively.

Note that the vector representation of time-velocity in spacetime is extended in spacekime to a second-order tensor representing the Jacobian matrix of the position-change (dx) relative to kime-change $(\boldsymbol{dk})$, i.e., $v = (v_{u,j}) = \frac{\boldsymbol{dx}(\boldsymbol{k})}{\boldsymbol{dk}}$, $u \in \{x, y, z\}$ and $j \in \{1, 2\}$. We can express the directional velocities in terms of $u \in \{x, y, z\}$ and $j \in \{1, 2\}$:

x–velocities: $\quad v_{x,1} = \frac{dx}{dk_1};\ v_{x,2} = \frac{dx}{dk_2} \Rightarrow dx = v_{x,1} dk_1 = v_{x,2} dk_2;$

y-velocities: $\quad v_{y,1} = \frac{dy}{dk_1};\ v_{y,2} = \frac{dy}{dk_2} \Rightarrow dy = v_{y,1} dk_1 = v_{y,2} dk_2;$

$$\text{z-velocities:} \qquad v_{z,1} = \frac{dz}{dk_1};\ v_{z,2} = \frac{dz}{dk_2} \Rightarrow dz = v_{z,1}dk_1 = v_{z,2}dk_2.$$

In a flat Euclidean kime manifold, the square of the *kime* change can be expressed in Cartesian coordinates as a sum of squares of the individual (directional) kime changes:

$$(dk)^2 = (dk_1)^2 + (dk_2)^2.$$

By definition, the total kime-velocity is

$$\mathbf{v}_k = \frac{d\mathbf{x}}{dk} = \underbrace{\frac{\sqrt{dx^2 + dy^2 + dz^2}}{dk}}_{\text{magnitude}} \underbrace{\mathbf{e}}_{\substack{\text{unit vector of spatial}\\ \text{direction change}}}$$

and the directional kime-velocities with respect to k_1 and k_2 can be expressed as

$$\mathbf{v}_1 = \frac{d\mathbf{x}}{dk_1} = \underbrace{\frac{\sqrt{dx^2 + dy^2 + dz^2}}{dk_1}}_{\text{magnitude}} \underbrace{\mathbf{e}}_{\substack{\text{unit vector of spatial}\\ \text{direction change}}},$$

$$\mathbf{v}_2 = \frac{d\mathbf{x}}{dk_2} = \underbrace{\frac{\sqrt{dx^2 + dy^2 + dz^2}}{dk_2}}_{\text{magnitude}} \underbrace{\mathbf{e}}_{\substack{\text{unit vector of spatial}\\ \text{direction change}}}.$$

Thus, we can connect the *total* kime velocity, $v_k = \|\mathbf{v}_k\|$, with the corresponding kime velocities:

$$\frac{1}{v_k^2} = \frac{(dk)^2}{dx^2 + dy^2 + dz^2} = \frac{(dk_1)^2 + (dk_2)^2}{dx^2 + dy^2 + dz^2} = \frac{(dk_1)^2}{dx^2 + dy^2 + dz^2} + \frac{(dk_2)^2}{dx^2 + dy^2 + dz^2} = \frac{1}{v_1^2} + \frac{1}{v_2^2}.$$

Therefore, $\frac{1}{v_k^2} = \frac{1}{v_1^2} + \frac{1}{v_2^2}$, where $v_1 = \|\mathbf{v}_1\|$ and $v_2 = \|\mathbf{v}_2\|$.

However, we can also express the relation between the standard time-velocity and the kime-velocity in polar coordinates. Let's consider a kime point $\boldsymbol{k}$. As complex-time (kime) is two-dimensional (2D), then the quantity $\boldsymbol{k}$ can be presented in vector form: $k = (k_1, k_2)$. In polar coordinates, $k_1 = t\ \cos\varphi$ and $k_2 = t\sin\varphi$, where the kime-magnitude t measures time and φ represents the phase direction of kime. Inverting these relations, we have:

$$t = \sqrt{k_1^2 + k_2^2}, \text{ and } \varphi = \operatorname{atan2}(k_2, k_1) =$$

$$\begin{cases} \arctan\left(\frac{k_2}{k_1}\right), \text{ if } k_1 > 0, \\ \arctan\left(\frac{k_2}{k_1}\right) + \pi, \text{ if } k_1 < 0 \cap k_2 \geq 0 \\ \arctan\left(\frac{k_2}{k_1}\right) - \pi, \text{ if } k_1 < 0 \cap k_2 < 0 \\ \frac{\pi}{2}, \text{ if } k_1 = 0 \cap k_2 > 0 \\ -\frac{\pi}{2}, \text{ if } k_1 = 0 \cap k_2 < 0 \\ \text{undefined, if } k_1 = 0 \cap k_2 = 0 \end{cases} = \begin{cases} 2\arctan\left(\frac{k_2}{k_1 + \sqrt{k_1^2 + k_2^2}}\right), \text{if } k_1 > 0, \\ 2\arctan\left(\frac{\sqrt{k_1^2 + k_2^2} - k_1}{k_2}\right), \text{if } k_1 \leq 0 \cap k_2 \neq 0 \cdot \\ -\pi, \text{if } k_1 < 0 \cap k_2 = 0 \\ \text{undefined, if } k_1 = 0 \cap k_2 = 0 \end{cases}$$

Therefore, $k = k(t, \varphi)$, i.e., $\mathbf{k}$ is a function of both *independent* variables t, φ. (In complex form: $k = k(t, \varphi) = te^{i\varphi}$.) Actually, $t = \sqrt{k_1^2 + k_2^2}$ is a scalar value and represents the length of the vector $\mathbf{k}$, i.e. $t = |\mathbf{k}|$. It is clear that t is the radius of the circle described with the equation $(k_1)^2 + (k_2)^2 = t^2$. The point $\mathbf{k} = (k_1, k_2)$ also lies on this circle. Therefore, $\mathbf{k}$ is only one of all possible points that lie on the circle centered at the origin of the coordinate system and a radius equal to t, **Figure 3.9**.

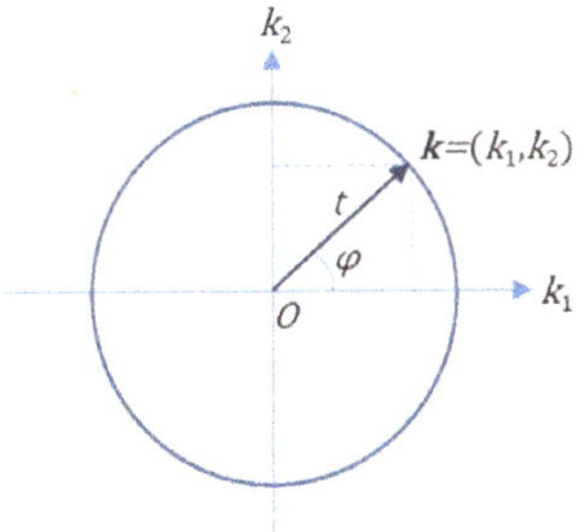

Figure 3.9: The relation between time (t) and kime ($\mathbf{k}$).

It may be worth clarifying that the atan2() function calculates the arctangent over the entire plane (in all four quadrants), whereas arctan() only computes in the first and fourth quadrants. For instance, when $\tan(\varphi)$ is positive, we cannot distinguish between the first and third quadrant, i.e., if $0 \leq \varphi \leq \frac{\pi}{2}$ or $-\pi \leq \varphi \leq -\frac{\pi}{2}$. Conversely, when $\tan(\varphi)$ is negative, the phase could be in either of the complementary, second or fourth, quadrants. The definition of arctan() yields a phase in the first or fourth quadrant, $-\frac{\pi}{2} \leq \varphi \leq \frac{\pi}{2}$. Whereas atan2() retrieves the full phase information, by separately utilizing the values of the sine and cosine functions, instead of their fraction $\left(\frac{\sin\varphi}{\cos\varphi}\right)$, to resolve all four quadrants by adding (or subtracting) π to the result of arctan() when the cosine denominator is negative.

The directional kime changes can be expressed as:

$$dk_1 = d(t\cos\varphi) = \cos\varphi dt + t\, d\cos\varphi = \cos\varphi dt - t\sin\varphi d\varphi,$$

$$dk_2 = d(t\sin\varphi) = \sin\varphi dt + t\, d\sin\varphi = \sin\varphi dt + t\cos\varphi d\varphi.$$

Substituting these in the formulas above we obtain the polar expressions for the directional kime derivatives v_1 and v_2:

$$\boldsymbol{v}_1 = \frac{\sqrt{dx^2+dy^2+dz^2}}{dk_1}\boldsymbol{e} = \frac{\sqrt{dx^2+dy^2+dz^2}}{\cos\varphi dt - t\sin\varphi d\varphi}\boldsymbol{e},$$

$$\boldsymbol{v}_2 = \frac{\sqrt{dx^2+dy^2+dz^2}}{dk_2}\boldsymbol{e} = \frac{\sqrt{dx^2+dy^2+dz^2}}{\sin\varphi dt + t\cos\varphi d\varphi}\boldsymbol{e}.$$

Therefore, the kime metric tensor in polar coordinates is:

$$(d\boldsymbol{k})^2 = (dk_1)^2 + (dk_2)^2 =$$
$$= (\cos\varphi\ dt - t\sin\varphi\ d\varphi)^2 + (\sin\varphi\ dt + t\cos\varphi\ d\varphi)^2 = (dt)^2 + t^2(d\varphi)^2.$$

In general, $d\boldsymbol{k} \neq dt$. Only if $d\varphi = 0$, then $\varphi = const$ and $d\boldsymbol{k} = dt$.

We can explore the relation between the (spacekime) kime-velocities $\left(\boldsymbol{v}_1 = \frac{dx}{d\kappa_1}\right.$ and $\left.\boldsymbol{v}_2 = \frac{dx}{d\kappa_2}\right)$ and the classical (spacetime) time-velocity $\left(v = v_t = \frac{\sqrt{dx^2+dy^2+dz^2}}{dt}\right)$. Let's start by considering the *path-based kime-velocity* defined on a curve in the kime manifold. In other words, suppose we have a parametric description of a path (curve, world line) in 2D where the kime order (magnitude) $t = t(s):[0,\infty) \to \mathbb{R}^+$ and the kime-phase $\varphi = \varphi(s):[0,\infty) \to [-\pi,\pi)$. The kime-velocity along the parametric curve is defined in terms of the change of radial distance $\left(t' = \frac{dt}{ds}\right)$ and the change of phase $\left(\varphi' = \frac{d\varphi}{ds}\right)$ with respect to the path parameter, s, which may correspond to the length of the curve up to the current location. Then, the square reciprocal of the classical time-velocity is

$$\frac{1}{v^2} = \frac{1}{v_t^2} = \frac{(dt)^2}{dx^2+dy^2+dz^2}.$$

On the other hand, the sum of the square reciprocals of the k_1 and k_2 directional kime-velocities is:

$$\frac{1}{v_k^2} = \frac{1}{v_1^2} + \frac{1}{v_2^2} = \frac{(\cos\varphi\, dt - t\sin\varphi\, d\varphi)^2}{dx^2+dy^2+dz^2} + \frac{(\sin\varphi\, dt + t\cos\varphi\, d\varphi)^2}{dx^2+dy^2+dz^2} = \frac{(dt)^2 + t^2(d\varphi)^2}{dx^2+dy^2+dz^2}.$$

Therefore,

$$\frac{1}{v_k^2} = \frac{1}{v_1^2} + \frac{1}{v_2^2} = \underbrace{\frac{(dt)^2}{dx^2+dy^2+dz^2}}_{(v_t^2)^{-1}} + t^2 \underbrace{\frac{(d\varphi)^2}{dx^2+dy^2+dz^2}}_{(v_\varphi^2)^{-1}} = \frac{1}{v_t^2} + t^2\frac{1}{v_\varphi^2}.$$

This relation provides a mechanism to solve for either the classical time-velocity ($v = v_t$), or the kime-velocities in either Cartesian (v_k, $v_1 = v_{\kappa_1}$ and $v_2 = v_{\kappa_2}$) or polar coordinates (v_t and v_φ). For instance, the time-velocity may be expressed as:

$$\frac{1}{v_t^2} = \frac{1}{v_1^2} + \frac{1}{v_2^2} - t^2 \frac{1}{v_\varphi^2} = \frac{1}{v_k^2} - t^2 \frac{1}{v_\varphi^2}.$$

We can also express the general kime-velocity in polar coordinates ($\boldsymbol{k} = (t, \varphi)$) using the Wirtinger derivative of the position with respect to kime:

$$v(\boldsymbol{k}) = \frac{\partial x}{\partial \boldsymbol{k}} = \frac{1}{2}\left(\cos\varphi \frac{\partial x}{\partial t} - \frac{1}{t}\sin\varphi \frac{\partial x}{\partial \varphi} - i\left(\sin\varphi \frac{\partial x}{\partial t} + \frac{1}{t}\cos\varphi \frac{\partial x}{\partial \varphi}\right)\right).$$

3.7.2.2 Newton's Equations of Motion in Kime

Newton's Equations of Motion	Derivation
$\boldsymbol{v}_1 = \boldsymbol{a}_1 k_1 + \boldsymbol{v}_{o1}$ $\boldsymbol{v}_2 = \boldsymbol{a}_2 k_2 + \boldsymbol{v}_{o2}$ $\boldsymbol{x} = \boldsymbol{x}_{o1} + \boldsymbol{v}_{o1} k_1 + \frac{1}{2}\boldsymbol{a}_1 k_1^2 =$ $= \boldsymbol{x}_{o2} + \boldsymbol{v}_{o2} k_2 + \frac{1}{2}\boldsymbol{a}_2 k_2^2,$ $2\boldsymbol{a}_1(\boldsymbol{x} - \boldsymbol{x}_{o1}) + \boldsymbol{v}_{o1}^2 = \boldsymbol{v}_1^2,$ $2\boldsymbol{a}_2(\boldsymbol{x} - \boldsymbol{x}_{o2}) + \boldsymbol{v}_{o2}^2 = \boldsymbol{v}_2^2,$	The equations on the left can again be derived from the definitions of velocity and acceleration: – Equ 1: As $\boldsymbol{a}_1 = \frac{d\boldsymbol{v}_1}{dk_1}$ and $\boldsymbol{a}_2 = \frac{d\boldsymbol{v}_2}{dk_2}$, integrating both sides yields $\int \boldsymbol{a}_1\, dk_1 = \int d\boldsymbol{v}_1$ and $\int \boldsymbol{a}_2\, dk_2 = \int d\boldsymbol{v}_2$. Since the acceleration is constant in kime, $\boldsymbol{v}_1 = \boldsymbol{a}_1 \int dk_1 = \boldsymbol{a}_1 k_1 + \boldsymbol{v}_{o1}$ and $\boldsymbol{v}_2 = \boldsymbol{a}_2 \int dk_2 = \boldsymbol{a}_2 k_2 + \boldsymbol{v}_{o2}$, where $\boldsymbol{v}_{o1}$ and $\boldsymbol{v}_{o2}$ are constants representing the initial k-velocities, defined in relation to the kime dimensions k_1 and k_2, respectively. – Equ 2: $\boldsymbol{v}_1 = \frac{d\boldsymbol{f}_1}{dk_1} = \boldsymbol{a}_1 k_1 + \boldsymbol{v}_{o1}$ and $\boldsymbol{v}_2 = \frac{d\boldsymbol{f}_2}{dk_2} = \boldsymbol{a}_2 k_2 + \boldsymbol{v}_{o2}$ (from equ 1), integrating we get $\int d\boldsymbol{f}_1 = \int \boldsymbol{a}_1 k_1 dk_1 + \boldsymbol{v}_{o1} \int dk_1$ and $\int d\boldsymbol{f}_2 = \int \boldsymbol{a}_2 k_2 dk_2 + \boldsymbol{v}_{o2} \int dk_2$. As $\boldsymbol{a}_1$ and $\boldsymbol{a}_2$ are constants, we have $\boldsymbol{x} = \boldsymbol{a}_1 \int d\frac{{k_1}^2}{2} + \boldsymbol{v}_{o1} k_1 = \boldsymbol{a}_1 \frac{{k_1}^2}{2} + \boldsymbol{v}_{o1} k_1 + \boldsymbol{C_1}$ and we can compute the constant $\boldsymbol{C_1} = \boldsymbol{x_{o1}}$ by setting $k_1 = 0$. Analogously, we will have $\boldsymbol{x} = \boldsymbol{a}_2 \int d\frac{{k_2}^2}{2} + \boldsymbol{v}_{o2} k_2 = \boldsymbol{a}_2 \frac{{k_2}^2}{2} + \boldsymbol{v}_{o2} k_2 + \boldsymbol{C_2}$, and we estimate the constant $\boldsymbol{C_2} = \boldsymbol{x_{o2}}$ by setting $k_2 = 0$. – Equ 3: $\boldsymbol{a}_1 = \frac{dv_1}{dk_1} = \frac{dv_1}{dx}\frac{dx}{dk_1} = \frac{dv_1}{dx}\boldsymbol{v}_1 = \boldsymbol{v}_1 \frac{dv_1}{dx}$. Again integrating, we get $\int \boldsymbol{a}_1 dx = \int \boldsymbol{v}_1 d\boldsymbol{v}_1$ and under the initial condition ($v_{o1} = v_1(0)$) this becomes $2\boldsymbol{a}_1(\boldsymbol{x} - \boldsymbol{x_{o1}}) + \boldsymbol{v}_{o1}^2 = \boldsymbol{v}_1^2$. Equ 4: Analogously, we will have $2\boldsymbol{a}_2(\boldsymbol{x} - \boldsymbol{x_{o2}}) + \boldsymbol{v}_{o2}^2 = \boldsymbol{v}_2^2$.

In a simple Euclidean spacekime manifold with a diagonal metric tensor, the *kime-motion* is a change in position of an object over kime. Suppose $\boldsymbol{x}_o$ and $\boldsymbol{x}$ are the particle's initial ($\boldsymbol{x}_o = \boldsymbol{x}(\boldsymbol{k} = \boldsymbol{0})$) and final ($\boldsymbol{x} = \boldsymbol{f}_1(k_1) = \boldsymbol{f}_2(k_2)$) *positions*, $\boldsymbol{v}_{o1}, \boldsymbol{v}_{o2}$ and $\boldsymbol{v}_1$, $\boldsymbol{v}_2$ are the particle's initial $\boldsymbol{v}_{o1} = \frac{d\boldsymbol{f}_1}{dk_1}\big|_{k_1=0}$, $\boldsymbol{v}_{o2} = \frac{d\boldsymbol{f}_2}{dk_2}\big|_{k_2=0}$ and final $\boldsymbol{v}_1 = \frac{d\boldsymbol{f}_1}{dk_1}$, $\boldsymbol{v}_2 = \frac{d\boldsymbol{f}_2}{dk_2}$ *velocities*, $\boldsymbol{a}_1 = \frac{d\boldsymbol{v}_1}{dk_1}$, $\boldsymbol{a}_2 = \frac{d\boldsymbol{v}_2}{dk_2}$ are the particle's *accelerations*, and kime is represented as

$\boldsymbol{k} = \underbrace{(k_1, k_2)}_{\text{Cartesian}}$ or $\boldsymbol{k} = \underbrace{re^{i\varphi}}_{\text{Polar}}$. Below, we introduce the kime-motion equations for a 3D particle moving linearly, along a straight line, with a constant kime acceleration.

Later, in **Chapter 5, Section 5.4** (*Uncertainty in 5D Spacekime*), we will also formulate the Lagrangian framework of the spacekime equations of motion, which describe particle trajectories as functions of kime ($\boldsymbol{k}$) and a kime-independent vector field (e.g., 4D spacetime momenta) relative to some initial kime ($\boldsymbol{k}_o$).

In the Eulerian framework, the 4-velocity field is represented as a function of the position x and time t, $\boldsymbol{v} = \frac{\partial x(t)}{\partial t} = \boldsymbol{v}(\boldsymbol{x}, t)$. The Lagrangian framework represents the particle motion by some (time-independent) vector field $\boldsymbol{x}_o$ at some initial time t_o, accounting for the possible changes of the trajectory curve over time and parameterizing the 4-velocity field. That is, $F(\boldsymbol{x}_o, t)$ expresses the position of the particle labeled $\boldsymbol{x}_o$ at time t.

Both the Eulerian and the Lagrangian representations describe the velocity of the particle labeled $\boldsymbol{x}_o$ at time t and are related by the Lagrange derivative:

$$\boldsymbol{v}(F(\boldsymbol{x}_o, t), t) = \frac{\partial F}{\partial t}(\boldsymbol{x}_o, t).$$

The pair of reference coordinate frames, $\boldsymbol{x}_o$ and $\boldsymbol{x}$, are also called the Lagrangian and the Eulerian coordinates of the flow motion. The total rate of change of the function F with respect to time may be computed by:

$$\underbrace{\frac{dF}{dt}}_{\substack{\text{total rate}\\ \text{of change}}} = \underbrace{\frac{\partial F}{\partial t}}_{\substack{\text{local rate}\\ \text{of change}}} + \underbrace{(\boldsymbol{v} \cdot \nabla)F}_{\substack{\text{convective}\\ \text{rate of change}}},$$

where ∇ is the gradient with respect to $\boldsymbol{x}$, and the inner-product $(\boldsymbol{v} \cdot \nabla)$ represents an operator applied to each component of F. The left-hand side represents the total rate of change of F as the particle moves through a flow field described by its Eulerian specification $\boldsymbol{v}(\boldsymbol{x}, t)$. The right-hand side represents the sum of the *local* (time) and the *convective* rates of change of F:

$$(\boldsymbol{v} \cdot \nabla)F = v_1 \frac{\partial F}{\partial x_1} + v_2 \frac{\partial F}{\partial x_2} + v_3 \frac{\partial F}{\partial x_3}.$$

This specific decomposition of the total rate of change follows from differentiating the composite function $F(\boldsymbol{x}(\boldsymbol{x}_o, t), t)$ with respect to time (t), which requires the use of the chain rule.

In their most general form and using Einstein summation indexing, the high-dimensional equations of motion of a particle are described in terms of the position x, $(n+2)$-velocity u, and $\mathcal{F}$ is the acceleration subject to some non-gravitational force [147–149]:

$$\left|\begin{array}{c} \mathcal{F}^B = u^A \nabla_A u^B \\ \kappa = g_{AB} u^\alpha u^\beta = u \cdot u = \langle u|u \rangle . \\ u^A = \dot{x}_A \end{array}\right.$$

The $(n+2)$-dimensional manifold $(\mathcal{M}, g_{AB})$ has a coordinate system $x \equiv \{x^A\}$ and tensor metric g_{AB}, with uppercase Latin indices $0 \le A \le n$. The velocity and the acceleration are orthogonal, $u \cdot \mathcal{F} = \langle u|\mathcal{F} \rangle = 0$, ∇_A is the covariant derivative on M defined with respect to the metric tensor g_{AB}. For a scalar function $l = l(x)$, the normal vector to the manifold foliation Σ_l (see **Chapter 5**) is given by $n_A = \epsilon \Phi \partial_A l$, where $\epsilon = +1$ when the fifth dimension is time-like (like in spacekime) or $\epsilon = -1$, when it's space-like, and the lapse function Φ is a scalar normalizing n^A, $\Phi^2 (\partial l)^2 = \langle n|n \rangle = \epsilon = +1$ (in spacekime). The symmetric projection metric tensor $h_{AB} = g_{AB} - \epsilon\, n_A n_B$ is orthogonal to n_A. Define the n-D coordinate system on the hypersurfaces Σ_l be $y \equiv \{y^\alpha\}$, with lowercase Greek indices $0 \le \alpha \le n-1$. Then the n-holonomic basis vectors will be orthogonal to n^A and tangent to the hypersurfaces Σ_l:

$$e_\alpha^A = \frac{\partial x^A}{\partial y^\alpha}, \quad n \cdot e_\alpha = \langle n|e_\alpha \rangle = 0.$$

Note that e_α^A allows us to project objects in $\mathcal{M}$ onto the hypersurfaces Σ_l, e.g., a 1-form T can be represented as $T_\alpha = e_\alpha^A T_A = \langle e_\alpha|T \rangle$, where T_α is the projection of T_A onto Σ_l. This induces a metric tensor $(h_{\alpha\beta})$ and its inverse $(h^{\alpha\gamma} h_{\gamma\beta} = \delta_\beta^\alpha)$ on Σ_l:

$$h_{\alpha\beta} = e_\alpha^A \, e_\beta^B \, g_{AB} = e_\alpha^A \, e_\beta^B \, h_{AB}.$$

These equations can be expanded to the following system of PDEs:

$$\left|\begin{array}{l} a^\beta(u) = -\epsilon\, u_n \left(K^{\alpha\beta} u_\alpha + e_B^\beta n^A \nabla_A u^B \right) + \mathcal{F}^\beta \\ \dot{u}_n = K_{\alpha\beta} u^\alpha u^\beta + \epsilon\, u_n n^A u^B \nabla_A n_B + \mathcal{F}_n \\ \kappa = h_{\alpha\beta} u^\alpha u^\beta + \epsilon\, u_n^2 \end{array}\right. .$$

This system of equations depends on: the acceleration, $a^\beta(u) = u^\alpha \nabla_\alpha u^\beta$; the tensor $g_{AB} = h_{\alpha\beta} e_A^\alpha e_B^\beta + \epsilon\, n^A n^B$; the index $\kappa = g_{AB} u^\alpha u^\beta = +1,\ 0,\ -1$ corresponding to spacelike, lightlike, and timelike intervals; the velocity $\dot{u}_n = \frac{du_n}{d\lambda} = u^A \nabla_A u_n$; the acceleration field projections $\mathcal{F}_n = \mathcal{F} \cdot n$; the curvature $K_{\alpha\beta} = \frac{1}{2\Phi} (\partial_l - \mathcal{L}_N) h_{\alpha\beta}$; the Lie derivative in the direction of the shift vector, $\mathcal{L}_N$; and the acceleration per unit mass, $\mathcal{F}^\beta$.

3.7.2.3 Wave Equation in Spacekime

The wave equation models the vibrations of strings (1D), thin plate membranes (2D), pressure or density in gasses, liquids or solids (3D), and so forth for higher dimensions [150]. In its most general form, the extension of the wave equation in higher dimensions is a natural generalization of its spacetime analogue and still represents a second-order linear PDE:

$$\underbrace{\Delta_{\boldsymbol{x}} u(\boldsymbol{x},\boldsymbol{\kappa})}_{\text{spatial Laplacian}} = \underbrace{\Delta_{\boldsymbol{\kappa}} u(\boldsymbol{x},\boldsymbol{\kappa})}_{\text{temporal Laplacian}} ,$$

$$\Delta_{\boldsymbol{x}} u = \sum_{i=1}^{d_s} \partial^2_{x_i} u; \ \Delta_{\boldsymbol{\kappa}} u = \sum_{i=1}^{d_t} \partial^2_{\kappa_i} u ,$$

where $\boldsymbol{x} = (x_1, x_2, \ldots, x_{d_s}) \in \mathbb{R}^{d_s}$ and $\boldsymbol{\kappa} = (\kappa_1, \kappa_2, \ldots, \kappa_{d_t}) \in \mathbb{R}^{d_t}$ are the Cartesian coordinates in the d_s space and d_t time dimensions. Of course, there may also be different weights, $\{\alpha_i\}_{i=1}^{d_s} \geq 0$ and $\{\beta_j\}_{j=1}^{d_t} \geq 0$, that can be introduced with each space or time dimension:

$$\underbrace{\sum_{i=1}^{d_s} \alpha_i \partial^2_{x_i} u}_{\Delta^{\alpha}_{\boldsymbol{x}} u(\boldsymbol{x},\boldsymbol{\kappa})} = \underbrace{\sum_{j=1}^{d_t} \beta_j \partial^2_{\kappa_j} u}_{\Delta^{\beta}_{\boldsymbol{\kappa}} u(\boldsymbol{x},\boldsymbol{\kappa})} .$$

Under the metric signature $(+,+,+,-,-)$, the special case involving the smallest flat 5D spacekime manifold has $d_s = 3$ spacelike (spatial) and $d_t = 2$ timelike (temporal) variables, $\boldsymbol{x} = (x_1, x_2, x_3) \in \mathbb{R}^3$ and $\boldsymbol{\kappa} = (\kappa_1, \kappa_2) \in \mathbb{C} \cong \mathbb{R}^2$, respectively.

Next, we will derive solutions to the wave equation in spacekime, or in higher dimensions, and explore their local and global existence, validity, and stability. Let's start with a simpler problem of functions defined with periodic boundary conditions. The d_s-dimensional spatial cube is

$$\boldsymbol{x} = (x_1, x_2, \ldots, x_{d_s}) \in D_s \equiv \underbrace{\left[-\frac{1}{2}, \frac{1}{2}\right] \times \left[-\frac{1}{2}, \frac{1}{2}\right] \times \ldots \times \left[-\frac{1}{2}, \frac{1}{2}\right]}_{d_s} \equiv \left[-\frac{1}{2}, \frac{1}{2}\right]^{d_s} \subset \mathbb{R}^{d_s}$$

and d_t-dimensional temporal hypercube is

$$\boldsymbol{t} = (\kappa_1, \kappa_2, \ldots, \kappa_{d_t}) \in D_t \equiv \underbrace{\left[-\frac{1}{2}, \frac{1}{2}\right] \times \left[-\frac{1}{2}, \frac{1}{2}\right] \times \ldots \times \left[-\frac{1}{2}, \frac{1}{2}\right]}_{d_t} \equiv \left[-\frac{1}{2}, \frac{1}{2}\right]^{d_t} \subset \mathbb{R}^{d_t}.$$

Depending on the exact normalization of the Fourier transform, a number of alternative configurations of the cubic domains are possible, e.g., $[-1, 1]^d$ or $[-\pi, \pi]^d$. Functions defined on such finite domains should be periodic, where the corresponding (spatial or temporal) frequencies are integer multiples of 2π, π, 1, etc.

Let $\boldsymbol{\eta} = (\eta_1, \eta_2, \ldots, \eta_{d_t})'$ and $\boldsymbol{\xi} = (\xi_1, \xi_2, \ldots, \xi_{d_s})'$ represent respectively the wavenumbers; frequencies vectors of *integers* corresponding to the temporal and spatial frequencies of the Fourier-transformed periodic solution of the wave equation. In general, when dealing with non-periodic functions, the spatial and temporal frequencies are real numbers, but for our periodic boundary condition case, the frequencies are integers. Assuming that the solution is twice differentiable, i.e., $u \in C^2\ (D_t \times D_s) \subseteq L^2$, we will use the Fourier transform, $\mathcal{F}: L^2 \to L^2$:

$$U(\boldsymbol{\eta},\ \boldsymbol{\xi}) = \mathcal{F}(u)(\boldsymbol{\eta},\ \boldsymbol{\xi}) = \int_{D_s,\ D_t} u(\boldsymbol{x}, \boldsymbol{\kappa}) \times e^{-2\pi i\langle \eta, \kappa\rangle} \times e^{-2\pi i\langle \xi, x\rangle} d\boldsymbol{x} d\boldsymbol{\kappa},$$

$$u(\boldsymbol{x}, \boldsymbol{\kappa}) = \mathcal{F}^{-1}(U)(\boldsymbol{x}, \boldsymbol{\kappa}) = \int_{D_\eta,\ D_\xi} U(\boldsymbol{\eta},\ \boldsymbol{\xi}) \times e^{2\pi i\langle \eta, \kappa\rangle} \times e^{2\pi i\langle \xi, x\rangle} d\boldsymbol{\eta} d\boldsymbol{\xi}.$$

In general, there is no direct algebraic relation between an arbitrary function $f = f(y)$ and its Laplacian $\Delta f = \sum_{i=1}^{d} \partial_{y_i}^2 f$. However, in the special case of a planar wave, $f(y) = e^{2\pi i\langle y, \xi\rangle}$, the Laplacian has the following interesting property:

$$\Delta f = \Delta e^{2\pi i\langle y, \xi\rangle} = \underbrace{-4\pi^2|\boldsymbol{\xi}|^2}_{\lambda} \times \underbrace{e^{2\pi i\langle \boldsymbol{y}, \boldsymbol{\xi}\rangle}}_{f}.$$

In other words, plane waves, $f(y) = e^{2\pi i\langle y, \xi\rangle}$, are *eigenfunctions* of the Laplacian operator, i.e., $\Delta f = \lambda f$, corresponding to the *eigenvalue* $\lambda = -4\pi^2|\boldsymbol{\xi}|^2$. Plane waves are base functions that allow us to represent any L^2 function as a superposition of plane waves, potentially infinitely many plane waves. As the Laplacian is a linear operator, any periodic square-integrable function with a smooth second derivative (necessary to justify integration by parts) $u(\boldsymbol{y}):[-1, 1]^d \to \mathbb{C}$, including $\Delta u(\boldsymbol{y})$, can be expressed in terms of its Fourier transform, $U(\boldsymbol{\xi}) = \mathcal{F}(u)(\boldsymbol{\xi})$:

$$\Delta_y u(\boldsymbol{y}) = \Delta u(\boldsymbol{y}) = \Delta\left(\underbrace{\int_D U(\boldsymbol{\xi}) \times e^{2\pi i\langle \boldsymbol{y}, \boldsymbol{\xi}\rangle} d\boldsymbol{\xi}}_{u(y) = \mathcal{F}^{-1}(U)(\boldsymbol{y})}\right) = \int_D U(\boldsymbol{\xi}) \Delta\left(e^{2\pi i\langle \boldsymbol{y}, \boldsymbol{\xi}\rangle}\right) d\boldsymbol{\xi} =$$

$$\int_D U(\boldsymbol{\xi}) \underbrace{\left[\left(-4\pi^2|\boldsymbol{\xi}|^2\right) e^{2\pi i\langle \boldsymbol{y}, \boldsymbol{\xi}\rangle}\right]}_{\Delta\left(e^{2\pi i\langle \boldsymbol{y}, \boldsymbol{\xi}\rangle}\right)} d\boldsymbol{\xi} = \int_D \left[\left(-4\pi^2|\boldsymbol{\xi}|^2\right) \times U(\boldsymbol{\xi})\right] e^{2\pi i\langle \boldsymbol{y}, \boldsymbol{\xi}\rangle} d\boldsymbol{\xi}.$$

In the above equation, the integrand $U(\xi)\Delta\left(e^{2\pi i\langle \boldsymbol{y}, \boldsymbol{\xi}\rangle}\right) \in L^1$ and we can interchange the integration and differentiation operators.

Shortly, we will prove Green's first identity that allows us to integrate by parts the Fourier transform of the potential function's partial derivatives. Suppose the twice differentiable potential function $u(\boldsymbol{x}, \boldsymbol{\kappa}) \in C^2(D_s \times D_t)$ is periodic in time, $\boldsymbol{\kappa} \in \boldsymbol{D}_t \equiv \left[-\frac{1}{2}, \frac{1}{2}\right]^{d_t}$. Then, $\forall 1 \le j \le d_t$ the Fourier transforms of the function's *temporal* partial derivatives are:

$$\mathcal{F}\left(\partial_{\kappa_j}u\right)(\boldsymbol{\eta},\boldsymbol{\xi}) = \int\limits_{D_s,\ D_t} \partial_{\kappa_j}u(\boldsymbol{x},\boldsymbol{\kappa}) \times e^{-2\pi i\langle\boldsymbol{\eta},\boldsymbol{\kappa}\rangle} \times e^{-2\pi i\langle\boldsymbol{\xi},\boldsymbol{x}\rangle}\,d\boldsymbol{x}d\boldsymbol{\kappa} =$$

$$\int\limits_{D_s}\int\limits_{D_t} \partial_{\kappa_j}u(\boldsymbol{x},\boldsymbol{\kappa}) \times e^{-2\pi i\langle\boldsymbol{\eta},\boldsymbol{\kappa}\rangle} \times e^{-2\pi i\langle\boldsymbol{\xi},\boldsymbol{x}\rangle}\,d\boldsymbol{x}d\boldsymbol{\kappa} \overbrace{=}^{\text{Green's identity}}$$

$$\int\limits_{D_s}\left[\underbrace{\int\limits_{\partial D_t} u(\boldsymbol{x},\boldsymbol{\kappa}) \times e^{-2\pi i\langle\boldsymbol{\eta},\boldsymbol{\kappa}\rangle} \times e^{-2\pi i\langle\boldsymbol{\xi},\boldsymbol{x}\rangle}\,d\boldsymbol{\kappa}}_{0,\ \text{periodic } u \text{ function on } \boldsymbol{\kappa}\in\boldsymbol{D}_t \equiv \left[-\frac{1}{2},\frac{1}{2}\right]^{d_t}} - \int\limits_{D_t} u(\boldsymbol{x},\boldsymbol{\kappa}) \times \partial_{\kappa_j} e^{-2\pi i\langle\boldsymbol{\eta},\boldsymbol{\kappa}\rangle} \times e^{-2\pi i\langle\boldsymbol{\xi},\boldsymbol{x}\rangle}\,d\boldsymbol{\kappa}\right] d\boldsymbol{x} =$$

$$\int\limits_{D_s}\left[-\int\limits_{D_t} u(\boldsymbol{x},\boldsymbol{\kappa}) \times \left(-2\pi\ i\ \eta_j\right) \times e^{-2\pi i\langle\boldsymbol{\eta},\boldsymbol{\kappa}\rangle} \times e^{-2\pi i\langle\boldsymbol{\xi},\boldsymbol{x}\rangle}\,d\boldsymbol{\kappa}\right] d\boldsymbol{x} =$$

$$2\pi\ i\ \eta_j \int\limits_{D_s,\ D_t} u(\boldsymbol{x},\boldsymbol{\kappa}) e^{-2\pi i\langle\boldsymbol{\eta},\boldsymbol{\kappa}\rangle} e^{-2\pi i\langle\boldsymbol{\xi},\boldsymbol{x}\rangle}\,d\boldsymbol{x}d\boldsymbol{\kappa} = 2\pi\ i\ \eta_j\ \mathcal{F}(u)(\boldsymbol{\eta},\boldsymbol{\xi}),\ \forall 1\le j\le d_t.$$

Similarly, the Fourier transform of the potential function's *spatial* partial derivatives are:

$$\mathcal{F}\left(\partial_{x_l}u\right)(\boldsymbol{\eta},\boldsymbol{\xi}) = 2\pi\ i\ \xi_l\ \mathcal{F}(u)(\boldsymbol{\eta},\boldsymbol{\xi}),\ \forall 1\le l\le d_s.$$

These relations between the Fourier transform and the first-order partial derivatives can be extended to the Fourier transform of the second-order partial derivatives $\partial^2_{x_l}u = \partial_{x_l}\left(\partial_{x_l}\right)(u)$ and $\partial^2_{\kappa_j}u = \partial_{\kappa_j}\left(\partial_{\kappa_j}\right)(u)$:

$$\mathcal{F}\left(\partial^2_{\kappa_j}u\right)(\boldsymbol{\eta},\boldsymbol{\xi}) = 2\pi i\ \eta_j\ \mathcal{F}\left(\partial_{\kappa_j}u\right)(\boldsymbol{\eta},\boldsymbol{\xi}) = \left(2\pi i\boldsymbol{\eta}_j\right)^2\ \mathcal{F}(u)(\boldsymbol{\eta},\boldsymbol{\xi}) = -\left(2\pi\,\eta_j\right)^2\ \mathcal{F}(u)(\boldsymbol{\eta},\boldsymbol{\xi}),\ \forall 1\le j\le d_t,$$

$$\mathcal{F}\left(\partial^2_{x_l}u\right)(\boldsymbol{\eta},\boldsymbol{\xi}) = 2\pi\ i\ \xi_l\ \mathcal{F}\left(\partial_{x_l}u\right)(\boldsymbol{\eta},\boldsymbol{\xi}) = \left(2\pi i\boldsymbol{\xi}_l\right)^2\ \mathcal{F}(u)(\boldsymbol{\eta},\boldsymbol{\xi}) = -\left(2\pi\ \xi_l\right)^2\ \mathcal{F}(u)(\boldsymbol{\eta},\boldsymbol{\xi}),\ \forall 1\le l\le d_s.$$

As the spatial and temporal Fourier transforms are linear, the Fourier transforms of the spatial and temporal Laplacians are:

$$\mathcal{F}(\Delta_x u)(\boldsymbol{\eta},\boldsymbol{\xi}) = \mathcal{F}\left(\sum_{l=1}^{d_s}\partial^2_{x_l}u\right)(\boldsymbol{\eta},\boldsymbol{\xi}) = \sum_{l=1}^{d_s}\mathcal{F}\left(\partial^2_{x_l}u\right)(\boldsymbol{\eta},\boldsymbol{\xi}) = -4\pi^2\ |\boldsymbol{\xi}|^2\ \mathcal{F}(u)(\boldsymbol{\eta},\boldsymbol{\xi}),$$

$$\mathcal{F}(\Delta_\kappa u)(\boldsymbol{\eta},\ \boldsymbol{\xi}) = \mathcal{F}\left(\sum_{j=1}^{d_t} \partial^2_{\kappa_j} u\right)(\boldsymbol{\eta},\ \boldsymbol{\xi}) = \sum_{j=1}^{d_t} \mathcal{F}\left(\partial^2_{\kappa_j} u\right)(\boldsymbol{\eta},\ \boldsymbol{\xi}) = -4\pi^2\ |\boldsymbol{\eta}|^2\ \mathcal{F}(u)(\boldsymbol{\eta},\ \boldsymbol{\xi}).$$

Therefore, $\underbrace{\mathcal{F}(\Delta u)(\boldsymbol{\xi})}_{\widehat{\Delta u}} = \underbrace{\left(-4\pi^2|\boldsymbol{\xi}|^2\right)}_{\text{multiplier}} \times \underbrace{\mathcal{F}(u)}_{U=\hat{u}}(\boldsymbol{\xi})$ and the Laplacian operator can be considered as a Fourier multiplier operator. That is, the Fourier transform of the Laplacian (Δu) at a frequency ξ is given by the Fourier transform of the original function (u) evaluated at the same frequency, multiplied by the value of the multiplier at that frequency, $-4\pi^2|\boldsymbol{\xi}|^2$. Utilizing the inverse Fourier transform:

$$\underbrace{\Delta u(\boldsymbol{y})}_{f(y)} = \mathcal{F}^{-1}\left(\underbrace{F}_{\mathcal{F}(f)}\right)(\boldsymbol{y}) = \int_D \underbrace{\mathcal{F}(\Delta u)(\boldsymbol{\xi})}_{\mathcal{F}(f)(\boldsymbol{\xi})} \times e^{2\pi i\langle \boldsymbol{y},\boldsymbol{\xi}\rangle} d\boldsymbol{\xi} = \int_D \underbrace{\left(-4\pi^2|\boldsymbol{\xi}|^2\right) \times U(\boldsymbol{\xi})}_{\mathcal{F}(\Delta u)(\boldsymbol{\xi})} e^{2\pi i\langle \boldsymbol{y},\boldsymbol{\xi}\rangle} d\boldsymbol{\xi}.$$

Of course, this general property for any (spatial and temporal) periodic function with continuous second derivatives holds in our case for (periodic) potential functions u of this type:

$$u(\boldsymbol{x},\boldsymbol{\kappa}) = e^{2\pi i\langle \boldsymbol{\eta},\boldsymbol{\kappa}\rangle} \times e^{2\pi i\langle \boldsymbol{x},\boldsymbol{\xi}\rangle} \in C^2(D_t \times D_s), \text{subject to} |\boldsymbol{\eta}|^2 \equiv |\boldsymbol{\xi}|^2,$$

which solves the wave equation $\Delta_{\boldsymbol{x}} u(\boldsymbol{x},\boldsymbol{\kappa}) = \Delta_{\boldsymbol{\kappa}} u(\boldsymbol{x},\boldsymbol{\kappa})$. Applying the Fourier transform to the wave equation $\Delta_{\boldsymbol{x}} u(\boldsymbol{x},\boldsymbol{\kappa}) = \Delta_{\boldsymbol{\kappa}} u(\boldsymbol{x},\boldsymbol{\kappa})$ yields:

$$-4\pi^2\ |\boldsymbol{\xi}|^2\ \mathcal{F}(u)(\boldsymbol{\eta},\ \boldsymbol{\xi}) = \mathcal{F}(\Delta_x u)(\boldsymbol{\eta},\ \boldsymbol{\xi}) \equiv \mathcal{F}(\Delta_{\boldsymbol{\kappa}} u)(\boldsymbol{\eta},\ \boldsymbol{\xi}) = -4\pi^2\ |\boldsymbol{\eta}|^2\ \mathcal{F}(u)(\boldsymbol{\eta},\ \boldsymbol{\xi}).$$

This suggests a non-local *necessary and sufficient wavenumbers condition* $|\boldsymbol{\xi}|^2 = |\boldsymbol{\eta}|^2$ for the relation between the integer spatial ($\boldsymbol{\xi}$) and integer temporal ($\boldsymbol{\eta}$) frequencies that guarantees the potential function $u(\boldsymbol{x},\boldsymbol{\kappa}) = e^{2\pi i\langle \boldsymbol{\eta},\boldsymbol{\kappa}\rangle} \times e^{2\pi i\langle \boldsymbol{x},\boldsymbol{\xi}\rangle}$ represents a wave equation solution.

Since the wave equation is a linear PDE, any finite linear combination of M such basic potential functions will also represent a (composite) solution:

$$u(\boldsymbol{x},\boldsymbol{\kappa}) = \sum_{\substack{m=1 \\ \left\{\boldsymbol{\xi}_m, \boldsymbol{\eta}_m \mid |\boldsymbol{\xi}_m|^2 = |\boldsymbol{\eta}_m|^2\right\}}}^{M} C_m \times e^{2\pi i\langle \boldsymbol{\eta}_m,\boldsymbol{\kappa}\rangle} \times e^{2\pi i\langle x,\xi_m\rangle}.$$

In spacekime, using polar coordinate representation of kime, the Laplacian defines a non-linear PDE for the wave equation:

$$\Delta u(\kappa_1,\kappa_2) = \frac{\partial^2 u}{\partial \kappa_1^2} + \frac{\partial^2 u}{\partial \kappa_2^2} = \frac{1}{r}\frac{\partial u}{\partial r}\left(r\frac{\partial u}{\partial r}\right) + \frac{1}{r^2}\frac{\partial^2 u}{\partial \varphi^2} = \frac{\partial^2 u}{\partial r^2} + \frac{1}{r}\frac{\partial u}{\partial r} + \frac{1}{r^2}\frac{\partial^2 u}{\partial \varphi^2} = \Delta u(r,\varphi).$$

In the simple case when $M = 1$, separable solutions of the wave equation can be expressed via:

$$\boldsymbol{\kappa} = te^{i\varphi} = t(\cos\varphi + i\sin\varphi) = \underbrace{(t\cos\varphi)}_{\kappa_1} + i\underbrace{(t\sin\varphi)}_{\kappa_2},$$

$$u(\mathbf{x}, \boldsymbol{\kappa}) = e^{2\pi i\langle \boldsymbol{\eta}, \boldsymbol{\kappa}\rangle} \times e^{2\pi i\langle \mathbf{x}, \boldsymbol{\xi}\rangle} = e^{2\pi i \sum_{j=1}^{d_t=2} \kappa_j \eta_j} \times e^{2\pi i \sum_{l=1}^{d_s=3} x_l \xi_l} =$$

$$e^{2\pi i(\eta_1 t\cos\varphi + \eta_2 t\sin\varphi)} \times e^{2\pi i(x_1\xi_1 + x_2\xi_2 + x_3\xi_3)}.$$

Figure 3.10 shows alternative views of one specific solution to the spacekime wave equation, which is given by:

$$u(\mathbf{x}, \boldsymbol{\kappa}) = u(x_1, x_2, x_3, t, \varphi) = e^{2\pi t\, i\,(-2\cos\varphi + 3\sin\varphi)} \times e^{2\pi i(-3x_1 + 2x_2)},$$

where $\boldsymbol{\eta} = (\eta_1, \eta_2) = (-2,\ 3)$ and $\boldsymbol{\xi} = (\xi_1,\ \xi_2, \xi_3) = (-3, 2, 0)$, $|\boldsymbol{\xi}|^2 = |\boldsymbol{\eta}|^2 = 13$.

Note the oscillatory patterns of the 2D kime dynamics, where the 2D spatial coordinates are along the horizontal (transverse) plane, the kime-phase (directions) and kime-magnitude (time) represent the vertical axis and the longitudinal snapshot of the dynamics, and different cross-sectional, volume and surface renderings depict the wave motion in the (reduced) 2D + 2D spacekime manifold.

A mathematical problem, like solving a PDE, is called *well-posed* when the following three properties are guaranteed: (1) existence, i.e., there exists at least one solution to the problem; (2) uniqueness, i.e., there is at most one solution; and (3) stability, i.e., the solution continuously depends on the data of the problem, e.g., Cauchy data for the wave equation, suggesting that a small change in the data (e.g., initial conditions) leads only to a small proportional change in the solution.

Non-temporally periodic solutions to the wave equation Cauchy initial value problem, correspond to non-integer (real) spatial and temporal frequency vectors that satisfy the same regularization condition, $|\boldsymbol{\xi}|^2 = |\boldsymbol{\eta}|^2$. This represents a strong assumption restricting the allowed wavenumbers that leads to a non-local constraint on the Cauchy initial data problem. An example of this non-local constraint ensuring the well-posedness of the initial value problem is:

$$\boldsymbol{\eta} = (\eta_1, \eta_2) = (\pi\sqrt{2},\ \pi\sqrt{2}) \text{ and } \boldsymbol{\xi} = (\xi_1,\ \xi_2, \xi_3) = (\pi,\ \pi\sqrt{3}, 0),\ |\boldsymbol{\xi}|^2 = |\boldsymbol{\eta}|^2 = 4\pi^2.$$

Let's dive deeper in the existence, stability, determinism, and uniqueness of local and global solutions to the wave equation with Cauchy initial data [151, 152]. The classical spacetime Cauchy initial value problem on co-dimension 1 hypersurfaces is well-posed and has global unique solutions in Sobolev spaces H^m. The Cauchy initial value problem, formulated on higher co-dimension hypersurfaces in terms of a finite number of derivatives of the data, is globally ill-posed and does not permit (global)

Solid cylinder space wave propagation in spacekime with backplane projections.

Cross-sectional view of the (x,y) space (horizontal plane) animated over kime-magnitude, with z-axis representing the kime-phase dynamics.

$\eta = (\eta_1, \eta_2) = (-2,3)$, $\xi = (\xi_1, \xi_2, \xi_3) = (-3, 2, 0)$ in a polar setting, $\varphi \in [-\pi, \pi]$, for $t = 1$.

Solid ball space wave propagation in spacekime

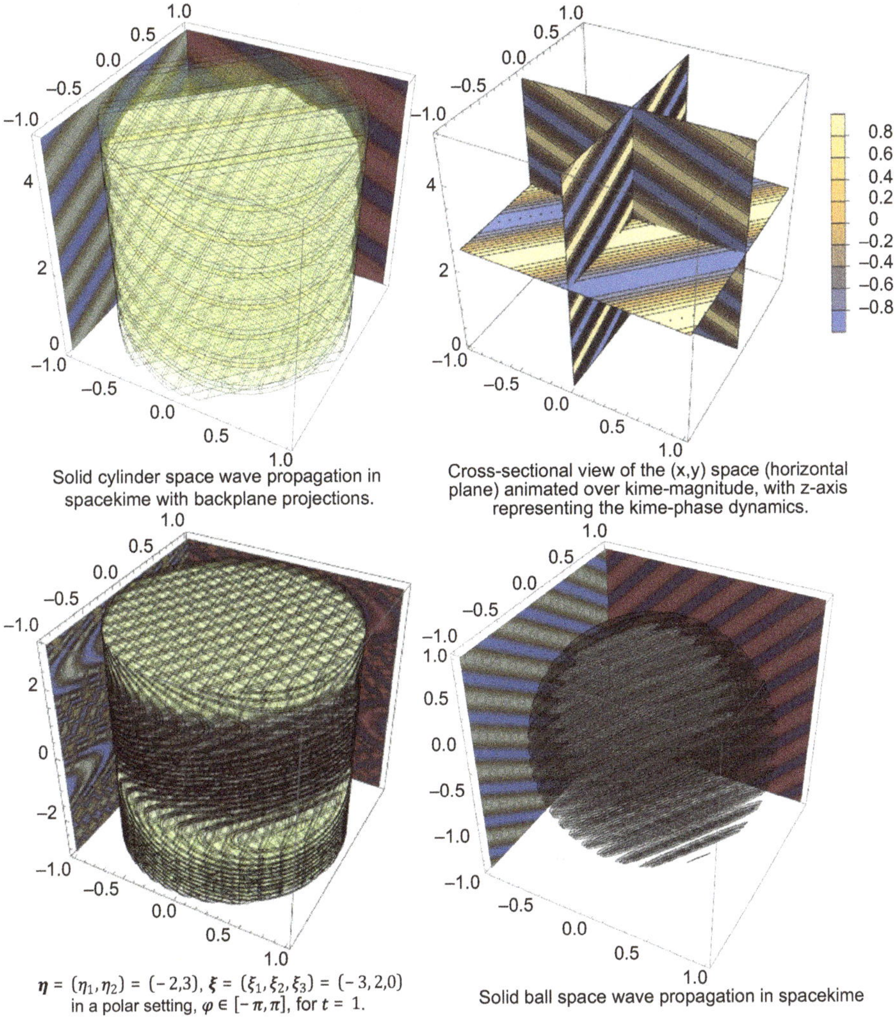

Figure 3.10: Example of the existence of a locally stable solution to the ultrahyperbolic wave equation in spacekime. The left and right columns illustrate alternative views and foliations of the 2D kime dynamics of the 4D (reduced) spacekime wave projected onto a flat 2D (x, y) spatial domain.

unique solutions. However, imposing some specific Cauchy-type constraints in terms of the initial data rectifies this problem and leads to a well-posed problem with locally unique solutions in neighborhoods of the initial hypersurfaces.

In essence, the general lack of global stability and uniqueness for the ultrahyperbolic wave equation, with Cauchy initial value formulation, can be resolved by imposing non-local constraints that arise naturally from the field equations. Such non-local constraints may preserve stability of the solutions but not their determinism or uniqueness. The Cauchy initial value problem associated with the ultrahyperbolic wave equation can be formulated as a linear constraint representing a hypersurface of co-dimension 1. Co-dimension 1 refers to a time variable (e.g., $\boldsymbol{\kappa} = (\kappa_1, \kappa_2) = te^{i\varphi}$) split of the temporal domain in two subspaces $\mathbb{R}^{d_t} \supseteq D_t = D_{t_1} \cup D_{t_{-1}}$.

The 1D subspace D_{t_1} represents κ_1, the dynamic evolution time dimension (e.g., t), and the complementary $(d_t - 1)$ dimensional subspace $D_{t_{-1}}$ consists of the remaining independent time-like dimensions (e.g., $\{\boldsymbol{\kappa}_{-1}\}$ or φ), where $\boldsymbol{\kappa} = \Bigg(\kappa_1, \underbrace{\kappa_2, \ldots, \kappa_{d_t}}_{\kappa_{-1}}\Bigg) \in D_t$.

Then, the ultrahyperbolic wave equation with Cauchy initial conditions and evolution along the first time-like coordinate, κ_1, is:

$$\sum_{i=1}^{d_s} \partial^2_{x_i} u \equiv \Delta_{\mathbf{x}} u = \Delta_{\boldsymbol{\kappa}} u \equiv \sum_{j=1}^{d_t} \partial^2_{\kappa_j} u,$$

$$\left| \begin{array}{l} u\Bigg(\underbrace{\mathbf{x}}_{x \in D_s}, \underbrace{0, \boldsymbol{\kappa}_{-1}}_{\kappa \in D_t}\Bigg) = f(\mathbf{x}, \boldsymbol{\kappa}_{-1}) \\ \partial_{\kappa_1} u(\mathbf{x}, 0, \boldsymbol{\kappa}_{-1}) = g(\mathbf{x}, \boldsymbol{\kappa}_{-1}) \end{array} \right. .$$

Typically, the initial constraints are formulated in terms of κ_1, a.k.a. the direction of (temporal) evolution, as restrictions over the neighborhood $N = \{(\mathbf{x}, \boldsymbol{\kappa}) \in D_s \times D_t | \kappa_1 = 0\}$ representing hypersurface subspaces of dimension one less than that of the entire space, $D_s \times D_t$. Higher co-dimensional constraints are defined analogously using two or more time dimensions to represent the temporal dynamics.

The Cauchy initial value problem depends on how much data (initial restrictions) are assumed or given a priori. For instance, one may fix the value of the potential function (zeroth derivative) and the first partial derivatives, or alternatively fix a finite number of partial derivatives of $u(x, \kappa)$, on the neighborhood N, and require compatibility of the imposed constraints with the general solutions of the wave equation, $\Delta_{\mathbf{x}} u = \Delta_{\boldsymbol{\kappa}} u$.

The standard *Sobolev space* of functions is defined as the closure, H^m, of the function space:

$$H^m = \text{closure}\left\{ f(\mathbf{x}, \boldsymbol{\kappa}_{-1}) \in C_o^{\infty}\left(\mathbb{R}^{d_s} \times \mathbb{R}^{d_t - 1}\right) \middle| \, \|f\|_m^2 \equiv \sum_{|\alpha| + |\beta| \le m} \int \left| \partial^{\alpha}_{\mathbf{x}} \partial^{\beta}_{\kappa_{-1}} f(\mathbf{x}, \boldsymbol{\kappa}_{-1}) \right|^2 d\mathbf{x} d\boldsymbol{\kappa}_{-1} \right\}.$$

As the spacekime wave equation can be expressed as $\partial^2_{\kappa_1} u = \Delta_{\mathbf{x}} u - \Delta_{\kappa_{-1}} u$, we can define the *energy functional*, associated with potential function solutions, as:

$$E(u) = \frac{1}{2}\int \left[\underbrace{\left|\partial_{\kappa_1} u\right|^2}_{\text{kinetic}} + \underbrace{\left|\nabla_x u\right|^2 - \left|\nabla_{\kappa_{-1}} u\right|^2}_{\text{potential}} \right] d\mathbf{x} d\boldsymbol{\kappa}_{-1},$$

where the *kinetic energy* term $\left|\partial_{\kappa_1} u\right|^2$ relates to the velocity in the dynamics temporal direction κ_1 and the *potential energy* term represents the spatio-temporal gradient $\left|\nabla_{\mathbf{x}} u\right|^2 - \left|\nabla_{\boldsymbol{\kappa}_{-1}} u\right|^2$, i.e., the spacekime displacement, reflecting the stored energy. This total energy indicates whether the wave-equation with Cauchy data (initial value problem) has unique solutions.

For non-temporally periodic potential functions that vanish at infinity, deriving their stability as solutions of the Cauchy initial value wave equation problem requires energy preservation, i.e., $\partial_{\kappa_1} E(u) = 0$. This can be accomplished by using the Gauss-Ostrogradsky theorem [153], which generalizes the univariate integration by parts to multiple dimensions. In particular, the Gauss-Ostrogradsky theorem relates a k-dimensional integral over a hypervolume to a $(k-1)$-dimensional integral over the surface boundary of the hypervolume domain.

Recall that if $U(y)$ and $V(y)$ are continuously differentiable scalar functions, integration by parts yields $\int UV'dy = UV - \int U'Vdy$. For a continuously differentiable vector field $F: V \to \mathbb{R}^n$, where $V = \{\mathbf{v} \in \mathbb{R}^n\}$ is compact with a piecewise smooth boundary ∂V, let's denote by $\vec{n}$ the outward pointing unit normal vector on the closed and orientable boundary manifold $S = \partial V$. For the given vector field (F), the *flux* $(F \cdot \vec{n})$ describes the magnitude of the flow over the closed surface boundary (∂V), and the *divergence* $(\nabla \cdot F)$ is a scalar field representing the aggregate of the input vector field. In Cartesian coordinates, the divergence of the field $F = (F_1, F_2, \ldots, F_n)$ at a given spatial position, $v = (x_1, x_2, \ldots, x_n) \in V$ represents the sum of the field partial derivatives at the location v:

$$\nabla \cdot F(v) = \langle \nabla \, | F(v) \rangle = \left\langle \left(\frac{\partial}{\partial x_1}, \frac{\partial}{\partial x_2}, \ldots, \frac{\partial}{\partial x_n} \right) \middle| (F_1, F_2, \ldots, F_n) \right\rangle = \sum_{i=1}^{n} \frac{\partial F_i(x_1, x_2, \ldots, x_n)}{\partial x_i}.$$

The flux of the field F across the boundary surface $S = \partial V$ measures the amount of flow passing through the boundary surface manifold per unit time. For a given small boundary patch with surface area ΔS, the *relative flux* is $F \cdot \vec{n} \Delta S$.

Connecting definite and indefinite integrals, the fundamental theorem of calculus states that integrating the derivative of a function over an interval can be expressed in terms of the functional values at the boundary of the interval. The higher-dimensional analog of the fundamental theorem of calculus is the Gauss-Ostrogradsky divergence theorem [153], which expresses the relation between the flux and divergence of the vector field F:

$$\int_V \underbrace{(\nabla \cdot F)}_{\text{divergence}} (\mathbf{v}) d\mathbf{v} = \underbrace{\oint_{\partial V} F \cdot \vec{n}\, ds}_{\text{flux}}.$$

For example, suppose $V \subseteq \mathbb{R}^3$, is a 3D solid circular cone with height $h = 1$ and base-radius $r = 1$. Let's define $F: V \to \mathbb{R}^3$ by $F = (F_1 = x - y,\ F_2 = x + z,\ F_3 = z - y)$. The positively oriented cone surface boundary is expressed by:

$$S = \partial V : \left\{ \begin{array}{c} \mathbf{v} = (x, y, z) \in \mathbb{R}^3 | \\ \{x^2 + y^2 - z^2 = 0\} \cap \{0 \le z \le 1\} \end{array} \right\}.$$

We can verify the divergence theorem by independently computing and comparing both hand sides representing the volume and surface integrals of the divergence and the flux, respectively.

Figure 3.11 shows a 3D rendering of the vector field $F = (x - y, x + z,\ z - y)$, the solid cone and its boundary, the surface normal ($\vec{n}$), the flux density (hot-metal color) ($F \cdot \vec{n}$), and the field magnitude (spectral color).

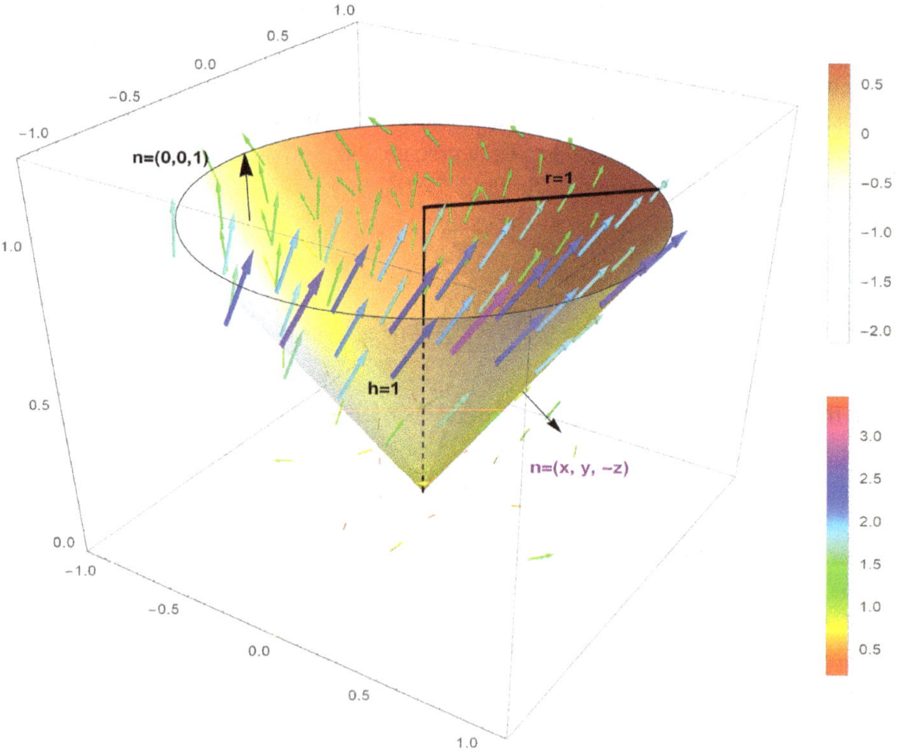

Figure 3.11: Conic domain flux density and field magnitude.

The divergence of this linear field at any point $v \in \mathbb{R}^3$ is constant:

$$\nabla \cdot F(v) = \left\langle \left(\frac{\partial}{\partial x}, \frac{\partial}{\partial y}, \frac{\partial}{\partial z}\right) | (x-y, x+z, z-y) \right\rangle (v) = 1 + 0 + 1 = 2.$$

The volume of a right circular cone of height h and base-radius r is obtained by integrating the area of a circle ($\pi\rho^2$) of radius $\rho = \frac{r}{h}x$, over $0 \le x \le h$:

$$\int_0^h \pi\rho^2 dx = \int_0^h \pi \frac{r^2}{h^2} x^2 dx = \pi \frac{r^2}{h^2} \int_0^h x^2 dx = \pi \frac{r^2}{h^2} \left[\frac{x^3}{3}\right]_0^h = \pi r^2 \frac{h}{3}.$$

Hence, the volume integral of the divergence over the solid unitary cone ($h = 1$, $r = 1$) is:

$$\int_V (\nabla \cdot F)(v) dv = \int_V 2\, dv = 2\left(\pi\, 1^2 \frac{1}{3}\right) = \frac{2\pi}{3}.$$

As the piecewise smooth cone surface boundary $S = \partial V$ consists of the base disc (S_d) and the attached cone surface (S_c), we can break the surface integral over the closed surface S in two pieces:

$$\oint_S F \cdot \vec{n} ds = \oint_{S_d \cup S_c} F \cdot \vec{n} ds = \oint_{S_d} F \cdot \vec{n} ds + \oint_{S_c} F \cdot \vec{n} ds.$$

To compute the first surface integral we will parameterize the disc surface (S_d):

$$r_d(\rho, v) = \begin{pmatrix} \rho \cos v \\ \rho \sin v \\ 1 \end{pmatrix}, \quad 0 \le \rho \le 1, \ 0 \le v \le 2\pi, \quad \frac{\partial r_d}{\partial \rho} = \begin{pmatrix} \cos v \\ \sin v \\ 0 \end{pmatrix}, \quad \frac{\partial r_d}{\partial v} = \begin{pmatrix} -\rho \sin v \\ \rho \cos v \\ 0 \end{pmatrix},$$

$$\left(\frac{\partial r_d}{\partial \rho} \times \frac{\partial r_d}{\partial v}\right) = \begin{pmatrix} \sin v \times 0 - 0 \times \rho \cos v \\ 0 \times (-\rho \sin v) - \cos v \times 0 \\ \cos v\, \rho \cos v - \sin v\,(-\rho \sin v) \end{pmatrix} = \begin{pmatrix} 0 \\ 0 \\ \rho \end{pmatrix},$$

$$\underbrace{\|J\|}_{\text{Jacobian}} = \left\| \frac{\partial r_d}{\partial \rho} \times \frac{\partial r_d}{\partial v} \right\|, \quad \vec{n} = \frac{\left(\frac{\partial r_d}{\partial \rho} \times \frac{\partial r_d}{\partial v}\right)}{\|J\|},$$

$$F = (x-y, x+z, z-y) = (\rho \cos v - \rho \sin v, \rho \cos v + 1, 1 - \rho \sin v).$$

Therefore, the flux over the disc S_d is

$$\oint_{S_d} F \cdot \vec{n} ds = \int_{\rho=0}^{1} \int_{v=0}^{2\pi} F(r_d(\rho, v)) \cdot \underbrace{\frac{\left(\frac{\partial r_d}{\partial \rho} \times \frac{\partial r_d}{\partial v}\right)}{\|J\|}}_{\vec{n}} dA =$$

$$\int_{\rho=0}^{1}\int_{v=0}^{2\pi}\underbrace{\begin{pmatrix}\rho\cos v-\rho\sin v\\ \rho\cos v+1\\ 1-\rho\sin v\end{pmatrix}'}_{F}\cdot\underbrace{\begin{pmatrix}0\\0\\ \rho\end{pmatrix}\frac{1}{\|J\|}}_{\vec{n}}\underbrace{\|J\|dv\,d\rho}_{dA}=$$

$$\int_{\rho=0}^{1}\int_{v=0}^{2\pi}\rho(1-\rho\sin v)dv\,d\rho=\int_{v=0}^{2\pi}\int_{\rho=0}^{1}(\rho-\rho^2\sin v)d\rho\,dv=\int_{v=0}^{2\pi}\left(\left(\frac{\rho^2}{2}-\frac{\rho^3}{3}\sin v\right)\Big|_{\rho=0}^{1}\right)dv=$$

$$\int_{v=0}^{2\pi}\left(\frac{1}{2}-\frac{1}{3}\sin v\right)\,dv=\frac{2\pi}{2}=\pi.$$

Similarly, we can parameterize the conic surface component (S_c) and compute the flux over the complementary surface. The only difference in this case is that the normal vector needs to point down (not up, as in the case of the flux over the disc). This is because $\vec{n}$ pointing up suggests an inward (not outward) flow. Thus, we use the flipped cross-product $\vec{n}=-\left(\frac{\partial r_c}{\partial\rho}\times\frac{\partial r_c}{\partial v}\right)=\left(\frac{\partial r_c}{\partial v}\times\frac{\partial r_c}{\partial\rho}\right)$.

$$r_c(\rho,v)=\begin{pmatrix}\rho\cos v\\ \rho\sin v\\ \rho\end{pmatrix},\quad 0\le\rho\le1,\quad 0\le v\le2\pi,\quad \frac{\partial r_c}{\partial\rho}=\begin{pmatrix}\cos v\\ \sin v\\ 1\end{pmatrix},\quad \frac{\partial r_c}{\partial v}=\begin{pmatrix}-\rho\sin v\\ \rho\cos v\\ 0\end{pmatrix},$$

$$\left(\frac{\partial r_c}{\partial\rho}\times\frac{\partial r_c}{\partial v}\right)=\begin{pmatrix}\sin v\times0-1\times\rho\cos v\\ 1\times(-\rho\sin v)-\cos v\times0\\ \cos v\,\rho\cos v-\sin v\,(-\rho\sin v)\end{pmatrix}=\begin{pmatrix}-\rho\cos v\\ -\rho\sin v\\ \rho\end{pmatrix},$$

$$-\left(\frac{\partial r_c}{\partial\rho}\times\frac{\partial r_c}{\partial v}\right)=\left(\frac{\partial r_c}{\partial v}\times\frac{\partial r_c}{\partial\rho}\right)=\begin{pmatrix}\rho\cos v\\ \rho\sin v\\ -\rho\end{pmatrix},\quad\underbrace{\|J\|}_{\text{Jacobian}}=\left\|\frac{\partial r_c}{\partial v}\times\frac{\partial r_c}{\partial\rho}\right\|,\ \vec{n}=\frac{\left(\frac{\partial r_c}{\partial v}\times\frac{\partial r_c}{\partial\rho}\right)}{\|J\|},$$

$$F=(x-y,x+z,z-y)=(\rho\cos v-\rho\sin v,\rho\cos v+\rho,\rho-\rho\sin v).$$

$$\oint_{S_c}F\cdot\vec{n}ds=\int_{\rho=0}^{1}\int_{v=0}^{2\pi}F(r_c(\rho,v))\cdot\underbrace{\frac{\left(\frac{\partial r_c}{\partial v}\times\frac{\partial r_c}{\partial\rho}\right)}{\|J\|}}_{\vec{n}}dA=$$

$$\int_{\rho=0}^{1}\int_{v=0}^{2\pi}\underbrace{\begin{pmatrix}\rho\cos v-\rho\sin v\\ \rho\cos v+\rho\\ \rho-\rho\sin v\end{pmatrix}'}_{F}\cdot\underbrace{\begin{pmatrix}\rho\cos v\\ \rho\sin v\\ -\rho\end{pmatrix}\frac{1}{\|J\|}}_{\vec{n}}\underbrace{\|J\|dv\,d\rho}_{dA}=$$

$$\int_{\rho=0}^{1}\int_{v=0}^{2\pi}[(\rho\cos v-\rho\sin v)(\rho\cos v)+(\rho\cos v+\rho)(\rho\sin v)-(\rho-\rho\sin v)\rho]dv\,d\rho=$$

$$-\int_{\rho=0}^{1}\int_{v=0}^{2\pi}\left(\rho^2\sin^2 v-2\rho^2\sin v\right)dv\,d\rho=-\frac{\pi}{3}.$$

This validates the divergence theorem:

$$\frac{2\pi}{3}=\pi-\frac{\pi}{3}=\oint_{S_d}F\cdot\vec{n}ds+\oint_{S_c}F\cdot\vec{n}ds\equiv\oint_{S}F\cdot\vec{n}ds\overbrace{=}^{?}\int_V(\nabla\cdot F)(\mathbf{v})d\mathbf{v}\equiv\int_V 2\,d\mathbf{v}=2\left(\pi\,1^2\frac{1}{3}\right)=\frac{2\pi}{3}.$$

Let's explore the special cases over the temporal domain $V=D_{t_{-1}}\subseteq\mathbb{R}^{d_{t_{-1}}}$ (excluding the first dynamics temporal variable, κ_1) and the spatial domain $V=D_x\subseteq\mathbb{R}^{d_x}$. For a pair of twice continuously differentiable scalar functions $\psi,\varphi:\mathbb{R}^d\to\mathbb{R}$, the relations between the flux and divergence of the vector field $F=\psi\nabla\varphi$ can be expressed in space and time domains by:

$$\int_{D_{t_{-1}}}\nabla_{\boldsymbol{\kappa}}\cdot\underbrace{\psi\nabla\varphi}_{F}\,d\boldsymbol{\kappa}_{-1}=\oint_{\mathbf{s}_{-1}\in\partial D_{t_{-1}}}\psi(\nabla\varphi\cdot\vec{n})d\mathbf{s}_{-1},\text{ and}$$

$$\int_{D_x}\nabla_{\mathbf{x}}\cdot\underbrace{\psi\nabla\varphi}_{F}\,d\mathbf{x}=\oint_{\mathbf{s}\in\partial D_x}\psi(\nabla\varphi\cdot\vec{n})d\mathbf{s}.$$

The multivariate product rule of differentiation (∇) allows us to simplify the following inner product in either the space ($\nabla_{\mathbf{x}}$) or the time $\left(\nabla_{\boldsymbol{\kappa}_{-1}}\right)$ domain:

$$\nabla\cdot F\equiv\nabla\cdot(\psi\nabla\varphi)=\nabla\psi\cdot(\nabla\varphi)+\psi\nabla\cdot(\nabla\varphi)=\nabla\psi\cdot(\nabla\varphi)+\psi\nabla^2\varphi=\nabla\psi\cdot(\nabla\varphi)+\psi\Delta\varphi.$$

This leads to the Green's first identity, which actually generalizes the univariate integration by parts to higher dimensions:

$$\int_V(\nabla\psi\cdot\nabla\varphi+\psi\Delta\varphi)d\mathbf{v}\equiv\int_V(\nabla\cdot F)(\mathbf{v})d\mathbf{v}\underbrace{=}_{\substack{\text{Gauss}\\\text{Ostrogradsky}}}\oint_{\partial V}F\cdot\vec{n}\,ds\equiv\oint_{\partial V}\psi\,(\nabla\varphi\cdot\hat{\mathbf{n}})d\mathbf{s}.$$

Therefore:

$$\text{(general)}\quad\int_V\psi\nabla^2\varphi\;d\mathbf{v}\equiv\int_V\psi\Delta\varphi\;d\mathbf{v}=\oint_{\partial V}\psi\;(\nabla\varphi\cdot\hat{\mathbf{n}})d\mathbf{s}-\int_V\nabla\psi\cdot\nabla\varphi\;d\mathbf{v},$$

$$\text{(spatial)}\quad\int_{D_{\mathbf{x}}}\psi\nabla_x^2\varphi\;d\mathbf{v}\equiv\int_{D_{\mathbf{x}}}\psi\Delta_x\varphi\;d\mathbf{v}=\oint_{\partial D_{\mathbf{x}}}\psi\;(\nabla_x\varphi\cdot\hat{\mathbf{n}})d\mathbf{s}-\int_{D_{\mathbf{x}}}\nabla_x\psi\cdot\nabla_x\varphi\;d\mathbf{v},$$

$$\text{(temporal)} \int_{D_{t_{-1}}} \psi \nabla^2_{\boldsymbol{\kappa}_{-1}} \varphi \, d\boldsymbol{\kappa}_{-1} \equiv \int_{D_{t_{-1}}} \psi \Delta_{\boldsymbol{\kappa}_{-1}} \varphi \, d\boldsymbol{\kappa}_{-1} = \oint_{\partial D_{t_{-1}}} \psi \left(\nabla_{\boldsymbol{\kappa}_{-1}} \varphi \cdot \hat{\boldsymbol{n}}\right) d\boldsymbol{s} - \int_{D_{t_{-1}}} \nabla_{\boldsymbol{\kappa}_{-1}} \psi \cdot \nabla_{\boldsymbol{\kappa}_{-1}} \varphi \, d\boldsymbol{\kappa}_{-1}.$$

Applying these equations to $\psi = \partial_{\kappa_1} u, \varphi = u$, we have:

$$\int \nabla_x u \cdot \partial_{\kappa_1} \nabla_x(u) \, d\boldsymbol{\kappa}_{-1} d\boldsymbol{x} = \int \nabla_x u \cdot \nabla_x(\partial_{\kappa_1} u) \, d\boldsymbol{\kappa}_{-1} d\boldsymbol{x} = \int_{D_{\boldsymbol{x}}} \int_{D_{t_{-1}}} \nabla_x(\partial_{\kappa_1} u) \cdot \nabla_x u \, d\boldsymbol{\kappa}_{-1} d\boldsymbol{x} =$$

$$\int_{D_{t_{-1}}} \underbrace{\left(\oint_{\partial D_{\boldsymbol{x}}} \partial_{\kappa_1} u \cdot (\nabla_x u \cdot \hat{\boldsymbol{n}}) ds \right)}_{0,\ u \text{ vanishes at infinity}} d\kappa_{-1} - \int_{D_{\boldsymbol{x}}} \int_{D_{t_{-1}}} \partial_{\kappa_1} u \cdot \nabla^2_x u \, d\boldsymbol{\kappa}_{-1} d\boldsymbol{x} =$$

$$= - \int_{D_{\boldsymbol{x}}} \int_{D_{t_{-1}}} \partial_{\kappa_1} u \cdot \Delta_x u \, d\boldsymbol{\kappa}_{-1} d\boldsymbol{x} = - \int \partial_{\kappa_1} u \cdot \Delta_x u \, d\boldsymbol{x} d\boldsymbol{\kappa}_{-1}.$$

Similarly,

$$\int \nabla_{\kappa_{-1}} u \cdot \partial_{\kappa_1} \nabla_{\kappa_{-1}}(u) \, d\boldsymbol{\kappa}_{-1} d\boldsymbol{x} = \int_{D_{\boldsymbol{x}}} \int_{D_{t_{-1}}} \nabla_{\kappa_{-1}}(\partial_{\kappa_1} u) \cdot \nabla_{\kappa_{-1}} u \, d\boldsymbol{\kappa}_{-1} d\boldsymbol{x} = - \int \partial_{\kappa_1} u \cdot \Delta_{\kappa_{-1}} u \, d\boldsymbol{x} d\boldsymbol{\kappa}_{-1}.$$

The initial conditions $\begin{pmatrix} u_o(\boldsymbol{x}, \boldsymbol{\kappa}_{-1}) \\ u_1(\boldsymbol{x}, \boldsymbol{\kappa}_{-1}) \end{pmatrix} = \begin{pmatrix} u\left(\boldsymbol{x}, \underbrace{0}_{\kappa_1}, \boldsymbol{\kappa}_{-1}\right) \\ \partial_{\kappa_1} u\left(\boldsymbol{x}, \underbrace{0}_{\kappa_1}, \boldsymbol{\kappa}_{-1}\right) \end{pmatrix}$ anchor the wave equation solution and its partial derivative at the starting value of the dynamics parameter (κ_1). When the dynamic evolution mapping $\kappa_1 \underset{\mathcal{M}}{\to} \begin{pmatrix} u(\boldsymbol{x}, \kappa_1, \boldsymbol{\kappa}_{-1}) \\ \partial_{\kappa_1} u(\boldsymbol{x}, \kappa_1, \boldsymbol{\kappa}_{-1}) \end{pmatrix}$ has a smooth first order derivative, i.e., $\mathcal{M} \in C^1\left(\mathbb{R}_{\kappa_1} \to H^1 \times H^o\right)$, then the energy functional is conserved along the solution path $u\left(\underbrace{--}_{x}, \kappa_1, \underbrace{--}_{\kappa_{-1}}\right)$ and the Cauchy initial value problem is well-posed $\left(\partial_{\kappa_1} E(u) = 0\right)$. This uses the divergence theorem (DT) as shown below:

$$\partial_{\kappa_1} E(u) = \frac{1}{2} \partial_{\kappa_1} \int \left[\left|\partial_{\kappa_1} u\right|^2 + |\nabla_x u|^2 - \left|\nabla_{\boldsymbol{\kappa}_{-1}} u\right|^2 \right] d\boldsymbol{x} d\boldsymbol{\kappa}_{-1} =$$

$$\frac{1}{2}\int 2\left[\partial_{\kappa_1}u\cdot\partial^2_{\kappa_1}u+\nabla_x u\cdot\partial_{\kappa_1}\nabla_x(u)-\nabla_{\boldsymbol{\kappa}_{-1}}u\cdot\partial_{\kappa_1}\nabla_{\boldsymbol{\kappa}_{-1}}(u)\right]d\boldsymbol{x}d\boldsymbol{\kappa}_{-1}=$$

$$\int\left(\left[\int\partial_{\kappa_1}u\cdot\partial^2_{\kappa_1}u\,d\boldsymbol{\kappa}_{-1}\right]+\left[\int\left[\nabla_x u\cdot\partial_{\kappa_1}\nabla_x(u)-\nabla_{\boldsymbol{\kappa}_{-1}}u\cdot\partial_{\kappa_1}\nabla_{\boldsymbol{\kappa}_{-1}}(u)\right]d\boldsymbol{\kappa}_{-1}\right]\right)d\boldsymbol{x}\underbrace{=}_{DT}$$

$$\int\left(\left[\int\partial_{\kappa_1}u\cdot\partial^2_{\kappa_1}u\,d\boldsymbol{\kappa}_{-1}\right]+\int\nabla^2_{\boldsymbol{\kappa}_{-1}}u\cdot\partial_{\kappa_1}(u)d\boldsymbol{\kappa}_{-1}-\int\nabla^2_x u\cdot\partial_{\kappa_1}(u)d\kappa_{-1}\right)d\boldsymbol{x}=$$

$$\int\left[\partial_{\kappa_1}u\cdot\partial^2_{\kappa_1}u+\left[\nabla^2_{\boldsymbol{\kappa}_{-1}}u-\nabla^2_x u\right]\cdot\partial_{\kappa_1}(u)\right]d\boldsymbol{\kappa}_{-1}d\boldsymbol{x}=$$

$$\int\left[\underbrace{\partial^2_{\kappa_1}u+\Delta_{\boldsymbol{\kappa}_{-1}}u-\Delta_x u}_{0,\ \text{wave equation}}\right]\cdot\partial_{\kappa_1}u\,d\boldsymbol{\kappa}_{-1}d\boldsymbol{x}=0.$$

Therefore, $E(u(\_\_,\kappa_1,\_\_))=E(u(\_\_,\ 0,\_\_))$, and the energy is preserved along the (univariate, κ_1) temporal evolution trajectory. However, by itself, energy preservation/conservation along κ_1 does not imply that the wave equation is well-posed in spacekime, since the negative sign of this component $-\left|\nabla_{\kappa_{-1}}u\right|$ in the energy allows for mixing unbounded positive and negative energy components while preserving the total constant energy. All three complementary energy components, $\left\{\left|\partial_{\kappa_1}u\right|,\ \left|\nabla_{\kappa_{-1}}u\right|,\ \left|\nabla_x u\right|\right\}$ may each be large while the energy still remains constant. Therefore, the above energy functional, $E(u)$, does not control the H^1-norm of the solution, $\|u\|^2_{m=1}$. In fact, the Cauchy initial value problem for the ultrahyperbolic wave equation is in general ill-posed [151].

However, these solution uniqueness and stability problems for the ultrahyperbolic wave equation, with Cauchy initial value formulation, can be resolved by imposing non-local constraints. For this Cauchy initial value problem, the Fourier transform maps the spatio-temporal variables $(\boldsymbol{x},\boldsymbol{\kappa}_{-1})\in D_s\times D_{t_{-1}}\subseteq\mathbb{R}^{d_s+d_t-1}$ to their frequency counterparts, the wavenumbers $(\boldsymbol{\xi},\boldsymbol{\eta}_{-1})\in\hat{D}_s\times\hat{D}_{t_{-1}}\subseteq\mathbb{R}^{d_s+d_t-1}$. We can examine the evolution operator defining the κ_1 dynamics subject to the initial constraints:

$$\begin{pmatrix}u_o(\boldsymbol{x},\boldsymbol{\kappa}_{-1})\\ u_1(\boldsymbol{x},\boldsymbol{\kappa}_{-1})\end{pmatrix}=\begin{pmatrix}u\left(\boldsymbol{x},\underbrace{0}_{\kappa_1},\boldsymbol{\kappa}_{-1}\right)\\ \partial_{\kappa_1}u\left(\boldsymbol{x},\underbrace{0}_{\kappa_1},\boldsymbol{\kappa}_{-1}\right)\end{pmatrix}\in H^{m+1}\times H^m,$$

$$\mathcal{F}\begin{pmatrix}u_o\\ u_1\end{pmatrix}=\widehat{\begin{pmatrix}u_o\\ u_1\end{pmatrix}}=\begin{pmatrix}\hat{u}_o(\boldsymbol{\xi},\boldsymbol{\eta}_{-1})\\ \hat{u}_1(\boldsymbol{\xi},\boldsymbol{\eta}_{-1})\end{pmatrix}=\int_{D_s,\ D_{t_{-1}}}\begin{pmatrix}u_o(\boldsymbol{x},\boldsymbol{\kappa}_{-1})\\ u_1(\boldsymbol{x},\kappa_{-1})\end{pmatrix}e^{-2\pi i\langle\boldsymbol{\eta}_{-1},\boldsymbol{\kappa}_{-1}\rangle}\times e^{-2\pi i\langle\boldsymbol{\xi},\boldsymbol{x}\rangle}d\boldsymbol{x}d\boldsymbol{\kappa}_{-1}.$$

Then, the wave equation becomes:

$$\sum_{i=1}^{d_s} \partial_{x_i}^2 u \equiv \Delta_{\mathbf{x}} u = \Delta_{\boldsymbol{\kappa}} u \equiv \sum_{i=1}^{d_t} \partial_{\kappa_i}^2 u,$$

$$\Delta_{\mathbf{x}} u - \Delta_{\boldsymbol{\kappa}_{-1}} u = \partial_{\kappa_1}^2 u,$$

$$\mathcal{F}\left(\Delta_{\mathbf{x}} u - \Delta_{\boldsymbol{\kappa}_{-1}} u\right) = \mathcal{F}\left(\partial_{\kappa_1}^2 u\right),$$

$$\mathcal{F}(\Delta_{\mathbf{x}}\, u)(\boldsymbol{\eta}_{-1}, \boldsymbol{\xi}) - \mathcal{F}(\Delta_{\boldsymbol{\kappa}_{-1}}\, u)(\boldsymbol{\eta}_{-1}, \boldsymbol{\xi}) \equiv \mathcal{F}\left(\partial_{\kappa_1}^2 u\right),$$

$$-4\pi^2\, |\boldsymbol{\xi}|^2\, \mathcal{F}(u)(\boldsymbol{\eta}_{-1}, \boldsymbol{\xi}) - \left(-4\pi^2\, |\boldsymbol{\eta}_{-1}|^2\, \mathcal{F}(u)(\boldsymbol{\eta}_{-1}, \boldsymbol{\xi})\right) = \partial_{\kappa_1}^2 \mathcal{F}(u)(\boldsymbol{\eta}_{-1}, \boldsymbol{\xi}),$$

$$-4\pi^2\, |\boldsymbol{\xi}|^2\, \hat{u}(\boldsymbol{\eta}_{-1}, \boldsymbol{\xi}) + 4\pi^2\, |\boldsymbol{\eta}_{-1}|^2\, \hat{u}(\boldsymbol{\eta}_{-1}, \boldsymbol{\xi}) = \partial_{\kappa_1}^2 \hat{u}(\boldsymbol{\eta}_{-1}, \boldsymbol{\xi}),$$

$$\underbrace{4\pi^2\left(-|\boldsymbol{\xi}|^2 + |\boldsymbol{\eta}_{-1}|^2\right)}_{\text{independent of } \kappa_1}\, \hat{u} = \partial_{\kappa_1}^2 \hat{u}.$$

Observe that the FT is applied to all spatial ($\mathbf{x}$) and all temporal ($\boldsymbol{\kappa}_{-1}$) variables except the time dynamics parameter (κ_1). This Fourier analysis allows us to convert the wave equation (PDE) to a simpler ordinary differential equation (ODE), in terms of κ_1, whose solution is the Fourier transform of the solution of the original PDE. By transforming the original Cauchy initial value problem to the Fourier domain, we get:

$$\begin{pmatrix} \hat{u}_o \\ \hat{u}_1 \end{pmatrix} = \begin{pmatrix} \hat{u}_o(\boldsymbol{\xi}, \boldsymbol{\eta}_{-1}) \\ \hat{u}_1(\boldsymbol{\xi}, \boldsymbol{\eta}_{-1}) \end{pmatrix} = \begin{pmatrix} \hat{u}(\boldsymbol{\xi}, \boldsymbol{\eta}_{-1}) \\ \partial_{\kappa_1} \hat{u}(\boldsymbol{\xi}, \boldsymbol{\eta}_{-1}) \end{pmatrix} \in H^{m+1} \times H^m.$$

Therefore, the Fourier space solution to the wave equation for Cauchy data is expressed as a piecewise smooth function:

$$\hat{u}(\boldsymbol{\xi}, \kappa_1, \boldsymbol{\eta}_{-1}) =$$

$$\begin{cases} \cos\left(2\pi\kappa_1\sqrt{|\boldsymbol{\xi}|^2 - |\boldsymbol{\eta}_{-1}|^2}\right)\hat{u}_o(\boldsymbol{\xi}, \boldsymbol{\eta}_{-1}) + \dfrac{\sin\left(2\pi\kappa_1\sqrt{|\xi|^2 - |\eta_{-1}|^2}\right)}{2\pi\sqrt{|\xi|^2 - |\eta_{-1}|^2}}\hat{u}_1(\boldsymbol{\xi}, \boldsymbol{\eta}_{-1}), & |\boldsymbol{\xi}| \geq |\boldsymbol{\eta}_{-1}| \\ \cosh\left(2\pi\kappa_1\sqrt{|\boldsymbol{\eta}_{-1}|^2 - |\boldsymbol{\xi}|^2}\right)\hat{u}_o(\boldsymbol{\xi}, \boldsymbol{\eta}_{-1}) + \dfrac{\sinh\left(2\pi\kappa_1\sqrt{|\eta_{-1}|^2 - |\xi|^2}\right)}{2\pi\sqrt{|\eta_{-1}|^2 - |\xi|^2}}\hat{u}_1(\boldsymbol{\xi}, \boldsymbol{\eta}_{-1}), & |\boldsymbol{\xi}| < |\boldsymbol{\eta}_{-1}| \end{cases}$$

This solution is easily confirmed by examining a simpler second-order ODE problem:

$$\partial_s^2 w(s) = a^2 w(s),$$

$$a = \underbrace{2\pi\sqrt{\left|\boldsymbol{\eta}_{-1}\right|^2 - |\boldsymbol{\xi}|^2}}_{\text{constant w.r.t. } s},\ a^2 = 4\pi^2\left(-|\boldsymbol{\xi}|^2 + \left|\boldsymbol{\eta}_{-1}\right|^2\right).$$

This ODE is solved by:

$$w(s) = c_1 e^{2\pi s\sqrt{\left|\boldsymbol{\eta}_{-1}\right|^2 - |\boldsymbol{\xi}|^2}} + c_2 e^{-2\pi s\sqrt{\left|\boldsymbol{\eta}_{-1}\right|^2 - |\boldsymbol{\xi}|^2}},\ c_1, c_2 \in \mathbb{C}.$$

In this problem simplification, s and w play the roles of κ_1 and $\hat{u}$ in the original ultrahyperbolic wave equation.

Since

$$\cos\phi = \frac{1}{2}\left(e^{i\phi} + e^{-i\phi}\right),\ \ \sin\phi = -\frac{i}{2}\left(e^{i\phi} - e^{-i\phi}\right),\ \ \text{and}$$

$$\cosh\phi = \frac{1}{2}\left(e^{\phi} + e^{-\phi}\right) = \cos(i\phi),\ \ \sinh\phi = \frac{1}{2}\left(e^{\phi} - e^{-\phi}\right) = -i\sin(i\phi),$$

the *circular* and *hyperbolic* functions $\hat{u}(\boldsymbol{\xi}, \boldsymbol{\eta}) = \hat{u}(\boldsymbol{\xi}, \kappa_1, \boldsymbol{\eta}_{-1})$ defined above indeed represent solutions to the $\partial^2_{\kappa_1}\hat{u} = \underbrace{4\pi^2\left(-|\boldsymbol{\xi}|^2 + \left|\boldsymbol{\eta}_{-1}\right|^2\right)}_{a^2}\hat{u}$ equation that obey the Cauchy initial value constraints:

$$\begin{pmatrix} \hat{u}\left(\boldsymbol{\xi}, \underbrace{0}_{\kappa_1}, \boldsymbol{\eta}_{-1}\right) \\ \partial_{\kappa_1}\hat{u}\left(\boldsymbol{\xi}, \underbrace{0}_{\kappa_1}, \boldsymbol{\eta}_{-1}\right) \end{pmatrix} = \underbrace{\begin{pmatrix} \hat{u}_o(\boldsymbol{\xi}, \boldsymbol{\eta}_{-1}) \\ \hat{u}_1(\boldsymbol{\xi}, \boldsymbol{\eta}_{-1}) \end{pmatrix}}_{\text{Cauchy data}}.$$

Case 1: For a Fourier frequency region defined by $|\boldsymbol{\xi}| \geq \left|\boldsymbol{\eta}_{-1}\right|$, the Cauchy data value problem $\begin{pmatrix} \hat{u}_o \\ \hat{u}_1 \end{pmatrix}$ is well-posed and yields a stable *local solution*, $\hat{u}$. Given also that the initial value problem $\begin{pmatrix} \hat{u}_o \\ \hat{u}_1 \end{pmatrix}$ is analytic and of exponential type, the solution is global and stable. This follows from the Paley-Wiener theorem [154], which states that the IFT of the Cauchy data $\begin{pmatrix} u_o \\ u_1 \end{pmatrix} = \mathcal{F}^{-1}\begin{pmatrix} \hat{u}_o \\ \hat{u}_1 \end{pmatrix}$ will be smooth and compactly supported if and only if its FT, $\mathcal{F}\begin{pmatrix} u_o \\ u_1 \end{pmatrix} = \widehat{\begin{pmatrix} u_o \\ u_1 \end{pmatrix}}$, is an *entire function*, i.e., analytic and of exponential type, rapidly decreasing with the frequency magnitude. In this case, for a pair (c_1, c_2) independent of κ_1, the general solution to the unconstrained ODE $\partial^2_{\kappa_1}\hat{u} = 4\pi^2\left(-|\boldsymbol{\xi}|^2 + \left|\boldsymbol{\eta}_{-1}\right|^2\right)\hat{u}$ is:

$$\hat{u} = c_1\cos\left(2\pi\,\kappa_1\sqrt{|\boldsymbol{\xi}|^2 - \left|\boldsymbol{\eta}_{-1}\right|^2}\right) + c_2\sin\left(2\pi\,\kappa_1\sqrt{|\boldsymbol{\xi}|^2 - \left|\boldsymbol{\eta}_{-1}\right|^2}\right).$$

This leads to a solution of the corresponding constrained problem (Cauchy data):

$$\hat{u}\left(\underbrace{\boldsymbol{\xi}, \kappa_1, \boldsymbol{\eta}_{-1}}_{\eta}\right) =$$

$$\cos\left(2\pi\kappa_1\sqrt{|\boldsymbol{\xi}|^2 - |\boldsymbol{\eta}_{-1}|^2}\right)\underbrace{\hat{u}_o(\boldsymbol{\xi}, \boldsymbol{\eta}_{-1})}_{c_1} + \sin\left(2\pi\kappa_1\sqrt{|\boldsymbol{\xi}|^2 - |\boldsymbol{\eta}_{-1}|^2}\right)\underbrace{\frac{\hat{u}_1(\boldsymbol{\xi}, \boldsymbol{\eta}_{-1})}{2\pi\sqrt{|\boldsymbol{\xi}|^2 - |\boldsymbol{\eta}_{-1}|^2}}}_{c_2}.$$

Case 2: However, over the complementary region, $|\boldsymbol{\xi}| < |\boldsymbol{\eta}_{-1}|$, given a non-trivial κ_1 value, the initial value problem $\begin{pmatrix} \hat{u}_o \\ \hat{u}_1 \end{pmatrix} = \begin{pmatrix} \hat{u}(\boldsymbol{\xi}, \kappa_1, \boldsymbol{\eta}_{-1}) \\ \partial_{\kappa_1}\hat{u}(\boldsymbol{\xi}, \kappa_1, \boldsymbol{\eta}_{-1}) \end{pmatrix} = \begin{pmatrix} \hat{u}_o(\boldsymbol{\xi}, \boldsymbol{\eta}_{-1}) \\ \hat{u}_1(\boldsymbol{\xi}, \boldsymbol{\eta}_{-1}) \end{pmatrix}$ $\in H^{m+1} \times H^m$ is ill-posed and the solutions are not bounded, unique, or stable with respect to the dynamic parameter κ_1, as the hyperbolic functions are not bandlimited. Thus, the spacekime-domain Cauchy initial value problem $\begin{pmatrix} u_o \\ u_1 \end{pmatrix} = \mathcal{F}^{-1}\begin{pmatrix} \hat{u}_o \\ \hat{u}_1 \end{pmatrix}$ would not be well-posed with respect to κ_1, and any spacekime solutions to the boundary value wave equation would not be well-defined, unique, or stable.

In this case, for a pair (c_3, c_4) independent of κ_1, the general solution of the unconstrained ODE $\partial^2_{\kappa_1}\hat{u} = 4\pi^2\left(-|\boldsymbol{\xi}|^2 + |\boldsymbol{\eta}_{-1}|^2\right)\hat{u}$, representing the Fourier dual to the original wave equation, is:

$$\hat{u} = c_3 e^{\left(2\pi\kappa_1\sqrt{|\boldsymbol{\eta}_{-1}|^2 - |\boldsymbol{\xi}|^2}\right)} + c_4 e^{\left(-2\pi\kappa_1\sqrt{|\boldsymbol{\eta}_{-1}|^2 - |\boldsymbol{\xi}|^2}\right)}.$$

The solution to the initial value problem is therefore:

$$\hat{u}(\boldsymbol{\xi}, \kappa_1, \boldsymbol{\eta}_{-1}) =$$

$$\cosh\left(2\pi\kappa_1\sqrt{|\boldsymbol{\eta}_{-1}|^2 - |\boldsymbol{\xi}|^2}\right)\underbrace{\hat{u}_o(\boldsymbol{\xi}, \boldsymbol{\eta}_{-1})}_{c_3} + \sinh\left(2\pi\kappa_1\sqrt{|\boldsymbol{\eta}_{-1}|^2 - |\boldsymbol{\xi}|^2}\right)\underbrace{\frac{\hat{u}_1(\boldsymbol{\xi}, \boldsymbol{\eta}_{-1})}{2\pi\sqrt{|\boldsymbol{\eta}_{-1}|^2 - |\boldsymbol{\xi}|^2}}}_{c_4}.$$

When $|\boldsymbol{\xi}| < |\boldsymbol{\eta}_{-1}|$, there are no globally stable solutions as cosh() and sinh() are unbounded and $\cosh\left(2\pi\kappa_1\sqrt{|\boldsymbol{\eta}_{-1}|^2 - |\boldsymbol{\xi}|^2}\right)$ and $\sinh\left(2\pi\kappa_1\sqrt{|\boldsymbol{\eta}_{-1}|^2 - |\boldsymbol{\xi}|^2}\right)$ grow exponentially with κ_1.

However, in the complementary domain, when $|\boldsymbol{\xi}| \geq |\boldsymbol{\eta}_{-1}|$, the magnitudes of their circular counterparts are bounded for any κ_1:

$$\left|\cos\left(2\pi\,\kappa_1\sqrt{|\boldsymbol{\xi}|^2-|\boldsymbol{\eta}_{-1}|^2}\right)\right|\le 1 \text{ and } \left|\frac{\sin\left(2\pi\,\kappa_1\sqrt{|\boldsymbol{\xi}|^2-|\boldsymbol{\eta}_{-1}|^2}\right)}{2\pi\sqrt{|\boldsymbol{\xi}|^2-|\boldsymbol{\eta}_{-1}|^2}}\right|\le 1,$$

which guarantees the global existence and stability of the solutions of the Cauchy initial value problem.

Fourier synthesis provides the connection between the Fourier solution, $\hat{u}(\boldsymbol{\xi},\boldsymbol{\eta})=\mathcal{F}(u)(\boldsymbol{\xi},\boldsymbol{\eta})$, of the ODE, and the corresponding spacekime solution, $u(\boldsymbol{x},\boldsymbol{\kappa})=\mathcal{F}^{-1}(\hat{u})(\boldsymbol{x},\boldsymbol{\kappa})$, under Cauchy initial value constraints. Since $\hat{u}(\boldsymbol{\xi},\kappa_1,\boldsymbol{\eta}_{-1})\in L^2(d\boldsymbol{\xi}, d\boldsymbol{\eta}_{-1})$, we can employ the inverse Fourier transform to obtain an analytic representation of the spacekime solution:

$$u\Bigg(\boldsymbol{x},\underbrace{\kappa_1,\boldsymbol{\kappa}_{-1}}_{\kappa}\Bigg)=\mathcal{F}^{-1}(\hat{u})(\boldsymbol{x},\boldsymbol{\kappa})=\int_{\hat{D}_s\times\hat{D}_{t_{-1}}}\hat{u}(\boldsymbol{\xi},\kappa_1,\boldsymbol{\eta}_{-1})\times e^{2\pi i\langle x,\xi\rangle}\times e^{2\pi i\langle \boldsymbol{\kappa}_{-1},\boldsymbol{\eta}_{-1}\rangle}d\boldsymbol{\xi}\,d\boldsymbol{\eta}_{-1}.$$

In the simple case where $|\eta_1|^2\equiv|\boldsymbol{\xi}|^2-|\boldsymbol{\eta}_{-1}|^2\ge 0$, the FT of the solution to the Cauchy initial value problem:

$$\hat{u}(\boldsymbol{\xi},\eta_1,\boldsymbol{\eta}_{-1})=\cos(2\pi|\eta_1|\,\kappa_1)\times\hat{u}_o(\boldsymbol{\xi},\boldsymbol{\eta}_{-1})+\frac{\sin(2\pi|\eta_1|\,\kappa_1)}{2\pi|\eta_1|}\times\hat{u}_1(\boldsymbol{\xi},\boldsymbol{\eta}_{-1})$$

can be represented explicitly as a sum of a pair of convolutions (*):

$$u(\boldsymbol{x},\kappa_1,\boldsymbol{\kappa}_{-1})=\mathcal{F}^{-1}(\hat{u})(\boldsymbol{x},\boldsymbol{\kappa})=C(\kappa_1)^{*}u_o(\boldsymbol{x},\boldsymbol{\kappa}_{-1})+S(\kappa_1)^{*}u_1(\boldsymbol{x},\boldsymbol{\kappa}_{-1}),$$

where

$$C(\kappa_1)=\mathcal{F}^{-1}_{(\xi,\eta_{-1})}(\cos(2\pi|\eta_1|\,\kappa_1)),$$

$$S(\kappa_1)=\mathcal{F}^{-1}_{(\xi,\eta_{-1})}\left(\frac{\sin(2\pi|\eta_1|\,\kappa_1)}{2\pi|\eta_1|}\right),$$

$$\frac{\partial}{\partial\kappa_1}S(\kappa_1)=\mathcal{F}^{-1}_{(\xi,\eta_{-1})}\left(\frac{\frac{\partial}{\partial\kappa_1}\sin(2\pi|\eta_1|\,\kappa_1)}{2\pi|\eta_1|}\right)=\mathcal{F}^{-1}_{(\xi,\eta_{-1})}\left(2\pi|\eta_1|\frac{\cos(2\pi|\eta_1|\,\kappa_1)}{2\pi|\eta_1|}\right)=C(\kappa_1),$$

$$|\eta_1|\equiv\sqrt{|\boldsymbol{\xi}|^2-|\boldsymbol{\eta}_{-1}|^2},\ |\boldsymbol{\xi}|^2\ge|\boldsymbol{\eta}_{-1}|^2.$$

Also note that the initial conditions (Cauchy data) may be specified by:

$$\underbrace{\begin{pmatrix}u_o(\boldsymbol{x},\boldsymbol{\kappa}_{-1})\\ u_1(\boldsymbol{x},\boldsymbol{\kappa}_{-1})\end{pmatrix}}_{\text{Cauchy data}}=\begin{pmatrix}u(\boldsymbol{x},\kappa_1=0,\ \boldsymbol{\kappa}_{-1})\\ \partial_{\kappa_1}u(\boldsymbol{x},\kappa_1=0,\ \boldsymbol{\kappa}_{-1})\end{pmatrix},$$

$$u\Big(\mathbf{x}, \underbrace{\kappa_1, \boldsymbol{\kappa}_{-1}}_{\boldsymbol{\kappa}}\Big) = e^{2\pi i\langle \boldsymbol{\eta}, \boldsymbol{\kappa}\rangle} \times e^{2\pi i\langle \mathbf{x}, \boldsymbol{\xi}\rangle}, \ |\boldsymbol{\xi}|^2 = |\eta_1|^2 + |\boldsymbol{\eta}_{-1}|^2.$$

The *Asgeirsson's mean value theorem* [155, 156] states that the average of the potential function over the spatial domain, for a fixed kime location, is the same as its average over the temporal domain, for a fixed spatial location:

$$\int_{|\mathbf{x}|=\rho} u(\mathbf{x}, \boldsymbol{\kappa}=0) dS(\mathbf{x}) = \int_{|\boldsymbol{\kappa}|=\rho} u(\mathbf{x}=0, \boldsymbol{\kappa}) dS(\boldsymbol{\kappa}),$$

where $\rho > 0$ is the radius of a Euclidean sphere in $\mathbb{R}^d$, $d \in \{d_s, d_t\}$, with surface area dS, and the potential function u represents a solution of the wave equation $\nabla_{\mathbf{x}}^2 u = \nabla_{\kappa}^2 u$ over a neighborhood $N = \left\{(\mathbf{x}, \boldsymbol{\kappa}) \in D_s \times D_t \subseteq \mathbb{R}^{d_s + d_t} \,||\mathbf{x}| + |\boldsymbol{\kappa}| \le \rho\right\}$. The mean value theorem suggests that the existence and stability of solutions of the ultrahyperbolic wave equation require some additional non-local constraints [151, 152]. The added restrictions may derived by Fourier synthesis (IFT) using the closed-form formulation of the Cauchy data problem in the Fourier frequency domain.

Earlier, we noted that the energy functional,
$E(u) = \frac{1}{2}\int \left[|\partial_{\kappa_1} u|^2 + |\nabla_{\mathbf{x}} u|^2 - |\nabla_{\boldsymbol{\kappa}_{-1}} u|^2\right] d\mathbf{x} d\kappa_{-1}$, is indefinite and does not lead to a proper norm on the Sobolev space of solutions of the Cauchy data problem.

However, the Cauchy initial value evolution operator $v = \begin{pmatrix} v_o \\ v_1 \end{pmatrix} \in \underbrace{X \equiv H^{m+1} \times H^m}_{\text{phase space}}$ may be used to naturally define a *modified energy norm* by parcellating the phase space:

$$\|v\|_X^2 \equiv \underbrace{\int_{|\boldsymbol{\eta}_{-1}| \le |\boldsymbol{\xi}|} \left[|\boldsymbol{\xi}|^2 - |\boldsymbol{\eta}_{-1}|^2\right] |\hat{v}_o(\boldsymbol{\xi}, \boldsymbol{\eta}_{-1})|^2 d\boldsymbol{\xi} d\boldsymbol{\eta}_{-1}}_{\ge 0} +$$

$$\underbrace{\int_{|\boldsymbol{\xi}| < |\boldsymbol{\eta}_{-1}|} \left[|\boldsymbol{\eta}_{-1}|^2 - |\boldsymbol{\xi}|^2\right] |\hat{v}_o(\boldsymbol{\xi}, \boldsymbol{\eta}_{-1})|^2 d\boldsymbol{\xi} d\boldsymbol{\eta}_{-1}}_{\text{condition } v \text{ to ensure } \ge 0} + \underbrace{\int |\hat{v}_1(\boldsymbol{\xi}, \boldsymbol{\eta}_{-1})|^2 d\boldsymbol{\xi} d\boldsymbol{\eta}_{-1}}_{\ge 0}.$$

This modified-energy norm, $\|v\|_X^2$, rectifies the problems with the original energy, and satisfies the three norm conditions; scalability, sub-additivity, and point-separability. Using this modified energy, we can split the phase space into three complementary components:

$$X \equiv H^{m+1} \times H^m = \underbrace{(X^\sigma \backslash X^\tau)}_{\text{center stable}} \cup \underbrace{(X^\tau \backslash X^\sigma)}_{\text{center unstable}} \cup \underbrace{(X^\tau \cap X^\sigma)}_{\text{center}}.$$

The entire domain of the evolution in terms of the temporal dynamics of $\kappa_1 \in \mathbb{R}^+$ is partitioned into three subspaces: *center stable* space, X^σ, the *center unstable* space, X^τ, and the *center* space, $X^c = X^\tau \cap X^\sigma$, which are defined by:

$$X^\sigma = \left\{ v = \begin{pmatrix} v_o \\ v_1 \end{pmatrix} \in X \;\middle|\; \frac{1}{2}\left(\hat{v}_o(\boldsymbol{\xi}, \boldsymbol{\eta}_{-1}) + \frac{\hat{v}_1(\boldsymbol{\xi}, \boldsymbol{\eta}_{-1})}{\sqrt{|\boldsymbol{\eta}_{-1}|^2 - |\boldsymbol{\xi}|^2}} \right) = 0, \forall |\boldsymbol{\xi}| < |\boldsymbol{\eta}_{-1}| \right\},$$

$$X^\tau = \left\{ v = \begin{pmatrix} v_o \\ v_1 \end{pmatrix} \in X \;\middle|\; \frac{1}{2}\left(\hat{v}_o(\boldsymbol{\xi}, \boldsymbol{\eta}_{-1}) - \frac{\hat{v}_1(\boldsymbol{\xi}, \boldsymbol{\eta}_{-1})}{\sqrt{|\boldsymbol{\eta}_{-1}|^2 - |\boldsymbol{\xi}|^2}} \right) = 0, \forall |\boldsymbol{\xi}| < |\boldsymbol{\eta}_{-1}| \right\},$$

$$X^c = \left\{ v = \begin{pmatrix} v_o \\ v_1 \end{pmatrix} \in X \;\middle|\; \text{support}\begin{pmatrix} \hat{v}_o \\ \hat{v}_1 \end{pmatrix} \subseteq \{ |\boldsymbol{\eta}_{-1}| \le |\boldsymbol{\xi}| \} \right\} = X^\sigma \cap X^\tau.$$

These three subdomains determine the corresponding three types of ultrahyperbolic wave equation solutions [152]: (1) For constraints in the central stable space, $v = \begin{pmatrix} v_o \\ v_1 \end{pmatrix} \in X^\sigma$, the Cauchy initial value problem has a *unique local solution* $u \in X$, for all $\kappa_1 \in \mathbb{R}^+$; (2) For constraints in the central unstable space, $v = \begin{pmatrix} v_o \\ v_1 \end{pmatrix} \in X^\tau$, the Cauchy initial value problem has a *unique solution* $u \in X$, only for all $\kappa_1 \in \mathbb{R}^-$; and (3) For constraints in the central space, $v = \begin{pmatrix} v_o \\ v_1 \end{pmatrix} \in X^c$, the Cauchy initial value problem has a *unique global solution* $u \in X$, for all $\kappa_1 \in \mathbb{R}$.

Let's next expand the derivation to non-periodic temporal functions. The rationale for considering spatially periodic ($\boldsymbol{x}$) and kime non-periodic temporal ($\boldsymbol{\kappa}$) domain is to generalize the very restrictive constraint for integer spatial and temporal frequencies, $|\boldsymbol{\xi}|^2 \equiv |\boldsymbol{\eta}|^2$, where $\boldsymbol{\xi} \in \mathbb{Z}^{d_s}$ and $\boldsymbol{\eta} \in \mathbb{R}^{d_t}$. Extending the spatio-temporal domain over a non-periodic temporal space $\boldsymbol{x} \in \left[-\frac{1}{2}, \frac{1}{2}\right]^{d_s}$ and $\boldsymbol{\kappa} = (\kappa_1, \boldsymbol{\kappa}_{-1}) \in \mathbb{R}^{d_t}$ allows for non-integer temporal frequencies, which yield a wider spectrum of solutions. To generalize the Fourier transform solution from the case of periodic temporal domain, we observe that linear combinations of solutions of the form:

$$e^{2\pi i \langle \boldsymbol{\eta}, \boldsymbol{\kappa} \rangle} \times e^{2\pi i \langle \boldsymbol{x}, \boldsymbol{\xi} \rangle}, \quad |\boldsymbol{\eta}|^2 = |\boldsymbol{\xi}|^2$$

still solve the wave equation, since:

$$\Delta_{\boldsymbol{\kappa}}\left(e^{2\pi i \langle \boldsymbol{\eta}, \boldsymbol{\kappa} \rangle} \times e^{2\pi i \langle \boldsymbol{x}, \boldsymbol{\xi} \rangle} \right) =$$

$$-4\pi^2 |\boldsymbol{\eta}|^2 e^{2\pi i \langle \boldsymbol{\eta}, \boldsymbol{\kappa} \rangle} e^{2\pi i \langle \boldsymbol{x}, \boldsymbol{\xi} \rangle} = -4\pi^2 |\boldsymbol{\xi}|^2 e^{2\pi i \langle \boldsymbol{\eta}, \boldsymbol{\kappa} \rangle} e^{2\pi i \langle \boldsymbol{x}, \boldsymbol{\xi} \rangle} = \Delta_{\boldsymbol{x}}\left(e^{2\pi i \langle \boldsymbol{\eta}, \boldsymbol{\kappa} \rangle} \times e^{2\pi i \langle \boldsymbol{x}, \boldsymbol{\xi} \rangle} \right).$$

The Cauchy initial value problem is well-posed when these solutions are in the central space, $v=\begin{pmatrix} v_o \\ v_1 \end{pmatrix} \in X^c$, and the solution $u \in X$ is *unique, global, and stable* for all $\kappa_1 \in \mathbb{R}$, since $|\boldsymbol{\eta}_{-1}|^2 \leq \underbrace{|\boldsymbol{\eta}_1|^2}_{|\kappa_1|^2} + |\boldsymbol{\eta}_{-1}|^2 = |\boldsymbol{\eta}|^2 = |\boldsymbol{\xi}|^2$, which implies that the support $\begin{pmatrix} \hat{v}_o \\ \hat{v}_1 \end{pmatrix} \subseteq \{|\boldsymbol{\eta}_{-1}| \leq |\boldsymbol{\xi}|\}$.

Conversely, given initial conditions $\begin{pmatrix} v_0 \\ v_1 \end{pmatrix} \in X^c$, the global and unique solutions of the well-posed ultrahyperbolic wave equation are derived by Fourier synthesis as follows:

$$\hat{u}(\boldsymbol{\xi}, \kappa_1, \boldsymbol{\eta}_{-1}) = \cos\left(2\pi\, \kappa_1 \sqrt{|\boldsymbol{\xi}|^2 - |\boldsymbol{\eta}_{-1}|^2}\right) \hat{v}_o(\boldsymbol{\xi}, \boldsymbol{\eta}_{-1})$$

$$+ \frac{\sin\left(2\pi\, \kappa_1 \sqrt{|\boldsymbol{\xi}|^2 - |\boldsymbol{\eta}_{-1}|^2}\right)}{2\pi \sqrt{|\boldsymbol{\xi}|^2 - |\boldsymbol{\eta}_{-1}|^2}} \hat{v}_1(\boldsymbol{\xi}, \boldsymbol{\eta}_{-1}).$$

Since $v=\begin{pmatrix} v_o \\ v_1 \end{pmatrix} \in X^c$, $2\pi\sqrt{|\boldsymbol{\xi}|^2 - |\boldsymbol{\eta}_{-1}|^2} \in \mathbb{R}$ and:

$$\left|\cos\left(2\pi\, \kappa_1 \sqrt{|\boldsymbol{\xi}|^2 - |\boldsymbol{\eta}_{-1}|^2}\right)\right| \leq 1 \text{ and } \left|\frac{\sin\left(2\pi\, \kappa_1 \sqrt{|\boldsymbol{\xi}|^2 - |\boldsymbol{\eta}_{-1}|^2}\right)}{2\pi \sqrt{|\boldsymbol{\xi}|^2 - |\boldsymbol{\eta}_{-1}|^2}}\right| \leq 1.$$

This implies that:

$$\|\hat{u}(\boldsymbol{\xi}, \kappa_1, \boldsymbol{\eta}_{-1})\|_{L^2} \leq \|\hat{v}_o(\boldsymbol{\xi}, \boldsymbol{\eta}_{-1})\|_{L^2} + \|\hat{v}_1(\boldsymbol{\xi}, \boldsymbol{\eta}_{-1})\|_{L^2} = \|v_0\|_{L^2} + \|v_1\|_{L^2},$$

where the last equality follows from Plancherel's Theorem. Therefore, we may invert the Fourier transform to obtain solution formula for $u(\boldsymbol{x},\ \kappa_1,\ \boldsymbol{\kappa}_{-1})$:

$$u(\boldsymbol{x},\ \kappa_1,\ \boldsymbol{\kappa}_{-1}) := \mathcal{F}^{-1}(\hat{u})(\boldsymbol{x}, \boldsymbol{\kappa}) = \sum_{\xi \in \mathbb{Z}^{d_s}} \int_{\mathbb{R}^{d_t - 1}} \hat{u}(\boldsymbol{\xi}, \kappa_1, \boldsymbol{\eta}_{-1})\, e^{2\pi i \langle \boldsymbol{\eta}_{-1}, \boldsymbol{\kappa}_{-1} \rangle} e^{2\pi i \langle \boldsymbol{x}, \boldsymbol{\xi} \rangle}\, d\boldsymbol{\eta}_{-1}.$$

The special solutions

$$u(\boldsymbol{x}, \boldsymbol{\kappa}) = \sum_{\substack{m=1 \\ \{\boldsymbol{\xi}_m, \boldsymbol{\eta}_m |\ |\boldsymbol{\xi}_m|^2 = |\boldsymbol{\eta}_m|^2\}}}^{M} \left(C_m \times e^{2\pi i \langle \boldsymbol{\eta}_m, \boldsymbol{\kappa} \rangle} \times e^{2\pi i \langle \boldsymbol{x}, \boldsymbol{\xi}_m \rangle}\right)$$

represent instances of the general solution formula where only finitely many coefficients in the infinite sum are non-trivial.

In the reduced spacekime domain (2D space), the basic (constrained) solutions can be explicated as:

$$u(x_1x_2, \kappa_1, \kappa_2) = e^{2\pi i(\eta_1\kappa_1 + \eta_2\kappa_2)} \times e^{2\pi i(\xi_1x_1 + \xi_2x_2)}, \ |\xi_1|^2 + |\xi_2|^2 = |\eta_1|^2 + |\eta_2|^2, \text{s.t.}$$

$$\underbrace{\begin{pmatrix} u_o(x_1x_2, \kappa_2) \\ u_1(x_1x_2, \kappa_2) \end{pmatrix}}_{\text{Cauchy data}} = \begin{pmatrix} u(x_1x_2, \kappa_1 = 0, \kappa_2) \\ \partial_{\kappa_1} u(x_1x_2, \kappa_1 = 0, \kappa_2) \end{pmatrix} = \begin{pmatrix} e^{2\pi i\eta_2\kappa_2} e^{2\pi i(\xi_1x_1 + \xi_2x_2)} \\ 2\pi i\eta_1 e^{2\pi i\eta_2\kappa_2} e^{2\pi i(\xi_1x_1 + \xi_2x_2)} \end{pmatrix}.$$

3.7.3 Lorentz Transformation in Spacekime

Let's consider the two inertial frames K and K' moving uniformly and rectilinearly to each other. We assume that the velocity of the frame K' against K, defined in relation to the first kime dimension k_1, is equal to the vector $\mathbf{v_1}$. Similarly, the velocity of the frame K' against K, defined in relation to the second kime dimension k_2, is equal to the vector $\mathbf{v_2}$. (Here $\mathbf{v_1}$ and $\mathbf{v_2}$ are the two components of the velocity of the frame K' against K.) Denote by x, y and z the axes of frame K and similarly, by x', y' and z' the axes of frame K'. The two kime dimensions, defined in the frame K, we will denote with k_1 and k_2 and in the frame K' with k_1' and k_2'. Let the point (p.) Q denote the origin of the spatial frame of reference K (i.e., $x = 0, \ y = 0, \ z = 0$) and p. Q' represents the origin of the second spatial frame of reference K' (i.e., $x' = 0, \ y' = 0, \ z' = 0$). The systems K and K' are chosen in such a way that p. Q' is moving along the axis x, in the direction of increasing the values along the axis x.

Further, we can choose the axes of the reference frames K and K' so that for an observer in K, the axis x coincides with the axis x', whereas the axes y and z are parallel to the axes y' and z' and all corresponding pairs of axes (e.g., x and x') have the same directions. At the initial moment, $k_1 = k_1' = 0$ and $k_2 = k_2' = 0$, as p. Q' coincides with p. Q ($p.Q' \equiv p.Q$). Under these conditions, the reference frames K and K' are in a *standard configuration*. Let's set: $\mathbf{v_1} = (v_1, 0, 0)$ and $\mathbf{v_2} = (v_2, 0, 0)$, where $v_1, 0, 0$ are the projections of the velocity vector $\mathbf{v_2}$ onto the axes x, y, z of frame K and similarly, $v_2, 0, 0$ are the projections of the velocity vector $\mathbf{v_2}$ onto the axes x, y, z. Also, assume that a particle's coordinates in K and K' are (x, y, z, k_1, k_2) and (x', y', z', k_1', k_2'), respectively.

For simplicity, we will denote: $x_1 = x$, $x_2 = y$, $x_3 = z$, $x_4 = ick_1$, $x_5 = ick_2$ and $x_1' = x'$, $x_2' = y'$, $x_3' = z'$, $x_4' = ick_1'$, $x_5' = ick_2'$ in the two reference frames, K and K'. In order to derive the transformations between K and K', we will use Lorentz boosting in an arbitrary direction. It generalizes the transformations between two inertial frames of reference, with parallel x,y and z axes and whose space-time origins coincide; these are the Lorentz transformations without rotation. First, let's consider a proper rotation angle α in the plane (x_4, x_5), while the other three dimensions (x_1, x_2, x_3) remain invariant. This transformation is described by a *circular rotation matrix* (rotating either the spatial or the temporal dimensions, but not both):

$$\mathbf{R} = \begin{pmatrix} 1 & 0 & 0 & 0 & 0 \\ 0 & 1 & 0 & 0 & 0 \\ 0 & 0 & 1 & 0 & 0 \\ 0 & 0 & 0 & \cos\alpha & \sin\alpha \\ 0 & 0 & 0 & -\sin\alpha & \cos\alpha \end{pmatrix}.$$

Let x_{4R} and x_{5R} denote the new axes, which arise from the rotation **R** of the x_4 and x_5 axes.

Then, an angle γ proper rotation in the plane (x_1, x_{4R}) preserving the other three dimensions (x_2, x_3, x_{5R}) is described by *hyperbolic (boosting) rotation matrix* (blending spatial and temporal dimensions):

$$\mathbf{L} = \begin{pmatrix} \cos\gamma & 0 & 0 & -\sin\gamma & 0 \\ 0 & 1 & 0 & 0 & 0 \\ 0 & 0 & 1 & 0 & 0 \\ \sin\gamma & 0 & 0 & \cos\gamma & 0 \\ 0 & 0 & 0 & 0 & 1 \end{pmatrix}.$$

To derive the transformations between K and K', we will sequentially apply the operations **R**, **L** and $\mathbf{R}^{-1}$, each of which preserves the spacetime interval ds^2, extending the classical Euclidean length. First, we will apply the **R**-transformation. As $\tan\alpha = \frac{x_5}{x_4}$, if $x_1' = 0$, then $x_1 = -i\frac{v_1}{c}x_4 = -i\frac{v_2}{c}x_5$ and thus $\tan\alpha = \frac{v_1}{v_2}$. Here the angle α is a real number. Denote $\beta = \frac{1}{\sqrt{\left(\frac{c}{v_1}\right)^2 + \left(\frac{c}{v_2}\right)^2}}$ and $\zeta = \frac{1}{\sqrt{1-\beta^2}}$. Observing that $0 \le \beta \le 1$ and $\zeta \ge 1$, we will then have:

$$\cos\alpha = \frac{c\beta}{v_1} \quad \text{and} \quad \sin\alpha = \frac{c\beta}{v_2}.$$

Next, we apply the transformation **L**. As $\tan\gamma = \frac{x_1}{x_{4R}} = \frac{x_1}{\sqrt{(x_4)^2 + (x_5)^2}}$, if $x_1' = 0$, then $x_1 = -i\frac{v_1}{c}x_4 = -i\frac{v_2}{c}x_5 > 0$. Therefore, the angle γ is an imaginary number and

$$\tan\gamma = -i\beta,$$

$$\sin\gamma = -i\beta\zeta,\ \cos\gamma = \zeta,\ \text{and}\ \tan\gamma = -i\beta.$$

The signs in the above expressions are chosen to ensure that when $v_1 \to 0$ and $v_2 \to 0$, we will have:

$$x' \to x,\ y' \to y,\ z' \to z,\ k_1' \to k_1,\ k_2' \to k_2.$$

Finally, we apply the transformation, $\mathbf{R}^{-1} = \begin{pmatrix} 1 & 0 & 0 & 0 & 0 \\ 0 & 1 & 0 & 0 & 0 \\ 0 & 0 & 1 & 0 & 0 \\ 0 & 0 & 0 & \cos\alpha & -\sin\alpha \\ 0 & 0 & 0 & \sin\alpha & \cos\alpha \end{pmatrix}$, the inverse of the initial transformation $\mathbf{R}$. The final transformation matrix from the coordinates x_1, x_2, x_3, x_4, x_5 in K onto $x_1', x_2', x_3', x_4', x_5'$ in K' can be represented as the product $\mathbf{R}^{-1} \times \mathbf{L} \times \mathbf{R}$:

$$\mathbf{R}^{-1} \times \mathbf{L} \times \mathbf{R} = \begin{pmatrix} \cos\gamma & 0 & 0 & -\cos\alpha\sin\gamma & -\sin\alpha\sin\gamma \\ 0 & 1 & 0 & 0 & 0 \\ 0 & 0 & 1 & 0 & 0 \\ \cos\alpha\sin\gamma & 0 & 0 & 1+(\cos\gamma-1)\cos^2\alpha & (\cos\gamma-1)\sin\alpha\cos\alpha \\ \sin\alpha\sin\gamma & 0 & 0 & (\cos\gamma-1)\sin\alpha\cos\alpha & 1+(\cos\gamma-1)\sin^2\alpha \end{pmatrix}.$$

Therefore,

$$\begin{pmatrix} x_1' \\ x_2' \\ x_3' \\ x_4' \\ x_5' \end{pmatrix} = \begin{pmatrix} \cos\gamma & 0 & 0 & -\cos\alpha\sin\gamma & -\sin\alpha\sin\gamma \\ 0 & 1 & 0 & 0 & 0 \\ 0 & 0 & 1 & 0 & 0 \\ \cos\alpha\sin\gamma & 0 & 0 & 1+(\cos\gamma-1)\cos^2\alpha & (\cos\gamma-1)\sin\alpha\cos\alpha \\ \sin\alpha\sin\gamma & 0 & 0 & (\cos\gamma-1)\sin\alpha\cos\alpha & 1+(\cos\gamma-1)\sin^2\alpha \end{pmatrix} \begin{pmatrix} x_1 \\ x_2 \\ x_3 \\ x_4 \\ x_5 \end{pmatrix}.$$

Using the trigonometric relations above to substitute $\sin\alpha$, $\cos\alpha$, $\sin\gamma$, $\cos\gamma$ by β and ζ we obtain the following spacekime Lorentz transformation:

$$\begin{pmatrix} x' \\ y' \\ z' \\ k_1' \\ k_2' \end{pmatrix} = \begin{pmatrix} \zeta & 0 & 0 & -\frac{c^2}{v_1}\beta^2\zeta & -\frac{c^2}{v_2}\beta^2\zeta \\ 0 & 0 & 0 & 0 & 0 \\ 0 & 0 & 1 & 0 & 0 \\ -\frac{1}{v_1}\beta^2\zeta & 0 & 0 & 1+(\zeta-1)\frac{c^2}{(v_1)^2}\beta^2 & (\zeta-1)\frac{c^2}{v_1v_2}\beta^2 \\ -\frac{1}{v_2}\beta^2\zeta & 0 & 0 & (\zeta-1)\frac{c^2}{v_1v_2}\beta^2 & 1+(\zeta-1)\frac{c^2}{(v_2)^2}\beta^2 \end{pmatrix} \begin{pmatrix} x \\ y \\ z \\ k_1 \\ k_2 \end{pmatrix}.$$

As $x_4' = ick_1'$ and $x_5' = ick_2'$, the kime transformation coefficients include an extra ic factor. If $v_1 = 0$ and $v_2 = 0$, then $\beta = 0$, $\zeta = 1$ and $x' = x, y' = y, z' = z$, $k_1' = k_1, k_2' = k_2$. These transformations belong to the group of proper orthochronous transformations, Lorentz group, $SO^+(1,3)$, that preserve both kime coordinates. The transformation is equivalent to its Minkowski spacetime counterpart, the 4D Lorentz transformation at $k_2 \to 0$ and accordingly $v_2 \to \pm\infty$. These transformations are applicable in the case of a Cartesian coordinate system.

The Lorentz transformation between the two inertial frames of reference,

$$(\boldsymbol{x},\boldsymbol{k}) = \Bigg(\underbrace{x,y,z}_{\boldsymbol{x}},\underbrace{k_1,k_2}_{\boldsymbol{k}}\Bigg) \text{ and } (\boldsymbol{x}',\boldsymbol{k}') = \Bigg(\underbrace{x',y',z'}_{\boldsymbol{x}'},\underbrace{k_1',k_2'}_{\boldsymbol{k}'}\Bigg) \in \mathbb{R}^3\times\mathbb{R}^2,$$

that are in a standard configuration (i.e., transformations without translations and/or rotations of the space-axis, in the space hyperplane, and of the kime-axis in the hyperplane of kime) can be expressed as follows:

$$\left|\begin{array}{l} k_1' = \left(1+(\zeta-1)\frac{c^2}{(v_1)^2}\beta^2\right)k_1 + (\zeta-1)\frac{c^2}{v_1v_2}\beta^2k_2 - \frac{1}{v_1}\beta^2\zeta x \\ k_2' = (\zeta-1)\frac{c^2}{v_1v_2}\beta^2k_1 + \left(1+(\zeta-1)\frac{c^2}{(v_2)^2}\beta^2\right)k_2 - \frac{1}{v_2}\beta^2\zeta x \\ x' = -c^2\beta^2\zeta\left(\frac{k_1}{v_1}+\frac{k_2}{v_2}\right) + \zeta x \\ y' = y \\ z' = z \text{ (} y \text{ and } z \text{ dimensions remain unchanged).} \end{array}\right. \tag{3.1}$$

In the above expression:

- $\boldsymbol{k} = (k_1,k_2)$, $\boldsymbol{x} = (x,y,z)$, and $\boldsymbol{k}' = (k_1',k_2')$, $\boldsymbol{x}' = (x',y',z')$ are the initial and the transformed spacekime coordinates,
- $\boldsymbol{v}_1 = (v_1,0,0)$ and $\boldsymbol{v}_2 = (v_2,0,0)$ are the vectors of the kime velocities of $\boldsymbol{x}'$ against $\boldsymbol{x}$, defined with respect to k_1, k_2,
- $\beta = \frac{1}{\sqrt{\left(\frac{c}{v_1}\right)^2+\left(\frac{c}{v_2}\right)^2}}$, and
- $\zeta = \frac{1}{\sqrt{1-\beta^2}}$.

These transformations are a generalization of the classical Lorentz transformation representing a *fixed space direction (x)* to the 5D spacekime manifold.

Using the equation (3.1) we will get:

$$\Delta k_1' = \left(1+(\zeta-1)\frac{c^2}{(v_1)^2}\beta^2\right)\Delta k_1 + (\zeta-1)\frac{c^2}{v_1v_2}\beta^2\Delta k_2 - \frac{1}{v_1}\beta^2\zeta\Delta x,$$

$$\Delta k_1' = (\zeta-1)\frac{c^2}{v_1v_2}\beta^2\Delta k_1 + \left(1+(\zeta-1)\frac{c^2}{(v_2)^2}\beta^2\right)\Delta k_2 - \frac{1}{v_2}\beta^2\zeta\Delta x$$

and

$$\Delta x' = -c^2\beta^2\zeta\left(\frac{\Delta k_1}{v_1} + \frac{\Delta k_2}{v_2}\right) + \zeta\Delta x.$$

For a pair of causally connected kevents, $\Delta s^2 = \Delta x^2 + \Delta y^2 + \Delta z^2 - c^2(\Delta k_1)^2 - c^2(\Delta k_2)^2 \le 0$ (see **Section 3.9**), equation (3.1) implies that if $\Delta k_1 = 0$, $\Delta x = 0$, $\Delta y = 0$, and $\Delta z = 0$, then $\Delta k_1' = (\zeta - 1)\frac{c^2}{v_1 v_2}\beta^2\Delta k_2$ and $\Delta x' = -\frac{c^2}{v_2}\beta^2\zeta\Delta k_2$.

Suppose two kevents are causally connected and in one inertial reference frame K their first four coordinates x, y, z, k_1 coincide, but the fifth (k_2) coordinates are distinct $\Delta k_2 \ne 0$. Then, in a different inertial frame K' the kevents coordinates x', k_1' must be distinct. In other words, when $\Delta k_2 \ne 0$ and $v_2 \ne \pm\infty$ (v_2 is finite), it is possible that $\Delta k_1' \ne 0$ and $\Delta x' \ne 0$.

In polar coordinates $\{(r, \varphi) \in \mathbb{R}^+ \times [-\pi, \pi)\}$, the transformations are given by the following equations:

- Frame of reference K:

$$k_1 = t\cos\varphi = \frac{r}{c}\cos\varphi,\ k_2 = t\sin\varphi = \frac{r}{c}\sin\varphi,$$

$$(r)^2 = (ct)^2 = (ck_1)^2 + (ck_2)^2,$$

$$\varphi = \operatorname{atan2}(k_2, k_1) = \begin{cases} 2\arctan\left(\frac{k_2}{k_1 + \sqrt{k_1^2 + k_2^2}}\right), & \text{if } k_1 > 0 \\ 2\arctan\left(\frac{\sqrt{k_1^2 + k_2^2} - k_1}{k_2}\right), & \text{if } k_1 \le 0 \cap k_2 \ne 0 \\ -\pi,\ k_1 < 0 \cap k_2 = 0 \\ \text{undefined},\ k_1 = 0 \cap k_2 = 0 \end{cases}.$$

- Frame of reference K':

$$(r')^2 = (ct')^2 = (ck_1')^2 + (ck_2')^2 =$$

$$= \left(\left(1 + (\zeta - 1)\frac{c^2}{(v_1)^2}\beta^2\right) r\cos\varphi + (\zeta - 1)\frac{c^2}{v_1 v_2}\beta^2 r\sin\varphi - \frac{c}{v_1}\beta^2\zeta x\right)^2$$

$$+ \left((\zeta - 1)\frac{c^2}{v_1 v_2}\beta^2 r\cos\varphi + \left(1 + (\zeta - 1)\frac{c^2}{(v_2)^2}\beta^2\right) r\sin\varphi - \frac{c}{v_2}\beta^2\zeta x\right)^2,$$

$$\varphi = \operatorname{atan2}(k_2', k_1') = \begin{cases} 2\arctan\left(\dfrac{k_2'}{k_1' + \sqrt{(k_1')^2 + (k_2')^2}}\right), & \mathit{if}\ k_1' > 0 \\ 2\arctan\left(\dfrac{\sqrt{(k_1')^2 + (k_2')^2} - k_1'}{k_2'}\right), & \mathit{if}\ k_1' \le 0 \cap k_2' \neq 0 \\ -\pi, \quad k_1' < 0 \cap k_2' = 0 \\ \text{undefined}, \quad k_1' = 0 \cap k_2' = 0 \end{cases}$$

3.7.4 Properties of the General Spacekime Transformations

Next, we are going to examine some of the properties of the general transformations between two reference frames, K and K'. Again, x^1, x^2, x^3 and x^4, x^5 denote the 3 space dimensions and the 2 kime dimensions, respectively, i.e., $x^1 = x, x^2 = y, x^3 = z$, $x^4 = ck_1, x^5 = ck_2$. Then, the Euclidean spacekime metric tensor dictates that the following equality is satisfied for the coordinates in the two frames:

$$\left(\sum_{\eta=1}^{3} (x^\eta)^2\right) - (x^4)^2 - (x^5)^2 = \left(\sum_{\eta=1}^{3} \left(x^{\eta'}\right)^2\right) - \left(x^{4'}\right)^2 - \left(x^{5'}\right)^2. \tag{3.2}$$

In general, this relation extends to the non-Euclidean spacekime manifold:

$$x^{\mu'} = a^\mu_\rho\, x^\rho + b^\mu, \tag{3.3}$$

where $\mu, \rho = 1, 2, 3, 4, 5$.[2] Here b^μ are five constant values, which are equal to the values of $x^{\mu'}$ for the case when $x^\mu = 0$ $(\mu = 1, 2, 3, 4, 5)$, i.e., b^μ is a 5-dimensional vector of translation (offset) in the spacekime. If the origins of both reference systems coincide, then $b^\mu = 0$ $(\mu = 1, 2, 3, 4, 5)$. Furthermore, we will examine spacekime affine transformations, excluding translations, i.e.,

$$x^{\mu'} = a^\mu_\rho\, x^\rho. \tag{3.4}$$

Let's introduce the following notation:

2 In this, and the following formulae, we use Einstein's summation convention for repeating indices.

$$g_{\mu\rho} = g_{\rho\mu} = \begin{cases} 1, & at\ \mu=\rho=1,2,3 \\ -1, & at\ \mu=\rho=4,5 \\ 0, & at\ \mu\neq\rho \end{cases}.$$

Then the condition (3.2) expressing the relation between two inertial frames can be presented in the form:

$$g_{\mu\rho}x^{\mu}x^{\rho} = g_{\mu\rho}x^{\mu'}x^{\rho'}. \tag{3.5}$$

If we substitute equality (3.4) on the right-hand side of the equation (3.5) and then compare the coefficients of x, we obtain:

$$g_{\mu\rho} = g_{\lambda\vartheta}\ a^{\lambda}_{\mu}\ a^{\vartheta}_{\rho}, \tag{3.6}$$

where $\lambda, \vartheta = 1,2,3,4,5$. Let us define a 5×5 matrix $(\boldsymbol{A})_{\mu\rho}$, with elements a^{μ}_{ρ}: $(\boldsymbol{A})_{\mu\rho} \equiv a^{\mu}_{\rho}$. Similarly, let us define the elements of the matrix $(\boldsymbol{G})_{\mu\rho} \equiv g_{\mu\rho}$. The matrix presentation in equation (3.6) can be written as $(\boldsymbol{G})_{\mu\rho} = (\boldsymbol{A}^t\boldsymbol{G}\boldsymbol{A})_{\mu\rho}$, and its determinant is:

$$\det(\mathbf{G}) = \det\left(\boldsymbol{A}^t\boldsymbol{G}\boldsymbol{A}\right) = \det\left(\boldsymbol{A}^t\right)\det(\boldsymbol{G})\det(\boldsymbol{A}).$$

As $\det(\boldsymbol{A}^t) = \det(\boldsymbol{A})$ and $\det(\boldsymbol{G}) = 1$, we obtain:

$$\det(\boldsymbol{A}) = \pm 1. \tag{3.7}$$

In equation (3.6), to solve for the coefficient a^4_4, we set $\mu=\rho=4$ to obtain:

$$-1 = \sum_{\eta=1}^{3}\left(a^{\eta}_{4}\right)^2 - \left(a^4_4\right)^2 - \left(a^5_4\right)^2, \qquad \text{i.e., } a^4_4 = \pm\sqrt{1+\sum_{\eta=1}^{3}\left(a^{\eta}_{4}\right)^2 - \left(a^5_4\right)^2}.$$

To solve for the coefficient a^5_5, we set $\mu=\rho=5$:

$$-1 = \sum_{\eta=1}^{3}\left(a^{\eta}_{5}\right)^2 - \left(a^4_5\right)^2 - \left(a^5_5\right)^2, \qquad \text{i.e., } a^5_5 = \pm\sqrt{1+\sum_{\eta=1}^{3}\left(a^{\eta}_{5}\right)^2 - \left(a^4_5\right)^2}.$$

Therefore, we can identify the following 9 possible situations:

$$\begin{aligned} &\mathrm{a}^4_4 > 0 \text{ and } \{\mathrm{a}^5_5 > 0 \text{ or } \mathrm{a}^5_5 < 0 \text{ or } \mathrm{a}^5_5 = 0\}; \\ &\mathrm{a}^4_4 < 0 \text{ and } \{\mathrm{a}^5_5 > 0 \text{ or } \mathrm{a}^5_5 < 0 \text{ or } \mathrm{a}^5_5 = 0\}; \\ &\mathrm{a}^4_4 = 0 \text{ and } \{\mathrm{a}^5_5 > 0 \text{ or } \mathrm{a}^5_5 < 0 \text{ or } \mathrm{a}^5_5 = 0\}; \end{aligned} \tag{3.8}$$

The additional binary condition (3.7) yields doubling of the possibilities, i.e., we have a total of 18 possible cases. In the case of two-dimensional kime, it is possible that $a^4_4 = 0$ and $a^5_5 = 0$. For instance, if $x^1 = x, x^2 = y,\ x^3 = z,\ x^4 = ck_1, x^5 = ck_2$ and the frame transformation is $x^{1'} = x, x^{2'} = y,\ x^{3'} = z,\ x^{4'} = ck_2, x^{5'} = -ck_1$, then $a^4_4 = 0$ and

$a_5^5 = 0$. The transformations corresponding with $a_4^4 \neq 0$ and $a_5^5 \neq 0$ are called *non-zero transformations*.

Now let's focus only on these non-zero transformations. In general, there are 8 non-zero transformations (see **Table 3.2**). The transformations that preserve the frame orientation ($\uparrow\rightarrow$) are called *proper orthochronous transformations*. In this case, there is one proper orthochronous transformation, denoted by $\Lambda_+^{\uparrow\rightarrow}$, corresponding to $\det(\boldsymbol{A}) = +1$, $a_4^4 > 0$ and $a_5^5 > 0$ $(+)$.

The more general transformations that just preserve the signs of k_1 and k_2, (e.g., $a_4^4 > 0$ and $a_5^5 > 0$) are called *orthochronous* in relation to k_1 and k_2, and are denoted by $\Lambda^{\uparrow\rightarrow}$, e.g., $\Lambda^{\uparrow\rightarrow} = \{\Lambda_+^{\uparrow\rightarrow} \cup \Lambda_-^{\uparrow\rightarrow}\}$.

Transformations preserving just the sign of k_1, (e.g., $a_4^4 > 0$) are called *orthochronous* in relation to k_1 and are denoted by $\Lambda^{\rightarrow} = \{\Lambda_+^{\uparrow\rightarrow} \cup \Lambda_+^{\downarrow\rightarrow} \cup \Lambda_-^{\uparrow\rightarrow} \cup \Lambda_-^{\downarrow\rightarrow}\}$. And similarly the sign-preserving transformations of k_2, (i.e., $a_5^5 > 0$), are called *orthochronous* in relation to k_2 and denoted by $\Lambda^{\uparrow} = \{\Lambda_+^{\uparrow\rightarrow} \cup \Lambda_+^{\uparrow\leftarrow} \cup \Lambda_-^{\uparrow\rightarrow} \cup \Lambda_-^{\uparrow\leftarrow}\}$.

The respective transformations that change the signs of the time dimensions, k_1 and k_2, are called *non-orthochronous* transformations and are denoted by $\Lambda^{\downarrow\leftarrow} = \{\Lambda_+^{\downarrow\leftarrow} \cup \Lambda_-^{\downarrow\leftarrow}\}$. It is clear that a transformation can be simultaneously orthochronous in relation to a given kime dimension and non-orthochronous in relation to the other kime dimension (e.g., the transformation $\Lambda_+^{\downarrow\rightarrow}$). Proper and non-proper transformations are denoted by $\Lambda_+ = \{\Lambda_+^{\uparrow\rightarrow} \cup \Lambda_+^{\downarrow\rightarrow} \cup \Lambda_+^{\uparrow\leftarrow} \cup \Lambda_+^{\downarrow\leftarrow}\}$, corresponding to $\det(A) = +1$, and $\Lambda_- = \{\Lambda_-^{\uparrow\rightarrow} \cup \Lambda_-^{\downarrow\rightarrow} \cup \Lambda_-^{\uparrow\leftarrow} \cup \Lambda_-^{\downarrow\leftarrow}\}$, corresponding to $\det(A) = -1$, respectively.

Next, we will define the following discrete operations that represent spatial or temporal reflection:

$$x^{\mu'} = \boldsymbol{\Pi}_\rho^\mu \, x^\rho, \; x^{\mu''} = \boldsymbol{\Gamma}_\rho^\mu x^\rho, \; x^{\mu'''} = \boldsymbol{\Omega}_\rho^\mu x^\rho, \; x^{\mu^{iv}} = \boldsymbol{\Xi}_\rho^\mu x^\rho,$$

where

$$\boldsymbol{\Pi}_\rho^\mu = \mathrm{diag}(-1, -1, -1, +1, +1), \;\; \boldsymbol{\Gamma}_\rho^\mu = \mathrm{diag}(+1, +1, +1, -1, -1),$$

$$\boldsymbol{\Omega}_\rho^\mu = \mathrm{diag}(+1, +1, +1, -1, +1), \;\; \boldsymbol{\Xi}_\rho^\mu = \mathrm{diag}(+1, +1, +1, +1, -1).$$

Note that scalar products are invariant under such spatial and temporal reflections in the full group of transformations. **Table 3.2** provides a mapping that connects spatial and temporal operations to the 8 non-zero transformations we described above, equation (3.8).

Table 3.2: Decomposition of the group of non-zero transformations.

Transformations between K and K'	det(A)	sign a_4^4	sign a_5^5	Operations applied on $\Lambda_+^{\uparrow\rightarrow}$:
$\Lambda_+^{\uparrow\rightarrow}$	+1	+1	+1	1 (unit matrix) $= \text{diag}(+1, +1, +1, +1, +1)$
$\Lambda_+^{\downarrow\rightarrow}$	+1	+1	−1	$\Pi\Xi = \text{diag}(-1, -1, -1, +1, -1)$
$\Lambda_+^{\uparrow\leftarrow}$	+1	−1	+1	$\Pi\Omega = \text{diag}(-1, -1, -1, -1, +1)$
$\Lambda_+^{\downarrow\leftarrow}$	+1	−1	−1	$\Gamma = \text{diag}(+1, +1, +1, -1, -1)$
$\Lambda_-^{\uparrow\rightarrow}$	−1	+1	+1	$\Pi = \text{diag}(-1, -1, -1, +1, +1)$
$\Lambda_-^{\downarrow\rightarrow}$	−1	+1	−1	$\Xi = \text{diag}(+1, +1, +1, +1, -1)$
$\Lambda_-^{\uparrow\leftarrow}$	−1	−1	+1	$\Omega = \text{diag}(+1, +1, +1, -1, +1)$
$\Lambda_-^{\downarrow\leftarrow}$	−1	−1	−1	$\Pi\Gamma$ (total inversion) $= \text{diag}(-1, -1, -1, -1, -1)$

3.7.5 Backwards Motion in the Two-Dimensional Kime

In the special theory of relativity, positive timelike intervals, $s^2 = x^2 + y^2 + z^2 - c^2t^2 < 0$, $t > 0$, cannot be mapped via proper orthochronous Lorentz transformations into negative timelike intervals, $s^2 < 0$, $t < 0$. If this was possible, then by continuity, this would also be true for $t = 0$, which yields a contradiction when $s^2 < 0$. Thus, relative to $t = 0$, the spacetime region $s^2 < 0$, $t > 0$ (inside of the positive light cone) is called "*absolute future*" and similarly the region $s^2 < 0$, $t < 0$ (the inside of the negative light cone) is called the "*absolute past*".

This property does not extend to the case of two-dimensional kime. For example, positive kimelike intervals $s^2 = x^2 + y^2 + z^2 - c^2(k_1)^2 - c^2(k_2)^2 < 0$, $k_1 > 0$, $k_2 > 0$ can be mapped through proper orthochronous transformations $\Lambda_+^{\uparrow\rightarrow}$ (see **Section 3.7.4** above) into negative or mixed kimelike intervals like $s^2 < 0$, $k_1 \leq 0$, $k_2 \leq 0$ or $s^2 < 0$, $k_1 \leq 0$, $k_2 \geq 0$ or $s^2 < 0, k_1 \geq 0, k_2 \leq 0$. Indeed, since k_1 and k_2 are independent components, it is possible to simultaneously satisfy the following conditions: $k_1 = 0$, $k_2 \neq 0$ and $x^2 + y^2 + z^2 - c^2(k_2)^2 < 0$. Thus, the equality $k_1 = 0$ does not contradict the inequality $s^2 < 0$. Similarly, there is no contradiction between the equality $k_2 = 0$ and the inequality $s^2 < 0$.

We will show that under certain conditions, applying proper orthochronous transformations $\Lambda_+^{\uparrow\rightarrow}$, see equation (3.1) and **Section 3.7.3**, leads to backwards movement in the kime manifold spanned by k_1 and k_2. Let us assume that $\Delta s^2 = \Delta x^2 + \Delta y^2 + \Delta z^2 - c^2(\Delta k_1)^2 - c^2(\Delta k_2)^2 \leq 0$, $\Delta x > 0$, $\Delta y = 0$, $\Delta z = 0$, $\Delta k_1 > 0$, and $\Delta k_2 \geq 0$. From equation (3.1), see **Section 3.7.3**, we have the following equality:

$$\Delta k_1' = \left(1 + (\zeta - 1)\frac{c^2}{(v_1)^2}\beta^2\right)\Delta k_1 + (\zeta - 1)\frac{c^2}{v_1 v_2}\beta^2 \Delta k_2 - \frac{1}{v_1}\beta^2\zeta\Delta x \tag{3.9}$$

Assuming that $v_1 > 0$, we can examine the conditions on the velocities v_1 and v_2 that guarantee $\Delta k_1' < 0$. That is, we will explore the kime-velocity (v_1 and v_2) conditions for backwards kime motion in the k_1 direction, which does not have a direct analogue in the standard theory of relativity.

First, take $\Delta k_2 = 0$ and set $\Delta x = c\Delta k_1$. Since $\Delta s^2 \le 0$, we have $\Delta x \le c\Delta k_1$. In this case, the inequality $\Delta k_1' < 0$ is equivalent to the inequality:

$$\frac{\Delta k_1'}{\Delta k_1} = 1 + (\zeta - 1)\frac{c^2}{(v_1)^2}\beta^2 - \frac{c}{v_1}\beta^2\zeta < 0. \tag{3.10}$$

If we set $p = \frac{c}{v_1}$, then we will obtain a quadratic inequality in relation to the parameter p. The expressions for β and ζ contain two independent variables $\frac{c}{v_1}$ and $\frac{c}{v_2}$, subject to a single restriction $\left(\frac{c}{v_1}\right)^2 + \left(\frac{c}{v_2}\right)^2 \ge 1$, see **Section 3.7.3** and **Section 3.11**. Let's set $\left(\frac{c}{v_1}\right)^2 + \left(\frac{c}{v_2}\right)^2 = \text{const} \ge 1$, $\beta = \text{const} \le 1$, and $\zeta = \text{const} \ge 1$. Computing the first and second derivatives of the function

$$f(p) = (\zeta - 1)\beta^2 p^2 - \beta^2 \zeta p + 1$$

allow us to compute the minimum of $f(p)$, which is attained at $p = \frac{\zeta}{2(\zeta - 1)}$ and $\zeta > 1$. Thus, for $1 \ge \beta > 0$ and $\zeta > 1$, the function $f(p)$ reaches its minimum at $p = \frac{\zeta}{2(\zeta - 1)}$. Therefore, setting $p = \frac{c}{v_1} = \frac{\zeta}{2(\zeta - 1)}$ and $\zeta > 1$ and using inequality (3.10) we obtain:

$$-\beta^2\zeta^2 + 4\zeta - 4 < 0. \tag{3.11}$$

As $\beta \to 1$, $\zeta = \frac{1}{\sqrt{1-\beta^2}} \to \infty$, $p = \frac{c}{v_1} = \frac{\zeta}{2(\zeta - 1)} \to \frac{1}{2}$, and $\frac{\Delta k_1'}{\Delta k_1} \to -\infty$. Further, if $\beta = \frac{2\sqrt{2}}{3}$ (i.e., $\zeta = 3$ and $p = \frac{3}{4}$), then the expression (3.11) becomes equal to zero. Hence, if $p \in \left[\frac{1}{2}, \frac{3}{4}\right)$, then condition (3.10) is satisfied and therefore $\frac{\Delta k_1'}{\Delta k_1} < 0$. In this case $\Delta k_1' < 0$, i.e., the particle will move backwards in kime dimension k_1', which we wanted to derive. At the two extremes, when $p = \frac{3}{4}$, then $\frac{\Delta k_1'}{\Delta k_1} = 0$, i.e., $\Delta k_1' = 0$, and when $p \to \frac{1}{2}$, then $\frac{\Delta k_1'}{\Delta k_1} \to -\infty$, i.e. $\Delta k_1' \to -\infty$. Using equation (3.1), see **Section 3.7.3**, we obtain the following equality:

$$\Delta k_2' = (\zeta - 1)\frac{c^2}{v_1 v_2}\beta^2 \Delta k_1 + \left(1 + (\zeta - 1)\frac{c^2}{(v_2)^2}\beta^2\right)\Delta k_2 - \frac{1}{v_2}\beta^2\zeta\Delta x. \tag{3.12}$$

Since $\Delta k_2 = 0$ and $x = c\Delta k_1$, we will have: $\Delta k_2' = \left(\left((\zeta - 1)\frac{c^2}{v_1 v_2}\beta^2 - \frac{c}{v_2}\beta^2\zeta\right)\right)\Delta k_1$. Let us assume that $v_2 > 0$ and $\frac{c}{v_1} = \frac{\zeta}{2(\zeta-1)}$. Then $(\zeta - 1)\frac{c^2}{v_1 v_2}\beta^2 - \frac{c}{v_2}\beta^2\zeta < 0$ and hence, $\Delta k_2' < 0$. If $\beta \to 1$, $\zeta \to \infty$, and $p \to \frac{1}{2}$, then $\frac{\Delta k_2'}{\Delta k_1} \to -\infty$, i.e., $\Delta k_2' \to -\infty$.

Suppose now that $\Delta k_2 > 0$, $\Delta k_1'$ is governed by the equation (3.9) and $\Delta k_2'$ by equation (3.12). The condition $\Delta s^2 \leq 0$ (and $\Delta y = 0$, $\Delta z = 0$) implies that $\Delta x \leq c\sqrt{(\Delta k_1)^2 + (\Delta k_2)^2}$. Let's set $\Delta x = \chi c\sqrt{(\Delta k_1)^2 + (\Delta k_1)^2}$, where $0 \leq \chi \leq 1$. We will examine the conditions on the kime velocities v_1 and v_2 that yield backwards kime motion in both kime dimensions $\Delta k_1' < 0$ and $\Delta k_2' < 0$.

Let $p = \frac{c}{v_1} = \frac{\zeta}{2(\zeta-1)}$ and $p \in \left[\frac{1}{2};\frac{3}{4}\right)$. Then, the following inequalities are satisfied

$$1 + (\zeta - 1)\frac{c^2}{(v_1)^2}\beta^2 - \frac{c}{v_1}\beta^2\zeta < 0 \text{ and } (\zeta - 1)\frac{c^2}{v_1 v_2}\beta^2 - \frac{c}{v_2}\beta^2\zeta < 0.$$

These two expressions tend to $(-\infty)$ as $\beta \to 1$ and accordingly $\zeta \to \infty$, $p \to \frac{1}{2}$. Since Δk_1, Δk_2 and Δx are independent spacekime characteristics, we can choose the value Δk_1 to be large enough, the value Δk_2 to be small enough, and the value χ to be close to 1. This yields that the expressions $(\zeta - 1)\frac{c^2}{v_1 v_2}\beta^2\frac{\Delta k_2}{\Delta k_1}$, $\frac{c}{v_1}\beta^2\zeta\left(1 - \chi\sqrt{1 + \left(\frac{\Delta k_2}{\Delta k_1}\right)^2}\right)$, $\left(1 + (\zeta - 1)\frac{c^2}{(v_2)^2}\beta^2\right)\frac{\Delta k_2}{\Delta k_1}$, and $\frac{c}{v_2}\beta^2\zeta\left(1 - \chi\sqrt{1 + \left(\frac{\Delta k_2}{\Delta k_1}\right)^2}\right)$ may by arbitrarily small.

When $p \in \left[\frac{1}{2};\frac{3}{4}\right)$, appropriate values of Δk_1, Δk_2, and x would guarantee the following inequalities:

$$\left(\left(1 + (\zeta - 1)\frac{c^2}{(v_1)^2}\beta^2\right) - \frac{c}{v_1}\beta^2\zeta\right) + (\zeta - 1)\frac{c^2}{v_1 v_2}\beta^2\frac{\Delta k_2}{\Delta k_1} + \frac{c}{v_1}\beta^2\zeta\left(1 - \chi\sqrt{1 + \left(\frac{\Delta k_2}{\Delta k_1}\right)^2}\right) < 0,$$

$$\left((\zeta - 1)\frac{c^2}{v_1 v_2}\beta^2 - \frac{c}{v_2}\beta^2\zeta\right) + \left(1 + (\zeta - 1)\frac{c^2}{(v_2)^2}\beta^2\right)\frac{\Delta k_2}{\Delta k_1} + \frac{c}{v_2}\beta^2\zeta\left(1 - \chi\sqrt{1 + \left(\frac{\Delta k_2}{\Delta k_1}\right)^2}\right) < 0.$$

In other words,

$$\Delta k_1' = \left(1 + (\zeta - 1)\frac{c^2}{(v_1)^2}\beta^2\right)\Delta k_1 + (\zeta - 1)\frac{c^2}{v_1 v_2}\beta^2\Delta k_2 - \frac{c}{v_1}\beta^2\zeta\chi\sqrt{(\Delta k_1)^2 + (\Delta k_1)^2} < 0 \text{ and}$$

$$\Delta k_2' = (\zeta - 1)\frac{c^2}{v_1 v_2}\beta^2\Delta k_1 + \left(1 + (\zeta - 1)\frac{c^2}{(v_2)^2}\beta^2\right)\Delta k_2 - \frac{c}{v_2}\beta^2\zeta\chi\sqrt{(\Delta k_1)^2 + (\Delta k_1)^2} < 0.$$

Examples of these situations are shown in **Table 3.3**.

Table 3.3: Some of the values of $\Delta k_1'$ and $\Delta k_2'$ provided that $\Delta k_1 = 1$, $\Delta k_2 = 0.3$, $\chi = 0.999$.

β	ζ	$p = \frac{c}{v_1}$	$\frac{c}{v_2}$	$\Delta k_1'$	$\Delta k_2'$
0.799	1.663	1.254	Imaginary number	Complex number	Complex number
0.800	1.667	1.250	0.000	0.276	0.300
0.840	1.843	1.093	0.472	0.320	0.007
0.841649	**1.852**	**1.087**	**0.480**	**0.320**	**0.000**
0.950	3.203	0.727	0.761	0.189	−0.549
0.960	3.571	**0.694**	0.776	**0.142**	−0.659
0.970	4.113	0.661	0.791	0.071	−0.813
0.976556	**4.645**	**0.637**	**0.802**	**0.000**	**−0.958**
0.980	5.025	0.624	0.807	−0.051	−1.060
0.990	7.089	0.582	0.825	−0.336	−1.594
0.999	22.366	0.523	0.853	−2.487	−5.384
0.999999	707.107	0.501	0.866	−99.434	−173.329

Table 3.3 illustrates that when $\beta < 0.800$, the value $\frac{c}{v_2}$ is purely an imaginary number and $\Delta k_1'$ and $\Delta k_2'$ are complex numbers. Similarly, when $\beta = 0.800$, $\Delta k_1' > 0$, $\Delta k_2' > 0$; when $\beta \approx 0.841649$, $\Delta k_1' > 0$, $\Delta k_2' = 0$; when $\beta = 0.950$, $\Delta k_1' > 0$, $\Delta k_2' < 0$; when $\beta \approx 0.976556$, $\Delta k_1' = 0$, $\Delta k_2' < 0$; and when $\beta = 0.999$, $\Delta k_1' < 0$, $\Delta k_2' < 0$.

These results for complex-time (2D kime) suggest that it is possible to simultaneously fulfill the following conditions $v_1 > c$ and $\Delta k_1' > 0$. For instance, if $p = \frac{c}{v_1} = 0.694$, then $\Delta k_1' = 0.142 > 0$. This is quite different from the special theory of relativity (STR) case corresponding to one-dimensional time.

3.7.6 Rotations in Kime and Space Hyperplanes

Suppose a particle moving in the 5D spacekime manifold is indexed by a pair of kime coordinates (k_1, k_2) and three spatial dimensions (x, y, z). We can separately and independently apply kime-axes rotations in the kime hyperplane, and space-axis rotations in the spatial hyperplane. These (passive) linear transformations can be expressed as changes of the kime or space bases in a frame of reference. The group of all proper and improper rotations in the kime hyperplane is isomorphic to the orthogonal group $O(2, \mathbb{R})$. Similarly, the group of all proper and improper rotations in the spatial hyperplane is isomorphic to the orthogonal group $O(3, \mathbb{R})$, where R denotes the field of real numbers. The kime and space intervals, $dk^2 =$

$(dk_1)^2+(dk_2)^2$ and $dx^2=dx^2+dy^2+dz^2$, as well as the 5D spacekime interval $ds^2=dx^2+dy^2+dz^2-c^2(dk_1)^2-c^2(dk_2)^2$, are invariant under such rotation operations.

Let's examine a kime hyperplane rotation that maps the (k_1,k_2) axes onto a pair of new kime axes, (k_1',k_2'). We'll denote the kime intervals by $\mathbf{dk}=(dk_1,dk_2)$ and $\mathbf{dk}'=(dk_1',dk_2')$, representing the vector components of dk (kime change) before and after applying of the rotation transformation. As rotations preserve distances, we have $(dk_1')^2+(dk_2')^2=(dk_1)^2+(dk_2)^2$. The transformation under consideration can be presented as an orthogonal matrix $\mathbf{A}=[a_{\mu\sigma}]_{2\times 2}$, a member of the orthogonal group $O(2,\mathbb{R})$, where $\mu,\sigma\in\{1,2\}$, $\mathbf{A}^t=\mathbf{A}^{-1}$, and $\det(\mathbf{A})=+1$, for proper rotation, $\mathbf{A}\in SO(2,\mathbb{R})$, or $\det(A)=-1$, for improper rotation. The relation between the components (dk_1',dk_2') and their counterparts (dk_1,dk_2) is given by the following equality:

$$\mathbf{dk}'=\mathbf{dk}\times\mathbf{A}.$$

More explicitly, $dk_\sigma'=dk_1a_{1\sigma}+dk_2a_{2\sigma}$, $\sigma\in\{1,2\}$. For example, in the new kime rotation space, the feasibility of $dk_1'=dk_2'=\frac{dt}{\sqrt{2}}>0$ is guaranteed by choosing an appropriate angle between the vector dk and the kime axes k_1',k_2' to be $\frac{\pi}{4}$.

Very similarly, we can define proper and improper spatial rotation of the space axes in the space hyperplane.

3.7.7 Velocity-addition Law

The velocity-addition law represents a 3D equation expressing the velocities of objects in different reference frames. The STR suggests that velocity-addition may not behave as a simple vector summation. If we have a transformation between two spacekime frames, K and K', we can derive the velocity-addition formulae. Let's denote the velocities in each spacekime frame by $V_{\sigma\eta}=\frac{dx_\eta}{dk_\sigma}$ and $V_{\sigma\eta}'=\frac{dx_\eta'}{dk_\sigma'}$, indexed by $\sigma=1,2$ (in kime) and $\eta=1,2,3$ (in space). Then, the velocity-addition formulae are given by:

$$V_{\sigma 1}'=\frac{V_{\sigma 1}\zeta\left(1-\beta^2\left(\frac{c^2}{v_1V_{11}}+\frac{c^2}{v_2V_{21}}\right)\right)}{1+\frac{V_{\sigma 1}}{v_\sigma}\beta^2\left((\zeta-1)\left(\frac{c^2}{v_1V_{11}}+\frac{c^2}{v_2V_{21}}\right)-\zeta\right)},$$

$$V_{\sigma\vartheta}'=\frac{V_{\sigma\vartheta}}{1+\frac{V_{\sigma 1}}{v_\sigma}\beta^2\left((\zeta-1)\left(\frac{c^2}{v_1V_{11}}+\frac{c^2}{v_2V_{21}}\right)-\zeta\right)}.$$

In the above expressions, the indices $\sigma=1,2$; $\vartheta=2,3$, $\mathbf{v_1}=(v_1,0,0)$ and $\mathbf{v_2}=(v_2,0,0)$ are the projections of the velocity vector $\mathbf{v_1}$ onto the spatial axes x, y, z of the frame K, and correspondingly, the velocity vector $\mathbf{v_2}$ onto the same spatial axes x, y, z of K. Also, we used the equalities $\frac{dk_\theta}{dk_\sigma}=\frac{dx_\eta}{dk_\sigma}\frac{dk_\theta}{dx_\eta}=\frac{V_{\sigma\eta}}{V_{\theta\eta}}$, $\frac{V_{1\eta}}{V_{1\pi}}=\frac{V_{2\eta}}{V_{2\pi}}$, where $\eta,\pi=1,2,3$, $\frac{V_{\sigma\eta}}{V_{\sigma\pi}}=\frac{dx_\eta}{dx_\pi}$, and $\sigma=1,2$.

3.7.8 Generalization of the Principle of Invariance of the Speed of Light

As a result of the 2D generalization of time in 5D spacekime, the *principle of invariance* of the *speed* of *light*, as stipulated in the Special Theory of Relativity, requires a new fresh look.

Let us assume that a particle moves in 5D spacekime relative to the kime dimensions, k_1, k_2, and space dimensions, x, y, z in a frame of reference K. The $(3+2)$-dimensional interval in this case is expressed as follows: $ds^2 = dx^2 + dy^2 + dz^2 - c^2dk_1^2 - c^2dk_2^2$. Let us set: $V_\theta = \frac{\sqrt{dx^2+dy^2+dz^2}}{|dk_\theta|}$, where $\theta = 1, 2$. Then, the total velocity in this case is equal to

$$u = \frac{\sqrt{dx^2+dy^2+dz^2}}{\sqrt{dk_1^2+dk_2^2}} = \frac{\sqrt{dx^2+dy^2+dz^2}}{|dk|} > 0.$$

For lightlike intervals, i.e., $ds^2 = 0$, we will have:

$$u = c, \text{i.e.}, \frac{c^2}{V_1^2} + \frac{c^2}{V_2^2} = 1. \tag{3.13}$$

Note that $V_1 = V_2$ corresponds to an appropriate rotation in the kime hyperplane, so that $dk_1 = dk_2$. This implies that

$$V_1 = V_2 = c\sqrt{2}. \tag{3.14}$$

Now consider the motion of the particle in the reference frame K', which is moving uniformly and rectilinearly in relation to K. The $(3+2)$-dimensional spacekime interval in this case is given by the expression: $(ds')^2 = (dx')^2 + (dy')^2 + (dz')^2 - c^2(dk_1')^2 - c^2(dk_2')^2$. We utilize analogous notation for the relative velocity in K':

$$V_\theta' = \frac{\sqrt{(dx')^2+(dy')^2+(dz')^2}}{|dk_\theta'|}, \ (\theta = 1, 2) \text{ and}$$

$$u' = \frac{\sqrt{(dx')^2+(dy')^2+(dz')^2}}{\sqrt{(dk_1')^2+(dk_2')^2}} = \frac{\sqrt{(dx')^2+(dy')^2+(dz')^2}}{|d\kappa'|}.$$

Since the interval ds is invariant, we have $ds = ds' = 0$, and therefore

$$u' = c, \text{ i.e.}, \frac{c^2}{(V_1')^2} + \frac{c^2}{(V_2')^2} = 1. \tag{3.15}$$

Again, an appropriate K' rotation in the kime hyperplane would yield $V_1' = V_2'$, $dk_1' = dk_2'$, and

$$V_1' = V_2' = c\sqrt{2}. \tag{3.16}$$

Consequently, there exist a pair of equivalence relations. The first one suggests that if the velocities u and V_θ of a particle in reference frame K satisfy equation (3.13), then the velocities u' and V_θ' of the particle in the other frame, K', would satisfy equation (3.15). The second equivalent relation is that when the velocities V_θ of a particle in K satisfy equation (3.14), then the corresponding velocities V_θ' of the particle in will satisfy equation (3.16).

3.7.9 Heisenberg's Uncertainty Principle

A freely moving microparticle in spacekime can be described with a flat wave of de Broglie [157]:

$$\psi(x) = \psi_{px} = \frac{1}{\sqrt{2\pi\hbar}} \exp\left[-\frac{i}{\hbar}(e_1 k_1 + e_2 k_2 - p_x x)\right],$$

where $e_1 = \hbar\omega_1$, $e_2 = \hbar\omega_2$ represent the particle energy, defined in relation to the kime dimensions k_1 and k_2, respectively, and $p_x = \frac{\hbar}{\gamma}$. The angular frequencies of the wave, ω_1 and ω_2, are defined in relation to the kime dimensions k_1 and k_2, respectively, and γ is the angular wavenumber. In **Chapter 5**, we will show that the classical Heisenberg 4D spacetime uncertainty may be explained as a reduction of Einstein-like 5D deterministic dynamics. In other words, the common spacetime uncertainty principle could be understood as a consequence of deterministic laws in 5D spacekime.

3.7.10 5D Spacekime Manifold Waves and the Doppler Effect

The d'Alembert wave operator represents the generalization of the Laplace operator to the 5D Minkowski spacekime

$$\square_{1,2} \equiv g^{\mu\rho}\frac{\partial}{\partial x^\mu}\frac{\partial}{\partial x^\rho} = \sum_{\eta=1}^{3}\left(\frac{\partial}{\partial x^\eta}\right)^2 - \left(\frac{\partial}{c\partial k_1}\right)^2 - \left(\frac{\partial}{c\partial k_2}\right)^2,$$

where $g^{\mu\rho} = \begin{cases} 1, & at\ \mu=\rho=1,2,3 \\ -1, & at\ \mu=\rho=4,5 \\ 0, & at\ \mu\neq\rho \end{cases}$, $x^4 = ck_1$, and $x^5 = ck_2$.

As the 5D Minkowski spacekime metric is invariant to Lorentz transformations, the d'Alembert wave operator $\square_{1,2}$ is a Lorentz scalar:

$$\square_{1,2} = \square_{1,2}' = \sum_{\eta=1}^{3}\left(\frac{\partial}{\partial x^{\eta\prime}}\right)^2 - \left(\frac{\partial}{c\partial k_1'}\right)^2 - \left(\frac{\partial}{c\partial k_2'}\right)^2.$$

Therefore, these expressions are valid for all standard 5D spacekime coordinates in every inertial frame.

For simplicity, let's denote $x^1 = x,\ x^2 = y,\ x^3 = z$. In the 5D spacekime manifold, the wave equation may be expressed as ultrahyperbolic partial differential equation

$$\frac{\partial^2 F}{\partial x^2} + \frac{\partial^2 F}{\partial y^2} + \frac{\partial^2 F}{\partial z^2} = \left(\frac{\gamma}{\omega_1}\right)^2 \frac{\partial^2 F}{\partial k_1^2} + \left(\frac{\gamma}{\omega_2}\right)^2 \frac{\partial^2 F}{\partial k_2^2},$$

where ω_1 and ω_2 are the angular frequencies of the wave defined in relation to the kime dimensions k_1 and k_2, respectively, γ is the angular wavenumber, and $u_1 = \frac{\omega_1}{\gamma}$ and $u_2 = \frac{\omega_2}{\gamma}$ represent the phase velocities relative to k_1 and k_2.

We will examine the Doppler effect in the 5D spacekime manifold where it is possible that certain waves have properties depending only on one or both of the two kime dimensions. We will consider the general case, when a wave is moving in the two kime dimensions k_1, k_2, as well as in the three spatial dimensions x, y, z.

Let ω_1 and ω_2 represent the angular frequencies of the wave in the frame K, indexed by the kime dimensions k_1 and k_2, and $\boldsymbol{\gamma} = \left(\gamma_x, \gamma_y, \gamma_z\right)$ be the wave vector of this wave in K. The phase of the wave in K will be given by the following expression:

$$\omega_1 k_1 + \omega_2 k_2 - \boldsymbol{\gamma}\mathbf{x}, \text{ where } \mathbf{x} = (x, y, z)^T.$$

Suppose ω_1' and ω_2' are the angular frequencies of the wave in another reference frame, K', indexed by k_1' and k_2', and $\boldsymbol{\gamma}' = \left(\gamma_x', \gamma_y', \gamma_z'\right)$ is the wave vector of the wave in K'. Then, the phase of the same wave in K' will be $\omega_1' k_1' + \omega_2' k_2' - \boldsymbol{\gamma}'\mathbf{x}'$, where $\mathbf{x}' = (x', y', z')^T$. To derive the Doppler effect for the wave, let's assume the wave phase is invariant, i.e.,

$$\omega_1 k_1 + \omega_2 k_2 - \boldsymbol{\gamma}\mathbf{x} = \omega_1' k_1' + \omega_2' k_2' - \boldsymbol{\gamma}'\mathbf{x}'.$$

We also label the angular wavenumber $\gamma' = \boldsymbol{\gamma}' > 0$, the phase velocities, relative to k_1' and k_2', respectively, $u_1' = \frac{\omega_1'}{\gamma'}$, $u_2' = \frac{\omega_1'}{\gamma'}$, and $\frac{\gamma_x'}{\gamma'} = \cos\phi'$. Applying the earlier transformations (3.1), from K to K' (see **Section 3.7.3**) we can express the relation between $\omega_1, \omega_2, \gamma_x, \gamma_y, \gamma_z$ and, $\omega_1', \omega_2', \gamma_x', \gamma_y', \gamma_z'$:

$$\omega_1 = \omega_1' \left(1 + (\zeta - 1)\frac{c^2}{v_1^2}\beta^2 + \frac{c^2}{v_1 u_1'}\beta^2 \zeta \cos\phi'\right) + \omega_2'(\zeta - 1)\frac{c^2}{v_1 v_2}\beta^2,$$

$$\omega_2 = \omega_2' \left(1 + (\zeta - 1)\frac{c^2}{v_2^2}\beta^2 + \frac{c^2}{v_2 u_2'}\beta^2 \zeta \cos\phi'\right) + \omega_2'(\zeta - 1)\frac{c^2}{v_1 v_2}\beta^2,$$

$$\gamma_x = \gamma_x' \zeta \left(1 + \frac{u_1'}{v_1 \cos\phi'}\beta^2\right) + \gamma_x' \zeta \beta^2 \frac{u_2'}{v_2 \cos\phi'},$$

$$\gamma_y = \gamma'_y,$$

$$\gamma_z = \gamma'_z.$$

These are derived from the invariance relation $\omega_1 k_1 + \omega_2 k_2 - \mathbf{y}\mathbf{x} = \omega'_1 k'_1 + \ \omega'_2 k'_2 - \mathbf{y}'\mathbf{x}' \Leftrightarrow$

$$\left(\omega_1, \omega_2, -\gamma_x, -\gamma_y, -\gamma_z\right)\begin{pmatrix} k_1 \\ k_2 \\ x \\ y \\ z \end{pmatrix} = \left(\omega'_1, \omega'_2, -\gamma'_x, -\gamma'_y, -\gamma'_z\right)\begin{pmatrix} k'_1 \\ k'_2 \\ x' \\ y' \\ z' \end{pmatrix} \underbrace{=}_{\substack{\text{transformation} \\ (3.1),\,\text{swap } 4^{th}, 5^{th} \\ \text{columns and rows} \\ \text{with } 1^{st}, 2^{nd}, 3^{rd}}}$$

$$\left(\omega'_1, \omega'_2, -\gamma'_x, -\gamma'_y, -\gamma'_z\right)\begin{pmatrix} 1+(\zeta-1)\frac{c^2}{(v_1)^2}\beta^2 & (\zeta-1)\frac{c^2}{v_1 v_2}\beta^2 & -\frac{1}{v_1}\beta^2\zeta & 0 & 0 \\ (\zeta-1)\frac{c^2}{v_1 v_2}\beta^2 & 1+(\zeta-1)\frac{c^2}{(v_2)^2}\beta^2 & -\frac{1}{v_2}\beta^2\zeta & 0 & 0 \\ -\frac{c^2}{v_1}\beta^2\zeta & -\frac{c^2}{v_2}\beta^2\zeta & \zeta & 0 & 0 \\ 0 & 0 & 0 & 1 & 0 \\ 0 & 0 & 0 & 0 & 1 \end{pmatrix}\begin{pmatrix} k_1 \\ k_2 \\ x \\ y \\ z \end{pmatrix},$$

$$\begin{pmatrix} x' \\ y' \\ z' \\ k'_1 \\ k'_2 \end{pmatrix} = \begin{pmatrix} \zeta & 0 & 0 & -\frac{c^2}{v_1}\beta^2\zeta & -\frac{c^2}{v_2}\beta^2\zeta \\ 0 & 1 & 0 & 0 & 0 \\ 0 & 0 & 1 & 0 & 0 \\ -\frac{1}{v_1}\beta^2\zeta & 0 & 0 & 1+(\zeta-1)\frac{c^2}{(v_1)^2}\beta^2 & (\zeta-1)\frac{c^2}{v_1 v_2}\beta^2 \\ -\frac{1}{v_2}\beta^2\zeta & 0 & 0 & (\zeta-1)\frac{c^2}{v_1 v_2}\beta^2 & 1+(\zeta-1)\frac{c^2}{(v_2)^2}\beta^2 \end{pmatrix}\begin{pmatrix} x \\ y \\ z \\ k_1 \\ k_2 \end{pmatrix},$$

from equation (3.1).

Therefore,

$$\omega_1 = \omega'_1\left(1+(\zeta-1)\frac{c^2}{{v_1}^2}\beta^2 + \frac{c^2}{v_1 u'_1}\beta^2\zeta\cos\phi'\right) + \omega'_2(\zeta-1)\frac{c^2}{v_1 v_2}\beta^2,$$

$$\omega_2 = \omega_2'\left(1 + (\zeta - 1)\frac{c^2}{{v_2}^2}\beta^2 + \frac{c^2}{v_2 u_2'}\beta^2\zeta\cos\phi'\right) + \omega_1'(\zeta - 1)\frac{c^2}{v_1 v_2}\beta^2,$$

$$\gamma_x = \gamma_x'\zeta\left(1 + \frac{u_1'}{v_1\cos\phi'}\beta^2\right) + \gamma_x'\zeta\beta^2\frac{u_2'}{v_2\cos\phi'},$$

$$\gamma_y = \gamma_y',$$

$$\gamma_z = \gamma_z'.$$

The expressions $\omega_2'(\zeta-1)\frac{c^2}{v_1 v_2}\beta^2$ and $\gamma_x'\zeta\beta^2\frac{u_2'}{v_2\cos\phi'}$ that are included in the angular frequency ω_1 and the wave vector γ_x, respectively, can be regarded as 2D kime-corrections of the Doppler effect formula in the spacekime manifold. Notice that setting $\omega_2' = 0$ yields a formula for the Doppler effect that reflects a wave moving only in one kime dimension k_1, which corresponds to the classical time Doppler effect. Still, in this case, the expressions for ω_1 and γ_x differ from the relativistic formulation of the Doppler effect in the Special Theory of Relativity.

3.7.11 Kime Calculus of Differentiation and Integration

Wirtinger derivatives are first-order partial differential operators that extend the ordinary univariate derivatives (defined with respect to a single real variable) to derivatives of differentiable functions defined on complex arguments. The Wirtinger derivative naturally leads to a differential calculus for functions defined on several complex variables that resembles the familiar ordinary differential calculus of univariate real functions [158, 160]. Below, we will derive the formulation of the first and second order Wirtinger derivatives.

Earlier in **Section 3.7.1**, we saw the *first* order Wirtinger derivative of a function of a complex variable defined in terms of the three common $\mathbb{C}$ plane parametrizations:

- in *Cartesian* coordinates: $f'(z) = \frac{\partial f(z)}{\partial z} = \frac{1}{2}\left(\frac{\partial f}{\partial x} - i\frac{\partial f}{\partial y}\right)$ and $f'(\bar{z}) = \frac{\partial f(\bar{z})}{\partial \bar{z}} = \frac{1}{2}\left(\frac{\partial f}{\partial x} + i\frac{\partial f}{\partial y}\right)$;
- in *conjugate-pair* coordinates: $df = \partial f + \bar{\partial} f = \frac{\partial f}{\partial z}dz + \frac{\partial f}{\partial \bar{z}}d\bar{z}$; and
- in *polar* kime coordinates:

$$f'(k) = \frac{\partial f(k)}{\partial k} = \frac{1}{2}\left(\cos\varphi\frac{\partial f}{\partial r} - \frac{1}{r}\sin\varphi\frac{\partial f}{\partial \varphi} - i\left(\sin\varphi\frac{\partial f}{\partial r} + \frac{1}{r}\cos\varphi\frac{\partial f}{\partial \varphi}\right)\right)$$

and

$$f'(\bar{k}) = \frac{\partial f(\bar{k})}{\partial \bar{k}} = \frac{1}{2}\left(\cos\varphi\frac{\partial f}{\partial r} - \frac{1}{r}\sin\varphi\frac{\partial f}{\partial \varphi} + i\left(\sin\varphi\frac{\partial f}{\partial r} + \frac{1}{r}\cos\varphi\frac{\partial f}{\partial \varphi}\right)\right).$$

The *second* order Wirtinger derivative with respect to kime ($k = (r, \varphi)$) is:

$$f''(k)=\frac{\partial}{\partial k}\left(\frac{\partial f(k)}{\partial k}\right)=\frac{1}{2}\,\frac{\partial}{\partial k}\left(\cos\varphi\frac{\partial f}{\partial r}-\frac{1}{r}\sin\varphi\frac{\partial f}{\partial\varphi}-i\left(\sin\varphi\frac{\partial f}{\partial r}+\frac{1}{r}\cos\varphi\frac{\partial f}{\partial\varphi}\right)\right)$$

$$=\frac{1}{4}\left(\cos\varphi\frac{\partial}{\partial r}\left(\cos\varphi\frac{\partial f}{\partial r}-\frac{1}{r}\sin\varphi\frac{\partial f}{\partial\varphi}-i\left(\sin\varphi\frac{\partial f}{\partial r}+\frac{1}{r}\cos\varphi\frac{\partial f}{\partial\varphi}\right)\right)\right.$$

$$-\frac{1}{r}\sin\varphi\frac{\partial}{\partial\varphi}\left(\left(\cos\varphi\frac{\partial f}{\partial r}-\frac{1}{r}\sin\varphi\frac{\partial f}{\partial\varphi}-i\left(\sin\varphi\frac{\partial f}{\partial r}+\frac{1}{r}\cos\varphi\frac{\partial f}{\partial\varphi}\right)\right)\right)$$

$$-i\left(\sin\varphi\frac{\partial}{\partial r}\left(\cos\varphi\frac{\partial f}{\partial r}-\frac{1}{r}\sin\varphi\frac{\partial f}{\partial\varphi}-i\left(\sin\varphi\frac{\partial f}{\partial r}+\frac{1}{r}\cos\varphi\frac{\partial f}{\partial\varphi}\right)\right)\right.$$

$$\left.\left.+\frac{1}{r}\cos\varphi\frac{\partial}{\partial\varphi}\left(\cos\varphi\frac{\partial f}{\partial r}-\frac{1}{r}\sin\varphi\frac{\partial f}{\partial\varphi}-i\left(\sin\varphi\frac{\partial f}{\partial r}+\frac{1}{r}\cos\varphi\frac{\partial f}{\partial\varphi}\right)\right)\right)\right)=$$

$$\frac{1}{4r^2}\left(\underbrace{\begin{pmatrix}4\cos\varphi\sin\varphi\frac{\partial f}{\partial\varphi}-\cos^2\varphi\frac{\partial^2 f}{\partial\varphi^2}+\sin^2\varphi\frac{\partial^2 f}{\partial\varphi^2}-r\cos^2\varphi\frac{\partial f}{\partial r}+r\sin^2\varphi\frac{\partial f}{\partial r}-\\ 4r\cos\varphi\sin\varphi\frac{\partial^2 f}{\partial r\partial\varphi}+r^2\cos^2\varphi\frac{\partial^2 f}{\partial r^2}-r^2\sin^2\varphi\frac{\partial^2 f}{\partial r^2}\end{pmatrix}}_{\text{Real part}}\right.$$
$$\left.\underbrace{+i\begin{pmatrix}2\cos^2\varphi\frac{\partial f}{\partial\varphi}-2\sin^2\varphi\frac{\partial f}{\partial\varphi}+2\cos\varphi\sin\varphi\frac{\partial^2 f}{\partial\varphi^2}+2r\cos\varphi\sin\varphi\frac{\partial f}{\partial r}-\\ 2r\cos^2\varphi\frac{\partial^2 f}{\partial r\partial\varphi}+2r\sin^2\varphi\frac{\partial^2 f}{\partial r\partial\varphi}-2r^2\cos\varphi\sin\varphi\frac{\partial^2 f}{\partial r^2}\end{pmatrix}}_{\text{Imaginary part}}\right).$$

Thus, the *kime-acceleration* (second order kime-derivative at $k=(r,\varphi)$) can be expressed more compactly as:

$$f''=\frac{1}{4r^2}\left((\cos\varphi-i\sin\varphi)^2\left(2i\frac{\partial f}{\partial\varphi}-\frac{\partial^2 f}{\partial\varphi^2}+r\left(-\frac{\partial f}{\partial r}-2i\frac{\partial^2 f}{\partial r\partial\varphi}+r\frac{\partial^2 f}{\partial r^2}\right)\right)\right).$$

In conjugate-pair coordinate space, the Laplacian is the divergence ($\nabla\cdot\equiv\langle\nabla|$) of the gradient ($\nabla$) operator:

$$\underbrace{\nabla\cdot\nabla}_{\substack{\text{inner}-\text{product}\\ \langle\nabla|\nabla\rangle\equiv\nabla'\nabla}}\equiv\nabla^2\equiv\Delta=4\frac{\partial}{dz}\frac{\partial}{d\bar{z}}\equiv4\frac{\partial}{d\bar{z}}\frac{\partial}{dz}.$$

Using first principles, we can derive that $\Delta\equiv4\frac{\partial}{\partial z}\frac{\partial}{\partial\bar{z}}=\frac{\partial^2}{\partial x^2}+\frac{\partial^2}{\partial y^2}$:

$$\frac{\partial}{\partial z}\frac{\partial}{\partial\bar{z}}f=\frac{\partial}{\partial z}\left(\frac{1}{2}\left(\frac{\partial f}{\partial x}+i\frac{\partial f}{\partial y}\right)\right)=\frac{1}{2}\left(\frac{1}{2}\frac{\partial}{\partial x}\left(\frac{\partial f}{\partial x}+i\frac{\partial f}{\partial y}\right)-\frac{1}{2}i\frac{\partial}{\partial y}\left(\frac{\partial f}{\partial x}+i\frac{\partial f}{\partial y}\right)\right)=$$

$$\frac{1}{4}\left(\frac{\partial^2 f}{\partial x^2}+\underbrace{i\frac{\partial^2 f}{\partial x\partial y}-i\frac{\partial^2 f}{\partial y\partial x}}_{0}+\frac{\partial^2 f}{\partial y^2}\right)=\frac{1}{4}\left(\frac{\partial^2}{\partial x^2}+\frac{\partial^2}{\partial y^2}\right)f.$$

Smooth complex-valued functions over open sets in $\mathbb{C}$ are called *analytic*, or *holomorphic*, if they have well-defined complex derivatives at each point in their domains. When such functions satisfy the Laplace equation, $\Delta f = 0$, they are called *harmonic* and can typically be expressed as mixtures of trigonometric sine and cosine functions.

These linear operators with constant coefficients may be extended to any space of generalized functions. The classical properties of differentiation apply to the Wirtinger derivatives, as well. For a pair of complex-valued multivariate functions, f and g, and two constants $u, v \in \mathbb{C}$, we have the following properties:

- Linearity:

$$\frac{\partial}{\partial \kappa}(uf + vg) = u\frac{\partial f}{\partial \kappa} + v\frac{\partial g}{\partial \kappa}.$$

- Product rule:

$$\frac{\partial}{\partial \kappa}(f \times g) = g\frac{\partial f}{\partial \kappa} + f\frac{\partial g}{\partial \kappa}.$$

- Chain rule:

$$\frac{\partial}{\partial \kappa}(f \circ g) = \left(\frac{\partial f}{\partial \kappa} \circ g\right)\frac{\partial g}{\partial \kappa} + \left(\frac{\partial f}{\partial \bar{\kappa}} \circ g\right)\frac{\partial \bar{g}}{\partial \kappa}.$$

- Conjugation:

$$\overline{\left(\frac{\partial f(\kappa)}{\partial \kappa}\right)} = \frac{\partial \bar{f}(\kappa)}{\partial \bar{\kappa}} = \frac{1}{2}\left(\underbrace{\left(\cos\varphi\frac{\partial \bar{f}}{\partial r} - \frac{1}{r}\sin\varphi\frac{\partial \bar{f}}{\partial \varphi}\right.}_{\text{real part}} + i\underbrace{\left(\sin\varphi\frac{\partial \bar{f}}{\partial r} + \frac{1}{r}\cos\varphi\frac{\partial \bar{f}}{\partial \varphi}\right)}_{\text{imaginary part}}\right).$$

It is common practice to pair calculus of differentiation and integration via the first fundamental theorem of calculus. For a real-valued function $g(x): \mathbb{R} \to \mathbb{R}$ the fundamental theorem of calculus states that if $G'(x) = g(x), \forall x \in [a, b] \subset \mathbb{R}$, i.e., G is the antiderivative of g, then

$$\int_a^b g(x)dx = G(b) - G(a).$$

Now that we have kime-differentiation, we can explore the corresponding kime path integration and the fundamental theorem of (kime) calculus. Let's define a function $f: \Omega \subset \mathbb{C} \to \mathbb{C}$ to be *holomorphic* (complex-analytic) in an open kime domain Ω if for any $\forall z_o \in \Omega$ the function can be expressed as an infinite Taylor series:

$$\sum_{k=0}^{\infty} \alpha_k (z - z_o)^k = \alpha_o + \alpha_1 (z - z_o) + \alpha_2 (z - z_o)^2 + \alpha_3 (z - z_o)^3 + \cdots + \alpha_n (z - z_o)^n + \cdots.$$

where the series coefficients $\alpha_k \in \mathbb{C}, \forall k$, and this expansion converges to $f(z)$ in a neighborhood of z_o, $N_{z_o} = \{z \in \Omega | \|z - z_o\| < \delta\}$. More specifically, a complex holomorphic function is infinitely differentiable and within a neighborhood of z_o its Taylor series expansion

$$T_n(z) = \sum_{k=0}^{n} \frac{f^{(k)}(z_o)}{k!} (z - z_o)^k$$

converges pointwise to the corresponding functional value, $T_n(z) \underset{n \to \infty}{\longrightarrow} f(z)$, $\forall z \in N_{z_o} \subset \Omega \subset \mathbb{C}$.

Next we will focus on functions of kime, $\kappa \in \Omega \subset \mathbb{C}$, over an open and simply connected kime domain Ω. Suppose $f : \Omega \to \mathbb{C}$ is an analytical holomorphic function and ζ is a rectifiable simple closed curve in Ω, i.e., the path ζ is not self-intersecting, permits a close approximation by a finite number of linear segments, and its starting and ending points coincide. Then the Cauchy-Goursat integral theorem [160] yields that the path integral of the holomorphic function over ζ is trivial:

$$\oint_{\zeta} f(\kappa) d\kappa = 0.$$

The proof of this resembles closely the classical complex analysis derivation in terms of kime, $k = re^{i\varphi}$ and $f(\kappa) = \underbrace{u(\kappa)}_{Re(f)} + i \underbrace{v(\kappa)}_{Im(f)}$, where the pair of Cauchy-Riemann equations:

$$\left| \begin{array}{l} \frac{\partial u}{\partial \kappa_1} = \frac{\partial v}{\partial \kappa_2} \\ \frac{\partial u}{\partial \kappa_2} = - \frac{\partial v}{\partial \kappa_1} \end{array} \right.$$

provide the necessary and sufficient conditions for f to be a complex differentiable (holomorphic) function:

$$\left| \begin{array}{l} \frac{\partial u}{\partial r} = \frac{1}{r} \frac{\partial v}{\partial \varphi} \\ \frac{\partial v}{\partial r} = - \frac{1}{r} \frac{\partial u}{\partial \varphi} \end{array} \right. \Leftrightarrow \left(\begin{array}{c} \text{Cartesian coord} \\ \text{change of variables} \\ \kappa_1 = r \cos \varphi \\ \kappa_2 = r \sin \varphi \end{array} \right) \Leftrightarrow \left| \begin{array}{cc} \frac{\partial \kappa_1}{\partial r} & \frac{\partial \kappa_1}{\partial \varphi} \\ \frac{\partial \kappa_2}{\partial r} & \frac{\partial \kappa_2}{\partial \varphi} \end{array} \right| = \underbrace{\left| \begin{array}{cc} \cos \varphi & -r \sin \varphi \\ \sin \varphi & r \cos \varphi \end{array} \right|}_{\text{Jacobian of Transform}} \Leftrightarrow$$

$$\left| \begin{array}{l} \frac{\partial u}{\partial \kappa_1} = \frac{\partial v}{\partial \kappa_2} \\ \frac{\partial v}{\partial \kappa_1} = - \frac{\partial u}{\partial \kappa_2} \end{array} \right. \Leftrightarrow \left| \begin{array}{l} \frac{\partial u}{\partial r} = \frac{\partial u}{\partial \kappa_1} \frac{\partial \kappa_1}{\partial r} + \frac{\partial u}{\partial \kappa_2} \frac{\partial \kappa_2}{\partial r} \\ \frac{\partial u}{\partial \varphi} = \frac{\partial u}{\partial \kappa_1} \frac{\partial \kappa_1}{\partial \varphi} + \frac{\partial u}{\partial \kappa_2} \frac{\partial \kappa_2}{\partial \varphi} \end{array} \right. .$$

Hence,

$$\oint_{\zeta} f(\kappa)d\kappa = \oint_{\zeta}(u(\kappa)+iv(\kappa))(d\kappa_1+i\,d\kappa_2) = \underbrace{\oint_{\zeta}(u\,d\kappa_1 - v\,d\kappa_2)}_{Re()} + i\underbrace{\oint_{\zeta}(v\,d\kappa_1 + u\,d\kappa_2)}_{Im()}$$

$$\underbrace{=}_{\text{Green's Theorem}} \underbrace{\iint_{\Omega}\left(-\frac{\partial v}{\partial\kappa_1}-\frac{\partial u}{\partial\kappa_2}\right)d\kappa_1 d\kappa_2}_{0,\ \text{since}\ \frac{\partial v}{\partial\kappa_1}=-\frac{\partial u}{\partial\kappa_2}} + i\underbrace{\iint_{\Omega}\left(\frac{\partial u}{\partial\kappa_1}-\frac{\partial v}{\partial\kappa_2}\right)d\kappa_1 d\kappa_2}_{0,\ \text{since}\ \frac{\partial v}{\partial\kappa_2}=\frac{\partial u}{\partial\kappa_1}} = 0.$$

Notice that the equality above uses the relation between double area integral and path integral calculations over simple closed curves (Green's theorem). The Cartesian coordinate formulation of the path integral representing the area inside the closed curve ζ, whose inside is the open simply connected set Ω, can first be transformed to kime coordinates and then related to the area integral by:

$$\text{Area} = \oint_{\zeta}\kappa_1 d\kappa_2 = \oint_{\zeta} r\cos\varphi d(r\sin\varphi) = \oint_{\zeta} r\cos\varphi(r\,d\sin\varphi + \sin\varphi\,dr)$$

$$= \oint_{\zeta}\underbrace{r^2\cos^2\varphi}_{U(r,\varphi)}\,d\varphi + \oint_{\zeta}\underbrace{r\cos\varphi\sin\varphi}_{V(r,\varphi)}\,dr = \oint_{\zeta} U(r,\varphi)d\varphi + V(r,\varphi)dr = (\text{Green's theorem})$$

$$= \iint_{\Omega}\left(\frac{\partial U}{\partial r}-\frac{\partial V}{\partial\varphi}\right)dr\,d\varphi = \iint_{\Omega}\left(\underbrace{2r\cos^2\varphi}_{\frac{\partial U}{\partial r}} + \underbrace{r\sin^2\varphi - r\cos^2\varphi}_{\frac{\partial V}{\partial\varphi}}\right)dr\,d\varphi = \iint_{\Omega} r dr\,d\varphi.$$

In the middle of this calculation, we used the standard Green's theorem, which is valid in any coordinate reference frame, including kime and Cartesian.

Therefore, integrals of holomorphic functions over paths in a simply connected domain are connected with the kime derivatives. A corollary of the Cauchy-Goursat theorem is an extension of the real-valued fundamental theorem of calculus to complex-valued functions. That is, over simply connected domains, path integrals of holomorphic complex-valued functions can be computed using antiderivatives. More precisely, let $\Omega \subset \mathbb{C}$ be an open and simply connected kime region, $f:\Omega \to \mathbb{C}$ be an analytical holomorphic function which is the kime-derivative of F, i.e., $f = F'$, and ζ be a piecewise continuously differentiable path (not necessarily closed) in the domain Ω that starts and ends at $\kappa_1, \kappa_2 \in \Omega$, i.e., $\zeta:[0,1] \to [\kappa_1, \kappa_2]$. Then,

$$\oint_{\zeta} f(\kappa)d\kappa = F(\kappa_2) - F(\kappa_1).$$

Path parameterization independence follows directly from Cauchy-Goursat integral theorem, as connecting in reverse the pair of paths, ζ_1 and ζ_2 that have the same starting and ending points, $\zeta_1, \zeta_2 : [0,1] \to [\kappa_1, \kappa_2]$ yields a closed curve $\zeta = \zeta_1 - \zeta_2$:

$$0 = \oint_{\zeta} f(\kappa)d\kappa = \oint_{\zeta_1 - \zeta_2} f(\kappa)d\kappa = \oint_{\zeta_1} f(\kappa)d\kappa - \oint_{\zeta_2} f(\kappa)d\kappa \Rightarrow \oint_{\zeta_1} f(\kappa)d\kappa = \oint_{\zeta_2} f(\kappa)d\kappa.$$

Thus, expanding the path integral over the (open) path ζ yields:

$$\oint_{\zeta} f(\kappa)d\kappa = \oint_{\zeta} F'(\kappa)d\kappa \underbrace{\quad = \quad}_{\text{path parameterization independence}} \int_0^1 F'(\zeta(t))d\zeta = \int_0^1 F'(\zeta(t))\,\zeta'(t)\,dt =$$

$$\int_0^1 (F \circ \zeta)'(t)dt = \int_0^1 d(F \circ \zeta) = (F \circ \zeta)(1) - (F \circ \zeta)(0) = F(\zeta(1)) - F(\zeta(0)) = F(\kappa_2) - F(\kappa_1).$$

It's important to note that the function needs to be holomorphic over an open and simply connected kime domain. If this requirement is violated (e.g., f is not differentiable or the kime region is closed or has holes), then this relation between the kime derivative and the kime path integral may be broken. Here is a simple counter example of a holomorphic function $f(\kappa) = \frac{1}{\kappa - 1}$ over the kime domain $\Omega = \{\kappa \in \mathbb{C} | 0 < \|\kappa - 1\| < 1\}$ (mind the singularity of f at $\kappa = 1$). We can define a simple closed path ζ to be just the unitary circle centered at $\kappa = 1$,

$$\zeta(t) = 1 + e^{i(2\pi t)} : [0,1] \to \Omega.$$

Then, the kime path integral is non-trivial despite the fact that the curve starts and ends at the same kime point, $\zeta(0) = \zeta(1) = 2 + i0 \in \mathbb{C}$:

$$\oint_{\zeta} f(\kappa)d\kappa = \int_0^1 \frac{1}{1 + e^{i(2\pi t)} - 1} d\left(1 + e^{i(2\pi t)}\right) = \int_0^1 \frac{(i2\pi)e^{i(2\pi t)}}{e^{i(2\pi t)}} dt = \int_0^1 i2\pi\ dt = i2\pi \neq 0.$$

3.8 The Copenhagen vs. Spacekime Interpretations

As shown earlier in **Section 3.6**, numerical Newtonian physics measurable quantities like the particle's position, momenta, spin, and energy, represent in quantum-mechanical sense silhouettes of their corresponding operators. For example, the total energy (Hamiltonian operator) $\hat{H} = \hat{T} + \hat{V}$ is a linear mixture of the *kinetic* energy operator, $\hat{T} = -\frac{\hbar^2}{2m}\nabla^2$, where $\nabla^2 = \frac{\partial^2}{\partial x^2} + \frac{\partial^2}{\partial y^2} + \frac{\partial^2}{\partial z^2}$ is the Laplacian, and the *potential* energy operator, $\hat{V}$, which depends on the specific type of force acting on the system, e.g., elastic, gravitational, Coulomb, and other conservative forces. In the special case of *spherically symmetric energy potentials*, the potential energy operator $\hat{V} = \hat{V}(r) = \frac{m}{2}\omega^2 r^2$

depends only the magnitude of the radius vector, r, the particle mass, m, the oscillator force constant, k, and the angular frequency of the oscillator, $\omega = \sqrt{\frac{k}{m}}$.

Thus, the eigenvalues, E_α, which capture all possible total energy states (observable energy values), and the eigenfunctions, ψ_α, which describe the particle's motion, represent the eigen solutions of the total energy Hamiltonian operator $\hat{H}\psi_\alpha = E_\alpha \psi_\alpha$.

The quantized *measurable values* including position and momenta represent the eigenvalues of the corresponding operator with which they are paired. Therefore, measuring the total energy of a particle would only yield numbers that can be eigenvalues of the Hamiltonian operator. To understand the bijective correspondence between eigenvalues and observable measures, we need to acknowledge that the particle wavefunctions are linear combinations of (base) eigenfunctions, and rarely a single wave eigenfunction. Even a base eigenfunction for one operator may actually be a linear combination of multiple wave eigenfunctions of another operator.

An interesting phenomenon arises in the process of taking a spacetime measurement and recording an observation. Despite the fact that the wavefunction is generally a combination of eigenfunctions, the actual measured value is always a singleton, and not a combination of eigenvalues. This apparent paradox is resolved by a quantum mechanics orthodoxy known as the *Copenhagen Interpretation* [161]. The latter states that a quantum particle doesn't exist in one unique state, but rather probabilistically occupies all of its possible states at the same time and the act of its observation captures a unique freeze-frame of its momentous state. This effectively answers axiomatically what happens to the eigenvalues in the linear combination that do not come out as measured values during the instantaneous snapshot of the system state:

> *An instant measurement causes the wavefunction Ψ to randomly collapse only into one of the eigenfunctions of the quantity that is being measured.*

Let's look at an example trying to measure the total energy of a particle. This instantaneous observation causes the wavefunction to collapse into one unique eigenfunction ψ_{α_0} of the total energy (Hamiltonian) operator. The observed total energy corresponds to an eigenvalue:

$$\left.\begin{array}{c}\overbrace{\Psi=\sum_\alpha c_\alpha\psi_\alpha(\text{wavefunction})}^{\text{Probabilistic state of the system}}\\ \underbrace{\text{Energy is unknown}}_{\text{Natural, intrinsic, fuzzy}}\end{array}\right\}\underbrace{\xrightarrow{\substack{\text{Copenhagen Interpretation}\\ \text{Measurement process}}}}_{\text{observe the total energy}}\left\{\begin{array}{c}\overbrace{\Psi=c_{\alpha_o}\psi_{\alpha_o},\ \text{for some index}\ \alpha_o,}^{\text{Wave function collapse}}\\ c_{\alpha_o}\in\mathbb{C}\\ \underbrace{\text{Total Energy}=E_{\alpha_o}}_{\text{Instance of the observed state of the system}}\end{array}\right. .$$

This is just a philosophical perspective discriminating between the (unobserved and theoretical) natural phenomena (prior to the act of observation) and an instantaneous measurement value generated by a device, which may itself be uncertain. Physical measurements are most commonly defined as interactions between macroscopic systems that produce the observed collapse of the native wavefunction into one unique eigenfunction.

Prior to a physical measurement, the wavefunction is a linear mixture of many different eigenfunctions, $\Psi=\sum_\alpha c_\alpha\psi_\alpha\in\mathbb{C}$ and $c_\alpha\in\mathbb{C}$. The determination about which eigenfunction projection will yield the observed measurement value is probabilistic, based on random sampling using the proportional square-magnitudes of the weights $|c_\alpha|^2$. For instance, if $|c_\alpha|^2=2|c_\beta|^2$, then a wavefunction collapse on the base eigenfunction ψ_α is expected to be twice as likely as a collapse on the other base eigenfunction ψ_β. In principle, these eigenfunction projections in spacetime may be random, non-random reflecting unobserved lurking variables, or they may be associated with higher dimensions that are hidden, undetectable, or non-traversable. However, Bell's theorem suggests that under certain conditions, quantum physics theory and experimental observations suggest that the eigenfunction projections must be random [85, 162].

There are several alternative *interpretations* of quantum mechanics (QM) reality including many-worlds, transactional, relational QM, ensemble, and de Broglie-Bohm theoretic interpretations [163, 164]. There are two philosophical constructs leading to different interpretations: (1) *ontology*, categorizing the existence of real things and world entities, and (2) *epistemology*, describing the state of knowledge, observable possibilities, and scope.

The *spacekime* generalization of spacetime leads to another interpretation of the dichotomy between the theoretical existence of wavefunctions as linear mixtures of base eigenfunctions and the reliable practical observation that real measurements always collapse the state of the system into only one (albeit random) base eigenfunction. The *Spacekime Interpretations* represent plausible elucidations of this eigenfunction collapse of the natural state of the system. Such more general interpretations may resolve the peculiar problem of the *choice* of one specific eigenfunction as a projection of the system state, i.e., explicate alternative views of the uncertainty of this projection-selection process.

By definition, spacekime measurements, samples, and observations are intrinsically instantaneous, anchored at specific spatial locations and kime moments. Such measurements represent discrete values, vectors, or tensors, limited simulations, finite experimental results, or countable interactions arising from known models, unknown distributions, or partially understood state-spaces. In a probabilistic sense, for any stochastic process, we can only observe discrete random samples or detect a finite number of instances. For some processes, it is mathematically feasible to (analytically) represent the entire (joint) probability process distribution holistically. However, in general, it's not possible to instantaneously sample (observationally) the entire probability distribution or measure the total wavefunction representing the complete state-space.

Some alternative *Spacekime Interpretations* include:

Option 1:

A fixed-time instantaneous measurement of the system at $t = t_o$, *yields a complete eigenfunction decomposition of the wavefunction* $\Psi(\kappa) = \Psi(t_o, \varphi)$ *where the kime-phase spans the range* $-\pi \le \varphi < \pi$. *However, as the entire distribution of kime phases* (Φ) *may not be directly, instantaneously, and holistically observed, the actual measurement only reflects a random phase draw, which yields a wavefunction value* $\Psi(\kappa) \equiv \Psi(t_o, \varphi) = \sum_\alpha c_\alpha \psi_\alpha(t_o, \varphi)$ *for some fixed random phase,* $\varphi = \varphi_o$. *Thus, each observation manifests as an immutable instantaneous measurement value,* $\Psi(t_o, \varphi_o) = \sum_\alpha c_\alpha \psi_\alpha(t_o, \varphi_o)$, *where* $\psi_\alpha(t_o) = \psi_\alpha(t_o, \varphi_o)$ *are just the classical spacetime eigenfunctions of the corresponding operator. In a measure-theoretic sense, a pair of repeated independent measurements of the exact same spacekime system would naturally yield two distinct observed values:* $\Psi' \equiv \Psi(t_o, \varphi'_o) = \sum_\alpha c_\alpha \psi_\alpha (t_o, \varphi'_o)$ *and* $\Psi'' \equiv \Psi(t_o, \varphi''_o) = \sum_\alpha c_\alpha \psi_\alpha(t_o, \varphi''_1)$, *where the two phases are independently sampled from the circular phase distribution, i.e.,* $\varphi'_o, \varphi''_1 \sim \Phi[-\pi, \pi)$.

Option 2:

For a fixed-time instantaneous measurement of the system at $t = t_o$, *the wavefunction, or inference-function,* $\Psi(\kappa) = \Psi(t_o, \varphi)$ *is naturally an aggregate measure over the entire kime-phase distribution with a range* $-\pi \le \varphi < \pi$. *However, as the entire kime phases distribution* (Φ) *may not be directly, holistically, and instantaneously observed, the actual measurement, or inference, only reflects a measurement* $\Psi(t_o, \varphi)$ *for one random phase,* φ_o. *In other words, the natural state of the system is theoretically described by a wavefunction, or inference-function,*

$$\underbrace{\Psi(t)}_{\substack{\text{Observed}\\ \text{Spacetime}\\ \text{wavefunction}}} = \int_{-\pi}^{\pi} \underbrace{\Psi(t, \varphi)}_{\substack{\text{Spacekime}\\ \text{wavefunction}}} \, d\Phi,$$

however, the actual observation reflects the value at a given time point (t_o) *for some fixed but randomly chosen phase,* $\varphi = \varphi_o$. *Thus, each observation manifests as an immutable instantaneous measurement value,* $\Psi(t_o) = \Psi(t_o, \varphi_o)$. *In a measure-*

theoretic sense, a pair of simultaneous $(t = t_o)$ *independent measurements of the exact same spacekime system would naturally yield two distinct observed values:* $\Psi' \equiv \Psi(t_o, \varphi'_o)$ *and* $\Psi'' \equiv \Psi(t_o, \varphi''_o)$, *where the two phases are independently sampled from the circular phase distribution, i.e.,* $\varphi'_o,\ \varphi''_o \sim \Phi[-\pi, \pi)$.

Note that both spacekime interpretations explain the common practice of taking independent and identically distributed (IID) samples of the same process (e.g., population polling, large sampling) to estimate a specific population characteristic, $\Psi(t_o)$, like the mean, dispersion, quantiles, etc. The implicit fix of the spatial location ($\boldsymbol{x}_o = (x_o,\ y_o,\ z_o)$) and temporal moment ($t_o$) coordinates specifically indicates the spacetime localization of this characterization. Suppose we measure repeatedly the state of the system under controlled environment (to ensure IID sampling) and obtain a series of observations, $\left\{\Psi^{(i)} \equiv \Psi\left(t_o, \varphi_o^{(i)}\right)\right\}_{i=1}^{n}$. By the law of large numbers (LLN) [165, 166], all sample-driven estimates would converge to their theoretical counterparts. For instance, if we want to estimate the population mean at time t_o, $\Psi(t_o) = \int_{-\pi}^{\pi} \underbrace{\Psi(t_o, \varphi)}_{\substack{\text{Spacekime}\\ \text{wavefunction}}}\ d\Phi$, we can compute the arithmetic average of the observed values,

$$\bar{\Psi}_n = \frac{1}{n}\sum_{i=1}^{n} \Psi^{(i)} \equiv \frac{1}{n}\sum_{i=1}^{n} \underbrace{\Psi\left(t_o, \varphi_o^{(i)}\right)}_{\substack{\text{spacetime}\\ \text{observables}}}.$$

By LLN, these sample-means will converge to the corresponding theoretical population mean, which is not directly observable (cf. Spacekime interpretation):

$$\underbrace{\bar{\Psi}_n}_{\substack{\text{spacetime}\\ \text{observables}}} \underset{n\to\infty}{\longrightarrow} \underbrace{\Psi(t_o)}_{\substack{\text{spacekime}\\ \text{unobservables}}}.$$

This process of random sampling and measure aggregation illustrates exactly how we cleverly interpret our limited 4D spacetime perceptions of reality to accurately represent the more enigmatic 5D spacekime, in which the obscure kime-phases may or may not be directly measurable.

Let's examine the extended notion of wavefunctions in spacekime. If $\psi(x, k)$ denotes the complex-valued amplitude and phase of a wave describing the spacekime motion of a particle, then the spatial 1D (special case) and 3D (general case) wave equations are:

$$1D\colon \frac{\partial^2 \psi(x,k)}{\partial k^2} = c^2 \frac{\partial^2}{\partial x^2}\psi(x,k),\ x \in \mathbb{R},\ k \in \mathbb{C}.$$

$$3D\colon \frac{\partial^2 \psi(\boldsymbol{x},k)}{\partial^2 k} = c^2\nabla^2\psi(\boldsymbol{x},k) = c^2\left(\frac{\partial^2}{\partial x^2}\psi(\boldsymbol{x},k) + \frac{\partial^2}{\partial y^2}\psi(\boldsymbol{x},k) + \frac{\partial^2}{\partial z^2}\psi(\boldsymbol{x},k)\right),\ \boldsymbol{x} \in \mathbb{R}^3,\ k \in \mathbb{C}.$$

The second order *Wirtinger* derivative of the wavefunction with respect to kime, $\frac{\partial^2 \psi(x,k)}{\partial^2 k}$, was defined earlier, see **Section 3.7.10**.

Next, we can try to explicate the Spacekime Interpretation for measuring the total energy of a particle whose motion is described by a 1D spatial spacekime wavefunction, representing a solution to this partial differential equation:

$$\frac{\partial^2}{\partial k^2}\Psi(x,k) = c^2\frac{\partial^2}{\partial x^2}\Psi(x,k), x \in \mathbb{R}, k \in \mathbb{C}.$$

In 4D Minkowski spacetime, an instantaneous ($t = t_o$) observation of the total energy of this system, E_φ, would represent an eigenvalue of corresponding Hamiltonian operator, $\hat{H}$, i.e.,

$$\hat{H}\ \underbrace{\Psi(\varphi)}_{\text{eigenfunction}} = \underbrace{E_\varphi}_{\text{eigenvalue}}\ \underbrace{\Psi(\varphi)}_{\text{eigenfunction}}.$$

In 5D Minkowski spacekime, the observed total energy corresponds to a wavefunction density value:

$$\left.\underbrace{\overbrace{\underbrace{\Psi(t)}_{\substack{\text{Spacetime}\\ \text{wavefunction}}} = \int_{-\pi}^{\pi} \underbrace{\Psi(t,\varphi)}_{\substack{\text{Spacekime}\\ \text{wavefunction}}}\, d\Phi}^{\text{Unknown Total Energy}}}_{\substack{\text{natural, intrinsic, fuzzy, probabilistic}\\ \text{state of the system}}}\right\} \underbrace{\xrightarrow[\text{Measurement process}]{\text{Spacekime Interpretation}}}_{\substack{\text{observe energy}\\ \text{at time } t_o}}$$

$$\left\{\underbrace{\begin{array}{c} \Psi(t_o) = \Psi\left(t_o, \varphi_o'\right) \equiv \Psi\left(\varphi_o'\right)\left\{\begin{array}{c}\text{wavefunction}\\ \text{density}\end{array}\right. \\ \varphi_o' \sim \Phi[-\pi,\pi)\left\{\begin{array}{c}\text{kime}-\text{phase}\\ \text{distribution}\end{array}\right. \\ \text{Observed Total Energy} = E_{\varphi'_o}, \text{which} \\ \text{still represents an } \hat{H} \text{ eigenvalue}\end{array}}_{\text{observable state of the system}}\right..$$

Note that the *observed* spacetime total energy can be expressed as a function of the square magnitude of the spacekime velocity. For instance, let's consider the simple case of spacekime energy of a *linear spring*, where the kime velocity is the Jacobian $v = \left(v_{x,j}\right) = \left(\frac{dx}{d\kappa_1}, \frac{dx}{d\kappa_2}\right)$.

The total energy of the particle moving in spacekime is defined by $\frac{m_0c^2}{\sqrt{1-\beta^2}}$, where m_0 is the particle mass at rest and $\beta = \frac{1}{\sqrt{\left(\frac{c}{v_1}\right)^2 + \left(\frac{c}{v_2}\right)^2}}$, see **Chapter 5, Section 5.7**. At low speeds, when $\beta \ll 1$, i.e., $v_1 \ll c$ or $v_2 \ll c$, then $E \approx m_0c^2\left(1+\frac{1}{2}\beta^2\right)$. Earlier in **Chapter 3, Section 3.7.2**, we showed for each *constant* $\varphi'_o \sim \Phi[-\pi, \pi)$, the relation between the classical time-velocity, $v = v_t$, and the directional kime-velocities, v_1, v_2,

$$\frac{1}{v_t^2} = \frac{1}{(v_1)^2} + \frac{1}{(v_2)^2} = \frac{\left(v_1^2+v_2^2\right)}{(v_1v_2)^2}.$$

Therefore,

$$E \approx m_0c^2 + \frac{m_0c^2}{2\left(\left(\frac{c}{v_1}\right)^2 + \left(\frac{c}{v_2}\right)^2\right)} = m_0c^2 + \frac{m_0}{2}\frac{(v_1v_2)^2}{\left(v_1^2+v_2^2\right)} = m_0c^2 + \frac{m_0}{2}\underbrace{v^2}_{v_t^2}.$$

This decomposes the total energy E in terms of the rest-mass (potential) energy, m_0c^2, and the low-speed kinetic energy $\frac{m_0}{2}\frac{(v_1v_2)^2}{\left(v_1^2+v_2^2\right)}$.

$$\underbrace{E}_{\substack{\text{spacetime}\\ \text{total energy}}} = \underbrace{\frac{1}{2}\underbrace{m_0}_{\text{rest mass}}\underbrace{v_t^2}_{\text{total time velocity}}}_{\text{kinetic energy}} + \underbrace{\frac{1}{2}\underbrace{k}_{\text{spring constant}}\underbrace{x^2}_{\text{distance}}}_{\text{potential energy}} =$$

$$= \frac{1}{2}\underbrace{m_0}_{\text{rest mass}}\underbrace{\frac{(v_1v_2)^2}{\left(v_1^2+v_2^2\right)}}_{\text{kime velocity}} + \frac{1}{2}kx^2 = \frac{1}{2}\underbrace{m_0}_{\text{rest mass}}\underbrace{v_\kappa^2}_{\text{kime velocity}} + \frac{1}{2}kx^2 = \underbrace{E}_{\substack{\text{spacekime}\\ \text{total energy}}}.$$

Using the relation between the time-velocity and the kime-velocity in polar coordinates

$$\frac{1}{v_\kappa^2} = \frac{1}{v_1^2} + \frac{1}{v_2^2} = \frac{(dt)^2}{dx^2+dy^2+dz^2} + \frac{t^2(d\varphi)^2}{dx^2+dy^2+dz^2} = \frac{1}{v_t^2} + t^2\underbrace{\frac{(d\varphi)^2}{dx^2+dy^2+dz^2}}_{=0,\ \text{as}\ \varphi=\varphi'_0\ \text{is constant}} = \frac{1}{v_t^2}.$$

Plugging in the spring energy equation we again see that the total spacetime observed energy equals the total spacekime energy

$$\underbrace{E}_{\substack{\text{spacetime}\\ \text{total energy}}} = \underbrace{\frac{1}{2}\underbrace{m}_{\text{mass}}\underbrace{v_t^2}_{\text{time velocity}}}_{\text{kinetic energy}} + \underbrace{\frac{1}{2}\underbrace{k}_{\text{spring constant}}\underbrace{x^2}_{\text{distance}}}_{\text{potential energy}} = \frac{1}{2}\underbrace{m}_{\text{rest mass}}v_\kappa^2 + \frac{1}{2}kx^2 = \underbrace{E_{\varphi'_0}}_{\substack{\text{spacekime}\\ \text{total energy}}}.$$

Open Problems

– *Ergodicity*: Let's look at the particle velocities in the 4D Minkowski spacetime (X), representing a measure space where gas particles move spatially and evolve longitudinally in time. Let $\mu = \mu_x$ be a measure on X, $f(\boldsymbol{x}, t) \in L^1(X, \mu)$ be an integrable function (e.g., velocity of a particle), and $T: X \to X$ be a measure-preserving transformation at position $\boldsymbol{x} \in \mathbb{R}^3$ at time $t \in \mathbb{R}^+$. Then, the pointwise ergodic theorem states that in a measure theoretic sense, the average of f over all particles in the gas system at a fixed time, $\bar{f} = E_t(f) = \int_{\mathbb{R}^3} f(\boldsymbol{x}, t) d\mu_x$, will be equal to the average velocity of just one particle estimated over the entire time span, $\hat{f} = \lim_{n \to \infty} \left(\frac{1}{n}\sum_{i=0}^{n} f(T^i x)\right)$ [167, 168]. That is, $\bar{f} \equiv \hat{f}$. In the above example, the spatial probability measure is denoted by μ_x and the transformation $T^i x$ represents the dynamics (time evolution) of the particle starting with an initial spatial location $T^o \boldsymbol{x} = \boldsymbol{x}$. Investigate the ergodic properties of various transformations in the 5D Minkowski spacekime. A starting point may be a formulation such as:

$$\underbrace{\bar{f} = E_\kappa(f) = \frac{1}{\mu_x(X)} \int f\left(\boldsymbol{x}, \underbrace{t, \phi}_{\kappa}\right) d\mu_x}_{\text{space averaging}} \overset{?}{=} \underbrace{\lim_{t \to \infty}\left(\frac{1}{t}\sum_{i=0}^{t}\left(\int_{-\pi}^{+\pi} f(T^i \boldsymbol{x}, t, \phi) d\Phi\right)\right)}_{\text{kime averaging}} = \hat{f}.$$

– *Duality*: It may be interesting to investigate deeper the Hamiltonian operator, $\hat{H}$, as a member of the dual space, as described below.

Spacekime wavefunctions describe particle states by complex-valued functions $\Psi: \mathbb{R}^3 \times \mathbb{C} \to \mathbb{C}$ of spatial $\boldsymbol{x} \in \mathbb{R}^3$ and kime $\boldsymbol{\kappa} \in \mathbb{C}$ arguments. Then, the wavefunction may be interpreted as a probability amplitude and the magnitude of its square modulus as $\rho(\boldsymbol{x}, \boldsymbol{\kappa})$:

$$|\Psi(\mathbf{x},\kappa)|^2 = \Psi^\dagger(\mathbf{x},\kappa)\Psi(\mathbf{x},\kappa) = \langle\Psi(\mathbf{x},\kappa) \mid \Psi(\mathbf{x},\kappa) = \rho(\mathbf{x},\kappa) \geq 0.$$

This magnitude is the 2D kime probability density of the particle at position $\mathbf{x}$. Effectively, the wavefunction describes the particle spatial probability distribution in kime, rather than a specific particle position in time. To quantify the likelihood that at a fixed time t, the particle is in a region $\mathbf{\Omega} \subseteq \mathbb{R}^3$, we can integrate the 2D kime probability density over this region:

$$P_{x\in\Omega}(t) = \int_{-\pi}^{\pi} P_{x\in\Omega}(\kappa)d\varphi = \int_{-\pi}^{\pi} P_{x\in\Omega}(t,\varphi)d\Phi = \int_{-\pi}^{\pi}\int_{\Omega}\langle\Psi|\Psi\rangle dxd\Phi =$$

$$\int_{\Omega}\int_{-\pi}^{\pi}\langle\Psi|\Psi\rangle d\Phi dx = \int_{\Omega}\Psi^\dagger\Psi\underbrace{\left(\int_{-\pi}^{\pi} d\Phi\right)}_{1} dx = \int_{\Omega}|\Psi(\mathbf{x},t)|^2 dx.$$

As spacekime wavefunctions form an infinite-dimensional space H, there is no finite set of base eigenfunctions for H. The norm in this Hilbert space of wavefunctions is derived by the inner product $\langle\cdot|\cdot\rangle$. Recall that the inner product of two wavefunctions Ψ_1 and Ψ_2 can be expressed using the bra-ket operator at kime $\kappa = (t, \varphi)$:

$$\langle\Psi_1(\mathbf{x},\kappa)|\Psi_2(\mathbf{x},\kappa)\rangle = \int_{\mathbb{R}^3}\Psi_1^\dagger(\mathbf{x},\kappa)\Psi_2(\mathbf{x},\kappa)dx.$$

In general, the inner product of a pair of wavefunctions is always a complex number, $\langle\Psi_1(\mathbf{x},\kappa)|\Psi_2(\mathbf{x},\kappa)\rangle \in \mathbb{C}$. However, for the special case of an inner product of a wavefunction Ψ with itself, it's a positive real number,

$$\Psi^\dagger(\mathbf{x},\kappa)\ \Psi(\mathbf{x},\kappa) = \langle\Psi(\mathbf{x},\kappa) \mid \Psi(\mathbf{x},\kappa)\rangle = \rho(\mathbf{x},\kappa) \geq 0.$$

At a fixed moment in time, t_o, the values of the spacekime wavefunction $\Psi(\mathbf{x},\kappa)$ represent uncountably many components of a vector in the infinite dimensional Hilbert state space:

$$|\Psi(t_o)\rangle = \int_{\mathbb{R}^3}\int_{-\pi}^{\pi}\Psi(\mathbf{x},\kappa)|x\rangle dxd\Phi = \int_{\mathbb{R}^3}\Psi(x,t_o)|x\rangle dx.$$

3.9 Space-Kime Formalism

In this section, we will summarize the special case of the Velev theory [169] of the causal structure of 5D spacekime. Specifically, we will present a description of particle motion in spacekime, formulate the Lorentz transformations between two inertial reference frames, as well as explain charge, parity, and time-reversal (CPT) symmetry, the invariance of the speed of light principle, the law of addition of velocities, and the energy-momentum conservation law. We will also derive the spacekime relations between energy, mass, and momentum of particles, determine the stability conditions for particles moving in space-kime, derive conditions for nonzero rest particle mass in space-kime, and examine the causal structure of space-kime by exploring causal regions.

3.9.1 Antiparticle in Spacekime

In the case of two-dimensional kime, the *CPT*-symmetry will be violated. Special relativity and the Lorentz covariance are in the base of the *CPT*-symmetry. Indeed, the even number of reflections of the coordinates in the Minkowski space-time is formally reduced to a rotation by an imaginary angle. Due to this fact, the existing physical theories, which are invariant relating to the Lorentz transformations (i.e. rotations in the Minkowski space-time) turn out to be automatically *CPT*-invariant. In the case of multidimensional time, the *CPT*-symmetry must be exchanged with another generalized symmetry.

If the number of the time dimensions where a particle is moving is greater than one, then this particle may have more than one antiparticle. If the number of the time dimensions is equal to m, then the number of the different antiparticles is equal to $(3^m - 2^m)$ [169]. For the case $m = 2$ we obtain $(3^2 - 2^2) = 5$ different antiparticles.

In the case of two-dimensional kime, the antiparticles can be defined as particles with a negative rest mass moving backward in at least one of the kime dimensions. **Figure 3.12** represents the different types of antiparticles of the particle M_{++}, moving forward in both kime dimensions k_1 and k_2. The different types of antiparticles in two-dimensional kime are marked with

$$A_{-+}\,,\; A_{-0}\,,\; A_{--}\,,\; A_{0-}\,,\; A_{+-}\,.$$

For example, the antiparticle A_{-+} is moving backwards in the kime dimension k_1, and forwards in the kime dimension k_2 (i.e., $dk_1 < 0$ and $dk_2 > 0$). The antiparticle A_{-0} is moving backwards in the kime dimension k_1 but does not move in the kime dimension k_2 (i.e., $dk_1 < 0$ and $dk_2 = 0$), etc.

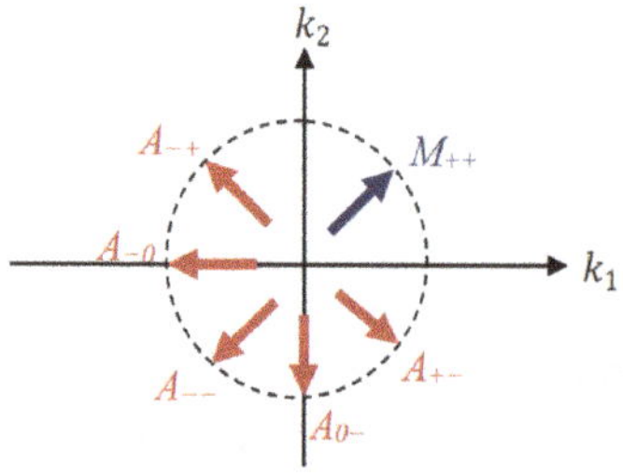

Figure 3.12: Antiparticles in two-dimensional kime.

If a rotation of the kime axes (k_1, k_2) is applied in the plane of kime, one particle can be converted to the corresponding type of antiparticle and vice-versa.

3.9.2 The Causal Structure of Spacekime

A number of authors discuss the issue of causality in higher dimensions [114, 115, 170]. Just like in the classical one-dimensional time, in the multidimensional time every localized object is moving along a one-dimensional time-like world line representing the kime-magnitude. Tegmark pointed out that when $m > 1$, there is no obvious reason why an observer could not perceive time as a one-dimensional construct to characterize and interpret reality as patterns of events ordered in 1D succession [115]. As all observers are intrinsically localized, the travel of each observer may be projected along an essentially one-dimensional (time-like) world line through the general $(n + m)$-dimensional[3] spacekime manifold. Therefore, even in two or more longitudinal dimensions, time may be perceived as one-dimensional because most observable physical processes have strong event-order characteristics allowing us to model them as linear consequences, which simplifies interpretation of reality. Regular clocks work in their usual manner in spacekime where every localized object will have one single "history" reflecting its kime-order. In this sense, the multidimensional kime notion does not differ philosophically from the well-known notion of time. However, in the case of multidimensional time there are problems concerning well-posed causality [115].

In higher-dimensional spacekime, it is always possible to construct closed kime-like curves [171].[4] Let us consider the case of two-dimensional kime and three-dimensional space. In the space-kime coordinates $\Big(\underbrace{x, y, z}_{\text{space}}, \underbrace{k_1, k_2}_{\text{kime}}\Big)$ we can consider a "motion" in the causal region in the plane (k_1, k_2), which begins and ends at the same kime location $(k_1 = k_1^o, k_2 = k_2^o)$. Assume $\Omega = const$; $\theta \in \Theta = [-\pi: \pi)$ and define $k_1 = \Omega \sin\theta, k_2 = \Omega(1 - \cos\theta)$, and x, y, $z = const$. Then, worldlines are kimelike everywhere [171] and we will have:

$$ds^2 = dx^2 + dy^2 + dz^2 - (cdk_1)^2 - (cdk_2)^2 = -c^2\Omega^2 d\theta^2 < 0.$$

The 2D-kime causal structure of a spacekime is different from the 1D-time causal structure in spacetime. Let us start with a 5D vector $\boldsymbol{A}_\mu$ in spacekime, $\mu = (n, m) = (3, 2)$. The scalar product of the vector $\boldsymbol{A}_\mu$ with itself will be[5]:

3 Here m is the number of the time dimensions; n is the number of the spatial dimensions.

4 This is different from the case of curved manifolds, where closed time-like curves may arise under certain circumstances.

5 In this formula, Einstein summation convention is used.

$$(\boldsymbol{A})^2 = \langle \boldsymbol{A}|\boldsymbol{A}\rangle = A^\mu A_\mu = g_{\mu\rho}A^\mu A^\rho = \sum_{\eta=1}^{3}(A^\eta)^2 - (A^4)^2 - (A^5)^2,$$

where $\mu, \rho = 1,2,3,4,5$, and $g_{\mu\rho} = \begin{cases} 1, \text{ at } \mu = \rho = 1,2,3 \\ -1, \text{ at } \mu = \rho = 4,5 \\ 0, \text{ at } \mu \neq \rho \end{cases}$.

If we denote $(\boldsymbol{A}_{3,1})^2 = \sum_{\eta=1}^{3}(A^\eta)^2 - (A^4)^2$, then we have $(\boldsymbol{A})^2 = (\boldsymbol{A}_{3,1})^2 - (A^5)^2$. While in the 4D Minkowski spacetime, $(n,m) = (3,1)$, the value $(\boldsymbol{A}_{3,1})^2$ is invariant, in the 5D spacekime manifold indexed by $(n,m) = (3,2)$, the value $(\boldsymbol{A}_{3,1})^2$ is not invariant, however, $\langle \boldsymbol{A}|\boldsymbol{A}\rangle = (\boldsymbol{A})^2$ is invariant.

For example, in spacekime, if the value $(\boldsymbol{A}_{3,1})^2$ is *spacelike* in one frame of reference K (i.e., $(\boldsymbol{A}_{3,1})^2 > 0$), in another frame of reference, K', it can also be *lightlike* (i.e., $(\boldsymbol{A}'_{3,1})^2 = 0$) or *kimelike* (i.e., $(\boldsymbol{A}'_{3,1})^2 < 0$). The spacekime causal region indexed by $(n,m) = (3,2)$ encompasses the region $(\boldsymbol{A})^2 \leq 0$, and the spacetime causal region indexed by $(n,m) = (3,1)$ encompasses the region $(\boldsymbol{A}_{3,1})^2 \leq 0$.

Let's set $A^1 = dx$, $A^2 = dy$, $A^3 = dz$, $A^4 = cdk_1$, $A^5 = cdk_2$, $dx^2 + dy^2 + dz^2 > 0$, $q_1 = \frac{\sqrt{dx^2+dy^2+dz^2}}{c|dk_1|} = \frac{V_1}{c} > 0$, $q_2 = \frac{\sqrt{dx^2+dy^2+dz^2}}{c|dk_2|} = \frac{V_2}{c} > 0$, $ds = \|\boldsymbol{A}\| = \sqrt{(\boldsymbol{A})^2}$. We will consider following three cases, depending on the inner product $\langle \boldsymbol{A}|\boldsymbol{A}\rangle = (\boldsymbol{A})^2$:

<u>*First (spacelike) case*</u>: $(\boldsymbol{A})^2 > 0$, i.e., $ds^2 > 0$. We have $(\boldsymbol{A}_{3,1})^2 > 0$ and it is not possible that $(\boldsymbol{A}_{3,1})^2 < 0$ or $(\boldsymbol{A}_{3,1})^2 = 0$. The condition $(\boldsymbol{A})^2 > 0$ is equivalent to the following inequalities:

$$(\boldsymbol{A})^2 = dx^2 + dy^2 + dz^2 - c^2dk_1^2 - c^2dk_2^2 > 0,$$

$$dx^2 + dy^2 + dz^2 > c^2dk_1^2 + c^2dk_2^2,$$

$$\frac{c^2dk_1^2}{dx^2+dy^2+dz^2} + \frac{c^2dk_2^2}{dx^2+dy^2+dz^2} < 1, \quad \text{i.e.,}$$

$$\frac{1}{q_1^2} + \frac{1}{q_2^2} < 1. \tag{3.17}$$

(Here $dx^2 + dy^2 + dz^2 \neq 0$.) The inequality (3.17) is satisfied if and only if the following inequalities are jointly satisfied:

$$q_1 > 1 \text{ and } q_2 > 1, \text{ (i.e., } V_1 > c \text{ and } V_2 > c).$$

<u>*Second (lightlike) case*</u>: $(\boldsymbol{A})^2 = 0$, i.e., $ds^2 = 0$. In this situation, $(\boldsymbol{A}_{3,1})^2 = c^2dk_2{}^2$ and it is therefore not possible that $(\boldsymbol{A}_{3,1})^2 < 0$. If $\boldsymbol{A}^5 = cdk_2 = 0$, then $(\boldsymbol{A}_{3,1})^2 = 0$. If $\boldsymbol{A}^5 = cdk_2 \neq 0$, then we have: $(\boldsymbol{A}_{3,1})^2 > 0$. The condition $(\boldsymbol{A})^2 = 0$ is equivalent to the following equality:

$$\frac{1}{q_1^2} + \frac{1}{q_2^2} = 1. \tag{3.18}$$

If $q_1 < 1$ (i.e., $V_1 < c$) or $q_2 < 1$ (i.e., $V_2 < c$), then equality (3.18) is violated. As $q_1 \to 1$ (i.e., $V_1 \to c$), then equality (3.18) is satisfied provided that $q_2 \to \infty$ (i.e., $V_2 \to \infty$).

This means that $dx^2 + dy^2 + dz^2 > 0$, $dk_1^2 > 0$, and $dk_2^2 = 0$, i.e., the particle moves in space and in kime dimension k_1, but its location in k_2 remains unchanged.

The same argument applies for the case $q_2 = 1$ (i.e., $V_2 = c$). Equation (3.18) yields the equality $\frac{c^2}{V_1^2} + \frac{c^2}{V_2^2} = 1$. Therefore, if the particle is moving with velocity $V_1 = \frac{c}{\sqrt{1 - \frac{c^2}{V_2^2}}}$, then

$(\boldsymbol{A})^2 = ds^2 = 0$. As we showed in **Section 3.7.8**, when the velocities V_1, V_2 of a particle in the frame of reference K satisfy $\frac{c^2}{V_1^2} + \frac{c^2}{V_2^2} = 1$, then for the corresponding velocities V_1', V_2' in the second reference frame K' will also satisfy this equation, $\frac{c^2}{V_1'^2} + \frac{c^2}{V_2'^2} = 1$.

Third (kimelike) case: $(\boldsymbol{A})^2 < 0$, i.e., $ds^2 < 0$. If $A^5 = cdk_2 \neq 0$, then in this case there are three possible situations: $(\boldsymbol{A}_{3,1})^2 < 0$, $(\boldsymbol{A}_{3,1})^2 = 0$, $(\boldsymbol{A}_{3,1})^2 > 0$. If $A^5 = cdk_2 = 0$, then $(\boldsymbol{A}_{3,1})^2 < 0$. Let us assume that in the frame of reference K, $A^5 = cdk_2 \neq 0$ and $(\boldsymbol{A}_{3,1})^2 > 0$. Then, it is clear that in the K' frame, which moves uniformly and rectilinearly in relation to K, we have $(\boldsymbol{A}')^2 = (\boldsymbol{A}'_{3,1})^2 - (A^{5'})^2 = (\boldsymbol{A})^2 < 0$.

If we assume that $A^{5'} = cdk_2' = 0$, then we have $(\boldsymbol{A}'_{3,1})^2 < 0$. Hence, we see that in the frame K, the value $\boldsymbol{A}_{3,1}$ is spacelike, $(\boldsymbol{A}_{3,1})^2 > 0$, and in frame K', the value $\boldsymbol{A}'_{3,1}$ is timelike, $(\boldsymbol{A}'_{3,1})^2 < 0$. Similar to case one above, we can show that the condition $(\boldsymbol{A})^2 < 0$ is equivalent to following inequality:

$$\frac{1}{q_1^2} + \frac{1}{q_2^2} > 1. \tag{3.19}$$

If $0 < q_1 \leq 1$ (i.e., $0 < V_1 \leq c$), then inequality (3.19) is satisfied for all real values of $q_2 \in \mathbb{R}$ (i.e., for all values of V_2). Recall that when referring to the velocities in K' relative to the first frame, K, we use lower-case letters (e.g., v_1, v_2) and when referring to particle velocities within K, we use capital-letters (e.g., V_1, V_2).

Similar arguments hold for the velocity V_2 in terms of q_2. Therefore, if the absolute value of the velocity of a particle, defined relative to one kime dimension, is less than or equal to the speed of light in vacuum, then the velocity of this particle, defined relative to the other kime dimension can be arbitrary large without violating the causality principle.

When $q_1 > 1$ and $q_2 > 1$, inequality (3.19) is satisfied for an appropriate choice of the parameters q_1 and q_2 (e.g., $q_1 = \frac{10}{9}$, $q_2 = 2$). It is clear that the kimelike condition $(\boldsymbol{A}_{3,1})^2 < 0$ is equivalent to the inequality $V_1 < c$, the lightlike condition $(\boldsymbol{A}_{3,1})^2 = 0$ is equivalent to the equality $V_1 = c$, and the spacelike condition $(\boldsymbol{A}_{3,1})^2 > 0$ is equivalent to the inequality $V_1 > c$.

This leads to the conclusion that the causal region of the 5D spacekime is larger and properly contains the 4D spacetime causal region.

Figure 3.13 shows the graph of the implicit function $\frac{c^2}{V_1^2} + \frac{c^2}{V_2^2} = 1$ only for non-negative values of V_1 and V_2.

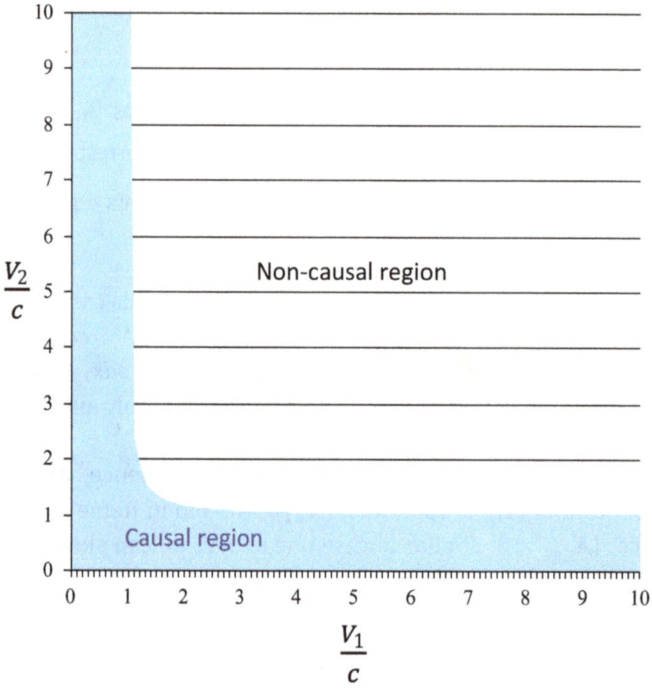

Figure 3.13: The values of the velocities V_1 and V_2 where the particle moves in the causal region of the spacekime (x, y, z, k_1, k_2).

Therefore, in 5D spacekime particles can move in the causal region with velocities that are greater than, less than, or equal to the speed of light in vacuum.

Let's assume a kevent E occurs at a point $O = (x, y = 0, z = 0, k_1 = 0, k_2 = 0)$ and an infinitesimal kime interval $dk > 0$ has passed since the kevent *E occurred*. Since the kime is two-dimensional, all possible combinations of coordinates ck_1, ck_2, for which the inequality $(ck_1)^2 + (ck_2)^2 \le c^2dk^2$ is valid, form a circle in the plane $ck_1 - ck_2$ with center the point O and radius equal to cdk. Let us add the spatial dimensions x, y, z in a way that at arbitrary values of x, y, z, the inequality $x^2 + y^2 + z^2 \le c^2dk^2$ is satisfied. The *causal region* includes all points (x, y, z, k_1, k_2), where one of these three conditions is satisfied:

$$\left\{x^2+y^2+z^2-(ck_1)^2-(ck_2)^2=0 \text{ and } x^2+y^2+z^2\le c^2dk^2\right\}, \text{or}$$

$$\left\{(ck_1)^2+(ck_2)^2=c^2dk^2 \text{ and } x^2+y^2+z^2\le c^2dk^2\right\}, \text{or}$$

$$\left\{x^2+y^2+z^2-(ck_1)^2-(ck_2)^2<0 \text{ and } x^2+y^2+z^2<c^2dk^2\right\}.$$

For simplicity, let's consider only one spatial dimension x and the two kime dimensions k_1 and k_2. The causal region connected to the event E includes the lateral surface of the right circular cone $(ck_1)^2+(ck_2)^2=\mathbf{x}^2\le c^2dk^2$, the lateral surface of the right circular cylinder $(ck_1)^2+(ck_2)^2=c^2dk^2$, $\mathbf{x}^2\le c^2dk^2$ and the inner region bounded by these surfaces $\{\mathbf{x}^2-(ck_1)^2-(ck_2)^2<0 \text{ and } \mathbf{x}^2<\mathrm{c}^2\mathrm{dk}^2\}$, **Figure 3.14**.

One can say that two-dimensional kime "flows" from the origin O in all directions in the kime plane (ck_1, ck_2) and yields the outer simultaneity circle shown in **Figure 3.14(b)**.

Let us look at three time moments $t_0<t_1<t_2$. In **Figure 3.15**, the present "is moving" from moment t_0 to t_1 and then towards t_2. At time t_0, t_0 is the present and t_1 and t_2 represent future moments; at time t_1, t_1 becomes the present, t_0 the past, and t_2 the future; and at time t_2, t_2 is the present with t_0 and t_1 both representing past moments.

Figure 3.15 resembles Barry Dainton's idea of the meta-time [172], which was proposed as a solution to the so-called overdetermination problem. This problem arises if we assume that the properties of the "running" time are not relational but rather inherent in the events which possess them. For example, the property of "presentness" may be inherent in a particular object at a fixed moment of time as well as a transient property, disappearing with the moment of "now" flying by, which leads to the paradoxical situation of overdetermination. If it is quite normal for an object to have two contrary properties at different times, it seems impossible for a given fixed time to combine (or the event at this moment of time to possess) contrary properties. A possible solution of overdetermination is the existence of an additional time dimension, the so-called meta-time, or kime-phase.

Dainton [172] identified two major objections against the hypothesis of meta-time. The first one is related to the fact that the concept of meta-time complicates the picture of the world by introducing an unobservable entity, i.e., kime-phase. The second objection comes down to the argument that the proposed two-dimensional model of time aims to save the dynamic aspect of time (the ever changing, moving present), but in fact the introduced two-dimensional model is entirely static. If meta-time is also flying by and the moment of "now" in it is moveable, then in order to rationalize the flow of meta-time requires introduction of a higher-order meta-meta-time. Clearly, this process leads into a continuously increasing time-dimensionality.

The concept of kime actually solves the overdetermination problem, circumventing the specified counter-arguments. **Figure 3.15** shows how one (ordinary) event time can have different inherent kime properties. The introduction of kime-phase

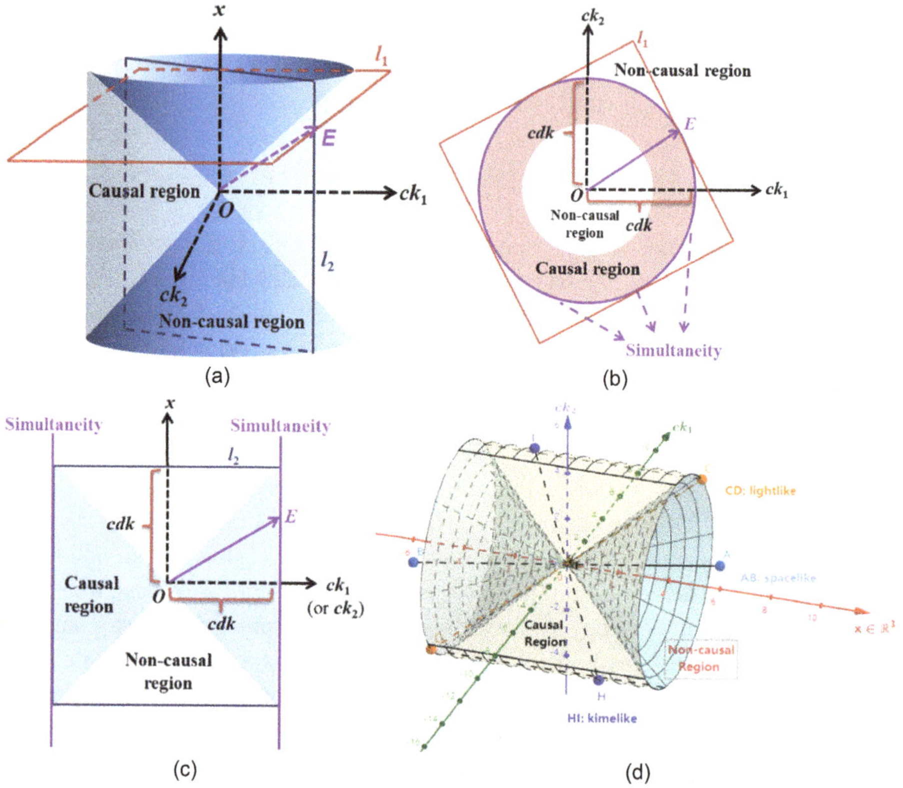

Figure 3.14: (a) Causal structure of a reduced spacekime manifold (1D space and 2D kime dimensions): (b) Causal region in the plane l_1; (c) causal region in the plane l_2; plane l_1 is parallel to the plane $c\kappa_1 - c\kappa_2$ and plane l_2 is perpendicular to the plane $c\kappa_1 - c\kappa_2$; and (d) shows the space-like, kime-like and light-like regions in spacekime, where 3D space is compressed into a 1D horizontal axis.

makes it possible for each single moment of which we are aware to be called "present," with the successive change of these moments, outlining the current present. The concept of kime is not an unnecessary complication of the world, and is connected with the explanation of experimentally observable things. The objection that the model of kime is also static and has to be constantly completed with additional measurements is valid only if it is (unjustifiably) assumed, that kime is again time and as such it also flies by. In fact, the model of kime is a useful instrument of clarifying the connection between the dynamic and static aspects of time.

In a universe with two-dimensional kime, other unexpected properties arise. Let us consider two non-relativistic observers, moving in different kime directions. In this case, the relativistic effects are neglected and observers can meet in spacekime and synchronize their clocks only if their directions of movements are crossing the same time foliation leaf (see **Chapter 5**). Then the observers may diverge again

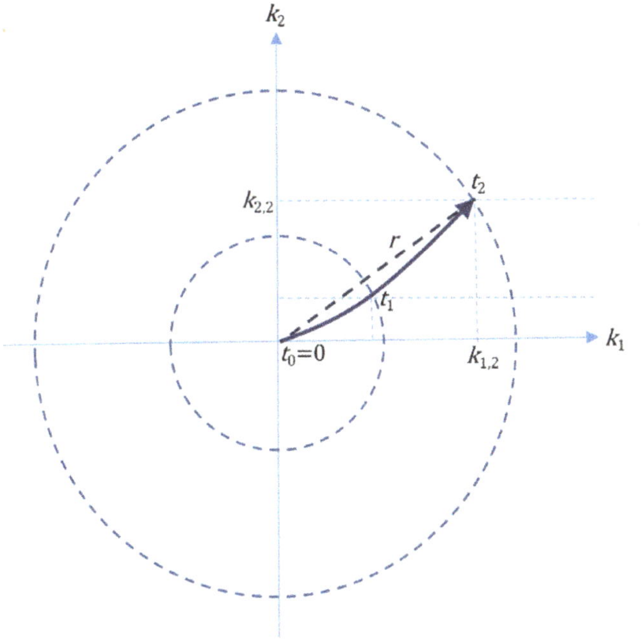

Figure 3.15: Pattern of "movement" ("running") of the present in kime. For instance, time is the radius of the outer circle and it equals $r = t_2 = \sqrt{\kappa_{1,2}^2 + \kappa_{2,2}^2}$.

and continue to traverse separate kime directions possibly without opportunities to converse and meet repeatedly [115].

When two observers are not moving in spacetime relative to one another, they may still move in different directions in the kime plane even if their time-clocks are synchronized. For instance, such pairs of observers may move in orthogonal or opposite kime-directions to each other, then they would naturally share the same kime foliation leaf and perceive each other as static with a common "nowness". However, when the observers move in spacetime relative to each other, then one observer may perceive the other as moving with unlimitedly large kime-velocity.

In that context, despite the theoretical possibility of time-travel in the spacekime manifold, time-travel is not feasible as it is as practical as matching a specific real number in a countable experiment of randomly choosing real numbers; the measure of the set of such successful experiments is trivial.

3.10 Kime Applications in Data Science

Now that we have established the mathematical foundations of spacekime, we can return to our interpretation of the 2D kime manifold as a Fourier dual. Effectively, we cannot directly observe kime in spacekime, however, we can measure quite accurately the event orders as kime-magnitudes (r) in k-space, which we call "*time*". To be able to reconstruct the 2D spatial structure of kime, we will use the same trick used by crystallographers when they resolve the structure of atomic particles by only observing the magnitudes of their X-ray diffraction pattern in k-space. This approach heavily relies on (1) prior information about the kime-phase directional orientation, which may be obtained from using similar datasets, provided by an oracle, or estimated via some analytical strategies, and (2) experimental reproducibility by repeated confirmations of the data analytic results using alternative (longitudinal datasets). **Figure 3.16** shows schematically the parallels between the experimental design space associated with planning an experiment and estimation (right) and the data science analytic space related with the scientific inference about a data-driven problem (left).

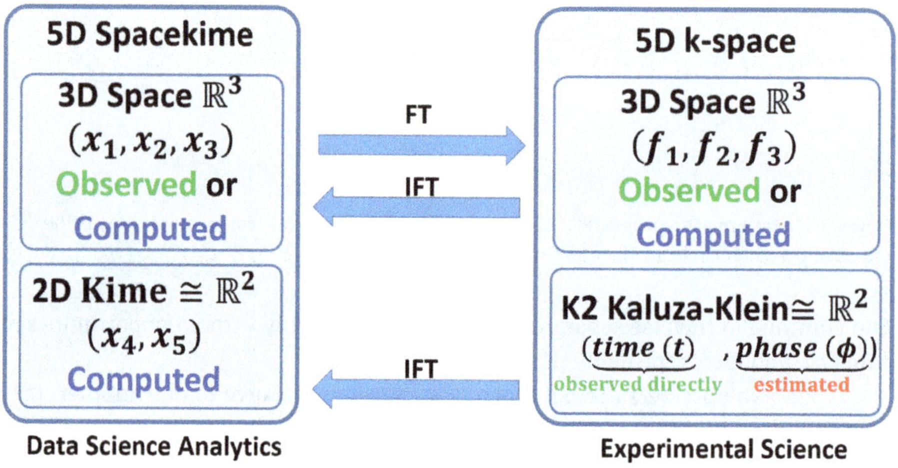

Figure 3.16: Diagrammatic depiction of the interplay between the data science analytic space and its experimental science counterpart.

Let's look at one specific example where the observed data can be symbolically represented as:

$$h(\boldsymbol{x}, \boldsymbol{\kappa}) = h\left(\underbrace{x, y}_{\text{space component}}, \underbrace{r = time, \varphi = phase}_{\text{kime component}}\right).$$

To keep this example realistic, let's assume the observed data is an fMRI time-series, which consists of measurements of hydrogen atom densities over a 2D lattice of spatial locations ($1 \le x, y \le 64$ pixels), about 3×3 millimeters^2 apart, recorded longitudinally over time ($1 \le t \le 180$) in increments of about 3 seconds, **Figure 3.17.** Albeit the general fMRI time-series are recorded on 3D spatial voxels, we will simplify the problem and consider just 2D spatial pixel intensities. This figure shows the temporal fMRI courses at some 2D pixels in the middle of the brain. These cylindrical tunnels represent the 2D spatial anatomical constraints of the fMRI intensities across time.

Assume only the three k-space frequency-time dimensions are observed $(x, y, r = t)$ and the time-direction (phase-angle, φ) is not measured or recorded. **Figure 3.17** illustrates the process of the 4D pseudo-spacekime reconstructing for this real dataset. Let's clarify again that for simplicity of the presentation and visualization, we are working in a lower dimensional 4D space, i.e., pseudo-spacekime, where there are only two space dimensions $x = (x, y)$, instead of the usual three spatial components $\mathbf{x} = (x, y, z)$ of the complete 5D spacekime. One can reduce the complexity of this experiment even further by only working with a single spatial component (x) and a time dimension, however, this may hide some of the intrinsic complexities of higher-dimensional multi-feature data.

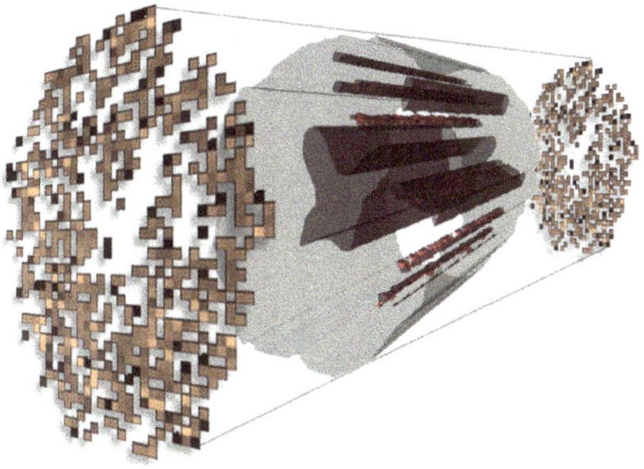

Figure 3.17: fMRI example: The 2D end-planes show the (thresholded) fMRI time-series at time = 1 and time = 180. In the middle, the perspective cylindrical view shows the temporal courses (intensities) at some pixels (or voxels). The colors illustrate the temporal intensities (fMRI blood oxygen-level dependent responses), which are constrained along cylindrical tunnels representing the 2D brain spatial anatomical structures.

Note the parallels and differences between the reconstructions of the data in spacekime using the correct *kime* = (*magnitude*, *phase*) and trivial phase-angle *kime* = (*magnitude*, 0) estimates, **Figure 3.18, panels C** and **D**. Indeed, working with the lower dimensional time, instead of kime, has the benefits of being more computationally efficient, but suffers from artifacts which may affect the final scientific inference based on the fMRI data reconstruction. The 5D spacekime data reconstructions that rely on correct kime-phase estimates provide more accurate representations of the underlying phenomena proxy-imaged by the fMRI data. However, the kime phase-angle directions may not always be directly available, accurately measured, or precisely estimated, which may lead to inaccurate result interpretation or complicated inference.

Many additional spacekime data-driven examples are provided in **Chapter 6** (Applications).

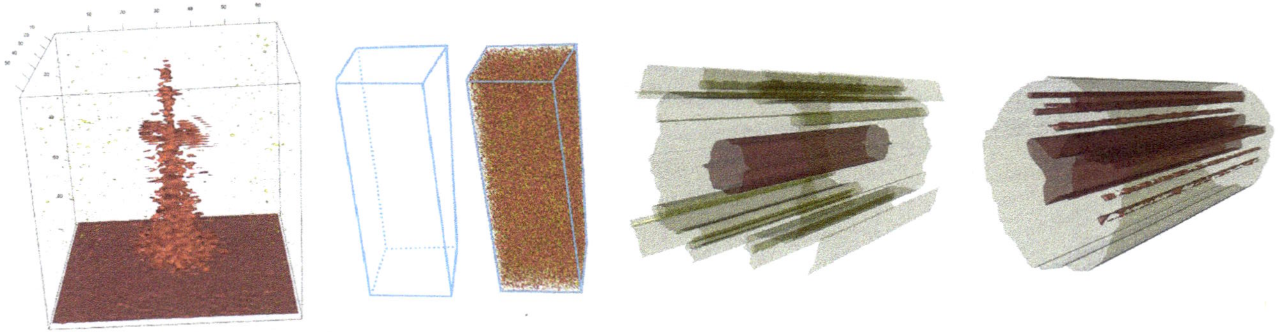

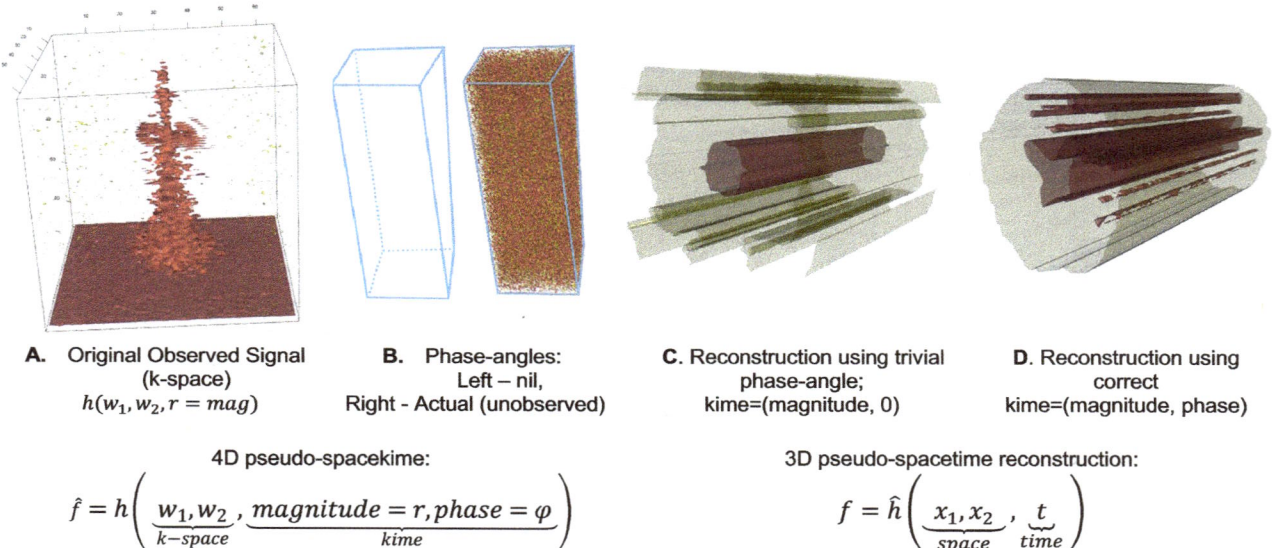

A. Original Observed Signal (k-space) $h(w_1, w_2, r = mag)$

B. Phase-angles: Left – nil, Right - Actual (unobserved)

C. Reconstruction using trivial phase-angle; kime=(magnitude, 0)

D. Reconstruction using correct kime=(magnitude, phase)

4D pseudo-spacekime:

$$\hat{f} = h\left(\underbrace{w_1, w_2}_{k-space}, \underbrace{magnitude = r, phase = \varphi}_{kime}\right)$$

3D pseudo-spacetime reconstruction:

$$f = \hat{h}\left(\underbrace{x_1, x_2}_{space}, \underbrace{t}_{time}\right)$$

Figure 3.18: An exemplary workflow illustrating the (pseudo) spacekime reconstruction of the k-space observed data. **Panel A** depicts the observed signal in k-space and **Panel B** illustrates the actual phase-angles, which we pretend are unknown, and the trivial (nil) phase direction. Notice the resemblance of the 3D renderings of **Panel C** (pretending the kime direction is unknown, or trivial) and **Panel D** (perfect spacekime reconstruction using the correct magnitude (r) and phase (φ)). In the absence of kime-phase knowledge, there is more artificial symmetry in the spacekime reconstruction. These artifacts impact the interpretation of the data and lead to fewer local details in spacekime (**Panel C**). This is due to the fact that magnitudes (time order) alone cannot perfectly represent the complexity of the 2D kime manifold. To correctly interpret the data, one needs kevent knowledge of both – time-order and phase-direction (**Panel D**). On the flip side, computing only using magnitudes (time) is more efficient, tractable and leads to simpler inference, whereas computing using order and direction (kime) is more intense, requires special handling, and leads to more elaborate inferential results that may require post-processing to yield specific decision or action.

3.11 Kime Philosophy

Just like an ant that lives in a 2D plane, or more generally in a 2D manifold like a Möbius band surface, cannot see or interpret the z-direction ($z\perp(x,y)\in\mathbb{R}^3$), it's impossible to directly perceive the universe beyond the 4D spacetime $\mathbb{R}^3\times\mathbb{R}^+$. The reader may find the appendix of the Data Science and Predictive Analytics textbook [10] useful, as it provides many interactive visualizations of 2-manifolds and 3-manifolds, see https://dspa.predictive.space, accessed January 29, 2021. We can discern differences between events that occur in exactly the same spacetime but have different directions (kime-phases), e.g., multiple independent factual and fictional reports about the same kevent encounters. Despite the unsettling modern contradictions of "alternative facts", there are multiple historical accounts of seemingly contradictory facts [173, 174]. The basic reason for these recorded variations of perceived reality is due to multiple alternative points of view, which represent different kime direction states experienced by multiple observers during and after the actual kevents. Various phenomena can be compared whether they co-occur in time or not, however, contrasting such observations needs to contextualize kime and take into account the phase-direction. For example, paleontological specimens, archaeological artifacts, and genetic sequencing endure natural decay, memory amnesia, and mutation. As such observations are indirect, circumstantial, or subjective, their reliable reproduction and use as evidence need to be supported by independently acquired information [175].

The event-related kime-order (r) and kime-direction (φ) characterizations are indirectly observed and interpreted. Initially, this may be unsettling since it suggests there may be alternative interpretations of complex events (kime states). However, we can illustrate many instances in which, we, as humans, experience alternative kime directions that may be complementary to other personal or collective interpretations of kevents, characterized by $(\boldsymbol{x},r,\varphi)$, that may correspond to the same spacetime events $(\boldsymbol{x},t)$. For instance, all of these experiences – watching movies, reading books, listening to science talks, or learning a new concept – yield rather different outcomes, interpretations, responses, and actions from different people. In addition, brain maturation, cultural norms, balance between individual and collective benefits, alter our perceptions of events (specifically their kime-phase characteristics).

The location and momentum of a particle in a quantum mechanical system is described by a probability density function. The Heisenberg's principle provides a lower limit on the precise identification of the particle's spatial position, energy, motion direction, and state. Similarly, for spacekime kevents, i.e., events described in 2D kime terms where kevent likelihoods are delineated as densities over kime. The event-analogue of Heisenberg's principle is that kevent characteristics may never be jointly, precisely, and accurately known. In other words, the more precisely we localize the event-related order (r) the less certain we are about its kime-direction (φ). Conversely, the more certain we are about the kime direction, the less accurate we can estimate its kime order. In principle, life, and by extension humans, have more affinity towards the event-related order, and much less intuition about kime direction. More details are provided later in **Chapter 5**.

Previously, we showed that there are several complementary representations of kime that provide alternative strategies, each with its specific advantages and disadvantages. These re-parameterizations simply illustrate different coordinate descriptions of the same two kime degrees of freedom, see **Figure 3.7**.

Appendix: Dirac Equation in Spacekime

Quantum mechanics is based on the correspondence principle, which associates physical quantities (e.g., particle measurements) to linear Hermitian operators that act as functions on the Hilbert space of wave functions. The wave equation in quantum mechanics is derived from the classical equation of motion of a given object by replacing the physical quantities with the corresponding operators and then treating the resulting expression as a differential operator on the wavefunction of the object.

There are multiple approaches to the generalization of the Dirac equation from four to five dimensions [176–187]. An extra dimension naturally emerges on the Dirac theory if we take into account the fact that the ordinary Minkowski complex Clifford algebra can be obtained as a real algebra with an extra time-like dimension [188]. Similar to the approaches of Dirac [189] and Bars and colleagues [190–192], Redington and Lodhi [193] chose the extra dimension to be timelike. To obtain Dirac equation in five dimensions (5D), the authors replaced the gamma matrices in the conventional Dirac equation by their spacetime algebra equivalents for the case with two time-like dimensions and three space-like dimensions. This leads to the so-called "Hestenes form" formulation of the Dirac equation [194]. Its solutions are eight-component spinors, which are interpreted as single-particle fermion wave functions in 4D Minkowski spacetime. Use of a "cylinder condition" allows the removal of explicit dependence on the fifth coordinate. This reduces each eight-component solution to a pair of degenerate four-component spinors obeying the classical Dirac equation [193]. When the cylinder condition is not applied or is treated like an approximation the dependence on the fifth dimension is retained. In this case, the number of spinor degrees of freedom is doubled, suggesting the possible existence of a new quantum number. This implies new physics beyond the classical Dirac equation.

We will apply a different approach to the generalization of the Dirac equation in 5D spacekime where the fifth dimension is also time-like and the energy of the particle has two components corresponding to each of the two time-like kime dimensions. We will use the Clifford algebra $C\ell_{3,2}(\mathbb{R})$ [195, 196], which is formed by the five gamma matrices: $\{\gamma^1, \gamma^2, \gamma^3, \gamma^4, \gamma^5\}$. In conventional settings, the matrix γ^5 is connected with the property of chirality. In the spacekime representation, the matrix γ^5 is connected with the existence of the second kime dimension and the second component of the particle's energy, respectively. Below, we suggest an approach to derive the Dirac equation in 5D spacekime.

The relativistic relation between energy and momentum is expressed by the following formula:

$$E^2 - \|\mathbf{p}\|^2 c^2 = m_0^2 c^4, \tag{3.20}$$

where E is the *total energy* of the particle, $\mathbf{p} = (p_x, p_y, p_z)$ is the *momentum* vector, m_0 is the rest *mass* of the particle, and c is the constant speed of light.

In spacekime, the total energy of the particle has two components and is represented as an energy magnitude:

$$E = \sqrt{(E_1)^2 + (E_2)^2}, \tag{3.21}$$

where E_1 and E_2 are the two energy components of the particle defined with respect to kime dimensions k_1 and k_2, respectively. Here $E_l = c^2 m_0 \frac{dk_l}{dk_0}$, $l \in \{1, 2\}$, $dk_0 = \|dk_0\| = \sqrt{(dk_{0,1})^2 + (dk_{0,2})^2} \geq 0$ (See **Chapter 5, Section 5.7.**)

Combining equations (3.20) and (3.21), we will obtain the relations between the total energy and the momentum in spacekime:

$$(E_1)^2 + (E_2)^2 - \|\mathbf{p}\|^2 c^2 - m_0^2 c^4 = 0. \tag{3.22}$$

Then, the spacekime energy and momentum operators are expressed as:

$$\underbrace{\hat{p}_1 = -i\hbar\frac{\partial}{\partial x}, \hat{p}_2 = -i\hbar\frac{\partial}{\partial y}, \hat{p}_3 = -i\hbar\frac{\partial}{\partial z}}_{\text{Momentum}}, \underbrace{\hat{p}_4 = \frac{i\hbar}{c}\frac{\partial}{\partial k_1}, \hat{p}_5 = \frac{i\hbar}{c}\frac{\partial}{\partial k_2}}_{\text{Energy}}, \tag{3.23}$$

where h is the reduced Planck constant and

$$\hat{p}_1\psi = p_x\psi, \quad \hat{p}_2\psi = p_y\psi, \quad \hat{p}_3\psi = p_z\psi, \quad \hat{p}_4\psi = \frac{E_1}{c}\psi, \quad \hat{p}_5\psi = \frac{E_2}{c}\psi. \tag{3.24}$$

Then, equation (3.22) may be expressed in operator form:

$$\left(\underbrace{-\hat{p}_1^2 - \hat{p}_2^2 - \hat{p}_3^2}_{\text{space}} + \underbrace{\hat{p}_4^2 + \hat{p}_5^2}_{\text{kime}}\right)\psi = m_0^2 c^2 \psi.$$

Following Dirac's approach, we can express this in terms of the gamma matrices:

$$\left(-\hat{p}_1^2 - \hat{p}_2^2 - \hat{p}_3^2 + \hat{p}_4^2 + \hat{p}_5^2\right) =$$

$$\left(\gamma^1\hat{p}_1 + \gamma^2\hat{p}_2 + \gamma^3\hat{p}_3 + \gamma^4\hat{p}_4 + \gamma^5\hat{p}_5\right)\left(\gamma^1\hat{p}_1 + \gamma^2\hat{p}_2 + \gamma^3\hat{p}_3 + \gamma^4\hat{p}_4 + \gamma^5\hat{p}_5\right),$$

where

$$\begin{cases} (\gamma^\mu)^2 = -1, & \mu = 1,2,3 \\ (\gamma^\mu)^2 = 1, & \mu = 4,5 \\ \gamma^\mu\gamma^\rho + \gamma^\rho\gamma^\mu = 0, & \mu \neq \rho \end{cases}.$$

In short, using Einstein summation convention, these conditions can be written in the following way:

$$\left(\gamma^\mu \hat{p}_\mu - m_0 c\right)\psi = 0 \text{ or } \left(\gamma^\mu \hat{p}_\mu + m_0 c\right)\psi = 0, \tag{3.25}$$

where

$$\{\gamma^\mu, \gamma^\rho\} = \gamma^\mu\gamma^\rho + \gamma^\rho\gamma^\mu = 2\eta^{\mu\rho} I_4. \tag{3.26}$$

Here $\{,\}$ is the *anticommutator*, $I_4 = \text{diag}(1,1,1,1)$ is the 4×4 identity matrix, and $\eta^{\mu\rho}$ is the metric tensor with components:

$$\eta^{\mu\rho} = \eta_{\mu\rho} = \begin{cases} -1, & \mu = \rho = 1,2,3 \\ 1, & \mu = \rho = 4,5 \\ 0, & \mu \neq \rho. \end{cases}$$

The anticommutation relations can be encoded using gamma matrices in terms of the Weyl (chiral) basis:

$$\text{space}\begin{cases} \gamma^1 = \begin{pmatrix} 0 & 0 & 0 & 1 \\ 0 & 0 & 1 & 0 \\ 0 & -1 & 0 & 0 \\ -1 & 0 & 0 & 0 \end{pmatrix} = \begin{pmatrix} 0 & \sigma_1 \\ -\sigma_1 & 0 \end{pmatrix}, \\ \gamma^2 = \begin{pmatrix} 0 & 0 & 0 & -i \\ 0 & 0 & i & 0 \\ 0 & i & 0 & 0 \\ -i & 0 & 0 & 0 \end{pmatrix} = \begin{pmatrix} 0 & \sigma_2 \\ -\sigma_2 & 0 \end{pmatrix}, \\ \gamma^3 = \begin{pmatrix} 0 & 0 & 1 & 0 \\ 0 & 0 & 0 & -1 \\ -1 & 0 & 0 & 0 \\ 0 & 1 & 0 & 0 \end{pmatrix} = \begin{pmatrix} 0 & \sigma_3 \\ -\sigma_3 & 0 \end{pmatrix}, \end{cases}$$

$$\underbrace{\gamma^4 = \begin{pmatrix} 0 & 0 & 1 & 0 \\ 0 & 0 & 0 & 1 \\ 1 & 0 & 0 & 0 \\ 0 & 1 & 0 & 0 \end{pmatrix} = \begin{pmatrix} 0 & I_2 \\ I_2 & 0 \end{pmatrix}, \gamma^5 = \begin{pmatrix} -1 & 0 & 0 & 0 \\ 0 & -1 & 0 & 0 \\ 0 & 0 & 1 & 0 \\ 0 & 0 & 0 & 1 \end{pmatrix} = \begin{pmatrix} -I_2 & 0 \\ 0 & I_2 \end{pmatrix}.}_{\text{kime}}$$

The chiral Weyl basis does not represent a unique set of matrices and more general gamma (Dirac) matrices may also be used.

Since 4×4 matrices operate on the wavefunction, it has to be four-component:

$$\psi = \begin{pmatrix} \psi_1 \\ \psi_2 \\ \psi_3 \\ \psi_4 \end{pmatrix} = \begin{pmatrix} \psi_A \\ \psi_B \end{pmatrix}, \text{where } \psi_A = \begin{pmatrix} \psi_1 \\ \psi_2 \end{pmatrix}, \ \psi_B = \begin{pmatrix} \psi_3 \\ \psi_4 \end{pmatrix}.$$

Again, employing Einstein summation convention, the γ^μ matrices operate on the wavefunction components $\psi_1, \ \psi_2, \ \psi_3, \ \psi_4$ by:

$$\gamma^\mu \psi_\theta = (\gamma^\mu)_{\theta\nu} \psi^\nu, \ \mu, \theta, \ \nu = 1, \ 2, \ 3, \ 4.$$

On the one hand, according to formula (3.26), the matrices $\{\gamma^1, \gamma^2, \gamma^3, \gamma^4, \gamma^5\}$ form a Clifford algebra $C\ell_{3,2}(\mathbb{R})$ in *spacekime* with signature $(- - - + +)$. On the other hand, in *spacetime* with signature $(- - - +)$, the matrix $\gamma^5 = i \ \gamma^1\gamma^2\gamma^3\gamma^4$ reflects chirality, a quantum-mechanical property in the Clifford algebra $C\ell_{3,1}(\mathbb{R})$.

Since the number of dimensions of spacetime is even, $(3+1)$D, the spinor space splits into two independent 2D spaces; one of which contains 'right-handed' objects, and the other 'left-handed' ones.

The chirality for Dirac fermion is determined through the operator γ^5, whose eigenvalues are ± 1, i.e., $(\gamma^5)^2 = 1$. Thus, a Dirac field can be projected onto its left-handed and right-handed components by:

$$\psi_L = \frac{1 - \gamma^5}{2} \psi = \begin{pmatrix} I_2 & 0 \\ 0 & 0 \end{pmatrix} \psi, \qquad \psi_R = \frac{1 + \gamma^5}{2} \psi = \begin{pmatrix} 0 & 0 \\ 0 & I_2 \end{pmatrix} \psi.$$

In fact, ψ_L and ψ_R represent the eigenvectors of γ^5:

$$\gamma^5 \psi_L = \frac{\gamma^5 - (\gamma^5)^2}{2} \psi = -\psi_L,$$

$$\gamma^5 \psi_R = \frac{\gamma^5 + (\gamma^5)^2}{2} \psi = \psi_R.$$

Therefore, there may be a deeper connection between the chirality property in $(3+1)$D spacetime and $(3+2)$D spacekime. The matrix γ^5 may represent a connecting link between the two chirality properties. If we look only at $(3+1)$D spacetime, the chirality property is defined by the matrix γ^5. Examining $(3+2)$D spacekime, the chirality can't be directly defined because of the odd number of dimensions (5). However, in $(3+2)$D spacekime, the matrix γ^5 appears as a left-hand-side multiple of the operator $\hat{p}_5 = \frac{i\hbar}{c}\frac{\partial}{\partial k_2}$, corresponding to the second kime dimension k_2.

Now, let's look at the solutions of the Dirac equation in spacekime in the absence of an external electromagnetic field. For a given free particle, the Dirac equation in spacekime is represented in equation (3.25). By analogy with the well-known case in spacetime, we focus on the expression involving a negative sign in front of the rest mass:

$$\left(i\hbar\gamma^{\mu}\hat{p}_{\mu} - m_0 c\right)\psi = 0. \tag{3.27}$$

We can examine the basic solutions of the plane-wave equation:

$$\psi = u \times e^{\left[\frac{i}{\hbar}\left(-E_1 k_1 - E_2 k_2 + \mathbf{p}.\boldsymbol{x}\right)\right]}, \tag{3.28}$$

where $\boldsymbol{x} = (x, y, z)$ and u is the constant amplitude that can be expressed as a column matrix:

$$u = \begin{pmatrix} u_A \\ u_B \end{pmatrix}, \qquad u_A = \begin{pmatrix} u_1 \\ u_2 \end{pmatrix}, \qquad u_B = \begin{pmatrix} u_3 \\ u_4 \end{pmatrix}, \tag{3.29}$$

where u_1, u_2, u_3, u_4 are constants. From equations (3.27), (3.28) and (3.29), we obtain:

$$\left[\begin{pmatrix} 0 & I_2 \\ I_2 & 0 \end{pmatrix} i\hbar\frac{\partial}{\partial k_1} + \begin{pmatrix} -I_2 & 0 \\ 0 & I_2 \end{pmatrix} i\hbar\frac{\partial}{\partial k_2} - c\begin{pmatrix} 0 & \vec{\sigma} \\ -\vec{\sigma} & 0 \end{pmatrix}\hat{\mathbf{p}} - \begin{pmatrix} I_2 & 0 \\ 0 & I_2 \end{pmatrix} m_0 c^2\right]_{4\times 4} \times \tag{3.30}$$

$$\begin{pmatrix} u_A \\ u_B \end{pmatrix} e^{\left[\frac{i}{\hbar}(-E_1 k_1 - E_2 k_2 + p.x)\right]} = \mathbf{0},$$

where $\vec{\sigma} \equiv (\sigma_1, \sigma_2, \sigma_3)$, $\hat{\mathbf{p}} \equiv (\hat{p}_1, \hat{p}_2, \hat{p}_3)$, and $I_2 = \text{diag}(1,1)$ is the 2×2 identity matrix.

Recall that the component corresponding, to the momentum operator $(\hat{\mathbf{p}})$ in equation (3.30) can be expanded to:

$$\begin{pmatrix} 0 & \vec{\sigma} \\ -\vec{\sigma} & 0 \end{pmatrix}\hat{\mathbf{p}} \equiv \sum_{\vartheta=1}^{3} \underbrace{\begin{pmatrix} 0 & \sigma_{\vartheta} \\ -\sigma_{\vartheta} & 0 \end{pmatrix}}_{4\times 4} \hat{p}_{\vartheta} =$$

$$\underbrace{\begin{pmatrix} 0 & 0 & 0 & \hat{p}_1 \\ 0 & 0 & \hat{p}_1 & 0 \\ 0 & -\hat{p}_1 & 0 & 0 \\ -\hat{p}_1 & 0 & 0 & 0 \end{pmatrix}}_{\hat{p}_1\gamma^1} + \underbrace{\begin{pmatrix} 0 & 0 & 0 & -i\hat{p}_2 \\ 0 & 0 & i\hat{p}_2 & 0 \\ 0 & i\hat{p}_2 & 0 & 0 \\ -i\hat{p}_2 & 0 & 0 & 0 \end{pmatrix}}_{\hat{p}_2\gamma^2} + \underbrace{\begin{pmatrix} 0 & 0 & \hat{p}_3 & 0 \\ 0 & 0 & 0 & -\hat{p}_3 \\ -\hat{p}_3 & 0 & 0 & 0 \\ 0 & \hat{p}_3 & 0 & 0 \end{pmatrix}}_{\hat{p}_3\gamma^3}.$$

Differentiating equation (3.30) with respect to the space and kime coordinates is equivalent to multiplying by the wavefunction and applying equations (3.24):

$$\left[\begin{pmatrix} 0 & I_2 \\ I_2 & 0 \end{pmatrix} E_1 + \begin{pmatrix} -I_2 & 0 \\ 0 & I_2 \end{pmatrix} E_2 - c\begin{pmatrix} 0 & \vec{\sigma} \\ -\vec{\sigma} & 0 \end{pmatrix}\mathbf{p} - \begin{pmatrix} I_2 & 0 \\ 0 & I_2 \end{pmatrix} m_0c^2\right]\begin{pmatrix} u_A \\ u_B \end{pmatrix} = 0.$$

Here,

$$\vec{\sigma}\mathbf{p} = \sum_{\vartheta=1}^{3} \sigma_\vartheta\,\hat{p}_\vartheta = \begin{pmatrix} 0 & \hat{p}_1 \\ \hat{p}_1 & 0 \end{pmatrix} + \begin{pmatrix} 0 & -i\hat{p}_2 \\ i\hat{p}_2 & 0 \end{pmatrix} + \begin{pmatrix} \hat{p}_3 & 0 \\ 0 & -\hat{p}_3 \end{pmatrix} = \begin{pmatrix} \hat{p}_3 & \hat{p}_1 - i\hat{p}_2 \\ \hat{p}_1 + i\hat{p}_2 & -\hat{p}_3 \end{pmatrix}.$$

Thus, we can express this in a closed-form as:

$$\underbrace{\begin{pmatrix} -(E_2 + m_0c^2)I_2 & E_1I_2 - c\vec{\sigma}\mathbf{p} \\ E_1I_2 + c\vec{\sigma}\mathbf{p} & (E_2 - m_0c^2)I_2 \end{pmatrix}}_{B}\begin{pmatrix} u_A \\ u_B \end{pmatrix} = B_{4\times4}\underbrace{\begin{pmatrix} u_A \\ u_B \end{pmatrix}}_{4\times1} = 0.$$

In the above equation, the 4×4 matrix B is written in terms of four 2×2 sub-matrices. This matrix equation can also be expressed as a pair of linear homogeneous equations in terms of the unknown two-dimensional constants u_A and u_B:

$$-(E_2 + m_0c^2)I_2u_A + (E_1I_2 - c\vec{\sigma}\mathbf{p})u_B = \mathbf{0}, \tag{3.31}$$

$$(E_1I_2 + c\vec{\sigma}\mathbf{p})u_A + (E_2 - m_0c^2)I_2u_B = \mathbf{0}. \tag{3.32}$$

A necessary and sufficient condition for nonzero solutions of this system of equations is the determinant in front of the unknown vectors u_A and u_B to be trivial. Then,

$$-(E_2 + m_0c^2)(E_2 - m_0c^2)I_2 - (E_1I_2 - c\vec{\sigma}\mathbf{p})(E_1I_2 + c\vec{\sigma}\mathbf{p}) = \mathbf{0}. \tag{3.33}$$

The properties of the Pauli matrices suggest that $\vec{\sigma}\mathbf{p}\vec{\sigma}\mathbf{p} = \mathbf{p}^2I_2$ and we can obtain the following connection: $(E_1)^2 + (E_2)^2 - \mathbf{p}^2c^2 - m_0^2c^4 = 0$ indicating that E_1 and E_2 can either be positive or negative values.

Equations (3.31), (3.32), and (3.33) show that u_A and u_B are not independent, since given a solution for u_A, we can calculate u_B, and vice-versa. Besides, comparing equations (3.31) and (3.32), we can see that u_A and u_B are symmetrical except for the reversals of the signs of the Pauli spin matrices and of the energy E_2. Thus, a particle described by the Dirac equation in spacekime has just two possible intrinsic states relative to a given basis, corresponding to the defined spin states and to the positive or negative energy values of E_2.

Let's first explore the simple solutions of equation (3.30) when $\mathbf{p} = 0$, i.e., when the particle is at rest:

$$\left[\begin{pmatrix} 0 & I_2 \\ I_2 & 0 \end{pmatrix} E_{0,1} + \begin{pmatrix} -I_2 & 0 \\ 0 & I_2 \end{pmatrix} E_{0,2}\right]\begin{pmatrix} u_A \\ u_B \end{pmatrix} = \begin{pmatrix} I_2 & 0 \\ 0 & I_2 \end{pmatrix} m_0 c^2 \begin{pmatrix} u_A \\ u_B \end{pmatrix}. \tag{3.34}$$

The energy of the particle at rest is expressed by the following formula:

$$(E_0)^2 = (E_{0,1})^2 + (E_{0,2})^2 = m_0^2 c^4,$$

where $E_{0,l} = c^2 m_0 \frac{dk_{0,l}}{dk_0}, l \in \{1,2\}$, see **Chapter 5, Section 5.7.**

Similarly to the approach in **Chapter 3, Section 3.7.6**, applying the appropriate kime hyperplane rotations of the kime axes $k_{0,1}, k_{0,2}$, we can identify the following cases:

$$dk_{0,1} = 0,\ dk_{0,2} > 0, \qquad E_{0,1} = 0,\ |E_{0,2}| > 0,$$

or

$$dk_{0,1} > 0,\ dk_{0,2} = 0, \qquad |E_{0,1}| > 0,\ E_{0,2} = 0,$$

or

$$dk_{0,1} = dk_{0,2} = \frac{dk_0}{\sqrt{2}} > 0, \qquad |E_{0,1}| = |E_{0,2}| = \frac{m_0 c^2}{\sqrt{2}}.$$

Let's explore each of these cases individually.

1) Let's assume that $E_{0,1} = 0$, then $E_{0,2} = \pm m_0 c^2$. Positive solutions $E_{0,2} = +m_0 c^2$ correspond to ordinary particles with a positive rest mass. From equation (3.34) we get:

$$m_0 c^2 \underbrace{\begin{pmatrix} -1 & 0 & 0 & 0 \\ 0 & -1 & 0 & 0 \\ 0 & 0 & 1 & 0 \\ 0 & 0 & 0 & 1 \end{pmatrix}\begin{pmatrix} u_1 \\ u_2 \\ u_3 \\ u_4 \end{pmatrix}}_{\gamma^5} = m_0 c^2 \underbrace{\begin{pmatrix} 1 & 0 & 0 & 0 \\ 0 & 1 & 0 & 0 \\ 0 & 0 & 1 & 0 \\ 0 & 0 & 0 & 1 \end{pmatrix}\begin{pmatrix} u_1 \\ u_2 \\ u_3 \\ u_4 \end{pmatrix}}_{I_{4\times 4}}.$$

This yields:

$$u_1 = 0,\ u_2 = 0,\ u = \begin{pmatrix} 0 \\ 0 \\ u_3 \\ u_4 \end{pmatrix} = \begin{pmatrix} 0 \\ u_B \end{pmatrix}.$$

Analogously, we can also consider the negative solutions corresponding to an antiparticle with negative rest mass: $E_{0,2} = -m_0c^2$:

$$u_3 = 0,\ u_4 = 0,\ u = \begin{pmatrix} u_1 \\ u_2 \\ 0 \\ 0 \end{pmatrix} = \begin{pmatrix} u_A \\ 0 \end{pmatrix}.$$

2) Next we consider $E_{0,2} = 0$ and $E_{0,1} = \pm m_0c^2$. First, the positive solutions for ordinary particles yield $E_{0,1} = +m_0c^2$. From equation (3.34) we get:

$$\underbrace{m_0c^2 \begin{pmatrix} 0 & 0 & 1 & 0 \\ 0 & 0 & 0 & 1 \\ 1 & 0 & 0 & 0 \\ 0 & 1 & 0 & 0 \end{pmatrix} \begin{pmatrix} u_1 \\ u_2 \\ u_3 \\ u_4 \end{pmatrix}}_{\gamma^4} = \underbrace{m_0c^2 \begin{pmatrix} 1 & 0 & 0 & 0 \\ 0 & 1 & 0 & 0 \\ 0 & 0 & 1 & 0 \\ 0 & 0 & 0 & 1 \end{pmatrix} \begin{pmatrix} u_1 \\ u_2 \\ u_3 \\ u_4 \end{pmatrix}}_{I_{4\times 4}},$$

i.e.,

$$\begin{pmatrix} m_0c^2u_3 \\ m_0c^2u_4 \\ m_0c^2u_1 \\ m_0c^2u_2 \end{pmatrix} = \begin{pmatrix} m_0c^2u_1 \\ m_0c^2u_2 \\ m_0c^2u_3 \\ m_0c^2u_4 \end{pmatrix}.$$

Therefore, in this case we obtain:

$$u_1 = u_3,\quad u_2 = u_4,\quad u = \begin{pmatrix} u_3 \\ u_4 \\ u_3 \\ u_4 \end{pmatrix}.$$

Second, the negative solutions similarly correspond to an antiparticle with negative rest mass: $E_{0,2} = -m_0c^2$:

$$u_3 = -u_1,\ u_4 = -u_2,\ u = \begin{pmatrix} u_1 \\ u_2 \\ -u_1 \\ -u_2 \end{pmatrix}.$$

Assuming $|E_{0,1}| = |E_{0,2}| = \frac{m_0c^2}{\sqrt{2}}$, there are four unique possibilities:

Case 1: $E_{0,1} > 0,\ E_{0,2} > 0$, where equation (3.34) yields:

$$\frac{m_0c^2}{\sqrt{2}}\underbrace{\begin{pmatrix} -1 & 0 & 1 & 0 \\ 0 & -1 & 0 & 1 \\ 1 & 0 & 1 & 0 \\ 0 & 1 & 0 & 1 \end{pmatrix}}_{\gamma^4+\gamma^5}\begin{pmatrix} u_1 \\ u_2 \\ u_3 \\ u_4 \end{pmatrix} = m_0c^2\underbrace{\begin{pmatrix} 1 & 0 & 0 & 0 \\ 0 & 1 & 0 & 0 \\ 0 & 0 & 1 & 0 \\ 0 & 0 & 0 & 1 \end{pmatrix}}_{I_{4\times 4}}\begin{pmatrix} u_1 \\ u_2 \\ u_3 \\ u_4 \end{pmatrix}.$$

In this case, $u_1 = u_3\left(\sqrt{2}-1\right),\ u_2 = u_4\left(\sqrt{2}-1\right)$.

Case 2: $E_{0,1} > 0,\ E_{0,2} < 0$, where equation (3.34) is reduced to:

$$\frac{m_0c^2}{\sqrt{2}}\underbrace{\begin{pmatrix} 1 & 0 & 1 & 0 \\ 0 & 1 & 0 & 1 \\ 1 & 0 & -1 & 0 \\ 0 & 1 & 0 & -1 \end{pmatrix}}_{\gamma^4-\gamma^5}\begin{pmatrix} u_1 \\ u_2 \\ u_3 \\ u_4 \end{pmatrix} = m_0c^2\underbrace{\begin{pmatrix} 1 & 0 & 0 & 0 \\ 0 & 1 & 0 & 0 \\ 0 & 0 & 1 & 0 \\ 0 & 0 & 0 & 1 \end{pmatrix}}_{I_{4\times 4}}\begin{pmatrix} u_1 \\ u_2 \\ u_3 \\ u_4 \end{pmatrix}.$$

Therefore, $u_3 = u_1\left(\sqrt{2}-1\right),\ u_4 = u_2\left(\sqrt{2}-1\right)$.

Case 3: $E_{0,1} < 0,\ E_{0,2} > 0$, where equation (3.34) simplifies to:

$$\frac{m_0c^2}{\sqrt{2}}\underbrace{\begin{pmatrix} -1 & 0 & -1 & 0 \\ 0 & -1 & 0 & -1 \\ -1 & 0 & 1 & 0 \\ 0 & -1 & 0 & 1 \end{pmatrix}}_{-\gamma^4+\gamma^5}\begin{pmatrix} u_1 \\ u_2 \\ u_3 \\ u_4 \end{pmatrix} = m_0c^2\underbrace{\begin{pmatrix} 1 & 0 & 0 & 0 \\ 0 & 1 & 0 & 0 \\ 0 & 0 & 1 & 0 \\ 0 & 0 & 0 & 1 \end{pmatrix}}_{I_{4\times 4}}\begin{pmatrix} u_1 \\ u_2 \\ u_3 \\ u_4 \end{pmatrix}.$$

Hence, $u_3 = -u_1\left(1+\sqrt{2}\right),\ u_4 = -u_2\left(1+\sqrt{2}\right)$.

Case 4: $E_{0,1} < 0,\ E_{0,2} < 0$, and equation (3.34) becomes:

$$\frac{m_0c^2}{\sqrt{2}}\underbrace{\begin{pmatrix} 1 & 0 & -1 & 0 \\ 0 & 1 & 0 & -1 \\ -1 & 0 & -1 & 0 \\ 0 & -1 & 0 & -1 \end{pmatrix}}_{-\gamma^4-\gamma^5}\begin{pmatrix} u_1 \\ u_2 \\ u_3 \\ u_4 \end{pmatrix} = m_0c^2\underbrace{\begin{pmatrix} 1 & 0 & 0 & 0 \\ 0 & 1 & 0 & 0 \\ 0 & 0 & 1 & 0 \\ 0 & 0 & 0 & 1 \end{pmatrix}}_{I_{4\times 4}}\begin{pmatrix} u_1 \\ u_2 \\ u_3 \\ u_4 \end{pmatrix}.$$

And therefore, $u_3 = -u_1\left(\sqrt{2}-1\right),\ u_4 = -u_2\left(\sqrt{2}-1\right)$.

By analogy with the case of $\mathbf{p} = 0$ and $E_{0,1} = 0$, we can identify u_3 and u_4 as solutions for particles with positive rest mass ($m_0 > 0$), and u_1 and u_2 as solutions for antiparticles with negative rest mass ($-m_0 < 0$). This follows analogously to the classical case of the Dirac equation in spacetime.

In the most general case, to obtain the solutions of the free Dirac equation for particles in spacekime ($m_0 > 0$), we can we express u_A in terms of u_B in the first equation (3.35):

$$u_A = \frac{1}{E_2 + m_0 c^2} \underbrace{(E_1 - c\vec{\sigma}\mathbf{p})u_B}_{2\times 1}.$$

Therefore,

$$\psi = \underbrace{\begin{pmatrix} \frac{1}{E_2 + m_0 c^2}(E_1 - c\vec{\sigma}\mathbf{p})u_B \\ u_B \end{pmatrix}}_{4\times 1} e^{\left[\frac{i}{\hbar}(-E_1 k_1 - E_2 k_2 + \mathbf{p}.\mathbf{x})\right]}. \tag{3.35}$$

Note that, $\mathbf{p} = 0$ and $E_{0,1} = 0$ correspond to the solutions we found for a particle at rest for $E_{0,2} > 0$. Here, u_B is a constant two-component vector characterizing the amplitudes of the two possible projections of the spin $\frac{\hbar}{2}$ and $-\frac{\hbar}{2}$ along the z axis. It can be expressed in terms of the eigenfunctions of the operator of the third projection of the spin:

$$u_B = \begin{pmatrix} u_3 \\ u_4 \end{pmatrix} = \begin{pmatrix} 1 \\ 0 \end{pmatrix} u_3 + \begin{pmatrix} 0 \\ 1 \end{pmatrix} u_4.$$

Therefore, we have two linearly independent solutions corresponding to the two projections of the spin along the axis z, where u_3 and u_4 are the amplitudes of the states with a different projection of the spin, respectively.

In the general case of antiparticles ($-m_0 < 0$), the solutions of the free Dirac equation in spacekime is derived from the second equation (3.32):

$$u_B = -\frac{1}{E_2 - m_0 c^2} \underbrace{(E_1 + c\vec{\sigma}\mathbf{p})u_A}_{2\times 1}.$$

In other words:

$$\psi = \underbrace{\begin{pmatrix} u_A \\ -\frac{1}{E_2 - m_0 c^2}(E_1 + c\vec{\sigma}\mathbf{p})u_A \end{pmatrix}}_{4\times 1} e^{\left[\frac{i}{\hbar}(-E_1 k_1 - E_2 k_2 + \mathbf{p}.\mathbf{x})\right]}. \tag{3.36}$$

Again, u_A is a constant two-component vector characterizing the amplitudes of the two possible projections of the spin angular momentum $\frac{\hbar}{2}$ and $-\frac{\hbar}{2}$ along the z axis. It can be expressed in terms of the eigenfunctions of the operator of the third projection of the spin:

$$u_A = \begin{pmatrix} u_1 \\ u_2 \end{pmatrix} = \begin{pmatrix} 1 \\ 0 \end{pmatrix} u_1 + \begin{pmatrix} 0 \\ 1 \end{pmatrix} u_2.$$

Thus, we have two linearly independent solutions that correspond to the two projections of the spin.

We define antiparticle states by just flipping the signs of E_1, E_2, and $\mathbf{p}$ following the Feynman-Stückelberg convention:

$$v_3(-E_1, -E_2, -\mathbf{p})e^{\left[-\frac{i}{\hbar}(-E_1k_1 - E_2k_2 + \mathbf{p.x})\right]} = u_2(E_1, E_2, \mathbf{p})e^{\left[\frac{i}{\hbar}(-E_1k_1 - E_2k_2 + \mathbf{p.x})\right]},$$

$$v_4(-E_1, -E_2, -\mathbf{p})e^{\left[-\frac{i}{\hbar}(-E_1k_1 - E_2k_2 + \mathbf{p.x})\right]} = u_1(E_1, E_2, \mathbf{p})e^{\left[\frac{i}{\hbar}(-E_1k_1 - E_2k_2 + \mathbf{p.x})\right]}.$$

As the spin matrix, $S_{antiparticle}$, for antiparticles is equal to $-S_{particle}$, v_3 and v_4 are the components of antiparticle spinors, associated with the u_2 and u_1, respectively. We can formulate Feynman-Stückelberg interpretation in the following way: *Particles with negative rest mass, moving backward in spacekime can be seen as antiparticles with positive rest mass, moving forward in spacekime.* Recami and Ziino [197, 198] formulated the so called strong CPT symmetry": the physical world is symmetric (i.e. the physical laws are invariant) during total 5-dimensional inversion of the axes x, y, z, ct_0 (where t_0 is the proper time).

It is possible to make a generalization of the strong CPT symmetry in the case of two dimensional kime: the physical world is symmetric (i.e. the physical laws are invariant) for inversion of the axes $x, y, z, ck_{0,1}, ck_{0,2}$ (where $k_{0,1}, k_{0,2}$ are the components of the proper kime $\boldsymbol{dk_0}$ – see **Chapter 5, Sections 5.7** and **5.10.1.**

Depending on the values of the components of energy E_1 and E_2 and the direction of motion in spacekime, we can distinguish 5 kinds of antiparticles:

- antiparticle A_{--} ($E_1 < 0$ and $E_2 < 0$);
- antiparticle A_{-+} ($E_1 < 0$ and $E_2 > 0$);
- antiparticle A_{+-} ($E_1 > 0$ and $E_2 < 0$);
- antiparticle A_{-0} ($E_1 < 0$ and $E_2 = 0$);
- antiparticle A_{0-} ($E_1 = 0$ and $E_2 < 0$).

We will consider the Dirac equation for a charged particle interacting with an electromagnetic field. In spacekime, the electric potential will not be a scalar, but a two-component vector $\phi = (\phi_4, \phi_5)$ defined with respect to kime dimensions k_1 and k_2.

By the substitution in the free Dirac equation (3.31), the spacekime Dirac equations for a particle with charge e, which interacts with an electromagnetic field in the standard way, may be expressed as:

$$\hat{\mathbf{p}}_k \to \hat{\mathbf{p}}_k - e\,\boldsymbol{\phi}(\boldsymbol{x}, k_1, k_2),$$

$$\hat{\mathbf{p}} \to \hat{\mathbf{p}} - \frac{e}{c}\boldsymbol{A}(\boldsymbol{x}, k_1, k_2).$$

This formulation relies on $\phi = (\phi_4, \phi_5)$, the electric potential vector, $\boldsymbol{A} = (A_1, A_2, A_3)$ the magnetic potential vector defined with respect to space dimensions x, y, z, and

$\widehat{p}_k = (\hat{p}_4, \hat{p}_5) = \left(\frac{i\hbar}{c}\frac{\partial}{\partial k_1}, \frac{i\hbar}{c}\frac{\partial}{\partial k_2}\right)$, the kime energy operator. Indeed, this formulation of Dirac equations agrees with the principle of gauge symmetry [71].

More explicitly, the Dirac equation in spacekime is:

$$\left[\vec{\gamma}^k(\hat{\mathbf{p}}_k - e\,\phi(\boldsymbol{x}, k_1, k_2)) - c\begin{pmatrix} 0 & \vec{\sigma} \\ -\vec{\sigma} & 0 \end{pmatrix}\left(\hat{\mathbf{p}} - \frac{e}{c}\mathbf{A}(\boldsymbol{x}, k_1, k_2)\right) - m_0c^2 I_{4\times 4}\right]\psi = 0, \tag{3.37}$$

where:

$$\vec{\gamma}^k = (\gamma^4, \gamma^5),$$

$$\vec{\gamma}^k(\hat{\mathbf{p}}_k - e\boldsymbol{\phi}(\boldsymbol{x}, k_1, k_2)) = \sum_{\tau=4}^{5} \underbrace{\gamma^\tau}_{4\times 4} (\hat{p}_\tau - e\phi_\tau(\boldsymbol{x}, k_1, k_2)).$$

$$\begin{pmatrix} 0 & \vec{\sigma} \\ -\vec{\sigma} & 0 \end{pmatrix}\left(\hat{\mathbf{p}} - \frac{e}{c}\mathbf{A}(\boldsymbol{x}, k_1, k_2)\right) = \sum_{\vartheta=1}^{3} \underbrace{\begin{pmatrix} 0 & \sigma_\vartheta \\ -\sigma_\vartheta & 0 \end{pmatrix}}_{4\times 4}\left(\hat{p}_\vartheta - \frac{e}{c}A_\vartheta(\boldsymbol{x}, k_1, k_2)\right).$$

Chapter 4
Kime-series Modeling and Spacekime Analytics

In this chapter, we examine strategies to statistically model kime-varying data, derive scientific information, or obtain statistical inference of spacekime signals. We start by reviewing the general formulation of functional magnetic resonance imaging (fMRI) activation analysis and the modern model-based likelihood-based inference technique. Then, we derive the magnitude-only inference on real-valued, longitudinal (time-varying) data, such as fMRI time-series defined on a 3D spatial lattice. We derive generalizations of analogous statistical inference for complex-valued spacetime signals and complex-valued spacekime data.

Various discrete and continuous strategies for mapping longitudinal time-varying data to kime-indexed kimesurfaces and spacekime analytics using tensor-based linear modeling are presented later in the chapter. Further alternative techniques, such as topological kime-series analysis, may need to be developed to enhance spacekime data modeling, inference, prediction, and forecasting.

4.1 General Formulation of fMRI Inference

Expanding some prior complex-valued longitudinal models [199–202], we can formulate some spacetime analytical models of complex-valued fMRI across time. Such spacetime models aim to understand, classify, and predict the state of an observed real- or complex-valued time-varying physical object $\rho(\boldsymbol{x},t)\colon \mathbb{R}^3 \times \mathbb{R}^+ \to \mathbb{C}$. In general, we start with an observed object $\rho(\boldsymbol{x},t)$ generated by measuring and transforming a 3D complex-valued signal $o_m(\boldsymbol{k})$, evaluated at 3D spatial frequency $\boldsymbol{k} = \left(k_x, k_y, k_z\right)$. In k-space, $o_m(\boldsymbol{k}) = \underbrace{o(\boldsymbol{k})}_{\text{true signal}} + \underbrace{e(\boldsymbol{k})}_{\text{noise}}$, where $o, e\colon \mathbb{R}^3 \to \mathbb{C}$ are respectively the (original) complex-valued true MRI signal and some random noise (error). The independent and identically distributed (IID) complex noise can be assumed to be normal. As both, the signal and the noise, are complex-valued, volume reconstructions in spacetime will utilize both their real and imaginary parts. The resulting volume reconstruction of the native Fourier space measurements will also be a complex-valued observed object mixed with some complex-valued noise.

Before we dive into the analytics in the kime domain indexed by *kappa*, $\kappa = (t, \varphi) \in \mathbb{K} \cong \mathbb{C} \ni \kappa = (\kappa_1, \kappa_2)$, we will examine the statistical inference using real- and complex-valued spacetime fMRI blood oxygenation level dependent (BOLD) signals indexed by time, t. Let's suppress the spatial indices and denote the complex-valued voxel intensity by ρ_t. This signal is defined for each time point by inverting the Fourier transform (FT) of the raw complex measurement $o_m(\boldsymbol{k})$ recorded in the Fourier

https://doi.org/10.1515/9783110697827-004

domain (k-space). Geometrically, the time-indexed intensities represent a complex-valued time-series that can be expressed as:

$$\rho_t = \underbrace{(\rho_{R,t} + e_{R,t})}_{Re(\rho_{m,t})} + \underbrace{i(\rho_{I,t} + e_{I,t})}_{Im(\rho_{m,t})}, \text{ or equivalently,}$$

$$\rho_t = \begin{pmatrix} \rho_{R,t} \\ \rho_{I,t} \end{pmatrix} + \begin{pmatrix} e_{R,t} \\ e_{I,t} \end{pmatrix}.$$

We denote the real and imaginary parts of the volume intensities by $(\rho_{R,t}, \rho_{I,t})'$, the bivariate normal distribution noise by $e = (e_{R,t}, e_{I,t})' \sim N(0, \Sigma)$, and the transpose of a vector or a matrix by $\square'$. The bivariate noise may be assumed to have a simpler scaled standard variance-covariance matrix, $\Sigma = \sigma^2 I$.

Following the notation in [199], we will express the fMRI time-series at a fixed voxel location as n complex-valued temporal intensities $\{\rho_t\}_t$ indexed by $1 \le t \le n$:

$$\rho = \underbrace{\left(\underbrace{b\mathbf{1}}_{\text{baseline}} + \underbrace{a\mathbf{r}}_{\text{activation}} \right)}_{\text{magnitude}} \times \underbrace{e^{i\theta}}_{\text{phase coupling}} + \underbrace{e}_{\text{error}}. \tag{4.1}$$

The *baseline component* ($b\mathbf{1}$) represents the longitudinal average of the time-series and is expressed as the product of the *unitary vector* of ones, $\mathbf{1} = (\underbrace{1, 1, \ldots, 1}_{n})'$, representing the static background signal at all spacetime locations, and the *baseline magnitude* ($b > 0$), a constant. A more flexible (non-constant) baseline model, e.g., polynomial, may also be used to account for drifts, trends, or habituation effects in the baseline.

The *activation component* ($a\mathbf{r}$) typically represents a periodically repeating experiment consisting of a *response vector* ($\mathbf{r}$) such as a known reference to the specific expected response characteristic function. For instance, in an "on-off" fMRI finger-tapping experiment, the response is an oscillatory step-wise characteristic function of the on-off stimulus paradigm. The second part of the activation is the *amplitude* (a), which characterizes the strength of the brain response, as measured by fMRI BOLD signal, due to the specific response function. Voxels that are not involved in brain function associated with specific stimulus will exhibit trivial activation amplitude, $a = 0$. Similarly, brain locations involved in processing a specific stimulus will have non-trivial activation responses, $a > 0$.

In equation (4.1), the parameter θ represents the *phase* of the complex-valued fMRI signal. The baseline ($b\mathbf{1}$) and activation ($a\mathbf{r}$) components of the native complex-valued fMRI data signal depend on a common phase (θ). This blend can be thought of as phase imperfection, or correction, of the raw magnitude (real-valued) signal. The final component in equation (4.1) represents the scaled additive model complex-valued

Gaussian *noise* vector (e). Later, we will see that this Gaussian noise model assumption can also be relaxed by incorporating Rician distribution noise (for modeling the magnitude of complex-valued MRI signal), Rayleigh distribution (for no-signal MRI background noise), or any other noise distribution [203, 204]. Clearly, all three components of model (4.1), magnitude, phase, and error, are indexed by time, $1 \le t \le n$. The phase component (θ) may be assumed to be fixed with time, but may still vary across space (voxel locations).

In model (4.1), both the baseline and activation signals share a common phase θ, i.e., both components of the intensity are phase-coupled. Experimental evidence supports the additive phase-coupling between the constant and response components [199]. Also, the parameters $\Theta = \{b,\ a,\ \theta,\ \sigma\}$ in model (4.1) are unknown and need to be estimated at all voxel spatial locations. Various assumptions about these parameters, e.g., b could be assumed to be constant across space and time, may lead to different models and subsequently to resulting inference variations.

A matrix representation of model (4.1) would facilitate the mathematical analysis and statistical formulation of test-statistics to identify spatial locations significantly associated with the stimulus conditions and specific brain activation. Unwinding the complex-valued representation from a wide (complex number) to a long (vector) format yields the following formulation of the fMRI intensity model:

$$\begin{aligned} \rho = \begin{pmatrix} \rho_R \\ \rho_I \end{pmatrix} &= \underbrace{\begin{pmatrix} \mathbf{1} & 0 \\ 0 & \mathbf{1} \end{pmatrix}}_{S} \underbrace{\begin{pmatrix} b\cos\theta \\ b\sin\theta \end{pmatrix}}_{\gamma} + \underbrace{\frac{a}{b}}_{q} \underbrace{\begin{pmatrix} \boldsymbol{r} & 0 \\ 0 & \boldsymbol{r} \end{pmatrix}}_{H} \underbrace{\begin{pmatrix} b\cos\theta \\ b\sin\theta \end{pmatrix}}_{\gamma} + \sigma e, \\ \rho &= \underbrace{(S + qH)}_{Y} \gamma + \sigma e, \end{aligned} \tag{4.2}$$

where $e \sim N(0, I)$, γ represents the complex baseline signal, and the activation to baseline ratio q relates to the signal-to-noise ratio (SNR) attributed to the actual fMRI activation. It may be informative to explicate the tensor dimensions of all terms in model (4.2):

$$\rho = \begin{pmatrix} \rho_R \\ \rho_I \end{pmatrix}_{2n\times 1} = S_{2n\times 2}\gamma_{2\times 1} + q_{1\times 1}H_{2n\times 2}\gamma_{2\times 1} + \sigma e_{2n\times 1} = \underbrace{(S + qH)}_{Y_{2n\times 2}} \gamma + \sigma e. \tag{4.3}$$

In the complex model (4.3), the multiplicative term ($qH\gamma$) represents the non-linear phase-coupling between the SNR (q) parameter and the complex baseline signal (γ). This is one of the substantive differences between the complex-valued model (4.3) and a classical linear regression model using magnitude-only (real-valued) fMRI signal where the baseline b and the activation a are assumed to be independent and linearly mixed:

$$\rho(\boldsymbol{x}) = \boldsymbol{b1} + \boldsymbol{ar} + \sigma\ e.$$

The spatial location, $\mathbf{x}$, will generally be suppressed. Matrix representation of orthogonal projections provides a useful contextualization of matrix manipulation for the purpose of orthogonal projection of an arbitrary vector $a \in \mathbb{R}^n$ onto a subspace $V \subseteq \mathbb{R}^n$. Let $\dim V = m \le n$ and the vectors $\{v_1, v_2, \ldots, v_m\} \in V$ represent any basis of V. If A is the matrix of column vectors $\{v_i\}_{i=1}^m$, we can find an orthonormal basis $\{u_1, u_2, \ldots, u_m\} \in V$, $u_i' u_j = \delta_{i,j}$, and the orthogonal projection onto V is represented by the $n \times n$ matrix:

$$P_V = \sum_i u_i u_i' = A(A'A)^{-1} A', \text{ where } A_{n\times m} = (v_1, v_2, \ldots, v_m).$$

Note that the square cross-product matrix $(A'A)_{m\times m}$ is invertible, since $\forall c = (c_1, c_2, \ldots, c_m) \in V \backslash \{0\}$, if we assume $c \in \text{kernel}(A'A)$, then $0 = A'Ac$ yields that $0 = c'A'Ac = (Ac)'(Ac) = \|Ac\|^2$ and $Ac = 0$. Since $\{v_i\}_{i=1}^m \in V$ represents any basis of V, then $0 = Ac = \sum_i c_i v_i$ introduces a trivial linear combination of the basis, which is only possible, when $c_i = 0, \forall i$. Therefore the kernel of $A'A$ is trivial, kernel $(A'A) = \{0\}$, and $A'A$ is invertible.

Next we can validate that the projection matrix $P = \sum_i u_i u_i' = A(A'A)^{-1} A'$ works as expected on all vectors $a \in \mathbb{R}^n$, for elements in V and $V^{\perp} = \mathbb{R}^n \backslash V$. Suppose $a = a_v + a_{\perp}$, where $a_v \in V$ and $a_{\perp} \in V^{\perp}$, first consider $a_{\perp}$, since $V = \text{Range}(A)$, $V^{\perp} = \text{Range}(A)^{\perp} = \text{kernel}(A')$, hence if $a_{\perp} \in V^{\perp}$, then $A'a_{\perp} = 0$. Next, consider $a_v \in V$, $\exists c = \{c_i\}_{i=1}^m$ such that a_v is a linear combination of the basis vectors $\{v_1, v_2, \ldots, v_m\} \in V$, and $a_v = Ac$.

Therefore:

$$P_V a = P_V a_v + P_V a_{\perp} = A(A'A)^{-1} A' a_v + A(A'A)^{-1} \underbrace{A' a_{\perp}}_{0} =$$

$$A(A'A)^{-1} A' \underbrace{Ac}_{a_v} + \underbrace{A(A'A)^{-1} 0}_{0} = A \underbrace{(A'A)^{-1}(A'A)}_{I} c = Ac = a_v.$$

This validates that the matrix operator $P_V = A(A'A)^{-1} A'$ orthogonally projects onto the subspace $V \subseteq \mathbb{R}^n$. Similarly, the matrix operator projecting onto $V^{\perp}$ is $P_{\perp} = \left(\boldsymbol{I} - A(A'A)^{-1} A'\right)$:

$$P_{\perp} a = P_{\perp} a_v + P_{\perp} a_{\perp} = a_{\perp}.$$

The *kernel* of the projection $P_{\perp} = \left(\boldsymbol{I} - A(A'A)^{-1} A'\right)$ and the *range* of the projection $P_V = A(A'A)^{-1} A'$ are both equal to V.

Without loss of generality, let's assume a trivial sum of the response vector components, $A'\boldsymbol{r} \equiv \mathbf{1}'\boldsymbol{r} = 0$. Otherwise, this property can be enforced by orthogonalizing the response states relative to the constant signal vector $A \equiv \mathbf{1}$ by projecting $\boldsymbol{r}$ into the complementary subspace spanned by the columns of $\mathbf{1}$. This corresponds to generating a new response:

$$\widehat{\boldsymbol{r}} \equiv P_{\perp} \boldsymbol{r} \equiv \left(\boldsymbol{I} - A(A'A)^{-1} A'\right)\boldsymbol{r} = \left(\boldsymbol{I} - \mathbf{1}(\mathbf{1}'\mathbf{1})^{-1} \mathbf{1}'\right)\boldsymbol{r},$$

where $\boldsymbol{I}$ is the identity matrix. Another observation is that we can always assume the normalization condition $\boldsymbol{r}'\boldsymbol{r} = n$ (the number of samples), otherwise, we can just renormalize the reference condition.

To account for varying, drifting, or nonstationary baseline, a more general model using non-constant baseline intensities component can also be fit. The multiple model parameters may still be estimated via linear or polynomial regression. In other words, we can extend the constant-baseline and activation model that relies on $\underbrace{b\mathbf{1}}_{\text{baseline}} + \underbrace{ar}_{\text{activation}}$ to a more flexible (non-constant-baseline) model using a design matrix, $X_{n\times(d+1)} = (\mathbf{1}\ x_1\ x_2 \ldots x_d)$, composed of d feature vectors $\{x_i\}_{i=1}^{d}$, each of which is assumed to be an observed time-series, and the baseline magnitude vector $\beta = (\beta_o, \beta_1, \ldots, \beta_d)'$:

$$\underbrace{X\beta}_{\text{baseline model}} = \underbrace{\beta_o\mathbf{1}}_{\beta_o} + \beta_1 x_1 + \beta_2 x_2 + \ldots + \beta_d x_d.$$

The observations for feature i across the n cases represent a vector $x_i = (x_{1,i}, x_{2,i}, \ldots, x_{n,i})'$, $\forall 1 \le i \le d$, and for a given case identifier (subject or time t), the observed feature values include $x_t' = (1, x_{t,1}, x_{t,2}, \ldots, x_{t,d})$, $\forall 1 \le t \le n$. In the special case of fMRI intensities, which indirectly measure the effective proton spin density at time t, the observations can be indexed by time $x_t' = (1, x_{t,1}, x_{t,2}, \ldots, x_{t,d})$, $\forall 1 \le t \le n$, and the fitted linear model-based value of the fMRI magnitude response at time t is represented as $\hat{y}_t = x_t'\beta$.

For instance, one can jointly fit a multi-term linear model that includes an intercept (β_o) and some multi-source components, e.g., a linear drift over time (β_1) representing brain habituation effect over time, a BOLD hemodynamic effect of a stimulus S_1 (β_2), a quadratic effect of S_1 or another stimulus S_2 (β_3), and so on.

Such a multivariate parameter model of the reference (baseline) signal incorporates fMRI signal uncertainties and can be used to account for other factors like the commonly observed habituation reflecting brain plasticity and adaptation to performing repetitive tasks. Thus, the multi-feature expansion of the complex-valued model (4.3) can be presented as:

$$\underbrace{\rho}_{2n\times 1} = \begin{pmatrix} \rho_R \\ \rho_I \end{pmatrix} = \underbrace{\overbrace{\begin{pmatrix} \mathbf{1} & 0 \\ 0 & \mathbf{1} \end{pmatrix}}^{S}}_{2n\times(2(d+1))} \underbrace{\underbrace{\begin{pmatrix} \beta & 0 \\ 0 & \beta \end{pmatrix}}_{(2(d+1))\times 2} \underbrace{\begin{pmatrix} \cos\theta \\ \sin\theta \end{pmatrix}}_{2\times 1}}_{\gamma} + \underbrace{\overbrace{M}^{qH}}_{2n\times 2(d+1)} \underbrace{\underbrace{\begin{pmatrix} \beta & 0 \\ 0 & \beta \end{pmatrix}}_{(2(d+1))\times 2} \underbrace{\begin{pmatrix} \cos\theta \\ \sin\theta \end{pmatrix}}_{2\times 1}}_{\gamma} \underbrace{e}_{2n\times 1} =$$

$$(S + qH)\gamma + \sigma e = \begin{pmatrix} X & 0 \\ 0 & X \end{pmatrix} \begin{pmatrix} \beta & 0 \\ 0 & \beta \end{pmatrix} \begin{pmatrix} \cos\theta \\ \sin\theta \end{pmatrix} + \begin{pmatrix} e_{R,t} \\ e_{I,t} \end{pmatrix}, \tag{4.4}$$

where the phase location parameter θ is assumed to be constant, the matrix $M_{2n\times 2(d+1)}$ depends on the SNR of activation to baseline and the response matrix H, the effect-size vector is $\beta=(\beta_o,\beta_1,\ldots,\beta_d)'$, and the design matrix X consists of the constant intercept, the baseline signal, and the expected fMRI BOLD response columns.

In general, if we define $A_{(2(d+1))\times 2}=\begin{pmatrix}\beta & 0\\ 0 & \beta\end{pmatrix}$, then the *left inverse* of A, i.e., $A^{-1}=A_L^{-1}$, is defined by:

$$A_L^{-1}=\underbrace{(A'A)^{-1}}_{2\times 2}\underbrace{A'}_{2\times(2(d+1))},$$

where $A'_{2\times(2(d+1))}=\begin{pmatrix}\beta' & 0\\ 0 & \beta'\end{pmatrix}$, $\underbrace{A'A}_{2\times 2}=\begin{pmatrix}\sum_{i=0}^{d}\beta_i^2 & 0\\ 0 & \sum_{i=0}^{d}\beta_i^2\end{pmatrix}$,

$$\underbrace{(A'A)^{-1}}_{2\times 2}=\begin{pmatrix}\left(\sum_{i=0}^{d}\beta_i^2\right)^{-1} & 0\\ 0 & \left(\sum_{i=0}^{d}\beta_i^2\right)^{-1}\end{pmatrix}, \text{ and } A^{-1}A=(A'A)^{-1}A'A=\begin{pmatrix}1 & 0\\ 0 & 1\end{pmatrix}=I_{2\times 2}.$$

Therefore,

$$a\underbrace{\begin{pmatrix}\boldsymbol{r} & 0\\ 0 & \boldsymbol{r}\end{pmatrix}}_{H}\begin{pmatrix}\cos\theta\\ \sin\theta\end{pmatrix}=a\underbrace{\begin{pmatrix}\boldsymbol{r} & 0\\ 0 & \boldsymbol{r}\end{pmatrix}}_{H}\underbrace{(A'A)^{-1}A'}_{A_L^{-1}}\underbrace{A\begin{pmatrix}\cos\theta\\ \sin\theta\end{pmatrix}}_{\gamma}=$$

$$\underbrace{a\underbrace{\begin{pmatrix}\boldsymbol{r} & 0\\ 0 & \boldsymbol{r}\end{pmatrix}}_{H}\underbrace{(A'A)^{-1}A'}_{A_L^{-1}}}_{M}\underbrace{\underbrace{\begin{pmatrix}\beta & 0\\ 0 & \beta\end{pmatrix}}_{(2(d+1))\times 2}\underbrace{\begin{pmatrix}\cos\theta\\ \sin\theta\end{pmatrix}}_{2\times 1}}_{\gamma}.$$

Analyzing the fMRI signal involves aggregating the repeated experimental condition measurements over time, e.g., event-related or task-driven performance design, and determining the "statistical significance" measuring some association between the observed fMRI intensity time course and the underlying experimenter-controlled reference function characterizing the stimulus conditions.

For instance, given a fixed spatial location $v=(x,y,z)$ and a specific *contrast* vector $c=(c_0,c_1,c_2,\ldots,c_d)'$, we can statistically analyze the likelihood of any linear association between the covariates (X) in terms of their effects (β). In other words, we can represent any concrete linear combination of the effects $\beta=(\beta_o,\beta_1,\ldots,\beta_d)'$ via their inner product with the contrast $c'\beta=\langle c|\beta\rangle$. This allows us to statistically evaluate the data-driven strength of the hypothesized linear association of the factors by contrasting a pair of hypotheses, e.g.,

null hypothesis $H_o: c'\beta=0$ vs. alternative *research hypothesis* $H_1: c'\beta\neq 0$.

The generalized likelihood ratio test (gLRT) [205] evaluates the strength of the data-driven evidence to reject the null hypothesis (H_o) and therefore accept the alternative research hypothesis (H_1). gLRT relies on maximizing the likelihood constrained by the null hypothesis.

4.2 Likelihood Based Inference

Next we will briefly introduce the method of maximum likelihood estimation (MLE) and the subsequent hypothesis-based inference using likelihood ratio testing (LRT). Given an a priori hypothesis in terms of a specific parameter vector, like the effect-size, β, e.g., $H_o : \beta = 0$ (no effect), the idea of likelihood inference involves three steps. It starts with computing the MLE parameter estimates, applying the central limit theorem (CLT) and the delta method to determine the sampling distributions for the statistics we are interested in, and finally using an appropriate statistical test (e.g., LRT) to quantify the data-driven evidence to reject the a priori hypothesis.

Having an observed data X and a set of candidate probability distribution models $\{P_\theta : \theta \in \Omega\}$ allows us to perform likelihood based inference using the likelihood function. In general, we start with a statistical model in which each probability distribution model (P_θ) is specified by a corresponding probability function f_θ. Assuming we have observed the data $X = x_o$, we can consider the *likelihood function* $L(\cdot|x_o) : \{\theta \in \Omega\} \to \mathbb{R}$ mapping the parameter space onto the reals, $L(\theta|x_o) = f_\theta(x_o)$, which explicitly depends on the data, x_o, and the set of probability models, $\{P_\theta\}$.

For any specific observation $X = x_o$, the value $L(\theta|x_o) = f_\theta(x_o)$ is the likelihood of the (unknown) parameter θ. In this setting, the usual roles of a fixed (known) parameter and unobserved (unknown) data are reversed, as the likelihood function considers the data to be known (fixed) and the model parameter to be unknown and requiring estimation. In essence, $f_\theta(x)$ quantifies the chance (probability) of observing the data $X = x$ given a known value of the parameter θ. Quantifying this probability imposes a natural *likely ordering* "$\underbrace{\theta_1 \ll \theta_2}_{\text{likely}}$" on the parameter set $\{\theta \in \Omega\}$, indicating the parameter θ_2 is more likely than the parameter θ_1 if and only if $f_{\theta_1}(x) < f_{\theta_2}(x)$. This *likely ordering* is different from comparing the actual *values* of the parameters, $\theta_1 \gtreqless \theta_2$. In other words, given the data $X = x$ is already observed, the parameter with higher likelihood is more likely to be associated with the source process of the observed data than another process associated with a different parameter whose likelihood is lower than the first. Of course, if $f_{\theta_1}(x) = f_{\theta_2}(x)$, then "$\underbrace{\theta_1 \equiv \theta_2}_{\text{likely}}$", but again, they may not necessarily be different as values. To reiterate, the likelihood value $L(\theta|x) = f_\theta(x)$ is the probability of observing the data $X = x$ given that θ is the true value of the process generating the data. Whereas probability density functions are restricted to be positive, $f_\theta(x) \geq 0$ and $\int f_\theta(x)dx = 1$, all probability values are within the range $[0, 1]$ and often the likelihood values may be very small over the entire parameter space $\{\theta \in \Omega\}$. In principle, it is not the *raw likelihood value* that is informative about the characteristics of the process we are studying. Of interest is the *relative likelihood value* contrasted to other likelihoods corresponding to other parameters in the model set $\{\theta \in \Omega\}$.

Let's start by examining one discrete and one continuous example of using LRT inference. The first example models an experiment of tossing a coin $n = 100$ times and observing $X = 60$ heads turn up. Solely based on this information, suppose the probability of getting a *Head* in a single toss is $P(H) = \theta$. Then, the Binomial distribution $Binomial(100, \theta)$ represents the appropriate statistical model for the entire experiment where $\{P_\theta = Binomial(100, \theta) : \theta \in \Omega = [0, 1]\}$. The Binomial likelihood function is:

$$L(\theta|X = 60) = f_\theta(60) = \binom{100}{60} \theta^{60} (1 - \theta)^{40}.$$

Let's try to infer if a fair coin ($\theta_1 = 0.50$) or a slightly biased coin, loaded for Heads ($\theta_2 = 0.55$), is more likely to have generated the observed $X = 60$ heads. Using the Distributome Binomial calculator (http://www.distributome.org/V3/calc/Binomial Calculator.html, accessed January 29, 2021), we can compute $L(\theta_1 = 0.5|X = 60) = 0.982 - 0.972 = 0.01$ and $L(\theta_2 = 0.55|X = 60) = 0.866 - 0.817 = 0.049$. This suggests that a coin loaded for heads (θ_2) is more likely to have generated the relatively larger number of heads, $X = 60$, compared to a fair coin (θ_1), which would be expected to yield 50 heads on average. A more relativized representation of the evidence in favor of a loaded coin is provided by the ratio of the pair of likelihoods:

$$\frac{L(\theta_2 = 0.55|X = 60)}{L(\theta_1 = 0.5|X = 60)} = \frac{0.049}{0.01} \cong 5.$$

Later, we will illustrate the use of the likelihood function, and it's close counterpart, the log-likelihood function ($ll(\theta|x) = \ln(L(\theta|x))$) to estimate unknown parameter vectors θ.

Let's look at another (continuous distribution) example based on a sample of n IID *normal* observations $\{x_i\}_{i=1}^n \in N(\theta, \sigma_o^2)$, over the parameter space $\{\theta \in \Omega = \mathbb{R}\}$ and assuming $\sigma_o^2 > 0$ is known. Denoting the sample mean and sample variance by $\bar{x}$ and s^2, we have:

$$(n-1)s^2 = \sum_{i=1}^n (x_i - \bar{x})^2 = \sum_{i=1}^n x_i^2 - 2\bar{x}\sum_{i=1}^n \bar{x} + n\bar{x}^2 = \sum_{i=1}^n x_i^2 - 2n\bar{x}^2 + n\bar{x}^2 = \sum_{i=1}^n x_i^2 - n\bar{x}^2,$$

$$\sum_{i=1}^{n}(x_i-\theta)^2=\sum_{i=1}^{n}x_i^2-2\theta\underbrace{\sum_{i=1}^{n}x_i}_{n\bar{x}}+n\theta^2=\sum_{i=1}^{n}x_i^2-2\theta n\bar{x}+n\theta^2=$$

$$\sum_{i=1}^{n}x_i^2\underbrace{-n\bar{x}^2+n\bar{x}^2}_{0}-2\theta n\bar{x}+n\theta^2=\underbrace{\sum_{i=1}^{n}x_i^2-n\bar{x}^2}_{(n-1)s^2}+n\underbrace{\left(\bar{x}^2-2\theta\bar{x}+\theta^2\right)}_{(\bar{x}-\theta)^2}=(n-1)s^2+n(\bar{x}-\theta)^2.$$

Then the likelihood function is:

$$L\left(\theta|X=\{x_i\}_{i=1}^{n}\right)\underbrace{=}_{\text{by independence}}\prod_{i=1}^{n}\underbrace{f_\theta(x_i)}_{\substack{\text{univariate}\\ \text{normal}\\ \text{density}}}=\frac{1}{\left(2\pi\sigma_o^2\right)^{\frac{n}{2}}}\prod_{i=1}^{n}e^{-\frac{(x_i-\theta)^2}{2\sigma_o^2}}=$$

$$\frac{1}{\left(2\pi\sigma_o^2\right)^{\frac{n}{2}}}e^{-\frac{\sum_{i=1}^{n}(x_i-\theta)^2}{2\sigma_o^2}}=\underbrace{\frac{1}{\left(2\pi\sigma_o^2\right)^{\frac{n}{2}}}e^{-\frac{(n-1)s^2}{2\sigma_o^2}}}_{\text{independent of }\theta}\times\underbrace{e^{-\frac{n(\bar{x}-\theta)^2}{2\sigma_o^2}}}_{\theta\text{ dependent}}\propto e^{-\frac{n(\bar{x}-\theta)^2}{2\sigma_o^2}}.$$

Remember that we are interested in the relative size of the likelihood function, ignoring any constant multipliers. Thus, we are using the "proportional" sign ($\propto$) to indicate the important relative-size component of the likelihood value. Then, the log-likelihood function $ll(\theta|X)$ is:

$$ll\left(\theta|X=\{x_i\}_{i=1}^{n}\right)=\log L\left(\theta|X=\{x_i\}_{i=1}^{n}\right)\propto-\frac{n(\bar{x}-\theta)^2}{2\sigma_o^2}.$$

Since the $\log(\cdot)$ function is monotonically increasing, $L(\theta|X)$ and $ll(\theta|X)$ share the same maxima and minima, which are obtained by setting to zero the first derivative(s) with respect to the unknown parameter θ:

$$\hat{\theta}=\arg\max_{\theta\in\Omega}L(\theta|X)=\arg\max_{\theta\in\Omega}ll(\theta|X)=\bar{x}.$$

Since the second derivative in negative, the *maximum likelihood* is attained for:

$$0=\underbrace{\frac{d}{d\theta}ll(\theta|X)}_{\text{score function}}\propto 2n(\bar{x}-\theta),\text{ or }\hat{\theta}=\bar{x},\text{ since }\frac{d^2}{d\theta^2}ll(\theta|X)\propto-2n<0.$$

In other words, the sample average $\bar{x}$ represents the MLE parameter estimate $\hat{\theta}$. The *score function* is the gradient of the log-likelihood with respect to the parameter. For multidimensional parameter vectors, the score function is the corresponding vector of partial derivatives of the log-likelihood. Evaluating the score at a particular parameter vector yields a score vector encoding the direction of change of the log-likelihood function with respect to changes of the values of the parameter vector.

If we are observing a normal distribution process with a known variance (σ_o^2) and unknown mean (μ), we can collect IID samples and estimate the most likely value of the unknown mean parameter μ. To formalize the inference process, we start with $\{P_\theta:\theta \in \Omega\} = \{\Phi_\theta:\theta = \mu \in \mathbb{R}\}$ and we take a random sample $\{x_i\}_{i=1}^n \in N(\mu, \sigma_o^2)$. We derived that the MLE of the population mean parameter is $\hat{\mu} = \bar{x}$. Now, we would like to understand the distribution of this estimate, i.e., the sampling distribution of the mean. The CLT provides a mechanism to track the asymptotics of the MLE estimate for large sample sizes [165, 206]. Clearly, prior to observing the $\{X_i = x_i\}_{i=1}^n$, the arithmetic average $\bar{X} = \frac{1}{n}\sum_{i=1}^n X_i$ is a random variable. The CLT provides explicit form of the sampling distribution of the mean:

$$\bar{X} \sim N\left(\mu, \frac{\sigma_o^2}{n}\right),\ E(\bar{X}) = \mu,\ MSE_\mu(\bar{X}) \equiv Var_\mu(\bar{X}) = \frac{\sigma_o^2}{n},\ SD_\mu(\bar{X}) = \frac{\sigma_o}{\sqrt{n}}.$$

A CLT generalization referred to as the "*delta method*" provides a mechanism to approximate the probability distribution for a function of an asymptotically normal statistical estimator given that the variance of that estimator is bounded [207]. As an example, we can think of asking what would be the distribution of a function of the sample mean. Again, we start with an IID sample $\{X_i\}_{i=1}^n \in \mathcal{D}$, from an arbitrary distribution $\mathcal{D}$. Assuming that θ and $\sigma > 0$ are well-defined constants and $\check{X} = \check{X}(X_1, X_2, \ldots, X_n)$ is an estimate of a parameter θ that is asymptotically normal:

$$\sqrt{n}\left(\check{X} - \theta\right) \xrightarrow{d} N(0, \sigma^2).$$

The notation $\xrightarrow{d}$ means convergence in distribution, with $n \to \infty$. Then, any smooth function $g()$ whose first derivative at θ exists and is non-trivial ($g'(\theta) \neq 0$), satisfies an analogous asymptotic property:

$$\sqrt{n}(g(\check{X}) - g(\theta)) \xrightarrow{d} N\left(0, \sigma^2 (g'(\theta))^2\right).$$

The delta method allows us to obtain closed-form analytical expressions for the asymptotic distributions of functions of parameter estimates. For instance, it is used to show that chi-square (χ^2_{df}) is the asymptotic distribution of the LRT statistic. More specifically, if we are trying to compare and contrast a pair of hypotheses, (null) $H_o:\theta = \theta_o$ and (alternative) $H_1:\theta \neq \theta_o$, the LRT statistics ($\lambda_n(X)$) is defined as:

$$\lambda_n(X) = \frac{\overbrace{\sup_{\theta = \theta_o} L(\theta|X)}^{\text{Restricted MLE}}}{\underbrace{\sup_{\theta} L(\theta|X)}_{\text{Global MLE}}} = \frac{L(\check{\theta}|X)}{L(\hat{\theta}|X)} = \frac{L(\theta_o|X)}{L(\hat{\theta}|X)}.$$

The asymptotic distribution of the log-likelihood ratio test statistics, $\Lambda_n(X)$, as $n \to \infty$, is $-2\log(\Lambda_n(X)) \xrightarrow{d} \chi^2_{df}$. The chi-square degrees of freedom (df) represents the difference between the number of free parameters in Ω and Ω_o, e.g., for our mean example above $\hat{\theta} = \bar{x}$, $df = 1$. This is because, $ll(\theta|X) := \log(L(\theta|X))$ and we can expand its Taylor series around $\hat{\theta}$:

$$\Lambda_n(X) \equiv -2\log \lambda_n(X) = 2\left(ll\left(\hat{\theta}|X\right) - ll(\theta_o|X)\right)$$

$$ll(\theta_o) \cong ll\left(\hat{\theta}\right) + \left(\hat{\theta} - \theta_o\right) ll'\left(\hat{\theta}\right) + \frac{1}{2}\left(\hat{\theta} - \theta_o\right)^2 ll''\left(\hat{\theta}\right).$$

Note that $ll\left(\hat{\theta}\right)$ does not depend upon θ_o and

$$\hat{\theta} = \arg\max_{\theta\in\Omega} L(\theta|X) = \arg\max_{\theta\in\Omega} \hat{\theta} = \arg\max_{\theta\in\Omega} L(\theta|X) = \arg\max_{\theta\in\Omega} ll(\theta|X)$$

is the extremum of the (log) likelihood, i.e., $ll'\left(\hat{\theta}\right) = 0$. Therefore,

$$\Lambda_n(X) \cong -\left(\hat{\theta} - \theta_o\right)^2 ll''\left(\hat{\theta}\right) = \left(\hat{\theta} - \theta_o\right)^2 J\left(\hat{\theta}\right) =$$

$$\left(\hat{\theta} - \theta_o\right)^2 I(\theta_o) \frac{J\left(\hat{\theta}\right)}{I(\theta_o)} = n\left(\hat{\theta} - \theta\right)^2 I(\theta_o) \times \frac{J\left(\hat{\theta}\right)}{n\, I(\theta_o)},$$

where $I(\theta) = -\mathbb{E}\left(\frac{d^2}{d\theta^2}\log f(X|\theta)\right)$ is the Fisher information, $J(\theta) = -ll''(\theta) \equiv -\frac{d^2}{d\theta^2} ll(\theta)$ is the observed information function, and $I(\theta) = \mathbb{E}\,(J(\theta))$.

By the delta method and the CLT, $\sqrt{n}\left(\hat{\theta} - \theta\right)\sqrt{I(\theta)} \xrightarrow{d} N(0, 1)$. Note the dependence of the parameter estimate on the sample size, $\hat{\theta} = \hat{\theta}_n$. Also by the law of large numbers, $\frac{J(\hat{\theta})}{n\,I(\theta)} \xrightarrow{p} 1$.

To understand this convergence in probability, $\xrightarrow{p}$, assume we are given a random (IID) sample, $\{x_i\}_{i=1}^n$, drawn from a probability distribution with a density function that is sufficiently regular [208, 209]. This regularity condition will guarantee smoothness, series convergence, and commutativity of differentiation and integration operators, i.e., we can interchange the expectation and the partial derivatives of the log-likelihood. Then, the average of the observed Fisher information across all terms converges to the expectation for a single term, which is just the Fisher information:

$$\lim_{n\to\infty}\left(\frac{J(\hat{\theta})}{n}\right) = -\lim_{n\to\infty}\left(\frac{1}{n}\sum_{i=1}^{n}\left(\frac{d^2}{d\theta^2}\log f(x_i|\theta)\right)\right) = -\mathbb{E}\left(\frac{d^2}{d\theta^2}\log f(x|\theta)\right) = I(\theta).$$

In other words, $\frac{J(\hat{\theta})}{n} \xrightarrow{p} I(\theta)$. Therefore, as convergence in probability implies convergence in distribution [210, 211], the distribution of the log-likelihood test statistics is:

$$\Lambda_n(X) \cong \underbrace{n\left(\hat{\theta} - \theta\right)^2 I(\theta)}_{\xrightarrow{d}\chi_1^2} \times \underbrace{\frac{J\left(\hat{\theta}\right)}{nI(\theta)}}_{\xrightarrow{p}1} \xrightarrow{d} \chi_1^2.$$

4.3 Magnitude-Only fMRI Intensity Inference

Let's first review the basics of the classical *magnitude-only fMRI inference*. For each fixed voxel location (spatial indexing is suppressed) and a given time t, $\rho_{R,t}$ and $\rho_{I,t}$ are the real and imaginary parts of the fMRI signal, and $\|\rho_t\|$ is the spacetime reconstruction of the real-valued magnitude-only fMRI intensity at time t:

$$\underbrace{\|\rho_t\|}_{\mathbb{R}\text{ value}} = \left[\left(\rho_{R,t}+e_{R,t}\right)^2+\left(\rho_{I,t}+e_{I,t}\right)^2\right]^{\frac{1}{2}} =$$

$$\left[\left(x_t{}'\beta\cos\theta+e_{R,t}\right)^2+\left(x_t{}'\beta\sin\theta+e_{I,t}\right)^2\right]^{\frac{1}{2}} = x_t{}'\beta\left[\left(\cos\theta+\frac{e_{R,t}}{x_t{}'\beta}\right)^2+\left(\sin\theta+\frac{e_{I,t}}{x_t{}'\beta}\right)^2\right]^{\frac{1}{2}} =$$

$$x_t{}'\beta\left[1+\frac{1}{\left(x_t{}'\beta\right)^2}\left(e_{R,t}^2+e_{I,t}^2\right)+\frac{2}{x_t{}'\beta}\left(\cos\theta e_{R,t}+\sin\theta e_{I,t}\right)\right]^{\frac{1}{2}} \approx x_t{}'\beta+e_t. \tag{4.5}$$

As $e_{R,t}^2+e_{I,t}^2 \sim \chi_2^2$ and $e_t = e_{R,t}\cos\theta+e_{I,t}\sin\theta \sim N(0,\sigma^2)$, the last approximation in equation (4.5) is valid for high SNR $\left(\frac{\left(x_t{}'\beta\right)^2}{e_{R,t}^2+e_{I,t}^2}\right)$ [199]. Therefore, the purely magnitude-based inference may be derived approximately from a linear model:

$$\underbrace{\|\rho\|}_{n\times 1} \cong \underbrace{X}_{n\times(d+1)}\ \underbrace{\beta}_{(d+1)\times 1} + \underbrace{e}_{n\times 1},$$

where $X = (\mathbf{1}\ x_1\ x_2 \cdots x_d)$ is the design matrix containing vectors of feature i observations denoted by $x_i = (x_{1,i}, x_{2,i}, \cdots, x_{n,i})', \forall 1\le i\le d$.

In general, without regard to the testable hypotheses, this *unconstrained* linear model, without any boundary conditions, can be solved by ordinary least squares (OLS) [212] and the solution is given by:

$$\underbrace{\hat{\beta}_{OLS}}_{(d+1)\times 1} = \underbrace{\underbrace{(X'X)^{-1}X'}}_{\underbrace{X_L^{-1}}_{(d+1)\times n}}\|\rho\|.$$

The corresponding variability of the OLS estimate is:

$$\hat{\sigma}_{OLS}^2 = \frac{1}{n}\left(\|\rho\|-X\hat{\beta}_{OLS}\right)'\left(\|\rho\|-X\hat{\beta}_{OLS}\right).$$

These estimates are derived using the matrix *differentiation operator* property [213]. Given a symmetric matrix $A = A'$, like the cross-product matrix $A = X'X$, and a column vector a, like $a = X'\|\rho\| = \left(\|\rho\|'X\right)'$, then $\frac{\partial a'\beta}{\partial\beta} = a$ and $\frac{\partial\beta'A\beta}{\partial\beta} = (A+A')\beta = 2A\beta$. These matrix-differentiation formulations are derived from the definition of the derivative (Jacobian matrix) of a vector transformation $y_{m\times 1} = \Psi(x_{n\times 1})$:

$$\frac{\partial y}{\partial x} = \begin{bmatrix} \frac{\partial y_1}{\partial x_1} & \cdots & \frac{\partial y_1}{\partial x_n} \\ \cdots & \frac{\partial y_k}{\partial x_l} & \cdots \\ \frac{\partial y_m}{\partial x_1} & \cdots & \frac{\partial y_m}{\partial x_n} \end{bmatrix}_{m \times n}.$$

For an affine transformation $x \xrightarrow{\Psi} y$, $y = \Psi(x) = Ax + b$, where $A_{m\times n}$ and b are the linear and offset parts of Ψ, $\frac{\partial y}{\partial x} = A = [a_{i,j}]$, as $x = (x_1, x_2, \cdots, x_n)'$, $y = (y_1, y_2, \cdots, y_m)'$, $y_k = \sum_{l=1}^{n} a_{k,l}x_l + b_k$, and $\frac{\partial y_k}{\partial x_l} = a_{k,l}$. Also, the derivative of a scalar ($\alpha \equiv \alpha'$) representing the vector inner product $\langle y|Ax\rangle \equiv y'Ax = \alpha \equiv \alpha' = (Ax)'y \equiv \langle Ax|y\rangle = \langle x|A'y\rangle \in \mathbb{R}$ is given by:

$$\frac{\partial \alpha}{\partial y} = (Ax)' = x'A', \quad \frac{\partial \alpha}{\partial x} = y'A = (A'y)',$$

since $\alpha = y'Ax = (y'A)x \equiv \alpha' = (y'Ax)' = x'A'y = (x'A')y$. The special case of $m = n$ and $x = y$ leads to square matrix $A_{n\times n}$ and a quadratic form

$$\alpha = x'Ax = \sum_{l=1}^{n} x_l \sum_{k=1}^{n} a_{l,k}x_k = \sum_{l=1}^{n}\sum_{k=1}^{n} a_{l,k}x_k x_l.$$

When $\alpha = x'Ax$, $\frac{\partial \alpha}{\partial x_r} = \underbrace{\sum_{k=1}^{n} a_{r,k}x_k}_{(Ax)_r} + \underbrace{\sum_{l=1}^{n} a_{l,r}x_l}_{(A'x)_r}$, $\forall 1 \le r \le n$, and $\frac{\partial \alpha}{\partial x} = Ax + A'x = (A + A')x$.

4.3.1 Likelihood Ratio Test (LRT) Inference

In repeated event-related fMRI, subject to a hemodynamic response function (HRF) delay, temporal intensity-magnitude measurements are associated with specific stimulations or tasks [214]. To identify the brain regions associated with processing the specific stimuli requires spatial voxel-by-voxel assessments of the statistical association between the observed fMRI magnitude and the corresponding stimuli presented as a characteristic reference function (e.g., *on* vs. *off* stimulation). As the fMRI intensity magnitude follows Rician distribution [201], under the general linear model in equation (4.5), the magnitude value and the corresponding probability density function (PDF) are defined by:

$$\underbrace{\|\rho_t\|}_{\text{magnitude}} = \left[\left(\underbrace{x_t'\beta\cos(\theta)}_{\rho_{R,t}} + e_{R,t}\right)^2 + \left(\underbrace{x_t'\beta\sin(\theta)}_{\rho_{I,t}} + e_{I,t}\right)^2\right]^{\frac{1}{2}},$$

$$\underbrace{p(\|\rho_t\| \,|\, x_t, \beta, \sigma^2)}_{\text{Rician PDF}(v = x_t'\beta, \sigma)} = \frac{\|\rho_t\|}{\sigma^2} e^{-\frac{1}{2\sigma^2}\left(\|\rho_t\|^2 + \left(\underbrace{x_t'\beta}_{v}\right)^2\right)} \times \underbrace{\frac{1}{2\pi}\int_{\theta_t=-\pi}^{\pi} e^{\frac{1}{\sigma^2}(x_t'\beta\|\rho_t\|\cos(\theta_t-\theta))} d\theta_t}_{\text{modified Bessel function } I_o\left(\frac{x_t'\beta\|\rho_t\|}{\sigma^2}\right)}. \tag{4.6}$$

The integral in equation (4.6) is the zeroth-order ($\alpha = 0$) modified *Bessel function* of the first kind [215], $I_o(z) = \frac{1}{2\pi}\int_{-\pi}^{\pi} e^{z\cos(\theta)} d\theta$, evaluated at $z = \frac{x_t'\beta\|\rho_t\|}{\sigma^2}$:

$$I_o\left(\frac{x_t'\beta\|\rho_t\|}{\sigma^2}\right) = \frac{1}{2\pi}\int_{\theta_t=-\pi}^{\pi} e^{\frac{1}{\sigma^2}(x_t'\beta\|\rho_t\|\cos(\theta_t-\theta))} d\theta_t \xrightarrow[x_t'\beta \to 0]{} 1.$$

The rationale for using a Rician distribution to model the fMRI data is based on the need to fuse the real and the imaginary parts of the observed signal into a univariate quantity, model the random noise using bivariate normal distribution, and represent the complex bivariate process $(\rho_{R,t} + e_{R,t}, \rho_{I,t} + e_{I,t})$ via a univariate distribution in terms of a magnitude $(r = \|\rho_t\|)$ and a phase $\left(\varphi = \theta_t = \arctan\left(\frac{\rho_{I,t}+e_{I,t}}{\rho_{R,t}+e_{R,t}}\right)\right)$. The role of the phase may be explicated by deriving the Rician distribution from bivariate normal model:

$$X \sim N(\mu = x_t'\beta\cos(\theta), \sigma^2), \quad Y \sim N(\mu = x_t'\beta\sin(\theta), \sigma^2).$$

Following a Cartesian-to-polar change of variables transformation, the bivariate probability density can be expressed as:

$$\underbrace{p_{r,\varphi}\left(\underbrace{r}_{\|\rho_t\|}, \underbrace{\varphi}_{\theta_t}\right)}_{\text{bivariate polar}} = \underbrace{\det\begin{vmatrix} \frac{\partial x}{\partial r} & \frac{\partial x}{\partial \varphi} \\ \frac{\partial y}{\partial r} & \frac{\partial y}{\partial \varphi}\end{vmatrix}}_{\text{Jacobian}} \underbrace{p_{XY}(x,y)}_{\substack{\text{normal} \\ \text{(Cartesian)}}} = r p_{XY}\left(\underbrace{x}_{r\cos(\varphi)}, \underbrace{y}_{r\sin(\varphi)}\right) =$$

$$\frac{r}{2\pi\sigma^2} e^{-\frac{1}{2\sigma^2}\left((x-\rho_{R,t})^2 + (y-\rho_{I,t})^2\right)} =$$

$$\frac{1}{2\pi\sigma^2}\underbrace{r}_{\|\rho_t\|} e^{-\frac{1}{2\sigma^2}\left(\underbrace{x^2+y^2}_{r^2} + \underbrace{\rho_{R,t}^2 + \rho_{I,t}^2}_{(x_t'\beta)^2} - 2r\cos\left(\underbrace{\varphi}_{\theta_t}\right)\rho_{R,t} - 2r\sin\left(\underbrace{\varphi}_{\theta_t}\right)\rho_{I,t}\right)} =$$

$$\frac{1}{2\pi\sigma^2}\|\rho_t\| e^{-\frac{1}{2\sigma^2}\left(\|\rho_t\|^2+\left(x_t'\beta\right)^2-2\|\rho_t\|x_t'\beta\cos(\varphi-\theta)\right)}.$$

Another way to explicate the distribution of the time-indexed magnitude-only (real-valued) fMRI signal is to use the linear model:

$$\underbrace{m}_{\text{model magnitude}} = x_t'\beta = \beta_o + \beta_1 x_{1,t} + \beta_2 x_{2,t} + \ldots + \beta_q x_{q,t}$$

and express the magnitude in terms of the complex-valued signal:

$$\rho_t = \left(\underbrace{x_t'\beta\cos\theta}_{\rho_{R,t}} + e_{R,t}\right) + i\left(\underbrace{x_t'\beta\sin\theta}_{\rho_{I,t}} + e_{I,t}\right) = \underbrace{\left(\rho_{R,t}+e_{R,t}\right)}_{y_R} + i\underbrace{\left(\rho_{I,t}+e_{I,t}\right)}_{y_I},$$

$$\underbrace{\|\rho_t\|}_{\text{magnitude}} = \sqrt{{y_R}^2+{y_I}^2} = \sqrt{\left(\rho_{R,t}+e_{R,t}\right)^2+\left(\rho_{I,t}+e_{I,t}\right)^2}.$$

Assume that at each voxel location the fixed phase imperfection is generally not known and needs to be estimated, subject to Gaussian noise, $(e_{R,t}, e_{I,t}) \sim N(0, \sigma^2 I_2)$. Note that the magnitude-only model, $\|\rho_t\|$, ignores the information contained in the complex intensity-phases at time t:

$$\varphi \equiv \theta_t = \arctan\left(\frac{\rho_{I,t}+e_{I,t}}{\rho_{R,t}+e_{R,t}}\right) = \arctan\left(\frac{y_I}{y_R}\right).$$

Again, we use Cartesian $(x = e_{R,t},\ y = e_{I,t})$ to polar coordinate (r, φ) transformations, $\mathcal{P}$ and $\mathcal{P}^{-1}$, on the range-space complex values:

$$\mathbb{R}^2 \ni (e_{R,t}, e_{I,t}) \underset{\mathcal{P}^{-1}}{\overset{\mathcal{P}}{\longleftrightarrow}} (r,\varphi) \in \mathbb{R}^+ \times [-\pi{:}\pi),\ \mathcal{P}{:}\left| \begin{array}{l} e_{R,t} = r\cos\varphi \\ e_{I,t} = r\sin\varphi \end{array}\right.,\ \mathcal{P}^{-1}{:}\left| \begin{array}{l} r = \sqrt{e_{R,t}^2 + e_{I,t}^2} \\ \varphi = \arctan\left(\frac{e_{I,t}}{e_{R,t}}\right) \end{array}\right..$$

As $\rho_t = \begin{pmatrix} y_R \\ y_I \end{pmatrix} = \begin{pmatrix} x_t'\beta\cos\theta \\ x_t'\beta\sin\theta \end{pmatrix} + \begin{pmatrix} e_{R,t} \\ e_{I,t} \end{pmatrix}$, and the error is assumed to be bivariate normally distributed, $e = \begin{pmatrix} e_{R,t} \\ e_{I,t} \end{pmatrix} = \begin{pmatrix} y_R - x_t'\beta\cos\theta \\ y_I - x_t'\beta\sin\theta \end{pmatrix} = \begin{pmatrix} y_R - m\cos\theta \\ y_I - m\sin\theta \end{pmatrix} \sim N(0, \sigma^2 I_2)$ and the likelihood function is:

$$p\left(\begin{pmatrix} e_{R,t} \\ e_{I,t} \end{pmatrix} \middle| m, \theta, \sigma^2\right) = \frac{1}{2\pi\sigma^2} e^{-\frac{e'e}{2\sigma^2}} = \frac{1}{2\pi\sigma^2} e^{-\frac{e_{R,t}^2 + e_{I,t}^2}{2\sigma^2}}.$$

Using the Jacobian of the polar transformation $J(\mathcal{P}) = r$, we can express the likelihood, i.e., the joint distribution of the complex-valued fMRI intensities, in polar coordinates:

$$\partial e_{R,t}\partial e_{I,t} = r\partial r\partial\varphi,$$

$$\langle e|e\rangle = e'e = e_{R,t}^2 + e_{I,t}^2 = \left(\rho_{R,t} - x_t{}'\beta\cos\theta\right)^2 + \left(\rho_{I,t} - x_t{}'\beta\sin\theta\right)^2 =$$

$$(r\cos\varphi - m\cos\theta)^2 + (r\sin\varphi - m\sin\theta)^2 =$$

$$r^2\left(\underbrace{\cos^2\varphi + \sin^2\varphi}_{1}\right) - 2rm\left(\underbrace{\cos\varphi\cos\theta + \sin\varphi\sin\theta}_{\cos(\varphi-\theta)}\right) + m^2\left(\underbrace{\cos^2\theta + \sin^2\theta}_{1}\right) =$$

$$r^2 + m^2 - 2\, r\, m\cos(\varphi - \theta).$$

For a general parameter space $\theta \in \Theta$, the Cartesian- and polar-based likelihoods are:

$$\underbrace{p(e_{R,t}, e_{I,t}|\Theta)}_{\text{Cartesian}} = \frac{1}{2\pi\sigma^2} e^{-\frac{e_{R,t}^2 + e_{I,t}^2}{2\sigma^2}} \partial e_{R,t}\partial e_{I,t} = \frac{r}{2\pi\sigma^2} e^{-\frac{1}{2\sigma^2}\left(r^2 + m^2 - 2r\, m\cos(\varphi-\theta)\right)} \partial r\partial\theta = \underbrace{p(r,\varphi|\Theta)}_{\text{polar}}.$$

In polar coordinates (bivariate kime), this estimate of the fMRI intensity distribution will be used in the section below to obtain the MLE estimate of the parameter vector $\theta = \left(\beta_o, \beta_1, \cdots, \beta_q, \sigma^2\right)$.

Integrating along the phase space yields the two-parameter univariate Rician distribution depending on the fMRI signal magnitude; $\|\rho_t\| \sim Rice(\upsilon, \sigma)$ where the parameter $\upsilon = x_t{}'\beta$ models the noise-free (deterministic) baseline fMRI magnitude:

$$\underbrace{p\left(\|\rho_t\|\,|x_t, \beta, \sigma^2\right)}_{\text{Rician PDF}\left(\upsilon = x_t'\beta, \sigma\right)} = p\left(\underbrace{r}_{\|\rho_t\|}\right) = \int_{-\pi}^{\pi} p_{r,\varphi}\left(\underbrace{r}_{\|\rho_t\|}, \varphi\right) d\varphi =$$

$$\int_{-\pi}^{\pi} \frac{1}{2\pi\sigma^2}\|\rho_t\| e^{-\frac{1}{2\sigma^2}\left(\|\rho_t\|^2 + (x_t'\beta)^2 - 2\|\rho_t\| x_t'\beta\cos(\varphi-\theta)\right)} d\varphi =$$

$$\frac{\|\rho_t\|}{\sigma^2} e^{-\frac{1}{2\sigma^2}\left(\|\rho_t\|^2 + \left(\underbrace{x_t'\beta}_{\upsilon}\right)^2\right)} \times \underbrace{\frac{1}{2\pi}\int_{-\pi}^{\pi} e^{\frac{1}{\sigma^2}\left(x_t'\beta\|\rho_t\|\cos(\varphi-\theta)\right)} d\varphi}_{\text{modified Bessel function } I_0\left(\frac{x_t'\beta\|\rho_t\|}{\sigma^2}\right)}.$$

Let's define the maximum likelihood estimates (MLE's) of β and σ for both the restricted null hypothesis (no activation) $H_o: C\beta = \tau$ (note that τ could be 0) and the complementary alternative hypotheses $H_1: C\beta \neq \tau$, where $C = C_{q\times(d+1)}$ is a full row-rank contrast matrix representing the constraint rows, and $\tau = \tau_{q\times 1}$ is a vector of size equal to the full rank of the contrast matrix C. The rows in the contrast matrix C correspond to all voxel locations in the spacetime domain. For a single voxel location, the matrix C is just a row vector (linear contrast) c'. Assuming temporal independence

$(1 \leq t \leq n)$, the likelihood (l) and the log-likelihood (ll) functions corresponding to model (4.6) are:

$$l = p(\|\rho\| \,|x, \beta, \sigma^2) = \prod_t p(\|\rho_t\| \,|x_t, \beta, \sigma^2) =$$

$$\left(\frac{1}{\sigma^2}\right)^n \times \left(\prod_t \|\rho_t\|\right) \times e^{-\frac{1}{2\sigma^2}\left(\sum_t \|\rho_t\|^2 + \sum_t (x_t{'}\beta)^2\right)} \times \prod_t I_o\left(\frac{x_t{'}\beta\|\rho_t\|}{\sigma^2}\right). \tag{4.7}$$

$$ll = \ln(l) = -n\ln(\sigma^2) + \sum_t \ln\|\rho_t\| - \frac{1}{2\sigma^2}\left(\sum_t \|\rho_t\|^2 + \sum_t (x_t{'}\beta)^2\right) + \sum_t \ln I_o\left(\frac{x_t{'}\beta\|\rho_t\|}{\sigma^2}\right).$$

By maximizing the log-likelihood function, ll, we find the optimal (MLE) β and σ parameter estimates. As shown above, the MLE parameter estimates, $\left(\hat{\beta}, \hat{\sigma}\right)$, of the unconstrained problem may be obtained by optimizing ll or by OLS.

Under the null hypothesis, $H_o{:}c'\beta = \tau$ (for single contrast) or $H_o{:}C\beta = \tau$ (for multiple contrast equations), we will denote the constrained-problem MLE estimates by $(\check{\beta}, \check{\sigma})$. These are obtained by maximizing the constrained problem and including an additional Lagrange-multiplier term $\left(\gamma'(c'\beta - \tau)\right)$ in the exponent term of the objective function. This additional term includes a vector of coefficients γ' and accounts for the joint optimization relative to the main parameters, β and σ, and the additional constraint-derived term, γ'.

Using equation (4.7), the LRT statistic for $H_o{:}c'\beta = \tau$ is the log-transformed fraction of the optimized constrained and unconstrained log-likelihoods:

$$LRT = -2\ln\left(\frac{\text{optimal constrained } ll}{\text{optimal unconstrained } ll}\right) = -2\ln\left(\frac{\max_{H_0} ll}{\max ll}\right) =$$

$$2\ln\left(\frac{\max ll}{\max_{H_0} ll}\right) = 2\ln\left(\frac{ll\left(\hat{\beta}, \hat{\sigma}\right)}{ll(\check{\beta}, \check{\sigma})}\right) = 2\ln\left(ll\left(\hat{\beta}, \hat{\sigma}\right)\right) - 2\ln\left(ll(\check{\beta}, \check{\sigma})\right) =$$

$$2n\ln\left(\frac{\check{\sigma}^2}{\hat{\sigma}^2}\right) + \frac{1}{\check{\sigma}^2}\left(\sum_t \|\rho_t\|^2 + \sum_t (x_t{'}\check{\beta})^2\right) - \frac{1}{\hat{\sigma}^2}\left(\sum_t \|\rho_t\|^2 + \sum_t \left(x_t{'}\hat{\beta}\right)^2\right) + \tag{4.8}$$

$$2\sum_t\left(\ln I_o\left(\frac{x_t{'}\hat{\beta}\|\rho_t\|}{\hat{\sigma}^2}\right) - \ln I_o\left(\frac{x_t{'}\check{\beta}\|\rho_t\|}{\check{\sigma}^2}\right)\right).$$

By Wilks' theorem [216], if the full row-rank of the contrast matrix C is df, then $LRT \sim \chi^2_{df}$, as $n \to \infty$. More generally, the contrast matrix C is specified by a null hypothesis $H_o : C\beta = \boldsymbol{\tau}$ and restricts the parameter region Ω_0 to be s dimensional, whereas the original unconstrained parameter region Ω is t dimensional. Then, $LRT \sim \chi^2_{t-s}$, as $n \to \infty$. In other words, for a full row-rank contrast matrix C with the number of independent constraint rows equal to $t - s = df$ the asymptotic distribution of the test statistics is $LRT \sim \chi^2_{df}$, as $n \to \infty$.

4.3.2 Rician and Normal Distributions of the Magnitude

Before we compute and compare the general and the constrained model parameter estimates, $\left(\hat{\beta}, \hat{\sigma}\right)$ and $(\breve{\beta}, \breve{\sigma})$, let's first derive the probability distribution density. We will show that asymptotically, for *large signal-to-noise ratio* (SNR), $v = x_t{}'\beta \to \infty$ and low noise corresponding to activated brain regions, the magnitude is Rician distributed, equation (4.6), can be approximated by normal distribution [201]:

$$p(\|\rho_t\| | x_t, \beta, \sigma^2) \sim \underbrace{N\left(\underbrace{x_t{}'\beta}_{\text{mean}}, \underbrace{\sigma^2}_{\text{variance}}\right)}_{\text{normal}}.$$

We will need the following two simple facts using the Bachmann-Landau asymptotic big-O function $O()$ notation [217]:

- A second-order Taylor-series expansion near the origin of the *cosine* function in the exponent of the Bessel function in equation (4.6) is given by:

$$\cos(\theta_t - \theta) = 1 - \frac{(\theta_t - \theta)^2}{2} + O\left((\theta_t - \theta)^4\right).$$

- A first-order Maclaurin series expansion of the *imaginary error function* (erfi) [218]:

$$\operatorname{erfi}(z) = -i \times \underbrace{\operatorname{erf}(iz)}_{\text{error function}} = -i \times \tfrac{1}{\sqrt{\pi}} \int_{-iz}^{iz} e^{-t^2} dt \cong \tfrac{2}{\sqrt{\pi}}\left(z + O(z^3)\right).$$

Let's examine the asymptotics of the zeroth-order modified Bessel function of the first kind as the argument increases, $x \to \infty$, which models the situation of a large SNR in fMRI signals. Over the real domain, $z \equiv x \in \mathbb{R}$, the Bessel function $(J_v(x))$ and modified Bessel function $(I_v(x))$ of the first kind and of order $v \in \mathbb{R}$ represent the canonical solutions $y = y(x)$ of the second-order (standard) Bessel differential equation and its modified counterpart, which naturally have two linearly independent solutions – the order v Bessel functions (J_v, Y_v) of the first and second kinds, respectively, which are paired with their *modified* Bessel function counterparts (I_v, K_v) of the first and second kinds [215, 219, 220]. Here, we are only concerned with the Bessel functions of the first kind that are solutions of:

$$(\text{Bessel equation})\; x^2 \frac{d^2y}{dx^2} + x\frac{dy}{dx} + (x^2 - v^2)y = 0,$$

$$(\text{modified Bessel equation})\; x^2 \frac{d^2y}{dx^2} + x\frac{dy}{dx} - (x^2 + v^2)y = 0.$$

For integer orders, $v \equiv n \in \mathbb{N}$, the Bessel functions of the first kind are reduced to:

$$J_n(z) = \frac{1}{2\pi}\int_{-\pi}^{\pi} e^{i(z\cdot\sin(\theta) - n\theta)}d\theta = \frac{1}{2\pi}\int_{-\pi}^{\pi}\cos(n\theta - z\sin(\theta))d\theta + \underbrace{\frac{i}{2\pi}\int_{-\pi}^{\pi}\sin(z\sin(\theta) - n\theta)d\theta}_{0} \equiv$$

$$\frac{1}{\pi}\int_{0}^{\pi}\cos(n\theta - z\sin(\theta))d\theta, \quad z \in \mathbb{C}, n \in \mathbb{N}.$$

Recall that when integrating a periodic function, shifting the limits of the definite integral by a multiple of the period does not alter the integral value. A special case of the general Bessel function of the first kind for integer orders $n \in \mathbb{N}$ and real arguments $n \in \mathbb{R}$ is the *modified* Bessel functions of the first kind:

$$(\text{modified})I_n(x) \equiv i^{-n}J_n\left(\underbrace{ix}_{z}\right) = \frac{e^{-\frac{\pi}{2}ni}}{2\pi}\int_{-\pi}^{\pi} e^{-x\cdot\sin(\theta) - in\theta}d\theta = \frac{1}{2\pi}\int_{-\pi}^{\pi} e^{-x\cdot\sin(\theta) - in\left(\theta + \frac{\pi}{2}\right)}d\theta =$$

$$\underbrace{\left\{\begin{array}{c}\theta + \frac{\pi}{2} = a \\ -\sin(\theta) = -\sin\left(a - \frac{\pi}{2}\right) = \cos(a)\end{array}\right\}}_{\text{integral change of variables}} = \frac{1}{2\pi}\int_{-\frac{\pi}{2}}^{\frac{3\pi}{2}} e^{x\cdot\cos(a) - ina}da =$$

$$\underbrace{\frac{1}{2\pi}\int_{-\frac{\pi}{2}}^{\frac{3\pi}{2}} e^{x\cdot\cos(a)}\cos(na)da}_{A} - \underbrace{\frac{i}{2\pi}\int_{-\frac{\pi}{2}}^{\frac{3\pi}{2}} e^{x\cdot\cos(a)}\sin(na)da}_{B} = \frac{1}{\pi}\int_{0}^{\pi} e^{x\cdot\cos(\theta)}\cos(n\theta)d\theta, x \in \mathbb{R}, n \in \mathbb{N}.$$

The periods of both integrands in A and B are 2π and the integral range can be split in half:

$$A = \frac{1}{2\pi}\int_{-\frac{\pi}{2}}^{\frac{3\pi}{2}} e^{x\cdot\cos(a)}\cos(na)da = \frac{1}{2\pi}\int_{-\frac{\pi}{2}}^{\frac{\pi}{2}} e^{x\cdot\cos(a)}\cos(na)da + \frac{1}{2\pi}\int_{\frac{\pi}{2}}^{\frac{3\pi}{2}} e^{x\cdot\cos(a)}\cos(na)da.$$

Note that $\forall x \in \mathbb{R}$, $\cos(a + \pi) = \cos(a - \pi)$, and the integrand $e^{x\cdot\cos(a)}\cos(na)$ is symmetric with respect to the mid-points 0 and π of the corresponding intervals $\left[-\frac{\pi}{2}, \frac{\pi}{2}\right]$ and $\left[\frac{\pi}{2}, \frac{3\pi}{2}\right]$. In other words:

$$\int_{-\frac{\pi}{2}}^{\frac{\pi}{2}} e^{x\cdot\cos(a)}\cos(na)da = 2\int_{0}^{\frac{\pi}{2}} e^{x\cdot\cos(a)}\cos(na)da \text{ and}$$

$$\int_{\frac{\pi}{2}}^{\frac{3\pi}{2}} e^{x\cdot\cos(a)}\cos(na)da = 2\int_{\frac{\pi}{2}}^{\pi} e^{x\cdot\cos(a)}\cos(na)da.$$

These equalities can be validated by mirroring the integration over $\left[\pi, \frac{3\pi}{2}\right]$, utilizing the 2π trigonometric periodicity, and showing that the integrals along the ranges $\left[\pi, \frac{3\pi}{2}\right]$ and $\left[\frac{\pi}{2}, \pi\right]$ are equal. Therefore:

$$A = \frac{1}{2\pi}\int_{-\frac{\pi}{2}}^{\frac{3\pi}{2}} e^{x\cdot\cos(a)}\cos(na)da =$$

$$\frac{1}{\pi}\int_{0}^{\frac{\pi}{2}} e^{x\cdot\cos(a)}\cos(na)da + \frac{1}{\pi}\int_{\frac{\pi}{2}}^{\pi} e^{x\cdot\cos(a)}\cos(na)da = \frac{1}{\pi}\int_{0}^{\pi} e^{x\cdot\cos(a)}\cos(na)da.$$

The second term, B, is trivial since

$$B = \int_{-\frac{\pi}{2}}^{\frac{3\pi}{2}} e^{x\cdot\cos(a)}\sin(na)da = \underbrace{\overbrace{\int_{-\frac{\pi}{2}}^{\frac{\pi}{2}} \overbrace{e^{x\cdot\cos(a)}\sin(na)}^{f(a)}\, da}^{0}}_{f(a)+f(-a)=0} + \underbrace{\int_{\frac{\pi}{2}}^{\frac{3\pi}{2}} e^{x\cdot\cos(a)}\sin(na)da}_{0} = 0.$$

The triviality of the second term can be derived similarly to the derivation of term A above:

$$\int_{\pi}^{\frac{3\pi}{2}} e^{x\cdot\cos(a)}\sin(na)da = -\int_{-\frac{3\pi}{2}}^{-\pi} e^{x\cdot\cos(a)}\sin(na)da = -\int_{\frac{\pi}{2}}^{\pi} e^{x\cdot\cos(a)}\sin(na)da.$$

Thus, for integer orders, $I_n(x) = A + B = \frac{1}{\pi}\int_0^{\pi} e^{x\cdot\cos(\theta)}\cos(n\theta)d\theta$ and the special case of the zeroth-order modified Bessel functions of the first kind is:

$$\text{(modified zeroth order Bessel function) } I_o(x) = \frac{1}{\pi}\int_0^{\pi} e^{x\cdot\cos(\theta)}d\theta.$$

An interesting property of the Bessel functions is that they are orthonormal:

$$\underbrace{\int_0^{\infty} J_n(x)dx = 1}_{\text{normalization}} \text{ and } \underbrace{\int_0^{a} J_n\left(a_{n,l}\frac{x}{a}\right) \times J_n\left(a_{n,k}\frac{x}{a}\right)xdx = \frac{1}{2}a^2[J_{n+1}(a_{n,l})]^2\delta_{l,k}}_{\text{orthogonality in}[0,a]},$$

where $a > 0$ is a constant, $a_{n,l}$ and $a_{n,k}$ are the lth and k^{th} zeros of J_n (there are infinitely many roots that extend to infinity) and the Kronecker delta $\delta_{l,k} = \begin{cases} 1, l = k \\ 0, l \neq k \end{cases}$.

The special case of the real-valued zeroth-order modified Bessel function is:

$$I_0(x) = \frac{1}{2\pi}\int_{-\pi}^{\pi} e^{x\cdot\cos(\theta)}d\theta \equiv \frac{1}{\pi}\int_{0}^{\pi} e^{x\cdot\cos(\theta)}d\theta = \sum_{k=0}^{\infty}\frac{x^{2k}}{4^k(k!)^2}, \forall x \in \mathbb{R}.$$

Recall that the Gauss integral provides the normalization coefficient of the normal distribution:

$$2\int_{0}^{\infty} e^{-a\theta^2}d\theta = \int_{-\infty}^{\infty} e^{-a\theta^2}d\theta = \sqrt{\frac{\pi}{a}}, \text{ or for } a = \frac{1}{2\sigma^2}, \ \frac{1}{\sqrt{2\pi\sigma^2}}\int_{-\infty}^{\infty} e^{-\frac{\theta^2}{2\sigma^2}}d\theta = 1.$$

We will provide three alternative derivations of the normal approximation to the Rician distribution of the fMRI intensities for high SNR by examining the asymptotic behavior of the zeroth-order modified Bessel function of the first kind when the argument increases, $x \to \infty$.

The *first approach* uses the *Laplace method* for analyzing integral convergence [221, 222]. For a pair of functions $g(\theta) > 0$ and $f(\theta) \in C^2((a,b) \subseteq \mathbb{R})$, if there is a unique maximum of f attained at $\theta_o \in (a,b)$, $f(\theta_o) = \max_{\theta\in(a,b)} f(\theta)$, and if $f''(\theta_o) < 0$, then the Laplace method allows us to approximate the integral as follows:

$$F(x) = \int_{a}^{b} g(\theta)e^{x\cdot f(\theta)}d\theta, \ \lim_{x\to\infty} F(x) \cong \lim_{x\to\infty}\left(\sqrt{\frac{2\pi}{x|f''(\theta_o)|}}g(\theta_o)e^{x\cdot f(\theta_o)}\right).$$

For such integrals, the Laplace method yields a valid approximation by considering a neighborhood where $f(\theta)$ attains its maximum.

In our specific case of the zeroth-order modified Bessel function, $I_0(x)$, we have $a = -\pi$, $b = \pi$, $g(\theta) = 1$, $f(\theta) = \cos(\theta)$, $f(\theta_o) = \max_{\theta\in(a,b)} f(\theta) = 1$, $\theta_o = 0$, and $f''(\theta_o) = -\cos 0 = -1 < 0$. Therefore, for large SNR ($x \to \infty$), the Laplace approximation method suggests:

$$\lim_{x\to\infty} I_0(x) = \frac{1}{2\pi}\lim_{x\to\infty}\int_{-\pi}^{\pi} e^{x\cdot\cos(\theta)}d\theta = \frac{1}{2\pi}\lim_{x\to\infty}\left(\sqrt{\frac{2\pi}{x}}e^{x\cdot\cos(\theta_o)}\right) = \lim_{x\to\infty}\left(\frac{e^x}{\sqrt{2\pi x}}\right).$$

This argument assumes that $f(\theta_o) \equiv \cos(\theta_o) = \max_{\theta\in(a,b)} f(\theta) = 1$, which is only true in a limiting sense, $\cos(\theta_o \equiv 0) = \sup_{\theta\in(-\pi,\pi)} \cos(\theta) = 1$.

The *second alternative derivation* proves the asymptotic behavior of $I_0(x)$ for large arguments using first principles. This approach examines the lower and upper bounds of the modified Bessel function as $x \to \infty$:

Case 1 (Lower Bound): Note that $\cos(\theta) \ge 1 - \frac{1}{2}\theta^2, \forall\theta \in [-\pi,\pi]$. This can be derived by letting $h(\theta) = \cos(\theta) - 1 + \frac{1}{2}\theta^2$ be the difference of the two hand sides and observing that h is a positive and monotonically increasing function that attains its minimum at $\theta = 0$:

$$h(0)=0,\ h''(\theta)=1-\cos(\theta)\geq 0.$$

Applying a change of variables transformation, $\left\{\begin{matrix}\sqrt{x}\theta=y\\ \sqrt{x}d\theta=dy\end{matrix}\right\}$, the modified Bessel function integral becomes:

$$\int_0^{\pi} e^{x\cdot\cos(\theta)}d\theta\geq\int_0^{\pi} e^{x\cdot\left(1-\frac{1}{2}\theta^2\right)}d\theta=e^x\int_0^{\pi} e^{-x\cdot\left(\frac{1}{2}\theta^2\right)}d\theta=e^x\frac{1}{\sqrt{x}}\int_0^{\sqrt{x}\pi} e^{-\left(\frac{1}{2}y^2\right)}dy.$$

Thus, the limit of the zeroth-order modified Bessel function of the first kind is bounded below:

$$\lim_{x\to\infty} I_0(x)=\frac{1}{2\pi}\lim_{x\to\infty}\int_{-\pi}^{\pi} e^{x\cdot\cos(\theta)}d\theta=\frac{1}{\pi}\lim_{x\to\infty}\int_0^{\pi} e^{x\cdot\cos(\theta)}d\theta\geq$$

$$\lim_{x\to\infty}\left(\frac{1}{\pi}e^x\frac{1}{\sqrt{x}}\int_0^{\infty} e^{-\left(\frac{1}{2}y^2\right)}dy\right)=\lim_{x\to\infty}\left(e^x\frac{1}{\pi\sqrt{x}}\sqrt{\frac{\pi}{2}}\right)=\lim_{x\to\infty}\left(\frac{e^x}{\sqrt{2\pi x}}\right).$$

Case 2 (Upper Bound): For each $\frac{\pi}{4}>|\varepsilon|>|\theta|>0$, $\cos(\theta)\leq 1-\frac{1}{2}\cos(\varepsilon)\theta^2,\ \forall|\theta|>0$. This can be derived again by examining the difference of the two terms, $h(\theta)=\cos(\theta)-1+\frac{1}{2}\cos(\varepsilon)\theta^2$, and observing that h is a negative monotonically decreasing function that attains its maximum at $\theta=0$:

$$h(0)=0,\ h''(\theta)=-\cos(\theta)+\cos(\varepsilon)\leq 0.$$

Therefore, modified Bessel function integral in this case is bounded above by:

$$\int_0^{\pi} e^{x\cdot\cos(\theta)}d\theta=\int_0^{\varepsilon} e^{x\cdot\cos(\theta)}d\theta+\int_{\varepsilon}^{\pi} e^{x\cdot\cos(\theta)}d\theta\leq$$

$$\int_0^{\varepsilon} e^{x\left(1-\frac{1}{2}\cos(\varepsilon)\theta^2\right)}d\theta+\pi e^{x(\cos(\varepsilon))}=$$

$$e^x\left(\int_0^{\varepsilon} e^{\left(-\frac{x}{2}\cos(\varepsilon)\theta^2\right)}d\theta+\pi e^{x(\cos(\varepsilon)-1)}\right)=\underbrace{\left\{\begin{matrix}\sqrt{x\cos(\varepsilon)}\theta=y\\ \sqrt{x\cos(\varepsilon)}d\theta=dy\end{matrix}\right\}}_{\text{integral change of variables}}=$$

$$\frac{e^x}{\sqrt{x\cos(\varepsilon)}}\left(\int_0^{\sqrt{x\cos(\varepsilon)}\varepsilon} e^{-\frac{y^2}{2}}dy+\pi e^{-x(1-\cos(\varepsilon))}\right).$$

Now we can put together these two cases, for any small $\varepsilon>0$, the limit of the modified Bessel function of the first kind is bounded by:

$$\lim_{x\to\infty}\left(\frac{e^x}{\sqrt{2\pi x}}\right)\underbrace{\le}_{\text{Case 1}}\lim_{x\to\infty} I_0(x)\underbrace{\le}_{\text{Case 2}}\lim_{x\to\infty}\left(\frac{1}{\pi}\frac{e^x}{\sqrt{x\cos(\varepsilon)}}\left(\underbrace{\int_0^{\sqrt{x\cos(\varepsilon)}\varepsilon} e^{-\frac{y^2}{2}}dy}_{\text{finite}}+\underbrace{\pi e^{\overbrace{-x\left(1-\cos(\varepsilon)\right)}^{\ge 0}}}_{\to 0}\right)\right)$$

$$\le \lim_{x\to\infty}\left(\frac{1}{\pi}\frac{e^x}{\sqrt{x\cos(\varepsilon)}}\underbrace{\left(\int_0^{\infty} e^{-\frac{y^2}{2}}dy\right)}_{\sqrt{\frac{\pi}{2}}}\right)=\lim_{x\to\infty}\left(\frac{e^x}{\sqrt{2\pi x\cos(\varepsilon)}}\right).$$

For large SNR (as $x\to\infty$), the dominating term of the limit on the right is effectively $\frac{e^x}{\sqrt{2\pi x}}$. Therefore, for large arguments, the zeroth-order modified Bessel function of the first kind can be approximated by:

$$I_0(x)=\frac{1}{2\pi}\int_{-\pi}^{\pi} e^{x\cdot\cos(\theta)}d\theta\equiv\frac{1}{\pi}\int_0^{\pi} e^{x\cdot\cos(\theta)}d\theta\cong\frac{e^x}{\sqrt{2\pi x}}.$$

Next, we recall that in equation (4.5), we have

$$\|\rho_t\|\approx x_t{}'\beta+N(0,\sigma^2),\ \ \frac{x_t{}'\beta}{\sigma}\to\infty\Rightarrow\sqrt{\frac{\|\rho_t\|}{x_t{}'\beta}}\xrightarrow{p}1,$$

$$p(\|\rho_t\|\ |x_t,\ \beta,\ \sigma^2)\sim N(x_t{}'\beta,\sigma^2).$$

To connect the Rician distribution of the fMRI magnitude to a normal distribution density in the case of large SNR, recall equation (4.6):

$$\underbrace{p(\|\rho_t\|\ |x_t,\beta,\sigma^2)}_{\text{Rician PDF}(\nu=x_t{}'\beta,\sigma)}$$

$$=\frac{\|\rho_t\|}{\sigma^2}e^{-\frac{1}{2\sigma^2}\left(\|\rho_t\|^2+\left(\underbrace{x_t{}'\beta}_{\nu}\right)^2\right)}\times\underbrace{\frac{1}{2\pi}\int_{\theta_t=-\pi}^{\pi} e^{\frac{1}{\sigma^2}\left(x_t{}'\beta\|\rho_t\|\cos(\theta_t-\theta)\right)}d\theta_t}_{\text{modified Bessel function } I_0\left(\frac{x_t{}'\beta\|\rho_t\|}{\sigma^2}\right)}.$$

$$p(\|\rho_t\|\ |x_t,\beta,\sigma^2)\approx\frac{\|\rho_t\|}{\sigma^2}e^{-\frac{1}{2\sigma^2}\left(\|\rho_t\|^2+\left(\underbrace{x_t{}'\beta}_{\nu}\right)^2\right)}\frac{1}{\sqrt{2\pi\frac{x_t{}'\beta\|\rho_t\|}{\sigma^2}}}e^{\frac{x_t{}'\beta\|\rho_t\|}{\sigma^2}}=$$

$$\frac{1}{\sqrt{\frac{(x_t'\beta)}{\|\rho_t\|}2\pi\sigma^2}} e^{-\frac{1}{2\sigma^2}\left(\|\rho_t\|^2 + \left(\underbrace{x_t'\beta}_{v}\right)^2\right)} e^{2\frac{x_t'\beta\|\rho_t\|}{2\sigma^2}} =$$

$$\frac{1}{\sqrt{\frac{(x_t'\beta)}{\|\rho_t\|}2\pi\sigma^2}} e^{-\frac{1}{2\sigma^2}\left(\|\rho_t\|^2 + \left(\underbrace{x_t'\beta}_{v}\right)^2 - 2x_t'\beta\|\rho_t\|\right)} =$$

$$\frac{1}{\sqrt{\frac{(x_t'\beta)}{\|\rho_t\|}2\pi\sigma^2}} e^{-\frac{1}{2\sigma^2}(\|\rho_t\| - x_t'\beta)^2} = \underbrace{\sqrt{\frac{\|\rho_t\|}{x_t'\beta}}}_{\substack{\cong 1 \\ \text{for high SNR,} \\ \text{due to equ (4.5)}}} \left(\underbrace{\frac{1}{\sqrt{2\pi\sigma^2}} e^{-\frac{1}{2\sigma^2}(\|\rho_t\| - x_t'\beta)^2}}_{N(x_t'\beta,\sigma^2)}\right).$$

The final *third approach* relies on algebraic derivation of the normal approximation to fMRI magnitude Rician distribution for large SNR. The zeroth-order modified Bessel Function of the first kind can be approximated as $x \to \infty$ by:

$$I_0(x) = e^{-x + O\left(\left(\frac{1}{x}\right)^2\right)}\left(\frac{i\sqrt{\frac{1}{x}}}{\sqrt{2\pi}} + O\left(\left(\frac{1}{z}\right)^{\frac{3}{2}}\right)\right) + e^{x + O\left(\left(\frac{1}{x}\right)^2\right)}\left(\frac{\sqrt{\frac{1}{x}}}{\sqrt{2\pi}} + O\left(\left(\frac{1}{x}\right)^{\frac{3}{2}}\right)\right).$$

We will derive this approximation up to a constant coefficient $\frac{1}{\sqrt{2\pi}}$. By definition, $I_0(x)$ is a solution of this second-order differential equation (where $v = 0$):

$$x^2\frac{d^2y}{dx^2} + x\frac{dy}{dx} - (x^2 + v^2)y = 0,$$

which is equivalent to:

$$\frac{d^2y}{dx^2} + \frac{1}{x}\frac{dy}{dx} - y = 0.$$

As x is a large positive real argument, we can make a change of variables transformation $u = u(x) \equiv y\sqrt{x}$ and rewrite the equation in terms of $u(x)$:

$$\frac{d^2u}{dx^2} = \left(1 - \frac{1}{4x^2}\right)u.$$

An additional variable transformation $u = v(x)e^x$ yields another second-order differential equation in terms of the new function $v = v(x) = ue^{-x}$:

$$\frac{d^2v}{dx^2} + 2\frac{dv}{dx} + \frac{v}{4x^2} = 0.$$

Let's assume a solution to this differential equation exists and can be expanded as an infinite series:

$$v = A\left(1 + \frac{c_1}{x} + \frac{c_2}{x^2} + \frac{c_3}{x^3} + \cdots\right),$$

for some global normalizing constant A, which in practice is $\frac{1}{\sqrt{2\pi}}$, and some series of constants $\{c_k\}_{k=1}^{\infty}$ as multipliers of the power terms in the series expansion.

Plugging in this infinite series representation of the solution into the differential equation, we obtain:

$$\left(2c_1 - \frac{1}{4}\right)\frac{1}{x^2} + \left(4c_2 - \left(\frac{3}{2}\right)^2 c_1\right)\frac{1}{x^3} + \left(6c_3 - \left(\frac{5}{2}\right)^2 c_2\right)\frac{1}{x^4} + \cdots = 0.$$

Note that the global constant A cancels out and we can set to zero the coefficients of all power terms $\left\{\frac{1}{x^k}\right\}_{k=2}^{\infty}$. These coefficients can be explicitly computed to be:

$$c_1 = \frac{1^2}{8}, c_2 = \frac{1^2 \cdot 3^2}{2!8^2}, c_3 = \frac{1^2 \cdot 3^2 \cdot 5^2}{3!8^3}, \cdots, c_k = \frac{\prod_{l=0}^{k-1}(2l+1)^2}{k!8^k}, \cdots.$$

Plugging the constant terms back into the series expansion of the solution $v(x)$, we obtain:

$$v = A\left(1 + \frac{1^2}{8}\frac{1}{x} + \frac{1^2 \cdot 3^2}{2!8^2}\frac{1}{x^2} + \frac{1^2 \cdot 3^2 \cdot 5^2}{3!8^3}\frac{1}{x^3} + \cdots + \frac{\prod_{l=0}^{k-1}(2l+1)^2}{k!8^k}\frac{1}{x^k} + \cdots\right).$$

By inverting the transformations, $y(x) = \frac{e^x}{\sqrt{x}}v(x)$ and letting $x \to \infty$, we can expand the zeroth-order modified Bessel function of the first kind as:

$$y(x) \equiv I_0(x) = \frac{Ae^x}{\sqrt{x}}\left(1 + \frac{1^2}{8}\frac{1}{x} + \frac{1^2 \cdot 3^2}{2!8^2}\frac{1}{x^2} + \frac{1^2 \cdot 3^2 \cdot 5^2}{3!8^3}\frac{1}{x^3} + \cdots + \frac{\prod_{l=0}^{k-1}(2l+1)^2}{k!8^k}\frac{1}{x^k} + \cdots\right). \tag{4.9}$$

To estimate the exact value of the global constant, $A = \frac{1}{\sqrt{2\pi}}$, one can use the Bessel function of the third kind (Hankel function) [223] and the integral expression of the modified Bessel function of the second kind. As the argument increases, $\frac{x_t{'}\beta\|\rho_t\|}{\sigma^2} \to \infty$, the dominant term in the function expansion (4.9) is $\frac{Ae^x}{\sqrt{x}}$ and therefore, $I_0\left(\frac{x_t{'}\beta\|\rho_t\|}{\sigma^2}\right) \cong \frac{1}{\sqrt{2\pi\frac{x_t{'}\beta\|\rho_t\|}{\sigma^2}}} e^{\frac{x_t{'}\beta\|\rho_t\|}{\sigma^2}}$.

The above derivations allow us to approximate the probability distribution of the magnitude and its likelihood function for *small*, *medium*, and *large* SNRs as follows:

1. For small SNR, $x_t'\beta \to 0$, $I_o\left(\frac{x_t'\beta\|\rho_t\|}{\sigma^2}\right) \to I_o(0) = 1$:

$$p(\|\rho_t\| \, |x_t, \beta, \sigma^2) \cong \underbrace{\frac{\|\rho_t\|}{\sigma^2} e^{-\frac{\|\rho_t\|^2}{2\sigma^2}}}_{\text{Rayleigh distribution}},$$

$$\underbrace{L(\|\rho\| \, |X, \beta, \sigma^2)}_{\text{likelihood}} = \prod_t p(\|\rho_t\| |x_t, \beta, \sigma^2) = \frac{1}{\sigma^{2n}} \left(\prod_t \|\rho_t\|\right) e^{-\frac{\sum_t (\|\rho_t\| - x_t'\beta)(\|\rho_t\| - x_t'\beta)}{2\sigma^2}},$$

$$ll = \ln L = -2n \ln \sigma + \sum_t \ln \|\rho_t\| - \frac{1}{2\sigma^2}(\|\rho\| - X\beta)'(\|\rho\| - X\beta) \cong$$

$$-n \ln \sigma^2 + \sum_t \ln\|\rho_t\| - \frac{1}{2\sigma^2}\left(\sum_t \|\rho_t\|^2 + \sum_t (x_t'\beta)^2\right) + \sum_t \ln I_o\left(\frac{x_t'\beta\|\rho_t\|}{\sigma^2}\right),$$

where the Rician density can be expressed as before

$$p(\|\rho_t\| |x_t, \beta, \sigma^2) \cong \frac{\|\rho_t\|}{\sigma^2} e^{-\frac{1}{2\sigma^2}\left(\|\rho_t\|^2 + \left(\underbrace{x_t'\beta}_{v}\right)^2\right)} \times I_o\left(\frac{x_t'\beta\|\rho_t\|}{\sigma^2}\right),$$

and $I_o\left(\frac{x_t'\beta\|\rho_t\|}{\sigma^2}\right) = \frac{1}{2\pi}\int_{-\pi}^{\pi} e^{-\frac{1}{\sigma^2}(x_t'\beta\|\rho_t\| \cos(\varphi - \theta))} d\varphi$ is a zeroth-order modified Bessel function of the first kind.

2. For intermediate SNR levels, we can use any order of the Taylor series in equation (4.9) and Markov Chain Monte Carlo techniques to numerically estimate the likelihood function.
3. For large SNR, $x_t'\beta \to \infty$, a first-order Taylor series approximation yields:

$$\underbrace{p(\|\rho_t\| \, |x_t, \beta, \sigma^2)}_{\text{Rician PDF}(v = x_t'\beta, \sigma)} =$$

$$\frac{\|\rho_t\|}{\sigma^2} e^{-\frac{1}{2\sigma^2}\left(\|\rho_t\|^2 + \left(\underbrace{x_t'\beta}_{v}\right)^2\right)} \times \underbrace{\frac{1}{2\pi}\int_{-\pi}^{\pi} e^{\frac{1}{\sigma^2}(x_t'\beta\|\rho_t\| \cos(\varphi - \theta))} d\varphi}_{\text{modified Bessel function } I_o\left(\frac{x_t'\beta\|\rho_t\|}{\sigma^2}\right)}.$$

$$p(\|\rho_t\| \, |x_t, \beta, \sigma^2) = \underbrace{\sqrt{\frac{\|\rho_t\|}{x_t'\beta}}}_{\cong 1} \left(\frac{1}{\sqrt{2\pi\sigma^2}} e^{-\frac{1}{2\sigma^2}(\|\rho_t\| - x_t'\beta)^2} \right) \cong \frac{1}{\sqrt{2\pi\sigma^2}} e^{-\frac{1}{2\sigma^2}(\|\rho_t\| - x_t'\beta)^2} \sim N(x_t'\beta, \sigma^2).$$

$$\underbrace{l(\| \rho \| \, |X, \beta, \sigma^2)}_{\text{likelihood}} = \prod_t p(\|\rho_t\| \, |x_t, \beta, \sigma^2) = \frac{1}{\sigma^n} \left(\frac{1}{2\pi} \right)^{\frac{n}{2}} e^{-\frac{(\|\rho\| - X\beta)'(\|\rho\| - X\beta)}{2\sigma^2}} =$$

$$\sigma^{-n} (2\pi)^{-\frac{n}{2}} e^{-\frac{(\|\rho\| - X\beta)'(\|\rho\| - X\beta)}{2\sigma^2}}, \tag{4.10}$$

$$\underbrace{ll}_{\text{log - likelihood}} = \ln l = -\frac{n}{2} \ln \sigma^2 - \underbrace{\frac{n}{2} \ln(2\pi)}_{\text{constant}} - \frac{(\| \rho \| - X\beta)'(\| \rho \| - X\beta)}{2\sigma^2} \propto$$

$$-n \ln \sigma^2 - \frac{1}{\sigma^2} \sum_t (\|\rho_t\| - x_t'\beta)(\|\rho_t\| - x_t'\beta).$$

In the next section, we will illustrate how to use this approximation of the log-likelihood function to compute the maximum likelihood estimates (MLE) of the parameters. For instance, we will show that the MLE of the variance, $\hat{\sigma}^2$, is obtained by setting to zero the partial derivative of the ll function w.r.t. σ^2:

$$0 = \frac{\partial ll}{\partial \sigma^2} = -\frac{n}{\sigma^2} + \frac{1}{2\sigma^4} \left(\|\rho\| - X\hat{\beta} \right)' \left(\|\rho\| - X\hat{\beta} \right) \Rightarrow \hat{\sigma}^2 = \frac{1}{2n} \left(\|\rho\| - X\hat{\beta} \right)' \left(\|\rho\| - X\hat{\beta} \right).$$

4.3.3 Parameter Estimation

In various brain regions, different SNR levels may correspond to different statistical inference. To estimate the parameters, β, σ, we can use alternative strategies such as least squares, the method of moments, or maximum likelihood estimation (MLE).

Assume we have a specific *contrast* matrix $C = C_{q\times(d+1)}$, where q = number of contract constraints (e.g., voxel locations) and d = number of covariate features in the linear model (e.g., linear and quadratic BOLD hemodynamic response effects, linear habituation effects). To represent a realistic inference problem, C should be a full row-rank and correspond to the boundary-constrained optimization under a null hypothesis $H_o : C\beta = \tau$ ($\tau_{q\times 1}$ could be 0) vs. an alternative $H_1 : C\beta \neq \tau$. It may be useful to keep in mind the following two pragmatic examples of constrained problems. First, assume that $C_{q\times(d+1)} = \begin{pmatrix} \boldsymbol{I}_{p\times p} & \boldsymbol{0}_{p\times r} \\ \boldsymbol{0}_{(q-p)\times p} & \boldsymbol{0}_{(q-p)\times r} \end{pmatrix}$, where $\boldsymbol{I}_{p\times p}$ is the identity matrix, $\boldsymbol{0}$ is a matrix of zeros, and $p + r = (d+1)$. Then the constraint $C_{q\times(d+1)}\beta_{(d+1)\times 1} = \boldsymbol{\tau}_{q\times 1}$ implies that p elements of the parameter vector β are preset to be exactly equal to the corresponding values provided in the constant vector $\boldsymbol{\tau}$, whereas the remaining $r = (d+1) - p$ elements of β are to be estimated. This constrained problem example simply reduces the dimension of the parameter estimation space. The second example involves a contrast matrix where $q = 1$, $\tau = 0$, and $C_{1\times(d+1)} = \mathbf{1}'$ is a row vector of 1's. This special constraint forces a model solution requiring a zero-sum of the regression parameters.

$$\underbrace{\begin{bmatrix} \begin{bmatrix} 1_{11} & \cdots & 0_{1p} \\ \vdots & \ddots & \vdots \\ 0_{p1} & \cdots & 1_{pp} \end{bmatrix} & \begin{bmatrix} 0_{1(p+1)} & \cdots & 0_{1(d+1)} \\ \vdots & \ddots & \vdots \\ 0_{p(p+1)} & \cdots & 0_{p(d+1)} \end{bmatrix} \\ \begin{bmatrix} 0_{(p+1)1} & \cdots & 0_{(p+1)p} \\ \vdots & \ddots & \vdots \\ 0_{q1} & \cdots & 0_{qp} \end{bmatrix} & \begin{bmatrix} 0_{(p+1)(p+1)} & \cdots & 0_{q(d+1)} \\ \vdots & \ddots & \vdots \\ 0_{q(p+1)} & \cdots & 0_{q(d+1)} \end{bmatrix} \end{bmatrix}}_{C_{q\times(d+1)}} \begin{bmatrix} \left.\begin{bmatrix}\beta_1\\ \vdots \\ \beta_p\end{bmatrix}\right\}\text{fixed} \\ \left.\begin{bmatrix}\beta_{p+1}\\ \vdots\\ \beta_{d+1}\end{bmatrix}\right\}\text{estimated}\end{bmatrix} = \begin{bmatrix}\tau_1\\ \vdots\\ \tau_p\\ \tau_{p+1}\\ \vdots\\ \tau_{d+1}\end{bmatrix}.$$

This *constrained* problem may be solved by maximizing the log-likelihood function (ll) and computing the ratio of the general and the constrained optimized likelihoods [224]. Again, $(\breve{\beta}, \breve{\sigma})$ and $(\hat{\beta}, \hat{\sigma})$ denote the corresponding $\arg\max(ll)$ of the constrained and unconstrained problems, respectively.

For large SNR, we can use equation (4.10) for unconstrained maximization of the likelihood function l with respect to the parameters, β, σ. This optimization is equivalent to minimization of the negative log-likelihood ll, which in turn is equivalent to minimizing the exponent term:

$$\hat{\beta} = \arg\min\left(\sum_t \left(\|\rho_t\| - x_t'\beta\right)^2\right) = \arg\min\left(\sum_t \left(\|\rho_t\| - x_t'\beta\right)\left(\|\rho_t\| - x_t'\beta\right)\right) = (X'X)^{-1}X'\|\rho\|$$

$$\hat{\sigma}^2 = \frac{1}{n}\left(\|\rho\| - X\hat{\beta}\right)'\left(\|\rho\| - X\hat{\beta}\right). \tag{4.11}$$

For the slightly more general model with non-identity (positive definite) variance-covariance matrix, W, the signal distribution is $p(\|\rho\| \,|\, X, \beta, \sigma^2) \sim N(X\beta, \sigma^2 W)$ and the least-square parameter estimates are:

$$\hat{\beta} = \left(X'W^{-1}X\right)^{-1}X'W^{-1}\|\rho\|$$

$$\hat{\sigma}^2 = \frac{1}{n}\left(\|\rho\| - X\hat{\beta}\right)'W^{-1}\left(\|\rho\| - X\hat{\beta}\right).$$

Recall that for a column vector a, like $a = X'\|\rho\| = \left(\|\rho\|'X\right)'$, and a symmetric matrix $A = A'$, like the cross-product matrix $A = X'X$, the *differentiation operator* has the following properties [213]:

$$\frac{\partial a'\beta}{\partial \beta} = a \text{ and } \frac{\partial \beta' A \beta}{\partial \beta} = (A + A')\beta = 2A\beta.$$

Under $H_o{:}C\beta=\tau$, the constrained problem is optimized using an additional Lagrange multiplier term [201, 224–226]. The parameter estimates are obtained by minimizing the residual sum square error (SSE):

$$\check{\beta}=\arg\min_{C\beta=\boldsymbol{\tau}} SSE(\beta)=\arg\min_{C\beta=\boldsymbol{\tau}}\left[(\|\rho\|-X\beta)'(\|\rho\|-X\beta)\right]=$$

$$\arg\min_{C\beta=\boldsymbol{\tau}}\left[\|\rho\|'\|\rho\|-2\|\rho\|'X\beta+\beta'X'X\beta\right],$$

$$\check{\beta}=\hat{\beta}-(X'X)^{-1}C'\left(C(X'X)^{-1}C'\right)^{-1}\left(C\hat{\beta}-\tau\right)= \tag{4.12}$$

$$\underbrace{\left[I-(X'X)^{-1}C'\left(C(X'X)^{-1}C'\right)^{-1}C\right]}_{D}\hat{\beta}+(X'X)^{-1}C'\left(C(X'X)^{-1}C'\right)^{-1}\tau,$$

$$\check{\sigma}^2=\frac{1}{n}(\|\rho\|-X\check{\beta})'(\|\rho\|-X\check{\beta}).$$

The Lagrange dual problem connects the solutions of the regular unconstrained and the constrained optimization problems:

$$\mathcal{L}(\beta,\lambda)=\|\rho\|'\|\rho\|\underbrace{-2\|\rho\|'X\beta+\beta'X'X\beta+\lambda'(C\beta-\tau)}_{\text{depend on }\beta}.$$

The solutions to the dual problem are obtained by setting to zero the partial derivative of the objective function

$$\frac{\partial}{\partial\beta}\mathcal{L}(\beta,\lambda)=\underbrace{-2X'\|\rho\|}_{\frac{\partial a'\beta}{\partial\beta}=a}+\underbrace{2X'X\beta}_{\frac{\partial\beta'A\beta}{\partial\beta}=2A\beta}+\underbrace{C'\lambda}_{\frac{\partial a'\beta}{\partial\beta}=a}=0$$

$$\Rightarrow\check{\beta}=\frac{1}{2}(X'X)^{-1}(2X'\|\rho\|-C'\lambda).$$

As the estimator $\check{\beta}$ is constrained, $C\check{\beta}=\tau$, the Lagrange multiplier estimate $\hat{\lambda}$ is computed by:

$$C\underbrace{\left[\frac{1}{2}(X'X)^{-1}(2X'\|\rho\|-C'\lambda)\right]}_{\check{\beta}}=C\left[\underbrace{(X'X)^{-1}X'\|\rho\|}_{\hat{\beta}}\right]-C\frac{1}{2}(X'X)^{-1}C'\lambda=\tau,$$

$$\Rightarrow\hat{\lambda}=2(C(X'X)C')^{-1}\left(C\hat{\beta}-\tau\right).$$

Using these two formulations of the optimization problem solutions, we can expend the constrained-parameter-estimate:

$$\check{\beta} = \frac{1}{2}(X'X)^{-1}(2X'\|\rho\| - C'\lambda) = \frac{1}{2}(X'X)^{-1}\left(2X'\|\rho\| - 2C'(C(X'X)C')^{-1}\left(C\hat{\beta} - \tau\right)\right) =$$

$$\hat{\beta} - (X'X)^{-1}C'\left(C(X'X)^{-1}C'\right)^{-1}\left(C\hat{\beta} - \tau\right).$$

In the original unconstrained OLS problem presented in equation (4.11), the estimation of the effect-size parameter vector is only valid when the variance-covariance matrix $X'X$ is invertible, i.e., a square matrix with non-trivial determinant. Constraining the problem by introducing the additional restriction, H_o: $C\beta = \boldsymbol{\tau}$, suggests a more generalized (perhaps not unique) solution to the optimization problem when the variance-covariance matrix $X'X$ is not necessarily invertible. In this case, the parameter vector β may be identifiable and its estimate computed using the following more general equation [224]:

$$\check{\beta} = R((XR)'XR)^{-1}(XR)'Y + \left[I_{(d+1)\times(d+1)} - R((XR)'XR)^{-1}(XR)'X\right]C'(CC')^{-1}\tau, \qquad (4.12A)$$

We can always define a new matrix $R_{(d+1)\times(d+1-q)}$ that may not be unique, as there may be multiple alternative solutions, which satisfies the following pair of conditions:

$$\det\left[C'_{(d+1)\times q},\ R_{(d+1)\times(d+1-q)}\right]_{(d+1)\times(d+1)} \neq 0, \text{ and } (CR)' = R'_{(d+1-q)\times(d+1)}C'_{(d+1)\times q} = 0.$$

Then, we can reformulate the linear model using the square invertible matrix A

$$A' = \left[C'_{(d+1)\times q},\ R_{(d+1)\times(d+1-q)}\ \right]_{(d+1)\times(d+1)} \text{ and } Z_{N\times(d+1)} = XA^{-1},$$

$$Y = X\beta + e = XA^{-1}\underbrace{A\beta}_{\gamma} + e = Z\gamma + e,$$

where the modified parameter vector γ can be partitioned into $\gamma = (\gamma_1', \gamma_2')'$, representing q specified constraints ($\gamma_1 = C\beta = \boldsymbol{\tau}$) and the remaining $(d+1-q)$ unspecified constraints ($\gamma_2 = R'\beta$), free parameters. Similar to the Lagrange multipliers approach for solving constrained optimization problems, this transformation $\begin{pmatrix} X \to Z_{N\times(d+1)} \equiv XA^{-1} \\ \beta \to \gamma \equiv A\beta \end{pmatrix}$ translates the initial problem of estimating $d+1$ parameters (β) subject to q constraints (C) to another optimization problem of estimating $(d+1-q)$ free parameters (γ_2) without any constraints (remember that the constant vector $\gamma_1 = \boldsymbol{\tau}$ is known).

The OLS estimate of γ_2 is:

$$\hat{\gamma}_2 = R'R((XR)'XR)^{-1}(XR)'\left(Y - XC'(CC')^{-1}\tau\right).$$

The invertibility of the variance-covariance $X'X$ condition of the original problem $(Y = X\beta + e)$ is replaced in the new model $(Y = Z\gamma + e)$ by invertibility of the *modified* covariance matrix $(XR)'XR$, whose non-singularity is ensured by the definition of $R_{(d+1)\times(d+1-q)}$. This parameter estimation formula, equation (4.12*A*) naturally agrees with the classical OLS formula when $X'X$ is invertible, equation (4.12).

For the constrained problem, the parameter estimates are obtained by minimizing the Lagrangian objective function:

$$\mathcal{L}(\beta, \sigma, \lambda) = \frac{1}{2} SSE(\beta, \sigma) + \lambda'(C\beta - \tau).$$

The minima of $\mathcal{L}(\beta, \sigma, \lambda)$ are obtained by solving the following system of equations:

$$\left| \begin{array}{l} \frac{\partial \mathcal{L}}{\partial \beta} = -X' \parallel \rho \parallel + X'X\beta + C'\lambda = 0 \\ \frac{\partial \mathcal{L}}{\partial \sigma^2} = -\frac{n}{\sigma^2} + \frac{(\|\rho\| - X\beta)'(\|\rho\| - X\beta)}{(\sigma^2)^2} = 0 \\ \frac{\partial \mathcal{L}}{\partial \lambda} = C\beta - \tau = 0 \end{array} \right. . \tag{4.13}$$

Left-multiplication by $C(X'X)^{-1}$ on both sides of the equation $\frac{\partial \mathcal{L}}{\partial \beta} = -X' \|\rho\| + X' X\beta + C'\lambda = 0$ yields:

$$-\underbrace{C(X'X)^{-1}X' \parallel \rho \parallel}_{\hat{\beta} = \hat{\beta}_{OLS}} + C\breve{\beta} + C(X'X)^{-1}C'\breve{\lambda} = -C\hat{\beta} + C\breve{\beta} + C(X'X)^{-1}C'\breve{\lambda} = 0,$$

where the unrestricted (OLS) estimates are denoted by $\hat{\square}$ and the restricted model estimates are denoted by $\breve{\square}$ (breve notation). Solving for $\breve{\lambda}$ using the third equation $C\breve{\beta} - \tau = 0$ we obtain $C(X'X)^{-1}C'\breve{\lambda} = C\hat{\beta} - \tau$, and therefore:

$$\breve{\lambda} = \left[C(X'X)^{-1}C'\right]^{-1}\left(C\hat{\beta} - \tau\right).$$

Note that the square matrix $R = C(X'X)^{-1}C'$ is positive definite, full rank, and invertible, i.e., $R > 0$ and $a'Ra > 0, \forall a > 0$.

Plugging in the unrestricted estimate $\breve{\lambda}$ in the first equation $\frac{\partial \mathcal{L}}{\partial \beta} = 0$ yields the unrestricted $\breve{\beta}$ estimate:

$$-X'\|\rho\| + X'X\breve{\beta} + C'\left[C(X'X)^{-1}C'\right]^{-1}\left(C\hat{\beta} - \tau\right) = 0,$$

$$X'X\breve{\beta} = X'\|\rho\| - C'\left[C(X'X)^{-1}C'\right]^{-1}\left(C\hat{\beta} - \tau\right),$$

$$\breve{\beta} = (X'X)^{-1}X'\|\rho\| - (X'X)^{-1}C'\left[C(X'X)^{-1}C'\right]^{-1}\left(C\hat{\beta} - \tau\right)$$

$$\breve{\beta} = \hat{\beta} - (X'X)^{-1}C'\left(C(X'X)^{-1}C'\right)^{-1}\left(C\hat{\beta} - \tau\right).$$

If $\tau = 0$, we have $\breve{\beta} = D\hat{\beta}_{OLS}$, where $D = I_{d+1} - (X'X)^{-1}C'\left(C(X'X)^{-1}C'\right)^{-1}C$, as shown in [201]. The last (variance) parameter estimate in equation (4.13) is obtained by solving the equation $\frac{\partial \mathcal{L}}{\partial \sigma^2} = 0$, which yields:

$$\breve{\sigma}^2 = \frac{1}{n}(\|\rho\| - X\breve{\beta})'(\|\rho\| - X\breve{\beta}).$$

Under the high SNR assumption, we have the log-likelihood approximation equation (4.10) and we can compute the maximum likelihood parameter estimates and their variability. Then, equation (4.8) suggests how we can conduct contrast-based magnitude-model inference by using the magnitude-only log-likelihood ratio test statistics (Λ_m):

$$\Lambda_m(X) \equiv -2\log \lambda_m(X) = n\log\left(\frac{\breve{\sigma}^2}{\hat{\sigma}^2_{OLS}}\right) \sim \chi^2_r,$$

where the degree of freedom of the chi-square distribution (χ^2_{df}) is $df = r$, the full rank of the contrast matrix C.

As an example, suppose we are fitting a three-term linear model using (i) an intercept, β_o, (ii) a longitudinal linear drift (habituation effect), β_1, and (iii) a contrast effect of a stimulus (BOLD hemodynamic response effect), β_2. Then, just testing for the effect of the stimulus at one spatial voxel location requires statistical quantification of the corresponding effect-size under a null hypothesis $H_o{:}C\beta = 0$. For instance, to solely test for a stimulus effect (β_2) in the general effect vector $\beta = (\beta_o, \beta_1, \beta_2)'$, we can use the contrast $C_{1\times(1+2)} = (0, 0, 1)$, since $0 = \langle C'|\beta\rangle = C\beta = \beta_2$. In this case, the LRT will asymptotically follow χ^2_1 distribution. Note that the chi-square degree of freedom $df = 1$ as we have just one additional alternative model parameter, β_2, compared to the trivial null model corresponding to the nil contrast $c = (0, 0, 0)'$. By Wilk's theorem [216], as the sample size (time) increases, i.e., $n \to \infty$, the LRT statistic $-2\log \Lambda_m$ approaches asymptotically χ^2_1, as the degrees of freedom reflect the difference in the parameters between the alternative and the research hypotheses. In general, if the full rank of the contrast C is r, then the distribution of the log LRT is asymptotically approximated by $\chi^2_{df=r}$ distribution.

In the simpler univariate case $H_o{:}\beta_2 = 0$ vs. $H_1{:}\beta_2 \neq 0$, and subject to common parametric assumptions, e.g., IID samples of a normal distribution population, the likelihood ratio may also be expressed approximately as:

$$\Lambda_m = \left(1 + \frac{t^2}{n-1}\right)^{-\frac{n}{2}},$$

where $t \sim t_{n-1}$, see [227]. This explains the relation with the familiar t test, the corresponding variance estimate $SE\left(\hat{\beta}_2\right) = \hat{\sigma}_{\hat{\beta}_2}$, and the test statistic:

$$t_2 = \frac{\hat{\beta}_2}{SE\left(\hat{\beta}_2\right)} \sim t_{df}.$$

4.3.4 Marginal Distribution of the Phase

For each variable, r (intensity-magnitude) and φ (intensity-phase), the corresponding marginal distributions are obtained by integrating the joint distribution over the range of the other variable. For instance, the marginal distribution of the magnitude (r) is computed by integrating the joint distribution over the intensity phase space (φ).

In the previous sections we estimated the *marginal distribution of the intensity magnitude*, r:

$$p(r|\Theta) \approx \begin{cases} \frac{r}{\sigma^2} e^{-\frac{r^2}{2\sigma^2}} \sim Rice(0,\sigma) \equiv Rayleigh\,(\sigma), \text{ for low SNR, under } H_o: \beta = 0, m = x_t{}'\beta = 0, \\ \frac{e^{-\frac{(r-m)^2}{2\sigma^2}}}{\sqrt{2\pi\sigma^2}} \sim N\,(m, \sigma^2), \text{ for high SNR, when } m = x_t'\beta \to \infty \end{cases} \tag{4.14}$$

This approximation relied on estimating the asymptotic behavior of the zeroth-order modified Bessel function of the first kind, $I_0(x)$ [228]. Using equations (4.10)–(4.13), we can denote the parameters maximizing the log-likelihood function, for the unconstrained and constrained (H_o:$C\beta = \tau$) problems by $(\hat{\beta}, \hat{\sigma})$ and $(\breve{\beta}, \breve{\sigma})$, respectively. Setting to zero the partial derivatives of the log-likelihoods, with and without constraining the magnitude-only problem, yields the maxima of the likelihood, which are attained at:

$$\left| \begin{array}{l} \hat{\beta} = (X'X)^{-1} X'r, \quad \hat{\sigma} = \frac{1}{n}\left(r - X\hat{\beta}\right)'\left(r - X\hat{\beta}\right) \\ \breve{\beta} = D(X'X)^{-1} X'r, \quad \breve{\sigma} = \frac{1}{n}(r - X\breve{\beta})'(r - X\breve{\beta}) \cdot \\ D = I_{d+1} - (X'X)^{-1} C'\left(C(X'X)^{-1} C'\right)^{-1} C \end{array} \right.$$

The *marginal distribution of the intensity-value phase*, φ, may be computed in terms of the error function $\text{erf}(z) = \frac{1}{\sqrt{\pi}} \int_{-z}^{z} e^{-\omega^2} d\omega$ by integrating the joint density over the polar radius space, $r \in [0{:}\ \infty)$:

$$p(\varphi|\Theta) = \int_0^\infty \frac{r}{2\pi\sigma^2}\, e^{-\frac{1}{2\sigma^2}\left(r^2 + m^2 - 2r\, m\cos(\varphi - \theta)\right)} dr =$$

$$\frac{e^{-\frac{m^2}{2\sigma^2}}}{4\pi\sigma^2}\left(2\sigma^2 + \sqrt{2\pi\sigma^2}\, m\cos(\varphi - \theta)\, e^{\frac{m^2\cos^2(\varphi - \theta)}{2\sigma^2}}\left(1 + \text{erf}\left(\frac{m\cos(\varphi - \theta)}{\sqrt{2\sigma^2}}\right)\right)\right).$$

To derive this equation, we let $a = m\cos(\varphi - \theta)$ and expand the marginal distribution:

$$p(\varphi|\Theta) = \int_0^\infty \frac{r}{2\pi\sigma^2} e^{-\frac{1}{2\sigma^2}\left(r^2+m^2-2r\,m\cos(\varphi-\theta)\right)} dr = \frac{e^{-\frac{m^2}{2\sigma^2}}}{2\pi\sigma^2} \int_0^\infty r e^{-\frac{1}{2\sigma^2}\left((r-a)^2-a^2\right)} dr =$$

$$\frac{e^{-\frac{m^2}{2\sigma^2}}}{2\pi\sigma^2} e^{\frac{a^2}{2\sigma^2}} \int_0^\infty (r-a+a) e^{-\frac{(r-a)^2}{2\sigma^2}} dr =$$

$$\frac{e^{-\frac{m^2}{2\sigma^2}}}{2\pi\sigma^2} e^{\frac{a^2}{2\sigma^2}} \int_0^\infty (r-a) e^{-\frac{(r-a)^2}{2\sigma^2}} dr + a \frac{e^{-\frac{m^2}{2\sigma^2}}}{2\pi\sigma^2} e^{\frac{a^2}{2\sigma^2}} \int_0^\infty e^{-\frac{(r-a)^2}{2\sigma^2}} dr.$$

As the integrand of the error function, $e^{-\omega^2}$, is an even function, we can rearrange the components in the first term and substitute $\frac{r-a}{\sqrt{2\sigma^2}} = \omega$ in the second term to obtain:

$$p(\varphi|\Theta) = \frac{e^{-\frac{m^2}{2\sigma^2}}}{2\pi} \underbrace{\left(e^{\frac{a^2}{2\sigma^2}} \int_0^\infty \frac{2(r-a)}{2\sigma^2} e^{-\frac{(r-a)^2}{2\sigma^2}} dr \right)}_{1} + a \frac{e^{-\frac{m^2}{2\sigma^2}}}{2\pi\sigma^2} e^{\frac{a^2}{2\sigma^2}} \sqrt{2\sigma^2} \int_{-\frac{a}{\sqrt{2\sigma^2}}}^{\infty} e^{-\omega^2} d\omega =$$

$$\frac{e^{-\frac{m^2}{2\sigma^2}}}{2\pi} + \frac{e^{-\frac{m^2}{2\sigma^2}}}{2\pi\sigma^2} e^{\frac{a^2}{2\sigma^2}} \sqrt{2\sigma^2} \left(a \int_{-\frac{a}{\sqrt{2\sigma^2}}}^{0} e^{-\omega^2} d\omega + a \int_0^\infty e^{-\omega^2} d\omega \right) =$$

$$\frac{e^{-\frac{m^2}{2\sigma^2}}}{2\pi} + \frac{e^{-\frac{m^2}{2\sigma^2}}}{2\pi\sigma^2} e^{\frac{a^2}{2\sigma^2}} \sqrt{2\sigma^2} \frac{1}{2} \left(\underbrace{a \int_{-\frac{a}{\sqrt{2\sigma^2}}}^{\frac{a}{\sqrt{2\sigma^2}}} e^{-k^2} d\omega}_{a\sqrt{\pi}\,\mathrm{erf}\left(\frac{a}{\sqrt{2\sigma^2}}\right)} + \underbrace{a \int_{-\infty}^{\infty} e^{-k^2} d\omega}_{a\sqrt{\pi}} \right) =$$

$$\frac{e^{-\frac{m^2}{2\sigma^2}}}{4\pi\sigma^2} \left(2\sigma^2 + \sqrt{2\pi\sigma^2}\, \underbrace{m\cos(\varphi-\theta)}_{a}\, e^{\frac{m^2\cos^2(\varphi-\theta)}{2\sigma^2}} \left(1 + \mathrm{erf}\left(\frac{m\cos(\varphi-\theta)}{\sqrt{2\sigma^2}} \right) \right) \right).$$

Therefore:

$$p(\varphi|\Theta) \approx \begin{cases} \int_0^\infty \frac{r}{2\pi\sigma^2} e^{-\frac{r^2}{2\sigma^2}} dr = -\frac{1}{2\pi}\int_0^\infty e^{-\frac{r^2}{2\sigma^2}} d\left(-\frac{r^2}{2\sigma^2}\right) = \frac{1}{2\pi} \sim \text{Uniform}(-\pi, \pi), \\ \quad \text{Under } H_o: \beta = 0, m = 0, \text{ for low SNR,} \\ \left(\frac{e^{-\frac{(\varphi-\theta)^2}{2\left(\frac{\sigma}{m}\right)^2}}}{\sqrt{2\pi\left(\frac{\sigma}{m}\right)^2}} \right) \sim N\left(\theta, \left(\frac{\sigma}{m}\right)^2\right), \text{ when } m = x_t{}'\beta \to \infty, \text{ for high SNR.} \end{cases} \tag{4.15}$$

Recall the Taylor series expansions of the trigonometric functions:

$$\sin^2(\varphi-\theta) = (\varphi-\theta)^2 - \frac{(\varphi-\theta)^4}{3} + \frac{2(\varphi-\theta)^6}{45} + O\left((\varphi-\theta)^8\right) \text{ and}$$

$$\cos(\varphi-\theta) = 1 - \frac{(\varphi-\theta)^2}{2} + \frac{(\varphi-\theta)^4}{24} + O\left((\varphi-\theta)^6\right).$$

The last approximation in equation (4.15) is valid asymptotically as the SNR increases, i.e., as $m = x_t{}'\beta \to \infty$, [228]. More specifically, for large SNR ($m \to \infty$), the first-order Taylor series expansion of $p(\vartheta|\Theta)$ at $m = \infty$ is:

$$p(\varphi|\Theta) \cong e^{-\left(\frac{m^2}{2\sigma^2} - O\left(\frac{1}{m}\right)^4\right)} \left(\frac{2\sigma^2}{4\pi m^2 \cos^2(\varphi-\theta)} + O\left(\frac{1}{m}\right)^3 \right) +$$

$$e^{-\left(\frac{\sin^2(\varphi-\theta)\, m^2}{2\sigma^2} + O\left(\frac{1}{m}\right)^4\right)} \left(\frac{m\cos(\varphi-\theta)}{\sqrt{2\pi\sigma^2}} + O\left(\frac{1}{m}\right)^3 \right).$$

Using only the first-order series expansions of sine and cosine near the origin, we can approximate the marginal phase distribution:

$$p(\varphi|\Theta) \cong \underbrace{e^{-\left(\frac{m^2}{2\sigma^2}\right)} \left(\frac{2\sigma^2}{4\pi m^2} \right)}_{\text{decays rapidly}} + \underbrace{e^{-\left(\frac{(\varphi-\theta)^2 m^2}{2\sigma^2}\right)} \left(\frac{m}{\sqrt{2\pi\sigma^2}} \right)}_{\text{dominating term}},$$

$$p(\varphi|\Theta) \cong e^{-\left(\frac{(\varphi-\theta)^2}{2\left(\frac{\sigma}{m}\right)^2}\right)} \left(\frac{1}{\sqrt{2\pi\left(\frac{\sigma}{m}\right)^2}} \right) \sim N\left(\theta, \left(\frac{\sigma}{m}\right)^2\right).$$

Recall the discussion in **Chapter 3** (see **Table 3.1**) about the symmetry of phase distributions of Fourier transformed real-valued functions. The derivations above show that in both cases (small or large SNR), the marginal phase distributions are symmetric. For small SNR, the uniform distribution is also zero-mean. Using the polar coordinate transformation, we often approximate the raw observed complex-valued fMRI signal using only the real part, which ignores the dependence on the phase imperfection (θ) of the original signal. Hence, in the case of large SNR, the normal approximation to the marginal phase distribution may potentially be non-zero-mean, although $\theta = 0$ corresponds to equivalence of the complex and real fMRI intensities, $m \equiv r\cos\theta + i\ r\sin\theta$.

4.4 Complex-Valued fMRI Time-Series Inference

Let's return now to the MLE solution of the *unconstrained likelihood* optimization problem involving complex-valued signals, see equations (4.1)–(4.3). To simplify the analytical expressions of the MLE solutions, we'll make the following short-hand notations:

$$\hat{\beta}_R := (X'X)^{-1}X'\rho_R,\ \hat{\beta}_I := (X'X)^{-1}X'\rho_I.$$

Below we will show that the MLE model parameter estimates are:

$$\hat{\theta} = \frac{1}{2}\text{artctan}\left(\frac{2\hat{\beta}_R{}'(X'X)\hat{\beta}_I}{\hat{\beta}_R{}'(X'X)\hat{\beta}_R - \hat{\beta}_I{}'(X'X)\hat{\beta}_I}\right), \tag{4.16}$$

$$\hat{\beta} = \hat{\beta}_R\cos\left(\hat{\theta}\right) + \hat{\beta}_I\sin\left(\hat{\theta}\right),\ \text{(weight-averaging real and imaginary parts)}$$

$$\hat{\sigma}^2 = \underbrace{\frac{1}{2n}\langle\epsilon|\epsilon\rangle}_{\text{MSE}} = \frac{1}{2n}\left(\rho - \begin{pmatrix} X & 0 \\ 0 & X\end{pmatrix}\begin{pmatrix}\hat{\beta} & 0 \\ 0 & \hat{\beta}\end{pmatrix}\begin{pmatrix}\cos\hat{\theta} \\ \sin\hat{\theta}\end{pmatrix}\right)'\left(\rho - \begin{pmatrix} X & 0 \\ 0 & X\end{pmatrix}\begin{pmatrix}\hat{\beta} & 0 \\ 0 & \hat{\beta}\end{pmatrix}\begin{pmatrix}\cos\hat{\theta} \\ \sin\hat{\theta}\end{pmatrix}\right).$$

Note that these estimates for the (bivariate normal) complex-valued model are different from the magnitude-only model we discussed earlier in **Section 4.3** where we used (univariate normal) approximation to Rician distribution for the magnitude.

The gLRT comparing $H_o: C\beta = 0$ against $H_a: C\beta \neq 0$ is computed by maximizing the *constrained likelihood* under H_o:

$$(\text{unchanged})\breve{\beta}_R \equiv \hat{\beta}_R = (X'X)^{-1}X'\rho_R,\ \breve{\beta}_I \equiv \hat{\beta}_I = (X'X)^{-1}X'\rho_I,$$

$$\underbrace{\mathbf{D}}_{(d+1)\times(d+1)} = I_{d+1} - (X'X)^{-1}C'\left(C(X'X)^{-1}C'\right)^{-1}C,$$

$$\breve{\theta} = \frac{1}{2}\text{artctan}\left(\frac{2\hat{\beta}_R{}'\mathbf{D}(X'X)\hat{\beta}_I}{\hat{\beta}_R{}'\mathbf{D}(X'X)\hat{\beta}_R - \hat{\beta}_I{}'\mathbf{D}(X'X)\hat{\beta}_I}\right), \tag{4.17}$$

$$\breve{\beta} = \mathbf{D}\left(\hat{\beta}_R{}'\cos\breve{\theta} + \hat{\beta}_I\sin\breve{\theta}\right),$$

$$\check{\sigma}^2 = \frac{1}{2n}\left(Y - \begin{pmatrix} X & 0 \\ 0 & X \end{pmatrix}\begin{pmatrix} \check{\beta} & X \\ X & \check{\beta} \end{pmatrix}\begin{pmatrix} \cos\check{\theta} \\ \sin\check{\theta} \end{pmatrix}\right)'\left(\underbrace{Y}_{2n\times 1} - \underbrace{\begin{pmatrix} X & 0 \\ 0 & X \end{pmatrix}}_{2n\times 2(d+1)}\underbrace{\begin{pmatrix} \check{\beta} & X \\ X & \check{\beta} \end{pmatrix}}_{2(d+1)\times 2}\underbrace{\begin{pmatrix} \cos\check{\theta} \\ \sin\check{\theta} \end{pmatrix}}_{2\times 1}\right).$$

Earlier, in the magnitude-only case, we ignored the complex-valued intensity phases and worked exclusively with the fMRI amplitudes $\|\rho_t\|$. Now, we consider *spacetime analytics based on the complex-valued fMRI signal* ($\rho \equiv \rho_t$). At time t, we now have a bivariate measurement with *real* and *imaginary* components:

$$\rho \equiv \rho_t = \begin{pmatrix} \rho_{R,t} \\ \rho_{I,t} \end{pmatrix} + \begin{pmatrix} \varepsilon_{R,t} \\ \varepsilon_{I,t} \end{pmatrix} = \begin{pmatrix} x_t'\beta\cos(\theta) \\ x_t'\beta\sin(\theta) \end{pmatrix} + \begin{pmatrix} \varepsilon_{R,t} \\ \varepsilon_{I,t} \end{pmatrix}, \quad 1 \le t \le n. \tag{4.18}$$

The observed (spacetime) complex-valued fMRI signal can be expressed in matrix form as:

$$\underbrace{\overbrace{\rho}^{Y}}_{2n\times 1} = \begin{pmatrix} \rho_R \\ \rho_I \end{pmatrix} = \underbrace{\begin{pmatrix} X & 0 \\ 0 & X \end{pmatrix}}_{2n\times 2(d+1)}\underbrace{\begin{pmatrix} \beta & 0 \\ 0 & \beta \end{pmatrix}}_{2(d+1)\times 2}\underbrace{\begin{pmatrix} \cos\theta \\ \sin\theta \end{pmatrix}}_{2\times 1} + \underbrace{\varepsilon}_{2n\times 1}, \varepsilon \sim N(\mu = 0, \Sigma = \sigma^2 I_{2n}).$$

Similar to the derivations in [199, 201], the *unconstrained* MLE estimates of the general parameter vector $\Theta = (\theta, \beta, \sigma)$, representing the intensity phase (θ), the effects ($\beta = (\beta_R, \beta_I)'$), and the variance ($\sigma$) may be derived as follows.

$$\underbrace{l(\rho|X, \theta, \beta, \sigma^2)}_{\text{likelihood}} = \prod_t p(\rho|x_t, \theta, \beta, \sigma^2) = \frac{1}{\sigma^{2n}(2\pi)^n}\left(\prod_t \begin{pmatrix} \rho_R \\ \rho_I \end{pmatrix}\right) \times$$

$$e^{-\frac{\left(\begin{pmatrix} \rho_R \\ \rho_I \end{pmatrix} - \begin{pmatrix} X & 0 \\ 0 & X \end{pmatrix}\begin{pmatrix} \beta & 0 \\ 0 & \beta \end{pmatrix}\begin{pmatrix} \cos\theta \\ \sin\theta \end{pmatrix}\right)'\left(\begin{pmatrix} \rho_R \\ \rho_I \end{pmatrix} - \begin{pmatrix} X & 0 \\ 0 & X \end{pmatrix}\begin{pmatrix} \beta & 0 \\ 0 & \beta \end{pmatrix}\begin{pmatrix} \cos\theta \\ \sin\theta \end{pmatrix}\right)}{2\sigma^2}},$$

$$ll(\rho) = \ln l \propto -n\ln\sigma^2$$

$$-\frac{\left(\begin{pmatrix} \rho_R \\ \rho_I \end{pmatrix} - \begin{pmatrix} X & 0 \\ 0 & X \end{pmatrix}\begin{pmatrix} \beta & 0 \\ 0 & \beta \end{pmatrix}\begin{pmatrix} \cos\theta \\ \sin\theta \end{pmatrix}\right)'\left(\begin{pmatrix} \rho_R \\ \rho_I \end{pmatrix} - \begin{pmatrix} X & 0 \\ 0 & X \end{pmatrix}\begin{pmatrix} \beta & 0 \\ 0 & \beta \end{pmatrix}\begin{pmatrix} \cos\theta \\ \sin\theta \end{pmatrix}\right)}{2\sigma^2}$$

$$= -n\ln\sigma^2 - \frac{1}{2\sigma^2}\left[\|\rho_R - X\beta\cos\theta\|^2 + \|\rho_I - X\beta\sin\theta\|^2\right].$$

Putting together these constrained and unconstrained likelihoods, we can estimate the LRT statistic for the (bivariate normal) complex-valued model in equation (4.4):

$$LRT = -2\log \Lambda_c = -2\ln\left(\frac{\sup_{\theta\in\Theta_o} p(\rho|\beta,\theta,\sigma^2,X)}{\sup_{\theta\in\Theta} p(\rho|\beta,\theta,\sigma^2,X)}\right) = -2\ln\left(\frac{p(\rho|\breve{\Theta},X)}{p\left(\rho|\hat{\Theta},X\right)}\right)$$

$$= 2\left(n\ln\breve{\sigma}^2 + \underbrace{\frac{1}{2\breve{\sigma}^2}\left[\|\rho_R - X\breve{\beta}\cos\breve{\theta}\|^2 + \|\rho_I - X\breve{\beta}\sin\breve{\theta}\|^2\right]}_{n,\ \text{since}\ \breve{\sigma}^2 = \breve{\sigma}^2_{MLE}} - n\ln\hat{\sigma}^2 \right.$$

$$\left. - \underbrace{\frac{1}{2\hat{\sigma}^2}\left[\|\rho_R - X\hat{\beta}\cos\hat{\theta}\|^2 + \|\rho_I - X\hat{\beta}\sin\hat{\theta}\|^2\right]}_{n,\ \text{since}\ \hat{\sigma}^2 = \hat{\sigma}^2_{MLE}} \right) = 2n\ln\left(\frac{\breve{\sigma}^2}{\hat{\sigma}^2}\right).$$

The MLE estimates for the three model parameters $(\beta,\ \theta,\ \sigma^2)$ are obtained by solving the score equations involving the partial derivatives of the log-likelihood:

$$0 = \tfrac{\partial}{\partial\beta} ll \propto \cos\theta X'(\rho_R - X\beta\cos\theta) + \sin\theta X'(\rho_I - X\beta\sin\theta) \Rightarrow X'X\beta = \cos\theta X'\rho_R + \sin\theta X'\rho_I$$

$$0 = \tfrac{\partial}{\partial\theta} ll \propto \sin\theta (X\beta)'\rho_R - \cos\theta (X\beta)'\rho_I$$

$$0 = \tfrac{\partial}{\partial\sigma} ll \propto -2n\tfrac{1}{\sigma} + \tfrac{1}{\sigma^3}\left(\begin{pmatrix}\rho_R\\ \rho_I\end{pmatrix} - \begin{pmatrix}X & 0\\ 0 & X\end{pmatrix}\begin{pmatrix}\beta & 0\\ 0 & \beta\end{pmatrix}\begin{pmatrix}\cos\theta\\ \sin\theta\end{pmatrix}\right)' \times \left(\begin{pmatrix}\rho_R\\ \rho_I\end{pmatrix} - \begin{pmatrix}X & 0\\ 0 & X\end{pmatrix}\begin{pmatrix}\beta & 0\\ 0 & \beta\end{pmatrix}\begin{pmatrix}\cos\theta\\ \sin\theta\end{pmatrix}\right).$$

$$0 = \tfrac{\partial}{\partial\sigma^2} ll \propto -n\tfrac{1}{\sigma^2} + \tfrac{1}{2\sigma^4}\underbrace{\left(\begin{pmatrix}\rho_R\\ \rho_I\end{pmatrix} - \begin{pmatrix}X & 0\\ 0 & X\end{pmatrix}\begin{pmatrix}\beta & 0\\ 0 & \beta\end{pmatrix}\begin{pmatrix}\cos\theta\\ \sin\theta\end{pmatrix}\right)' \times \left(\begin{pmatrix}\rho_R\\ \rho_I\end{pmatrix} - \begin{pmatrix}X & 0\\ 0 & X\end{pmatrix}\begin{pmatrix}\beta & 0\\ 0 & \beta\end{pmatrix}\begin{pmatrix}\cos\theta\\ \sin\theta\end{pmatrix}\right)}_{\left[\|\rho_R - X\beta\cos\theta\|^2 + \|\rho_I - X\beta\sin\theta\|^2\right]}$$

From the first score function equation:

$$\hat{\beta} = \hat{\beta}_{MLE} = \underbrace{(X'X)^{-1}X'\rho_R}_{\hat{\beta}_R}\cos\theta + \underbrace{(X'X)^{-1}X'\rho_I}_{\hat{\beta}_I}\sin\theta = \hat{\beta}_R\cos\theta + \hat{\beta}_I\sin\theta.$$

Plugging this $\hat{\beta}$ estimate in the second score equation leads to the MLE of the parameter θ:

$$0 = \sin\theta\left(X\hat{\beta}\right)'\rho_R - \cos\theta\left(X\hat{\beta}\right)'\rho_I =$$

$$\sin\theta\left(X\hat{\beta}_R\cos\theta+X\hat{\beta}_I\sin\theta\right)'\rho_R-\cos\theta\left(X\hat{\beta}_R\cos\theta+X\hat{\beta}_I\sin\theta\right)'\rho_I.$$

Therefore,

$$0=0\times\underbrace{\frac{1}{\cos^2\theta}}_{\neq 0}=$$

$$\left(\sin\theta\cos\theta\left(X\hat{\beta}_R\right)'\rho_R+\sin^2\theta\left(X\hat{\beta}_I\right)'\rho_R-\cos^2\theta\left(X\hat{\beta}_R\right)'\rho_I-\sin\theta\cos\theta\left(X\hat{\beta}_I\right)'\rho_I\right)\times\underbrace{\frac{1}{\cos^2\theta}}_{\neq 0}=$$

$$\frac{\sin\theta}{\cos\theta}\left(X\hat{\beta}_R\right)'\rho_R+\frac{\sin^2\theta}{\cos^2\theta}\left(X\hat{\beta}_I\right)'\rho_R-\left(X\hat{\beta}_R\right)'\rho_I-\frac{\sin\theta}{\cos\theta}\left(X\hat{\beta}_I\right)'\rho_I=$$

$$\tan\theta\left(\left(X\hat{\beta}_R\right)'\rho_R-\left(X\hat{\beta}_I\right)'\rho_I\right)+\tan^2\theta\left(X\hat{\beta}_I\right)'\rho_R-\left(X\hat{\beta}_R\right)'\rho_I.$$

As $\hat{\beta}_R=(X'X)^{-1}X'\rho_R$ and $\hat{\beta}_I=(X'X)^{-1}X'\rho_I$ we have $(X'X)\hat{\beta}_R=X'\rho_R$ and $(X'X)\hat{\beta}_I=X'\rho_I$:

$$0=\tan\theta\left(\left(X\hat{\beta}_R\right)'\rho_R-\left(X\hat{\beta}_I\right)'\rho_I\right)+\tan^2\theta\left(X\hat{\beta}_I\right)'\rho_R-\left(X\hat{\beta}_R\right)'\rho_I=$$

$$\tan\theta\left(\hat{\beta}_R'(X'X)\hat{\beta}_R-\hat{\beta}_I'(X'X)\hat{\beta}_I\right)+\tan^2\theta\,\hat{\beta}_I'(X'X)\hat{\beta}_R-\underbrace{\hat{\beta}_R'(X'X)\hat{\beta}_I}_{(\hat{\beta}_R'(X'X)\hat{\beta}_I)'}=$$

$$\tan\theta\left(\hat{\beta}_R'(X'X)\hat{\beta}_R-\hat{\beta}_I'(X'X)\hat{\beta}_I\right)+(\tan^2\theta-1)\;\hat{\beta}_I'(X'X)\hat{\beta}_R=$$

$$\tan\theta\left(\hat{\beta}_R'(X'X)\hat{\beta}_R-\hat{\beta}_I'(X'X)\hat{\beta}_I\right)+\left[\tan^2\theta-1\right]\hat{\beta}_R'(X'X)\hat{\beta}_I.$$

As $\tan 2\theta=\frac{2\tan\theta}{1-\tan^2\theta}$,

$$\frac{\tan 2\theta}{2}=\frac{\tan\theta}{1-\tan^2\theta}=\frac{\hat{\beta}_R'(X'X)\hat{\beta}_I}{\hat{\beta}_R'(X'X)\hat{\beta}_R-\hat{\beta}_I'(X'X)\hat{\beta}_I}.$$

Hence, we can solve for the parameter θ by using the inverse tangent function arctan():

$$\hat{\theta}=\hat{\theta}_{MLE}=\frac{1}{2}\arctan\left(\frac{2\hat{\beta}_R'(X'X)\hat{\beta}_I}{\hat{\beta}_R'(X'X)\hat{\beta}_R-\hat{\beta}_I'(X'X)\hat{\beta}_I}\right).$$

To summarize

$$\hat{\beta}_R=(X'X)^{-1}X'\rho_R,\;\hat{\beta}_I=(X'X)^{-1}X'\rho_I,$$

$$\hat{\theta} = \frac{1}{2} \arctan\left(\frac{2\hat{\beta}_R{}'(X'X)\hat{\beta}_I}{\hat{\beta}_R{}'(X'X)\hat{\beta}_R - \hat{\beta}_I{}'(X'X)\hat{\beta}_I} \right),$$

$$\hat{\beta} = \hat{\beta}_R \cos\left(\hat{\theta}\right) + \hat{\beta}_I \sin\left(\hat{\theta}\right), \text{ (weight-averaging the real and imaginary parts)}$$

$$\underbrace{\hat{\sigma}^2 = \frac{1}{2n}\langle \epsilon | \epsilon \rangle}_{\text{MSE}} = \frac{1}{2n} \underbrace{\left(\rho - \begin{pmatrix} X & 0 \\ 0 & X \end{pmatrix} \begin{pmatrix} \hat{\beta} & 0 \\ 0 & \hat{\beta} \end{pmatrix} \begin{pmatrix} \cos\hat{\theta} \\ \sin\hat{\theta} \end{pmatrix}\right)' \left(\rho - \begin{pmatrix} X & 0 \\ 0 & X \end{pmatrix} \begin{pmatrix} \hat{\beta} & 0 \\ 0 & \hat{\beta} \end{pmatrix} \begin{pmatrix} \cos\hat{\theta} \\ \sin\hat{\theta} \end{pmatrix}\right)}_{\left[\|\rho_R - X\hat{\beta}\cos\hat{\theta}\|^2 + \|\rho_I - X\hat{\beta}\sin\hat{\theta}\|^2 \right]}. \tag{4.19}$$

The variance estimate, $\hat{\sigma}^2$, is slightly biased since $\mathbb{E}\left(\hat{\sigma}^2\right) = \frac{2n-(d+1)}{2n}\sigma^2 < \sigma^2$. Under the bivariate normal model, equation (4.18), we have:

$$\varepsilon \sim N\left(\mu = 0, \Sigma = \sigma^2 I_{2n}\right), \; \rho \sim N\left(\begin{pmatrix} X & 0 \\ 0 & X \end{pmatrix} \begin{pmatrix} \beta\cos\theta \\ \beta\sin\theta \end{pmatrix}, \Sigma = \sigma^2 I_{2n} \right) \Rightarrow$$

$$\frac{1}{\sigma^2}\langle \epsilon | \epsilon \rangle \sim \chi^2_{df = 2n - (d+1)} \Rightarrow$$

$$\underbrace{\mathbb{E}(\langle \epsilon | \epsilon \rangle)}_{\substack{\text{expectation of} \\ \text{residual } SSE}} = (2n - d - 1)\sigma^2 \Rightarrow \mathbb{E}(MSE) \equiv \mathbb{E}\left(\hat{\sigma}^2\right) = \mathbb{E}\left(\frac{1}{2n}\langle \epsilon | \epsilon \rangle\right) = \frac{(2n-d-1)}{2n}\sigma^2.$$

This can be explicated as follows:

$$\hat{\rho} = \begin{pmatrix} \hat{\rho}_R \\ \hat{\rho}_I \end{pmatrix} = \begin{pmatrix} X & \hat{\beta}_R \\ X & \hat{\beta}_I \end{pmatrix} = \begin{pmatrix} X & 0 \\ 0 & X \end{pmatrix} \begin{pmatrix} \hat{\beta}\cos\hat{\theta} \\ \hat{\beta}\sin\hat{\theta} \end{pmatrix},$$

$$\begin{pmatrix} X'X\hat{\beta}\cos\hat{\theta} \\ X'X\hat{\beta}\sin\hat{\theta} \end{pmatrix} = \begin{pmatrix} X'\hat{\rho}_R \\ X'\hat{\rho}_I \end{pmatrix},$$

$$\epsilon = (\rho - \hat{\rho}) = \begin{pmatrix} \rho_R - \hat{\rho}_R \\ \rho_I - \hat{\rho}_I \end{pmatrix} = \begin{pmatrix} \rho_R - X\hat{\beta}_R \\ \rho_I - X\hat{\beta}_I \end{pmatrix} = \begin{pmatrix} \rho_R \\ \rho_I \end{pmatrix} - \begin{pmatrix} X(X'X)^{-1}X' & 0 \\ 0 & X(X'X)^{-1}X' \end{pmatrix} \begin{pmatrix} \rho_R \\ \rho_I \end{pmatrix} =$$

$$\underbrace{\left(I_{2n} - \underbrace{\begin{pmatrix} X(X'X)^{-1}X' & 0 \\ 0 & X(X'X)^{-1}X' \end{pmatrix}}_{\text{hat matrix, } H} \right)}_{I-H} \begin{pmatrix} \rho_R \\ \rho_I \end{pmatrix},$$

$$\langle \epsilon|\epsilon\rangle = \begin{pmatrix} \rho_R - X\hat{\beta}_R \\ \rho_I - X\hat{\beta}_I \end{pmatrix}' \begin{pmatrix} \rho_R - X\hat{\beta}_R \\ \rho_I - X\hat{\beta}_I \end{pmatrix} = \begin{pmatrix} \rho_R \\ \rho_I \end{pmatrix}' (I_{2n}-H)(I_{2n}-H) \begin{pmatrix} \rho_R \\ \rho_I \end{pmatrix} = \underbrace{\begin{pmatrix} \rho_R \\ \rho_I \end{pmatrix}'}_{\rho'} (I_{2n}-H) \underbrace{\begin{pmatrix} \rho_R \\ \rho_I \end{pmatrix}}_{\rho},$$

where the symmetric ($H = H'$) and idempotent ($H'H = HH = H$) *hat* matrix is

$$H = \begin{pmatrix} X(X'X)^{-1}X' & 0 \\ 0 & X(X'X)^{-1}X' \end{pmatrix}.$$

Note that $(I_{2n} - H)$ is also idempotent, $(I_{2n} - H)(I_{2n} - H) = (I_{2n} - H)$, and $(I_{2n}-\mathrm{H}) \begin{pmatrix} X & 0 \\ 0 & X \end{pmatrix} = \begin{pmatrix} 0 & 0 \\ 0 & 0 \end{pmatrix}$. Since the linear model is

$$\rho = \begin{pmatrix} X & 0 \\ 0 & X \end{pmatrix} \begin{pmatrix} \beta & 0 \\ 0 & \beta \end{pmatrix} \begin{pmatrix} \cos\theta \\ \sin\theta \end{pmatrix} + \varepsilon,\ \varepsilon \sim N(\mu = 0, \Sigma = \sigma^2 I_{2n}) \Rightarrow \rho'(I_{2n} - H)\rho \sim \sigma^2 \chi^2_{df = 2n-(d+1)}.$$

Hence, $\mathbb{E}(\langle\epsilon|\epsilon\rangle) = \mathbb{E}(\rho'(I_{2n} - H)\rho) = (2n - d - 1)\sigma^2$ and $\mathbb{E}\left(\hat{\sigma}^2\right) = \frac{2n-(d+1)}{2n}\sigma^2 < \sigma^2$.

In practice, this slight bias is not critical as the time (n) is generally much larger than $d + 1$, the number of multi-source components (an intercept and feature covariates). Hence, the variance estimate $\hat{\sigma}^2$ may only be slightly *underestimating* the unknown theoretical variance σ^2.

The likelihood function of the complex-valued model is:

$$p(\rho|X,\beta,\theta,\sigma^2) = \frac{1}{(2\pi\sigma^2)^n} e^{-\frac{1}{2\sigma^2}\left(Y - \begin{pmatrix} X & 0 \\ 0 & X \end{pmatrix} \begin{pmatrix} \beta & 0 \\ 0 & \beta \end{pmatrix} \begin{pmatrix} \cos\theta \\ \sin\theta \end{pmatrix}\right)' \left(Y - \begin{pmatrix} X & 0 \\ 0 & X \end{pmatrix} \begin{pmatrix} \beta & 0 \\ 0 & \beta \end{pmatrix} \begin{pmatrix} \cos\theta \\ \sin\theta \end{pmatrix}\right)}.$$

The effects ($\hat{\beta}$) are estimated by phase-dependent weight-averaging of the real and imaginary parts of the classical regression coefficients ($\hat{\beta}_R, \hat{\beta}_I$), see equation (4.19).

Similar to the first case of inference based on real-valued magnitude-only signals, under the null hypothesis, $H_o : C\beta = 0$, the corresponding *constrained* MLE estimates can be expressed by:

$$(\text{unchanged})\ \check{\beta}_R = (X'X)^{-1} X'\rho_R = \hat{\beta}_R,\ \check{\beta}_1 = (X'X)\, X'\rho_I \equiv \hat{\beta}_R,$$

$$\underbrace{\mathbf{D}}_{(d+1)\times(d+1)} = I_{d+1} - (X'X)^{-1}C'\left(C(X'X)^{-1}C'\right)^{-1}C,$$

$$\check{\theta} = \frac{1}{2}\arctan\left(\frac{2\hat{\beta}_R{}'\mathbf{D}(X'X)\hat{\beta}_I}{\hat{\beta}_R{}'\mathbf{D}(X'X)\hat{\beta}_R - \hat{\beta}_I{}'\mathbf{D}(X'X)\hat{\beta}_I}\right), \tag{4.20}$$

$$\breve{\beta} = \boldsymbol{D}\left(\hat{\beta}_R \cos\left(\breve{\theta}\right) + \hat{\beta}_I \sin\left(\breve{\theta}\right)\right),$$

$$\breve{\sigma}^2 = \underbrace{\frac{1}{2n}\left(\rho - \begin{pmatrix} X & 0 \\ 0 & X \end{pmatrix}\begin{pmatrix} \breve{\beta} & 0 \\ 0 & \breve{\beta} \end{pmatrix}\begin{pmatrix} \cos\breve{\theta} \\ \sin\breve{\theta} \end{pmatrix}\right)'\left(\underbrace{\rho}_{2n\times 1} - \underbrace{\begin{pmatrix} X & 0 \\ 0 & X \end{pmatrix}}_{2n\times 2(d+1)}\underbrace{\begin{pmatrix} \breve{\beta} & 0 \\ 0 & \breve{\beta} \end{pmatrix}}_{2(d+1)\times 2}\underbrace{\begin{pmatrix} \cos\breve{\theta} \\ \sin\breve{\theta} \end{pmatrix}}_{2\times 1}\right)}_{\left[\|\rho_R - X\breve{\beta}\cos\breve{\theta}\|^2 + \|\rho_I - X\breve{\beta}\sin\breve{\theta}\|^2\right]}.$$

Under the same parametric assumptions as in the real-valued case, these MLE estimates naturally lead to the gLRT for spacetime complex-valued fMRI signal activation:

$$-2\log \Lambda_c = 2n\log\left(\frac{\breve{\sigma}^2}{\hat{\sigma}^2}\right) \sim \chi_r^2,$$

where r is the full rank of the contrast vector c specifying the null hypothesis.

4.5 Complex-Valued Kime-Indexed fMRI Kintensity Inference

Finally, let's consider an extension of the prior spacetime analytics approach, based on the complex-valued fMRI signals, to spacekime inference.

Spacekime analytics depend jointly on kime domain indexing, $\kappa = (t,\varphi) \in \mathbb{K} \cong \mathbb{C} \ni \kappa = (\kappa_1, \kappa_2)$, and spatial voxel indexing, $v = (x, y, z) \in \mathbb{R}^3$. For simplicity, at a given voxel we will still consider fMRI signals whose spacekime complex-intensities (*kintensities*) depend on the kime-magnitude ($r = t$) and kime-phase (φ) and are defined for each kime by inverting the FT of the complex measurement $o_m(\boldsymbol{k})$. Geometrically, at each spatial location, ρ_κ are the observed complex-valued kime-indexed intensities (kintensities) representing a kimesurface that can be expressed as:

$$\rho_\kappa = \underbrace{\left(\rho_{R,\kappa} + e_{R,\kappa}\right)}_{Re(\rho_\kappa)} + \underbrace{i\left(\rho_{I,\kappa} + e_{I,\kappa}\right)}_{Im(\rho_\kappa)},\ \text{or equivalently,}$$

$$\rho_\kappa = \begin{pmatrix} \rho_{R,\ \kappa} \\ \rho_{I,\ \kappa} \end{pmatrix} + \begin{pmatrix} e_{R,\kappa} \\ e_{I,\kappa} \end{pmatrix} \equiv \begin{pmatrix} \rho_{R,\ \kappa_1,\kappa_2} \\ \rho_{I,\ \kappa_1,\kappa_2} \end{pmatrix} + \begin{pmatrix} e_{R,\kappa_1,\kappa_2} \\ e_{I,\kappa_1,\kappa_2} \end{pmatrix},$$

where $\kappa = (\kappa_1, \kappa_2)$, and the real and imaginary parts of the volume kintensities are $\left(\rho_{R,\kappa}, \rho_{I,\kappa}\right)'$. While more elaborate and higher-order models can certainly be explored, we will assume multivariate normal distribution noise $(e_{R,\kappa}, e_{I,\kappa})' \sim N(0, \Sigma)$ with a scaled standard variance-covariance matrix, $\Sigma = \sigma^2 I$, where $I = I_{2\times 2}$ is the identity matrix of rank 2, and $'$ denotes the transpose of a vector or a matrix. A more explicit representation of the residual error of the linear model is:

$$e = \left(e_{R,\kappa}, e_{I,\kappa}\right)' \sim N\left(0, \Sigma \otimes I_{n^2 \times n^2}\right),$$

where $\otimes$ denotes the matrix tensor product and $\Sigma = \sigma^2 I_{2\times 2}$. In **Chapter 5**, we will prove that the observed uncertainty in 4D Minkowski spacetime is the result of a natural one degree of freedom resulting from the projection (lossy-compression) of the 5D spacekime process into spacetime. Repeated random sampling in spacetime is analogous to coupling a single spacetime observation with a prior kime-phase distribution. However, the spacekime error term may not be completely eliminated since some errors may be unavoidable, e.g., measurement errors and discrepancy between the true process and a particular modeling strategy. That is why, spacekime linear modeling may still need to include a residual error term $e \sim N(0, \Sigma)$.

The simplest model has variance-covariance tensor expressed as:

$$\Sigma \otimes I_{n^2 \times n^2} = \begin{pmatrix} \sigma^2 & \\ & \sigma^2 \end{pmatrix}_{2\times 2} \otimes \begin{pmatrix} 1 & \cdots & \\ \vdots & \ddots & \vdots \\ & \cdots & 1 \end{pmatrix}_{n^2 \times n^2} =$$

$$\begin{pmatrix} \sigma^2 \begin{pmatrix} 1 & \\ & 1 \end{pmatrix} & & & \\ & \cdots & & \\ & & \cdots & \\ & & & \sigma^2 \begin{pmatrix} 1 & \\ & 1 \end{pmatrix} \end{pmatrix}_{2n^2 \times 2n^2}.$$

In certain situations, it may be appropriate or necessary to introduce and derive more advanced models accounting for different centrality, variation, and/or correlation in the kime-magnitude and kime-phase subspaces.

It's worth clarifying some of the notations. First, note that there is a difference between (*kappa*) $\kappa = (t, \varphi) \in \mathbb{K}$, which represents the complex-time (kime) domain indexing, and the wave spatial-frequency vector $\boldsymbol{k} \in \mathbb{R}^3$, which represents the Fourier domain (k-space) indexing of the fMRI signal kintensities. The second clarification is about the pair of independent phases playing complementary roles in this model – *kime-phases* (domain indexing), φ, and *range-phases* (related to the complex values of the kintensities), θ. Although we will discuss "tensors" in the next section, it is important to realize that spacekime observations, or data, are intrinsically higher-dimensional manifolds that are represented as complex-valued kimesurfaces. Additional details will be provided later. For the time being, the components of the kintensities linear model are defined on the 2D kime domain, $1 \le \kappa_1, \ \kappa_2 \le n$, and can be represented as fourth-order tensors ($\kappa_1 \times \kappa_2 \times \text{effects} \times \text{value}$). The *effects* encode the specific linear modeling terms, e.g., intercepts, BOLD signal, linear trends, quadratic trends, etc., and *value* corresponds to the real and imaginary (range) components of the outcome. For instance, the design tensor for the tensor-based linear model may be represented as a $2n^2 \times (d+1)$ matrix, where rows are the possible kime domain grid vertices for both the real and imaginary range values and the columns represent the effects in the linear model.

For complex-time, we need to extend the fMRI signal from time-series longitudinal indexing (t) to the corresponding (complex kime) kime-series, or kimesurfaces, which are parameterized by $(t, \varphi) \in \mathbb{K} \cong \mathbb{C} \ni \kappa = (\kappa_1, \kappa_2)$. Note that in spacetime, there is only one unique longitudinal event order (the time index runs from small to large). However, spacekime observations represent kime-surfaces, not just time-series, which allow for many different paths, trajectories, or simple closed contours throughout the complex-plane.

For a fixed voxel location, the *spacetime* analytical models of complex-valued fMRI signals, equations (4.4), (4.5), and (4.6) can be translated to their *spacekime* counterparts as follows:

$$\underbrace{\rho_\kappa}_{(2n^2)\times 1} = \underbrace{\begin{pmatrix} \rho_R \\ \rho_I \end{pmatrix}}_{(2n^2)\times 1} = \underbrace{\begin{pmatrix} I_{2\times 2} \otimes \underbrace{X_\kappa}_{n^2\times(d+1)} \end{pmatrix}}_{(2n^2)\times(2(d+1))} \times \underbrace{(I_2 \otimes \beta)}_{(2(d+1))\times 2} \times \begin{pmatrix} \cos\theta \\ \sin\theta \end{pmatrix}_{2\times 1} + e =$$

$$\underbrace{\begin{pmatrix} X_\kappa & 0 \\ 0 & X_\kappa \end{pmatrix}}_{(2n^2)\times(2(d+1))} \underbrace{\begin{pmatrix} \beta & 0 \\ 0 & \beta \end{pmatrix}}_{(2(d+1))\times 2} \underbrace{\begin{pmatrix} \cos\theta \\ \sin\theta \end{pmatrix}}_{2\times 1} + \underbrace{\begin{pmatrix} e_R \\ e_I \end{pmatrix}}_{(2n^2)\times 1}, \tag{4.21}$$

where $\beta \equiv \beta_\kappa = \begin{pmatrix} \beta_R \\ \beta_I \end{pmatrix}$ is the effect vector of length $(d+1)$ defined over the kime domain, and X_κ represents the design-matrix, $2n^2 \times (d+1)$, as a matricized version of the third-order tensor of dimension $(2 \times \underbrace{n \times n}_{\text{kime domain}}) \times (d+1)$.

If $E^2 = e_{R,\kappa}^2 + e_{I,\kappa}^2 \sim \chi_{df=2n^2}^2$, using the Taylor series expansion of $\sqrt{x} \approx \frac{1+x}{2}$ around $x = 1$, we can approximate the kime-indexed real-valued and complex-valued magnitude-only kintensities, $\|\rho_\kappa\|$. The most general case of complex-valued kintensities defined over the kime domain requires the more general *tensor-linear modeling* approach [16, 229, 230]. However, the unwind long matrix representations (concatenating rows of κ_1 indexed values corresponding to independent κ_2 values) would be analogous to equations (4.16) and (4.17):

$$\hat{\beta}_R := (X_\kappa' X_\kappa)^{-1} X_\kappa' \rho_R, \ \hat{\beta}_I := (X_\kappa' X_\kappa)^{-1} X_\kappa' \rho_I.$$

Then, the MLE model parameter estimates are:

$$\hat{\theta}=\frac{1}{2}\text{artctan}\left(\frac{2\hat{\beta}_R{}'(X_\kappa{}'X_\kappa)\hat{\beta}_I}{\hat{\beta}_R{}'(X_\kappa{}'X_\kappa)\hat{\beta}_R-\hat{\beta}_I{}'(X_\kappa{}'X_\kappa)\hat{\beta}_I}\right),$$

$\hat{\beta}=\hat{\beta}_R\cos\left(\hat{\theta}\right)+\hat{\beta}_I\sin\left(\hat{\theta}\right)$, (weight-averaging real and imaginary parts)

$$\hat{\sigma}^2=\underbrace{\frac{1}{2n^2}\langle e|e\rangle}_{\text{MSE}}=\frac{1}{2n^2}\left(\rho_\kappa-\begin{pmatrix}X_\kappa & 0\\ 0 & X_\kappa\end{pmatrix}\begin{pmatrix}\hat{\beta} & 0\\ 0 & \hat{\beta}\end{pmatrix}\begin{pmatrix}\cos\hat{\theta}\\ \sin\hat{\theta}\end{pmatrix}\right)'\left(\rho_\kappa-\begin{pmatrix}X_\kappa & 0\\ 0 & X_\kappa\end{pmatrix}\begin{pmatrix}\hat{\beta} & 0\\ 0 & \hat{\beta}\end{pmatrix}\begin{pmatrix}\cos\hat{\theta}\\ \sin\hat{\theta}\end{pmatrix}\right).$$

4.6 Spacekime Tensor-Based Linear Modeling

Now we will formulate spacekime inference using tensor representation of the data. Following a brief review of the basic tensor definitions, representation, and operations, we will define tensor-based linear models for inputs and outputs and illustrate how these enable linear modeling, statistical inference, and more general analytics.

4.6.1 Tensor Representation and Basic Operations

We will follow the vector, matrix, and tensor notation and the corresponding product definitions presented in [229, 231, 232], to define a *tensor of order Q* as a real (or complex) valued multi-dimensional array with dimension lengths given by $J_j\in\mathbb{N}, \forall 1\le j\le Q$:

$$A=\left(a_{j_1,j_1,\ldots,j_Q}\right)\in\mathbb{R}^{J_1\times J_2\times\cdots\times J_Q}.$$

Tensors are multidimensional arrays extending to higher dimensions the classical vector-based outcomes and two-way (design) input data matrices (e.g., samples by features) [16, 233]. For example, a multi-subject study of aging and dementia using 3D MRI neuroimaging may be represented as a multi-way array of dimensions storing subjects, time points, brain voxel locations, regions of interest, clinical phenotypes, and so on. Another example involves observing a collection of 13,000 labeled facial images, faces in a large database [234, 235], that contains RGB image intensities on a 90×90 spatial pixel grid. This dataset can be represented as a fourth-order tensor indexed by $FaceLabel\times\underbrace{X\times Y}_{\text{pixel}}\times RGBColor$.

In general, given N training data observations, inference, forecasting, or prediction of an outcome tensor Y of order M, $\dim(Y)=N\times D_1^Y\times D_2^Y\times\cdots\times D_M^Y$, may be derived under certain conditions using an observed input (design) tensor X of order L, $\dim(X)=N\times D_1^{X+}\times D_2^{X+}\times\cdots\times D_L^{X+}$. Linear modeling of the input-output tensor relation involves estimating a parameter coefficient tensor B of order:

$$\dim(B) = \underbrace{D_1^{X+} \times D_2^{X+} \times \cdots \times D_L^{X+}}_{\dim(X)} \times \underbrace{D_1^{Y} \times D_2^{Y} \times \cdots \times D_M^{Y}}_{\dim(Y)}.$$

This linear tensor model may be represented as $Y = \langle X, B\rangle + E$, where the order M error (residual) tensor E has $\dim(E) = N \times D_1^Y \times D_2^Y \times \cdots \times D_M^Y$ and the tensor product $\langle X, B\rangle$ is defined below. The *plus dimension notation*, D_1^{X+}, represents $D_1^{X+} = D_1^X$, for purely linear model with no intercepts, and $D_1^{X+} = D_1^{X+1}$ for affine model with intercepts. For simplicity, we often write D_1^X for D_1^{X+}. Having the model tensor estimates (effects) $\hat{B}$, the linear model ($Y = \langle X, B\rangle + E$) predicts outcomes as vectors indexed by $(d_1^Y, d_2^Y, \cdots, d_M^Y)$, $\hat{Y} = \langle X, \hat{B}\rangle$, by:

$$\hat{Y}(N, d_1^Y, d_2^Y, \cdots, d_M^Y) = \sum_{d_1^X=1}^{D_1^{X+}} \sum_{d_2^X=1}^{D_2^{X+}} \cdots \sum_{d_L^X=1}^{D_L^{X+}} \left[X(N, d_1^X, d_2^X, \cdots, d_L^X)\, \hat{B}(d_1^X, d_2^X, \cdots, d_L^X; d_1^Y, d_2^Y, \cdots, d_M^Y)\right].$$

The upper limit indexing of each sum $\left(D_j^{X+}\right)$ reflects the presence or absence of intercepts, which may be necessary when the input X and output Y tensors are not centered to be zero-mean. The special case of a magnitude only (real-valued) univariate model with *spacekime indexing* corresponds to $M = 1, L = 2$, as we have two kime degrees of freedom, one for the kime magnitude (time) and one for the kime direction (phase).

An Mth-order tensor A is an array of dimensions $D_1^A \times D_2^A \times \cdots \times D_M^A$, where D_j^A is the size of the jth *mode dimension*, $1 \le j \le M$. The tensor elements are indexed as $A[i_1, i_2, \cdots, i_M]$, $1 \le i_j \le D_j^A, \forall 1 \le j \le M$. For a pair of vectors, i.e., first order tensors, $a = (a_{j_1}) \in \mathbb{R}^{J_1}$ and $b = (b_{j_2}) \in \mathbb{R}^{J_2}$, the *outer product* is defined as a $J_1 \times J_2$ matrix M:

$$M = a^\circ b = (m_{j_1, j_2}) \in \mathbb{R}^{J_1 \times J_2},$$

whose entries $m_{j_1, j_2} = a_{j_1} \times b_{j_2}$.

Each tensor A may be expressed as an *outer product* ($^\circ$) of vectors $(a_1, a_2, \cdots, a_M)$ that are each of length $D_1^A, D_2^A, \ldots, D_M^A$, respectively:

$$a^\circ b = \left\{(a^\circ b)_{i,j} = a_i b_j\right\} \equiv ab',$$

$$A = a_1{}^\circ a_2{}^\circ \cdots {}^\circ a_M.$$

Then, the tensor elements are $A[i_1, i_2, \cdots, i_M] = \prod_{j=1}^{M} a_j[i_j]$, where $a_j[i_j]$ is the i_j^{th} element of the jth vector in the outer product defining the tensor A.

In general, the *inner product* of a pair of tensors:

$$A = \left(a_{j_1, j_2, \ldots, j_Q}\right) \text{ and } B = \left(b_{j_1, j_2, \ldots, j_Q}\right) \in \mathbb{R}^{J_1 \times J_2 \times \cdots \times J_Q}$$

of order Q is defined by:

$$A.B = \langle A, B\rangle = \sum_{j_1=1}^{J_1}\sum_{j_2=1}^{J_2}\cdots\sum_{j_Q=1}^{J_Q}\left(\underbrace{a_{j_1,j_2,\ldots,j_Q}\times b_{j_1,j_2,\ldots,j_Q}}_{\text{numeric multiplication}}\right).$$

Of course, the inner product of matrices, i.e., second-order tensors, is just a special case corresponding to $Q=2$. If $A_{m\times n}$ and $B_{k\times l}$ are a pair of matrices (not necessarily of congruent dimensions), then the *Kronecker product* is defined as a $km\times ln$ matrix:

$$(A\otimes B)_{km\times ln} = \begin{bmatrix} a_{1,1}B & a_{1,2}B & \ldots & a_{1,n}B \\ a_{2,1}B & a_{2,2}B & \ldots & a_{2,n}B \\ \ldots & \ldots & \ldots & \ldots \\ a_{m,1}B & a_{m,2}B & \ldots & a_{m,n}B \end{bmatrix}.$$

An order Q tensor $A=\left(a_{j_1,j_2,\ldots,j_Q}\right)\in\mathbb{R}^{J_1\times J_2\times\cdots\times J_Q}$ is called *rank-one tensor* when it has the special property that it can be expressed as an outer product of Q order 1 tensors, i.e., vectors, $a^{(j)}\in\mathbb{R}^{J_j}, \forall 1\le j\le Q$:

$$A = a^{(1)}\circ a^{(2)}\circ\ldots\circ a^{(Q)}.$$

Not every tensor is a rank-one tensor, however, every tensor may be factorized via canonical polyadic decomposition (*CP decomposition*) into a weighted sum of R rank-one tensors. Therefore, each tensor A can modeled as the weighted sum of rank-one tensors plus some additive residual (error) term:

$$A = \sum_{r=1}^{R}\left(\lambda_r\underbrace{a_r^{(1)}\circ a_r^{(2)}\circ\ldots\circ a_r^{(Q)}}_{\text{rank}-\text{one tensors}}\right)+\epsilon.$$

In this representation, the *tensor rank* is $R\in\mathbb{N}$, $\{\lambda_r\}_{r=1}^{R}$ are the scalar weights of the different rank-one tensor components $a_r^{(j)}=\left(a_{i_j,r}^{(j)}\right)\in\mathbb{R}^{J_j}, 1\le j\le Q, 1\le r\le R$, and the residual tensor ϵ has the same size (order and dimension lengths) as the original tensor A. **Figure 4.1** shows an illustration of the $R=4$ CP decomposition of a third-order tensor, which may represent the intensities of a 3D sMRI volume, or the 2D kime indexed 1D line of an fMRI series, or a location by time by phenotype dataset.

For computational purposes, we can transform long vectors into tensors and vice-versa, and high-order tensors into second-order tensors (matrices) or linear vectors. This is accomplished via the vectorization operator, which allows us to unfold a tensor into long vectors. An example to keep in mind is storing third-order tensors representing 3D sMRI volumes in computer memory as linear arrays or linked lists. These linear memory vectors can be conversely used to reconstruct the native 3D volume structures that can be visualized, **Figure 4.2**:

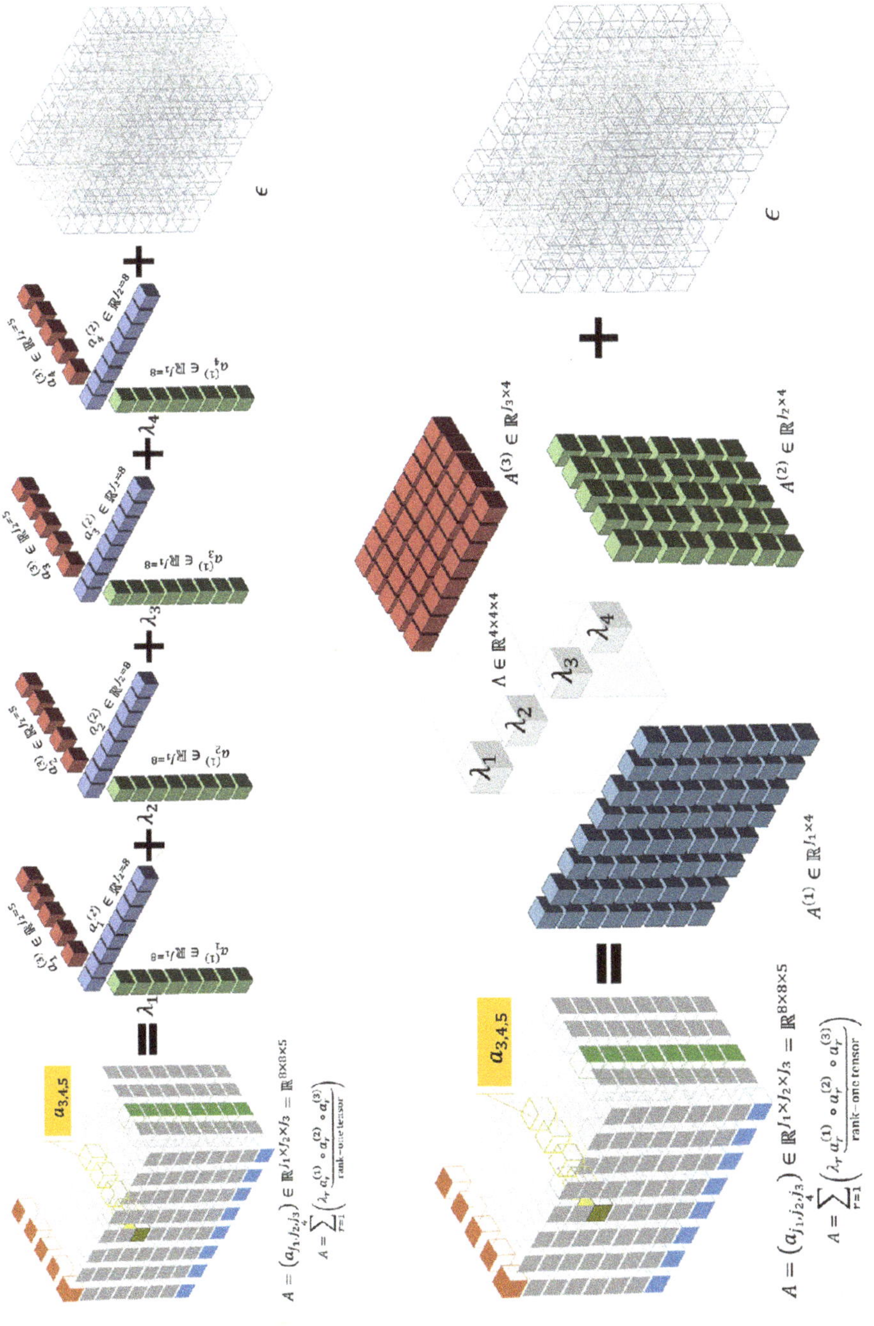

Figure 4.1: Schematic of a CP decomposition of a third-order tensor A using $r = 4$ components. In this CP decomposition, the rank-one tensor components, $a_r^{(j)} = \left(a_{i_j,r}^{(j)}\right) \in \mathbb{R}^{J_j}, 1 \le j \le 3, 1 \le r \le 4$, represent the rth column of the factor matrix $A^{(j)} = \left[a_1^{(j)}, a_2^{(j)}, a_3^{(j)}, a_4^{(j)}\right] \in \mathbb{R}^{J_j \times R}$ from the jth mode dimension.

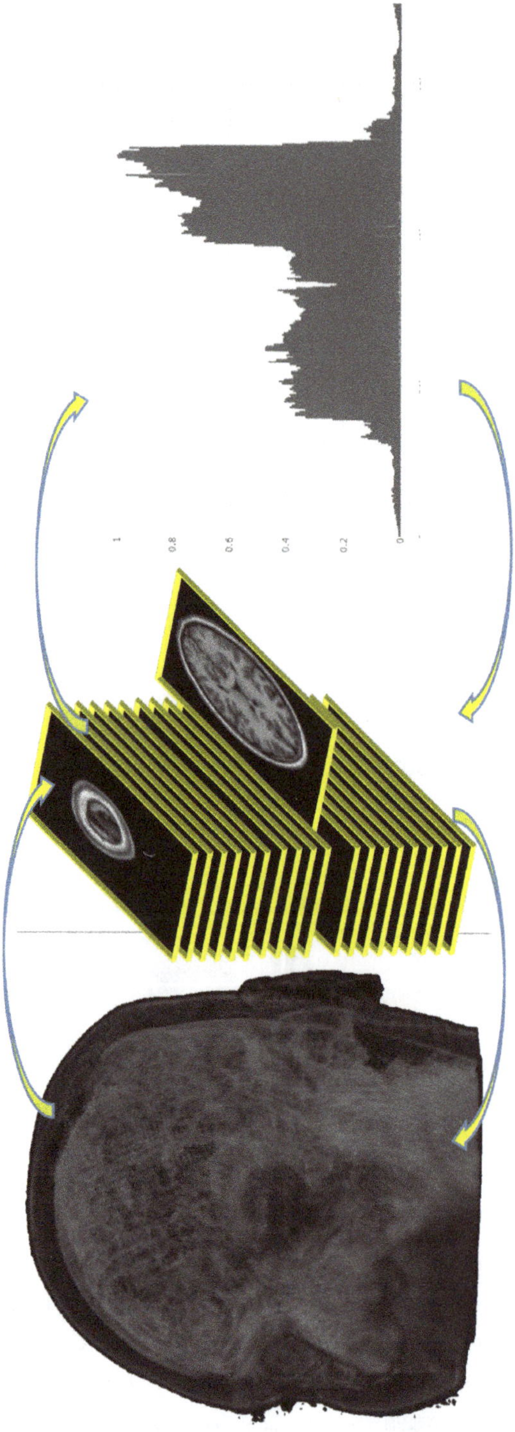

Figure 4.2: 3D sMRI illustration of tensor (*A*) vectorization (*i*) and the reversed tensorization of the corresponding linear array representation of the same sMRI data. Assuming the following dimension indexing $0 \le x < D_x$, $0 \le y < D_y$, $0 \le z < D_z$ and the *modulo* (remainder of the division) function is *mod*, then the forward and reverse tensor-to-vector indexing transformations are.

$$\underbrace{A[\_,\_,\_] \rightarrow vec(A)[\_]}_{\text{forward:}3D \rightarrow 1D} : A[x,y,z] \rightarrow vec(A)\left[\underbrace{D_x \times D_y \times z + D_x \times y + x}_{i}\right],$$

$$\underbrace{vec(A)[\_] \rightarrow A[\_,\_,\_]}_{\text{reverse:}1D \rightarrow 3D} : vec(A)[i] \rightarrow A[x,y,z], \text{ where } \left| \begin{array}{l} x = i \, mod \, D_x \\ y = \left(\frac{i-x}{D_x}\right) mod(D_y). \\ z = \left(\frac{i-x-D_x y}{D_x \times D_y}\right) \end{array} \right.$$

In general, the tensor vectorization operator $\mathbb{V}(\cdot) = vec(\cdot)$ transforming a higher-order tensor A to a first-order tensor (vector) is defined by:

$$\underbrace{\mathbb{V}(A)\left[i_1 + \sum_{j=2}^{M}\left(\left(\prod_{l=1}^{j-1} D_l^A\right)(i_j - 1)\right)\right]}_{\text{vectorization index}} = \underbrace{A[i_1, i_2, \cdots, i_M]}_{\text{tensor element}}.$$

Similarly, it's often useful to represent high-order tensors as matrices (second-order tensors). This tensor-to-matrix (*tenmat*) operation unfolds a tensor along a given mode dimension (k_o) and allows the application of classical matrix manipulations to the resulting matrix. For instance, denoting the rows of the resulting tenmat representation of the tensor A by $A^{(k)}$, these rows will unfold to give the vectorized versions of each subarray (lower-order sub-tensor) in the k_o^{th} mode dimension:

$$\dim(tenmat(A, k_o)) = \underbrace{D_{k_o}^A}_{\text{rows}} \times \underbrace{\left(\prod_{j \neq k_o} D_j^A\right)}_{\text{columns}}.$$

The *contracted tensor product* [230, 234, 235] extends the usual matrix multiplication to a higher-order tensor product. For a pair of tensors U and V,

$$\dim(U) = \underbrace{D_1^U \times D_2^U \times \cdots \times D_M^U}_{U \text{ specific}} \times \underbrace{D_1 \times D_2 \times \cdots \times D_L}_{U \,\&\, V \text{ common}}$$

and

$$\dim(V) = \underbrace{D_1 \times D_2 \times \cdots \times D_L}_{U \,\&\, V \text{ common}} \times \underbrace{D_1^V \times D_2^V \times \cdots \times D_K^V}_{V \text{ specific}}.$$

Then, *contracted tensor product* (L) of dimension:

$$\dim(\langle U, V\rangle_L) = \underbrace{D_1^U \times D_2^U \times \cdots \times D_M^U}_{U \text{ specific}} \times \underbrace{D_1^V \times D_2^V \times \cdots \times D_K^V}_{V \text{ specific}}$$

is defined by:

$$\langle U,V\rangle_L\left[\underbrace{i_1,i_2,\cdots,i_M}_{U\text{ specific}},\underbrace{j_1,j_2,\cdots,j_K}_{V\text{ specific}}\right]=$$

$$\sum_{d_1=1}^{D_1}\sum_{d_2=1}^{D_2}\cdots\sum_{d_L=1}^{D_L}\left(U\left[\underbrace{i_1,i_2,\cdots,i_M}_{U\text{ specific}},\underbrace{d_1,d_2,\cdots,d_L}_{\text{common}}\right]\times V\left[\underbrace{d_1,d_2,\cdots,d_L}_{\text{common}},\underbrace{j_1,j_2,\cdots,j_K}_{V\text{ specific}}\right]\right).$$

Note that the contracted tensor product agrees with classical matrix multiplication for the special case of second-order tensors (matrices), i.e., $M=L=K=1$. Suppose $U_{D_1^U\times D}$ and $V_{D\times D_1^V}$, then:

$$\left\langle U_{D_1^U\times D},V_{D\times D_1^V}\right\rangle_{L=1}[i,j]=\sum_{d=1}^{D}\underbrace{(U[i,d]V[d,j])}_{\text{row by column}}=UV[i,j],$$

$$\left\langle U_{D_1^U\times D},\ V_{D\times D_1^V}\right\rangle_{L=1}=U\ V.$$

It may be helpful to illustrate one simple and specific formulation of tensor-based linear modeling for fMRI data. Let's denote all the fMRI brain spatial locations by the voxels

$$v=(x,y,z)\in[1{:}64,1{:}64,1{:}40]\subset\mathbb{R}^3.$$

Typical MRI scans have (higher) isotropic resolution within the axial (transverse) plane (for a fixed $z=z_o$) and the between plane slices are thicker (lower resolution). The observed (complex) *outcome* that will be modeled will be a tensor $Y_{(t,x,y,z,2)}$ indexed by time $t\in[1{:}160]\subset\mathbb{R}^1$, voxel $v=(x,y,z)$, and the Real (1) and Imaginary (2) indices corresponding to the observed complex-valued fMRI signal at (v,t). In this simple example, the tensor-model *design matrix, X*, representing the experimental design, may include three components: (a) intercept vector of ones, (b) the HRF, which transforms the original stimulus (ON/OFF) into the observed fMRI BOLD signal, and (c) the linear model signal (brain habituation linear trend):

$$X\left[\underbrace{X_o=1}_{\text{intercept}},\underbrace{X_1}_{\text{BOLD signal}},\underbrace{X_2}_{\text{linear signal}}\right]_{160\times3\times1}$$

If the effect-size tensor is $B[feature,1,x,y,z,2]$ and the residual error tensor is $e_{(t,x,y,z,2)}$, then the tensor-based linear model is represented as:

$$Y_{(160,64,64,40,2)}=X_{(160,3,1)}\times B_{(3,1,64,64,40,2)}+e_{(160,64,64,40,2)}.$$

Often it is convenient to represent a tensor X of an arbitrary order as a second-order tensor, or a matrix. These lower-order representations are not unique and there are multiple ways to arrange the tensor elements into matrix format. A *mode-d dimension matricization* of X [231] is formulated by:

$$X_{(d)} = \underbrace{A^{(d)}}_{J_d\times R} \underbrace{\Lambda}_{R\times R} \underbrace{\left(A^{(L)} \odot A^{(L-1)} \odot \cdots \odot A^{(d+1)} \odot A^{(d-1)} \odot \cdots \odot A^{(2)} \odot A^{(1)}\right)'}_{R\times \prod_{i\neq d} J_i},$$

where the *Khatri-Rao dot product* $\odot$ of $A=[a_1,a_2,\cdots,a_J]\in\mathbb{R}^{I\times J}$ and $B=[b_1,b_2,\cdots,b_J]\in\mathbb{R}^{K\times J}$ is defined as column-wise Kronecker product of matrices $A\odot B=[a_1\otimes b_1,a_2\otimes b_2,\cdots,a_J\otimes b_J]\in\mathbb{R}^{IK\times J}$, $A^{(d)}$ represents the R components of the corresponding d^{th} component rank-one tensor stacked together, $\Lambda=\text{diag}\ (\lambda)$, $\lambda=(\lambda_1,\lambda_2,\cdots,\lambda_R)\in\mathbb{R}^R$ is a vector of normalization weights that ensures that the CP-decomposition factor matrices $A^{(l)}$ are composed of unitary column vectors, and R is the tensor rank. In the special case of the third-order fMRI BOLD tensor $X\in\mathbb{R}^{160\times 3\times 1}$, the CP decomposition factorizes the fMRI tensor into a sum of rank-one tensor components (which in this case are vectors, $x_o=[x_{o,1}\ x_{o,2}\cdots\ x_{o,R}], x_1=[x_{1,1}\ x_{1,2}\cdots\ x_{1,R}],\ x_2=[x_{2,1}\ x_{2,2}\cdots\ x_{2,R}]$):

$$X=\sum_{r=1}^{R} x_{o,r}{}^{o}x_{1,r}{}^{o}x_{2,r}=\left(x_{i,j,k}\right)=\left(\sum_{r=1}^{R} x_{o,i,r}{}^{o}x_{1,j,r}{}^{o}x_{2,k,r}\right), \forall 1\le i\le 160,\ 1\le j\le 3,\ 1\le k\le 1.$$

Thus, there are alternative matricized representations (one for each of the 3 modes) of the fMRI tensor:

$$X_{(1)}=\mathbf{x}_o(\mathbf{x}_1\odot\mathbf{x}_2)',\ X_{(2)}=\mathbf{x}_1(\mathbf{x}_2\odot\mathbf{x}_o)',\ X_{(3)}=\mathbf{x}_2(\mathbf{x}_1\odot\mathbf{x}_o)'.$$

By introducing the weights $\Lambda=\text{diag}\ (\lambda)$, $\lambda=(\lambda_1,\lambda_2,\ldots,\lambda_R)\in\mathbb{R}^R$, we can assure the CP-decomposition factor matrices $A^{(l)}=\mathbf{x}_l$ are composed of unitary column vectors:

$$X=[\![\lambda;\mathbf{x}_o,\mathbf{x}_1,\mathbf{x}_2]\!]=\sum_{r=1}^{R}\lambda_r x_{o,r}{}^{o}x_{1,r}{}^{o}x_{2,r}.$$

Finally, tensor inner, outer, and contracted products lead to the Frobenius tensor norm.

1. *Inner tensor product* of a pair of congruent (same orders and dimension sizes) tensors $A,B\in\mathbb{R}^{J_1\times J_2\times\ldots\times J_D}$ is a scalar defined by:

$$\langle A,B\rangle=\langle vec(A),vec(B)\rangle\in\mathbb{R},$$

$$\langle A,B\rangle=\sum_{j_1=1}^{J_1}\sum_{j_2=1}^{J_2}\cdots\sum_{j_D=1}^{J_D} a_{j_1,j_2,\ldots,j_D}\times b_{j_1,j_2,\ldots,j_D}.$$

2. *Outer tensor product* of a general pair of tensors $A\in\mathbb{R}^{J_1\times J_2\times\cdots\times J_P}$ and $B\in\mathbb{R}^{K_1\times K_2\times\cdots\times K_Q}$ is defined by:

$$A^{o}B = \left(z_{j_1,j_2,\ldots,j_P,k_1,k_2,\ldots,k_O}\right) = \left(a_{j_1,j_2,\ldots,j_P} \times b_{k_1,k_2,\ldots,k_O}\right) \in \mathbb{R}^{J_1 \times J_2 \times \ldots \times J_P \times K_1 \times K_2 \times \ldots \times K_O}.$$

3. *Frobenius tensor norm* (derived from *Frobenius* inner product) of a general tensor $A \in \mathbb{R}^{J_1 \times J_2 \times \ldots \times J_P}$ is defined by:

$$||A|| = ||A||_F = \sqrt{\langle A, A\rangle} = \sqrt{\sum_{j_1=1}^{J_1} \sum_{j_2=1}^{J_2} \cdots \sum_{j_P=1}^{J_P} A^2_{j_1,j_2,\ldots,j_P}}.$$

Observe that the *tensor inner product* is a special case of the *tensor contracted product* when all the modes are in common between the two tensor multipliers, i.e., if $A, B \in \mathbb{R}^{J_1 \times J_2 \times \ldots \times J_D}$, then the constant $z = \langle A, B\rangle = \langle A, B\rangle_{\{j_1,j_2,\ldots,j_D;\, j_1,j_2,\ldots,j_D\}}$. At the other extreme, when there are no common indices between two tensors ($L = 0$), the contracted tensor product is the tensor outer product, i.e., $Z = A^{o}B = \langle A, B\rangle_{\{0;0\}}$.

4.6.2 Linear Modeling Using Tensors

Tensor algebra naturally leads to linear modeling and least-squares parameter estimation for high-order tensors. In the most general setting, the dimension of the *outcome tensor* is $\dim(Y) = N \times D_1^Y \times D_2^Y \times \cdots \times D_M^Y$, the dimension of the observed *input (design) tensor* is $\dim(X) = N \times D_1^{X+} \times D_2^{X+} \times \cdots \times D_L^{X+}$, the dimension of the model effects is $\dim(B) = D_1^{X+} \times D_2^{X+} \times \cdots \times D_L^{X+} \times D_1^Y \times D_2^Y \times \cdots \times D_M^Y$, and the residual (error) tensor dimension is congruent to that of the output, i.e., $\dim(E) = N \times D_1^Y \times D_2^Y \times \cdots \times D_M^Y$. Note that for notation brevity, we will write D_l^X for D_l^{X+}. Then the *tensor linear model* is:

$$Y = \langle X,\ B\rangle_L + E.$$

And the model-driven predictions are:

$$\hat{Y}\left(n, d_1^Y, d_2^Y, \cdots, d_M^Y\right) = \langle X, \hat{B}\rangle_L =$$

$$\sum_{d_1^X=1}^{D_1^{X+}} \sum_{d_2^X=1}^{D_2^{X+}} \cdots \sum_{d_L^X=1}^{D_L^{X+}} \left[X\left(N, d_1^X, d_2^X, \cdots, d_L^X\right)\hat{B}\left(d_1^X, d_2^X, \cdots, d_L^X; d_1^Y, d_2^Y, \cdots, d_M^Y\right)\right].$$

Denote by N the total number of observations, $P = \prod_{l=1}^{L} D_l^X$ the total number of predictors for each outcome, and $O = \prod_{m=1}^{M} D_m^Y$ the total number of outcomes for each observation. Then the tensor-linear model is equivalent to a classical matrix formulation of a linear model:

$$Y^{(1)} = X^{(1)}B^{(1)} + E^{(1)}.$$

This vectorized tenmat representation of the linear model relies on unfolding each of the tensors along the first mode dimension, $\left(Y^{(1)}, X^{(1)}, B^{(1)}, E^{(1)}\right)$, where the resulting matrix dimensions are $\dim\left(Y^{(1)}\right) = N \times O$, $\dim\left(X^{(1)}\right) = N \times P$, $\dim\left(B^{(1)}\right) = P \times O$, and $\dim\left(E^{(1)}\right) = N \times O$. Explicitly, the relation between the tensor effects and their vectorized tenmat representations is:

$$\operatorname{tenmat}(B)\left[\underbrace{d_1^X + \sum_{j=2}^{L}\left(\left(\prod_{l=1}^{j-1} D_l^X\right)\left(d_l^X - 1\right)\right)}_{\text{row index}}, \underbrace{d_1^Y + \sum_{j=2}^{M}\left(\left(\prod_{l=1}^{j-1} D_l^Y\right)\left(d_l^Y - 1\right)\right)}_{\text{column index}}\right] =$$

$$B\left[\underbrace{\underbrace{d_1^X, d_2^X, \cdots, d_L^X}, \underbrace{d_1^Y, d_2^Y, \cdots, d_M^Y}}_{\text{tensor indexing}}\right].$$

Having the (*tenmat*) matrix representation of general tensors suggests a strategy to estimate the parameter coefficient tensor B by minimizing the squared residuals based on the *Frobenius* norm:

$$\hat{B} = \arg\min_{\text{B}} \|Y - \langle X, B\rangle_L\|^2 = \arg\min_{\text{B}} \|Y - \langle X, B\rangle\|_F^2.$$

The unrestricted solution for B over the entire (high-dimensional) parameter space may be computed by OLS regressions for each of the $O = \prod_{i=1}^{M} D_i^Y$ outcomes constituting the multivariate output Y. The input design matrix $X^{(1)}$ corresponds to the first of M mode dimensions and the B columns are individually estimated by OLS regressions of $X^{(1)}$ on each column of $Y^{(1)}$. The existence of this unrestricted problem solution requires that $P \le N$ and $X^{(1)}$ be a full column rank matrix. This OLS solution *may not be optimal* as the unfolding of the tensor into a matrix operation ignores the multidimensional (multi-way) structure of the tensors X and Y. In practice, the OLS parameter estimation of the coefficient tensor B fits a large number of parameters, $PO = \left(\prod_{l=1}^{L} D_l^X\right)\left(\prod_{m=1}^{M} D_m^Y\right)$. Schematically, N is the number of linear equations and P is the number of features whose effects are the parameters we need to estimate in the matrix $B^{(1)}$:

$$N\left\{\underbrace{\begin{bmatrix} \square & \square & \square \\ \square & X^{(1)} & \square \\ \square & \square & \square \end{bmatrix}}_{P}\right., P\left\{\underbrace{\begin{bmatrix} \square & \square & \square \\ \square & B^{(1)} & \square \\ \square & \square & \square \end{bmatrix}}_{O}\right..$$

Regularized tensor linear modeling using orthogonal matching pursuit (OMP), least absolute shrinkage and selection operator (LASSO), ridge, or elastic net [10, 238] may be employed to tamper this problem of extremely large number of parameters. Regularization properties include (1) blending the coefficient estimation with feature selection, (2) preserving some of the multidimensional tensor structure into the tenmat representation, and (3) reducing the chance for overfitting. The regularized linear tensor model can be expressed as:

$$\breve{B} = \arg \min_{\text{rank}(B) \leq R} \left(\underbrace{\left\| Y - \langle X, B \rangle_L \right\|^2}_{\text{fidelity}} + \underbrace{\lambda_o \, \|B\|_o + \lambda_1 \, \|B\|_1 + \lambda_2 \, \|B\|_2^2}_{\text{regularizer}} \right),$$

where R is the desired solution tensor rank, for $l \in \{0,1,2\}$, $\{\lambda_l\}$ are the weights of the elastic net regularization penalty, and the lth norm $\|B\|_l$ corresponds to $\|B\|_o =$ the number of non-trivial components (OMP), $\|B\|_1 = \sum_{j=1}^{P} |b_j|$, for LASSO, and $\|B\|_2^2 = \sum_{j=1}^{P} |b_j|^2$, for ridge regularization, respectively. Some special cases of this model for various (O, P) combinations are discussed in [230]. Solutions to the unregularized OLS model and regularized ridge, LASSO and elastic net regularized models are provided in [230, 239–241]. Below, we will illustrate some special cases of specific tensor linear models and their solutions.

4.6.3 Unregularized OLS Model Solutions

The general alternating least squares (ALS) approach for model estimation relies on fixing the outcome Y and optimizing X followed by fixing X and optimizing Y, iteratively repeating the process until convergence is achieved. Similar to ALS, tensor-based linear model solutions may be obtained by an iterative algorithm sequentially updating each vector component of $B\left[d_1^X, d_2^X, \ldots, d_L^X; d_1^Y, d_2^Y, \ldots, d_M^Y\right]$, while keeping the remaining components fixed. This is applicable for

$$\dim B = \underbrace{R}_{\text{tensor rank }(B)} \times \left(\underbrace{D_1^X + D_2^X + \ldots + D_L^X}_{\substack{\text{shared dimensions} \\ \text{with predictor tensor } X}} + \underbrace{D_1^Y + D_2^Y + \ldots + D_M^Y}_{\substack{\text{shared dimensions} \\ \text{with outcome tensor } Y}} \right),$$

$$B = \sum_{i=1}^{R} \left[\overbrace{\left(\underbrace{u_{1,i}}_{D_1^X} \circ \underbrace{u_{2,i}}_{D_2^X} \circ \cdots \circ \underbrace{u_{L,i}}_{D_L^X} \right) \circ \left(\underbrace{v_{1,i}}_{D_1^Y} \circ \underbrace{v_{2,i}}_{D_2^Y} \circ \cdots \circ \underbrace{v_{M,i}}_{D_M^Y} \right)}^{\text{rank 1 tensor}} \right].$$

As the objective function is invariant to indexing permutations of the L dimension modes of the design matrix tensor, the protocol for estimating the first component $\left(D_1^X\right)$ with the other components remaining fixed, provides a generic solution for estimating all components of the effects B. The coefficient estimation problem permits solutions under certain conditions on the *identifiability of the B effects*, e.g., the resulting components can be uniquely factorized using tensor rank canonical polyadic decomposition (CP) [242–244].

This iterative estimation protocol uses the (aggregate) matrix $C = [C_1, C_2, \cdots, C_R]$, $\dim C = (N \times O) \times \left(R \times D_1^X\right)$, composed of C_r defined as the contracted tensor products of the input design matrix X and the rth component of the CP tensor factorization (except the first one, which is being currently estimated) in each mode dimension k. Then, $\dim C_r = (N \times O) \times \left(D_1^X\right)$, $\forall 1 \le r \le R$ and C_r can be expressed as:

$$C_r = \langle X,\ u_{2,r} \circ u_{3,r} \circ \cdots \circ u_{L,r} \circ v_{1,r} \circ v_{1,r} \circ \cdots \circ v_{M,r} \rangle_{L-1}.$$

By unfolding the CP factorization component in the dimension corresponding to P_1, we can obtain a tenmat input design matrix, C_r, that can be used to predict the vectorized outcome $(Y^{(1)})$ corresponding to the rth column of the first mode direction (D_1^X). Aggregating all C_r matrices $\forall 1 \le r \le R$, yields the master design input matrix $C = [C_1, C_2, \ldots, C_R]$, for all elements of the first component, $U^{(1)}$. Specifically, the vectorized estimates of the effects in the first component D_1^X can be obtained using:

$$vec\left(U^{(1)}\right) = (C'C)^{-1} C'\ vec\left(Y^{(1)}\right).$$

A more explicit formulation of the CP tensor factorization can be formulated by flattening higher order tensors as vectors and matrices:

$$C_r = \overbrace{\langle X, u_{2,i} \circ \cdots \circ u_{L,i} \circ v_{1,i} \circ \cdots \circ v_{M,i} \rangle_{L-1}}^{N \times D_1^X \times \overbrace{D_1^Y \times D_2^Y \times \cdots \times D_M^Y}^{O}},\quad 1 \le r \le R,$$

$$\underbrace{\begin{bmatrix} Y_1 \\ \vdots \\ Y_{NO} \end{bmatrix}}_{Y^{(1)}} = \underbrace{\begin{bmatrix} \underbrace{\begin{bmatrix} C_{1,(1,1)} & \cdots & C_{1,\left(1,D_1^X\right)} \\ \vdots & \vdots & \vdots \\ C_{1,(NO,1)} & \cdots & C_{1,\left(NO,D_1^X\right)} \end{bmatrix}}_{C_1} \underbrace{\begin{bmatrix} C_{2,(1,1)} & \cdots & C_{2,\left(1,D_1^X\right)} \\ \vdots & \vdots & \vdots \\ C_{2,(NO,1)} & \cdots & C_{2,\left(NO,D_1^X\right)} \end{bmatrix}}_{C_2} \cdots \underbrace{\begin{bmatrix} C_{R,(1,1)} & \cdots & C_{R,\left(1,D_1^X\right)} \\ \vdots & \vdots & \vdots \\ C_{R,(NO,1)} & \cdots & C_{R,\left(NO,D_1^X\right)} \end{bmatrix}}_{C_R} \end{bmatrix}}_{C} \underbrace{\underbrace{\begin{bmatrix} \begin{bmatrix} \vdots \\ u_{1,1} \\ \vdots \end{bmatrix} \\ \begin{bmatrix} \vdots \\ u_{2,1} \\ \vdots \end{bmatrix} \\ \vdots \\ \begin{bmatrix} \vdots \\ u_{R,1} \\ \vdots \end{bmatrix} \end{bmatrix}}_{U^{(1)}}}_{\left(D_1^X R\right)\times 1}$$

In matrix equation form, the tensor linear model may be expressed by:

$$\underbrace{vec\left(Y^{(1)}\right)}_{NO\times 1} = \underbrace{C}_{NO\times D_1^X R} \times \underbrace{vec\left(U^{(1)}\right)}_{D_1^X R\times 1} \Rightarrow vec\left(U^{(1)}\right) = (C'C)^{-1}C'vec\left(Y^{(1)}\right).$$

This can be derived from first principles as follows. For a given data, X, the prediction from a tensor linear model

$$Y = \langle X, B\rangle_L + E$$

is expressed as

$$\hat{Y}\left(N, d_1^Y, d_2^Y, \cdots, d_M^Y\right) = \sum_{d_1^X=1}^{D_1^{X+}} \sum_{d_2^X=1}^{D_2^{X+}} \cdots \sum_{d_L^X=1}^{D_L^{X+}} \left[X\left(N, d_1^X, d_2^X, \cdots, d_M^X\right)\hat{B}\left(d_1^X, d_2^X, \cdots, d_M^X; d_1^Y, d_2^Y, \cdots, d_M^Y\right)\right].$$

Assuming that the effect tensor B is of rank R, it can be expanded as a sum of R simple tensors B_i that are completely factorizable in terms of vector outer products:

$$B = \sum_{i=1}^{R} \underbrace{\left(u_{1,i}{}^{\circ}u_{2,i}{}^{\circ}\cdots{}^{\circ}u_{L,i}\right){}^{\circ}\left(v_{1,i}{}^{\circ}v_{2,i}{}^{\circ}\cdots{}^{\circ}v_{M,i}\right)}_{B_i}.$$

For a fixed index i, the R vector components, $\{u_{l,i}\}_{l=1}^{L}$ and $\{v_{m,i}\}_{m=1}^{M}$, correspond to stacking the R dimension to form the $U^{(j)}$ and $V^{(j)}$ matrices. Specifically,

$$U^{(j)} = \begin{bmatrix} \vdots & \vdots & \vdots \\ u_{j,1} & \cdots & u_{j,R} \\ \vdots & \vdots & \vdots \end{bmatrix}, \quad V^{(j)} = \begin{bmatrix} \vdots & \vdots & \vdots \\ v_{j,1} & \cdots & v_{j,R} \\ \vdots & \vdots & \vdots \end{bmatrix}.$$

Expanding the tensor linear model in terms of the outer products, we obtain:

$$Y\left(\overbrace{N,\ \underbrace{d_1^Y, d_2^Y, \cdots, d_M^Y}_{\substack{\text{index mapping to} \\ \text{a single number } d^Y}}}^{\text{flattening on } N}\right) = Y\left(\underbrace{N}_{\substack{\text{row}\\ \text{index}}},\ \underbrace{d^Y}_{\substack{\text{column}\\ \text{index}}}\right) =$$

$$\underbrace{\sum_{i=1}^{R} \left\langle X,\ \underbrace{u_{1,i} \circ u_{2,i} \circ \cdots \circ u_{L,i} \circ v_{1,i} \circ \cdots \circ v_{M,i}}_{B_i} \right\rangle}_{\langle X, B\rangle_L} =$$

$$\sum_{i=1}^{R} \sum_{d_1^X=1}^{D_1^X} \sum_{d_2^X=1}^{D_2^X} \cdots \sum_{d_L^X=1}^{D_L^X} \left[X\left(N, d_1^X, d_2^X, \cdots, d_M^X\right) B_i \left(d_1^X, d_2^X, \cdots, d_M^X;\ \underbrace{d_1^Y, d_2^Y, \cdots, d_M^Y}_{\substack{\text{index mapping to} \\ \text{a single number } d^Y}} \right) \right] =$$

$$\sum_{i=1}^{R} \sum_{d_1^X=1}^{D_1^X} \sum_{d_2^X=1}^{D_2^X} \cdots \sum_{d_L^X=1}^{D_L^X} \left[X\left(N, d_1^X,\ \underbrace{d_2^X, \cdots, d_M^X}_{\substack{\text{index mapping to} \\ \text{a single number } d_{-1}^X}} \right) B_i \left(d_1^X,\ \underbrace{d_2^X, \cdots, d_M^X}_{\substack{\text{index mapping to} \\ \text{a single number } d_{-1}^X}}\ ; d^Y \right) \right].$$

Multiplying the numerator and the denominator by $u_{1,i}\left(d_1^X\right)$ we get:

$$Y\left(\underbrace{N}_{\substack{\text{row}\\ \text{index}}},\ \underbrace{d^Y}_{\substack{\text{column}\\ \text{index}}}\right) = \sum_{i=1}^{R} \sum_{d_1^X=1}^{D_1^X} \sum_{d_{-1}^X=1}^{\prod_{l=2}^{L} D_l^X} \left[X\left(N, d_1^X, d_{-1}^X\right) u_{1,i}\left(d_1^X\right) \frac{B_i\left(d_1^X, d_{-1}^X; d^Y\right)}{u_{1,i}\left(d_1^X\right)} \right] =$$

$$\sum_{i=1}^{R} \sum_{d_1^X=1}^{D_1^X} u_{1,i}\left(d_1^X\right) \sum_{d_{-1}^X=1}^{\prod_{l=2}^{L} D_l^X} \left[X\left(N, d_1^X, d_{-1}^X\right) \frac{B_i\left(d_1^X, d_{-1}^X; d^Y\right)}{u_{1,i}\left(d_1^X\right)} \right] =$$

$$\sum_{i=1}^{R} \underbrace{\left[\underbrace{\sum_{d_{-1}^X=1}^{\prod_{l=2}^{L} D_l^X} \left[X\left(N, 1, d_{-1}^X\right) \frac{B_i\left(1, d_{-1}^X; d^Y\right)}{u_{1,i}(1)} \right]}_{\text{scalar}}, \ldots, \underbrace{\sum_{d_{-1}^X=1}^{\prod_{l=2}^{L} D_l^X} \left[X\left(N, D_1^X, d_{-1}^X\right) \frac{B_i\left(D_1^X, d_{-1}^X; d^Y\right)}{u_{1,i}\left(D_1^X\right)} \right]}_{\text{scalar}} \right]}_{C_i \text{ with row specified indexing}(N, d^Y)} \underbrace{\begin{bmatrix} u_{1,i}(1) \\ u_{1,i}(2) \\ \vdots \\ u_{1,i}\left(D_1^X\right) \end{bmatrix}}_{\left(D_1^X\right) \times 1}.$$

If we stack in $N \times O$ rows as a column vector the predicted outcome, $Y^{(1)}$, and horizontally stack the R matrices, we obtain the matrix C (this can be thought of as a design tenmat object) and the vectorization of the outcome prediction of the tensor linear model becomes:

$$\underbrace{vec\left(Y^{(1)}\right)}_{NO\times 1} = \underbrace{\begin{bmatrix} Y_1 \\ \vdots \\ Y_{NO} \end{bmatrix}}_{Y^{(1)}} = \underbrace{C}_{NO\times D_1^X R} \times \underbrace{vec\left(U^{(1)}\right)}_{D_1^X R\times 1} \Rightarrow vec\left(U^{(1)}\right) = (C'C)^{-1} C' vec\left(Y^{(1)}\right).$$

As the objective cost function is independent of the ordering of the M mode dimensions of the outcome Y, the computational procedure to update the D^Y components ($v_{i,r}$) is similar to the one above for estimating the D^X components $u_{i,\ r}$. Having fixed all components $\left[D_1^X, D_2^X, \cdots, D_L^X, D_1^Y, D_2^Y, \cdots, D_{M-1}^Y\right]$, except the last one, D_M^Y, we can update the components of the last outcome mode dimension. If we unfold the outcome tensor Y along the mode dimension corresponding to D_M^Y and denote it by Y_M, then $\dim Y_M = D_M^Y \times N \times \prod_{l=1}^{M-1} D_l^Y$. Let D be the contracted tensor product of the input data X and the R CP factorization component excluding D_M^Y:

$$\dim D = N \times \prod_{l=1}^{M-1} \left(D_l^Y \times R\right),\ D = [d_1, d_2, \cdots, d_R]$$

$$d_r = \text{vec}\left(\left\langle X, u_{1,r} \circ u_{2,r} \circ \cdots \circ u_{L,r} \circ v_{1,r} \circ v_{2,r} \circ \cdots \circ v_{(M-1),r} \right\rangle_L\right).$$

Then, the D_M^Y elements V can be estimated as a vector of independent OLS estimations of size D_M^Y:

$$\underbrace{V_M^Y}_{R\times D_M^Y} = \left(\underbrace{D'D}_{R\times R}\right)^{-1} \times \underbrace{D'}_{R\times N\prod_{l=1}^{M-1} D_l^Y} \times \underbrace{Y_M{}'}_{\left(N\prod_{l=1}^{M-1} D_l^Y\right)\times D_M^Y}.$$

Having the matrix estimates of U and V, we can reconstruct the parameter-tensor, i.e., effect-tensor, B, using the recursively computed $U^{(1)}, U^{(2)}, \cdots, U^{(L)}$ and $V_1^Y = V^{(1)}, V_2^Y = V^{(2)}, \cdots, V_M^Y = V^{(M)}$ matrices:

$$\hat{B} = \sum_{i=1}^{R} \left[\overbrace{\left(\underbrace{u_{1,i}}_{D_1^X} \circ \underbrace{u_{2,i}}_{D_2^X} \circ \cdots \circ \underbrace{u_{L,i}}_{D_L^X}\right) \circ \left(\underbrace{v_{1,i}}_{D_1^Y} \circ \underbrace{v_{2,i}}_{D_2^Y} \circ \cdots \circ \underbrace{v_{M,i}}_{D_M^Y}\right)}^{\text{rank 1 tensor}}\right].$$

To obtain the d_r and C_r estimates

$$C_r = \langle X,\ u_{2,r} {}^{\circ} u_{3,r} {}^{\circ} \cdots {}^{\circ} u_{L,r} {}^{\circ} v_{1,r} {}^{\circ} v_{1,r} {}^{\circ} \cdots {}^{\circ} v_{M,r} \rangle_{L-1},$$

$$d_r = \text{vec}\left(\langle X,\ u_{1,r} {}^{\circ} u_{2,r} {}^{\circ} \cdots {}^{\circ} u_{L,r} {}^{\circ} v_{1,r} {}^{\circ} v_{2,r} {}^{\circ} \cdots {}^{\circ} v_{(M-1),r} \rangle_L \right)$$

we use the vectorizing operator on $\langle X,\ u_{1,r} {}^{\circ} u_{2,r} {}^{\circ} \cdots {}^{\circ} u_{L,r} {}^{\circ} v_{1,r} {}^{\circ} v_{2,r} {}^{\circ} \cdots {}^{\circ} v_{(M-1),r} \rangle_L$, where all the entries following the contraction are all unique to Y. Contrast this to the general inner-product $\langle X,\ u_{2,r} {}^{\circ} u_{3,r} {}^{\circ} \cdots {}^{\circ} u_{L,r} {}^{\circ} v_{1,r} {}^{\circ} v_{1,r} {}^{\circ} \cdots {}^{\circ} v_{M,r} \rangle_{L-1}$ where the last term has dimension that belongs to X. We address this inability to flatten out this exclusive dimension by flattening the predicted outcome tensor $\text{vec}\left(U^{(1)}\right) = \left(C'C\right)^{-1} C' vec\left(Y^{(1)}\right)$.

4.6.4 Regularized Linear Model Solutions

Analogous solutions to regularized tensor-based linear models may be derived using alternative penalty terms, e.g., OMP, LASSO, and ridge. These solutions can be explicated by completing the squares and transforming the regularized model to a dual unregularized optimization problem with modified input and output tensors, $\tilde{X}, \tilde{Y}$. For instance, the ridge penalty model can be expressed as:

$$\breve{B} = \arg\min \left(\underbrace{\| Y - \langle X, B \rangle_L \|^2}_{\text{fidelity}} + \underbrace{\lambda_1 \| B \|_1^2}_{\text{regularizer}} \right) = \arg \min_{\text{rank}(B) \le R} \left(\| \tilde{Y} - \langle \tilde{X}, B \rangle_L \|^2 \right),$$

where $\tilde{X}$ is the concatenated tensor product of X with a tensor where each slice of dimension $D_1^X \times D_2^X \times \cdots \times D_L^X$ has a single entry equal to $\sqrt{\lambda}$, and all other entries are trivial (0). And similarly, $\tilde{Y}$ is the concatenated tensor products of Y with a tensor of dimension $P \times O$, where $P = \prod_{l=1}^{L} D_l^X$ the total number of predictors for each outcome and $O = \prod_{m=1}^{M} D_m^Y$. To explicate these derived input and output tensors, we can unfold them along the first dimension and express them as tenmats:

$$\tilde{X}^{(1)} = \begin{bmatrix} X^{(1)} \\ \sqrt{\lambda} I_{P \times P} \end{bmatrix} \text{and } \tilde{Y}^{(1)} = \begin{bmatrix} Y^{(1)} \\ 0_{P \times O} \end{bmatrix}.$$

Thus, the ridge regularized optimization problem (left-hand size) may be solved by alternating least squares and swapping $\tilde{X}$ and $\tilde{Y}$ for X and Y, respectively, and solving the non-regularized linear model problem (right-hand side of the ridge regularized tensor linear model above).

In practice, for high-dimensional data, the tenmat representations $\tilde{X}$ and $\tilde{Y}$ may be quite large and computationally intractable. Some algorithmic approaches have been proposed to make these calculations feasible by modifying the OLS updating steps for $vec\left(U^{(1)}\right)$ and V_M^Y:

$$\text{Unregularized: } vec\left(U^{(1)}\right) = \left(C'C\right)^{-1} C'\text{vec}(Y) \Rightarrow$$

$$\text{Regularized: } vec\left(U^{(1)}\right) =$$

$$\left(C'C + \lambda\left(U^{(2)\prime}U^{(2)} \odot U^{(3)\prime}U^{(3)} \odot \ldots \odot U^{(L)\prime}U^{(L)} \odot V_1'V_1 \odot V_2'V_2 \odot \ldots \odot V'_{M-1}V_{M-1}\right)\right.$$
$$\left. \otimes I_{P_1 \times P_1}\right)^{-1} C'\text{vec}(Y),$$

$$\text{Unregularized: } V_M^Y = \left(D'D\right)^{-1} D'\ Y'_M \Rightarrow$$

$$\text{Regularized: } V_M^Y =$$

$$\left(D'D + \lambda\left(U^{(1)\prime}U^{(1)} \odot U^{(2)\prime}U^{(2)} \odot \ldots \odot U^{(L)\prime}U^{(L)} \odot V_1'V_1 \odot V_2'V_2 \odot \ldots \odot V'_{M-1}V_{M-1}\right)\right)^{-1} D'\,Y'_M,$$

where the Khatri-Rao dot product and the Kronecker product are denoted by $\odot$ and $\otimes$ [229, 232].

This iterative updating algorithm sequentially improves the regularized least squares estimates by step-wise minimizing the objective function. Even though the objective function and the parameter space are both convex, the full space of low-rank tensors is not a convex space in this iteratively optimization process. Thus, there are no guarantees of convergence to a global minimum and the algorithm may be trapped in local minima [245]. In practice, the algorithm generally tends to converge to coordinate-wise minima where the solution cannot be improved by iterative unidimensional perturbations of the parameters. As with all regularized linear models, the regularization penalty weight (λ) affects the speed of convergence [239].

This generic tensor-based linear model formulation can be tailored for the specific 2D kime indexing of the fMRI data ($L = 2$). This may lead to effective algorithmic implementations to obtain linear model based inference using the corresponding spacekime signal reconstructions – kimesurfaces.

4.7 Mapping Longitudinal Data to Kimesurfaces

There are many alternative strategies to estimate the missing kime-phases and translate longitudinally indexed time-series observations to kime-indexed kimesurfaces. Some of these approaches involve using kime-phase estimates from analogous problems or prior studies, utilizing phase-aggregators based on various statistics (e.g., mean, median, order, etc.) that yield unique values or a finite number of phases, random sampling from a prior kime-phase model-distribution, or kime-magnitude and kime-phase indexing operators. The properties of different phase-estimation techniques are not yet fully understood, e.g., their optimality, asymptotics, bias, and precision need to be investigated further.

Let's try to explicate one specific strategy (*direct-mapping*) to translating the observed spacetime fMRI series into spacekime fMRI kime-surface. In this direct-mapping approach, the longitudinal (time) indexing, $x(t)$, in the spacekime model (9) only includes one epoch of the event-related fMRI design. This is illustrated by the kime-magnitude time-indices $x(t) \in \{1, 2, 3, \ldots, 10\}$ shown in **Figure 4.3**. This indexing map notation, $x(t)$, should not be confused with the previous spacetime indexing notation, (x, y, z, t).

Also, the number of epochs of one specific stimulus (e.g., finger-tapping) indicates the number of kime-phases that need to be drawn (or estimated) from the appropriate kime-phase distribution for the specific stimulus condition. In the same **Figure 4.3** example, this corresponds to the kime-magnitude epoch-based phase-indices $x(\varphi) \in \{1, 2, 3, \ldots, 8\}$.

Note that the most general kintensities, corresponding to complex-valued and kime-indexed data, are somewhat difficult to visualize as they are intrinsically 4-dimensional structures. However, there are 3D complex-plot methods rendering the kimesurfaces as 2-manifolds embedded in 3D. At each voxel spatial location, the triangulated kime surface intensity-height and surface-color correspond respectively to the magnitude and the phase of the complex value.

There is a need to clarify a couple of important points:

1. There is a difference between the kime-phases (φ), which are indirectly estimated, and the phases of the complex-valued fMRI signal (θ), which are observed and obtained via the IFT of the k-space acquired complex-valued BOLD signal.
2. Effectively, we may consider each repeated stimulus (e.g., activation or rest state) of the (temporally long) longitudinal fMRI time-series, as an instance of a (temporally short) kimesurface over one time-epoch and across the kime-phase spectrum.

Let's apply spacekime inference to a complex-valued fMRI dataset representing a dichotomous finger-tapping "*on-off*" (activation vs. rest) event-related experiment [246, 247]. A k-space signal over a period of 8-minutes tracking the functional BOLD signal of a normal volunteer was acquired performing a standard finger-tapping "on-off" task. The data was Fourier transformed into spacetime and stored as a 4D double-precision array of complex-valued intensities. The size of the data is about ½ GB (gigabytes) and the dimensions of the data are $\underbrace{64x \times 64y \times 40z}_{\text{voxel}, v} \times \underbrace{160t}_{\text{time}}$. The

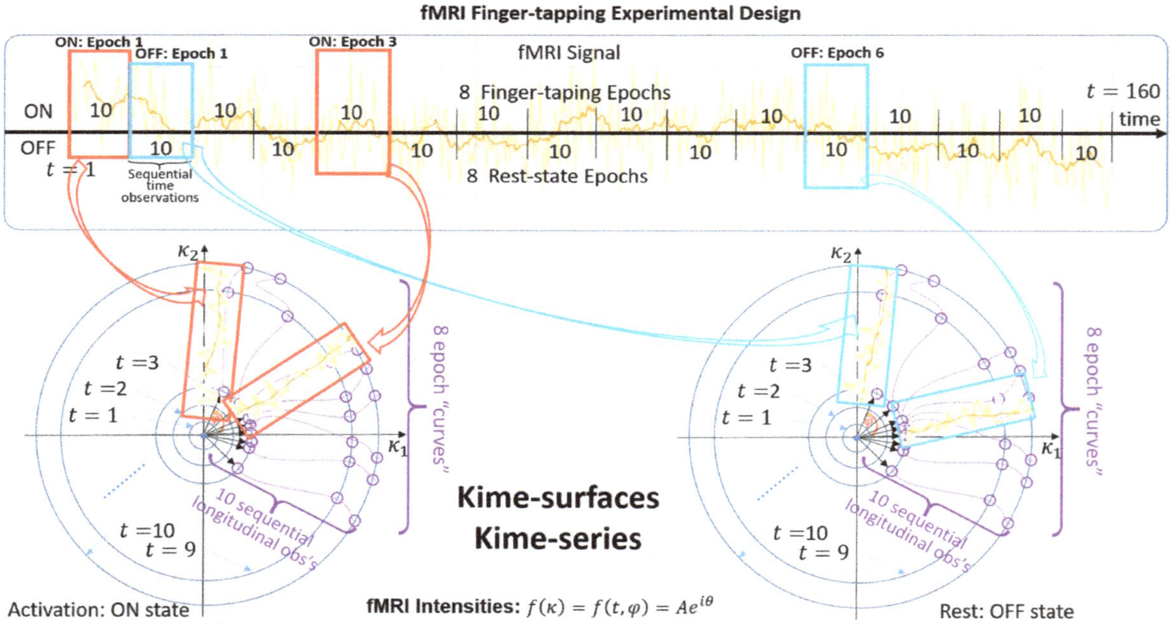

Figure 4.3: Direct-mapping of a typical spacetime-reconstructed event-related fMRI time-series into a kime-surface. The top portion shows the original (magnitude-only) fMRI time-source and the 8 repeated alterations (epochs) of stimulus (finger-tapping) and rest-state paradigms. As the kime-phases are not observed, they need to be estimated, randomly simulated, or approximated. Assuming different epochs of the fMRI time-series span the space of kime-phases, the schematic on the bottom depicts one strategy of reconstructing the 2D kime-surfaces by stacking the epoch curves and generating an approximate piecewise parametric surface indexed over kime with kintensities derived from the corresponding complex-valued fMRI signal. The complex value of the function $f(t,\varphi) = z = A\, e^{i\theta}$ is in the range-space and has magnitude $A = A(t,\varphi)$ and (range) phase $\theta = \theta(t,\varphi)$ that correspond to the polar kime coordinates (t,φ).

longitudinal (time) finger-tapping task associated with the data has a basic pattern of 10 ON (activation) time-points followed by 10 OFF (rest) time-points. The ON and OFF epochs of 10 are intertwined and repeated 8 times for a total of 160 time-points, each of about 3 seconds. A more accurate representation of the task activation vector may need to account for the natural hemodynamic delay, which is about 5–8 seconds, or approximately two time instances.

Standard fMRI data preprocessing can be applied, e.g., using the R package *fmri*, to motion-correct, skull-strip, register, and normalize the fMRI signal [248, 249]. **Figure 4.4** shows some of the raw complex-valued fMRI data, a model of the spacekime kime-series, and reconstructions of the real spacetime time-series as spacekime kimesurfaces.

Figure 4.5 shows some of the raw data and the corresponding statistical maps identifying the voxel locations highly associated with the underlying event-related on-off stimulus paradigm.

In **Chapter 6**, we will apply spacekime tensor-based linear modeling, which we presented in the previous **Section 4.6.4**, to expand this fMRI data analytics example and shown 3D voxel-based as well as region-of-interest based statistical maps.

4.8 The Laplace Transform of Longitudinal Data: Analytic Duality of Time-Series and Kimesurfaces

By using the Laplace transformation, we can explore interesting properties between the space-time and space-kime representations of longitudinal data. In **Chapter 3**, we showed that the Fourier transformation is a separable linear operator that maps complex-valued functions of real variables (e.g., space and time domains) to complex-valued functions of real harmonic variables in k-space (e.g., spatial and angular frequencies). The Laplace transform is similar, however, it sends complex-valued functions of positive real variables (e.g., time) to complex-valued functions defined on complex variables (e.g., kime). The linearity and separability properties of the Laplace transform ensure that it can be applied to linear mixtures of higher dimensional signals by applying it to each component and each dimension separately and then reconstructing the multidimensional linear combination.

More specifically, the *Laplace transform* (LT), $\mathcal{L}$, of a time-based complex-valued function $f(t)\colon \mathbb{R}^+ \to \mathbb{C}$, is another complex-valued function $F(z)\colon \mathbb{C} \to \mathbb{C}$ defined by:

$$\mathcal{L}(f)(z) \equiv F(z) = \int_0^\infty f(t) e^{-zt} dt.$$

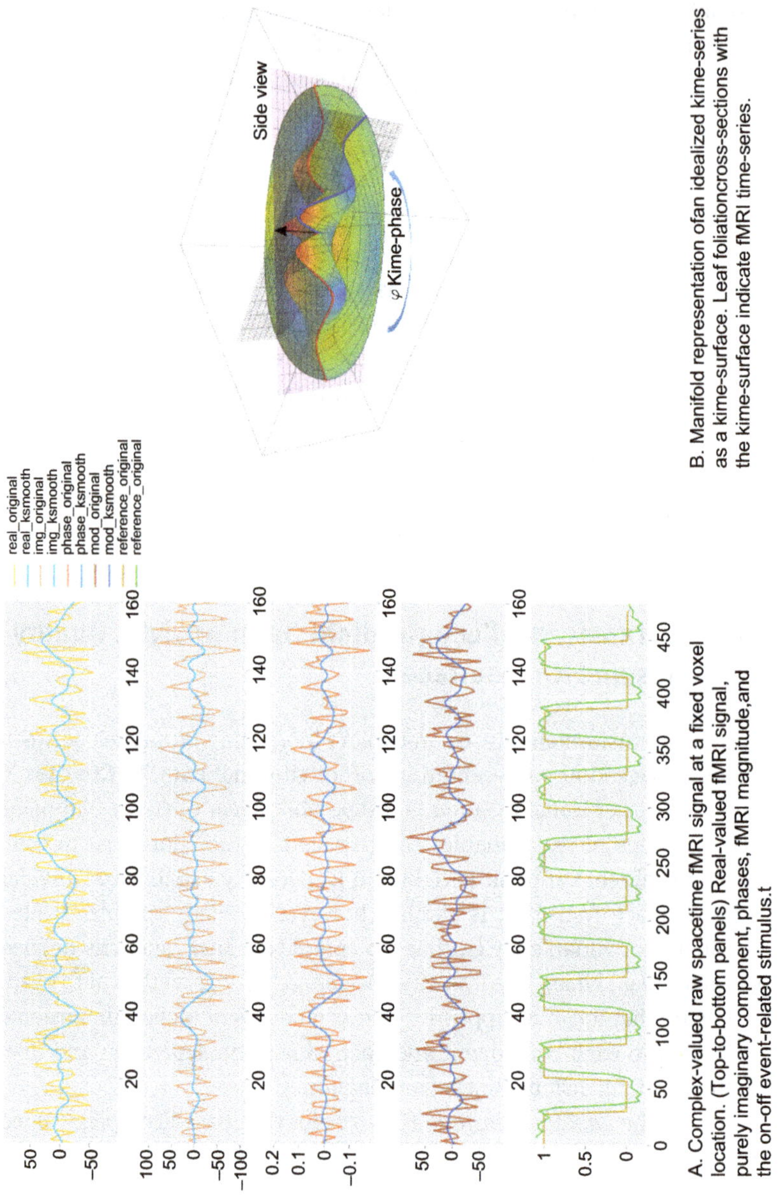
real_original
real_ksmooth
img_original
img_ksmooth
phase_original
phase_ksmooth
mod_original
mod_ksmooth
reference_original
reference_original
Side view
φ Kime-phase
A. Complex-valued raw spacetime fMRI signal at a fixed voxel location. (Top-to-bottom panels) Real-valued fMRI signal, purely imaginary component, phases, fMRI magnitude,and the on-off event-related stimulus.t
B. Manifold representation ofan idealized kime-series as a kime-surface. Leaf foliationcross-sections with the kime-surface indicate fMRI time-series.

Figure 4.4: fMRI signal. Panel A: Raw complex-valued fMRI data (real, imaginary, phase, magnitude, stimulus characteristics function, and HRF); Panel B: Kime-surface model of an idealized spacekime kime-series; Panel C: Reconstructions of the real spacetime time-series as spacekime kimesurfaces and ON-OFF kime-surface differences; Panel D: A couple of 3D scenes showing the ON-OFF fMRI kime-surface, which is ultimately used to identify statistical significance associated with the event-related fMRI activation paradigm.

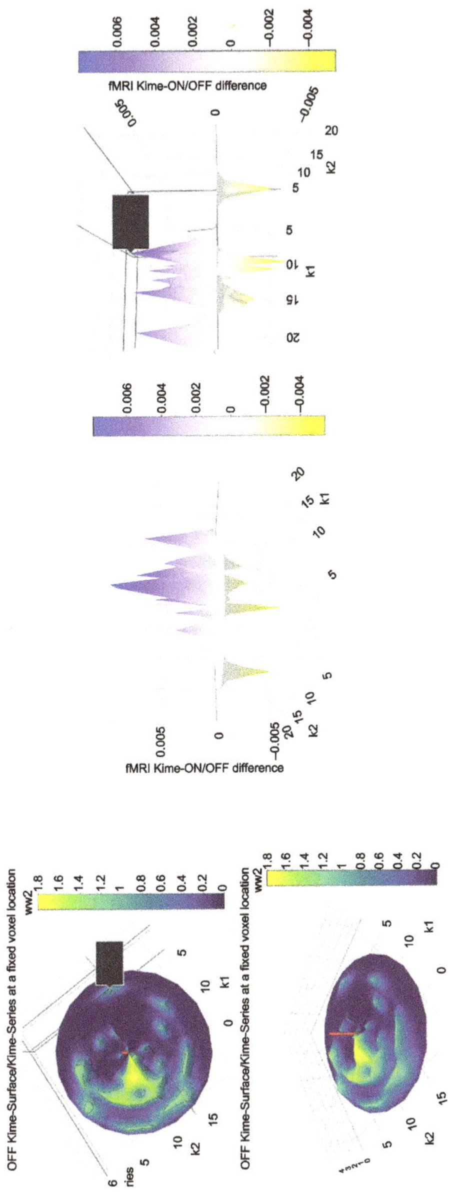

C. Top and side views of the kime-surface reconstructions using observed fMRI time-series.

D. Two views of the ON-OFF fMRI kimesurface differences. Blue and yellow colors indicate positive and negative kimesurface differences, respectively.

Figure 4.4 (continued)

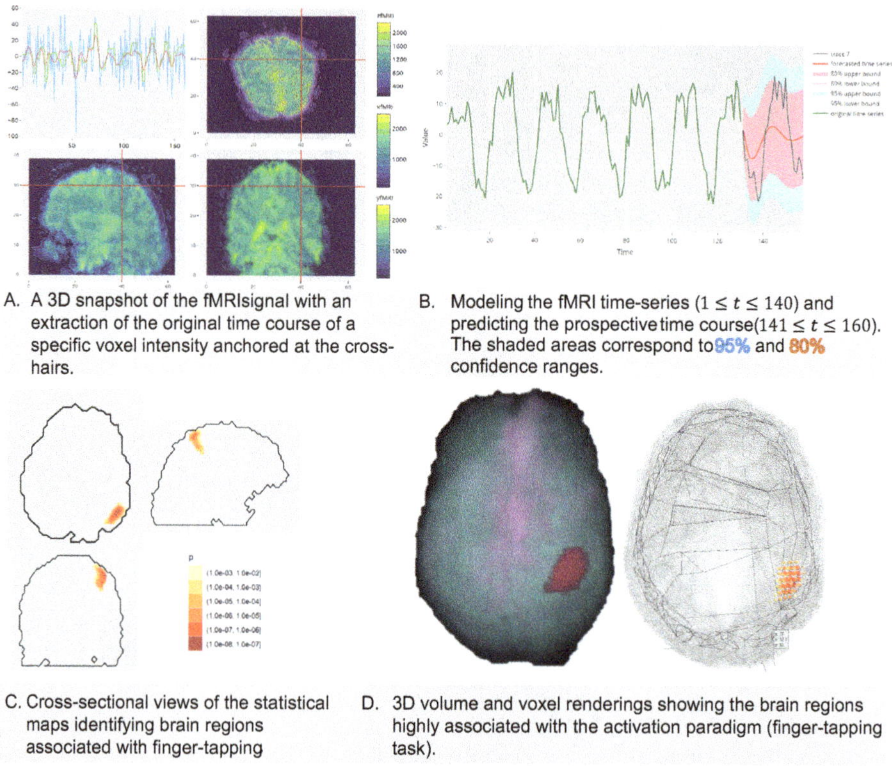
A. A 3D snapshot of the fMRIsignal with an extraction of the original time course of a specific voxel intensity anchored at the cross-hairs.
B. Modeling the fMRI time-series (1 ≤ t ≤ 140) and predicting the prospectivetime course(141 ≤ t ≤ 160). The shaded areas correspond to95% and 80% confidence ranges.
Time
Value
C. Cross-sectional views of the statistical maps identifying brain regions associated with finger-tapping
D. 3D volume and voxel renderings showing the brain regions highly associated with the activation paradigm (finger-tapping task).

Figure 4.5: Spacekime fMRI analysis. Panel A: Stereotaxic representation of the data; Panel B: Longitudinal modeling and prediction of the fMRI time-course; Panel C: Statistical maps illustrating the highly significant voxel locations associated with the on-off finger-tapping task; Panel D: 3D scenes showing the fMRI statistical maps.

4.8.1 Continuous Laplace Transform

Just like the FT is invertible and acts as a linear operator on functions, the Laplace transform has linear properties and has an inverse, $\mathcal{L}^{-1}$. In the most general case, the LT is invertible in a measure-theoretic sense. That is, if the LTs of a pair of integrable functions f and g are identical, i.e., $\mathcal{L}(f) = \mathcal{L}(g)$, then the Lebesgue measure of the set $\{t : f(t) \neq g(t)\}$ is zero. Thus, the inverse Laplace transform (ILT) is well defined and $\mathcal{L}(\mathcal{L}^{-1}) \equiv \mathcal{L}^{-1}(\mathcal{L}) \equiv I$, the identity operator. Also, the result of the Laplace transform, i.e., the LT image, is the space of the complex analytic (holomorphic) functions. For a complex holomorphic function, F (typically $F = \mathcal{L}(f)$), the ILT of F is defined as a complex path integral:

$$f(t) = \mathcal{L}^{-1}(F)(t) = \frac{1}{2\pi i} \lim_{T\to\infty} \oint_{\gamma - i\,T}^{\gamma + i\,T} e^{zt} F(z)dz.$$

In the last equation, γ is any real number ensuring that the contour path of integration $(\gamma - i\ T \to \gamma + i\ T)$ is entirely within the region of convergence of the initial complex function $F(z)$.

The LT and ILT have many connections and applications to probability and statistics. For instance, the LT of a random variable X, with a given probability density function f, is the expectation ($\mathbb{E}$), which directly relates to the moment generating function (MGF) of the process:

$$\mathcal{L}(f)(z) = \mathbb{E}(e^{-zX}) = MGF(-z) = \int_0^\infty e^{-zt} f(t)dt.$$

Also, the ILT of the scaled expectation allows us to compute the *cumulative distribution function* (CDF) of the process, $\mathcal{L}(CDF)(z) = \frac{1}{z}\mathcal{L}(f)(z)$, i.e.,

$$CDF(t) = \int_{-\infty}^{t} f(x)dx = \mathcal{L}^{-1}\left(\frac{1}{z}\mathbb{E}(e^{-zX})\right)(t) = \mathcal{L}^{-1}\left(\frac{1}{z}\mathcal{L}(f)(z)\right)(t).$$

For some random variables and their corresponding probability density functions, this direct connection between the CDF, MGF, and ILT provides an indirect method for estimating the cumulative distributions or the quantile functions. The above equation is valid over the region of convergence of the density f and over the half plane $Re(z) > 0$. These conditions ensure that the 2D domain $\{(t,\tau)|t \in \mathbb{R} \cap \tau < t\}$ can be reparameterized as $\{(t,\tau)|t \in \mathbb{R} \cap t > \tau\}$ and the double *integral converges absolutely*:

$$\int_{\mathbb{R}} \left| e^{-zt} \int_{-\infty}^{t} f(\tau)d\tau \right| dt < \infty.$$

To explicate that

$$\underbrace{\int_{\mathbb{R}} e^{-zt} F(t)dt}_{\mathcal{L}(CDF)(z)} \equiv \int_{\mathbb{R}} \left(e^{-zt} \int_{-\infty}^{t} f(\tau)d\tau \right) dt \overset{?}{=} \underbrace{\int_{\mathbb{R}} f(\tau) \left(\int_{\tau}^{-\infty} e^{-zt} dt \right) d\tau}_{\frac{1}{z}\mathcal{L}(f)(z)},$$

we expand the left-hand side

$$\mathcal{L}(CDF)(z) = \int_{\mathbb{R}} \left(e^{-zt} \int_{-\infty}^{t} f(\tau)d\tau \right) dt = -\frac{1}{z}\int_{\mathbb{R}} \left(\underbrace{\left(-ze^{-zt}\right)}_{v'(t)} \underbrace{\int_{-\infty}^{t} f(\tau)d\tau}_{u(t)} \right) dt \underbrace{=}_{\text{integration by parts}}$$

$$-\frac{1}{z}\left(\underbrace{e^{-zt}}_{v(t)} \underbrace{\int_{-\infty}^{t} f(\tau)d\tau}_{u(t)} \right)\Bigg|_{-\infty}^{\infty} + \frac{1}{z}\int_{\mathbb{R}} \underbrace{e^{-zt}}_{v(t)} \underbrace{f(t)}_{u'(t)} dt =$$

$$\frac{1}{z}\lim_{t\to-\infty} \left(\underbrace{e^{-zt} \int_{-\infty}^{t} f(\tau)d\tau}_{A(t,z)} \right) - \underbrace{\frac{1}{z}\lim_{t\to\infty} \left(e^{-zt} \overbrace{\int_{-\infty}^{t} f(\tau)d\tau}^{CDF \le 1} \right)}_{0} + \frac{1}{z}\int_{\mathbb{R}} f(t)e^{-zt}dt = \frac{L(f)(z)}{z}.$$

To verify that the first term is trivial, $\lim_{t\to-\infty}(A(t,z)) = 0$, let's denote the CDF of the random variable X by

$$F(t) = \int_{-\infty}^{t} f(\tau)d\tau : \mathbb{R} \to [0,1],$$

and let $z \in \{u + iv \mid u = Re(z) > 0\} \subset \mathbb{C}$. A proof by contrapositive shows that the limit of $A(t,z)$ as $t \to -\infty$ is trivial. Assume that $A(t,z)$ does not tend to zero as t decreases. Then, there exists $\varepsilon > 0$ such that $\forall t \in \mathbb{R}$:

$$|A(t,z)| = \left|e^{-zt}F(t)\right| = \left| e^{-zt} \int_{-\infty}^{t} f(\tau)d\tau \right| > \varepsilon.$$

Then for a fixed $z_o = u_o + iv_o$, $\exists t' < t'' < t$ such that

$$\int_{t'}^{t''} \left|e^{-z_o t}F(t)\right|dt \ge \int_{t'}^{t''} \varepsilon\left|e^{-z_o(t-t')}\right|dt = \varepsilon\int_{t'}^{t''} e^{-u_o(t-t')}dt = \frac{\varepsilon}{u_o}\left(1 - e^{-u_o(t''-t')}\right).$$

There exists a monotonically decreasing sequence $\{a_k\}_{k=1}^{\infty}$, such that $a_o = 0$, $a_{k+1} < a_k - 1, \forall k \ge 0$, with $\lim_{k\to\infty} a_k = -\infty$ for which

$$\left|e^{-z_o a_k}F(a_k)\right| > \varepsilon.$$

This leads to

$$\int_{a_{n+1}}^{a_o} \left|e^{-z_o t}F(t)\right|dt = \sum_{k=0}^{n} \int_{a_{k+1}}^{a_k} \left|e^{-z_o t}F(t)\right|dt \geq \sum_{k=0}^{n} \frac{\varepsilon}{u_o}\left(1-e^{-u_o(a_k - a_{k+1})}\right) \geq$$

$$\sum_{k=0}^{n} \frac{\varepsilon}{u_o}(1-e^{-u_o}) = \frac{\varepsilon(n+1)}{u_o}(1-e^{-u_o}) \underset{n\to\infty}{\longrightarrow} \infty.$$

Because $\lim_{n\to\infty} \int_{a_{n+1}}^{a_o} |e^{-z_o t}F(t)|dt \leq \int_{-\infty}^{\infty} |e^{-z_o t}F(t)|dt$, we obtain a contradiction with the *a priori* assumption of the absolute convergence of the double integral

$$\int_{-\infty}^{\infty} \left|e^{-z_o t}F(t)\right|dt \equiv \int_{\mathbb{R}} \left|e^{-z_o t} \int_{-\infty}^{t} f(\tau)d\tau\right| dt < \infty.$$

This implies that there is no such lower limit $|A(t,z)| > \varepsilon > 0$, and hence,

$$\lim_{t\to-\infty} \left(\underbrace{e^{-zt} \int_{-\infty}^{t} f(\tau)d\tau}_{A(t,z)} \right) = 0.$$

We are mostly interested in the LT and ILT to map between *time-domain data*, represented as classical longitudinal time-series processes, and *kime-domain functions*. This mapping transformation results in kimesurfaces that can be subsequently space-kime analyzed using appropriate exploratory, classification, regression, clustering, ensemble, or other AI methods.

Let's start by cataloguing some of the LT properties. The following notation will be useful, lowercase letters denote functions of time (f, g), whereas capital letters (F, G) denote their LTs, which are functions of complex time (kime):

$$\mathcal{L}(f)(z) \equiv F(z), \; f(t) = \mathcal{L}^{-1}(F)(t),$$

$$\mathcal{L}(g)(z) \equiv G(z), \; g(t) = \mathcal{L}^{-1}(G)(t).$$

Many of the LT properties listed in **Table 4.1** are derived similarly to this derivation of the LT differentiation property [250, 251]:

Table 4.1: Some properties of the forward and inverse Laplace transforms, where $H(\cdot)$ is the heaviside step function.

LT property	Time domain function	Kime-domain LT function
Linearity	$a\,f(t)+b\,g(t)$	$\mathcal{L}(a\,f+b\,g)(z)=a\mathcal{L}(f)(z)+b\mathcal{L}(g)(z)=a\,F(z)+b\,G(z)$
Time to Kime Magnitude	$f(t)=t$ (time, identity)	$\mathcal{L}(f)(z)=\frac{1}{z^2}$ (reciprocal square kime magnitude)
Unitary Time to Reciprocal Kime	$f(t)=1$ (constant)	$\mathcal{L}(f)(z)=\frac{1}{z}$ (kime reciprocal)
First Derivative	$f'(t)$	$\mathcal{L}\left(f'\right)(z)=z\mathcal{L}(f)(z)-f(0)=zF(z)-f(0)$
Second Derivative	$f''(t)$	$\mathcal{L}\left(f''\right)(z)=z^2F(z)-zf(0)-f'(0)$
General Derivative	$f^{(n)}(t)$	$\mathcal{L}(f^{(n)})(z)=z^nF(z)-\sum_{k=1}^{n}z^{n-k}f^{(k-1)}(0)$
Kime-domain Derivative	$t^nf(t)$	$(-1)^nF^{(n)}(z)$
Time Integration	$\int\limits_0^t f(\tau)d\tau$	$\frac{1}{z}F(z)$
Time-domain Multiplication	$f(t)g(t)$	$\frac{1}{2\pi i}\lim_{T\to\infty}\oint_{a-iT}^{a+iT}F(\kappa)G(z-\kappa)d\kappa$, the integral is along the vertical line $Re(\kappa)=a$ within the kime-space region of convergence of F, for a real constant $a\in\mathbb{R}$
Time Shift	$f(t-a)H(t-a),\ t>a$	$e^{-az}F(z)$
Time Scale	$f(at)$	$\frac{1}{a}F\left(\frac{z}{a}\right)$
Power Function	$f(t)=t^n,\ n>-1$	$F(z)=\begin{cases}\frac{n!}{z^{n+1}}, & n=integer\\ \frac{\Gamma(n+1)}{z^{n+1}}, & n\in(-1,\ \infty)\end{cases}$

Periodic Function	$[f(t) \equiv f(t+T), \forall t > 0]$ $f(t)H(t)$	$\frac{1}{1-e^{-Tz}} \int_0^T e^{-zt} f(t)dt$
Time-domain Convolution	$(f^*g)(t)$	$\mathcal{L}(f^*g)(z) = F(z)\ G(z)$
Trigonometric Functions	$f_1(t) = \sin(wt)\,H(t),\ t \geq 0$ $f_2(t) = \cos(wt)\,H(t),\ t \geq 0$	$F_1(z) = \mathcal{L}(f_1)(z) = \frac{w}{z^2+w^2}$ $F_2(z) = \mathcal{L}(f_2)(z) = \frac{z}{z^2+w^2}$
Exponential Decay	$f(t) = e^{-wt}H(t),\ t \geq 0$	$F(z) = \mathcal{L}(f)(z) = \frac{1}{z+w}$

$$\mathcal{L}(f)(z) = \int_0^\infty f(t)e^{-zt}dt = \int_0^\infty \frac{1}{-z} f(t)de^{-zt} \underbrace{=}_{\substack{\text{integration}\\ \text{by parts}}}$$

$$\left[\frac{f(t)e^{-zt}}{-z}\right]_{t=0}^{\infty} - \int_0^\infty \frac{1}{-z} e^{-zt} f'(t)dt = -\frac{f(0)}{-z} + \frac{1}{z}\,\mathcal{L}(f')(z).$$

Therefore, $z\ \mathcal{L}(f)(z) = f(0) + \mathcal{L}(f')(z)$ and

$$\mathcal{L}(f')(z) = z\ \mathcal{L}(f)(z) - f(0) = zF(z) - f(0).$$

For example, a complex-valued fMRI time-series signal $f(t)\colon \mathbb{R}^+ \to \mathbb{C}$ can be represented as an ordered (discrete) sequence of complex fMRI BOLD intensities:

$$\{f(t) = a_t + i\ b_t\colon 1 \le t \le T\}.$$

Then, the discrete LT of f will be represented as an infinite series corresponding to the discretized LT integral above. For a specified step-size $\eta > 0$, the discrete Laplace transform $L(f)$ is defined by:

$$\mathcal{L}_\eta(f)(z) = \eta \sum_{k=0}^{\infty} e^{-zk\eta} f(k\eta).$$

There is a difference between the discrete Laplace and the discrete FTs. In the case of the fMRI time-series, the discrete FT takes a finite sequence of (real or complex) numbers and returns another finite sequence of complex numbers. The discrete LT evaluates the input function, f, at an infinite number of time points ($k \in \mathbb{R}^+$), and derives another analytic (holographic) function, $F = \mathcal{L}_\eta(f)$, as an infinite series.

Let's look at three examples, **Figure 4.6**. In the first case, applying the LT we will explicitly reconstruct the Laplace-dual kimesurface, $\hat{F} = \mathcal{L}(f)$, that corresponds to the observed time-series $f(t) = \sin(w\ t)\ H(t)$, where $H(t)$ is the Heaviside function defined below. In the second case, applying the ILT we will recover the Laplace-dual time-series, $\hat{f} = \mathcal{L}^{-1}(F)$, that corresponds to the original kimesurface,

$$F(z) = \frac{1}{z+1} + \left(\frac{1}{z^2+1}\right) \times \left(\frac{z}{z^2+1}\right) + \frac{1}{z^2}.$$

The last third example shows the result of applying the Laplace transform to a smoothed version, $\tilde{f}(t)$, of a (noisy) real fMRI time-series, $f(t)$. Gaussian convolution smoothness may be required for some observed fMRI signals to ensure the numerical estimates (e.g., integrals, derivatives, limits, convergence) are stable despite the presence of significant white noise or for non-smooth functions. In this situation, the Laplace-dual kimesurface, $\hat{F} = \mathcal{L}(f)$, is a complex-valued function over kime. The kimesurface can be rendered in 3D as a 2D manifold whose altitude (height in the z-axis) and color (RGBA value) may represent the *Real, Imaginary, Magnitude,* or *Phase* values of the kimesurface kintensity, $\hat{F}(z)=\hat{F}(\kappa_1,\kappa_2)=\check{F}(t,\varphi)$.

The first example starts with a known time-series:

$$f(t)=\sin(w\ t)\ H(t),\ \text{where } w=2 \text{ and } H(t)=\begin{cases}0, t<0\\ \frac{1}{2},\ t=0\\ 1,\ t>0\end{cases} \text{ is the Heaviside step function.}$$

For this specific model-based time-series, the Laplace-dual has a closed form analytical expression, **Table 4.1**:

$$F(z)=\mathcal{L}(f)(z)=\frac{w}{z^2+w^2}.$$

The second example starts a known kimesurface $F(z)=\frac{1}{z+1}+\left(\frac{1}{z^2+1}\right)\left(\frac{z}{z^2+1}\right)+\frac{1}{z^2}$. We can compute the Laplace-dual time-series, $f(t)=\mathcal{L}^{-1}(F)(t)$ by decomposing the original kimesurface into its building blocks and using the linear properties of the Laplace transform:

$$F(z)=\mathcal{L}(f)=\underbrace{\frac{1}{z+1}}_{F_1(z)=\mathcal{L}(f_1(t)=e^{-t})}+\underbrace{\frac{1}{z^2+1}}_{F_2(z)=\mathcal{L}(f_2(t)=\sin(t))}\times\underbrace{\frac{z}{z^2+1}}_{F_3(z)=\mathcal{L}(f_3(t)=\cos(t))}+\underbrace{\frac{1}{z^2}}_{F_4(z)=\mathcal{L}(f_4(t)=t)}.$$

Decomposition of compound (*time* or *kime* domain) functions in terms their building blocks provides a mechanism to use the linearity and convolution-to-product properties of the Laplace transform to analytically simplify the corresponding duals:

$$F(z)=F_1(z)+F_2(z)\times F_3(z)+F_4(z)$$

Therefore,

$$f(t)=\mathcal{L}^{-1}(F)=\mathcal{L}^{-1}(F_1+F_2\times F_3+F_4))=$$

$$\mathcal{L}^{-1}(F_1)+\left(\underbrace{\mathcal{L}^{-1}(F_2)^*\mathcal{L}^{-1}(F_3)}_{\text{convolution}}\right)(t)+\mathcal{L}^{-1}(F_4)=$$

$$\mathcal{L}^{-1}(\mathcal{L}(f_1))(t)+\left(\mathcal{L}^{-1}(\mathcal{L}(f_2))^*\mathcal{L}^{-1}(\mathcal{L}(f_3))\right)(t)+\mathcal{L}^{-1}(\mathcal{L}(f_4))(t).$$

Finally,

$$f(t)=\mathcal{L}^{-1}(F)(t)=f_1(t)+(f_2{}^*f_3)(t)+f_4(t)=e^{-t}+\int_0^t\sin(\tau)\times\cos(t-\tau)d\tau+t=$$

$$t+e^{-t}+\frac{t\sin(t)}{2}.$$

Many other examples of analytical function can be similarly transformed into kime-surfaces, e.g.,

$$f_1(t) = t \Leftrightarrow F_1(z) = \mathcal{L}(f_1)(z) = \frac{1}{z^2} \text{ and } f_2(t) = e^{-5t} \Leftrightarrow F_2(z) = \mathcal{L}(f_2)(z) = \frac{1}{z+5}.$$

The third example shows a raw fMRI time-series, $f_{(x_o,y_o,z_o)}(t)$, at a fixed spatial voxel location, $v_o = (x_o, y_o, z_o) \in \mathbb{R}^3$, a corresponding spline smoothed time curve, $\hat{f}_{(x_o,y_o,z_o)}(t)$, and the Laplace-dual, $F_{(x_o,y_o,z_o)}(z) = \mathcal{L}(\hat{f}_{(x_o,y_o,z_o)})(z)$. The intermediate smoothing operation is applied to temper the extreme noise of the original fMRI time-series, which hampers function calculations involving numerical integration, differentiation, and optimization. The resulting kimesurface is a complex-valued analytic (holomorphic) function over complex-time.

Laplace Transform Duality	Time-series ($\mathbb{R}^+$ domain)	Kimesurfaces ($\mathbb{C}$ domain)
Example 1 (analytic function) special case, $w = 2$	$f(t) = \sin(w\,t)\,H(t)$	(Laplace-dual) $\hat{F}(z) = \mathcal{L}(f)(z) \equiv \dfrac{w}{z^2 + w^2}$
	(Laplace-dual) $f(t) = \mathcal{L}^{-1}(F)(t) = t + e^{-t} + \dfrac{t\sin(t)}{2}$	$F(z) = \dfrac{1}{z+1} + (\dfrac{1}{z^2+1}) \times (\dfrac{z}{z^2+1}) + \dfrac{1}{z^2}$ (surface height=Real part, color=Imaginary part)

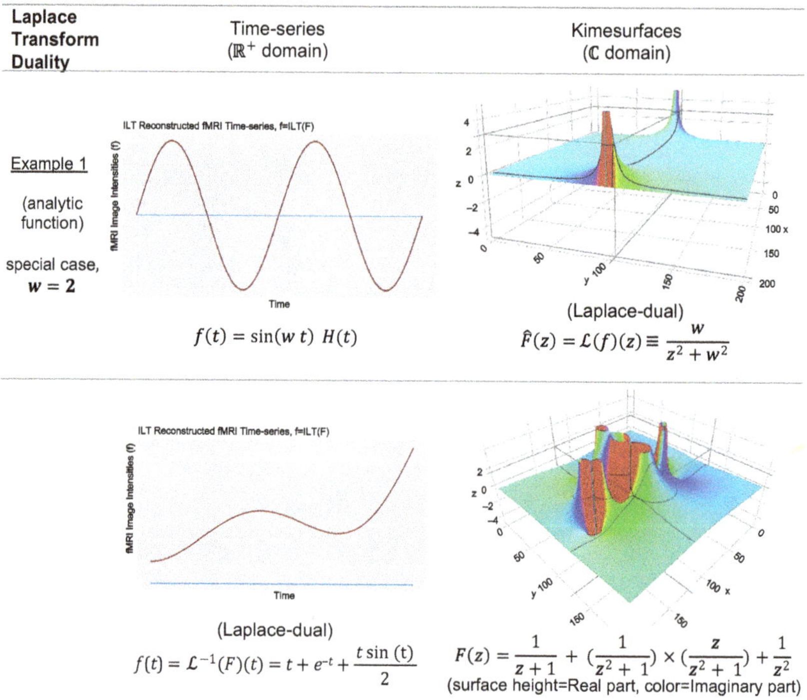

Figure 4.6: Examples of Laplace transform duality between the time and kime domain representations of complex-valued signals. These three examples illustrate the direct analytical correspondence between complex-valued, signals defined over $\mathbb{R}^+$ (time), and their counterparts, defined over the complex plane (kime). The first two examples start with either knowing the time-series (example one) or the kimesurface (example 2) and then reconstructing its counterpart as a Laplace dual. The surface-height and the surface-color correspond to the real and the imaginary parts, respectively, of the complex-valued kimesurfaces. For the second example, the kimesurface in the bottom shows the *magnitude* and the *phase* of the kintensity as height and color, respectively. Example 3 shows the duality between a real fMRI time-series and its corresponding kime-surface.

Example 2

(analytic function)

$$F(z) = \frac{1}{z+1} + \left(\frac{1}{z^2+1}\right) \times \left(\frac{z}{z^2+1}\right) + \frac{1}{z^2}$$

(surface height=magnitude, color=phase)

Example 3

(fMRI)

Original fMRI Time-series f(t) and Smoothed curve

fMRI Image Intensities (f and f)

Time

(original and smooth fMRI signal)

(Laplace-dual)

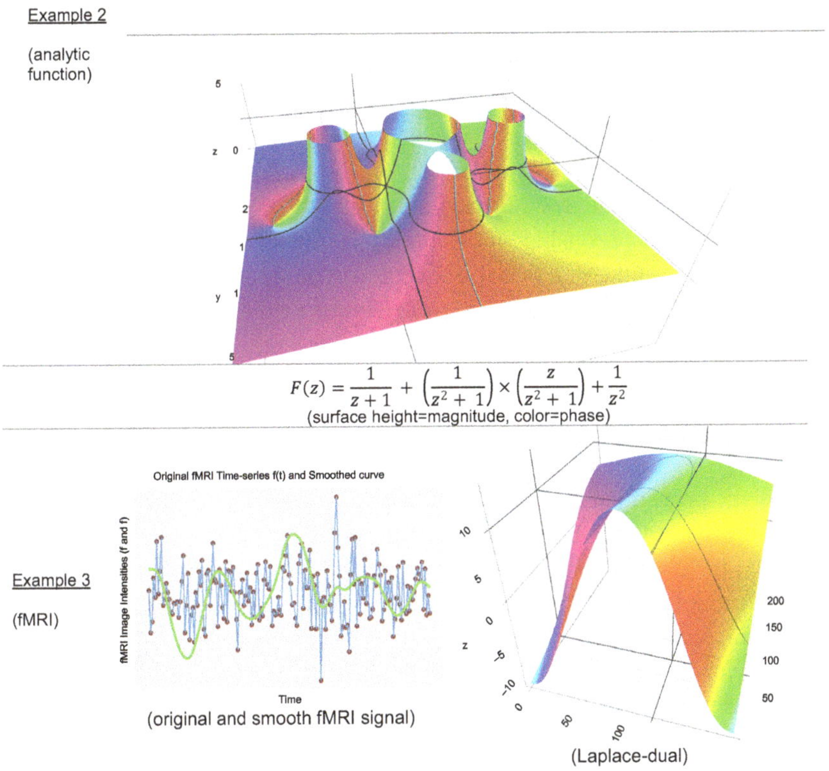

Figure 4.6 (continued)

Figure 4.6 illustrates the three pairs of time-series $\leftrightarrow$ kimesurfaces dualities via the Laplace transformation. The first two examples involve analytic functions whose Laplace-duals can be directly computed in closed-form using the LT and ILT formulations. The last example involves a discretely sampled fMRI time-series whose Laplace dual kimesurface is numerically estimated using a discrete implementation of the forward Laplace transform.

The spacekime representation of complex-valued and kime-indexed longitudinal data as kimesurfaces has some advantages to their real-valued and time-indexed time-series counterparts. *Analyticity* of kimesurfaces makes them theoretically more interesting than their 1D time-series counterparts. All first-order differentiable complex-valued functions $f(\kappa)$ are analytic. They have local representations as convergent power series (Taylor series) and are infinitely differentiable. More specifically, for all $\kappa_o \in \mathbb{K} \subset \mathbb{C}$, in some κ_o neighborhood $N(\kappa_o)$, their Taylor series expansions, $T(\kappa)$, are convergent to the functional value $f(\kappa_o)$:

$$T(\kappa) = \lim_{N \to \infty} \sum_{n=0}^{N} \frac{f^n(\kappa_o)}{n!} (\kappa - \kappa_o)^n \underbrace{\rightarrow}_{\forall \kappa \in N(\kappa_o) \subset \mathbb{K}} f(\kappa).$$

As kimesurfaces are complex-valued functions defined on complex domains (open kime-space regions, e.g., $\mathbb{K} = \{\kappa| \, ||\kappa|| < 1\} \subset \mathbb{C}$), if a kimesurface is differentiable on $\mathbb{K}$, then it is holomorphic (infinitely differentiable) on $\mathbb{K}$ and analytic. This analytic property is unique for all complex-valued functions, including kimesurfaces; however, it is not true for real-valued time-series [154]. For complex-valued functions, analyticity results from the much stronger condition of existence of the first-order derivative, $f'(\kappa_o) = \lim_{\underbrace{\kappa \rightarrow \kappa_o}_{\{\kappa_y\}}} \left(\frac{f(\kappa) - f(\kappa_o)}{\kappa - \kappa_o}\right)$, where the difference quotient value must approach the same complex number, for all paths $\kappa \underbrace{\rightarrow}_{\{\kappa_y\}} \kappa_o, \kappa_y, \kappa_o \in \mathbb{K}$ in the complex plane.

This analytic property of complex-valued kimesurfaces has much stronger implications than real-valued time-series differentiability. For instance, all holomorphic kimesurfaces are infinitely differentiable, whereas for time-series, the existence of the nth order derivative does not guarantee the existence of the next $(n+1)^{th}$ order derivative. In addition, at every kime point the kimesurface may be locally represented as a convergent Taylor series. This highly accurate polynomial approximation of kimesurfaces in every kime neighborhood could suggest powerful inferential strategies, which have no parallels in the general real-valued time-series. For instance, even if a time-series is known to be infinitely differentiable, it's not guaranteed to be analytic anywhere.

4.8.2 Discrete Signal Laplace Transform

In the previous section we considered the Laplace transform for analytical and continuous signals. In practice, many observed signals represent windowed versions of *discrete processes* with restricted domains (finite support). Such discrete signals can be approximated by continuous analytical functions, splines, polynomials, Taylor, or Fourier series expansions. Some of these interpolations are based on power functions like $f_n(t) = t^n$.

In such analytical expansions, reducing the support of power functions is necessary to control the (model or base function) extreme growth, or decay, as the argument tends to infinity. The characteristic function, a.k.a. indicator function, of a half-open real interval $[a, b)$, $\chi_{[a,\,b)}(t)$, may be expressed in terms of the Heaviside function, $H_a(t)$:

$$\underbrace{\chi_{[a,b)}}_{\text{indicator}}(t) = H(t-a) - H(t-(b-a)) = \begin{cases} 0, t < a \\ 1, 0 \leq t < b, \\ 0, t \geq b \end{cases}$$

$$\underbrace{H}_{\text{Heaviside}}(t) = \begin{cases} 0, t < 0 \\ 1, t \geq 0 \end{cases}.$$

For instance, to constrain the support of the power function $f_n(t) = t^n$ to the interval $[0, 2\pi)$, we can multiply it by $\chi_{[0,\ 2\pi)}(t)$, i.e.,

$$f_{n,\ supp[0,2\pi)}(t) = f_n(t) \times \chi_{[0,\ 2\pi)}(t) = \begin{cases} f_n(t) = t^n,\ 0 \leq t < 2\pi \\ 0,\ \text{otherwise} \end{cases}.$$

As a linear operator, the Laplace transform of the power function, compactly supported over the interval $[a, b)$, will be:

$$\mathcal{L}\left(f_{n,supp[a,b)}\right)(z) = \mathcal{L}\left(f_n\chi_{[a,b)}\right)(z) = \underbrace{\mathcal{L}(f_nH(t-a))(z)}_{A_{a,n}(z)} - \underbrace{\mathcal{L}(f_nH(t-b))(z)}_{A_{b,n}(z)}.$$

We need three derivations to explicate this Laplace transform of the windowed power function.

First, let's validate that the Laplace transform of the power function is $\mathcal{L}(f_n(\cdot)H(\cdot))(z) = \frac{n!}{z^{n+1}}$. This is based on Cauchy-Goursat theorem [252], which states that for a simply connected open set $D \subseteq \mathbb{C}$ with a boundary ∂D representing the closed contour bounding D, a function $f{:}D \to \mathbb{C}$ that is holomorphic everywhere in D:

$$\oint_{\partial D} f(z)dz = 0\ .$$

This is the case since all holomorphic functions

$$\left| \begin{array}{c} f(z) = u(z) + iv(z) \\ z = x + iy \\ dz = dx + idy \end{array} \right.$$

satisfy the Cauchy-Riemann equations:

$$\left| \begin{array}{l} \frac{\partial v}{\partial x} + \frac{\partial u}{\partial x} = 0 \\ \frac{\partial v}{\partial x} - \frac{\partial u}{\partial x} = 0 \end{array} \right. .$$

Thus,

$$\oint_{\partial D} f(z)dz = \oint_{\partial D}(u+iv)(dx+idy) = \oint_{\partial D}(udx - vdy) + i\oint_{\partial D}(vdx+udy) \underbrace{=}_{\text{Green's theorem}}$$

$$\iint_D \left(-\frac{\partial v}{\partial x} - \frac{\partial u}{\partial x}\right)dxdy + \iint_D \left(\frac{\partial u}{\partial x} - \frac{\partial v}{\partial x}\right)dxdy = 0.$$

Let's use the following change of variables transformation to show that $\mathcal{L}(f_n(\cdot)H(\cdot))(z) = \frac{n!}{z^{n+1}}$.

$$\left| \begin{array}{c} u = zt = (x+iy)t \\ \epsilon \le t = \frac{u}{z} \le L \\ zdt = du \end{array} \right. .$$

Then,

$$\int_{\epsilon}^{L} t^n e^{-zt}dt = \oint_{C_3}\left(\frac{u}{z}\right)^n e^{-u} d\frac{u}{z} =$$

$$\oint_{x\epsilon+iy}^{xL+iyL}\left(\frac{u}{z}\right)^n e^{-u} d\frac{u}{z} = \frac{1}{z^{n+1}}\oint_{x\epsilon+iy}^{xL+iyL} u^n e^{-u}du.$$

Let's denote the general path line integral in the complex plane by $I_C = \int_C u^n e^{-u}du$ and consider the closed (boundary) contour $C = C_1 + C_2 - C_3 + C_4$, as in **Figure 4.7.** We will apply the Cauchy-Goursat theorem over the entire boundary $\partial D = C_1 + C_2 - C_3 + C_4$ of the open set D.

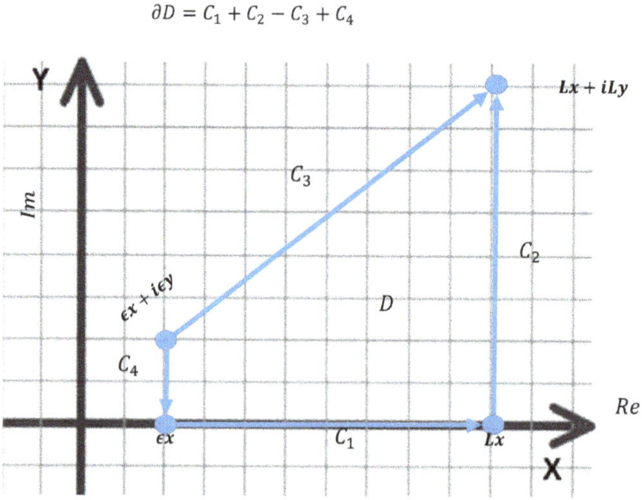

Figure 4.7: Line path integral avoiding the singularity at the origin.

$$\underbrace{\oint_C u^n e^{-u} du}_{I_C} = \underbrace{\oint_{C_1} u^n e^{-u} du}_{I_{C_1}} + \underbrace{\oint_{C_2} u^n e^{-u} du}_{I_{C_2}} - \underbrace{\oint_{C_3} u^n e^{-u} du}_{I_{C_3}} + \underbrace{\oint_{C_4} u^n e^{-u} du}_{I_{C_4}} = 0.$$

Let's show that $\lim\limits_{L\to\infty} I_{C_2} = \lim\limits_{\epsilon\to 0} I_{C_4} = 0$:

$$|I_{C_2}| = \left|\oint_{C_2} u^n e^{-u} du\right| = \left|\oint_{Lx}^{Lx+iLy} u^n e^{-u} du\right| \leq \max_{u\in C_2} L|u^n e^{-u}| = \max_{u\in C_2} L|u^n| e^{-Re(u)} \cong L|L^n| e^{-L} \underset{L\to\infty}{\longrightarrow} 0.$$

$$|I_{C_4}| = \left|\oint_{C_4} u^n e^{-u} du\right| = \left|\oint_{\epsilon x + i\epsilon y}^{x} u^n e^{-u} du\right| \leq \max_{u\in C_4} |u^n e^{-u}| \epsilon = \max_{u\in C_4} |u^n| e^{-Re(u)} \epsilon \cong |\epsilon^n| e^{-\epsilon} \underset{\epsilon\to 0}{\longrightarrow} 0.$$

Hence, in the limit, $I_C = I_{C_1} - I_{C_3} = 0$, $I_{C_1} = I_{C_3}$, and

$$\lim_{\epsilon\to 0^+} \lim_{L\to\infty} I_{C_1}(imn, L) = \lim_{\epsilon\to 0^+} \lim_{L\to\infty} \oint_{e\, imx}^{Lx} u^n e^{-u} du \underbrace{=}_{I_{C_1} = I_{C_3}} \int_0^{\infty} u^n e^{-u} du \underbrace{=}_{zt=u} \frac{1}{z^{n+1}} \oint_0^{\to +\infty} t^{(n+1)-1} e^{-t} dt.$$

Therefore,

$$\mathcal{L}(f_n(\cdot)H(\cdot))(z) = \lim_{\epsilon\to 0^+} \lim_{L\to\infty} I(\epsilon, L) = \frac{1}{z^{n+1}} \int_0^{\to +\infty} t^{(n+1)-1} e^{-t} dt = \frac{\Gamma(n+1)}{z^{n+1}} = \begin{cases} \frac{n!}{z^{n+1}}, n = \text{integer} \\ \frac{\Gamma(n+1)}{z^{n+1}}, n > 0 \end{cases}.$$

Next, to estimate the term $A_{a,n}(z) = \mathcal{L}(f_n(t)H(t-a))(z)$, we will make a transformation:

$$s = t - a,$$

$$A_{a,n}(z) = \mathcal{L}(f_n(t)H(t-a))(z) = e^{-az} \mathcal{L}(f_n(s+a)H(s))(z) = e^{-az} \mathcal{L}((s+a)^n H(s))(z) =$$

$$e^{-az} \int_0^{\infty} (s+a)^n e^{-zs} ds = e^{-az} \int_0^{\infty} \left(\sum_{k=1}^{n} \binom{n}{k} s^k a^{n-k} \right) e^{-zs} ds =$$

$$e^{-az} \sum_{k=1}^{n} \left[\binom{n}{k} a^{n-k} \int_0^{\infty} s^n e^{-zs} ds \right] = e^{-az} \sum_{k=0}^{n} \left[\binom{n}{k} a^{n-k} \underbrace{\frac{k!}{z^{k+1}}}_{\mathcal{L}(s^n H(s))(z)} \right].$$

Finally, following the same argument as we used above to compute $A_{a,n}(z)$, we can compute

$$A_{b,n}(z) = \mathcal{L}(f_n H(t-b))(z) = e^{-bz} \sum_{k=0}^{n} \left[\binom{n}{k} b^{n-k} \frac{k!}{z^{k+1}} \right].$$

Therefore, the windowed he Laplace transform power function is:

$$\mathcal{L}(f_{n,supp[a,b)})(z) = \mathcal{L}\left(f_n\, \chi_{[a,b)}\right)(z) = A_{a,n}(z) - A_{b,n}(z) =$$

$$e^{-az}\sum_{k=0}^{n}\left[\binom{n}{k}a^{n-k}\frac{k!}{z^{k+1}}\right] - e^{-bz}\sum_{k=0}^{n}\left[\binom{n}{k}b^{n-k}\frac{k!}{z^{k+1}}\right].$$

In practice, integration by parts

$$\int t^n e^{-zt}dt = -\frac{t^n}{z}e^{-zt} + \frac{n}{z}\int t^{n-1}e^{-zt}dt$$

may be used to implement the Laplace transform in a closed-form analytical expression:

$$\int t^n e^{-zt}dt = \frac{1}{z^{n+1}}e^{-zt}\left(\sum_{k=0}^{n} c_k t^k\right),$$

where we need to estimate the coefficients c_k and we can transform the indefinite to definite integral over the support $[a, b)$.

Now, suppose we are dealing with discrete data sampled over a finite time interval $[a, b]$. One strategy to compute the Laplace transform of the discrete signal is by direct numerical integration. Alternatively, we can first approximate the data using some base-functions that permit a closed-form Laplace transform analytical representations, see **Table 4.1**. Then, we can aggregate the piecewise linear transforms of all the base functions to derive an analytical approximation to the Laplace transform of the complete discrete dataset without explicit integration. The Inverse Laplace transform may be computed similarly by transforming and aggregating the analytical counterparts of the base functions.

For instance, suppose we decide to preprocess the discrete signal by using a fourth-order spline interpolation of the discrete dataset to model the observations. Thus, we will numerically obtain a finitely supported spline model $f(t) = f_{n=4,supp[a,b)}(t)$ of the data over the time interval $t \in [a,\ b)$ [253]. The Laplace transform for this spline model may be computed exactly as an analytical function, by integrating it against the exponential term:

$$\mathcal{L}(f_{n=4,supp[a,b)})(z) = \int_0^\infty f_{n=4,supp[a,b)}(t)\ e^{-zt}dt = \int_a^b f_{n=4}(t)\ e^{-zt}dt.$$

For simplicity, assume that the fourth-order spline model smoothly concatenates d 4-order polynomial functions, $f_{n=4,\ supp[a,b)} \equiv \left\{f_{k,\ n=4,supp[a_k,a_{k+1})}\right\}_{k=0}^{d-1}$, defined over the domain support on the corresponding partition:

$$\underbrace{[a,b)}_{\text{support}} = \bigcup_{k=0}^{d} \overbrace{\underbrace{[a_k, a_{k+1})}_{\text{partition interval}}}^{\substack{\text{support of} \\ f_{k,n=4,supp[a_k,a_{k+1})}}}, a \equiv a_o < a_1 < \cdots < a_d \equiv b,$$

$$\underbrace{f_{k,n=4,supp[a_k,a_{k+1})}}_{\text{spline model component}} = \sum_{n=0}^{4} q_{n,k} t^n \chi_{[a_k,a_{k+1})}(t), \forall 0 \le k \le d-1,$$

$$\underbrace{f_{n=4,supp[a,b)}}_{\text{spline model}} = \sum_{k=0}^{d-1} \left[\underbrace{\sum_{n=0}^{4} q_{n,k} t^n \chi_{[a_k,a_{k+1})}(t)}_{f_{k,n=4,supp[a_k,a_{k+1})}} \right].$$

Then, the closed-form analytical expression of the Laplace transform of the (approximate) dataset can be derived:

$$\mathcal{L}(\text{dataset})(z) \cong \mathcal{L}\left(f_{n=4,supp[a,b)}\right)(z) = \int_0^\infty f_{n=4,supp[a,b)}(t) e^{-zt} dt =$$

$$\underbrace{\sum_{k=0}^{d-1}}_{\text{spline}} \left[\underbrace{\sum_{n=0}^{4}}_{\text{polynomial}} \left(\underbrace{q_{n,k}}_{\text{poly coef}} \underbrace{\int_{a_k}^{a_{k+1}} t^n e^{-zt} dt}_{\mathcal{L}\left(f_{n,supp[a_k,a_{k+1})}\right)(z)} \right) \right] = \sum_{k=0}^{d-1} \left[\sum_{n=0}^{4} \left(q_{n,k} \left(A_{a_k,n}(z) - A_{a_{k+1},n}(z) \right) \right) \right],$$

where

$$A_{a_k,n}(z) - A_{a_{k+1},n}(z) = e^{-a_k z} \sum_{l=0}^{n} \left[\binom{n}{l} a_k^{n-l} \frac{l!}{z^{l+1}} \right] - e^{-a_{k+1} z} \sum_{l=0}^{n} \left[\binom{n}{l} a_{k+1}^{n-l} \frac{l!}{z^{l+1}} \right].$$

Hence, the strategy of first spline-interpolating the discrete dataset, followed by exact integration of these analytical functions, represents an alternative approach to numerically integrating the (possibly noisy) observed signals. This method circumvents potential instabilities and extremely time-consuming numerical integrations. To obtain the Laplace transform of the dataset, we effectively sum up the integrals of the Laplace transformed power functions, across the range of polynomial powers and across the spline partition of the signal time-domain support.

For time-varying functions, $f(t)H(t)\colon \mathbb{R}^+ \to \mathbb{C}$, there is also a direct connection between the Laplace and the Fourier transforms:

$$\mathcal{L}(f(\cdot)H(\cdot))(z) \equiv \int_0^\infty f(t)H(t)e^{-zt}dt \underbrace{=}_{\substack{z=i\omega \\ -iz=\omega}} \int_0^\infty f(t)H(t)e^{-i\omega t}dt \equiv \mathcal{F}(f(\cdot)H(\cdot))(-iz).$$

Suppose $w \in \mathbb{C}$ and we consider the Laplace transform of the exponential decay function

$$f(t) = e^{-wt}H(t),\ t \geq 0,$$

$$\mathcal{L}(f)(z) = \mathcal{L}\left(e^{-wt}H(t)\right)(z) = \frac{1}{z+w}.$$

The Laplace transform of the shifted function $f(t) = e^{-w(t+a)}$ will be

$$\mathcal{L}(f)(z) = \mathcal{L}\left(e^{-w(t+a)}H(t)\right)(z) = e^{-wa}\mathcal{L}\left(e^{-wt}H(t)\right)(z) = \frac{e^{-wa}}{z+w}.$$

Alternative Laplace transform approximations may be obtained for discrete data using other base functions. For instance, the Laplace transforms of the windowed trigonometric functions may be obtained as follows.

$$\mathcal{L}(\sin(wt)H(t))\ (z) = \frac{w}{z^2+w^2},\ \mathcal{L}(\cos(wt)\ H(t))(z) = \frac{z}{z^2+w^2},$$

$$\mathcal{L}(\sin(w(t+a))H(t))(z) = \sin(wa)\frac{z}{z^2+w^2} + \cos(wa)\frac{w}{z^2+w^2},$$

$$\mathcal{L}(\sin(wt)\ H(t-a))(z) = e^{-az}\ \frac{z\sin(wa) + w\cos(wa)}{z^2+w^2},$$

$$\mathcal{L}(\cos(wt)\ H(t-a))(z) = e^{-az}\ \frac{z\cos(wa) + w\sin(wa)}{z^2+w^2},$$

$$\mathcal{L}\left(\sin(wt)_{supp[a,b)}\right)(z) = \mathcal{L}(\sin(wt)\ H(t-a))(z) - \mathcal{L}(\sin(wt)\ H(t-b))(z).$$

Open Problems

1. Expand the derivations of the magnitude-only model in equation (4.5) for a more general non-constant baseline component in equation (4.1).
2. Derive the distribution, likelihood, and log-likelihood functions for the magnitude-only model (4.6) under *intermediate SNR* conditions? Perhaps use higher order Taylor or Maclaurin series expansions.
3. Expand the complex-valued and Kime-indexed fMRI kintensity inference. For instance, what assumptions may be necessary to derive a more accurate estimate of the magnitude-only intensity? For the fully complex-valued kintensities, derive explicit and tractable closed-form representation of the constrained and unconstrained log-likelihood functions. In both cases, determine the appropriate conditions to estimate or approximate the LRT statistics.
4. Expand the tensor-based spacekime analytics formulation presented in **Section 4.6.**
5. Explore alternative strategies for transforming time-series to kimesurfaces.
6. Propose effective approaches to estimate the enigmatic kime phases.

Chapter 5
Inferential Uncertainty

In the mid-1920's, the German physicist Werner Heisenberg was working on the mathematical foundations of quantum mechanics. Heisenberg realized that atomic particles do not travel on static orbits, which leads to the discrepancy between the classical description of motion and the quantum level motion. As we saw in **Chapter 2**, observables in all physical experiments are measurable quantities associated with self-adjoint linear operators on the state space (Hilbert space). Eigenvectors of operators represent results of specific physical measurements. What led to the discovery of one of the most important scientific principles was the understanding that *some* pairs of observables can be simultaneously measured with infinite precision and some cannot. In technical terms, if the commutator of a pair of functional operators has a zero-dimensional null-space, then the operators cannot share eigenvectors. Heisenberg formulated the uncertainty principle asserting a fundamental limit to the precision of knowledge of pairs of complementary physical properties, such as position, momentum, and energy.

As we pointed out earlier in **Chapter 2**, in the field of data science, the physical concepts of observables, states, and wavefunctions correspond to data, features, and inference functions, respectively. We will start by defining some important terminologies and then illustrate their practical applications in data analytics. Then, we will go deeper into understanding uncertainty by considering an embedding of classical 4D spacetime into the 5D spacekime manifold. This Campbell embedding suggests a coupling between 4D spacetime and fifth spacekime dimension as well as the existence of an extra force that is parallel to the 4-velocity. The core of this derivation is based on extending the 4D solutions of Einstein‘s equations to their corresponding 5D solutions of the Ricci-flat field equations and explicating the Ricci tensor.

In this chapter, we draw parallels between quantum mechanics concepts and their corresponding data science counterparts. After extending velocity, momentum, and Lorentz transformations in 5D spacekime, we discuss synergies between random sampling in spacetime and spacekime and present a Bayesian formulation of spacekime analytics. Alternative representations of uncertainty in data science are shown at the end. The extensive chapter appendix includes bra-ket formulations of multivariate random vectors, time-varying processes, conditional probability, linear modeling, and the derivation of the cosmological constant (Λ) in the 5D Ricci-flat spacekime.

https://doi.org/10.1515/9783110697827-005

5.1 Core Definitions

5.1.1 Observables (Datasets)

In data science, *observables* are dynamic datasets consisting of collections of measurable features. For instance, clinical observables may include various measurements of human health that are acquired by clinicians, physicians, or technicians. Imaging observables are sets of raw time, space, spacetime, or hyper-volume records representing the anatomical, functional, or physiological states of living organisms, molecules, cells, organs, systems, networks, or other biological specimens. There are also multiple endogenous (e.g., omics, biome, hereditary), exogenous (e.g., environmental, epigenetics, diet), and stochastic (e.g., known and unknown variability) observables that are recorded under various experimental conditions.

Most often, observables are structured real-valued functions on the combinations of all possible system states. That is, for a given set of data features, measurable observables represent data collections (datasets) that include instances of data elements (e.g., numbers, strings, class labels, intensities, "NA", etc.) for each feature and each case. Observables may also be thought of as operators, or gauges, where properties of the data state can be determined by some sequences of such operations. For example, these operations might involve applying physical, chemical, biological, mechanical, or other processes to eventually obtain, or read, the values for all features (i.e., the set of observables) in the specific problem scope.

Problem-context meaningful observables must satisfy *transformation laws*, which relate datasets recorded by different investigators (observers) at different frames of reference (e.g., site, time, location, condition). These transformation laws are automorphisms of the state space suggesting that they transform bijectively the inferential states while preserving some specific mathematical properties.

5.1.2 Inference Function

In data science, an *inference function* represents a mathematical description of the outcome states of a specific data analytic system (problem). These *inferential (outcome) states* are paired with a specific analytical strategy and a specific probability distribution for the corresponding inference estimates based on all observables (instances of datasets or data archives). The resulting inference outcomes (states) correspond to each possible measurement of the process phenomenon (i.e., dataset). Knowledge of the inference states, their likelihoods, and the rules for the system's time evolution (data augmentation, aggregation, expansion, etc.), provides complete data science description of the analytical problems and shows what and how it can be modeled, predicted, inferred, or concluded about the system's behavior based on input data.

In general, as a mathematical description of a solution to a specific data analytic system (problem), the inference function is a real multivariate function quantifying likelihoods, kevent probabilities, odds of discrete classes or categorical labels, or yielding regression quantization. A problem inference represents the possible outcome conclusions that can be obtained, or derived, from the system measurements, i.e., observed datasets. To keep notation synergistic with quantum physics' wavefunctions, we denote inference functions by Greek letters like ψ, ϕ, Ψ, or Φ.

In physics, the state of a particle is always identical to the system wavefunction, which represents the probability amplitude of finding the system in that state. However, in data science, for a given data system, the *inference function does not have the same relation to the dataset*. Albeit the inference always depends on the (input) data, the resulting scientific conclusion may substantially change or differ across different (model-based and model-free) analytical strategies, and often may be computed using estimation that relies on stochastic optimization algorithms.

Example: To clarify the relations between these concepts, it helps to keep some simple examples in mind. For instance, let's review a couple of inference functions on this specific data system (problem):

- *Context*: Predict the binary disease state of patients using clinical data.
- *Observables (Data)*: predictors = $\{X_{i,j}\}_{i=1,j=1}^{n,m}$, n = number of patients we have data for, and m = number of features, and the clinical outcomes Y_i represent binary labels, e.g., 0 (alive, control, or baseline) or 1 (dead, patient, or treatment).
- *Transformation laws*: Two examples of transformation laws include (1) the logit transformation of the outcome (Y), which may be used to transform between binary outcomes and probability values, and (2) variance-stabilizing transforms, which may be employed to "normalize" or standardize the predictors (X).
- *Analytical strategy*: There are model-based and model-free approaches that can be used in this analytical problem setting. Let's look at some examples of each type of inference.

A linear model ($Y \cong X\beta + \varepsilon$) may be fitted on the observables ($O = \{X, Y\}$) to predict the outcome (Y) using the independent features (X). Alternatively, a model-free machine learning technique may be employed to generate estimates $\left(\hat{Y} = \hat{Y}(X)\right)$ of the likelihood, probability, or state of the outcome based on the observed covariates. If the outcome is binary, we will be suppressing some of the technical details of the formulation of the generalized linear model and the corresponding logit parameter estimation as an equation root-finding problem $0 = X^T(Y - X\beta)$. In this situation, maximizing the likelihood of the linear logistic regression model is equivalent to obtaining solutions of the score (vector) equation:

$$0 = \frac{\partial}{\partial \beta} \log L(\beta, X, Y) = X^T\left(Y - g^{-1}(X\beta)\right),$$

where $Z = g(\mu) = \log \frac{\mu}{1-\mu}$ is the *link function*, its inverse is the likelihood, and for $Y = 1$, $g^{-1}(Z) = \frac{e^Z}{1+e^Z} = \frac{1}{1+e^{-Z}} = P(Y = 1|X, \beta)$, $Z = X\beta$ is the linear predictor of the binary outcome vector $Y = (y_1, y_2, \ldots, y_n)^T$, and the expected value of Y_i is the likelihood for $y_i = 1$, $g^{-1}(x_i\beta) = \mu_i$. As the logit model can be expressed in terms of a multivariate linear model:

$$\operatorname{logit}\left(\mathbb{E}\left[y_i \middle| \underbrace{x_{i,1}, x_{i,2}, \ldots, x_{i,m}}_{x_i}\right]\right) = y_i = \beta_o + \sum_{j=1}^{m} x_{i,j}\beta_j = x_i\beta,$$

we can assume that we are working with a general linear model of a continuous outcome Y.

As the binary independent and identically distributed (IID) observations represent a Bernoulli process, the likelihood and the log-likelihood functions are expressed in terms of the observations ($\{y_i\}$) and the parameter vector (β):

$$L(\beta, X, Y) = \prod_{i=1}^{n} \underbrace{\mu_i^{y_i}(1-\mu_i)^{1-y_i}}_{\text{Bernoulli}(\mu_i)}, \qquad \mu_i = g^{-1}(x_i\beta),$$

$$ll = \log L = \sum_{i=1}^{n}\left[\log\left(1 - \underbrace{g^{-1}(x_i\beta)}_{\mu_i}\right) + y_i \log\left(\frac{g^{-1}(x_i\beta)}{1-g^{-1}(x_i\beta)}\right)\right].$$

As $\mu_i \equiv g^{-1}(x_i\beta)$, its partial derivative with respect to the parameter β_j is:

$$\mu'_{i,j} = \frac{\partial}{\partial\beta_j} g^{-1}\left(\underbrace{x_{i,\mu}\beta_\mu}_{z_i}\right) = \frac{\partial}{\partial z_i}\left(g^{-1}\left(\underbrace{x_{i,\mu}\beta_\mu}_{z_i}\right)\right)\frac{\partial}{\partial\beta_j} z_i =$$

$$\frac{\partial}{\partial z_i}\left(\frac{1}{1+e^{-z_i}}\right)x_{i,j} = \frac{e^{-z_i}}{(1+e^{-z_i})^2}x_{i,j} = g^{-1}(z_i)(1-g^{-1}(z_i))x_{i,j} = \mu_i(1-\mu_i)x_{i,j}.$$

Then, the partial derivative of the log-likelihood function is:

$$\underbrace{\frac{\partial}{\partial\beta_j} ll}_{1\times 1} = \frac{\partial}{\partial\beta_j}\log L = \sum_{i=1}^{n}\left(-\frac{\mu'_{i,j}}{1-\mu_i} + \frac{y_i\mu'_{i,j}}{\mu_i(1-\mu_i)}\right) =$$

$$\sum_{i=1}^{n}\left((y_i - \mu_i)\underbrace{\frac{\mu'_{i,j}}{\mu_i(1-\mu_i)}}_{x_{i,j}}\right) = \underbrace{s_j^T}_{1\times n}\left(Y - \underbrace{g^{-1}(X\beta)}_{n\times 1}\right),$$

where $Y = (y_1, y_2, \ldots, y_n)^T$, $g^{-1}(X\beta) = (g^{-1}(x_1\beta), g^{-1}(x_2\beta), \ldots, g^{-1}(x_n\beta))^T$, $X_{n\times m} = (s_1, s_2, \ldots, s_m)$, and the m column vectors s_j represent the observed features, as vectors of dimension $n\times 1$. Therefore, the score vector is:

$$\underbrace{\frac{\partial}{\partial\beta} ll}_{m\times 1} = \left(\frac{\partial}{\partial\beta_1}, \frac{\partial}{\partial\beta_2}, \ldots, \frac{\partial}{\partial\beta_m}\right)^T = \underbrace{X^T}_{m\times n}\underbrace{(Y - g^{-1}(X\beta))}_{n\times 1}.$$

To illustrate an example of an inference function, let's first focus on the simpler logit model where we have a closed-form analytical solution (via *least squares*) to the problem of predicting the outcome (Y) using the independent features (X).

Specifically, the inference function, $\psi = \psi\left(X, Y \middle| \underbrace{\text{linear model}}_{\text{analytical strategy}}\right)$, quantifies the effects of all independent features (X) on the outcome (Y). The linear model has a closed-form analytical representation where the predicted outcomes are expressed as $\hat{Y} = X\hat{\beta}$. Ordinary least squares (OLS) can be used to compute an estimate of the parameter vector:

$$\hat{\beta} = \hat{\beta}^{OLS} = \langle X|X\rangle^{-1}\langle X|Y\rangle \equiv (X^TX)^{-1}X^TY.$$

The resulting *inference* includes three complementary components:

(1) the actual effect-size estimates $\left(\hat{\beta} = \left(\hat{\beta}_1, \hat{\beta}_2, \hat{\beta}_3, \ldots, \hat{\beta}_m\right)'\right)$,
(2) their variability ($\sigma^2_{\hat{\beta}_i}$), and
(3) their associated probabilities relative to some *a priori* null and research hypotheses, e.g.,

$$\underbrace{P\left(|t_{n-m-1}| < \left|t^{\beta}_{i,o}\right|\right)}_{\text{tail probability}}, \quad \text{where} \quad \underbrace{t^{\beta}_{i,o}}_{\substack{\text{test}\\ \text{statistics}}} = \frac{\hat{\beta}_i - \beta_i}{\sigma_{\hat{\beta}_i}\sqrt{\left((X'X)^{-1}\right)_{i,i}}} \sim \underbrace{\overbrace{t_{n-m-1}}^{df=n-m-1}}_{\substack{\text{Student's t}\\ \text{distribution}}}.$$

Thus, the basic model-based inference for the *generalized linear model* (GLM) represents solutions to a prediction inference problem. The corresponding inference function, which quantifies the effects for all independent features (X) on the dependent outcome (Y), can be expressed as:

$$\psi(O) = \psi(X, Y) = \hat{\beta} = \hat{\beta}^{OLS} = \langle X|X\rangle^{-1}\langle X|Y\rangle = \left(X^T X\right)^{-1} X^T Y.$$

For the same data system example, support vector machines (SVM) classification provides an alternative example of a non-parametric, non-linear inference.

Assume that $\psi_x \in H$ is the lifting function $\psi: \mathbb{R}^n \to \mathbb{R}^d$ ($\psi: x \in \mathbb{R}^n \to \tilde{x} = \psi_x \in H$), where $n \ll d$, the kernel $\psi_x(y) = \langle x|y\rangle : O \times O \to \mathbb{R}$, and the observed data $O_i = \{x_i, y_i\}$ are lifted to ψ_{O_i}. Then, the SVM prediction operator is the weighted sum of the kernel functions at ψ_{O_i}, which generalizes the notion of inner product similarity:

$$\langle \psi_O | \beta^* \rangle_H = \sum_{i=1}^{n} p_i^* \langle \psi_O | \psi_{Oi} \rangle_H,$$

where β^* is a solution to the regularized optimization problem:

$$\min_{\beta \in H}\left(\underbrace{L\left(\tilde{X}\beta, Y\right)}_{\text{fidelity}} + \frac{\lambda}{2}\underbrace{\|\beta\|_H^2}_{\text{regularizer}}\right),$$

$\tilde{X} = \left\{\psi_{x_i}\right\}_{i=1}^{n}$ are the lifted features, L is the objective (loss) function (e.g., least-squares cost) assessing the fidelity of the classifier, and the coefficients in this linear combination, p_i^*, are the dual weights that are multiplied by the label corresponding to each training instance, $\{y_i\}$. Our oversimplified binary classification example aims to predict the disease state of participants, e.g., patient vs. control. Supervised SVM classification represents one analytical strategy that outputs concrete decision-making inference. For instance, SVM can classify new data points (testing case participants), z, that have homologous features to the training case participants, x. Explicitly, SVM classification labeling of testing data is based on computing the (binary) sign ($\pm$) of the transformation

$$z \to \text{sign}\left(\left[\sum_{j=1}^{n} c_j y_j k(x_j, z)\right] - b\right),$$

where the pair of parameters, *offset* $b = w.\psi_i - y_i = \left(\sum_{j=1}^{n} c_j y_j k(x_j, x_i)\right) - y_i$, and *normal vector*, w, to the separation plane are estimated using the training data (O_i consisting of the observed features, x_i, and their corresponding labels, y_i):

$$y_i = \begin{cases} -1, & \text{if } w^t x_i + b \le -1 \\ +1, & \text{if } w^t x_i + b \ge 1 \end{cases} \underset{\text{equivalent to}}{\Longleftrightarrow} y_i\left(w^t x_i + b\right) \ge 1.$$

Also, note that the coefficients, c_j, are learned on the training data, $O_i = \{x_i, y_i\}$, using the Lagrangian dual optimization problem, and the kernel depends on the lifting function ψ:

$$k\left(\underbrace{x_i}_{\substack{\text{training}\\ \text{data}}}, \underbrace{z}_{\substack{\text{testing}\\ \text{data}}}\right) = \langle \psi_x | \psi_z \rangle_H.$$

By the *representer theorem* [254], all predictions of SVM models, as well as other models resulting from kernel methods, can be expressed as a linear combination of kernel evaluations, inner products, $\langle \cdot | \cdot \rangle_H$, between some training instances (the support vectors) and the testing instances.

Note that the data points ($O_i = \{x_i, y_i\}$) only appear within the inner product, which can be calculated in the feature space. Thus, we do not need an explicit definition of the mapping (lifting function). Therefore, a solution, $\beta^* \in H$, to the regularized optimization problem only lives in the lower (n) dimensional space spanned by the lifted observed features, ψ_{x_i}. The resulting predictions can be expressed as a non-linear function of x and ψ_x:

$$\langle \tilde{x} | \beta^* \rangle_H = \sum_{i=1}^{n} p_i^* \langle \psi_x | \psi_{x_i} \rangle_H = \sum_{i=1}^{n} p_i^* k(x_i, x),$$

where the kernel (k) is defined $\forall x, y \in X$ by $k(x, y) = \langle \psi_x | \psi_y \rangle_H$.

This prediction representation scheme avoids the problem of computing in the higher (d) dimensional space H, which may actually be infinite dimensional. In practice, the process is actually reversed. Designing a kernel is much easier than explicating the lifting ψ operator. Hence, we will directly focus on the kernel function $k(x, y)$. Examples of commonly used SVM kernels include:

$$k(x,y) = \begin{cases} \text{Linear}, & k(x,y) = \langle x|y \rangle = x'y \\ \text{Polynomial}, & k(x,y) = (\langle x|y \rangle + c)^a, a \in \mathbb{Z}^+, c > 0 \\ \text{Sigmoid}, & k(x,y) = \tanh(\langle ax|y \rangle + b) \\ \text{Gaussian (Radial Base)}, & \begin{cases} k(x,y) = \exp\left(\frac{-\langle x-y|x-y \rangle}{2\sigma^2}\right) = \\ \quad = \exp\left(\frac{-\|x-y\|^2}{2\sigma^2}\right) \end{cases} \\ \text{etc.}, & \ldots \end{cases}$$

When defining a kernel, we need to consider its geometric properties, e.g., how does the kernel respond to addition. The kernel must correspond to a lifting function, $\psi : x \in \mathbb{R}^n \to \tilde{x} = \psi_x = \psi(x) \in H$, where H is an implicit higher dimensional feature space (Hilbert space) with dimension $\dim H = d$, and all choices of kernels must be positive definite and symmetric, $k(x_i, x_j) = k(x_j, x_i)$ for any pair of sampling points x_j, x_i. This requirement is equivalent to the existence of a lifting function $\psi_x(y) = \langle x|y \rangle = x'y$ that generates the kernel, which can be defined on flat (Euclidean) or curved (non-Euclidean) spaces.

For *linear* kernels, this one-to-one correspondence between kernels, lifting functions, and inner products, $k(x, y) = \psi_x(y) = \langle x|y \rangle$, is associated with a trivial lifting, $\psi_x = \psi(x) = x$.

For homogeneous ($c = 0$) and inhomogeneous ($c \neq 0$) *polynomial* kernels, $k(x, y) = (\langle x|y \rangle + c)^a$, the function ψ corresponds to lifting the data points to a finite dimensional space, H. For instance, if the observable process includes three features ($p = 3$) and the kernel power exponent is $a = 2$, then $\dim(H) = 10$. To see this, we will expand the polynomial kernel and complete the symbolic calculations:

$$k(x, y) = \langle \psi(x) | \psi(y) \rangle = (\langle x|y \rangle + c)^2 = (x_1 y_1 + x_2 y_2 + x_3 y_3 + c)^2.$$

To work this case explicitly, we can simplify the notation by denoting

$$\left| \begin{array}{l} \mathbf{x} = (x_1, x_2, x_3) \\ \mathbf{y} = (y_1, y_2, y_3) \end{array} \right. .$$

Let's expand the polynomial kernel:

$$k(\mathbf{x}, \mathbf{y}) = \langle \psi(\mathbf{x}) | \psi(\mathbf{y}) \rangle = (\langle \mathbf{x} | \mathbf{y} \rangle + c)^2 = (x_1y_1 + x_2y_2 + x_3y_3 + c)^2 =$$

$$c^2 + 2c\ x_1y_1 + 2c\ x_2y_2 + 2c\ x_3y_3 + x_1^2y_1^2 + x_2^2y_2^2 + x_3^2y_3^2 +$$

$$2x_1x_2y_1y_2 + \ 2x_1x_3y_1y_3 + \ 2x_2x_3y_2y_3.$$

Therefore,

$$k(\mathbf{x}, \mathbf{y}) = \langle \psi(\mathbf{x}) | \psi(\mathbf{y}) \rangle = \underbrace{(\psi(\mathbf{x}))^{\dagger}}_{\substack{\text{adjoint} \\ \text{(conjugate transpose)}}} \cdot \psi(\mathbf{y}) =$$

$$= \left(c, \sqrt{2c}\, x_1, \sqrt{2c}\, x_2, \sqrt{2c}\, x_3, x_1^2, x_2^2, x_3^2, \sqrt{2c}\, x_1x_2, \sqrt{2c}\, x_1x_3, \sqrt{2c}\, x_2x_3\right) \times \begin{pmatrix} c \\ \sqrt{2c}\, y_1 \\ \sqrt{2c}\, y_2 \\ \sqrt{2c}\, y_3 \\ y_1^2 \\ y_2^2 \\ y_3^2 \\ \sqrt{2c}\, y_1y_2 \\ \sqrt{2c}\, y_1y_3 \\ \sqrt{2c}\, y_2y_3 \end{pmatrix},$$

where the lift function, $\psi : \mathbf{x} \in \mathbb{R}^{n=3} \to \mathbb{R}^{d=10} \supseteq H \ni \psi(\mathbf{x}) =$

$$\psi(x_1, x_2, x_3) = \left(c, \sqrt{2c}\, x_1, \sqrt{2c}\, x_2, \sqrt{2c}\, x_3, x_1^2, x_2^2, x_3^2, \sqrt{2c}\, x_1x_2, \sqrt{2c}\, x_1x_3, \sqrt{2c}\, x_2x_3\right)' \in \mathbb{R}^{10},$$

and the kernel is defined in terms of the inner product, $k(\mathbf{x}, \mathbf{y}) = \langle \psi(\mathbf{x}) | \psi(\mathbf{y}) \rangle$.

Non-linear kernels correspond to infinite dimensional lifting functions. For instance, the Gaussian kernel is:

$$\psi : \mathbf{x} \in \mathbb{R}^{n} \to \psi(\mathbf{x}) \in L^2(\mathbb{R}^{n}),\ \textit{i.e.}, \text{square integrable linear functions, } \mathbb{R}^{n} \text{ dual,}$$

where

$$k(\mathbf{x}, \mathbf{y}) = \exp\left(\frac{-\langle \mathbf{x} - \mathbf{y} | \mathbf{x} - \mathbf{y} \rangle}{2\sigma^2}\right) = \exp\left(\frac{-\|\mathbf{x} - \mathbf{y}\|^2}{2\sigma^2}\right).$$

In practice, cross-validation may be used to estimate (or tune) the parameter $\sigma > 0$ that controls the bandwidth of the model.

Many alternative kernels may be used. For instance, the Laplacian kernel resembles a Gaussian, but relies instead on the L^1 norm: $k(\mathbf{x}, \mathbf{y}) = \exp\left(\frac{-\|\mathbf{x} - \mathbf{y}\|_1}{\sigma}\right)$, and variations of string and graph kernels that measure graph similarity are based on an inner product defined on the graphs (or trees) [255].

Finally, any linear combination of kernels, $k(\boldsymbol{x},\boldsymbol{y}) = \sum_{i=1}^{m} \lambda_i k_i(\boldsymbol{x},\boldsymbol{y})$, $\lambda_i \geq 0$, will itself be another kernel that can be used. Note that we can use statistical internal cross-validation to estimate the kernel weight components, λ_i, that yield an *optimal* mixed kernel, $k(\boldsymbol{x},\boldsymbol{y})$.

The positive definite requirements for all translation-invariant kernels $k(\boldsymbol{x},\boldsymbol{y}) = k(\boldsymbol{x}-\boldsymbol{y})$on $\mathbb{R}^\eta$ is essentially equivalent to $\hat{k} > 0$, where $\hat{k}(\omega)$ is the Fourier transform of the kernel $k(x)$. As $k(\boldsymbol{x},\boldsymbol{y}) = \langle \psi(\boldsymbol{x}) | \psi(\boldsymbol{y}) \rangle$, the lifting function, ψ, corresponding to the kernel, $k(\boldsymbol{x},\boldsymbol{y}) = k(\boldsymbol{x}-\boldsymbol{y})$, may be computed by inverting the Fourier transform of $\hat{h} = \sqrt{\hat{k}}$. In other words, all lifting operations that correspond to translation-invariant kernels k are defined by:

$$\psi : \boldsymbol{x} \in \mathbb{R}^\eta \rightarrow \psi_{\boldsymbol{x}} = h_{\boldsymbol{x}} \in L^2(\mathbb{R}^\eta), \text{where } h = \hat{\hat{h}} \text{ and } h_{\boldsymbol{x}}(\boldsymbol{y}) = h(\boldsymbol{x}-\boldsymbol{y}).$$

Recall that for each function, $k \in L^1(\mathbb{R})$, up to an appropriate normalization factor, it's Fourier transform is defined by:

$$\forall \omega \in \mathbb{R},\ \hat{k}(\omega) = \int_{\mathbb{R}} k(x) e^{-i x \omega} dx,$$

and the inverse Fourier transform is

$$\forall x \in \mathbb{R},\ k(x) = \hat{\hat{k}}(x) = \frac{1}{2\pi} \int_{\mathbb{R}} \hat{k}(\omega) e^{i x \omega} d\omega.$$

Both the forward and inverse Fourier transforms are linear transformations that satisfy $\hat{\hat{h}} = h$ and $\widehat{h * g} = \hat{h} \times \hat{g}$, where the *convolution operator* (*), defined by $(f * g)(t) \equiv \int_{\mathbb{R}} f(\tau) g(t-\tau) d\tau = \int_{\mathbb{R}} f(t-\tau)\, g(\tau) d\tau$, is the Fourier transform of the *product operator* ($\times$), and vise-versa.

Most of the time, we work with inference functions that are symmetric. That is, the order of listing features does not matter in computing the inference as a function of the total degrees of freedom (e.g., $\eta + \kappa - 1$), which corresponds to some maximum set of commuting observables (*features*). Once such a representation is chosen, the inference function can be derived from the specific model formulation, the computational strategy, the analytical approach, and the specific state (instance) of the observed data.

For a given data science challenge, the choice of which commuting observables to use is not unique. In fact, it is a really hard problem to identify the smallest subset of the most salient features that can be reliably, reproducibly, consistently, and efficiently used to obtain the retrodiction, prediction, or forecasting inference.

The domain of the inference function is not uniquely defined either. For instance, in spacelike cross-sectional data analytics, the observables are recorded at one fixed time point and the inference function (*inference*) may act only on cross-sectional data elements. Alternatively, in timelike or kimelike analytics, the inference may act on longitudinal data, e.g., time-series analysis.

Of course, sometimes data preprocessing may be either optional, necessary, or required. In such cases, the raw or transformed data may be used in the data analytic process to compute the inference. For instance, magnetic resonance imaging (MRI) data is naturally recorded in the Fourier domain (frequency, k-space). However, radiographic reading of MRI data, as well as its clinical interpretation by physicians, requires the imaging data to be inverted into spacetime. In this case, the forward and inverse Fourier transforms provide a specific relation between the corresponding observables (features) in the time (image) domain and their corresponding counterparts in the native Fourier (frequency) space.

5.2 Inner Product

The **inner product** between two inference functions, $\langle \psi | \ \phi \rangle \equiv \langle \psi, \phi \rangle$, measures the level of inference overlap, result consistency, agreement or synergies between their corresponding inferential states. The inner product provides the foundation for a probabilistic interpretation of data science inference in terms of transition probabilities.

The squared modulus of an inference function, $\langle \psi | \psi \rangle = \| \psi \|^2$, represents the probability density that allows us to measure specific inferential outcomes for a given set of observables. To facilitate probability interpretation, the law of total probability requires the normalization condition, i.e., $1 = \int \| \psi \|^2$. Let's illustrate the modulus in the scope of the above logistic example with real-value data (X, Y) where matrix Hermitian conjugation corresponds to just the matrix transposition, $X^* = X^T$. In this case, the square modulus of the inference function is:

$$\| \psi \|^2 = \langle \psi | \psi \rangle = \langle \psi(X, Y) | \psi(X, Y) \rangle = \left\langle \hat{\beta}^{OLS} \middle| \hat{\beta}^{OLS} \right\rangle =$$

$$= \langle (X^T X)^{-1} X^T Y | (X^T X)^{-1} X^T Y \rangle = \left((X^T X)^{-1} X^T Y \right)^T (X^T X)^{-1} X^T Y =$$

$$= Y^T X (X^T X)^{-1} (X^T X)^{-1} X^T Y = Y^T \underbrace{X (X^T X)^{-2} X^T}_{D} Y = Y^T D Y = \left\langle \left(D^{\frac{1}{2}} \right)^T Y \middle| \left(D^{\frac{1}{2}} \right) Y \right\rangle = \| Y \|_D^2.$$

Open Problems

(1) What would be the effect of exploring the use of the matrix D as a constant normalization factor $\left(\boldsymbol{D}^{\frac{1}{2}} \right)$?

(2) More research is needed to really define an appropriate coherence metric that captures the agreement, or overlap, between a pair of complementary inference functions or data analytic strategies. Ideally, some inference consistency measures can be derived that are analogous to:

$$\text{Coherence} = \frac{\langle \psi | \phi \rangle}{\sqrt{\langle \psi | \psi \rangle \times \langle \phi | \phi \rangle}} = \frac{\langle \psi | \phi \rangle}{\| \psi \| \| \phi \|}.$$

(3) Alternatively, as the data represent random variables (vectors, or tensors) and the specific data-analytic strategy yields the inference function, we may explore **mutual information of operators,** which may also be interpreted as linear or non-linear operator acting on the data:

$$I(\psi;\phi) = \sum_i \sum_j \langle \psi_i | \phi_j \rangle \log \left(\frac{\langle \psi_i | \phi_j \rangle}{\| \psi_i \| \| \phi_j \|} \right),$$

where the inference states ψ_i and ϕ_i are the **eigenfunctions** corresponding to the eigenvalues O_i, see the eigenspectra section below.

In the most general complex-valued data, any observable dataset that can be collected via simulation, prospectively designed experiments, retrospectively as secondary data, or recorded real time should be associated with a **self-adjoint linear operator.** For instance, suppose our observed dataset consists of complex-valued measurements of $\kappa + 1$ features ($O = \{\{X_i\}_{i=1}^{\kappa} \cup Y\}$) for η cases, participants, or units:

$$\langle X| = (X_1, X_2, \ldots, X_\kappa) = \begin{bmatrix} \underbrace{\begin{matrix} x_{1,1} \\ \vdots \\ x_{\eta,1} \end{matrix}}_{X_1} & \begin{matrix} \cdots \\ \ddots \\ \cdots \end{matrix} & \underbrace{\begin{matrix} x_{1,\kappa} \\ \vdots \\ x_{\eta,\kappa} \end{matrix}}_{X_\kappa} \end{bmatrix} \text{ and } |\beta\rangle = \begin{pmatrix} \beta_1 \\ \beta_2 \\ \cdots \\ \beta_\eta \end{pmatrix}.$$

This observable, O, corresponds to the linear operator $\hat{O} = \langle X|$, which acts on (linear model-based) inference, ψ and generates predictions, e.g., fitted values $\hat{Y} = \langle X|\hat{\beta}\rangle = |X^*\hat{\beta}\rangle$.

Then, the expectation of the linear operator $\hat{O}$ can be defined by:

$$\langle \hat{O} \rangle = \langle \psi(Y) | \hat{O} | \psi(Y) \rangle = \int dX \, (\psi(Y))^* \, \hat{O} \, \psi(Y).$$

5.3 Eigenspectra (Eigenvalues and Eigenfunctions)

Suppose the observable O is measured multiple times (r) under identical conditions, i.e., we observe several instances of the data under identical conditions, using the same features and cases. Symbolically, the result will be a multi-instance dataset $\{O_1, O_2, \ldots, O_r\}$. The expected value of the observed data, $\langle O \rangle$, represents the asymptotic limit of the average of its multiple instances $\lim_{r\to\infty} \left[\frac{1}{r} \sum_{i=1}^{r} O_i \right]$. Note the synergy with the law of large numbers [164]. All of the r different observed data instances, O_i, represent possible outcomes of the same data-generating experiment. More importantly, the O_i datasets are actually the *eigenvalues* of the linear operator $\hat{O}$, and the corresponding inference states ψ_i are the *eigenfunctions* associated with the eigenvalues O_i. That is,

$$\hat{O}\psi_i = O_i\psi_i.$$

Note that each eigenfunction, ψ_i represents the inference function of a state in which the data (observed measurement) O yields the value O_i with probability 1. The eigenfunctions may also be normalized, $\langle\psi_i|\psi_j\rangle = \delta_{i,j} = \begin{cases} 1, i = j \\ 0, i \neq j \end{cases}$. This fact can be validated by computing the variance of O in the state ψ_i:

$$Var_i(O) = \langle\hat{O}^2\rangle - \langle\hat{O}\rangle^2 = \langle\psi_i|\hat{O}^2|\psi_i\rangle - \langle\psi_i|\hat{O}|\psi_i\rangle^2 =$$
$$= O_i^2\langle\psi_i|\psi_i\rangle - (O_i\langle\psi_i|\psi_i\rangle)^2 = 0.$$

These linear operators have real data as eigenvalues, since these values result from specific observable scientific experiments. Real eigenvalues are associated with Hermitian matrices where the probability of each eigenvalue is related to the projection of the physical state on the subspace spanned by the corresponding eigenfunction related to that eigenvalue.

5.4 Uncertainty in 5D Spacekime

Since the 1990's, an international team of scholars (the 5D Space-Time-Matter Consortium, http://5dstm.org, accessed January 29, 2021) published a series of papers examining the implications of a 5D world with an extra time dimension [62, 69, 147, 256–259]. One of the results of this monumental work was the conclusion that in $(3+2)$ space + time dimensions, the interpretation of particle spacetime motion may be slightly modified by an extra force to produce a correlation between the momentum and position similar to the uncertainty relation in quantum mechanics. This induced-matter (aka space-time-matter) theory led to a derivation of the classical 4D spacetime Heisenberg uncertainty as a reduction of Einstein-like 5D deterministic dynamics. In the higher-dimensional spacekime extension, the common spacetime uncertainty principle could be understood in 5D as consequences of deterministic laws. As Paul Wesson stated "*Heisenberg was right in 4D, because Einstein was right in 5D*" and "*God does not play dice in a higher-dimensional world*" [260]. It's probably useful to draw a distinction between Hermann Weyl's space-time-matter (*Raum-Zeit-Materie*) formulation [261] and the 5D space-time-matter proposed by Wesson and colleagues [69]. In the 1920's, Weyl described matter as a substance composed of elementary quanta with invariable mass and a definitive charge that are observed in specific spatial locations at certain times.

This section presents an approach for realizing the 4D Heisenberg uncertainty as a silhouette of 5D Einstein deterministic dynamics. Let's start by expressing the original Heisenberg's uncertainty relation between the momentum and the position using Einstein summation indexing convention:

$$\underbrace{dp^{\mu}}_{\substack{\text{increment in the}\\ 4-\text{momentum}}} \quad \underbrace{dx_{\mu}}_{\substack{\text{increment in the}\\ 4-\text{position}}} \sim h.$$

We can divide both sides of this equation by two increments in the proper time s, which represents the time measured within the internal coordinate reference frame:

$$\frac{dp^{\mu}}{ds}\frac{dx_{\mu}}{ds} = F^{\mu}\ u_{\mu} \sim \frac{h}{ds^2}.$$

In the limit, this suggests that there is a force acting parallel to the velocity, whose inner product with velocity is non-trivial. However, this contradicts the well-known orthogonality condition in Einstein's 4D theory of relativity.

For completeness, let's demonstrate the orthogonality condition in 4D spacetime. The path of an object moving relative to a particular reference frame is defined by four coordinate functions $x^{\mu}(s)$, where μ is a spacetime index with $\mu = 0$ for the time-like component and $\mu = 1, 2, 3$ for the spacelike coordinates. The zeroth component, $x^0 = ct$, is defined as the time coordinate multiplied by c, the speed of light, and each coordinate function depends on the proper time, s.

The *4-velocity* is the tangent 4-vector of a timelike world line, which can be defined at any world line point, $x^{\mu}(s)$ by:

$$u^{\mu} = \frac{dx^{\mu}}{ds}.$$

More explicitly, for $\mu = 0$, the temporal component of the 4-velocity is the derivative of the x^0 position relative to proper time (s):

$$u^0 = \frac{dx^0}{ds} = \frac{dx^0}{dt}\frac{dt}{ds} = c\frac{dt}{ds} = c\ \gamma(v).$$

Similarly for the three spatial components of the 4-velocity, indexed by $\mu = 1, 2, 3$:

$$u^{\mu} = \frac{dx^{\mu}}{ds} = \frac{dx^{\mu}}{dt}\frac{dt}{ds} = \frac{dx^{\mu}}{dt}\gamma(v) = v^{\mu}\gamma(v),$$

where the Lorentz factor, $\gamma(v) = \frac{dt}{ds} = \frac{1}{\sqrt{1-\frac{v^2}{c^2}}} = \frac{1}{\sqrt{1-\beta^2}}$, is a function of the square Euclidean norm of the *3D velocity vector*, v between the two inertial frames of reference (K and K'):

$$v = \|\mathbf{v}\| = \sqrt{(v^1)^2 + (v^2)^2 + (v^3)^2},$$

$$v^{\mu} = \frac{dx^{\mu}}{dt} \quad (\mu = 1, 2, 3).$$

Thus, the 4-velocity is $u = (u^{\mu})_{4\times1} = \gamma(c, \mathbf{v}) = (\gamma c, \gamma \mathbf{v})$. This is an explicit relation between a particle's 3-velocity and its 4-velocity.

Let's examine the relation between the 4-acceleration (a^{μ}) and the classical 3-acceleration $\left(\frac{dv}{dt}\right)$:

$$a = (a^{\mu})_{4\times1} = \frac{du}{ds} = \left(\frac{du^{\mu}}{ds}\right)_{4\times1} = \gamma\left(\frac{du^{\mu}}{dt}\right) = \gamma\frac{d}{dt}(\gamma c, \gamma \mathbf{v}) = \gamma\left(\underbrace{c\frac{d\gamma}{dt}}_{\text{temporal}}, \underbrace{\frac{d\gamma}{dt}\mathbf{v} + \gamma\frac{d\mathbf{v}}{dt}}_{\text{spatial}}\right).$$

In the particle's internal resting frame, $\mathbf{v} = \mathbf{0}$, hence,

$$\gamma = 1, \frac{d\gamma}{dt} = \frac{d}{dt}\left(\frac{1}{\sqrt{1-\frac{v^2}{c^2}}}\right) = -\frac{1}{2\left(1-\frac{v^2}{c^2}\right)^{\frac{3}{2}}}\left(-\frac{2v}{c^2}\right)\frac{dv}{dt} = 0,$$

and the 4-velocity $u \equiv (u^{\mu})_{4\times1} = (c, \mathbf{0})$, and the 4-acceleration $a \equiv (a^{\mu})_{4\times1} = \left(0, \frac{d\mathbf{v}}{dt}\right)$.

Remember that this is relative to the particle's instantaneous resting frame. After an infinitesimal time change, dt, the particle is no longer in this instantaneous resting frame – it will be in a new instantaneous resting frame. Of course, the particle's velocity in the new rest frame is still $\mathbf{0}$, but in the old rest frame, its velocity has now changed and $\frac{d\mathbf{v}}{dt}$ is not trivial. As a result, the inner product of the 4-velocity $u \equiv (u^{\mu})_{4\times1} = (c, \mathbf{0})$ and the 4-acceleration $a \equiv (a^{\mu})_{4\times1} = \left(0, \frac{d\mathbf{v}}{dt}\right)$ is:

$$\langle u|a\rangle = u^{\mu}a_{\mu} = (c, \mathbf{0}) \cdot \left(0, \frac{d\mathbf{v}}{dt}\right) = 0.$$

This explains the classical 4D relativity orthogonality condition that the inner product of the 4-acceleration and the 4-velocity is trivial.

We can also derive this using first principles by explicitly computing the square magnitude of the 4-velocity using the metric tensor $g = (g_{\alpha\beta})$:

$$\|\mathbf{u}\|^2 = \langle u|u\rangle = \sum_{\alpha=0}^{3}\sum_{\beta=0}^{3} g_{\alpha\beta}u^{\alpha}u^{\beta} = \sum_{\mu=0}^{3} g_{\mu\mu}u^{\mu}u^{\mu}$$

$$\underbrace{=}_{u^{\mu}=\gamma(c,\mathbf{v})} \pm(\gamma(v))^2\left(\underbrace{c^2-(v^1)^2-(v^2)^2-(v^3)^2}_{\langle(c,\mathbf{v})|(c,\mathbf{v})\rangle}\right) = \pm\frac{c^2-v^2}{1-\frac{v^2}{c^2}} = \pm c^2.$$

The positive or negative sign in the equation above depends on the choice of metric signature:

$$g = diag(+1, -1, -1, -1) \text{ or } g = diag(-1, +1, +1, +1).$$

Therefore, $\|\mathbf{u}\|^2$ is a constant and its derivative with respect to the proper time is trivial:

$$\frac{d\|\boldsymbol{u}\|^2}{ds} = \frac{d}{ds}(\boldsymbol{u}\cdot\boldsymbol{u}) = 2\boldsymbol{u}\cdot\frac{d\boldsymbol{u}}{ds} = 0 \Rightarrow$$

$$u^{\mu}a_{\mu} = u^{\mu}\frac{du_{\mu}}{ds} = 0, \quad a_{\mu} = g_{\mu\gamma}a^{\gamma},$$

and we have that the 4-acceleration vector (a) is orthogonal to 4-velocity (u).

Next, we will try to show that extending 4D spacetime to 5D spacekime suggests the existence of an extra force. One component of this additional force is parallel to the 4-velocity and explains the intrinsic Heisenberg uncertainty relation in the lower 4D spacetime embedding.

Campbell's theorem [257] provides necessary conditions for an N-dimensional Riemannian manifold to be embedded as a hypersurface leaf in a similar $(N+1)$-dimensional manifold. This result suggests an interpretation of 5D spacekime as a Ricci-flat space where the 4D Minkowski spacetime can be embedded using the *canonical metric*, which provides a smooth embedding extending any 4D metric space into a 5D manifold. More details of this embedding and a historical perspective is available in Wesson's 2010 manuscript [262].

Let's first investigate the properties of this embedding and the canonical metric. Consider a D-dimensional "generating" space that we *foliate* by a family of $(D-1)$ hypersurfaces [263]. Foliation, also called slicing, represents a decomposition of a higher-dimensional manifold, e.g., Minkowski 5D spacekime, into hypersurfaces of lower dimension, e.g., Minkowski 4D spacetime leaves. This embedding also requires the existence of a regular smooth scalar field with non-trivial gradient, whose level-surfaces represent the foliation leaves or hypersurfaces. Since the scalar field is regular, the hypersurfaces are non-intersecting. Let's assume that the manifold is globally hyperbolic, all hypersurfaces are spacelike, and the foliation covers the whole manifold. Each hypersurface is called a leaf or a slice of the foliation. **Figure 5.1** shows

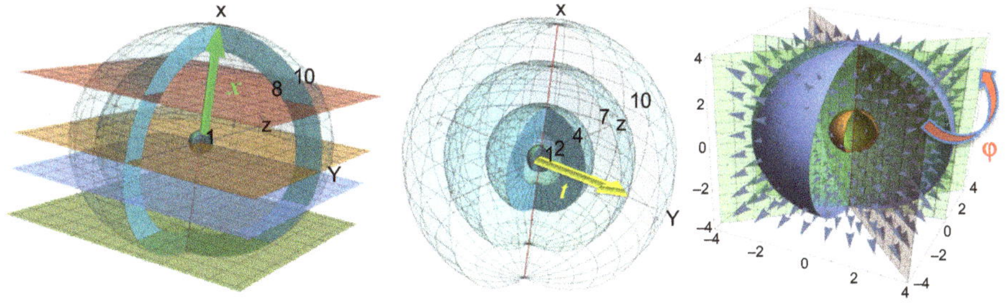

Space ($\boldsymbol{x}$) Foliation of Spacekime | (Radial, t) Time-Foliation of Spacekime | (Angular, φ) Phase-Foliation of *Kime*

Figure 5.1: Examples of manifold foliations that are relevant in the lifting of spacetime analytics to spacekime.

schematics of simple alternative manifold foliations, which play important roles in the generalization of classical spacetime to spacekime analytics, see **Chapter 6.**

In normal coordinates, the spacekime interval (dS^2) in $D \equiv 5$ dimensions is defined in terms of the metric tensor (g_{AB}) by:

$$dS^2 = \sum_0^{D-1} \sum_0^{D-1} g_{AB} dx^A dx^B.$$

And according to the canonical 5D spacekime metric, the interval can be expressed as:

$$dS^2 = \sum_0^{D-2} \sum_0^{D-2} g_{\alpha\beta}(x^\mu, l) dx^\alpha dx^\beta + \epsilon \, dl^2,$$

where x^μ is the $(D-1)$ spacetime location and l is the extra kime dimension. When using metric signature $(+, -, -, -)$, the sign factor ϵ is taken to be $+1$ or -1 corresponding to timelike or spacelike hypersurfaces. The 5D spacekime metric is conformally related to the induced Minkowski 4D metric by:

$$dS^2 = \Omega \sum_0^{D-2} \sum_0^{D-2} g_{\alpha\beta}(x^\mu, l) dx^\alpha dx^\beta + \epsilon dl^2.$$

The warping conformal factor is the constant Ω and $g_{\alpha\beta}(x^\mu, l)$ is the physical metric on the embedded hypersurface of one lower dimension (foliating leaf). For example, using metric signature $(+, -, -, -)$, the *flat* 4D Minkowski spacetime metric relative to the coordinates $x^\mu = (ct, x, y, z)$ is:

$$g = \begin{bmatrix} 1 & 0 & 0 & 0 \\ 0 & -1 & 0 & 0 \\ 0 & 0 & -1 & 0 \\ 0 & 0 & 0 & -1 \end{bmatrix}.$$

Another example is the Schwarzschild metric, which describes the spacetime around a heavy spherically-symmetric body, such as a planet or a black hole. Using spherical coordinates $x^\mu = (ct, r, \theta, \varphi)$, we can write the Schwarzschild metric as:

$$g = \begin{bmatrix} \left(1 - \frac{2GM}{rc^2}\right) & 0 & 0 & 0 \\ 0 & -\left(1 - \frac{2GM}{rc^2}\right)^{-1} & 0 & 0 \\ 0 & 0 & -r^2 & 0 \\ 0 & 0 & 0 & -r^2 \sin^2\theta \end{bmatrix},$$

where G is the gravitational constant, M represents the total mass-energy content of the central object, and (θ, φ) represent the two rotational degrees of freedom.

In 5D space-time-matter theory, $\Omega = \left(\frac{l}{L}\right)^2$ is the conformal factor, and L is a constant length defined in terms of the cosmological constant $\Lambda = -\epsilon \frac{3}{L^2}$, where, in the metric signature $(+, -, -, -)$, $\Lambda > 0$ for a spacelike extra coordinate and $\Lambda < 0$ for a timelike extra fifth coordinate. Thus, the *canonical metric* can be expressed as:

$$dS^2 = \frac{l^2}{L^2} \sum_0^{D-2} \sum_0^{D-2} g_{\alpha\beta}(x^\mu, l) dx^\alpha dx^\beta + \epsilon dl^2.$$

In simpler terms, 5D spacekime relativity is similar to the standard 4D spacetime relativity, except for an extra quadratic factor in the second kime dimension which augments the spacetime components. This extra factor is orthogonal to spacetime and determines the embedding of the classical Minkowski spacetime into spacekime.

This *canonical metric* is often used to represent the motion of particles. It suggests the existence of an extra force acting on the particles that can be measured by a foliation observer in a hypersurface leaf of one lower dimension.

Let's derive this extra force using the canonical metric. Consider a line element and metric in 5D given by:

$$dS^2 = \sum_{A=0}^{4} \sum_{B=0}^{4} g_{AB} dx^A dx^B,$$

which contains a 4D subspace counterpart:

$$ds^2 = \sum_{\alpha=0}^{3} \sum_{\beta=0}^{3} g_{\alpha\beta} dx^\alpha dx^\beta.$$

In general, the 5D and 4D metric tensors, g_{AB} and $g_{\alpha\beta}$ are functions of all five coordinates, which we label as $x^0 = t$ (time), $x^{1,2,3} = x, y, z$ (space), and $x^4 = l$ (the extra kime-dimension), but act on 5D and 4D objects, respectively. Normalizing the square interval, we use measuring units that absorb the speed of light and assume unitary mass, i.e., $c = 1,\ m = 1$.

Assuming that the extra dimension is timelike ($\epsilon = +1$), the 5D interval in the 5D canonical metric is:

$$dS^2 = \frac{l^2}{L^2} \sum_0^{3} \sum_0^{3} g_{\alpha\beta}(x^\mu, l) dx^\alpha dx^\beta + dl^2, \tag{5.1}$$

where L is a constant length related to the 4D cosmological constant.

Then, the induced 4D spacetime interval is:

$$ds^2 = \sum_0^{3} \sum_0^{3} g_{\alpha\beta}(x^\mu, l) dx^\alpha dx^\beta.$$

Hence, we can express the relation between the 5D canonical metric and the induced 4D metric by:

$$dS^2 = \frac{l^2}{L^2} ds^2 + dl^2.$$

This implies that the 5D coordinates can be constructed by taking a 4D hypersurface in the 5D manifold and regarding the lines normal to this hypersurface as the extra coordinate. These lines will be geodesics curves with proper length equal to the value of the fifth coordinate $x^4 = l$.

This allows a continuously dense foliation of spacekime by 4D spacetime hypersurfaces that depend on the choice of the initial 4D hypersurface embedded in the 5D spacekime manifold.

Dividing the equation $ds^2 = \sum_0^3 \sum_0^3 g_{\alpha\beta}(x^\mu, l) dx^\alpha dx^\beta$ by the interval ds^2 will normalize the 4-velocities $u^\mu \equiv \frac{dx^\mu}{ds}$:

$$\sum_0^3 \sum_0^3 g_{\alpha\beta}\, u^\alpha\, u^\beta \equiv g_{\alpha\beta}\, u^\alpha\, u^\beta = 1.$$

Mind the somewhat awkward notation where s represents proper time, whereas ds^2 represents the Lorentz invariant spacetime interval.

Recall from **Chapter 3, Section 3.7.2** (*Kime Motion Equations*) the discussion of the Newtonian equations of motion, the Lagrangian and Eulerian frames of reference equations of motion, and the most general spacekime motion equations. The kime equations of motion in **Chapter 3** generalized the Newtonian equations of motion in the Special Theory of Relativity (STR) into 5D spacekime.

We will now derive the Lagrange-framework equations of motion by maximizing the distance (integral along the geodesic path) between two points in 5D via $d\left[\int dS\right] = 0$. If y is an arbitrary affine parameter along the geodesic path (change of variables setting), this relation can be written as $d\left[\int \mathcal{L}\, dy\right] = 0$. Using equation (5.1), the quantity $\mathcal{L}$ which is commonly referred to as the *Lagrangian* is:

$$\mathcal{L} \equiv \frac{dS}{dy} = \left[\frac{l^2}{L^2} \sum_0^3 \sum_0^3 g_{\alpha\beta} \frac{dx^\alpha}{dy}\frac{dx^\beta}{dy} + \left(\frac{dl}{dy}\right)^2\right]^{\frac{1}{2}} = \left[\frac{l^2}{L^2} g_{\alpha\beta} \frac{dx^\alpha}{dy}\frac{dx^\beta}{dy} + \left(\frac{dl}{dy}\right)^2\right]^{\frac{1}{2}}.$$

With respect to the new parameter variable y, a 5D path is described by $x^\alpha = x^\alpha(y)$, $l = l(y)$.

In a Lagrangian framework, the momenta of 4D spacetime ($\alpha = 0, 1, 2, 3$) and the extra kime-dimension (l) are:

$$p_\alpha = \frac{\partial \mathcal{L}}{\partial\left(\frac{dx^\alpha}{dy}\right)}, \quad p_l = \frac{\partial \mathcal{L}}{\partial\left(\frac{dl}{dy}\right)}.$$

And the equations of motion in a Lagrangian framework are given as usual by:

$$\frac{\partial \mathcal{L}}{\partial x^\alpha} \underbrace{=}_{\substack{\text{Euler – Lagrange}\\ \text{equation}}} \frac{d}{dy}\underbrace{\frac{\partial \mathcal{L}}{\partial\left(\frac{dx^\alpha}{dy}\right)}}_{p_\alpha} = \frac{dp_\alpha}{dy}, \tag{5.2}$$

$$\frac{\partial \mathcal{L}}{\partial l} \underbrace{=}_{\substack{\text{Euler – Lagrange}\\ \text{equation}}} \frac{d}{dy}\underbrace{\frac{\partial \mathcal{L}}{\partial\left(\frac{dl}{dy}\right)}}_{p_l} = \frac{dp_l}{dy} \tag{5.3}$$

Let's define $u^\alpha \equiv \frac{dx^\alpha}{dy}$ $(\alpha = 0, 1, 2, 3)$, $u^l \equiv \frac{dl}{dy}$, and $\theta \equiv \mathcal{L}^2 = \frac{l^2}{L^2}\sum_0^3\sum_0^3 g_{\beta\gamma}u^\beta u^\gamma + \left(u^l\right)^2$. Note that $u^\beta = \sum_{\alpha=0}^3 \delta_\alpha^\beta u^\alpha = \sum_{\gamma=0}^3 \delta_\gamma^\beta u^\gamma$, $u^\gamma = \sum_{\alpha=0}^3 \delta_\alpha^\gamma u^\alpha = \sum_{\beta=0}^3 \delta_\beta^\gamma u^\beta$, $\frac{\partial u^\beta}{\partial u^\alpha} = \delta_\alpha^\beta$, and we can express the momenta p_α and p_l as:

$$p_\alpha = \frac{\partial \mathcal{L}}{\partial\left(\frac{dx^\alpha}{dy}\right)} = \frac{\partial}{\partial u^\alpha}\theta^{\frac{1}{2}} = \left(\frac{1}{2\theta^{\frac{1}{2}}}\right)\frac{\partial\theta}{\partial u^\alpha} = \left(\frac{1}{2\theta^{\frac{1}{2}}}\right)\left(\frac{l^2}{L^2}\right)\left(\sum_{\beta=0}^3\sum_{\gamma=0}^3 g_{\beta\gamma}\underbrace{\frac{\partial u^\beta}{\partial u^\alpha}}_{\delta_\alpha^\beta}u^\gamma + \sum_{\beta=0}^3\sum_{\gamma=0}^3 g_{\beta\gamma}u^\beta\underbrace{\frac{\partial u^\gamma}{\partial u^\alpha}}_{\delta_\alpha^\gamma}\right) =$$

$$\left(\frac{1}{2\theta^{\frac{1}{2}}}\right)\left(\frac{l^2}{L^2}\right)\left(\sum_{\beta=0}^3\sum_{\gamma=0}^3 g_{\beta\gamma}\delta_\alpha^\beta u^\gamma + \sum_{\beta=0}^3\sum_{\gamma=0}^3 g_{\beta\gamma}\delta_\alpha^\gamma u^\beta\right) =$$

$$\left(\frac{1}{2\theta^{\frac{1}{2}}}\right)\left(\frac{l^2}{L^2}\right)\left(\sum_{\gamma=0}^3\left(\underbrace{\sum_{\beta=0}^3 g_{\beta\gamma}\delta_\alpha^\beta}_{g_{\alpha\gamma}}\right)u^\gamma + \sum_{\beta=0}^3\left(\underbrace{\sum_{\gamma=0}^3 g_{\beta\gamma}\delta_\alpha^\gamma}_{g_{\beta\alpha}}\right)u^\beta\right) =$$

$$\left(\frac{1}{2\theta^{\frac{1}{2}}}\right)\left(\frac{l^2}{L^2}\right)\left(\sum_{\beta=0}^3 g_{\beta\alpha}u^\beta + \sum_{\gamma=0}^3 g_{\gamma\alpha}u^\gamma\right) \underbrace{=}_{\substack{\beta \text{ and } \gamma\\ \text{index symmetries}}}$$

$$\left(\frac{1}{2\theta^{\frac{1}{2}}}\right)\left(\frac{l^2}{L^2}\right)\left(2\sum_{\beta=0}^3 g_{\beta\alpha}u^\beta\right) \underbrace{=}_{g_{\beta\alpha}=g_{\alpha\beta}} \left(\frac{1}{\theta^{\frac{1}{2}}}\right)\left(\frac{l^2}{L^2}\right)\left(\sum_{\beta=0}^3 g_{\alpha\beta}u^\beta\right), 1 \le \alpha \le 3 \tag{5.4}$$

and

$$p_l = \frac{\partial \mathcal{L}}{\partial\left(\frac{dl}{dy}\right)} = \frac{\partial}{\partial u^l}\theta^{\frac{1}{2}} = \left(\frac{1}{2\theta^{\frac{1}{2}}}\right)\frac{\partial\theta}{\partial u^l} = \left(\frac{1}{2\theta^{\frac{1}{2}}}\right)\left(2u^l\right) = \frac{u^l}{\theta^{\frac{1}{2}}}. \tag{5.5}$$

Substituting the expression (5.4) for p_α in the right hand side of equation (5.2) we get:

$$\frac{dp_\alpha}{dy} = \left(\frac{1}{\theta^{\frac{1}{2}}}\right)\left(\frac{l^2}{L^2}\right)\frac{d}{dy}\left(\sum_{\beta=0}^{3} g_{\alpha\beta}u^\beta\right) + \left(\frac{1}{\theta^{\frac{1}{2}}}\right)\frac{d}{dy}\left(\frac{l^2}{L^2}\right)\sum_{\beta=0}^{3} g_{\alpha\beta}u^\beta + \frac{d}{dy}\left(\frac{1}{\theta^{\frac{1}{2}}}\right)\left(\frac{l^2}{L^2}\right)\sum_{\beta=0}^{3} g_{\alpha\beta}u^\beta =$$

$$\left(\frac{1}{\theta^{\frac{1}{2}}}\right)\left(\frac{l^2}{L^2}\right)\frac{d}{dy}\left(\sum_{\beta=0}^{3} g_{\alpha\beta}u^\beta\right) + \left(\frac{1}{\theta^{\frac{1}{2}}}\right)\frac{2l}{L^2}\frac{dl}{dy}\sum_{\beta=0}^{3} g_{\alpha\beta}u^\beta - \frac{1}{2\theta^{\frac{3}{2}}}\frac{d\theta}{dy}\left(\frac{l^2}{L^2}\right)\sum_{\beta=0}^{3} g_{\alpha\beta}u^\beta.$$

As $\theta \equiv \mathcal{L}^2$, the left hand side of equation (5.2) is:

$$\frac{\partial\mathcal{L}}{\partial x^\alpha} = \frac{\partial}{\partial x^\alpha}\theta^{\frac{1}{2}} = \left(\frac{1}{2\theta^{\frac{1}{2}}}\right)\frac{\partial\theta}{\partial x^\alpha} = \left(\frac{1}{2\theta^{\frac{1}{2}}}\right)\left(\frac{l^2}{L^2}\right)\sum_{\beta=0}^{3}\sum_{\gamma=0}^{3}\frac{\partial g_{\beta\gamma}}{\partial x^\alpha}u^\beta u^\gamma.$$

Dividing both sides of equation (5.2) by $\left(\frac{1}{\theta^{\frac{1}{2}}}\right)\left(\frac{l^2}{L^2}\right)$ yields:

$$\frac{d}{dy}\left(\sum_{\beta=0}^{3} g_{\alpha\beta}u^\beta\right) + \frac{2}{l}\left(\frac{dl}{dy}\right)\sum_{\beta=0}^{3} g_{\alpha\beta}u^\beta - \left(\frac{1}{2\theta}\right)\frac{d\theta}{dy}\sum_{\beta=0}^{3} g_{\alpha\beta}u^\beta = \frac{1}{2}\sum_{\beta=0}^{3}\sum_{\gamma=0}^{3}\frac{\partial g_{\beta\gamma}}{\partial x^\alpha}u^\beta u^\gamma, \tag{5.6}$$

which represent the *4D spacetime equations of motion* ($\alpha = 0, 1, 2, 3$).

We can do the same operations for the equations of motion in the extra time dimension (l). Let's substitute the expression (5.5) for p_l in the right hand side of equation (5.3):

$$\frac{dp_l}{dy} = \frac{1}{\theta^{\frac{1}{2}}}\frac{du^l}{dy} + u^l\frac{d}{dy}\left(\frac{1}{\theta^{\frac{1}{2}}}\right) = \frac{1}{\theta^{\frac{1}{2}}}\frac{d^2l}{dy^2} - \frac{1}{2\theta^{\frac{3}{2}}}\frac{d\theta}{dy}\frac{dl}{dy}.$$

Since $\theta \equiv \mathcal{L}^2$, the left hand side of equation (5.3) is:

$$\frac{\partial\mathcal{L}}{\partial l} = \frac{\partial}{\partial l}\theta^{\frac{1}{2}} = \left(\frac{1}{2\theta^{\frac{1}{2}}}\right)\frac{\partial\theta}{\partial l} = \left(\frac{1}{2\theta^{\frac{1}{2}}}\right)\left(\frac{2l}{L^2}\sum_{\alpha=0}^{3}\sum_{\beta=0}^{3} g_{\alpha\beta}u^\alpha u^\beta + \frac{l^2}{L^2}\sum_{\alpha=0}^{3}\sum_{\beta=0}^{3}\frac{\partial g_{\alpha\beta}}{\partial l}u^\alpha u^\beta\right).$$

Again, dividing both sides of equation (5.3) by $\frac{1}{\theta^{\frac{1}{2}}}$ yields:

$$\frac{d^2l}{dy^2} - \left(\frac{1}{2\theta}\right)\frac{d\theta}{dy}\frac{dl}{dy} = \frac{l}{L^2}\sum_{\alpha=0}^{3}\sum_{\beta=0}^{3} g_{\alpha\beta}u^\alpha u^\beta + \frac{1}{2}\left(\frac{l^2}{L^2}\right)\sum_{\alpha=0}^{3}\sum_{\beta=0}^{3}\frac{\partial g_{\alpha\beta}}{\partial l}u^\alpha u^\beta, \tag{5.7}$$

which is the *equation of motion of the extra kime dimension l.*

In proper time, $y = s$ and using $g_{\alpha\beta} = g_{\alpha\beta}(x^\mu, l)$, the first term on the left hand side of equation (5.6) becomes:

$$\frac{d}{ds}\left(\sum_{\beta=0}^{3} g_{\alpha\beta}u^\beta\right) = \sum_{\beta=0}^{3}\frac{dg_{\alpha\beta}}{ds}u^\beta + \sum_{\beta=0}^{3} g_{\alpha\beta}\frac{du^\beta}{ds} =$$

$$\sum_{\gamma=0}^{3}\sum_{\beta=0}^{3}\frac{\partial g_{\alpha\beta}}{\partial x^\gamma}\frac{dx^\gamma}{ds}u^\beta + \sum_{\beta=0}^{3}\frac{\partial g_{\alpha\beta}}{\partial l}\frac{dl}{ds}u^\beta + \sum_{\beta=0}^{3} g_{\alpha\beta}\frac{du^\beta}{ds} =$$

$$\sum_{\gamma=0}^{3}\sum_{\beta=0}^{3}\frac{\partial g_{\alpha\beta}}{\partial x^{\gamma}}u^{\gamma}u^{\beta}+\sum_{\beta=0}^{3}\frac{\partial g_{\alpha\beta}}{\partial l}\frac{dl}{ds}u^{\beta}+\sum_{\beta=0}^{3}g_{\alpha\beta}\frac{du^{\beta}}{ds}.$$

Let's substitute this expression in equation (5.6), simplify by multiplying both sides by $g^{\mu\alpha}$, and sum up over $\alpha=0,1,2,3$:

$$\sum_{\alpha=0}^{3}\sum_{\gamma=0}^{3}\sum_{\beta=0}^{3}g^{\mu\alpha}\frac{\partial g_{\alpha\beta}}{\partial x^{\gamma}}u^{\gamma}u^{\beta}+\sum_{\alpha=0}^{3}\sum_{\beta=0}^{3}g^{\mu\alpha}\frac{\partial g_{\alpha\beta}}{\partial l}\frac{dl}{ds}u^{\beta}+\sum_{\alpha=0}^{3}g^{\mu\alpha}\sum_{\beta=0}^{3}g_{\alpha\beta}\frac{du^{\beta}}{ds}-$$
$$\left[\frac{1}{2\theta}\frac{d\theta}{ds}-\frac{2}{l}\frac{dl}{ds}\right]\sum_{\alpha=0}^{3}g^{\mu\alpha}\sum_{\beta=0}^{3}g_{\alpha\beta}u^{\beta}=\frac{1}{2}\sum_{\alpha=0}^{3}\sum_{\gamma=0}^{3}\sum_{\beta=0}^{3}g^{\mu\alpha}\frac{\partial g_{\beta\gamma}}{\partial x^{\alpha}}u^{\beta}u^{\gamma}.$$

Rearranging the terms, we obtain:

$$\sum_{\alpha=0}^{3}g^{\mu\alpha}\sum_{\beta=0}^{3}g_{\alpha\beta}\frac{du^{\beta}}{ds}+\sum_{\alpha=0}^{3}\sum_{\gamma=0}^{3}\sum_{\beta=0}^{3}g^{\mu\alpha}\left(\frac{\partial g_{\alpha\beta}}{\partial x^{\gamma}}-\frac{1}{2}\frac{\partial g_{\beta\gamma}}{\partial x^{\alpha}}\right)u^{\gamma}u^{\beta}+\sum_{\alpha=0}^{3}\sum_{\beta=0}^{3}g^{\mu\alpha}\frac{\partial g_{\alpha\beta}}{\partial l}\frac{dl}{ds}u^{\beta}=$$
$$\left[\frac{1}{2\theta}\frac{d\theta}{ds}-\frac{2}{l}\frac{dl}{ds}\right]\sum_{\alpha=0}^{3}g^{\mu\alpha}\sum_{\beta=0}^{3}g_{\alpha\beta}u^{\beta}. \tag{5.8}$$

Using symmetries due to index exchange of β and γ, we have $\sum_{\alpha=0}^{3}\sum_{\gamma=0}^{3}\sum_{\beta=0}^{3}\frac{\partial g_{\alpha\beta}}{\partial x^{\gamma}}=\{\beta\to\gamma,\gamma\to\beta\}=\sum_{\alpha=0}^{3}\sum_{\beta=0}^{3}\sum_{\gamma=0}^{3}\frac{\partial g_{\alpha\gamma}}{\partial x^{\beta}}$. Then, the second term on the left hand side of equation (5.8), $\left(\frac{\partial g_{\alpha\beta}}{\partial x^{\gamma}}-\frac{1}{2}\frac{\partial g_{\beta\gamma}}{\partial x^{\alpha}}\right)$, becomes:

$$\left(\frac{1}{2}\frac{\partial g_{\alpha\beta}}{\partial x^{\gamma}}+\frac{1}{2}\frac{\partial g_{\alpha\beta}}{\partial x^{\gamma}}-\frac{1}{2}\frac{\partial g_{\beta\gamma}}{\partial x^{\alpha}}\right)=\frac{1}{2}\left(\frac{\partial g_{\alpha\beta}}{\partial x^{\gamma}}+\frac{\partial g_{\alpha\gamma}}{\partial x^{\beta}}-\frac{\partial g_{\beta\gamma}}{\partial x^{\alpha}}\right).$$

Recall that the classical 4D Christoffel symbols of the second kind are defined by:

$$\Gamma^{\mu}_{\beta\gamma}=\sum_{\alpha=0}^{3}\frac{1}{2}g^{\mu\alpha}\left(\frac{\partial g_{\alpha\beta}}{\partial x^{\gamma}}+\frac{\partial g_{\alpha\gamma}}{\partial x^{\beta}}-\frac{\partial g_{\beta\gamma}}{\partial x^{\alpha}}\right),\ \forall 0\le\mu,\beta,\gamma\le 3.$$

Therefore, the second term on the left hand side of equation (5.8) is:

$$\sum_{\gamma=0}^{3}\sum_{\beta=0}^{3}\left[\sum_{\alpha=0}^{3}g^{\mu\alpha}\left(\frac{\partial g_{\alpha\beta}}{\partial x^{\gamma}}-\frac{1}{2}\frac{\partial g_{\beta\gamma}}{\partial x^{\alpha}}\right)\right]u^{\gamma}u^{\beta}=$$
$$\sum_{\gamma=0}^{3}\sum_{\beta=0}^{3}\left[\sum_{\alpha=0}^{3}\frac{1}{2}g^{\mu\alpha}\left(\frac{\partial g_{\alpha\beta}}{\partial x^{\gamma}}+\frac{\partial g_{\alpha\gamma}}{\partial x^{\beta}}-\frac{\partial g_{\beta\gamma}}{\partial x^{\alpha}}\right)\right]u^{\gamma}u^{\beta}=\sum_{\gamma=0}^{3}\sum_{\beta=0}^{3}\Gamma^{\mu}_{\beta\gamma}u^{\gamma}u^{\beta}.$$

From the definition of the 4-velocity, $u^{\mu}=\frac{dx^{\mu}}{ds}$ and the corresponding dual operator $\langle u^{\mu}|\cdot\rangle=\sum_{\beta=0}^{3}g_{\mu\beta}u^{\beta}=u_{\mu}$, we can simplify the right hand side of equation (5.8):

$$g^{\mu\alpha}g_{\alpha\beta}u^{\beta}=\sum_{\alpha=0}^{3}g^{\mu\alpha}\sum_{\beta=0}^{3}g_{\alpha\beta}u^{\beta}=\sum_{\alpha=0}^{3}g^{\mu\alpha}u_{\alpha}=u^{\mu}.$$

And similarly, the first term of the left hand side of equation (5.8) is:

$$\sum_{\alpha=0}^{3} g^{\mu\alpha} \sum_{\beta=0}^{3} g_{\alpha\beta} \frac{du^{\beta}}{ds} = \sum_{\alpha=0}^{3} g^{\mu\alpha} \frac{du_{\alpha}}{ds} = \frac{du^{\mu}}{ds}.$$

These simplifications and transformations allow us to rewrite equation (5.8) as:

$$\frac{du^{\mu}}{ds} + \sum_{\beta=0}^{3} \sum_{\gamma=0}^{3} \Gamma^{\mu}_{\beta\gamma} u^{\beta} u^{\gamma} = \underbrace{- \sum_{\alpha=0}^{3} \sum_{\beta=0}^{3} g^{\mu\alpha} \frac{\partial g_{\alpha\beta}}{\partial l} \frac{dl}{ds} u^{\beta} + \left[\frac{1}{2\theta} \frac{d\theta}{ds} - \frac{2}{l} \frac{dl}{ds} \right] u^{\mu}}_{\text{fifth force: } f^{\mu}}. \tag{5.9}$$

This contrasts the 4D scenario, where

$$\frac{du^{\mu}}{ds} + \sum_{\beta=0}^{3} \sum_{\gamma=0}^{3} \Gamma^{\mu}_{\beta\gamma} u^{\beta} u^{\gamma} = 0.$$

In equation (5.7), we can substitute proper time, s, for the general geodesic path parameter, y, and employ the velocity normalization condition, $\sum_0^3 \sum_0^3 g_{\alpha\beta} u^{\alpha} u^{\beta} = 1$, to obtain:

$$\frac{d^2 l}{ds^2} - \left(\frac{1}{2\theta}\right) \frac{d\theta}{ds} \frac{dl}{ds} = \frac{l}{L^2} + \frac{1}{2} \left(\frac{l^2}{L^2}\right) \sum_{\alpha=0}^{3} \sum_{\beta=0}^{3} \frac{\partial g_{\alpha\beta}}{\partial l} u^{\alpha} u^{\beta}. \tag{5.10}$$

Therefore, when using 4D proper time, s, as the parameter along the integral path, we have

$\theta = \left(\frac{l^2}{L^2}\right) + \left(\frac{dl}{ds}\right)^2$, and equation (5.10) can be written as:

$$\frac{d^2 l}{ds^2} - \left(\frac{1}{2\theta}\right) \left(\frac{2l}{L^2} \frac{dl}{ds} + 2 \frac{dl}{ds} \frac{d^2 l}{ds^2}\right) \frac{dl}{ds} = \frac{l^2}{L^2} \left[\frac{1}{l} + \frac{1}{2} \sum_{\alpha=0}^{3} \sum_{\beta=0}^{3} \frac{\partial g_{\alpha\beta}}{\partial l} u^{\alpha} u^{\beta} \right],$$

$$\frac{d^2 l}{ds^2} - \frac{1}{\theta} \left[\frac{l}{L^2} + \frac{d^2 l}{ds^2} \right] \left(\frac{dl}{ds}\right)^2 = \frac{l^2}{L^2} \left[\frac{1}{l} + \frac{1}{2} \sum_{\alpha=0}^{3} \sum_{\beta=0}^{3} \frac{\partial g_{\alpha\beta}}{\partial l} u^{\alpha} u^{\beta} \right].$$

Multiplying both sides by $\theta = \left(\frac{l^2}{L^2}\right) + \left(\frac{dl}{ds}\right)^2$, we obtain:

$$\frac{d^2 l}{ds^2} \left[\left(\frac{l^2}{L^2}\right) + \left(\frac{dl}{ds}\right)^2 \right] - \left[\frac{l}{L^2} + \frac{^2 l}{ds^2} \right] \left(\frac{dl}{ds}\right)^2 = \frac{l^2}{L^2} \left[\frac{l^2}{L^2} + \left(\frac{dl}{ds}\right)^2 \right] \left[\frac{1}{l} + \frac{1}{2} \sum_{\alpha=0}^{3} \sum_{\beta=0}^{3} \frac{\partial g_{\alpha\beta}}{\partial l} u^{\alpha} u^{\beta} \right],$$

$$\frac{l^2}{L^2} \frac{d^2 l}{ds^2} - \frac{l}{L^2} \left(\frac{dl}{ds}\right)^2 = \frac{l^2}{L^2} \left[\frac{l^2}{L^2} + \left(\frac{dl}{ds}\right)^2 \right] \left[\frac{1}{l} + \frac{1}{2} \sum_{\alpha=0}^{3} \sum_{\beta=0}^{3} \frac{\partial g_{\alpha\beta}}{\partial l} u^{\alpha} u^{\beta} \right].$$

And dividing both sides by $\frac{l^2}{L^2}$ yields:

$$\frac{d^2l}{ds^2}-\frac{1}{l}\left(\frac{dl}{ds}\right)^2=\left[\frac{l^2}{L^2}+\left(\frac{dl}{ds}\right)^2\right]\left[\frac{1}{l}+\frac{1}{2}\sum_{\alpha=0}^{3}\sum_{\beta=0}^{3}\frac{\partial g_{\alpha\beta}}{\partial l}u^\alpha u^\beta\right]. \tag{5.11}$$

We can now substitute θ in the second term on the right hand side of equation (5.9) to obtain:

$$\left[\frac{1}{2\theta}\frac{d\theta}{ds}-\frac{2}{l}\frac{dl}{ds}\right]u^\mu=\left[\frac{1}{2\theta}\left(\frac{2l}{L^2}\frac{dl}{ds}+2\frac{dl}{ds}\frac{d^2l}{ds^2}\right)-\frac{2}{l}\frac{dl}{ds}\right]u^\mu=\left[\frac{1}{\theta}\left(\frac{l}{L^2}+\frac{d^2l}{ds^2}\right)-\frac{2}{l}\right]\frac{dl}{ds}u^\mu.$$

Using equation (5.11), we get:

$$\frac{\frac{d^2l}{ds^2}-\frac{1}{l}\left(\frac{dl}{ds}\right)^2}{\theta}=\frac{1}{l}+\frac{1}{2}\sum_{\alpha=0}^{3}\sum_{\beta=0}^{3}\frac{\partial g_{\alpha\beta}}{\partial l}u^\alpha u^\beta,\quad \text{(divide both sides by } \theta)$$

$$\frac{\left[\frac{d^2l}{ds^2}+\frac{l}{L^2}\right]-\left[\frac{l}{L^2}+\frac{1}{l}\left(\frac{dl}{ds}\right)^2\right]}{\theta}-\frac{1}{l}=\frac{1}{2}\sum_{\alpha=0}^{3}\sum_{\beta=0}^{3}\frac{\partial g_{\alpha\beta}}{\partial l}u^\alpha u^\beta,$$

$$\frac{1}{\theta}\left(\frac{l}{L^2}+\frac{d^2l}{ds^2}\right)-\frac{2}{l}=\frac{1}{2}\sum_{\alpha=0}^{3}\sum_{\beta=0}^{3}\frac{\partial g_{\alpha\beta}}{\partial l}u^\alpha u^\beta.$$

Therefore,

$$\left[\frac{1}{2\theta}\frac{d\theta}{ds}-\frac{2}{l}\frac{dl}{ds}\right]u^\mu=\frac{1}{2}\frac{dl}{ds}u^\mu\sum_{\alpha=0}^{3}\sum_{\beta=0}^{3}\frac{\partial g_{\alpha\beta}}{\partial l}u^\alpha u^\beta=\sum_{\alpha=0}^{3}\sum_{\beta=0}^{3}\frac{1}{2}u^\mu u^\alpha\frac{dl}{ds}\frac{dx^\beta}{ds}\frac{\partial g_{\alpha\beta}}{\partial l}.$$

Now we can derive the second-order PDE equations of motions by substituting this expression in equation (5.9). Note that the 4D components of the spacekime equations of motion can be written explicitly in terms of the fifth force f^μ measured in units of inertia mass, i.e., assuming $m=1$:

$$\frac{du^\mu}{ds}+\sum_0^3\sum_0^3\Gamma^\mu_{\beta\gamma}\,u^\beta\,u^\gamma=f^\mu,$$

$$f^\mu\equiv\sum_{\alpha=0}^{3}\sum_{\beta=0}^{3}\left(-g^{\mu\alpha}+\frac{1}{2}u^\mu u^\alpha\right)\frac{dl}{ds}\frac{dx^\beta}{ds}\frac{\partial g_{\alpha\beta}}{\partial l}. \tag{5.12}$$

And the 5D component of the spacekime equation of motion (5.11) can be rewritten as a second order differential equation:

$$\frac{d^2l}{ds^2}-\frac{2}{l}\left(\frac{dl}{ds}\right)^2-\frac{l}{L^2}=\frac{1}{2}\left[\frac{l^2}{L^2}+\left(\frac{dl}{ds}\right)^2\right]\sum_{\alpha=0}^{3}\sum_{\beta=0}^{3}u^\alpha u^\beta\frac{\partial g_{\alpha\beta}}{\partial l}. \tag{5.13}$$

Equation (5.12) suggests that in 5D spacekime, the conventional geodesic motion is perturbed by an extra force f^μ. We can decompose this extra force into two parts

$f^\mu = f_\perp^\mu + f_\parallel^\mu$, where $f_\perp^\mu$ is normal to the 4-velocity and $f_\parallel^\mu$ is parallel to the 4-velocity, u_μ. The normal component, $f_\perp^\mu$, is similar to other conventional forces and obeys the usual orthogonality condition $f_\perp^\mu\ u_\mu = 0$. However, the parallel component $f_\parallel^\mu$ has no analog in 4D spacetime. In general, it has a non-trivial inner product with the 4-velocity u^μ, $f_\parallel^\mu u_\mu \neq 0$.

Let's demonstrate this observation by computing the inner product $f_\parallel^\mu u_\mu$. From equation (5.12), we can express the inner product of the extra force and the 4-velocity as:

$$f_\parallel^\mu u_\mu \overset{f_\perp^\mu u_\mu = 0}{\widehat{=}} f^\mu u_\mu = \sum_{\alpha=0}^{3}\sum_{\beta=0}^{3}\left(-g^{\mu\alpha}u_\mu + \frac{1}{2}u^\mu u_\mu u^\alpha\right)\frac{dl}{ds}\frac{dx^\beta}{ds}\frac{\partial g_{\alpha\beta}}{\partial l} =$$

$$\sum_{\alpha=0}^{3}\sum_{\beta=0}^{3}\left(-u^\alpha + \frac{1}{2}u^\alpha\right)\frac{dl}{ds}\frac{dx^\beta}{ds}\frac{\partial g_{\alpha\beta}}{\partial l} = -\frac{1}{2}\sum_0^3\sum_0^3\left(\frac{\partial g_{\alpha\beta}}{\partial l}u^\alpha u^\beta\right)\frac{dl}{ds}. \tag{5.14}$$

Multiplying both hand sides by u^μ and using the normalization condition $u^\mu u_\mu = 1$, the parallel force component can be expressed as:

$$f_\parallel^\mu = -\frac{1}{2}u^\mu\sum_0^3\sum_0^3\left(\frac{\partial g_{\alpha\beta}}{\partial l}u^\alpha u^\beta\right)\frac{dl}{ds}. \tag{5.15}$$

Contextualizing in terms of Heisenberg's uncertainty, we will now directly derive the inner product of the fifth force and 4-velocity $f^\mu u_\mu$ without using the spacekime equations of motion.

First, differentiate $\sum_0^3\sum_0^3 g_{\alpha\beta}(x^\mu, l)u^\alpha u^\beta = 1$ with respect to proper time, s:

$$\sum_{\alpha=0}^{3}\sum_{\beta=0}^{3}\frac{\partial g_{\alpha\beta}}{\partial s}u^\alpha u^\beta + \sum_{\alpha=0}^{3}\sum_{\beta=0}^{3}g_{\alpha\beta}\frac{du^\alpha}{ds}u^\beta + \sum_{\alpha=0}^{3}\sum_{\beta=0}^{3}g_{\alpha\beta}u^\alpha\frac{du^\beta}{ds} = 0,$$

$$\sum_{\gamma=0}^{3}\sum_{\alpha=0}^{3}\sum_{\beta=0}^{3}\frac{\partial g_{\alpha\beta}}{\partial x^\gamma}u^\gamma u^\alpha u^\beta + \sum_{\alpha=0}^{3}\sum_{\beta=0}^{3}\frac{\partial g_{\alpha\beta}}{\partial l}\frac{dl}{ds}u^\alpha u^\beta + 2\sum_{\alpha=0}^{3}\sum_{\mu=0}^{3}g_{\alpha\mu}\frac{du^\mu}{ds}u^\alpha = 0. \tag{5.16}$$

(using symmetry of interchanging the indices α and β)

Again using the α and β index symmetry, we have $\sum_{\alpha=0}^3\sum_{\gamma=0}^3\sum_{\beta=0}^3\frac{\partial g_{\alpha\beta}}{\partial x^\gamma} = \sum_{\alpha=0}^3\sum_{\beta=0}^3\sum_{\gamma=0}^3\frac{\partial g_{\alpha\gamma}}{\partial x^\beta}$. The first term on the left hand side can be rewritten as:

$$\sum_{\gamma=0}^{3}\sum_{\alpha=0}^{3}\sum_{\beta=0}^{3}\frac{\partial g_{\alpha\beta}}{\partial x^\gamma}u^\gamma u^\alpha u^\beta = \sum_{\gamma=0}^{3}\sum_{\alpha=0}^{3}\sum_{\beta=0}^{3}\left(\frac{\partial g_{\alpha\beta}}{\partial x^\gamma} + \frac{\partial g_{\alpha\gamma}}{\partial x^\beta} - \frac{\partial g_{\beta\gamma}}{\partial x^\alpha}\right)u^\gamma u^\alpha u^\beta.$$

By definition, we have:

$$\left(\frac{\partial g_{\alpha\beta}}{\partial x^\gamma} + \frac{\partial g_{\alpha\gamma}}{\partial x^\beta} - \frac{\partial g_{\beta\gamma}}{\partial x^\alpha}\right) = 2g_{\alpha\mu}\Gamma^\mu_{\beta\gamma}.$$

It follows that equation (5.16) can then be expressed as:

$$2\sum_{\alpha=0}^{3}\sum_{\mu=0}^{3} g_{\alpha\mu}u^{\alpha}\underbrace{\left(\frac{du^{\mu}}{ds}+\sum_{\beta=0}^{3}\sum_{\gamma=0}^{3}\Gamma^{\mu}_{\beta\gamma}u^{\beta}u^{\gamma}\right)}_{\text{fifth force}}+\sum_{\alpha=0}^{3}\sum_{\beta=0}^{3}\frac{\partial g_{\alpha\beta}}{\partial l}\frac{\partial l}{\partial s}u^{\alpha}u^{\beta}=0.$$

Then, we can compute the inner product:

$$f^{\mu}_{\parallel}u_{\mu}=f^{\mu}u_{\mu}=\sum_{\alpha=0}^{3}\sum_{\mu=0}^{3} g_{\alpha\mu}u^{\alpha}\underbrace{\left(\frac{du^{\mu}}{ds}+\sum_{\beta=0}^{3}\sum_{\gamma=0}^{3}\Gamma^{\mu}_{\beta\gamma}u^{\beta}u^{\gamma}\right)}_{\text{fifth force}}=-\frac{1}{2}\sum_{0}^{3}\sum_{0}^{3}\left(\frac{\partial g_{\alpha\beta}}{\partial l}u^{\alpha}u^{\beta}\right)\frac{dl}{ds}.$$

This is the same expression as equation (5.14). However, irrespective of the coordinate system reference, the nature of this alternative strategy of deriving the inner product $f^{\mu}_{\parallel}u_{\mu}$ does not involve the canonical metric and yields the general form of the fifth force. In other words, $f^{\mu}_{\parallel}$ is a 4-vector whose spacetime transformations do not involve the fifth coordinate and can be expressed by equation (5.14) generally on the 4D part of 5D spacekime.

If there is no coupling between 4D spacetime and the fifth dimension (kime-phase), the metric tensor will be independent of the extra time coordinate:

$$\frac{\partial g_{\alpha\beta}}{\partial l}=0,\ \forall\alpha,\beta\in\{0,1,2,3\},\ \text{and } f^{\mu}_{\parallel}\text{ is zero.}$$

This is called the *pure-canonical metric* [262], where $g_{\alpha\beta}(x^{\mu},l)=g_{\alpha\beta}(x^{\mu}\text{ only})$, $\mu=0,1,2,3$, and the extra time dimension only enters in 5D through the quadratic warp factor in the extra kime coordinate attached to the 4D spacetime [262].

However, this situation assumes that the 4-velocities are normalized without the conformal factor:

$$\langle u|u\rangle=\sum_{0}^{3}\sum_{0}^{3} g_{\alpha\beta}\ u^{\alpha}\ u^{\beta}=1.$$

The fact that the pure-canonical metric may coincide with the conventional geodesic motion depends on a special choice of the coordinates where $g_{\alpha\beta}(x^{\mu},l)\equiv g_{\alpha\beta}(x^{\mu}\text{ only})$, $\mu=0,1,2,3$.

In the general case, where $g_{\alpha\beta}(x^{\mu},l)$ depends on l, the parallel force $f^{\mu}_{\parallel}$ will be expressed via an unfactorized l-dependent tensor metric, $g'_{\alpha\beta}$ by applying shifting along the extra kime axis, $l\to(l-l_0)$. We can use that kime-shift to transform the pure-canonical metric:

$$dS^2=\frac{l^2}{L^2}ds^2+dl^2,$$

$$ds^2 = \sum_0^3 \sum_0^3 g_{\alpha\beta}(x^\mu) dx^\alpha dx^\beta,$$

$$\Lambda = -\frac{3}{L^2}.$$

Note that the $(+, -\,-\,-, +)$ metric signature assumes the fifth dimension is timelike and the cosmological constant has a negative sign. The *transformed metric* and its corresponding cosmological constant (Λ) may be written as:

$$dS^2 = \frac{(l - l_0)^2}{L^2} ds^2 + dl^2,$$

$$ds^2 = \sum_0^3 \sum_0^3 g_{\alpha\beta}(x^\mu) dx^\alpha dx^\beta,$$

$$\Lambda = -\frac{3}{L^2}\left(\frac{l}{l - l_0}\right)^2.$$

These derivations are based on the Campbell-Wesson embedding theorem and the details are illustrated in Appendix 5.10.

Note that the 5D cosmological constant agrees with its 4D counterpart when $l_0 = 0$, however in the general 4D → 5D embedding, Λ depends on the value of the fifth coordinate, l. The complete details of the derivation extending the 4D solutions of Einstein's equations to their corresponding 5D solutions of the Ricci-flat field equations are provided in Mashhoon and Wesson's paper [264]. The derivation of the *transformed metric* and its corresponding cosmological constant (Λ) in Appendix 5.10 represents a special case of the Campbell 4D → 5D embedding theorem [265–268]. Campbell's embedding theorem implies that an analytic n-dimensional Riemannian space $V_n(s, t)$, where $n = s + t$, s is the number of spacelike and t the number of timelike dimensions, can be locally embedded in a corresponding $(n+1)$-dimensional Ricci-flat Riemannian space $V_{n+1}{}'(s', t')$, where $(s', t') \equiv (s, t+1)$ or $(s', t') = (s+1, t)$. This result suggests that solutions to the n-dimensional Einstein field equations with arbitrary energy-momentum tensor can be locally embedded as solutions in a higher $(n+1)$-dimensional space-kime-like Ricci-flat space [257, 266].

Then the 5D distance interval can be expressed in terms of the unfactorized metric tensor:

$$dS^2 = \sum_0^3 \sum_0^3 g'_{\alpha\beta}(x^\mu, l) dx^\alpha dx^\beta + \epsilon dl^2,$$

where

$$g'_{\alpha\beta}(x^\mu, l) = \frac{(l-l_0)^2}{L^2} g_{\alpha\beta}(x^\mu), \quad \underbrace{\epsilon = \pm 1}_{\text{metric signature}} .$$

Using the shifted pure-canonical metric, the normalization condition of the 4-velocities is:

$$\langle u|u\rangle = \sum_0^3 \sum_0^3 g'_{\alpha\beta}(x^\mu, l) u^\alpha u^\beta = 1.$$

Then, the fifth force has the same form as equation (5.15), with the pure-canonical metric $g_{\alpha\beta}(x^\mu, l)$ replaced by the shifted pure-canonical metric $g'_{\alpha\beta}(x^\mu, l)$:

$$f_\parallel^\mu = -\frac{1}{2} \sum_0^3 \sum_0^3 \left(\frac{\partial g'_{\alpha\beta}}{\partial l} u^\alpha u^\beta \right) \frac{dl}{ds} u^\mu .$$

The particle-wave duality of electrons in 4D spacetime depends on distribution theory and involves uncertainty. Now, the lift of 4D spacetime to 5D spacekime provides a more structured explanation by applying the null-path conditions (i.e., wave traveling along the membrane, following lightlike geodesics, $dS^2 = 0 \Rightarrow \frac{dl}{l-l_0} \approx \frac{ds}{L}$) to the shifted pure-canonical metric:

$$dS^2 = \frac{(l-l_0)^2}{L^2} ds^2 + \epsilon dl^2, \text{where } \epsilon = +1 \text{ for timelike and}$$
$$\epsilon = -1 \text{ for spacelike fifth dimension.}$$

This yields

$$l \,\text{spacelike} \Rightarrow \frac{(l-l_0)^2}{L^2} ds^2 - dl^2 = 0 \Rightarrow ds = \pm \frac{L}{l-l_0} dl \Rightarrow l = l_0 + l_* e^{\pm \frac{s}{L}}, \tag{5.17}$$

$$l \,\text{timelike} \Rightarrow \frac{(l-l_0)^2}{L^2} ds^2 + dl^2 = 0 \Rightarrow ds = \pm i \frac{L}{l-l_0} dl \Rightarrow l = l_0 + l_* e^{\pm \frac{is}{L}} . \tag{5.18}$$

The path $l = l(s)$ is defined by the shift constant l_0, the constant amplitude of the wave l_*, and the wavelength L. For spacelike l, the path *moves away* from an l-hypersurface, $l = l_0 + l_* e^{\pm \frac{s}{L}}$, and the motion is monotonic. For timelike l, the path *oscillates* around a 4D spacetime hypersurface, $l = l_0 + l_* e^{\pm \frac{is}{L}}$. Equation (5.17) is particlelike while equation (5.18) is wavelike and the distinction between both reflects free monotonic motion vs. confined oscillatory motion.

In 5D spacekime, the extra kime coordinate is timelike, which results in the presence of wave oscillations in the vacuum that have similar properties as de Broglie waves associated with particles of mass m and Compton wavelength $L = \frac{h}{mc}$. The extra fifth kime coordinate causes the appearance of a hypersurface at $l = l_0$ where the energy density of the vacuum formally diverges.

For waves traveling along the hypersurface membrane (lightlike cone), where $dS^2 = 0$, i.e., $\frac{dl}{(l-l_0)} \sim \frac{ds}{L}$, the scalar coupling inner product term in $f_\parallel^\mu$ is:

$$\sum_0^3\sum_0^3\left(\frac{\partial g'_{\alpha\beta}}{\partial l}u^\alpha u^\beta\right)=\sum_0^3\sum_0^3 2(l-l_0)L^{-2}g_{\alpha\beta}(x^\mu)u^\alpha u^\beta=\frac{2}{l-l_0}\sum_0^3\sum_0^3 g'_{\alpha\beta}(x^\mu,l)u^\alpha u^\beta=\frac{2}{l-l_0}.$$

Therefore, using equation (5.15), the parallel component of the fifth force can be expressed as:

$$f_\parallel^\mu=-\frac{1}{2}u^\mu\underbrace{\sum_0^3\sum_0^3\left(\frac{\partial g'_{\alpha\beta}(x^\mu,l)}{\partial l}u^\alpha u^\beta\right)}_{\frac{2}{l-l_0}}\underbrace{\frac{dl}{ds}}_{\frac{dl}{ds}\sim\frac{(l-l_0)}{L}}=-\frac{u^\mu}{l-l_0}\frac{(l-l_0)}{L}=-\frac{u^\mu}{L}=-\frac{1}{L}\frac{dx^\mu}{ds}.$$

Recall that $\frac{du^\mu}{ds}+\sum_0^3\sum_0^3\Gamma^\mu_{\beta\gamma}u^\beta u^\gamma=f^\mu\equiv f_\parallel^\mu+f_\perp^\mu$, $\frac{dp^\mu}{ds}=f_\parallel^\mu$, $f_\perp^\mu\underbrace{dx_\mu}_{u_\mu}=0$, and we have,

$$\frac{dp^\mu}{ds}=-\frac{1}{L}\frac{dx^\mu}{ds},$$

$$dp^\mu=-\frac{1}{L}dx^\mu.$$

Then, the inner product,

$$\langle dp|dx\rangle=dp^\mu\ dx_\mu,\ \frac{dx^\mu dx_\mu}{L}=\frac{ds^2}{L}.$$

Therefore, since the fifth force f^μ is measured in unitary inertia mass ($m=1$) and unitary speed of light ($c=1$) near the leaf membrane hypersurface, we have

$$\langle dp|dx\rangle=\sum_{\mu=0}^3 dp^\mu\ dx_\mu=\pm L\left(\frac{dl}{l-l_0}\right)^2=\pm\frac{h}{m\ c}\left(\frac{dl}{l-l_0}\right)^2\sim\pm h.$$

This relation is analogous to the quantum mechanics uncertainty principle in 4D Minkowski spacetime; however, it is derived from 5D Einstein deterministic dynamics. In other words, in spacetime, Heisenberg's uncertainty principle manifests simply because of lack of sufficient information about the second kime dimension, l. In Minkowski 4D spacetime, the lack of kime-phase information naturally leaves one degree of freedom in the system, which manifests as Heisenberg's uncertainty.

This argument about a higher-dimensional deterministic formulation of the 4D spacetime observation of the Heisenberg's principle also supports the de Broglie–Bohm theory [269, 270], which provides an explicit deterministic model of a system configuration and its corresponding wavefunction. It suggests that in a higher-dimensional extension of the observed probabilistic spacetime, the universe may be represented by objective deterministic principles, which add degrees of freedom, hidden variables, or unobservable characteristics to measurements and experiences in the lower-dimensional spacetime [271]. Bell's theorem [84] suggests that any deterministic hidden-variable theory, which is consistent with quantum mechanics predictions, has to be non-local. In other words, valid hidden variable theories must be non-local, permit quantum entanglement, and allow particles to be directly influenced by both neighboring surroundings and spatially-remote objects. This implies the existence of instantaneous, faster than the speed of light, interactions between particles that are significantly separated in 3D space (non-local relations) [272].

5.5 Fundamental Law of Data Science Inference

Let's try now to formulate a **fundamental law of data science inference** that delineates the *data-to-inference* duality. The inference state of a data science problem (a system/question) is described by an inference function ψ, which depends on an observed dataset ($O = \{X, Y\}$) and a specific analytical strategy (ζ):

$$\text{data state} \sim \underbrace{\psi_\zeta(O) \equiv \psi_\zeta(X, Y) \equiv \psi(X, Y|\zeta)}_{\text{inference}} .$$

The inference function does not directly yield a final analytical decision or suggest a specific practical action based on the data. Yet, it encodes the entire distribution of the decision space, i.e., it represents the problem system in a probabilistic sense. Thus, data science can only report *probabilistically* the inference outcome of an analytic experiment. In principle, this likelihood can be computed from the inference function; however, sometimes, when the analytical scheme is complex, this calculation may be complicated or even intractable. Other times, when the analytical approach permits a quick, efficient, and robust solution (e.g., parametric linear models), this probability may be easily obtained.

For instance, the inference square-modulus $\|\psi_\zeta(X, Y)\|^2 dX$ is the probability that a data measurement (X, Y) and a specific analytical strategy (ζ) yield a resulting inference (e.g., forecast) in the interval $X \to X + dX$. Thus, $\|\psi_\zeta(X, Y)\|^2$ represents a probability per unit length, i.e., a probability density function. The total probability of finding something about the data, i.e., some inference, may need to be restricted to be unitary:

$$\|\psi_\zeta\|^2 = \int \|\psi_\zeta(X, Y)^2\| dX = 1.$$

Recall that any square integrable function ($\psi_\zeta \in L^2$) can be normalized by multiplying it by an appropriate (normalization) constant, $c = \frac{1}{\|\psi_\zeta\|^2}$. In practice, two inference functions that differ by an arbitrary factor $c \in \mathbb{C}$ may in fact describe the same data science analytical system.

5.6 Superposition Principle

The inference superposition principle suggests that linear combinations of inference-state vectors for a specific problem yield new inference states that are also admissible, i.e., inference functions represent a space that is closed under linear combinations.

Open Problem

In data science, the **superposition principle** may or may not be valid. This concept may also relate to the notion of data value metrics (DVMs) [273]. DVMs try to quantify the energy of a dataset by uniquely decomposing the inference errors (e.g., discrepancies between theoretical population parameters and their statistical, sample-driven, analytical strategy estimates) into three independent components [274]. For example, suppose, θ and $\hat{\theta}$ represent a theoretical characteristic of interest (e.g., population mean) and its sample-based parameter estimate (e.g., sample arithmetic average), respectively. Then, the canonical decomposition of the inference error may be expressed as:

$$\theta - \hat{\theta} = \underbrace{A}_{\substack{\text{data}\\ \text{quality}}} + \underbrace{B}_{\substack{\text{data}\\ \text{quantity}}} + \underbrace{C}_{\substack{\text{inference problem}\\ \text{complexity}}} .$$

If the superposition principle is valid for inference functions, it may suggest that linear operations (addition, multiplication, differentiation, integration, etc.) act linearly on inference functions.

A simple 1D system provides an example that may help us better understand how the information about a data science analytical system is encoded in the inference function. The inference function describes the outcome states of a specific data analytic problem. Suppose the problem (system) is to use the data and classify, or predict, patients into six (ordinal) discretized disease states labeled:

$$0\,(control),\ 1\,(mild),\ 2\,(moderate),\ 3(elevated),\ 4(high),\ 5\,(severe).$$

Let's denote by ϵ the lattice spacing representing the distance between any two of these ordinal state labels.

For a specific analytical strategy (ζ), the probabilistic interpretation of the inference function suggests that given the observed data $\{X, Y\}$, the probability of a patient to be classified (or predicted) to be in a certain clinical state $0 \le S \le 5$ is:

$$\left|\psi_\zeta(S, X, Y)\right|^2 \times \epsilon.$$

The multiplicative factor ϵ in the inference function is needed because the square of the function modulus is the probability density *per unit length*. Hence, to find the probability, we need to multiply it by the distance between two neighboring point states. To simplify, we can redefine $\psi_{\zeta,S} = \sqrt{\epsilon}\psi_\zeta(S, X, Y)$, $0 \le S \le 5$, and the complete information about this analytical inference problem (system) will be encoded in the 6D ket as a complex inference-state vector:

$$|\psi_{\zeta,S}\rangle = (\psi_0, \psi_1, \psi_2, \psi_3, \psi_4, \psi_5).$$

Of course, there is no significance in the fact that we have chosen an example with six inference outcome states. The same approach applies to (binary) classification or (continuous) regression problems. Note that there is no explicit X dependence in the ket vector, $|\psi_{\zeta,S}\rangle$. The values of $\psi_{\zeta,S}$ at different cross-sectional observations, i.e., spatial points, are the components of the state vector. As necessary, we can also use $|\psi_\zeta(S)\rangle$ to indicate the explicit inference dependence of the outcome states.

Let's examine the inference-state vector, $|2\rangle = (0, 0, 1, 0, 0, 0)$, which represents the inference for case x_2. An inferential label at x_2 embodies the probability $= 1$ classification inference of being a label x_2, and a probability $= 0$ for any other class label. Therefore, $|2\rangle = (0, 0, 1, 0, 0, 0)$. Moreover, any state of the form $|2\rangle = (0, 0, z, 0, 0, 0)$, $|z|^2 = 1$ indicates a localized inferential solution. In this simple discrete inference system, we can see explicitly that *the state of the system is represented by a 6D vector*, which is an important concept that generalizes to higher dimensions.

The normalization condition encodes the fact that the total probability of inferring the class label of a patient, somewhere between the six label states, must integrate to one, i.e.,

$$\langle\psi|\psi\rangle = \sum_{i=0}^{5} |\psi_i|^2 = 1.$$

This illustrates the fact that $|\psi\rangle$ is a complex vector with unitary norm. In general, the coherence between a pair of inference functions may be computed as $\langle\phi|\psi\rangle = \sum_{i=0}^{5} \phi_i^* \psi_i$. The discrete case of inference functions generalizes to inference functions for continuous systems (e.g., regression problems).

Let's now return to the general discussion of the superposition principle, which may provide clues to linearly combining inference-state vectors to obtain new admissible inference states. If ψ_1 and ψ_2 are a pair of inference states, and if any linear mixture also represents a possible inference state:

$$\psi(x,t) = c_1\psi_1(x,t) + c_2\psi_2(x,t), \; c_1, c_2 \in \mathbb{C},$$

then, the space of possible inference states is a closed vector space. The state superposition principle may imply that longitudinal (or even kime) evolution of an inference system may be determined by a linear equation $\mathcal{L}(\psi) = 0$, where the linear operator $\mathcal{L}$ satisfies these linearity conditions:

$$\mathcal{L}(c_1\psi_1(x,t) + c_2\psi_2(x,t)) = c_1\mathcal{L}(\psi_1(x,t)) + c_2\mathcal{L}(\psi_2(x,t)).$$

There is a synergy between inference superposition and various machine learning ensemble methods like sequential boosting, bagging, or parallel random forests. There is a strong empirical evidence that ensemble machine learning techniques do improve the inference results by combining several (typically weaker) models, learners, or classifiers. As aggregating methods, ensemble approaches tend to produce better predictive performance compared to single model learners.

5.7 Terminology

Table 5.1 shows some of the parallels between concepts in quantum mechanics and their corresponding data science counterparts.

Table 5.1: Terminology parallels between data science and quantum physics.

Data science	Physics
An *object* is something that exists by itself, actually or potentially, concretely or abstractly, physically or incorporeal (e.g., person, subject, thing, entity, case, unit, etc.)	A *particle* is a small localized object that permits observations and characterization of its physical or chemical properties
A *feature* is a dynamic *variable* or an attribute about an object that can be measured	An *observable* is a dynamic *variable* about particles that can be measured
Datum is an observed quantitative or qualitative value, an instantiation, of a feature	Particle *state* is an observable particle characteristic (e.g., position, momentum)
Problem, aka *Data system*, is a collection of questions, which may or may not be associated with some *a priori* hypotheses	Particle *system* is a collection of independent particles and observable characteristics, in a closed system
Spacekime (*SK*) Complex-events (*kevents*) are points or states in the spacekime manifold. – *Spacelike* intervals correspond to positive spacetime intervals, $\boldsymbol{ds^2 > 0}$, where an inertial frame can be found such that two kevents $\boldsymbol{a}, \boldsymbol{b} \in \boldsymbol{SK}$ occur simultaneously. An object cannot be present at two kevents which are separated by a spacelike interval – *Lightlike* intervals correspond to $\boldsymbol{ds^2 = 0}$. If two events are on the line of a photon, then they are separated by a lightlike interval and a ray of light could travel between the two events – *Kimelike* intervals correspond to $\boldsymbol{ds^2 < 0}$. An object can be present at two different kevents, which are separated by a kimelike interval Lightlike and kimelike intervals between events suggest that they are causal, i.e., one event influences the other, but spacelike events are not causal	**(Minkowski) Spacetime** *Events* are instantaneous physical situations or occurrences associated with points in spacetime. All events are referenced by their spacetime coordinates (x, t). The Minkowski metric tensor facilitates determining the causal structure of spacetime. Spacetime interval between events are categorized as *spacelike*, *lightlike* and *kimelike* separations

(continued)

Table 5.1 (continued)

Data science	Physics
Dataset (*data*) is an observed instance of a set of datum elements about the problem system, $\boldsymbol{O} = \{\boldsymbol{X}, \boldsymbol{Y}\}$: – The dataset need not be complete, error-free, uni-source, single-scale, homogeneous, or directly computable – Supervised inference requires a clearly defined outcome (**Y**), $\boldsymbol{O} = \{\boldsymbol{X}, \boldsymbol{Y}\}$ – Unsupervised inference assumes no explicit outcome $\boldsymbol{O} = \{\boldsymbol{X}\}$	*State of the system* is an observed measurement of all particles. The *state* is identical to the *wavefunction*, defined below, which represents the probability amplitude of finding the system in that state
Computable data object is a very special representation of a dataset, which allows direct application of computational processing, modeling, analytics, or inference based on the observed dataset	A particle system is *computable* if (1) the entire system is logical, consistent, and complete, and (2) the unknown internal states of the system don't influence the computation (wavefunction, intervals, probabilities, etc.)
Inference function: mathematical description of a solution to a specific data analytic system (problem), which leads to rational evidence-based decision-making. For example, – A linear (e.g., GLM) model represents a solution of a parametric prediction inference problem where the inference function quantifies the effects for all independent features (**X**) on the dependent outcome (**Y**): $\boldsymbol{\psi}(\boldsymbol{O}) = \boldsymbol{\psi}(\boldsymbol{X}, \boldsymbol{Y}) = \widehat{\boldsymbol{\beta}} = \widehat{\boldsymbol{\beta}}^{\boldsymbol{OLS}} =$ $= \langle \boldsymbol{X} \vert \boldsymbol{X}^{-1} \rangle \langle \boldsymbol{X} \vert \boldsymbol{Y} \rangle = \left(\boldsymbol{X}^T \boldsymbol{X}\right)^{-1} \boldsymbol{X}^T \boldsymbol{Y}.$	*Wavefunction*: $\psi(x, t) = Ae^{i(kx - wt)}$ represents a wave traveling in the positive x direction and a corresponding wave traveling in the opposite direction, giving rise to standing waves. Note for $\boldsymbol{x} = (x_1, x_2, \ldots, x_l)$ that: $\left(\frac{\partial^2}{\partial x^2} - \frac{1}{v^2}\frac{\partial^2}{\partial t}\right)\psi(x, t) = 0.$ A complex solution $\psi(x, t) = Ae^{i(kx - wt)}$ represents a traveling wave, where k is the wave number and w is the angular frequency

– A non-parametric, <u>non-linear</u>, alternative inference is SVM classification. If $\boldsymbol{\psi}_{\boldsymbol{x}} \in \boldsymbol{H}$ is the lifting function $\boldsymbol{\psi}:\mathbb{R}^{\eta} \to \mathbb{R}^{\boldsymbol{d}}$ ($\boldsymbol{\psi}:\boldsymbol{x} \in \mathbb{R}^{\eta} \to \tilde{\boldsymbol{x}} = \boldsymbol{\psi}(\boldsymbol{x}) \in \boldsymbol{H}$), where $\eta \ll \boldsymbol{d}$ and the kernel $\psi_{\boldsymbol{x}}(\boldsymbol{y}) = \langle \boldsymbol{x} \vert \boldsymbol{y} \rangle : \boldsymbol{O} \times \boldsymbol{O} \to \mathbb{R}$, the observed data $\boldsymbol{O}_i = \{\boldsymbol{x}_i, \boldsymbol{y}_i\}$ are lifted to $\psi_{\boldsymbol{O}_i}$. Then, the SVM prediction operator is the weighted sum of the kernel functions at $\psi_{\boldsymbol{O}_i}$: $$\langle \boldsymbol{\psi_O} \vert \boldsymbol{\beta}^* \rangle_H = \sum_{i=1}^{n} \boldsymbol{p}_i^* \left\langle \boldsymbol{\psi_O} \vert \boldsymbol{\psi}_{\boldsymbol{O}_i} \right\rangle_{\boldsymbol{H}},$$ see the "*Inference Function*" section above for complete details of this example. The inference always depends on the (input) data, however, it does not have a one-to-one bijective correspondence with the data, as the inference function quantifies the predictions in a probabilistic sense	
A data system is *separable* when the data features $\{f_1, f_2\}$ are independent: $$\boldsymbol{\psi}\left(\boldsymbol{O} = \left\{\boldsymbol{O}_{f_1}, \boldsymbol{O}_{f_2}\right\}\right) = \widehat{\boldsymbol{\beta}} = \begin{pmatrix} \widehat{\boldsymbol{\beta}}_{f_1} \\ \widehat{\boldsymbol{\beta}}_{f_2} \end{pmatrix}$$ Separable data systems are *complete*, i.e., any data system can be represented as a linear combination of separable data. This simplifies the data representation, calculations, and inference. However, separable system representations may not be unique, easy to obtain, or simple to interpret	The wavefunction (particle system $(\{x_1, x_2\})$ is *separable* when: $\psi(x_1, x_2) = \psi(x_1)\psi(x_2)$

(continued)

Table 5.1 (continued)

Data science	Physics
Data are *entangled* when their features are dependent, i.e., the opposite property of separability: $$\psi(\boldsymbol{O}=\{\boldsymbol{O}_{f_1},\boldsymbol{O}_{f_2}\})=\widehat{\beta}(\boldsymbol{f_1},\boldsymbol{f_2})\neq\begin{pmatrix}\widehat{\boldsymbol{\beta}}_{f_1}\\ \widehat{\boldsymbol{\beta}}_{f_2}\end{pmatrix}$$ Although most multi-feature inference functions are not separable, it's common practice to focus on studying separable inference functions because they are mathematically simpler and are complete, i.e., any other inference function may be expressed as a linear superposition of separable inference functions	When a particle-system wavefunction is not separable, the particles are entangled. That is, we can "learn" something about particle x_2 by performing a measurement only on x_1
The total *energy* of a data analytic problem still needs to be defined. It's likely this formulation may involve some of the following components: – The *eigenspectra*, $$\hat{O}\psi_i = O_i\psi_i$$ – Various *data value metrics (DVMs)* [273]	Total energy of a system of κ particles is represented by the Hamiltonian: $$H=\underbrace{-\sum_{i=1}^{\kappa}\frac{\hbar^2}{2m_i}\frac{\partial^2}{\partial x_i^2}}_{\text{Kinetic energy}}+\underbrace{V(x_i,\ldots,x_\kappa)}_{\text{Potential energy}},$$ where m_i represents the mass of the ith particle, and $\hbar$ is the reduced Planck constant

Identical features have the same distributions, i.e., same statistical moments (mean, variance, skewness, kurtosis, etc.) Note that swapping identical features may still have a non-trivial effect on the resulting problem decision obtained via the inference function. This is simply because, randomly rearranging the order of a salient feature will preserve the probability distribution of the feature. However, this reordering of the feature may in fact have significant impact on the analytical strategy leading to the final problem decision	*Identical particles* have the same intrinsic properties (mass, charge, momentum, position, orientation, etc.) Quantum mechanics (QM) permits multiple particles to be identical, i.e., they have all the same intrinsic properties and their joint particle probability density is unchanged under the operation of interchanging a pair of its arguments: $\lvert\psi(x_1, \ldots, x_i, \ldots, x_j, \ldots, x_\eta)\rvert^2 = \lvert\psi(x_1, \ldots, x_j, \ldots, x_i, \ldots, x_\eta)\rvert^2$ Then, two particles, x_i, x_j, cannot be distinguished by their locations, velocities, or by the likelihood of being found at various locations or with various velocities. If the particles are identical and their wavefunction is interchangeable at one particular time, then the time-dependent Schrödinger equation yields that the wavefunction is interchangeable for all possible times. In QM, there are two different ways that interchangeability can be realized: For bosons: $\psi(x_1, \ldots, x_i, \ldots, x_j, \ldots, x_\eta) = +\psi(x_1, \ldots, x_j, \ldots, x_i, \ldots, x_\eta)$. Alternatively, for fermions: $\psi(x_1, \ldots, x_i, \ldots, x_j, \ldots, x_\eta) = -\psi(x_1, \ldots, x_j, \ldots, x_i, \ldots, x_\eta)$
Inference analytics facilitates the computational processing of the inference function For example, suppose we want to predict the class labels for a specific supervised classification problem, given a concrete inference function (e.g., kNN classifier), an objective function, some parameter settings, and a concrete optimization strategy. Under these assumptions, the inference analytics would yield a specific inference result, e.g., forecasting label assignments based on the training observations $\boldsymbol{O} = \{\boldsymbol{X}, \boldsymbol{Y}\}$	*Particle analytics* allow us to quantify aggregate particle states. For instance, given the wavefunction description of the state $\lvert\psi\rangle$, the probability of finding the particle in a region A of volume $d\boldsymbol{x} \subseteq \mathbb{R}^3$ would be: $\int_A \lvert\langle \boldsymbol{x} \vert \psi\rangle\rvert^2 d\boldsymbol{x}$

(continued)

Table 5.1 (continued)

Data science	Physics
In the <u>5D spacekime</u> (SK) manifold, an *interval* $(\Delta s)^2$ is the generalization of the 4D interval, which remains <u>invariant</u> for all inertial reference frames. Because of this invariance, observers in two different inertial frames will always measure identical time- and interval-distances between any two kevents separated in kime by $\Delta\kappa = \kappa - \kappa'$ and in space by distance $\Delta x = x - x'$. Kevents are characterized by their spacekime coordinates $(x, \kappa) = \left(\underbrace{x_1, x_2, x_3}_{space}, \underbrace{\kappa_1, \kappa_2}_{kime}\right) \in SK$. An invariant interval between two kevents, $a, b \in SK$, is defined by: $(\Delta s)^2 = (x_1 - x_1')^2 + (x_2 - x_2')^2 + (x_3 - x_3')^2 - (ic\kappa_1 - ic\kappa_1')^2 - (ic\kappa_2 - ic\kappa_2')^2,$ $ds^2 = (\Delta x_1)^2 + (\Delta x_2)^2 + (\Delta x_3)^2 - c^2(\Delta\kappa_1)^2 - c^2(\Delta\kappa_2)^2.$ Or simply: $ds^2 = (x_1 - x_1')^2 + (x_2 - x_2')^2 + (x_3 - x_3')^2 - c^2(\kappa_1 - \kappa_1')^2 - c^2(\kappa_2 - \kappa_2')^2,$ $ds^2 = (dx_1)^2 + (dx_2)^2 + (dx_3)^2 - c^2(d\kappa_1)^2 - c^2(d\kappa_2)^2,$ where each term represents a classical 1D distance-interval in $\mathbb{R}^1$. These intervals are invariant because another observer using a different coordinate system $(\tilde{x}_1, \tilde{x}_2, \tilde{x}_3, \tilde{\kappa}_1, \tilde{\kappa}_2)\|$, but the same metric, would measure the same interval independent of the coordinate system: $ds^2 = (dx_1)^2 + (dx_2)^2 + (dx_3)^2 - c^2(d\kappa_1)^2 - c^2(d\kappa_2)^2.$ <u>Metric distance nomenclature</u>: There are two sign conventions: Either $\left(\underbrace{+\,+}_{kime}\ \underbrace{-\,-\,-}_{space}\right)$: $ds^2 = c^2(d\kappa_1)^2 + c^2(d\kappa_2)^2 - (dx_1)^2 - (dx_2)^2 - (dx_3)^2.$ Or $\left(\underbrace{+\,+\,+}_{space}\ \underbrace{-\,-}_{kime}\right) ds^2 = (dx_1)^2 + (dx_2)^2 + (dx_3)^2 - c^2(d\kappa_1)^2 - c^2(d\kappa_2)^2.$	**Manifold Metric: Spacetime Interval** In <u>3D, the Euclidean distance</u> Δd between two points $x = (x_1, x_2, x_3)$ and $x' = (x'_1, x'_2, x'_3)$ is defined using the Pythagorean theorem: $(\Delta d)^2 = (\Delta x)^2 + (\Delta y)^2 + (\Delta z)^2 = = (x_1 - x_1')^2 + (x_2 - x_2')^2 + (x_3 - x_3')^2$ In Minkowski <u>4D spacetime</u>, an *interval* $(\Delta s)^2$ is the generalization of 3D distance, which is <u>invariant</u> for all inertial frames. In other words, observers in two different inertial frames will always measure identical time and interval-distance between any two events separated in time by $\Delta t = t - t'$ and in space by distance $\Delta x = x - x'$. The spacetime interval $(\Delta s)^2$ between two events separated by a distance Δx in space and by $\Delta\tau = \Delta ict = ic\Delta t$ in the $\Delta\tau$ coordinate is: $(\Delta s)^2 = (\Delta x)^2 + (\Delta\tau)^2 = (\Delta x)^2 + (\Delta y)^2 + (\Delta z)^2 + (\Delta ict)^2 =$ $(\Delta x)^2 + (\Delta y)^2 + (\Delta z)^2 - (\Delta ct)^2,$ where the constant speed of light, c, converts the time measured units (seconds) into space-distance measured units (meters)

Open Problems

There are potentially many other spacekime concepts that still need to be extended to the field of data science. Some of these examples are included below:

1. Reference Frames

The 4D notion of Minkowski spacetime reference frame, K, in terms of (x, y, z, t) may be extended to a spacekime k-frame of reference K^* in terms of $(x_1, x_2, x_3, k_1, k_2)$.

The 5D transformation between two k-frames, K and K^*, is provided by:

$$\left| \begin{array}{l} x_1 = x \\ x_2 = y \\ x_3 = z \\ k_1 = t\cos(\varphi) \\ k_2 = t\sin(\varphi) \end{array} \right.$$

2. Galilean Transformations

The Galilean transformations between the two inertial k-frames of reference $K(x_1, x_2, x_3, k_1, k_2)$ and $K'(x_1', x_2', x_3', k_1', k_2')$, which are in a standard configuration (i.e., transformations without translations and/or rotations of the space axis in the hyperplane of space and of the time axis in the hyperplane of time), are given as follows:

$$k_1' = k_1$$

$$k_2' = k_2$$

$$x_1' = x_1 - v_1 k_1 = x_1 - v_2 k_2$$

$$x_2' = x_2$$

$$x_3' = x_3$$

where:

$\mathbf{v}_1 = (v_1, 0, 0)$ and $\mathbf{v}_2 = (v_2, 0, 0)$

are the vectors of the velocities of x' against x, defined accordingly in relation to the two kime dimensions.

3. Lorentz Transformations

The spacekime Lorentz transformations between the two inertial k-frames of reference $K(x_1, x_2, x_3, k_1, k_2)$ and $K'(x_1', x_2', x_3', k_1', k_2')$, which are in a standard configuration (i.e., transformations without translations and/or rotations of the space axis in the hyperplane of space and of the kime axis in the hyperplane of kime), are given as follows:

$$k_1' = \left(1 + (\zeta - 1)\frac{c^2}{(v_1)^2}\beta^2\right)k_1 + (\zeta - 1)\frac{c^2}{v_1 v_2}\beta^2 k_2 - \frac{1}{v_1}\beta^2\zeta x_1,$$

$$k_2' = (\zeta - 1)\frac{c^2}{v_1 v_2}\beta^2 k_1 + \left(1 + (\zeta - 1)\frac{c^2}{(v_2)^2}\beta^2\right)k_2 - \frac{1}{v_2}\beta^2\zeta x_1,$$

$$x_1' = -c^2\beta^2\zeta\left(\frac{k_1}{v_1} + \frac{k_2}{v_2}\right) + \zeta x_1,$$

$$x_2' = x_2$$

$$x_3' = x_3$$

where:

$$\mathbf{v}_1 = (v_1, 0, 0) \text{ and } \mathbf{v}_2 = (v_2, 0, 0)$$

are the vectors of the velocities of $\boldsymbol{x}'$ against $\boldsymbol{x}$, defined accordingly in relation to the two kime dimensions k_1, k_2, $\beta = \frac{1}{\sqrt{\frac{c^2}{v_1} + \frac{c^2}{v_2}}}$, $\zeta = \frac{1}{\sqrt{1-\beta^2}}$.

4. Mass, Energy, Velocity, and Momentum

Assume that a particle is moving in the 5D spacekime plane, where the pair of kime dimensions are denoted by k_1, k_2 and the three spatial dimensions are denoted by x_1, x_2, x_3. We can extend the STR concept of *proper time* to *proper kime*, k_0, to reflect the kime measured in the proper frame of reference K_0, where the particle is at rest. Note that the *proper kime interval* is commonly the quantity of interest, since proper kime itself is fixed only up to an arbitrary complex constant, which specifies the initial setting of the kime clock at some point in the 2D kime space. The proper kime interval between two kevents depends not only on the kevents themselves but also on the planar path (kevent trajectory) connecting the pair of kevents. Hence, the *proper kime interval* is a function of the kevent trajectory as a path integral over the kime manifold.

Since the kime manifold is 2D, the proper kime interval $\boldsymbol{dk_0} = (dk_{0,1}, dk_{0,2})$ will be a 2D vector, where $dk_{0,j}$ represents the projections of the proper kime on the pair of kime axes, $1 \le j \le 2$, and the magnitude of the kime interval is

$$dk_0 = ||\boldsymbol{dk_0}|| = \sqrt{(dk_{0,1})^2 + (dk_{0,2})^2} \ge 0.$$

We can similarly extend the STR velocity u to spacekime. Suppose a particle is observed in a pair of (3 + 2)D k-frames, K_0 and K, in positions $\boldsymbol{r_0} = (x_{0,1}, x_{0,2}, x_{0,3}, ck_{0,1}, ck_{0,2})$ and $\boldsymbol{r} = (x_1, x_2, x_3, ck_1, ck_2)$, respectively.

As the particle moves in spacekime, the dynamics of its new locations in K_o and K are represented by $\boldsymbol{r_0} + \boldsymbol{dr_0} = (x_{0,1}, x_{0,2}, x_{0,3}, ck_{0,1} + cdk_{0,1}, ck_{0,2} + cdk_{0,2})$ and $\boldsymbol{r} + \boldsymbol{dr} = (x_1 + dx_1, x_2 + dx_2, x_3 + dx_3, ck_1 + cdk_1, ck_2 + cdk_2)$, respectively.

Let us consider the motion of a particle in relation to the reference frame K. The invariant (3 + 2)-dimensional interval is given through the expression

$$ds^2 = (dx_1)^2 + (dx_2)^2 + (dx_3)^2 - c^2(dk_1)^2 - c^2(dk_2)^2.$$

Since the particle moves in the causal region, this interval is kimelike or lightlike, that is, $ds^2 \le 0$ (see **Chapter 3, Section 3.9.2**). By analogy with STR, we have to assume that the proper kime is invariant. If we consider the quotient of the two invariant quantities ds^2 and c, then we will obtain an invariant value having physical dimension of time. We assume that this value is equal to the length of the vector $\boldsymbol{dk_0}$, an invariant scalar quantity, $\boldsymbol{dk_0} \equiv ||\boldsymbol{dk_0}|| = \frac{\sqrt{-ds^2}}{c}$. Also,

$$ds^2 = -c^2\left(dk_1^2 + dk_2^2\right)\left(1 - \beta^2\right) = -c^2 dk_0^2,$$

where $\beta = \frac{1}{\sqrt{\frac{c^2}{v_1^2} + \frac{c^2}{v_2^2}}}$. The scalar values $dk_{0,1}, dk_{0,2}$ are projections of the proper kime $\boldsymbol{dk_0}$ along the kime axes $k_{0,1}, k_{0,2}$ in the K_0 frame.

In the 5D spacekime, we have $\boldsymbol{dk_0} \times \boldsymbol{u} = \boldsymbol{dr}$, where $\boldsymbol{dr} = (dx_1, dx_2, dx_3, cdk_1, cdk_2)$, and the generalized velocity $\boldsymbol{u} = (u_{\theta,\mu})$ is a second-order tensor, a 2×5 matrix, with elements indexed by $\theta \in \{1, 2\}$, $\mu \in \{1, 2, 3, 4, 5\}$. If $\boldsymbol{u}_{\cdot,\mu} = \begin{pmatrix} u_{1,\mu} \\ u_{2,\mu} \end{pmatrix}$ $\forall 1 \le \mu \le 5$, then $\langle \boldsymbol{dk_0} | \boldsymbol{u}_{\cdot,\eta} \rangle = dx_\eta$, $\forall \eta \in \{1, 2, 3\}$ and $\langle \boldsymbol{dk_0} | \boldsymbol{u}_{\cdot,(\sigma+3)} \rangle = cdk_\sigma$, $\sigma \in \{1, 2\}$, that is,

$$\langle \boldsymbol{dk_0} | \boldsymbol{u}_{\cdot,\eta} \rangle = \frac{dk_{0,1}}{dx_\eta} u_{1,\eta} + \frac{dk_{0,2}}{dx_\eta} u_{2,\eta} = 1,$$

$$\langle \boldsymbol{dk_0} | \boldsymbol{u}_{\cdot,(\sigma+3)} \rangle = \frac{dk_{0,1}}{cdk_\sigma} u_{1,(\sigma+3)} + \frac{dk_{0,2}}{cdk_\sigma} u_{2,(\sigma+3)} = 1.$$

Likewise, a spacekime energy–momentum tensor (2×5 matrix) is $\boldsymbol{P} = m_0 \boldsymbol{u} = (p_{\theta,\mu})_{2\times 5}$, where $m_0 \ge 0$ is the proper mass of the particle at rest ($\boldsymbol{v_1} = \boldsymbol{v_2} = 0$), $\boldsymbol{u} = (u_{\theta,\mu})_{2\times 5}$ is the generalized kime velocity, $u_{\theta,\eta} = \frac{dx_\eta}{dk_{0,\theta}}$, $\eta \in \{1,2,3\}$, and $u_{\theta,(\sigma+3)} = \frac{cdk_\sigma}{dk_{0,\theta}}$, $\sigma, \theta \in \{1,2\}$. Let us consider the momentum components $\boldsymbol{p}_\mu = m_0 \boldsymbol{u}_{\cdot,\mu}$, $p_{\theta,\mu} = m_0 u_{\theta,\mu}$, $\theta \in \{1,2\}$, and $\mu \in \{1,2,3,4,5\}$.

Following the classical STR recipe, multiplying the momentum components $p_{\theta,(\sigma+3)}$, $\sigma, \theta \in \{1,2\}$ by the speed of light in vacuum, c, yields the corresponding components of the particle energy $e_{\theta,\sigma}$. More specifically, for each $\sigma \in \{1,2\}$, the particle energy in kime dimension k_σ is a two-component vector, $\boldsymbol{E}_\sigma = \begin{pmatrix} e_{1,\sigma} \\ e_{2,\sigma} \end{pmatrix}$, where $e_{\theta,\sigma} = c m_0 u_{\theta,(\sigma+3)} = c^2 m_0 \frac{dk_\sigma}{dk_{0,\theta}}$, $\sigma, \theta \in \{1,2\}$.

Thus, the *aggregate energy* in kime dimension k_σ is $E_\sigma \equiv ||\boldsymbol{E}_\sigma|| = c^2 m_0 \frac{dk_\sigma}{dk_0}$, that is,

$$E_\sigma = c^2 m_0 \frac{dk_\sigma}{\sqrt{(dk_{0,1})^2 + (dk_{0,2})^2}}.$$

Therefore, for $\sigma \in \{1,2\}$,

$$\frac{1}{(E_\sigma)^2} = \frac{(dk_{0,1})^2 + (dk_{0,2})^2}{(c^2 m_0)^2 (dk_\sigma)^2} = \frac{1}{\left(\frac{c^2 m_0 dk_\sigma}{dk_{0,1}}\right)^2} + \frac{1}{\left(\frac{c^2 m_0 dk_\sigma}{dk_{0,2}}\right)^2} = \frac{1}{(e_{1,\sigma})^2} + \frac{1}{(e_{2,\sigma})^2} \Rightarrow$$

$$E_\sigma = \frac{1}{\sqrt{\frac{1}{(e_{1,\sigma})^2} + \frac{1}{(e_{2,\sigma})^2}}}.$$

The total energy of the particle moving in spacekime is defined as the scalar $E \equiv ||\boldsymbol{E}|| = c^2 m_0 \frac{dk}{dk_0} = \frac{m_0 c^2}{\sqrt{1-\beta^2}}$, where $dk \equiv ||\boldsymbol{dk}|| = \sqrt{(dk_1)^2 + (dk_2)^2} \geq 0$. Hence,

$$E \equiv ||\boldsymbol{E}|| = \sqrt{(E_1)^2 + (E_2)^2} = \sqrt{\frac{1}{\frac{1}{(e_{1,1})^2} + \frac{1}{(e_{2,1})^2}} + \frac{1}{\frac{1}{(e_{1,2})^2} + \frac{1}{(e_{2,2})^2}}}.$$

In the case of 2D kime, the momentum of the particle, in space the direction x_η, $\eta \in \{1,2,3\}$, has two components $\boldsymbol{p}_\eta = \begin{pmatrix} p_{1,\eta} \\ p_{2,\eta} \end{pmatrix}$, where $p_{\theta,\eta} = m_0 u_{\theta,\eta}$, $\theta \in \{1,2\}$. Hence, in spatial dimension x_η, the magnitude of the particle momentum is a scalar $p_\eta \equiv ||\boldsymbol{p}||_\eta = m_0 \frac{dx_\eta}{dk_0}$, and we have the relation:

$$p_\eta = m_0 \frac{dx_\eta}{\sqrt{(dk_{0,1})^2 + (dk_{0,2})^2}}.$$

So, for each spatial index, $\eta \in \{1, 2, 3\}$,

$$\frac{1}{(p_\eta)^2} = \frac{(dk_{0,1})^2 + (dk_{0,2})^2}{m_0^2 (dx_\eta)^2} = \frac{1}{\left(\frac{m_0 dx_\eta}{dk_{0,1}}\right)^2} + \frac{1}{\left(\frac{m_0 dx_\eta}{dk_{0,2}}\right)^2} = \frac{1}{(p_{1,\eta})^2} + \frac{1}{(p_{2,\eta})^2},$$

which is just

$$p_\eta = \sqrt{\frac{1}{\frac{1}{(p_{1,\eta})^2} + \frac{1}{(p_{2,\eta})^2}}}.$$

The magnitude of the total momentum of the particle moving in the 2D kime plane is a scalar

$$p \equiv ||\boldsymbol{p}|| = m_0 \frac{dx}{dk_0} = \frac{m_0 c\beta}{\sqrt{1-\beta^2}},$$

where $dx = \sqrt{(dx_1)^2 + (dx_2)^2 + (dx_3)^2}$. Thus, we have

$$(dx_1)^2 + (dx_2)^2 + (dx_3)^2 = c^2(dk_1)^2 + c^2(dk_2)^2 - c^2 dk_0^2$$

and

$$\frac{dx}{dk_0} = \frac{\sqrt{(dx_1)^2 + (dx_2)^2 + (dx_3)^2}}{dk_0} = \sqrt{\frac{c^2(dk_1)^2 + c^2(dk_2)^2 - c^2 dk_0^2}{dk_0^2}} = \frac{c\beta}{\sqrt{1-\beta^2}}.$$

In other words,

$$p = ||\boldsymbol{p}|| = \sqrt{\sum_{\eta=1}^{3} (p_\eta)^2} = \sqrt{\sum_{\eta=1}^{3} \left(\frac{1}{\frac{1}{(p_{1,\eta})^2} + \frac{1}{(p_{2,\eta})^2}} \right)}.$$

Let us explicate the relations between the energy tensor $\boldsymbol{E}$, its vector components $\boldsymbol{E}_\sigma$, and its scalar magnitude $||\boldsymbol{E}||$, as well as the connection between the momentum tensor $\boldsymbol{p}$, its vector components $\boldsymbol{p}_\eta$, and its scalar magnitude $||\boldsymbol{p}||$. The generalized velocity in spacekime is

$$\boldsymbol{u} = (u_{\theta,u})_{2\times 5} = \begin{pmatrix} \frac{dx_1}{dk_{0,1}} & \frac{dx_2}{dk_{0,1}} & \frac{dx_3}{dk_{0,1}} & \frac{cdk_1}{dk_{0,1}} & \frac{cdk_2}{dk_{0,1}} \\ \underbrace{\frac{dx_1}{dk_{0,2}} \quad \frac{dx_2}{dk_{0,2}} \quad \frac{dx_3}{dk_{0,2}}}_{\text{Space dimensions}} & & & \underbrace{\frac{cdk_1}{dk_{0,2}} \quad \frac{cdk_2}{dk_{0,2}}}_{\text{Kime dimensions}} & \end{pmatrix}$$

The total energy–momentum matrix in spacekime can be expressed as

$$\boldsymbol{P} = \left(\boldsymbol{p}_1, \boldsymbol{p}_2, \boldsymbol{p}_3, \frac{\boldsymbol{E}_1}{c}, \frac{\boldsymbol{E}_2}{c}\right)_{2\times 5} = \begin{pmatrix} p_{1,1} \; p_{1,2} \; p_{1,3} \; \frac{e_{1,1}}{c} \; \frac{e_{1,2}}{c} \\ p_{2,1} \; p_{2,2} \; p_{2,3} \; \frac{e_{2,1}}{c} \; \frac{e_{2,2}}{c} \end{pmatrix}.$$

And the total energy matrix in spacekime is

$$\boldsymbol{E} = \underbrace{(\boldsymbol{E_1}, \boldsymbol{E_2})}_{\text{Kime dimensions}} = \begin{pmatrix} e_{1,1} \; e_{1,2} \\ e_{2,1} \; e_{2,2} \end{pmatrix}_{2\times 2}.$$

The scalar aggregate energy in kime dimension k_σ is

$$E_\sigma \equiv ||\boldsymbol{E}_\sigma|| = \frac{1}{\sqrt{\frac{1}{(e_{1,\sigma})^2} + \frac{1}{(e_{2,\sigma})^2}}}, \quad \sigma \in \{1, 2\}.$$

Thus, the magnitude of the total energy is

$$E \equiv ||\boldsymbol{E}|| = \sqrt{(E_1)^2 + (E_2)^2}\,.$$

and the total momentum matrix in spacekime is

$$\boldsymbol{p} = \underbrace{(\boldsymbol{p_1}, \boldsymbol{p_2}, \boldsymbol{p_3})}_{\text{Space dimensions}} = \begin{pmatrix} p_{1,1} \; p_{1,2} \; p_{1,3} \\ p_{2,1} \; p_{2,2} \; p_{2,3} \end{pmatrix}_{2\times 3}.$$

The scalar momentum in space dimension x_η is

$$p_\eta \equiv ||\boldsymbol{p}_\eta|| = \sqrt{\frac{1}{\frac{1}{(p_{1,\eta})^2} + \frac{1}{(p_{2,\eta})^2}}}, \quad \eta \in \{1, 2, 3\},$$

and the magnitude of the total momentum is

$$p = ||p|| = \sqrt{(p_1)^2 + (p_2)^2 + (p_3)^2}.$$

Summarizing, this formulation of energy–momentum yields the following relations:

$$\frac{cdk}{dk_0} = \frac{c}{\sqrt{1-\beta^2}} = \frac{E}{m_0 c},$$

$$\frac{dx}{dk_0} = \frac{c\beta}{\sqrt{1-\beta^2}} = \frac{p}{m_0},$$

$$dx^2 - c^2 dk^2 = ds^2 = -c^2 dk_0^2.$$

Using these equations, we obtain the following important equality expressing the relation between particle total energy E, total momentum p, and rest mass m_0:

$$E^2 - p^2 c^2 = m_0^2 c^4.$$

In STR, the *energy–momentum conservation law* is derived as a consequence of the continuous spacetime symmetry (Noether's theorem) [275]. For example, let us consider the process of a particle decay, where the energy–momentum conservation law states that the vector sum of the energy–momentum four-vector of the decay products should equal the corresponding energy–momentum four-vector of the original particle. Some aspects of the problem for conservation of energy and momentum in the multidimensional time are discussed by J. Dorling [113].

One may assume that in the case of 2D kime, the energy may be represented as a 2D vector. However, in the 5D spacekime manifold, the energy is a 2×2 tensor and the momentum is a 2×3 tensor. For particle decay processes in complex time, this leads to the following formulation of the energy–momentum conservation law.

The matrix entrywise sum of the second-order energy–momentum tensor (2×5 matrix) of the decay products is equal to the corresponding energy–momentum tensor (also a 2×5 matrix) of the original particle. Specifically, the matrix entrywise sum of the 2×2 energy matrix of the decay products is equal to the 2×2 energy matrix of the original particle. And similarly, the matrix entrywise sum of the 2×3 momentum matrix of the decay products is equal to the 2×3 momentum matrix of the original particle.

For example, if $\boldsymbol{P}^H = \left(p^H_{\theta,\mu}\right)_{2\times 5}$ denotes the energy–momentum tensor of the original particle (particle H) and accordingly $\boldsymbol{P}^A = \left(p^A_{\theta,\mu}\right)_{2\times 5}, \ldots, \boldsymbol{P}^D = \left(p^D_{\theta,\mu}\right)_{2\times 5}$ are the energy–momentum tensors of the decay by-products (particles $A, \ldots, D$), then we will have:

$$\boldsymbol{P}^H = \boldsymbol{P}^A + \cdots + \boldsymbol{P}^D, \text{ i.e.,}$$

$$p^H_{\theta,\mu} = p^A_{\theta,\mu} + \cdots + p^D_{\theta,\mu} \quad \forall \theta \in \{1,2\}; \quad \mu \in \{1,2,\ldots,5\}.$$

More specifically, the separate energy and momentum conservation laws can be expressed as follows:

$$e^H_{\theta,\sigma} = e^A_{\theta,\sigma} + \cdots + e^D_{\theta,\sigma} \ \forall \theta \in \{1,2\}, \ \sigma \in \{1,2\},$$

$$p^H_{\theta,\eta} = p^A_{\theta,\eta} + \cdots + p^D_{\theta,\eta} \ \forall \theta \in \{1,2\}, \ \eta \in \{1,\ 2,3\}.$$

Thus, the *magnitudes* of the energy and momentum tensors of particle H are

$$E^H = \sqrt{(E^H)_1^2 + (E^H)_2^2} = \sqrt{\frac{1}{\frac{1}{\left(e_{1,1}^A + \cdots + e_{1,1}^D\right)^2} + \frac{1}{\left(e_{2,1}^A + \cdots + e_{2,1}^D\right)^2}} + \frac{1}{\frac{1}{\left(e_{1,2}^A + \cdots + e_{1,2}^D\right)^2} + \frac{1}{\left(e_{2,2}^A + \cdots + e_{2,2}^D\right)^2}}},$$

$$p^H = \sqrt{(p^H)_1^2 + (p^H)_2^2} = \sqrt{\sum_{\eta=1}^{3}\left(\frac{1}{\frac{1}{\left(p_{1,\eta}^A + \cdots + p_{1,\eta}^D\right)^2} + \frac{1}{\left(p_{2,\eta}^A + \cdots + p_{2,\eta}^D\right)^2}}\right)}.$$

Recall that for each of the particles $A, \ldots, D, H$, we have

$$(E^A)^2 - (p^A)^2 c^2 = (m_0^A)^2 c^4$$

$$\vdots$$

$$(E^D)^2 - (p^D)^2 c^2 = (m_0^D)^2 c^4$$

$$(E^H)^2 - (p^H)^2 c^2 = (m_0^H)^2 c^4.$$

Hence, as the particle H decays into sub-particle by-products, $A, \ldots, D$, its energy and rest mass satisfy the following inequalities:

$$\sqrt{(E^H)^2 - (p^H)^2 c^2} \geq \sqrt{(E^A)^2 - (p^A)^2 c^2} + \underbrace{\cdots}_{\forall \text{ sub-particles}} + \sqrt{(E^D)^2 - (p^D)^2 c^2}.$$

More accurate relations, approximations, and computational bounds on the energies may be derived by expanding the energy terms further:

$$(E^H)^2 - (p^H)^2 c^2 =$$

$$\frac{1}{\frac{1}{\left(e_{1,1}^A + \cdots + e_{1,1}^D\right)^2} + \frac{1}{\left(e_{2,1}^A + \cdots + e_{2,1}^D\right)^2}} + \frac{1}{\frac{1}{\left(e_{1,2}^A + \cdots + e_{1,2}^D\right)^2} + \frac{1}{\left(e_{2,2}^A + \cdots + e_{2,2}^D\right)^2}} -$$

$$\sum_{\eta=1}^{3}\left(\frac{1}{\frac{1}{\left(p_{1,\eta}^A + \cdots + p_{1,\eta}^D\right)^2} + \frac{1}{\left(p_{2,\eta}^A + \cdots + p_{2,\eta}^D\right)^2}}\right) c^2 \geq$$

$$\frac{1}{\frac{1}{\left(e_{1,1}^{A}\right)^{2}}+\frac{1}{\left(e_{2,1}^{A}\right)^{2}}}+\frac{1}{\frac{1}{\left(e_{1,2}^{A}\right)^{2}}+\frac{1}{\left(e_{2,2}^{A}\right)^{2}}}-\sum_{\eta=1}^{3}\left(\frac{1}{\frac{1}{\left(p_{1,\eta}^{A}\right)^{2}}+\frac{1}{\left(p_{2,\eta}^{A}\right)^{2}}}\right)c^{2}+\underbrace{\cdots}_{\forall\ \text{sub-particles}}+$$

$$\frac{1}{\frac{1}{\left(e_{1,1}^{D}\right)^{2}}+\frac{1}{\left(e_{2,1}^{D}\right)^{2}}}+\frac{1}{\frac{1}{\left(e_{1,2}^{D}\right)^{2}}+\frac{1}{\left(e_{2,2}^{D}\right)^{2}}}-\sum_{\eta=1}^{3}\left(\frac{1}{\frac{1}{\left(p_{1,\eta}^{D}\right)^{2}}+\frac{1}{\left(p_{2,\eta}^{D}\right)^{2}}}\right)c^{2}+\sum_{i\neq j\in\{A,\ldots,D\}}2m_{0}^{i}m_{0}^{j}c^{4}.$$

Therefore, in the case of complex time, we can derive the counterpart of the classical STR rest mass inequality, stating that the rest mass of the decaying particle exceeds the aggregate sum of the sub-particle rest masses, which guarantees the stability of the particles:

$$m_{0}^{H}\geq m_{0}^{A}+\cdots+m_{0}^{D}.$$

5.8 Spacetime IID vs. Spacekime Sampling

For all natural spacetime processes, various population characteristics like the mean, variance, range, and quantiles can be estimated by collecting IID samples. These samples represent observed data that is traditionally used as a process proxy to obtain sample-driven parameter estimates of specific population characteristics via standard formulas like the *sample* arithmetic average, variance, range, quantiles, etc. The latter approximate their population counterparts and form the basis for classical parametric, non-parametric, and Bayesian statistical inference.

Typically, reliable *spacetime* statistical inference depends on the characteristics of the phenomenon, the distribution of the native process, and the choice of the sample-size. By clever estimation of the (unobserved) process kime-phases, we will demonstrate *spacekime* analytics that rely on a single spacetime observation, instead of a large IID sample. Thus, spacekime representation facilitates effective and reliable inference that resembles the classical spacetime analytics, but relaxes the condition requiring samples with a large number of independent observations.

Without loss of generality, suppose we have a pair of cohorts A and B and we obtain a series of measurements $\{X_{A,i}\}_{i=1}^{n_A}$ and $\{X_{B,i}\}_{i=1}^{n_B}$, respectively. Obviously the relations between the cohorts could widely vary, from being samples of the same process, to representing loosely related or completely independent processes.

There are two extreme cases of cohort pairing: (1) independent and differently distributed cohorts (A and B) or (2) independent and identically distributed (IID) cohorts (A, C, and D). An example of the latter case is a random split of the first cohort (A) into complementary *training* (C) and *testing* (D) groups. This design allows us to compare the classical spacetime-derived population characteristics of cohort A to their spacekime-reconstructed counterparts obtained using a single random kime-magnitude measurement from A and kime-phase estimates derived from cohorts B, C or D.

We can regard the Fourier representation of each dataset X as complex-valued function $\hat{X}=\hat{X}(\omega)$ of the variable $\omega=re^{i\varphi}\in\mathbb{C}$, representing the kime frequency $\boldsymbol{\omega}=(\omega_1,\omega_2,\omega_3,\ldots,\omega_{n_X})$, kime magnitude $\boldsymbol{r}=(r_1,r_2,r_3,\ldots,r_{n_X})$, and kime phase $\boldsymbol{\varphi}_X=(\varphi_1,\varphi_2,\varphi_3,\ldots,\varphi_{n_X})$. We will manipulate the kime magnitudes (r) and the kime-phases (φ), reconstruct the signal in spacetime $\left(\hat{\hat{X}}\right)$, and compare the distributions of the original (X) and spacekime reconstructed data $\left(\hat{\hat{X}}\right)$. The specific spacekime operations include replacing the original kime-magnitudes (r_j) by $r_j'=r_{j_o}$, $1\le j\le n_X$, for some randomly chosen r_{j_o}, and swapping the X signal kime phases θ_j by the phases of another dataset Y, $\boldsymbol{\varphi}_Y'=(\varphi_1',\varphi_2',\varphi_3',\ldots,\varphi_{n_Y}')$. Spacekime analytics based on a single spacetime observation corresponding to a unique kime-magnitude r_{j_o}, use prior knowledge about the kime-phases that can be estimated from another dataset Y. In our simulation case, this prior knowledge is represented by using the kime-phases derived from cohorts B, C, and D. Each of the three corresponding spacetime reconstructions are obtained by the inverse Fourier transform $\hat{\hat{X}}=IFT\left(\hat{X}(\boldsymbol{\omega}')\right)$, where $\boldsymbol{\omega}_j'=r_j'e^{i\varphi_Y'}$, $1\le j\le n_X$.

As shown in the previous chapters, the Fourier transform represents just one example illustrating synthetic phase generation. Phase distribution priors, Laplace transform, and empirical phase-estimation represent alternative strategies that can be employed to derive spacekime reconstructions of temporal longitudinal data.

5.9 Bayesian Formulation of Spacekime Analytics

Spacekime analytics can be formulated in Bayesian inference terms. Suppose we have a single spacetime observation $X=\{x_{i_o}\}\sim p(x|\gamma)$ and $\gamma\sim p(\gamma|\varphi)$ is a process parameter (could be a vector) that we are trying to estimate. The focus of spacekime analytics is to make appropriate inference about the process X. Note that the *sampling distribution*, $p(x|\gamma)$, is the distribution of the observed data X conditional on the parameter γ and the *prior distribution*, $p(\gamma|\varphi)$, is the distribution of the parameter γ before the data X is observed.

Also assume that the hyperparameter (vector) φ representing the kime-phase estimates for the process can be estimated by $\hat{\varphi}=\varphi'$. These estimates may be obtained from an oracle, approximated using similar datasets, acquired as phases from samples of analogous processes, or derived via some phase-aggregation strategy. Then, the posterior distribution $p(\gamma|X,\varphi')$ of the parameter γ given the observed data $X=\{x_{i_o}\}$ and the kime-phase hyperparameter vector φ of the process parameter distribution, $\gamma\sim p(\gamma|\varphi)$, formulates spacekime inference as a Bayesian parameter estimation problem:

$$\underbrace{p(\gamma|X,\varphi')}_{\text{posterior distribution}} = \frac{p(\gamma,X,\varphi')}{p(X,\varphi')} = \frac{p(X|\gamma,\varphi')\times p(\gamma,\varphi')}{p(X,\varphi')} = \frac{p(X|\gamma,\varphi')\times p(\gamma,\varphi')}{p(X|\varphi')\times p(\varphi')} =$$

$$\frac{p(X|\gamma,\varphi')}{p(X|\varphi')}\times\frac{p(\gamma,\varphi')}{p(\varphi')} = \underbrace{\frac{p(X|\gamma,\varphi')\times p(\gamma|\varphi')}{p(X|\varphi')}}_{\text{observed evidence}} \propto \underbrace{p(X|\gamma,\varphi')}_{\text{likelihood}} \times \underbrace{p(\gamma|\varphi')}_{\text{prior}}.$$

In Bayesian terms, the posterior probability distribution of the unknown parameter γ is proportional to the product of the likelihood and the prior. In probability terms, the posterior = likelihood times prior, divided by the observed evidence, in this case, a single spacetime data point, x_{i_o}.

Spacekime analytics based on a single spacetime observation x_{i_o} can be thought of as a type of Bayesian prior or posterior *predictive distribution* estimation problem. For instance,

- *Posterior predictive distribution* of a new data point x_{j_o}, marginalized over the *posterior* (i.e., the sampling distribution $p(x_{j_o}|\gamma)$, weight-averaged by the posterior distribution):

$$p(x_{j_o}|x_{i_o},\varphi') = \int p(x_{j_o}|\gamma)\times \underbrace{p(\gamma|x_{i_o},\varphi')}_{\text{posterior distribution}}\, d\gamma.$$

- *Prior predictive distribution* of a new data point x_{j_o}, marginalized over the *prior* (i.e., the sampling distribution $p(x_{j_o}|\gamma)$, weight-averaged by the pure prior distribution):

$$p(x_{j_o}|\varphi') = \int p(x_{j_o}|\gamma)\times \underbrace{p(\gamma|\varphi')}_{\text{prior distribution}}\, d\gamma.$$

The difference between these two predictive distributions is that the *posterior* predictive distribution is updated by the observation $X=\{x_{i_o}\}$ and the hyperparameter, φ, whereas the *prior* predictive distribution only relies on the values of the hyperparameters that appear in the prior distribution.

This framework allows formulating spacekime analytics as Bayesian predictive inference where the posterior predictive distribution may be used to sample or forecast the *distribution* of a prospective, yet unobserved, data point x_{j_o}. Rather than predicting a specific outcome point, the posterior predictive distribution provides the holistic distribution of all possible observable states, i.e., outcome points. Clearly, the posterior predictive distribution spans the entire parameter state-space (Domain(γ)), much like the wavefunction represents the distribution of particle positions over the complete particle state-space.

Using maximum likelihood or maximum a posteriori estimation, we can also estimate an individual parameter point-estimate, γ_o. In this frequentist approach, the point estimate may be plugged into the formula for the distribution of a data point, $p(x|\gamma_o)$, which enables drawing IID samples or individual outcome values. This approach has the drawback that it does not account for the intrinsic uncertainty of the parameter, γ, which may lead to an underestimation of the variance of the predictive distribution.

The following simulation example generates two random samples drawn from mixture distributions each of $n_A = n_B = 10,000$ observations: $\{X_{A,i}\}_{i=1}^{n_A}$, where $X_{A,i} = 0.3U_i + 0.7V_i$, $U_i \sim N(0,1)$ and $V_i \sim N(5,3)$, and $\{X_{B,i}\}_{i=1}^{n_B}$, where $X_{B,i} = 0.4P_i + 0.6Q_i$, $P_i \sim N(20,20)$ and $Q_i \sim N(100,30)$. **Figure 5.2** illustrates the density plots of the pair of processes ($\{X_A\}$ and $\{X_B\}$).

Clearly, the intensities of cohorts A and B are independent and follow different mixture distributions. We'll split the first cohort (A) into *training* (C) and *testing* (D) subgroups, and then perform the following four steps:

- Transform all four cohorts into Fourier k-space,
- Iteratively randomly sample single observations from cohort C,
- Reconstruct three separate versions of the process in spacetime by using a single kime-magnitude value and three alternative kime-phase estimates, each derived separately from cohorts B, C, and D, and

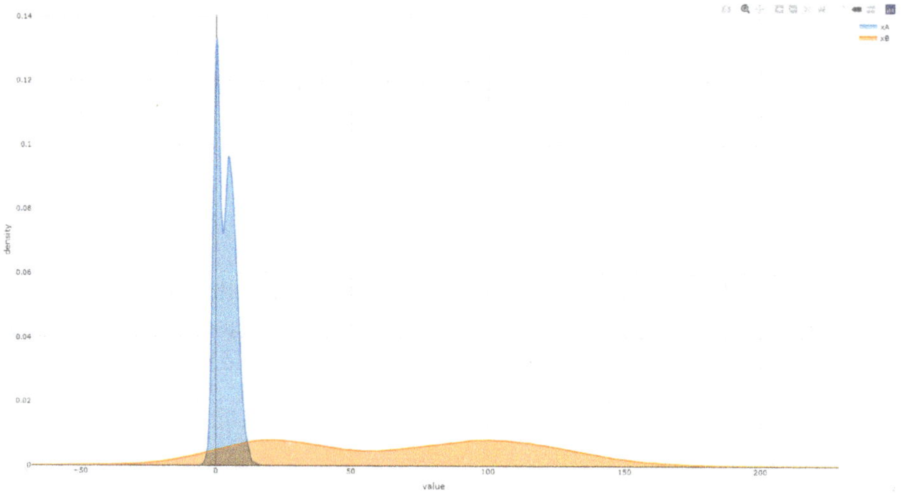

Figure 5.2: Spacetime simulation. The two processes $\{X_A\}$ and $\{X_B\}$ represent bimodal mixture distributions.

- Compute the *classical spacetime-derived* population characteristics of cohort A and compare them to their spacekime counterparts obtained using a single C kime-magnitude paired with B, C, or D kime-phases.

Table 5.2 shows some of the summary statistics for the original process (cohort A) and the corresponding values of their counterparts computed using the spacekime reconstructed signals based on kime-phases of cohorts, C, and D. The estimates for the latter three cohorts correspond to reconstructions using a single spacetime observation (i.e., single kime-magnitude) and alternative kime-phases (in this case, kime-phases derived from cohorts B, C, and D). The results show strong agreement between the original data and its three spacekime reconstructions, as measured by the seven summary statistics (rows in the table). Reconstructions (C) and (D) match the original signal better than reconstruction (B).

Table 5.2: Mixture distribution modeling. Summary statistics for the scaled original and spacekime reconstructed process distributions. Cohorts *A* and *B*, represent the pair of original mixture distributions, and cohorts *C* and *D* represent the training and testing split of cohort *A*.

	Spacetime	**Spacekime reconstructions (single kime-magnitude)**		
Summaries	(*A*) Original	(*C*) Phase=True	(*B*) Phase=Diff. Process	(*D*) Phase=Independent
Min	−2.38798	−2.98116	−3.798440	−2.69808
1st Quartile	−0.89359	−0.76765	−0.636799	−0.76453
Median	0.03311	−0.05982	0.009279	−0.08329
Mean	0.00000	0.00000	0.000000	0.00000
3 Quartile	0.75772	0.72795	0.645119	0.69889
Max	3.61346	3.64800	3.986702	3.22987
Skewness	0.348269	0.2372526	0.001021943	0.31398
Kurtosis	−0.68176	−0.4452207	0.2149918	−0.3270084

Additional measures quantifying the differences between these distributions (original signal and spacekime reconstructions) include the two-sample Kolmogorov-Smirnov (KS) test and the correlation coefficient. The value of the KS test statistic (D) represents the maximum distance between the estimated cumulative distribution functions. Its corresponding p-value is the probability of seeing a test statistic as high or higher than the one observed given that (under the default null hypothesis) the two samples were drawn from the same distribution. In our case, comparing the distributions of the original data (A) and its reconstruction using a single kime magnitude and the correct kime-phases (C) yields a KS statistics $D = 0.053875$, and $p_{value} = 1.647 \times e^{-10}$. This provides a very strong statistical evidence (mostly due to the large sample size) but a marginal practical difference between the real and reconstructed signals (low value of the test statistics). To calibrate these practical and statistical significance values, we can compare a pair of reconstructions using a single kime-magnitude value and two independent kime-phase estimates (cohorts C and D), which yields KS test statistics $D = 0.017375$ and $p_{value} = 0.1786$.

The correlation between the original data (A) and its reconstruction using a single kime magnitude and the correct kime-phases (C) is $\rho(A, C) = 0.89$, **Figure 5.3**. This relatively large correlation suggests that a substantial part of the A process energy can be recovered using only a single observation. In this case, to reconstruct the signal back into spacetime and compute the corresponding correlation, we used a single kime-magnitude (sample-size=1) and kime-phase estimates derived from process C.

Figure 5.4 shows graphically the similarities and differences between the (scaled) distributions of the original process (A) and three spacekime reconstructions (B, C, D) using alternative strategies for estimating the kime-phases. Notice the bimodal shape of the original data (**A=blue**) and the spacekime reconstructions (C=orange and D=green), which are based on cohorts C and D. These bimodal shapes are indeed related to the mixture distributions of the corresponding underlying processes. This plot also shows that the right-modes of the reconstructions agree perfectly with the corresponding right-mode in the original data. However, the imperfection of the spacekime reconstructions of the data is evident by the positive offsets of the left-modes of the signal reconstructions. The last reconstruction (B=red), whose distribution has a unimodal shape, is obtained using the kime-phases of a completely different process and hence represents a poor estimate of the original signal via a single kime-magnitude measurement paired with kime-phases derived from process B.

Let's demonstrate the Bayesian inference corresponding to this spacekime data analytic problem using the following simulated bimodal experiment: $X_A = 0.3U + 0.7V$, where $U \sim N(0, 1)$ and $V \sim N(5, 3)$. Specifically, we will illustrate the Bayesian inference using repeated single spacetime observations from cohort A, $X = \{x_{i_o}\}$, and varying kime-phase priors obtained from cohorts B, C, or D, which represent the posterior predictive distribution. **Figure 5.5** depicts relations between the empirical data distribution (**dark blue**) and samples from the posterior predictive distribution, Bayesian simulated spacekime reconstructions (light-blue). Although the derived Bayesian estimates do not perfectly match the empirical distribution of the simulated data, there is clearly information encoding that is captured by the spacekime data reconstructions. This signal compression can be exploited by subsequent model-based or model-free data analytic strategies for retrospective prediction, prospective forecasting, classification, clustering, and other spacekime inference.

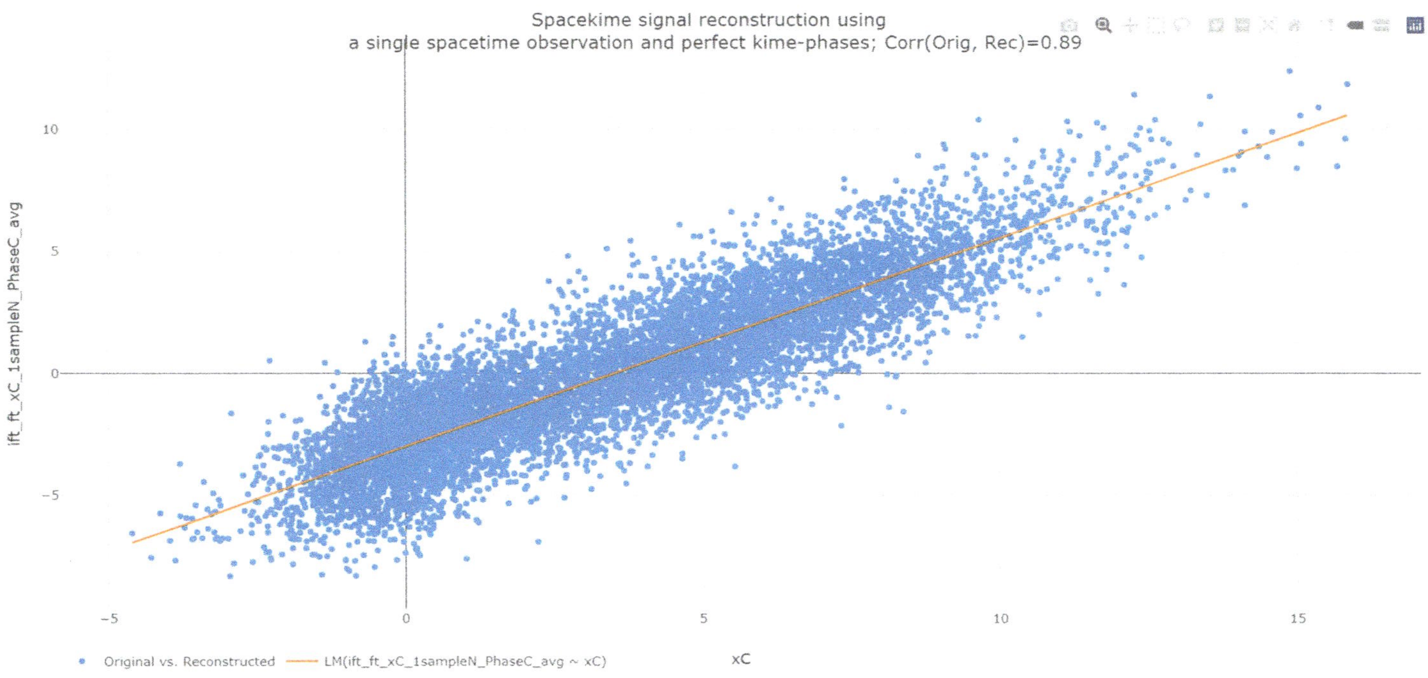

Figure 5.3: Scatterplot and correlation between original spacetime IID sample (*A*) and a spacekime reconstruction based on a single kime-magnitude and kime-phase estimates derived from process *C*.

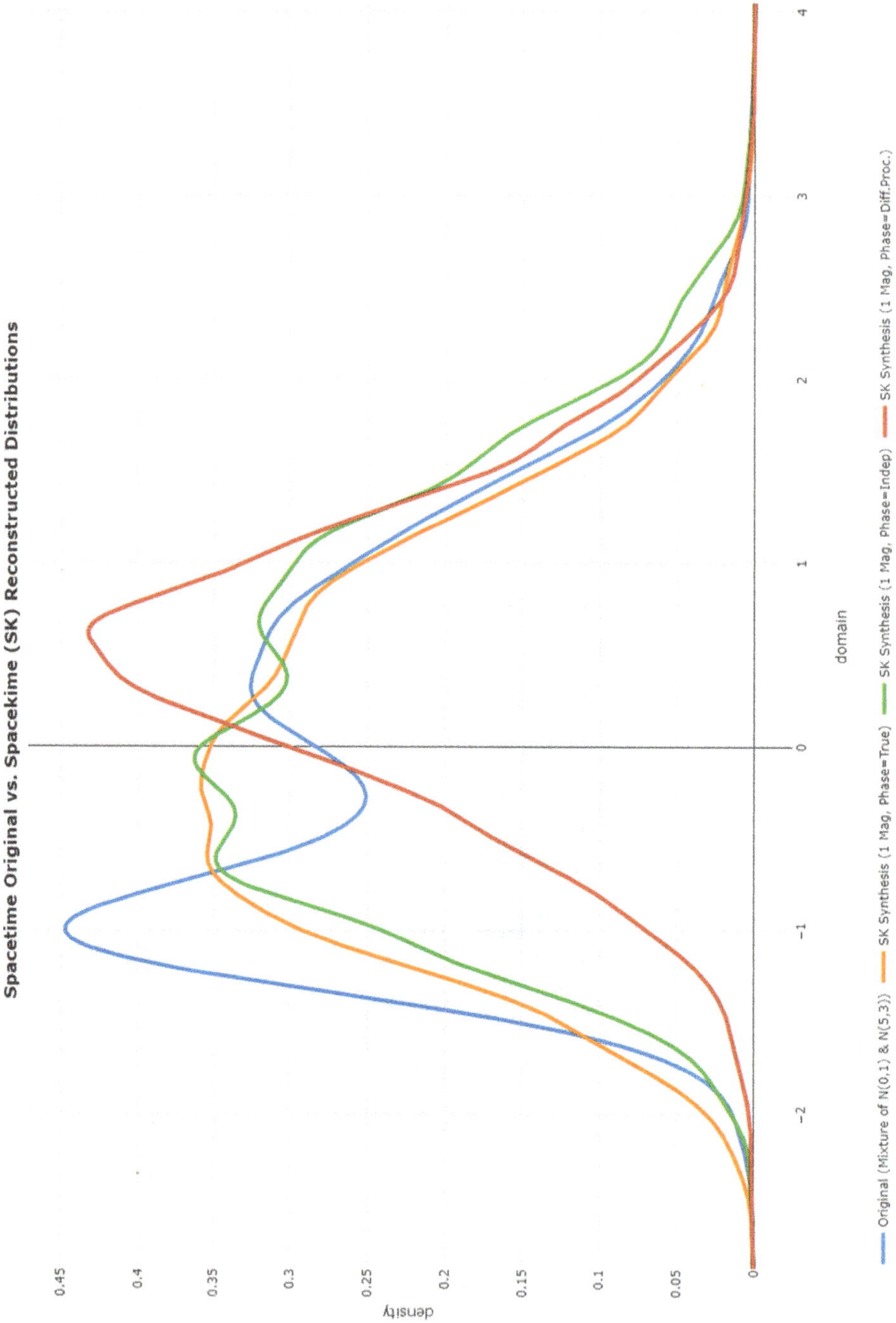

Figure 5.4: Distributions of the original spacetime IID sample (*A*) and three corresponding spacekime reconstructions based on a single kime-magnitude and kime-phase estimates derived from processes *B*, *C*, and *D*.

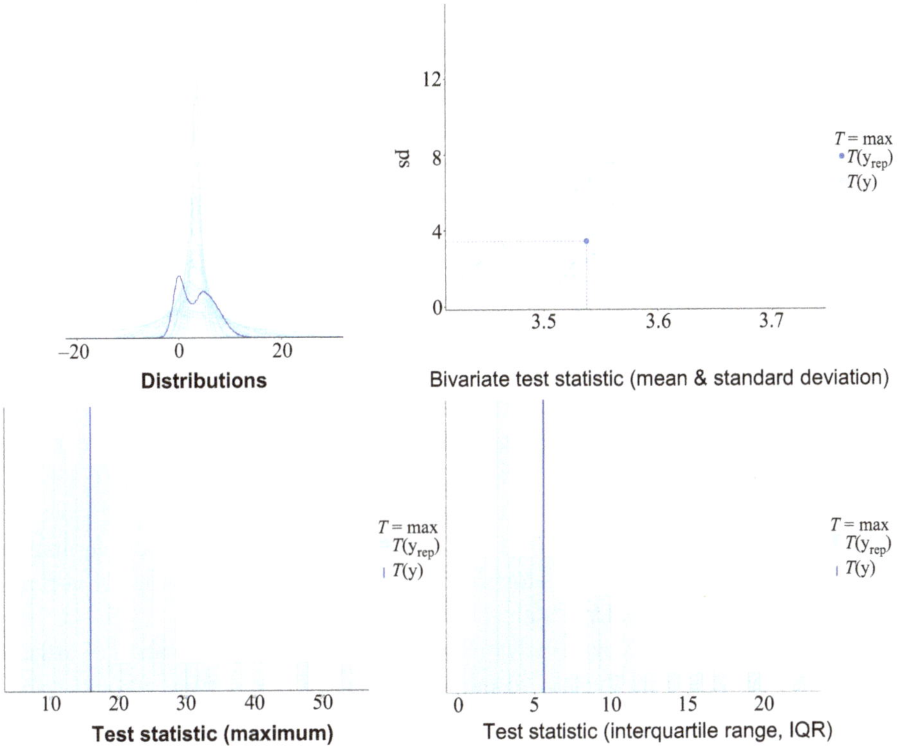

Figure 5.5: Relations between the empirical data distribution (dark blue) and samples from the posterior predictive distribution, Bayesian simulated spacekime reconstructions (light-blue).

5.10 Uncertainty in Data Science

The Heisenberg uncertainty principle may have a corresponding analogue in data science. Its oversimplified data science formulation can be expressed as follows. Observational data-driven inference cannot be simultaneously completely generic (complete coverage) and, at the same time, reliably accurate (consistency). The quantum mechanical space-frequency duality, expressed via the Fourier transformation, may be replaced in data science by the coupling of inferential scope (indicating the generalizability of the inference) and the inferential precision (affirming the veracity and robustness of the inference).

In essence, the data science uncertainty principle suggests that we are always in one of two states, but never both: Either we get perfectly robust inference, but with limited analytical application scope, i.e., analytical models under strict assumptions and using small well-managed data archives. Alternatively, at the opposite extreme, when using big data, such as the entire population or census, without many restrictions like a priori assumptions, we may compromise the precision, timeliness, or reproducibility of the resulting inference. At the two extremes are (1) the classical statistical inference framework, which is narrowly applicable but generates (nearly) optimal, reproducible, and robust inferential models for idealized situations (e.g., parametric assumptions, linear models, small IID datasets, resulting in best linear unbiased estimators, etc.), and (2) the modern semi-supervised machine learning approaches, which certainly are broadly applicable and allow handling of large, complex, and multisource datasets with limited *a priori* assumptions, but may be intrinsically dynamic or unstable, possibly trading off the result validity, reproducibility, or reliability.

Most scholars involved in statistical computing using non-trivial datasets can attest that as the sources, heterogeneity, size, scale, and missingness in the data increase, the complexity of the data-driven scientific inference grows exponentially [276, 277]. This is due to many complementary factors like (1) the increased costs for data management and preprocessing, (2) the quality of the data is usually compromised with the increase of the sample-size, sampling rate, and the number of recording devices, (3) for complex datasets, *a priori* model assumptions may be violated, and there are challenges with the data representation, processing, and information extraction, and (4) various estimators (functions) and estimates (parameter vectors) may become computationally unstable, intractable, or expensive.

The impact of technological advances, IT services, and digitalization of human experience over the past decade all provided effective mechanisms for collecting, tracking, processing, and interpreting enormous amounts of information. The Heisenberg information duality in data science relates to the management of the data (i.e., position) and derived inference (i.e., energy of the information). Specifically, there appears to be non-commuting properties associated with most big data science applications. We cannot perfectly represent and handle "all observed data" *and* at the same time derive truly consistent, timely, and complete inference of the holistic information contained in the entire dataset. Why is this interesting and important?

The limit of the duality between the properties of perfectly managing and precisely interpreting the information content of a big data case-study is currently unknown [40]. However, the tradeoffs between an ideal data representation and an impeccable inference are clear. Let's look at a concrete example using a hypothetical clinical decision support inference problem.

Jane Doe is a physician treating elderly patients with memory problems. Mr. Real is a patient in Dr. Doe's clinic who presents with cognitive symptoms of impaired recollection, i.e., memory formation and/or retrieval. Dr. Doe suspects the patient may suffer from Alzheimer's disease, a progressive neurodegenerative disorder. To investigate her hunch, Dr. Doe orders and inspects Mr. Real's neuroimaging, clinical tests, cognitive assessment, genetic phenotyping, and various biospecimen tests. These data will allow her to confirm the diagnosis, suggest an appropriate treatment regimen, and provide prognosis for the patient and his family. Dr. Doe also has access to a large data archive of asymptomatic controls and patients that match Mr. Real's demographic phenotype and include participants at various stages of dementia. In an ideal world, Dr. Doe has access to a clinical decision support system that provides the quintessential dual functionality – perfect data management and optimal scientific inference. The automated decision support system ingests the patient's data and holistically compares Mr. Real's records to the normative information contained in the available large cohort database. Essentially, the system facilitates precision health assessment and personalized medical treatment allowing the physician to map the patient onto a comprehensive Alzheimer's disease atlas. The inferential component of this automated clinical decision support system will provide direct quantification of the likelihood that Mr. Real has Alzheimer's, suggest the most beneficial treatment, and predict the clinical outcome and cognitive state of the patient in the future.

In practice, however, this ideal situation is not attainable for several reasons. The *first* one is the lack of accessible large-scale, well curated, complete and representative normative data that can be used to compute precisely the necessary statistical models forecasting the disease progression. The *second* issue is a lack of general tools that would provide optimal, reliable, and consistent scientific inference by jointly modeling all available information (big data). The data science uncertainty principle suggests that there is a limit to how accurate *and* broad this automated decision support system could be, in general. Solving one of the two challenges (either perfect big data handling or flawless analytical inference) precisely will compromise the efficacy of the other.

Below we provide some background and ideas of mathematical descriptions of the uncertainty principle in relation to data science.

5.10.1 Quantum Mechanics Formulation

Quantum physics rules allow us to convert *classical* physics equations into their *quantum* mechanical counterparts. For instance, classical physics *energy* is converted to a quantum partial differential operator: $E \mapsto i\hbar\frac{\partial}{\partial t}$; and classical physics *momentum* is mapped to another partial differential operator: $p_x \mapsto -i\hbar\frac{\partial}{\partial x}$. The commutator algebra of differential operators identifies pairs of physical observables that either can or cannot be measured simultaneously with infinite precision. Consider the commutator of the momentum and position operators:

$$\left[-i\hbar\frac{\partial}{\partial x}, x\right]\psi(x,t).$$

The general mathematical representation of the uncertainty principle can be expressed in terms of a particular inequality:

$$\|D_x u\| \, \|xu\| \geq \frac{\hbar}{2} \|u\|^2,$$

where the norm is defined in terms of the bra-ket inner product notation, $\frac{\hbar}{2}\|u\|^2 = \langle \frac{\hbar}{i}\partial_x u | ixu \rangle$, and the non-commutation of the unbounded operators $D_x = \frac{\hbar}{i}\partial_x$ and x, (i.e., multiplication by x).

In general, the identity operator on a Hilbert space can never be expressed as the commutator of bounded operators [278]. Thus, the non-commutation of bounded operators $A = \frac{1}{i}\partial_x$ and $B = ix$ (e.g., finite matrices) could never lead to $[A, B] = I$, since the trace of the commutator is 0. Therefore, quantization requires necessarily infinite dimensional unbounded operators.

In quantum mechanics, this specific non-commutation property leads to various uncertainty principle inequalities. The algebra of operators is in general non-commutative, which explains the common natural occurrence of the uncertainty principle. If unknown quantities (measurable, random variables) do not obey a commutative algebra, an uncertainty principle is likely in play. In classical mechanics, measurable quantities, like position x and momentum p, belong to a commutative algebra, as functions on the phase space. However, in quantum mechanics the unknown quantities represent quantizations (operators) of observables, e.g., $\langle D_x|$, momentum operator, and $|x\rangle$, position operator. In quantum mechanics and beyond, the uncertainty principle is a signature of non-commutativity.

5.10.2 Statistics Formulation

In a classical statistical sense, the data science uncertainty principle is realized via the sample-size – power-analysis framework [279]. It's well known that the key factors playing roles in the inverse relation of size vs. power include sensitivity (power), sample size, inter-individual variability, effect-size (magnitude of the response), significance level, and the formulation of the alternative hypothesis. Power analysis demonstrates one simple view of the pragmatics of data science uncertainty. By trading off statistical power, sample size and resources (energy), we can get either high power or real-time maximum efficiency, but not both. At one extreme, getting high power requires lots of resources to manage (relatively) larger sample sizes. On the other extreme, extremely large sample-sizes present enormous non-sampling problems and require substantial allocations of resources. These tradeoffs directly relate to quality control, computational complexity, resource allocation, and model limitations. Non-sampling errors are prevalent in large samples. These represent deviations from the true parameter estimates that are not due to the sample selection. Non-sampling errors are difficult to quantify and include systematic errors, random errors, coverage errors, biases in population representation, inconsistencies of observed information about all sample cases, response errors, deliberate data adulteration, mistakes due to human and machine failures, consistent coding to standard classifications, errors of collection, non-random non-response bias, inconsistent preprocessing, etc.

More formally, suppose we are performing a hypothesis test comparing the mean of a continuous outcome variable in a single population to a known mean. We can express the data science uncertainty principle in terms of a statistical inference framework. Let's consider a simple parametric inference problem (hypothesis testing) where the null is $H_o : \mu = \mu_o$, the research hypothesis is $H_1 : \mu \neq \mu_o$, and μ_o represents a known mean value. Under common parametric assumptions, the relation between sample-size (η) and statistical-power ($\pi = 1 - \beta$), i.e., sensitivity, is given by:

$$\eta = \left(\frac{z_{1-\alpha/2} + z_{1-\beta}}{ES}\right)^2,$$

where α is the level of significance; z_γ is a critical value of the distribution of a specific parametric test-statistic, e.g., the standard normal distribution with a right-tail probability γ; and ES is the effect size, e.g., $ES = \frac{|\mu_1 - \mu_o|}{\sigma}$; μ_1 is the mean under the alternative hypothesis, H_1; and σ is the standard deviation of the outcome of interest.

Of course, the *size-power* relation will change depending on the model assumptions and the specific study design. In fact, closed-form analytical expressions for the power are not always known. In general, the statistical-power of the test is a function $\pi = \pi(\eta, \theta, \alpha, \ldots)$, where θ represents the parameter vector of the inference testing, η is the sample size, and there are other additional parameters that play roles in this association.

In the above quadratic *size-power* relation, increasing η can certainly drive the power $(1 - \beta)$ up and the false-negative rate β down, however, this is not the only way to get that effect. Stating this in Heisenberg's uncertainty principle terms, $\beta \times \eta \geq \delta > 0$, where the constant δ depends on all other inferential factors.

When we increase the resolution, or work in a fine scale, we tradeoff precision (variability) and coverage (generality). A lot of factors, such as model assumptions, time effects, test interpretation, inference understanding, variable and parameter inter-dependencies, and estimates variability, are all entangled in the process of obtaining scientific inference from raw observations (data). In high-resolution and fine-scale, uncertainty is ubiquitous. The effects of uncertainty in large-scale statistical computing are also well-documented [280].

5.10.3 Decision Science Formulation

In decision science, Heisenberg's uncertainty principle may be formulated for any multiple-criteria decision-making models [281] for solving problems that can be represented in the *criterion space* or the *decision space*. If different criteria are combined by a weighted linear function, it is also possible to represent the problem in the weight space. The *decision space* corresponds to the set of possible decisions that are allowed. The *criterion space* represents the specific consequences of the decisions we make.

Suppose a decision model is represented as a set of criteria that rank a few alternatives differently according to some preference aggregation function. Decision aggregating functions accumulate decision weights, whether additively, as in a linear model, or non-additively, as in higher order models.

When evaluating, reviewing, assessing, or refining decision criteria of all possible alternatives, we can consider few or many more criteria depending on the desired model complexity, even possibly considering rather unlikely events that can only be observed in big data studies. This flexibility allows us to increase the discriminative power on the *criterion space*, however, it comes at the expense of losing precision on the decision side (too many alternative options). This also ties in to the Gödel's incompleteness theorem which argues a system that obeys arithmetic rules can never be both complete and consistent [23].

In decision space representation, the ideal situation is to have just one criterion, according to which we can make an action recommendation based on how the options all rank against each other. When we have many alternatives, some decisions will be best according to some criteria and some will be best according to another, thus making the actual decision is somewhat inhibited. In these situations, we need to determine specific protocols weights ranking all criteria according to their importance and then aggregating the final ensemble decision. When we have a large number of criteria, each alternative will represent the optimal decision according to some of the criteria, but probably not all. In some models, this subjectivity of criteria weights construction may yield erroneous decisions or equally competitive alternatives. In essence, expanding a decision-making model with more criteria is guaranteed to increase the model discriminative power, which comes with a negative consequence of more dispersed resulting action. There appears to be a lower limit on jointly minimizing the uncertainty attached to a decision making model and maximizing the utility to act on the decision.

5.10.4 Information Theoretic Formulation

Mathematical representation and quantification of uncertainty and unpredictability typically relies on estimations of two classes of functions – dispersion or entropy – that both are computed using random observations and measure the intrinsic process uncertainty. Information theoretic representations of the data science uncertainty principle can be formulated in terms of Shannon's information and channel capacity limit [282, 283] or in terms of Fisher information.

5.10.4.1 Shannon's Information and Channel Capacity Limit

Suppose the bandwidth of a communication channel is $B\,Hz$, the signal-to-noise ratio is $\frac{S}{N}$, and Shannon's maximum channel capacity C is $C = B \times \log_2\left(1 + \frac{S}{N}\right)$. Increasing the levels of a signal increases the probability of an error occurring, in other words it reduces the reliability of the system. Let's define the overall maximal capacity of any communication channel by $\nu = C_{max}$. This effectively allows us to do arithmetic on capacity, bandwidth, and transfer rates, much like fixing the speed of light allows us to compute on the independent space and time in the joint 4D Minkowski spacetime manifold [284].

Thus, we can express a modified version of Shannon's channel capacity as:

$$\Delta C = \Delta\left(B \times \log_2\left(1 + \frac{S}{N}\right)\right).$$

Therefore, the relative variance of the mean energy, i.e., the maximal capacity variance, is:

$$\frac{\Delta E}{h} = \Delta \nu \geq \Delta C = \Delta\left(B \times \log_2\left(1 + \frac{S}{N}\right)\right).$$

Denoting by h and $\hbar$ the Planck and the reduced Planck constants, respectively, we can express the uncertainty as $\Delta E \Delta t \geq \frac{\hbar}{2}$. Thus,

$$h = \Delta\left(B \times \log_2\left(1 + \frac{S}{N}\right)\right) \times \Delta t \geq \hbar,$$

$$\Delta\left(B \times \log_2\left(1 + \frac{S}{N}\right)\right) \times \Delta t \geq \frac{1}{2\pi}.$$

If n is the number of bits transferred via the channel with a constant bandwidth B, we can assume $n = B \times \Delta t$, then we obtain an information-theoretic representation of the uncertainty principle: $n \times \log_2\left(1 + \frac{S}{N}\right) \geq \frac{1}{2\pi}$.

A useful tool to quantify unpredictability in information theory is the Shannon's entropy measure, which explicates the intrinsic uncertainty of the states by mapping the distribution of random variables, or quantum states, to real numbers. For instance, flipping a fair coin, $P(Head) = \frac{1}{2}$, represents a random process embedding more uncertainty about the outcome compared to an alternative biased coin with $P(Head) = \frac{9}{10}$. Computing and contrasting the entropy measures for both coins (discrete processes) yields:

$$\underbrace{H\left(\text{Coin } 1 \mid P(Head) = \frac{1}{2}\right)}_{\text{higher uncertainty}} \equiv -\sum_{i=1}^{2} p_i \log_2 p_i = \frac{1}{2} + \frac{1}{2} = 1 > 0.469 =$$

$$-0.9\log_2(0.9) - 0.1\log_2(0.1) \equiv \underbrace{H\left(\text{Coin } 2 \mid P(Head) = \frac{9}{10}\right)}_{\text{lower uncertainty}}.$$

Shannon's *entropy* can also be used just like the deviation (or standard deviation) to quantify distribution structure. For instance, for any $f \in L^2$ that is normalized to $\int |f(x)|^2 dx = \int \left|\hat{f}(x)\right|^2 dx = 1$, i.e., $|f(x)|^2 = \varphi(x)$ is a probability distribution function, the entropy may represent a better measure of dispersion than the standard deviation. For

$$FT(\varphi)(\omega) = \hat{\varphi}(\omega) = \int_R \varphi(x) e^{-2\pi i x \omega} dx \text{ and } IFT(\hat{\varphi})(x) = \widehat{\hat{\varphi}}(x) = \varphi(x) = \int_R \hat{\varphi}(\omega) e^{2\pi i x \omega} d\omega,$$

the total entropy [285] $H_x + H_\omega$ represents the aggregate sum of the spatial and spectral Shannon differential entropies, which naturally extend the (discrete Shannon entropy measure):

$$H_x = H(\varphi) = -\int \varphi(x) \log(\varphi(x)) dx \text{ (space)},$$

and

$$H_\omega = H(\hat{\varphi}) = -\int \hat{\varphi}(\omega) \log(\hat{\varphi}(\omega)) d\omega \text{ (spectral)},$$

and satisfies the inequality,

$$H_x + H_\omega \geq \log\left(\frac{e}{2}\right).$$

The remaining part of this section is agnostic to the base of the logarithmic function log(), which easily converts between the natural-logarithm ln() and a logarithm of an arbitrary base $b \in \mathbb{R}^+ \backslash \{1\}$, $\log_b(x) = \frac{1}{\ln b} \ln x$, $\ln x = \frac{1}{\log_b e} \log_b x$. The Shannon entropy is a special case of the more general Rényi entropy uncertainty [286] and this total entropy lower bound is attained for normal density functions. The Rényi entropy is defined by:

$$H_\alpha(\varphi) = \frac{1}{1-\alpha} \log \int (\varphi(x))^\alpha dx.$$

The limit $\lim_{\alpha \to 1} \underbrace{H_\alpha(\varphi)}_{\text{Rényi}} = \underbrace{H(\varphi)}_{\text{Shannon}} \equiv -\int \varphi(x) \log(\varphi(x)) dx$ [286] can be derived using l'Hopital's theorem for the limit of a quotient of functions:

$$\lim_{x \to b} \frac{f(x)}{g(x)} = \lim_{x \to b} \frac{f'(x)}{g'(x)}.$$

In this case, $\alpha \to 1$, $f(\alpha) = \log \int (\varphi(x))^\alpha dx$, and $g(\alpha) = 1 - \alpha$, i.e., $H_\alpha(\varphi) = \frac{f(\alpha)}{g(\alpha)}$. The derivative of the denominator is a constant, $g'(\alpha) = -1$, and the derivative of f is computed by the chain-rule for differentiation:

$$\frac{d}{d\alpha} f(\alpha) = \frac{d}{d\alpha}\left(\log \int (\varphi(x))^\alpha dx\right) =$$

$$\frac{1}{\int (\varphi(x))^\alpha dx}\left(\int \frac{d}{d\alpha}(\varphi(x))^\alpha dx\right) = \frac{1}{\int (\varphi(x))^\alpha dx}\left(\int \frac{d}{d\alpha}\left(e^{\alpha \ln \varphi(x)}\right) dx\right) =$$

$$\frac{1}{\int (\varphi(x))^\alpha dx}\left(\int e^{\alpha \ln \varphi(x)} \frac{d}{d\alpha}(\alpha \ln \varphi(x)) dx\right) = \frac{1}{\int (\varphi(x))^\alpha dx}\left(\int e^{\alpha \ln \varphi(x)} (\ln \varphi(x)) dx\right) =$$

$$\frac{1}{\int (\varphi(x))^\alpha dx}\left(\int (\varphi(x))^\alpha (\ln \varphi(x)) dx\right).$$

Hence, the limit

$$\lim_{\alpha \to 1} \tfrac{d}{d\alpha} f(\alpha) = \frac{1}{\underbrace{\int \varphi(x) dx}_{1}} \left(\int \varphi(x)(\ln \varphi(x)) dx\right) = \int \varphi(x)(\ln \varphi(x)) dx.$$

Assuming we are working with natural log, since the limit of the denominator is a constant $\lim_{\alpha \to 1} g'(\alpha) = \lim_{\alpha \to 1}(-1) = -1$, the Rényi entropy tends to the Shannon entropy as $\alpha \to 1$:

$$\lim_{\alpha \to 1} H_\alpha(\varphi) = \lim_{x \to b} \frac{f(x)}{g(x)} = \lim_{x \to b} \frac{f'(x)}{g'(x)} = -\int \varphi(x)(\ln \varphi(x)) dx = H(\varphi).$$

For a pair of positive non-unitary numbers $\alpha, \beta \in \left\{\mathbb{R}^+ \setminus \{1\} \mid \frac{1}{\alpha} + \frac{1}{\beta} = 2\right\}$, the Babenko–Beckner inequality [285] for the Rényi entropy is:

$$H_\alpha(\varphi) + H_\beta(\hat{\varphi}) \ge \frac{1}{2}\left(\frac{\log \alpha}{\alpha - 1} + \frac{\log \beta}{\beta - 1}\right) - \log 2.$$

Taking the limit of the sum of the Rényi entropies of the (general) probability distribution φ and its Fourier transform $\hat{\varphi}$, as α, β approach 1, yields:

$$\underbrace{\lim_{\substack{\alpha \to 1 \\ \beta \to 1}} \left(H_\alpha(\varphi) + H_\beta(\hat{\varphi})\right)}_{H_X + H_\omega = H_X(\varphi) + H_\omega(\varphi)} \ge \lim_{\substack{\alpha \to 1 \\ \beta \to 1}} \left(\frac{1}{2}\left(\frac{\log \alpha}{\alpha - 1} + \frac{\log \beta}{\beta - 1}\right) - \log 2\right) = \frac{1}{2}(1+1) - \log 2 = \log \frac{e}{2}.$$

The normal probability distribution has many properties that make it unique. One of these properties characterizes the relationship between entropy and variance. One way to state the information-theoretic uncertainty principle is as follows: Given a fixed distribution variance, the normal distribution is the one that maximizes the entropy measure, and conversely, given a fixed entropy, normal distribution yields the minimal possible variance.

To make this entropy-variance relation more explicit for distribution functions, suppose we are given a random variable $X \sim \varphi(x)$ with a general real-valued probability density function φ with a finite process variance $\sigma = Var(X)$. Then, the relation between the Shannon's entropy and the process variance for this density is:

$$H(\varphi) \leq \log \sqrt{2\pi e \sigma^2}.$$

In this relation, the equality is achieved only for a normal distribution density, $g(x) \equiv \frac{1}{\sqrt{2\pi\sigma^2}} e^{-\frac{(x-\mu)^2}{2\sigma^2}} \sim N(\mu, \sigma^2)$. To derive this upper-bound of the entropy by the variance, let's plug in $g(x)$ in the entropy definition:

$$H(g) = -\int g(x) \log(g(x))dx = -\int \frac{1}{\sqrt{2\pi\sigma^2}} e^{-\frac{(x-\mu)^2}{2\sigma^2}} \log\left(\frac{1}{\sqrt{2\pi\sigma^2}} e^{-\frac{(x-\mu)^2}{2\sigma^2}}\right) dx =$$

$$-\int \frac{1}{\sqrt{2\pi\sigma^2}} e^{-\frac{(x-\mu)^2}{2\sigma^2}} \left[\log\left(\frac{1}{\sqrt{2\pi\sigma^2}}\right) - \frac{(x-\mu)^2}{2\sigma^2}\right] dx =$$

$$-\log\left(\frac{1}{\sqrt{2\pi\sigma^2}}\right) \underbrace{\int \frac{1}{\sqrt{2\pi\sigma^2}} e^{-\frac{(x-\mu)^2}{2\sigma^2}} dx}_{1} + \frac{1}{2\sigma^2} \underbrace{\int \frac{1}{\sqrt{2\pi\sigma^2}} e^{-\frac{(x-\mu)^2}{2\sigma^2}} (x-\mu)^2 dx}_{Var(g)} =$$

$$\frac{1}{2}\log(2\pi\sigma^2) + \frac{\sigma^2}{2\sigma^2} = \frac{1}{2}\log(2\pi e\sigma^2) = \log\left(\sqrt{2\pi e\sigma^2}\right).$$

For any general distribution, φ, we can assume that both densities g, φ have equal variances, σ; otherwise, we can set the variance of the normal distribution g equal to the actual variance of the general distribution φ. The Kullback–Leibler divergence between the two distributions is:

$$0 \leq D_{KL}(\varphi \| g) = \int \varphi(x) \log\left(\frac{\varphi(x)}{g(x)}\right) dx = \underbrace{\int \varphi(x) \log(\varphi(x))dx}_{-H(\varphi)} - \int \varphi(x) \log(g(x))dx =$$

$$-H(\varphi) - \int \varphi(x) \log\left(\frac{1}{\sqrt{2\pi\sigma^2}} e^{-\frac{(x-\mu)^2}{2\sigma^2}}\right) dx =$$

$$-H(\varphi) - \int \varphi(x) \log\left(\frac{1}{\sqrt{2\pi\sigma^2}}\right) dx - \log(e) \int \varphi(x) \left(-\frac{(x-\mu)^2}{2\sigma^2}\right) dx =$$

$$-H(\varphi) + \log\left(\sqrt{2\pi\sigma^2}\right) + \log(e)\frac{\sigma^2}{2\sigma^2} = -H(\varphi) + \tfrac{1}{2}(\log(2\pi\sigma^2) + \log(e)) =$$

$$-H(\varphi) + \underbrace{\frac{1}{2}\log(2\pi e\sigma^2)}_{H(\text{Normal})} = -H(\varphi) + H(g).$$

Thus, for any probability distribution φ, $H(\varphi) \le H(g)$ and the normal distribution g maximizes the entropy. The result of this is the following upper bound on the entropy of any distribution:

$$H(\varphi) \le \log \sqrt{2\pi e \sigma^2}.$$

Note that the normal (Gaussian) probability $g \sim N(\mu = 0, \sigma^2)$ has the additional property that its Fourier transform $\varphi(\omega) \equiv \hat{g}(\omega) \equiv FT(g)(\omega) \sim N\left(\hat{\mu} = 0, \hat{\sigma}^2 = \frac{1}{\sigma^2}\right)$ and their squares, g^2, $\hat{g}^2$, are all Gaussian with correspondingly modified variances.

In the general case for an arbitrary distribution, φ, we can combine the exponentiated lower bound,

$$\log\left(\frac{e}{2}\right) \le H_x(\varphi) + H_\omega(\hat{\varphi})$$

with the upper bound,

$$H_\omega(\hat{\varphi}) + H_x(\varphi) \le \log \sqrt{2\pi e \hat{\sigma}^2} + \log \sqrt{2\pi e \sigma^2} = \log\left(2\pi e \sqrt{\hat{\sigma}^2 \sigma^2}\right).$$

This yields another representation of the uncertainty in terms of a product of variances that can't be simultaneously very small:

$$\frac{1}{4\pi} \le \frac{e^{[H(\hat{\varphi}) + H(\varphi)]}}{2\pi e} \le \sqrt{Var(\hat{\varphi}) Var(\varphi)}.$$

The lower bound may be just $\frac{1}{2}$ depending on the normalization of the Fourier transform.

5.10.4.2 Fisher Information

An alternative form of information-theoretic uncertainty can be formulated in terms of Fisher information and Cramér–Rao lower bound of the product of two variances [286, 287]. We saw the Fisher information earlier, in **Chapter 4**, when we discussed log-likelihood based inference.

Given an observable random variable $X \sim f_X(x|\theta)$, where θ is some unknown parameter, the *Fisher information* captured the amount of information about θ that is embedded in the random process X. When the density (or mass) function f_X chances slowly (flat curve) or rapidly (peaked curve) with respect to θ, X carries little or a lot of information, respectively, to correctly identify the parameter θ based on observed data. Recall that the *score function* is the partial derivative of the log-likelihood function with respect to the parameter θ, i.e., $\frac{\partial}{\partial\theta} \log f_X(x|\theta)$. In general, the *expected value of the score* is trivial:

$$\mathbb{E}\left(\frac{\partial}{\partial\theta} \log f_X(x|\theta)|\theta\right) = \int \left(\frac{\partial}{\partial\theta} \log f_X(x|\theta)\right) f_X(x|\theta) dx =$$

$$\int \left(\frac{\frac{\partial}{\partial \theta} f_X(x|\theta)}{f_X(x|\theta)} \right) f_X(x|\theta) dx = \int \frac{\partial}{\partial \theta} f_X(x|\theta) dx = \frac{\partial}{\partial \theta} \underbrace{\int f_X(x|\theta) dx}_{1} = 0.$$

Therefore, the *Fisher information*, $I(\theta)$, defined as the *variance of the score* is:

$$I(\theta) = Var\left(\frac{\partial}{\partial \theta} \log f_X(x|\theta) \right) = \mathbb{E}\left(\left(\frac{\partial}{\partial \theta} \log f_X(x|\theta) \right)^2 | \theta \right) = \int \left(\frac{\partial}{\partial \theta} \log f_X(x|\theta) \right)^2 f_X(x|\theta) dx \geq 0.$$

Random variables corresponding to low or high Fisher information indicate low or high absolute values of the score function. When the log-likelihood function is regular and has a smooth second derivative with respect to θ, we have:

$$\frac{\partial^2}{\partial \theta^2} \log f_X(x|\theta) = \frac{\partial}{\partial \theta} \left(\frac{\frac{\partial}{\partial \theta} f_X(x|\theta)}{f_X(x|\theta)} \right) = \frac{\frac{\partial^2}{\partial \theta^2} f_X(x|\theta)}{f_X(x|\theta)} - \underbrace{\left(\frac{\frac{\partial}{\partial \theta} f_X(x|\theta)}{f_X(x|\theta)} \right)^2}_{\left(\frac{\partial}{\partial \theta} \log f_X(x|\theta) \right)^2} =$$

$$\frac{\frac{\partial^2}{\partial \theta^2} f_X(x|\theta)}{f_X(x|\theta)} - \left(\frac{\partial}{\partial \theta} \log f_X(x|\theta) \right)^2,$$

$$\mathbb{E}\left(\frac{\frac{\partial^2}{\partial \theta^2} f_X(x|\theta)}{f_X(x|\theta)} \bigg| \theta \right) = \int \left(\frac{\frac{\partial^2}{\partial \theta^2} f_X(x|\theta)}{f_X(x|\theta)} \right) f_X(x|\theta) dx = \frac{\partial^2}{\partial \theta^2} \underbrace{\int f_X(x|\theta) dx}_{1} = 0,$$

$$\mathbb{E}\left(\frac{\partial^2}{\partial \theta^2} \log f_X(x|\theta) \right) = -\mathbb{E}\left(\left(\frac{\partial}{\partial \theta} \log f_X(x|\theta) \right)^2 \right).$$

Thus, taking the expectations of both hand sides of the equation above yields the Fisher information:

$$I(\theta) = \underbrace{\int \left(\frac{\partial}{\partial \theta} \log f_X(x|\theta) \right)^2 f_X(x|\theta) dx}_{\mathbb{E}\left(\left(\frac{\partial}{\partial \theta} \log f_X(x|\theta) \right)^2 | \theta \right)} = - \underbrace{\int \left(\frac{\partial^2}{\partial \theta^2} \log f_X(x|\theta) \right) f_X(x|\theta) dx}_{\frac{\partial^2}{\partial \theta^2} \mathbb{E}\left(\log f_X(x|\theta) | \theta \right)}.$$

Suppose now that we have an independent sample from the process distribution, $f_X(x|\theta)$, and $\hat{\theta} = \hat{\theta}(X)$ is an *unbiased* estimate of the unknown population parameter θ, i.e., $\mathbb{E}\left(\hat{\theta} \right) = \int \hat{\theta} f_X(x|\theta) dx = \theta$. The Cramér–Rao lower bound represents a minimal value of the variance of $\hat{\theta}$ expressed in terms of the Fisher information. In other words, there is a lower bound on the product of the score function variance (data-derived information about the unknown parameter) and the estimator variance (precision).

The lack of assumed bias ensures that $\forall\theta \in \Omega$:

$$\mathbb{E}\left(\hat{\theta}(X) - \theta|\theta\right) = \int\left(\hat{\theta}(x) - \theta\right) f_X(x|\theta)dx = 0.$$

Differentiating both sides yields,

$$0 = \frac{\partial}{\partial\theta}\mathbb{E}\left(\hat{\theta}(X) - \theta|\theta\right) = \frac{\partial}{\partial\theta}\int\left(\hat{\theta}(x) - \theta\right) f_X(x|\theta)dx = \int\left(\hat{\theta}(x) - \theta\right)\frac{\partial}{\partial\theta} f_X(x|\theta)dx - \underbrace{\int f_X(x|\theta)dx}_{1},$$

$$\int\left(\hat{\theta}(x) - \theta\right)\tfrac{\partial}{\partial\theta} f_X(x|\theta)dx = 1.$$

Let's consider a sample of n independent identically distributed observations, $X = \{X_i\}_{i=1}^n$. Then:

$$\int\left(\sum_{i=1}^n \frac{\partial}{\partial\theta}\log f_{X_i}(x_i|\theta)\sqrt{f_{X_1,X_2,\ldots,X_n}(x_1,x_2,\ldots,x_n|\theta)}\right)^2 dx_1dx_2\ldots dx_n =$$

$$\underbrace{\int\sum_{i=1}^n\left(\frac{\partial}{\partial\theta}\log f_{X_i}(x_i|\theta)\sqrt{f_{X_i}(x_i|\theta)}\prod_{j\neq i}\sqrt{f_{X_j}(x_j|\theta)}\right)^2 dx_1dx_2\ldots dx_n}_{\text{IID}, n\int\left(\frac{\partial}{\partial\theta}\log f_X(x|\theta)\right)^2 f_X(x|\theta)dx} +$$

$$\underbrace{2\int\sum_{1\le i<j\le n}\left(\left[\tfrac{\partial}{\partial\theta}\log f_{X_i}(x_i|\theta)\sqrt{f_{X_1,X_2,\ldots,X_n}(x_1,x_2,\ldots,x_n|\theta)}\right]\times \left[\tfrac{\partial}{\partial\theta}\log f_{X_j}(x_j|\theta)\sqrt{f_{X_1,X_2,\ldots,X_n}(x_1,x_2,\ldots,x_n|\theta)}\right]\right)dx_1dx_2\ldots dx_n}_{\text{vanishes as } \forall i\neq j,\ X_i \text{ and } X_j \text{ are IID.}}.$$

$$\int\sum_{i=1}^n\left(\frac{\partial}{\partial\theta}\log f_{X_i}(x_i|\theta)\sqrt{f_{X_i}(x_i|\theta)}\prod_{j\neq i}\sqrt{f_{X_j}(x_j|\theta)}\right)^2 dx_1dx_2\ldots dx_n =$$

$$\sum_{i=1}^n\int\left(\tfrac{\partial}{\partial\theta}\log f_{X_i}(x_i|\theta)\sqrt{f_{X_i}(x_i|\theta)}\right)^2 dx_i$$

since:

$$\int\left(\prod_{j\neq i}\sqrt{f_{X_j}(x_j|\theta)}\right)^2 \underbrace{dx_1dx_2\ldots dx_{i-1}}\underbrace{dx_{i+1}\ldots dx_n} = 1, \forall i.$$

Note that for any pair of indices $i \neq j$,

$$\int \left[\frac{\partial}{\partial\theta}\log f_{X_i}(x_i|\theta)\sqrt{f_{X_1,X_2,\ldots,X_n}(x_1,x_2,\ldots,x_n|\theta)}\right] \times$$

$$\left[\frac{\partial}{\partial\theta}\log f_{X_j}(x_j|\theta)\sqrt{f_{X_1,X_2,\ldots,X_n}(x_1,x_2,\ldots,x_n|\theta)}\right] dx_1dx_2\ldots dx_n =$$

$$\int \frac{\partial}{\partial\theta}\log f_{X_i}(x_i|\theta)\,\frac{\partial}{\partial\theta}\log f_{X_j}(x_j|\theta)\, f_{X_1,X_2,\ldots,X_n}(x_1,x_2,\ldots,x_n|\theta)\, dx_1dx_2\ldots dx_n =$$

$$\int \left(\frac{\partial}{\partial\theta}\log f_{X_i}(x_i|\theta)\,\frac{\partial}{\partial\theta}\log f_{X_j}(x_j|\theta)\right)\prod_{l=1}^{n} f_{X_l}(x_l|\theta)\, dx_1dx_2\ldots dx_n =$$

$$\underbrace{\left(\int \frac{\partial}{\partial\theta}\log f_{X_i}(x_i|\theta)\, f_{X_i}(x_i|\theta)dx_i\right)}_{\mathbb{E}\left(\frac{\partial}{\partial\theta}\log f_X(x|\theta)|\theta\right)=0} \times \left(\int \frac{\partial}{\partial\theta}\log f_{X_j}(x_j|\theta)\prod_{l\neq i} f_{X_l}(x_l|\theta)dx_{-i}\right) = 0.$$

Therefore, using the Cauchy–Schwarz inequality, $\left|\int f(x)\overline{g(x)}dx\right|^2 \leq \left(\int |f(x)|^2dx\right)\left(\int |g(x)|^2dx\right)$, and the fact that

$$f_X(x|\theta)\,\tfrac{\partial}{\partial\theta}\log f_X(x|\theta) = f_X(x|\theta)\,\tfrac{1}{f_X(x|\theta)}\,\tfrac{\partial}{\partial\theta}f_X(x|\theta) = \tfrac{\partial}{\partial\theta}f_X(x|\theta),$$

we have:

$$1 = \int\left(\hat{\theta}(\boldsymbol{x}) - \theta\right)\frac{\partial}{\partial\theta}f_{\boldsymbol{X}}(\boldsymbol{x}|\theta)d\boldsymbol{x} = \int\left(\hat{\theta}(\boldsymbol{x}) - \theta\right)\left(\frac{\partial}{\partial\theta}\log f_{\boldsymbol{X}}(\boldsymbol{x}|\theta)\right)(f_{\boldsymbol{X}}(\boldsymbol{x}|\theta))d\boldsymbol{x} =$$

$$\int\left(\hat{\theta}(x_1,x_2,\ldots,x_n) - \theta\right)f_{X_1,X_2,\ldots,X_n}(x_1,x_2,\ldots,x_n|\theta)\times$$

$$\frac{\partial}{\partial\theta}\log f_{X_1,X_2,\ldots,X_n}(x_1,x_2,\ldots,x_n|\theta)dx_1dx_2\ldots dx_n =$$

$$\left(\int\left(\hat{\theta}(x_1,x_2,\ldots,x_n)-\theta\right)f_{X_1,X_2,\ldots,X_n}(x_1,x_2,\ldots,x_n|\theta)\frac{\partial}{\partial\theta}\left(\log\left(\prod_{i=1}^{n}f_{X_i}(x_i|\theta)\right)\right)dx_1dx_2\ldots dx_n\right)^2 =$$

$$\left(\int\left(\hat{\theta}(x_1,x_2,\ldots,x_n)-\theta\right)\left(\sum_{i=1}^{n}\frac{\partial}{\partial\theta}\log f_{X_i}(x_i|\theta)\right)f_{X_1,X_2,\ldots,X_n}(x_1,x_2,\ldots,x_n|\theta)dx_1dx_2\ldots dx_n\right)^2 \leq$$

$$\left(\int\left(\left(\hat{\theta}(x_1,x_2,\ldots,x_n) - \theta\right)\sqrt{f_{X_1,X_2,\ldots,X_n}(x_1,x_2,\ldots,x_n|\theta)}\right)^2 dx_1dx_2\ldots dx_n\right)\times$$

$$\left(\int\left(\sum_{i=1}^{n}\frac{\partial}{\partial\theta}\log f_{X_i}(x_i|\theta)\sqrt{f_{X_1,X_2,\ldots,X_n}(x_1,x_2,\ldots,x_n|\theta)}\right)^2 dx_1dx_2\ldots dx_n\right) =$$

$$\left(\int\left(\left(\hat{\theta}(x_1,x_2,\ldots,x_n)-\theta\right)\right)^2 f_{X_1,X_2,\ldots,X_n}(x_1,x_2,\ldots,x_n|\theta)dx_1dx_2\ldots dx_n\right)\times$$

$$\underbrace{\sum_{i=1}^{n}\int\left(\frac{\partial}{\partial\theta}\log f_{X_i}(x_i|\theta)\right)^2 f_{X_i}(x_i|\theta)dx_i}_{\text{IID},\, n\int\left(\frac{\partial}{\partial\theta}\log f_X(x|\theta)\right)^2 f_X(x|\theta)dx}.$$

Hence, an alternative information-theoretic formulation of the data science uncertainty principle is:

$$\int\left(\left(\hat{\theta}(x_1,x_2,\ldots,x_n)-\theta\right)\right)^2 f_{X_1,X_2,\ldots,X_n}(x_1,x_2,\ldots,x_n|\theta)dx_1dx_2\ldots dx_n\times$$

$$\int\left(\frac{\partial}{\partial\theta}\log f_X(x|\theta)\right)^2 f_X(x|\theta)dx\geq\frac{1}{n}.$$

In other words,

$$\underbrace{\int\left(\left(\hat{\theta}(\mathbf{x})-\theta\right)\right)^2 f_X(\mathbf{x}|\theta)d\mathbf{x}}_{\text{Variance of the estimate }\hat{\theta}}\times\underbrace{\int\left(\frac{\partial}{\partial\theta}\log f_X(x|\theta)\right)^2 f_X(x|\theta)dx}_{\substack{\text{Variance of the score,}\\ \text{Fisher Information, I}(\theta)}}\geq\frac{1}{n}.$$

For the more general case with possible estimator bias, the Cramér–Rao lower bound can be similarly derived in terms of the bias expectation, $\mathbb{E}\left(\hat{\theta}(X)-\theta|\theta\right)=\underbrace{\mathbb{E}\left(\hat{\theta}(X)|\theta\right)}_{\psi(\theta)}-\theta$:

$$\underbrace{\int\left(\left(\hat{\theta}(\mathbf{x})-\theta\right)\right)^2 f_X(\mathbf{x}|\theta)d\mathbf{x}}_{\text{Variance of the estimate }\hat{\theta}}\times\underbrace{\int\left(\frac{\partial}{\partial\theta}\log f_X(x|\theta)\right)^2 f_X(x|\theta)dx}_{\substack{\text{Variance of the score,}\\ \{\text{Fisher Information, I}(\theta)}}\geq\frac{\left(\psi'(\theta)\right)^2}{n}=$$

$$\frac{1}{n}\left(\frac{\partial}{\partial\theta}\psi(\theta)\right)^2>0.$$

The Cramér–Rao inequality effectively states that for a given sample size, the observed data can't yield parameter estimates that have simultaneously low score function variance (amount of information about the unknown parameter contained in the observables) *and* low estimator variance (precision). A decrease in one of these variances drives the other proportionately higher. That is, for a given sample size, we cannot guarantee that the estimator is arbitrary precise while also extracting a lot of reliable information about the parameter of interest. This delicate balance between precision and information is reminiscent of Gödel's completeness and consistency theorem [23] and the Heisenberg's uncertainty principle.

5.10.4.3 Normal Model Estimation Example

To illustrate the information-theoretic uncertainty principle, suppose we are modeling a normal distribution process using a sample of n random observations $\{x_k\}_{k=1}^{n} \sim N(\mu, \sigma^2)$. To estimate the unknown parameter vector $\theta = (\mu, \sigma^2)$, we will use the unbiased estimator vector

$$\hat{\theta} = \left(\hat{\mu} = \frac{1}{n}\sum_{k=1}^{n} x_k, \hat{\sigma}^2 = \frac{1}{n-1}\sum_{k=1}^{n}(x_k - \hat{\mu})^2\right)'.$$

Recall that for any distribution, the following parameter vector

$$\hat{\tau} = \left(\hat{\mu} = \tfrac{1}{n}\sum_{k=1}^{n} x_k,\ \hat{\sigma}^2 = \tfrac{1}{n-1}\sum_{k=1}^{n}(x_k - \hat{\mu})^2,\ \hat{\mu}_3 = \tfrac{1}{n}\sum_{k=1}^{n}(x_k - \hat{\mu})^3\right)'$$

represents an unbiased estimate for the vector of the first three moments, mean, variance, and third-moment, i.e., the skewness scaled by a factor $\hat{\sigma}^3$:

$$\tau = \left(\mu = \mathbb{E}(X),\ \sigma^2 = \mathbb{E}(X-\mu)^2,\ \mu_3 = \mathbb{E}(X-\mu)^3\right).$$

A necessary and sufficient condition for the mutual independence of the sample mean and sample variance is that the underlying distribution is normal [288]. The fact that the sample mean and sample variance are always uncorrelated $\left(Cov\left(\hat{\mu}, \hat{\sigma}^2\right) = 0\right)$, is true for any symmetric distribution ($\mu_3 = 0$). This is because

$$\mathbb{E}\left((x_k-\mu)(x_l-\mu)^2\right) = \underbrace{\mathbb{E}(x_k-\mu)}_{0} \times \mathbb{E}\left((x_l-\mu)^2\right) = 0,\ \forall k \neq l\ (\text{independence}),$$

$$Cov\left(\hat{\mu}, \hat{\sigma}^2\right) = \mathbb{E}\left(\hat{\mu} \times \hat{\sigma}^2\right) - \mathbb{E}(\hat{\mu}) \times \mathbb{E}\left(\hat{\sigma}^2\right) = \mathbb{E}\left((\hat{\mu} - \mu + \mu) \times \hat{\sigma}^2\right) - \mu\mathbb{E}\left(\hat{\sigma}^2\right) = \mathbb{E}\left((\hat{\mu} - \mu)\hat{\sigma}^2\right) =$$

$$\mathbb{E}\left(\underbrace{\left(\frac{1}{n}\sum_{k=1}^{n}(x_k-\mu)\right)}_{(\hat{\mu}-\mu)} \underbrace{\left(\frac{1}{n-1}\left[\sum_{k=1}^{n}(x_k-\mu)^2 - \frac{1}{n}\left(\sum_{k=1}^{n}(x_k-\mu)\right)^2\right]\right)}_{\hat{\sigma}^2}\right) =$$

$$\frac{1}{n(n-1)}\mathbb{E}\left(\left(\sum_{k=1}^{n}(x_k-\mu)\right)\left(\sum_{l=1}^{n}(x_l-\mu)^2\right) - \frac{1}{n}\left(\sum_{k=1}^{n}(x_k-\mu)\right)\left(\sum_{k=1}^{n}(x_k-\mu)\right)^2\right) =$$

$$\frac{1}{n(n-1)}\left[\underbrace{\mathbb{E}\left(\sum_{l=1}^{n}(x_l-\mu)^3\right)}_{n\mu_3} - \frac{1}{n}\mathbb{E}\left(\left(\sum_{k=1}^{n}(x_k-\mu)\right)\left(\sum_{k=1}^{n}(x_k-\mu)\right)^2\right)\right] =$$

$$\frac{1}{n(n-1)}\left[n\mu_3-\frac{1}{n}\mathbb{E}\left(\left(\sum_{k=1}^{n}(x_k-\mu)\right)\left(\sum_{k=1}^{n}(x_k-\mu)^2+\sum_{\substack{k,l=1\\ l\neq k}}^{n}(x_k-\mu)(x_l-\mu)\right)\right)\right]=$$

$$\frac{1}{n(n-1)}\left[n\mu_3-\mathbb{E}\left(\frac{1}{n}\sum_{k=1}^{n}(x_k-\mu)^3\right)\right]=\frac{1}{n(n-1)}\left[n\mu_3-\mu_3\right]=\frac{(n-1)\mu_3}{n(n-1)}=\frac{\mu_3}{n}.$$

Even though all symmetric distributions ($\mu_3\equiv 0$) have uncorrelated sample mean and variance, i.e., $Cov\left(\hat{\mu},\hat{\sigma}^2\right)=0$, the normal distribution is special since it is the only one that guarantees that the sample mean and variance are also mutually independent (a much stronger condition than being uncorrelated).

In our example, the density of the normal distribution model is $f_X(x|\theta)=\frac{1}{\sqrt{2\pi\sigma^2}}e^{-\frac{(x-\mu)^2}{2\sigma^2}}$ and given the observations $\{x_k\}_{k=1}^{n}$, the likelihood of the parameter vector θ is:

$$l(\theta|x_1,x_2,\ldots,x_n)=\prod_{k=1}^{n}\frac{1}{\sqrt{2\pi\sigma^2}}e^{-\frac{(x-\mu)^2}{2\sigma^2}}=\frac{1}{\left(\sqrt{2\pi\sigma^2}\right)^n}e^{-\sum_{k=1}^{n}\frac{(x_k-\mu)^2}{2\sigma^2}}.$$

The log likelihood, ll, and it's first and second order partial derivatives evaluated with respect to the parameter vector $\theta=\Big(\underbrace{\mu}_{\theta_1},\underbrace{\sigma^2}_{\theta_2}\Big)'$ are:

$$ll(\theta|x_1,x_2,\ldots,x_n)=-\frac{n}{2}\log\left(2\pi\sigma^2\right)-\sum_{k=1}^{n}\frac{(x_k-\mu)^2}{2\sigma^2},$$

$$\frac{\partial}{\partial\theta_1}ll(\theta)=\sum_{k=1}^{n}\frac{2(x_k-\mu)}{2\sigma^2}=\sum_{k=1}^{n}\frac{(x_k-\mu)}{\sigma^2},$$

$$\frac{\partial}{\partial\theta_2}ll(\theta)=-\frac{n}{2}\frac{2\pi}{2\pi\sigma^2}+\sum_{k=1}^{n}\frac{(x_k-\mu)^2}{2\sigma^4}=-\frac{n}{2\sigma^2}+\sum_{k=1}^{n}\frac{(x_k-\mu)^2}{2\sigma^4},$$

$$\frac{\partial^2}{\partial\theta_2\partial\theta_1}ll(\theta)\equiv\frac{\partial^2}{\partial\theta_1\partial\theta_2}ll(\theta)=-\sum_{k=1}^{n}\frac{x_k-\mu}{\sigma^4},$$

$$\frac{\partial^2}{\partial\theta_1^2}ll(\theta)=-\sum_{k=1}^{n}\frac{1}{\sigma^2}=-\frac{n}{\sigma^2},$$

$$\frac{\partial^2}{\partial\theta_2^2}ll(\theta)=\frac{n}{2\sigma^4}-\sum_{k=1}^{n}\frac{(x_k-\mu)^2}{(\sigma^2)^3}=\frac{n}{2\sigma^4}-\sum_{k=1}^{n}\frac{(x_k-\mu)^2}{\sigma^6}.$$

As we are trying to estimate the parameter vector $\hat{\theta} = (\theta_1, \theta_2)' = \left(\hat{\mu}, \hat{\sigma}^2\right)'$, the Fisher information will be a symmetric 2×2 matrix that captures the relative slope of the log-likelihood function with respect to each of the 2 parameters:

$$I(\theta) = (I_{i,j}(\theta)) = \mathbb{E}_\theta(|\nabla_\theta \log f_X\rangle \langle \nabla_\theta \log f_X|) = \mathbb{E}_\theta((\nabla_\theta \log f_X)(\nabla_\theta \log f_X)') =$$

$$\begin{bmatrix} \mathbb{E}_\theta\left(\frac{\partial}{\partial\theta_1}\log f_X \frac{\partial}{\partial\theta_1}\log f_X\right) & \mathbb{E}_\theta\left(\frac{\partial}{\partial\theta_1}\log f_X \frac{\partial}{\partial\theta_2}\log f_X\right) \\ \mathbb{E}_\theta\left(\frac{\partial}{\partial\theta_2}\log f_X \frac{\partial}{\partial\theta_1}\log f_X\right) & \mathbb{E}_\theta\left(\frac{\partial}{\partial\theta_2}\log f_X \frac{\partial}{\partial\theta_2}\log f_X\right) \end{bmatrix},$$

where:

$$\nabla_\theta g(\theta) = \left(\frac{\partial}{\partial\theta_1} g, \frac{\partial}{\partial\theta_2} g\right)',$$

$$I_{i,j}(\theta) = \mathbb{E}_\theta\left(\frac{\partial}{\partial\theta_i}\log f_X(x|\theta) \frac{\partial}{\partial\theta_j}\log f_X(x|\theta)\right) =$$

$$\int \left(\frac{\partial}{\partial\theta_i}\log f_X(x|\theta)\right)\left(\frac{\partial}{\partial\theta_j}\log f_X(x|\theta)\right) f_X(x|\theta)\, dx, \forall 1 \le i, j \le 2.$$

Following the previous discussion,

$$\mathbb{E}_\theta\left(\frac{\partial}{\partial\theta_i}\log f_X(x|\theta)\right) = \int \left(\frac{\partial}{\partial\theta_i}\log f_X(x|\theta)\right) f_X(x|\theta)\, dx =$$

$$\int \left(\frac{\frac{\partial}{\partial\theta_i} f_X(x|\theta)}{f_X(x|\theta)}\right) f_X(x|\theta) dx = \frac{\partial}{\partial\theta_i} \underbrace{\int f_X(x|\theta) dx}_{1} = 0.$$

Similarly,

$$\mathbb{E}_\theta\left(\frac{\frac{\partial^2}{\partial\theta_i\partial\theta_j} f_X(x|\theta)}{f_X(x|\theta)}\right) = \int \left(\frac{\frac{\partial^2}{\partial\theta_i\partial\theta_j} f_X(x|\theta)}{f_X(x|\theta)}\right) f_X(x|\theta)\, dx = \int \frac{\partial^2}{\partial\theta_i\partial\theta_j} f_X(x|\theta)\, dx =$$

$$\frac{\partial^2}{\partial\theta_i\partial\theta_j} \underbrace{\int f_X(x|\theta) dx}_{1} = 0.$$

Thus, $I_{i,j}(\theta) = Cov_\theta\left(\frac{\partial}{\partial\theta_i}\log f_X(x|\theta), \frac{\partial}{\partial\theta_j}\log f_X(x|\theta)\right) = \left(\mathbb{E}_\theta((\nabla_\theta \log f_X)(\nabla_\theta \log f_X)')\right)_{i,j}$. To show that the Fisher information matrix is a semi-positive definite covariance matrix, let's set $\xi_i \equiv \frac{\partial}{\partial\theta_i}\log f_X(x|\theta)$. Then, $\forall v = (v_1, v_2) \in \mathbb{R}^2$ we have:

$$0 \leq Var\left(\sum_i v_i \xi_i\right) = \mathbb{E}\left(\sum_i v_i \xi_i (\xi_i - \mathbb{E}(\xi_i))\right)^2 = \sum_{i,j} \mathbb{E}\left(v_i v_j (\xi_i - \mathbb{E}(\xi_i))(\xi_i - \mathbb{E}(\xi_i))\right) =$$

$$\sum_{i,j} \left[v_i \mathbb{E}(\xi_i - \mathbb{E}(\xi_i))(\xi_i - \mathbb{E}(\xi_i)) v_j\right] = \sum_{i,j} v_i \, Cov\left(\xi_i, \xi_j\right) v_j = \sum_{i,j} v_i I_{i,j}(\theta) v_j = v' \underbrace{I(\theta)}_{2\times 2} v.$$

Under regularity conditions, the second order derivatives exist and integration and differentiation can be swapped as the support of the integrand does not depend the parameter we have:

$$\frac{\partial^2}{\partial\theta_i \partial\theta_j} \underbrace{\log f_X(x|\theta)}_{ll} = \frac{\frac{\partial^2}{\partial\theta_i \partial\theta_j} f_X(x|\theta)}{f_X(x|\theta)} - \frac{\frac{\partial}{\partial\theta_i} f_X(x|\theta)}{f_X(x|\theta)} \times \frac{\frac{\partial}{\partial\theta_j} f_X(x|\theta)}{f_X(x|\theta)},$$

$$-\mathbb{E}_\theta\left(\frac{\partial^2}{\partial\theta_i \partial\theta_j} \underbrace{\log f_X(x|\theta)}_{ll}\right) = \mathbb{E}_\theta\left(\frac{\partial}{\partial\theta_i} \log f_X(x|\theta) \frac{\partial}{\partial\theta_j} \log f_X(x|\theta)\right) \equiv I_{i,j}(\theta).$$

Returning to our problem, we are trying to estimate the 2-parameter vector $\theta = (\theta_1, \theta_2)'$. The Fisher information matrix is:

$$I(\theta) = \begin{bmatrix} \mathbb{E}_\theta\left(\frac{\partial}{\partial\theta_1} \log f_X \frac{\partial}{\partial\theta_1} \log f_X\right) & \mathbb{E}_\theta\left(\frac{\partial}{\partial\theta_1} \log f_X \frac{\partial}{\partial\theta_2} \log f_X\right) \\ \mathbb{E}_\theta\left(\frac{\partial}{\partial\theta_2} \log f_X \frac{\partial}{\partial\theta_1} \log f_X\right) & \mathbb{E}_\theta\left(\frac{\partial}{\partial\theta_2} \log f_X \frac{\partial}{\partial\theta_2} \log f_X\right) \end{bmatrix}$$

$$= \begin{bmatrix} I_{1,1} & I_{1,2} \\ I_{2,1} & I_{2,2} \end{bmatrix}.$$

The four components of the Fisher information matrix, $I(\theta)_{2\times 2}$, track the aggregate amount of information about the parameter vector, $\theta = (\theta_1, \theta_2)' \equiv (\mu, \sigma^2)'$, that is captured by the random process in terms of the observed data $\{x_k\}_{k=1}^n$.

$$I_{1,1} = -\mathbb{E}_{\theta_1,\theta_1}\left(\frac{\partial^2}{\partial\theta_1^2} ll(\theta)\right) = -\mathbb{E}_{\theta_1,\theta_1}\left(\frac{\partial}{\partial\mu} \sum_{k=1}^n \frac{(x_k - \mu)}{\sigma^2}\right) = \frac{n}{\sigma^2}$$

$$I_{2,2} = -\mathbb{E}_{\theta_2,\theta_2}\left(\frac{\partial^2}{\partial\theta_2^2} ll(\theta)\right) = -\mathbb{E}_{\theta_2,\theta_2}\left(\frac{\partial}{\partial\theta_2} \sum_{k=1}^n -\frac{1}{2\sigma^2} + \frac{(x_k - \mu)^2}{2(\sigma^2)^2}\right) =$$

$$-\mathbb{E}_{\sigma^2,\sigma^2}\left(\frac{n}{2\sigma^4} - \sum_{k=1}^n \frac{(x_k - \mu)^2}{(\sigma^2)^3}\right) = -\frac{n}{2\sigma^4} + n\frac{\sigma^2}{(\sigma^2)^3} = \frac{n}{2\sigma^4}$$

$$I_{1,2} = I_{2,1} = -\mathbb{E}_{\sigma^2,\mu}\left(\frac{\partial^2}{\partial\theta_2 \partial\theta_1} ll(\theta)\right) = \mathbb{E}_{\sigma^2,\mu}\left(\sum_{k=1}^n \frac{x_k - \mu}{\sigma^4}\right) = \frac{1}{\sigma^4} \mathbb{E}_{\sigma^2,\mu}\left(\sum_{k=1}^n (x_k - \mu)\right) = 0.$$

Therefore, the Fisher information matrix and its inverse are

$$I(\theta) = \begin{bmatrix} I_{1,1} I_{1,2} \\ I_{2,1} I_{2,2} \end{bmatrix} = - \begin{bmatrix} \mathbb{E}_{\theta_1,\theta_2}\left(\frac{\partial^2}{\partial\theta_1^2} ll(\theta)\right) \mathbb{E}_{\theta_1,\theta_2}\left(\frac{\partial^2}{\partial\theta_1\partial\theta_2} ll(\theta)\right) \\ \mathbb{E}_{\theta_2,\theta_1}\left(\frac{\partial^2}{\partial\theta_2\partial\theta_1} ll(\theta)\right) \mathbb{E}_{\theta_2,\theta_2}\left(\frac{\partial^2}{\partial\theta_2^2} ll(\theta)\right) \end{bmatrix} = \begin{bmatrix} \frac{n}{\sigma^2} & 0 \\ 0 & \frac{n}{2\sigma^4} \end{bmatrix},$$

$$I^{-1}(\theta) = \begin{bmatrix} \frac{n}{\sigma^2} & 0 \\ 0 & \frac{n}{2\sigma^4} \end{bmatrix}^{-1} = \begin{bmatrix} \frac{\sigma^2}{n} & 0 \\ 0 & \frac{2\sigma^4}{n} \end{bmatrix}.$$

As the estimator vector

$$\hat{\theta} = \hat{\theta}(x_1, x_2, \ldots, x_n) = \left(\hat{\theta}_1, \hat{\theta}_2\right)' = \left(\hat{\mu} = \frac{1}{n}\sum_{k=1}^{n} x_k,\ \hat{\sigma}^2 = \frac{1}{n-1}\sum_{k=1}^{n}(x_k - \hat{\mu})^2\right)'$$

is unbiased,

$$\beta'(\theta) = \left(\frac{\partial}{\partial\theta_1} \underbrace{\mathbb{E}_\theta\left(\hat{\theta}\right)}_{\begin{pmatrix}\theta_1\\ \theta_2\end{pmatrix} = \begin{pmatrix}\mu\\ \sigma^2\end{pmatrix}},\ \frac{\partial}{\partial\theta_2} \underbrace{\mathbb{E}_\theta\left(\hat{\theta}\right)}_{\begin{pmatrix}\theta_1\\ \theta_2\end{pmatrix} = \begin{pmatrix}\mu\\ \sigma^2\end{pmatrix}} \right) = \begin{pmatrix} \frac{\partial}{\partial\mu}\mu & \frac{\partial}{\partial\sigma^2}\mu \\ \frac{\partial}{\partial\mu}\sigma^2 & \frac{\partial}{\partial\sigma^2}\sigma^2 \end{pmatrix} = \begin{pmatrix} 1 & 0 \\ 0 & 1 \end{pmatrix} = I_{2\times 2}.$$

Therefore, the information-theoretic form of the uncertainty principle manifests as semi-positive-definiteness of a quadratic form $Cov_\theta\left(\hat{\theta}\right) \geq I^{-1}(\theta)$, i.e.,

$$Cov_\theta\left(\hat{\theta}(x_1, x_2, \ldots, x_n)\right) \geq \beta'(\theta)\, I^{-1}(\theta)\, \beta(\theta) = \begin{pmatrix} 1 & 0 \\ 0 & 1 \end{pmatrix} I^{-1}(\theta) \begin{pmatrix} 1 & 0 \\ 0 & 1 \end{pmatrix} = I^{-1}(\theta).$$

Recall that for any distribution $Cov\left(\hat{\mu}, \hat{\sigma}^2\right) = \frac{\mu_3}{n}$ and for symmetric distributions $\mu_3 = 0$, i.e., the covariance of the sample mean and sample variance is trivial. Also, since $\hat{\mu} \sim N\left(\mu, \frac{\sigma^2}{n}\right)$ and $\frac{(n-1)\hat{\sigma}^2}{\sigma^2} \sim \chi^2_{\underbrace{n-1}_{\text{df}}} \equiv \Gamma\left(\underbrace{\alpha = \frac{n-1}{2}}_{\text{scale}}, \underbrace{\beta = 2}_{\text{shape}}\right)$, the variance of the parameter estimate vector is:

$$Var\left(\hat{\theta}\right) = \begin{pmatrix} Var(\hat{\mu}) \\ Var\left(\hat{\sigma}^2\right) \end{pmatrix} = \begin{pmatrix} \frac{\sigma^2}{n} \\ \frac{2\sigma^4}{n-1} \end{pmatrix}.$$

This follows from:

$$Var(\hat{\mu}) = \frac{\sigma^2}{n},$$

$$Var\left(\frac{(n-1)\hat{\sigma}^2}{\sigma^2} \sim \chi^2_{n-1}\right) = Var\left(\Gamma\left(\alpha = \frac{n-1}{2}, \beta = 2\right)\right) = \alpha\beta^2 = 2(n-1),$$

$$Var\left(\hat{\sigma}^2\right) = Var(S^2) = 2(n-1)\frac{\sigma^4}{(n-1)^2} = \frac{2\sigma^4}{n-1}.$$

$$Cov_\theta\left(\hat{\theta}\right) = \begin{bmatrix} Cov(\hat{\mu},\hat{\mu}) & Cov\left(\hat{\mu},\hat{\sigma}^2\right) \\ Cov\left(\hat{\mu},\hat{\sigma}^2\right) & Cov\left(\hat{\sigma}^2,\hat{\sigma}^2\right) \end{bmatrix} = \begin{bmatrix} Var(\hat{\mu}) & \frac{\mu_3}{n} \\ \frac{\mu_3}{n} & Var\left(\hat{\sigma}^2\right) \end{bmatrix} = \begin{bmatrix} \frac{\sigma^2}{n} & 0 \\ 0 & \frac{2\sigma^4}{n-1} \end{bmatrix} \geq I^{-1}(\theta) \equiv \begin{bmatrix} \frac{\sigma^2}{n} & 0 \\ 0 & \frac{2\sigma^4}{n-1} \end{bmatrix}.$$

Finally, the representation of the uncertainty principle in terms of a semi-positive-definite quadratic form becomes:

$$M \equiv Cov_\theta\left(\hat{\theta}\right) - I^{-1}(\theta) > 0,$$

$$M = \begin{bmatrix} \frac{\sigma^2}{n} & 0 \\ 0 & \frac{2\sigma^4}{n-1} \end{bmatrix} - \begin{bmatrix} \frac{\sigma^2}{n} & 0 \\ 0 & \frac{2\sigma^4}{n-1} \end{bmatrix} = \begin{bmatrix} 0 & 0 \\ 0 & \frac{2\sigma^4}{n(n-1)} \end{bmatrix} > 0,$$

$$v' \underbrace{M}_{2\times 2} v > 0, \quad \forall v \in \{\mathbb{R}\backslash 0\} \times \{\mathbb{R}\backslash 0\}.$$

In other words, for non-trivial processes, there is a lower limit to simultaneously reducing the variability of the estimate vector at the same time as we reduce the variance of the score matrix.

5.10.4.4 Single-tone Example

Another demonstration of the information-theoretic uncertainly principle in the field of data science is based on a complex-valued single-tone process [289, 290]:

$$Y(t|\theta) = m(t,\theta) \equiv A\, e^{i(\omega t + \varphi)} \in \mathbb{C},$$

where the parameter vector $\theta = (\theta_1, \theta_2, \theta_3) = (A, \varphi, \omega)$ specifies the signal amplitude, phase, and frequency. This is a special case of a more general multi-frequency tone process $Y(t|\theta) = m(t,\theta) \equiv \sum_{l=1}^{L} A_l e^{i(\omega_l t + \varphi_l)}$ that superimposes a number of single-tones into a more nuanced pattern.

Suppose we record a sequence of random observations from this process, $\{Y_t\}_{t=0}^{n-1}$, and model the complex-valued process using additive Gaussian noise,

$$Y_t = m_t(\theta) + \epsilon_t, \qquad \epsilon_t \sim N\left(0, \Sigma = \sigma^2 I_{2\times 2}\right).$$

Assume that the process expectation is $E_\theta(Y_t) = m_t(\theta) = \begin{pmatrix} \mu(\theta) \\ \nu(\theta) \end{pmatrix} \in \mathbb{C}$, the (real and imaginary) process variance σ^2 is known, and we are aiming to estimate the time (t) invariant parameter vector $\theta = (\theta_1, \theta_2, \theta_3) = (A, \varphi, \omega)$ consisting of the process amplitude (A), phase angle (φ), and angular frequency (ω).

Solutions to this single-tone estimation problem have various applications in power systems, communication-system carrier-recovery, object localization using radar and sonar systems, estimation of prenatal heart rates, to name a few [291, 292].

Suppose the estimation of the exponential single tone complex-valued model is based on n discrete time IID observations:

$$Y_t = \underbrace{A\, e^{i(\omega(t\cdot T + t_0) + \varphi)}}_{\text{True signal}} + \underbrace{\epsilon_t}_{\text{Noise}}, \forall 0 \le t \le n-1,$$

where the amplitude $A > 0$, the phase $-\pi \le \varphi < \pi$, and the frequency $0 < \omega < \pi$ are all non-random model parameters. Assume the data sampling rate is $\frac{1}{T}$ (recording 1 observation in time T) and the initial (starting) time point, t_0, can be specified. The real and imaginary parts of the signal noise (error) ϵ_t, which potentially corrupts the original signal, are modeled as independent zero-mean Gaussian processes with variance σ^2. Given the parameter vector, $\theta = (\theta_1, \theta_2, \theta_3) = (A, \varphi, \omega)$, the probability distribution of this process may be expressed in terms of the real, $Re(Y_t)$, and imaginary, $Im(Y_t)$, parts of the observations:

$$f(Y_0, \ldots, Y_{n-1}|\theta) = \prod_{t=0}^{n-1} \frac{1}{\sqrt{2\pi\sigma^2}} e^{-\frac{\epsilon'_t \epsilon_t}{2\sigma^2}} =$$

$$\prod_{t=0}^{n-1} \frac{1}{\sqrt{2\pi\sigma^2}} e^{-\frac{1}{2\sigma^2}\left(\left(Re(Y_t) - \underbrace{Re\left(A\, e^{i(\omega(t\cdot T + t_0)+\varphi)}\right)}_{\mu_t}\right)^2 + \left(Im(Y_t) - \underbrace{Im\left(A\, e^{i(\omega(t\cdot T + t_0)+\varphi)}\right)}_{\nu_t}\right)^2\right)}.$$

We can estimate the parameter vector $\hat{\theta} = \left(\hat{A}, \hat{\varphi}, \hat{\omega}\right)$ in terms of the process distribution of Y, $f_Y = f_Y(Y|\theta)$. For a multivariate normal distribution, $Y \sim N(m, \Sigma)$, with a mean vector is $m = m(\theta)$ and a variance-covariance matrix Σ independent of θ, the elements of the 3×3 Fisher information are expressed as $I_{i,j} = \frac{\partial m'}{\partial \theta_i} \Sigma^{-1} \frac{\partial m}{\partial \theta_j}$.

As Y is complex-valued, the n samples yield $2n$ observations (independent real and imaginary parts) and the density of the multivariate normal distribution is:

$$f(y|m, \Sigma) = (2\pi)^{-n} \underbrace{|\Sigma|}_{\det \Sigma}{}^{-\frac{n}{2}} e^{-\frac{1}{2}(y-m)'\Sigma^{-1}(y-m)},$$

with a corresponding log-likelihood function:

$$l(m,\Sigma|x) = \underbrace{-n\log(2\pi) - \frac{n}{2}\log|\Sigma|}_{\text{independent of }\theta} - \tfrac{1}{2}(x-m)'\Sigma^{-1}(x-m) \propto -\tfrac{1}{2}(x-m)'\Sigma^{-1}(x-m).$$

$$I_{i,j} = \mathbb{E}\left[\left(\frac{\partial l}{\partial\theta_i}\right)'\frac{\partial l}{\partial\theta_j}\right] = \mathbb{E}\left[\left(\underbrace{\frac{\partial l}{\partial(x-m)}}_{\substack{\text{quadratic form}\\ \text{derivative}}}\frac{\partial(x-m)}{\partial\theta_i}\right)'\frac{\partial l}{\partial(x-m)}\frac{\partial(x-m)}{\partial\theta_j}\right] =$$

$$\mathbb{E}\left[\left((x-m)'\Sigma^{-1}\frac{\partial m}{\partial\theta_i}\right)'(x-m)'\Sigma^{-1}\frac{\partial m}{\partial\theta_j}\right] = \frac{\partial m'}{\partial\theta_i}(\Sigma^{-1})'\underbrace{\mathbb{E}[(x-m)(x-m)']}_{\Sigma}\Sigma^{-1}\frac{\partial m}{\partial\theta_j} =$$

$$\frac{\partial m'}{\partial\theta_i}\Sigma^{-1}\Sigma\Sigma^{-1}\frac{\partial m}{\partial\theta_j} = \frac{\partial m'}{\partial\theta_i}\Sigma^{-1}\frac{\partial m}{\partial\theta_j}.$$

In our $(2n)$-variate case, $Y \sim N\left(m = \begin{pmatrix}\mu(\theta)\\ \nu(\theta)\end{pmatrix}, \sigma^2 \boldsymbol{I}_{2\times 2}\right)$:

$$I_{i,j} = \frac{\partial}{\partial\theta_i}\begin{pmatrix}\mu\\ \nu\end{pmatrix}'\Sigma^{-1}\frac{\partial}{\partial\theta_j}\begin{pmatrix}\mu\\ \nu\end{pmatrix} = \frac{1}{\sigma^2}\sum_{t=0}^{n-1}\left[\frac{\partial\mu_t}{\partial\theta_i}\frac{\partial\mu_t}{\partial\theta_j} + \frac{\partial\nu_t}{\partial\theta_i}\frac{\partial\nu_t}{\partial\theta_j}\right].$$

Let's denote by $n_0 = \frac{t_0}{T}$ the number of missing initial observations prior to the start of data acquisition, and also denote the sum of the indices and the sum of the squared indices by:

$$P = \sum_{l=0}^{n-1} l = \frac{n(n-1)}{2} \text{ and } Q = \sum_{j=0}^{n-1} j^2 = \frac{n(n-1)(2n-1)}{6}.$$

Then, we can explicate the term $I_{3,3}$ in the 3×3 Fisher information matrix:

$$I_{3,3} = \frac{1}{\sigma^2}\sum_{t=0}^{n-1}\left[\frac{\partial(A\cos(\omega(t\cdot T+t_0)+\varphi))}{\partial\omega}\times\frac{\partial(A\cos(\omega(t\cdot T+t_0)+\varphi))}{\partial\omega} + \frac{\partial(A\sin(\omega(t\cdot T+t_0)+\varphi))}{\partial\omega}\times\frac{\partial(A\sin(\omega(t\cdot T+t_0)+\varphi))}{\partial\omega}\right] =$$

$$\frac{1}{\sigma^2}\sum_{t=0}^{n-1}\left[(-A(t\cdot T+t_0)\sin(\omega(t\cdot T+t_0)+\varphi))^2 + (-A(t\cdot T+t_0)\cos(\omega(t\cdot T+t_0)+\varphi))^2\right] =$$

$$\frac{1}{\sigma^2}\sum_{t=0}^{n-1}(A(t\cdot T+t_0))^2 = \frac{1}{\sigma^2}\sum_{t=0}^{n-1}(A(t\cdot T+n_0T))^2 =$$

$$\frac{1}{\sigma^2}\sum_{t=0}^{n-1}(A(t\cdot T+n_0T))^2=\frac{A^2T^2}{\sigma^2}\left(n_0^2n+2n_0P+Q\right).$$

Similarly, we can explicate the other 5 terms in the symmetric Fisher information matrix of the parameter vector $\theta=(A,\varphi,\omega)$:

$$I(\theta)=\left(I_{i,j}(\theta)\right)=\frac{1}{\sigma^2}\begin{bmatrix} n & 0 & 0 \\ 0 & A^2n & A^2T(n_0n+P) \\ 0 & A^2T(n_0n+P) & A^2T^2\left(n_0^2n+2n_0P+Q\right)\end{bmatrix}.$$

We can invert the Fisher information matrix provided we have:

$$A^2n\cdot A^2T^2\left(n_0^2n+2n_0P+Q\right)-\left(A^2T(n_0n+P)\right)^2\neq 0\Leftrightarrow nQ-P^2\neq 0\Leftrightarrow n^2\frac{n^2-1}{12}\neq 0.$$

The corresponding inverse of the Fisher information matrix is:

$$I(\theta)^{-1}=\sigma^2\begin{bmatrix}\frac{1}{n} & 0 & 0 \\ 0 & \frac{n_0^2n+2n_0P+Q}{A^2\left(nQ-P^2\right)} & \frac{n_0n+P}{A^2T\left(P^2-nQ\right)} \\ 0 & \frac{n_0n+P}{A^2T\left(P^2-nQ\right)} & \frac{n}{A^2T^2\left(nQ-P^2\right)}\end{bmatrix}=\sigma^2\begin{bmatrix}\frac{1}{n} & 0 & 0 \\ 0 & \frac{12\left(n_0^2n+2n_0P+Q\right)}{A^2n^2\left(n^2-1\right)} & -\frac{12(n_0n+P)}{A^2Tn^2\left(n^2-1\right)} \\ 0 & -\frac{12(n_0n+P)}{A^2Tn^2\left(n^2-1\right)} & \frac{12n}{A^2T^2n\left(n^2-1\right)}\end{bmatrix}.$$

Recall that the Cramér–Rao lower bound (CRLB) is a lower limit on the variance of unbiased estimators and the Fisher Information measures the amount of information that an observable random variable X carries about an unknown parameter θ associated with the distribution that models the process X [287, 293, 294].

Observe that in our case:

1. The Cramér–Rao lower bound (CRLB) for the magnitude ($\theta_1=A$), i.e., $I_{1,1}$, is independent of the other variables, whereas the CRLBs for the phase (φ), i.e., $I_{2,2}$, and the frequency (ω), i.e., $I_{3,3}$, are dependent on the magnitude (A).
2. When the sampling starting point is too late (t_0 is large), the reliability of the inference for the *phase* ($\theta_2=\varphi$) may be compromised. This problem can be rectified by increasing the subsequent sample size n. This ties to the well-known law of large numbers [164, 295].
3. The correlation between phase and frequency estimation may be disentangled only when $n_0n+P=0$. This choice also minimizes the CRLB for the phase (φ), as $\frac{\partial}{\partial n_0}\left[I(\theta)^{-1}\right]_{2,2}=0\Leftrightarrow n_0n+P=0$.

Thus far, we derived the *Fisher information* of the parameter vector $\theta=(A,\varphi,\omega)$. Next, we can examine the *variance* of the parameter estimate $\hat{\theta}=\left(\hat{A},\hat{\varphi},\hat{\omega}\right)$ for the observed process (Y) and the corresponding theoretical single-tone model (m):

$$Y_t=A\,e^{i(\omega t\cdot T+\varphi)}+\epsilon_t,\ t=n_0,\ n_0+1,\ \ldots,\ n_0+(n-1),$$

$$m_t = A\, e^{i(\omega t \cdot T + \varphi)},$$

$$Y_t = \left(1 + \underbrace{\frac{\epsilon_t}{m_t}}_{g_t} \right) m_t,$$

where, $g_t \equiv \frac{\epsilon_t}{m_t} = Re\left(\frac{\epsilon_t}{m_t}\right) + i\, Im\left(\frac{\epsilon_t}{m_t}\right)$ is intrinsically complex and $\epsilon_t \sim N(0, \sigma^2 \boldsymbol{I}_{2\times 2})$. As ϵ_t is invariant under orthogonal transformation, the rotation factor $e^{-i(\omega t \cdot T + \varphi)}$ will not affect its distribution and $Im\left(\frac{\epsilon_t}{m_t}\right) \sim N\left(0, \frac{\sigma^2}{A^2}\right)$. That is,

$$Var\left(Im\left(\frac{\epsilon_t}{m_t} \right) \right) = \frac{1}{A^2} \sigma^2 = \frac{1}{SNR},$$

$$1 + \frac{\epsilon_t}{m_t} = \underbrace{\sqrt{\left(1 + Re\left(\frac{\epsilon_t}{m_t}\right)\right)^2 + Im\left(\frac{\epsilon_t}{m_t}\right)^2}}_{\text{magnitude}} \times e^{\overbrace{i \arctan\left(\frac{Im\left(\frac{\epsilon_t}{m_t}\right)}{1 + Re\left(\frac{\epsilon_t}{m_t}\right)} \right)}^{\text{phase}}}.$$

Let's consider high signal-to-noise ratio (SNR), i.e., $1 \gg Re(g_t) \equiv Re\left(\frac{\epsilon_t}{m_t}\right)$ and $1 \gg Im(g_t)$. Then,

$$1 + \frac{\epsilon_t}{m_t} = \sqrt{\underbrace{\left(1 + Re\left(\frac{\epsilon_t}{m_t}\right)\right)^2 + Im\left(\frac{\epsilon_t}{m_t}\right)^2}_{\simeq 1}} \times \underbrace{e^{i \arctan\left(\frac{Im\left(\frac{\epsilon_t}{m_t}\right)}{1 + Re\left(\frac{\epsilon_t}{m_t}\right)} \right)}}_{\simeq e^{i \arctan Im\left(\frac{\epsilon_t}{m_t}\right)} \simeq e^{i Im\left(\frac{\epsilon_t}{m_t}\right)}} \simeq e^{i Im\left(\frac{\epsilon_t}{m_t}\right)}.$$

The phase uncertainty (error) is an additive Gaussian noise:

$$Y_t = A\, e^{i\left(\omega t \cdot T + \varphi + \underbrace{Im\left(\frac{\epsilon_t}{m_t}\right)}_{\substack{\text{additive phase}\\ \text{uncertainty}}} \right)}, \quad t = n_0, n_0 + 1, \ldots, n_0 + (n-1).$$

We can compute the Least Squares unbiased estimate of the *reduced* parameter vector $(\theta_2, \theta_3) = (\varphi, \omega)$:

$$\underbrace{E}_{\substack{\text{least squares}\\ \text{error}}} = \sum_{t=n_0}^{n_0 + (n-1)} \left(\underbrace{\omega t \cdot T + \varphi + Im\left(\frac{\epsilon_t}{m_t}\right)}_{\phi[t]} - \hat{\omega} t \cdot T - \hat{\varphi} \right)^2.$$

Setting to zero the partial derivatives with respect to $\hat{\varphi}$ and $\hat{\omega}$, we obtain:

$$0=\frac{\partial E}{\partial \hat{\varphi}}=2\sum_{t=n_0}^{n_0+(n-1)}(\phi[t]-\hat{\omega}t\cdot T-\hat{\varphi})\Leftrightarrow\sum_{t=n_0}^{n_0+(n-1)}\phi[t]=\hat{\omega}(n_0n+P)\cdot T+n\hat{\varphi},$$

$$0=\frac{\partial E}{\partial \hat{\omega}}=2\sum_{t=n_0}^{n_0+(n-1)}(\phi[t]-\hat{\omega}t\cdot T-\hat{\varphi})t\cdot T\Leftrightarrow$$

$$\sum_{t=n_0}^{n_0+(n-1)}tT\phi[t]=\hat{\omega}T^2\sum_{t=n_0}^{n_0+(n-1)}t^2+\hat{\varphi}T\sum_{t=n_0}^{n_0+(n-1)}t=\hat{\omega}T^2\sum_{t=0}^{n-1}(n_0+t)^2+\hat{\varphi}T\sum_{t=0}^{n-1}(n_0+t)=$$

$$\hat{\omega}T^2\left(n_0{}^2n+2n_0P+Q\right)+\hat{\varphi}T(n_0n+P).$$

Summarizing these estimations in a more concise matrix form,

$$\begin{bmatrix} n & T(n_0n+P) \\ T(n_0n+P) & T^2(n_0{}^2n+2n_0P+Q)\end{bmatrix}\begin{bmatrix}\hat{\varphi}\\ \hat{\omega}\end{bmatrix}=\begin{bmatrix}\sum_{t=n_0}^{n_0+n-1}\phi[t]\\ \sum_{t=n_0}^{n_0+n-1}tT\phi[t]\end{bmatrix}.$$

The matrix on the left can be inverted by observing that,

$$\underbrace{\begin{bmatrix} n & T(n_0n+P) \\ T(n_0n+P) & T^2(n_0{}^2n+2n_0P+Q)\end{bmatrix}}_{\text{(reduced) original Fisher information matrix}}\times\underbrace{\begin{bmatrix} T^2(n_0{}^2n+2n_0P+Q) & -T(n_0n+P) \\ -T(n_0n+P) & n\end{bmatrix}}_{\text{inverse matrix (up to a constant)}}=$$

$$T^2\left(nQ-P^2\right)\begin{bmatrix}1&0\\0&1\end{bmatrix}=n^2T^2\frac{n^2-1}{12}\begin{bmatrix}1&0\\0&1\end{bmatrix}.$$

Therefore,

$$\underbrace{\begin{bmatrix} A^2n & A^2T(n_0n+P) \\ A^2T(n_0n+P) & A^2T^2(n_0{}^2n+2n_0P+Q)\end{bmatrix}^{-1}}_{\text{original Fisher information submatrix}}=$$

$$\underbrace{\begin{bmatrix} \frac{1}{A^2}T^2(n_0{}^2n+2n_0P+Q)\frac{1}{T^2\left(nQ-P^2\right)} & \frac{1}{A^2}(-T(n_0n+P))\frac{1}{T^2\left(nQ-P^2\right)} \\ \frac{1}{A^2}(-T(n_0n+P))\frac{1}{T^2\left(nQ-P^2\right)} & \frac{1}{A^2}n\frac{1}{T^2\left(nQ-P^2\right)}\end{bmatrix}}_{\text{inverted submatrix}}.$$

The reduced parameter estimate for $(\theta_2,\theta_3)=(\varphi,\,\omega)$ becomes:

$$\begin{bmatrix}\hat{\varphi}\\ \hat{\omega}\end{bmatrix}=\frac{1}{T^2(nQ-P^2)}\begin{bmatrix} T^2(n_0{}^2n+2n_0P+Q) & -T(n_0n+P) \\ -T(n_0n+P) & n\end{bmatrix}\begin{bmatrix}\sum_{t=n_0}^{n_0+n-1}\phi[t]\\ \sum_{t=n_0}^{n_0+n-1}tT\phi[t]\end{bmatrix},$$

and the frequency and phase parameter estimates become:

$$\hat{\varphi}=\frac{1}{T^2(nQ-P^2)}\left[-T(n_0 n+P)\sum_{t=n_0}^{n_0+(n-1)} tT(\omega tT+\varphi+Im(g_t))+\right.$$

$$\left.T^2\left(n_0{}^2 n+2n_0P+Q\right)\sum_{t=n_0}^{n_0+(n-1)}\omega tT+\varphi+Im(g_t)\right],$$

$$\hat{\omega}=\frac{1}{T^2(nQ-P^2)}\times$$

$$\left(n\sum_{t=n_0}^{n_0+(n-1)} tT(\omega tT+\varphi+Im(g_t))-T(n_0 n+P)\sum_{t=n_0}^{n_0+(n-1)}(\omega tT+\varphi+Im(g_t))\right).$$

To compute the variability of the frequency and phase estimates, we first look at their dispersions:

$$\hat{\varphi}-\mathbb{E}[\hat{\varphi}]=\frac{1}{T^2(nQ-P^2)}\left(-T(n_0 n+P)\sum_{t=n_0}^{n_0+(n-1)} tTIm(g_t)\right.$$

$$\left.+T^2\left(n_0{}^2 n+2n_0P+Q\right)\sum_{t=n_0}^{n_0+(n-1)} Im(g_t)\right),$$

$$\hat{\omega}-\mathbb{E}[\hat{\omega}]=\frac{1}{T^2(nQ-P^2)}\left(nT\sum_{t=n_0}^{n_0+(n-1)} tTIm(g_t)-T(n_0 n+P)\sum_{t=n_0}^{n_0+(n-1)} Im(g_t)\right).$$

Let's examine some of the expectations that play roles in the estimation of the variance-covariance matrix:

$$\mathbb{E}[(Im(g_t))(Im(g_l))]=\frac{1}{A^2}\sigma^2\delta_{t,l}=\frac{1}{SNR}\delta_{t,l},$$

$$\mathbb{E}\left[\left(\sum_{t=n_0}^{n_0+(n-1)} t\,Im(g_t)\right)^2\right]=\frac{1}{SNR}\sum_{t=n_0}^{n_0+(n-1)} t^2=\frac{1}{SNR}\left(n_0{}^2 n+2n_0P+Q\right),$$

$$\mathbb{E}\left[\left(\sum_{t=n_0}^{n_0+(n-1)} Im(g_t)\right)^2\right]=\sum_{t=n_0}^{n_0+(n-1)}\sum_{l=n_0}^{n_0+(n-1)}\mathbb{E}[(Im(g_t))(Im(g_l))]=\frac{n}{SNR},$$

$$\mathbb{E}\left[\left(\sum_{t=n_0}^{n_0+(n-1)} t\,Im(g_t)\right)\left(\sum_{t=n_0}^{n_0+(n-1)} Im(g_t)\right)\right]=\frac{1}{SNR}\sum_{t=n_0}^{n_0+(n-1)} t=\frac{1}{SNR}(n_0 n+P).$$

Therefore, the three elements of the variance-covariance matrix corresponding to the frequency and phase estimates are:

$$Cov[\hat{\varphi},\hat{\varphi}] = \mathbb{E}[(\hat{\varphi}-\mathbb{E}[\hat{\varphi}])(\hat{\varphi}-\mathbb{E}[\hat{\varphi}])] =$$

$$\left(\frac{1}{T^2(nQ-P^2)}\right)^2 \frac{1}{SNR} \times$$

$$\left(T^4(n_0n+P)^2(n_0{}^2n+2n_0P+Q)+nT^4(n_0{}^2n+2n_0P+Q)^2-2T^4(n_0{}^2n+2n_0P+Q)(n_0n+P)^2\right)=$$

$$\left(\frac{1}{T^2(nQ-P^2)}\right)^2 \frac{1}{SNR}(T^4(n_0{}^2n+2n_0P+Q)(nQ-P^2)) =$$

$$\frac{1}{T^2(nQ-P^2)}\frac{1}{SNR}T^2(n_0{}^2n+2n_0P+Q).$$

$$Cov[\hat{\omega},\hat{\omega}] = \mathbb{E}[(\hat{\omega}-\mathbb{E}[\hat{\omega}])(\hat{\omega}-\mathbb{E}[\hat{\omega}])] =$$

$$\left(\frac{1}{T^2(nQ-P^2)}\right)^2\left(n^2T^2\frac{1}{SNR}\left[\sum_{t=n_0}^{n_0+(n-1)} t^2\right]+T^2(n_0n+P)^2n\frac{1}{SNR}-\frac{1}{SNR}2nT^2(n_0n+P)^2\right)=$$

$$\left(\frac{1}{T^2(nQ-P^2)}\right)^2 \frac{1}{SNR}\left((n_0{}^2n+2n_0P+Q)n^2T^2-T^2(n_0n+P)^2n\right)=$$

$$\left(\frac{1}{T^2(nQ-P^2)}\right)^2 \frac{1}{SNR}nT^2(nQ-P^2) = \frac{1}{T^2(nQ-P^2)}\frac{1}{SNR}n.$$

$$Cov[\hat{\varphi},\hat{\omega}] = \mathbb{E}[(\hat{\varphi}-\mathbb{E}[\hat{\varphi}])(\hat{\omega}-\mathbb{E}[\hat{\omega}])] =$$

$$\left(\frac{1}{T^2(nQ-P^2)}\right)^2 \frac{1}{SNR} \times$$

$$[-nT^3(n_0n+P)(n_0{}^2n+2n_0P+Q)+nT^3(n_0n+P)(n_0{}^2n+2n_0P+Q)+$$

$$T^3(n_0n+P)^3-nT^3(n_0n+P)(n_0{}^2n+2n_0P+Q)]=$$

$$\left(\frac{1}{T^2(nQ-P^2)}\right)^2 \frac{1}{SNR}T^3(n_0n+P)\left((n_0n+P)^2-n(n_0{}^2n+2n_0P+Q)\right)=$$

$$-\frac{1}{T^2(nQ-P^2)}\frac{1}{SNR}T(n_0n+P).$$

And the phase-frequency covariance matrix is:

$$Cov\begin{bmatrix}\hat{\varphi}\\ \hat{\omega}\end{bmatrix} = \frac{1}{T^2(nQ-P^2)}\underbrace{\frac{\sigma^2}{A^2}}_{\frac{1}{SNR}}\begin{bmatrix}T^2(n_0^2n+2n_0P+Q) & -T(n_0n+P)\\ -T(n_0n+P) & n\end{bmatrix}=$$

$$\sigma^2\begin{bmatrix} \frac{T^2(n_0^2n+2n_0P+Q)}{T^2(nQ-P^2)A^2} & -\frac{T(n_0n+P)}{T^2(nQ-P^2)A^2} \\ -\frac{T(n_0n+P)}{T^2(nQ-P^2)A^2} & \frac{n}{A^2T^2(nQ-P^2)} \end{bmatrix} = \sigma^2\begin{bmatrix} \frac{(n_0^2n+2n_0P+Q)}{(nQ-P^2)A^2} & -\frac{(n_0n+P)}{T(nQ-P^2)A^2} \\ -\frac{(n_0n+P)}{T(nQ-P^2)A^2} & \frac{n}{A^2T^2(nQ-P^2)} \end{bmatrix}.$$

Earlier we computed the Fisher information matrix and its inverse:

$$I(\theta) = (I_{i,j}(\theta)) = \frac{1}{\sigma^2}\begin{bmatrix} n & 0 & 0 \\ 0 & A^2n & A^2T(n_0n+P) \\ 0 & A^2T(n_0n+P) & A^2T^2(n_0{}^2n+2n_0P+Q) \end{bmatrix},$$

$$I(\theta)^{-1} = \sigma^2\begin{bmatrix} \frac{1}{n} & 0 & 0 \\ 0 & \frac{n_0^2n+2n_0P+Q}{A^2(nQ-P^2)} & -\frac{n_0n+P}{A^2T(P^2-nQ)} \\ 0 & -\frac{n_0n+P}{A^2T(P^2-nQ)} & \frac{n}{A^2T^2(nQ-P^2)} \end{bmatrix}.$$

Therefore, the difference matrix of the reduced parameter variance matrix of the phase and frequency and the reduced inverse Fisher information matrix corresponding to the phase frequency is trivial:

$$Cov\begin{bmatrix}\hat{\varphi}\\ \hat{\omega}\end{bmatrix} - \left(I\begin{bmatrix}\hat{\varphi}\\ \hat{\omega}\end{bmatrix}\right)^{-1} = \begin{bmatrix}0 & 0\\ 0 & 0\end{bmatrix}.$$

This implies that for *efficient estimates* [296], which attain the CRLB, there is no residual uncertainty. However, for *non-efficient estimates*, the difference matrix may be strictly positive-definite, $Cov\begin{bmatrix}\hat{\varphi}\\ \hat{\omega}\end{bmatrix} - \left(I\begin{bmatrix}\hat{\varphi}\\ \hat{\omega}\end{bmatrix}\right)^{-1} > 0$, implying some residual uncertainty in the estimator. Recall that the matrix notation $A \geq B$ denotes semi-positive definiteness of the corresponding difference matrix $A - B \geq 0$, i.e., $u'(A-B)u = \langle u|u\rangle_{A-B} \geq 0, \forall u \in \mathbb{R}^3 \setminus \{0\}$.

In the previous example of the normal distribution, the difference matrix $Cov\begin{bmatrix}\hat{\varphi}\\ \hat{\omega}\end{bmatrix} - \left(I\begin{bmatrix}\hat{\varphi}\\ \hat{\omega}\end{bmatrix}\right)^{-1} > 0$ is always positive definite. However, this single-tone example with high signal-to-noise ratio illustrates a case where the least square estimate yields a "trivial" difference matrix

$$Cov\begin{bmatrix}\hat{\varphi}\\ \hat{\omega}\end{bmatrix} - \left(I\begin{bmatrix}\hat{\varphi}\\ \hat{\omega}\end{bmatrix}\right)^{-1} = \mathbf{0}_{2\times 2}.$$

The high signal-to-noise ratio assumption is essential as we are able to shift the additive Gaussian frequency noise onto an additive phase noise. Without this assumption, the matrix on the left hand size, which represents the difference of the covariance matrix of the phase and frequency estimator and its corresponding inverse Fisher information matrix, won't be strictly positive definite.

A governing principle in quantifying the uncertainty for unbiased estimators in data science utilizes the positive definiteness of the matrix difference between the parameter covariance matrix and inverse Fisher information matrix, i.e., $u'\left(Var\left(\hat{\theta}\right) - I\left(\hat{\theta}\right)^{-1}\right)u \geq 0, \forall u$. This single-tone parameter estimation problem has many natural applications and illustrates an example of resolving the uncertainty principle in data science problems where efficient estimates exist.

Additional modern generalizations of entropic uncertainty relations in finite- and infinite-dimensional spaces, wave-particle quantum duality, communication security, and cryptography applications are presented in [297].

5.10.5 Data Science Formulation

Let's start by examining $L^2(\mathbb{R})$ uncertainty. By Schwartz-Paley-Wiener theorem [298], it is impossible for $f \in L^2$ and $\hat{f}$ to both decrease extremely rapidly. If both have rapidly decreasing tails: $|f(x)| \leq C(1+|x|)^n e^{-a\pi x^2}$ and $\left|\hat{f}(\omega)\right| \leq C(1+|\omega|)^n e^{-b\pi\omega^2}$, for some constant C, polynomial power n, and $a, b \in \mathbb{R}$, then $f = 0$ (when $ab > 1$); $f(x) = P_k(x)e^{-a\pi x^2}$ and $\hat{f}(\omega) = \hat{P}_k(\omega) * \left(\frac{1}{\sqrt{a}} e^{-\frac{\omega^2}{4\pi a}}\right)$, where $\deg(P_k) \leq n$ (when $ab = 1$); or $\hat{f} = 0$ (when $ab < 1$) [299]. Effectively, this suggests that a function and its Fourier transform can't be simultaneously localized in their respective space and frequency domains.

Another way to formulate the data science uncertainty principle builds on the information-theoretic approach. As the data size increases, $n \to \infty$, we must have low signal energy, $\frac{S}{N} \to 0$. And vice-versa, for high energy signals, if $\frac{S}{N} > \epsilon > 0$, then we must have $n < N_\epsilon$. Less formally, this direct statement translates into an uncertainty principle dictating a limit on the joint generality and scope of inference against the precision and robustness of big data analytics.

Similar to the thermodynamic limitation, the uncertainty principle in data science suggests a duality between data size and information energy. In other words, there is a limitation on the increase of the information value (energy) of the data and the complexity and size of the data.

Many big data science applications represent complex data as graphs. Related to the information-theoretic formulation above, a graph-theoretic explication of the uncertainty principle can also be obtained based on the graph Fourier transform [300]. Big data graph representation also facilitates a direct relation between uncertainty principle for performance of signal recovery, sampling rate, and sample location [301].

Data science research is very broad, deep, and transdisciplinary. Two specific issues that commonly arise in many big data applications involve access to valuable datasets and data representation as computable objects. For instance, most of the health data cannot be readily shared because they contain personal, protected, or sensitive information. Findable, accessible, interoperable, and reusable (FAIR) data-sharing and open-science principles clearly demonstrate the potential of big data to expand our knowledge and rapidly translate science into practice [38, 302]. The main problem is that there are no mechanisms in place to enable sharing of, and collaboration with, big data without risking inappropriate use of sensitive information. Methods like ϵ-differential privacy [303], DataSifter statistical obfuscation, and homomorphic encryption [304] enable the desensitization of large, complex and heterogeneous data archives by statistically obfuscating the data, yet preserving their joint distribution. Such techniques allow physicians from different clinics, or data governors, to share their data and enable the construction of normative databases needed by Dr. Doe, and others, to use for diagnosis, treatment, and prognosis for patients like Mr. Real.

There are also some early approaches addressing the big data challenge of modeling, interrogating and scientific inference. For instance, tensor mining [305] and Compressive Big Data Analytics (CBDA) [15, 128] allow the modeling and inference based on large, heterogeneous, incomplete, multiscale and multi-source datasets. Such techniques can form the basis for building an automated clinical decision support system that clinicians, like Dr. Doe, can utilize to obtain data-driven evidence-based actionable knowledge necessary to identify the clinical state, suggest medical treatment, and communicate longitudinal forecasting for her patients.

This discussion leads to two conclusions. First, albeit the exact interplays of data quality, size, complexity, and energy may be widely varying across studies and applications, there are also some specific data science limitations. For big data that heavily depend on the specific representation and meta-data interpretation, these limitations relate to inferential generalizability, scope, precision, and reliability.-Second, the scientific, computational, and engineering communities need to join forces in laying the mathematical foundations for representation, inference, and interpretation of big datasets. This will provide a framework to track more precisely limits on the data science uncertainty principle within the scope of each specific data representation scheme.

Appendix

1 Probabilistic Bra-Ket Notation and Data Science Synergies with Quantum Mechanics

The classical quantum mechanics parallels between measurable quantities and operators translate in data science to *data* and *meta-data* specifications (measurables) and *inference* (operators). For instance, given a dataset $X_{\eta\times\kappa}\sim\varrho$, descriptors such as *data size* η and *feature size* κ remain unaffected when they are transformed to (inference) operators, as these are generally known, however the *data distribution* ϱ often is unknown and requires data-driven estimation:

$$\left| \begin{array}{l} \eta \to \hat{\eta} \\ \kappa \to \hat{\kappa}\,. \\ \varrho \to \hat{\varrho} \end{array} \right. \tag{1}$$

There may be multiple strategies to translate operator ($\widehat{\ }$) quantization in physics to data science *linear-model inference* operators ($\langle \cdot \mid \cdot \rangle$) that ensure consistent, reliable, and model-based prospective state-prediction:

$$\left| \begin{array}{l} E \to i\hbar\frac{\partial}{\partial t}\ \text{(quantum mechanics)} \\ p_{\eta,\kappa} \to \langle \cdot \mid \cdot \rangle\ \text{(data science inference)} \end{array} \right. . \tag{2}$$

Data science forecasting is the process of estimating a function $f(X)$ that predicts the typical output value Y corresponding to specific input X. *Model-based inference* is the dual problem of inductively learning the relationship between the observed covariates X and the outcome Y that explicates the relation between combinations of X and particular changes of Y. In other words, model inference aims to understand the (model-based) underlying mechanistic association, e.g., physical, biological, social, environmental, between X and Y.

The *conditional expectation* of Y given X, defined by $f(X)=\mathbb{E}(Y|X)$, can be interpreted in two complementary ways. First, as a deterministic function of X, reflecting a large sample size (big data with significant number of observed (X, Y) pairs) from their joint distribution, and averaging the output Y values that are coupled with their corresponding X value, i.e.,

$$\mathbb{E}(Y|X=x)=\int_{\Omega} yp(y|x)dy.$$

Alternatively, the conditional expectation may be interpreted as a scalar random variable representing a realization obtained by sampling X from its marginal distribution and plugging this value into the deterministic function $\mathbb{E}(Y|X=x)$.

Using linear models to represent relations between independent variables

$$\langle X'| = (X_1,\ X_2,\ \ldots,\ X_\kappa)' = \begin{bmatrix} x_{1,1} & \cdots & x_{1,\kappa}\}X_1 \\ \vdots & \ddots & \vdots \\ x_{\eta,1} & \cdots & x_{\eta,\kappa}\}X_\eta \end{bmatrix}$$

and a dependent variable $|Y\rangle = \begin{pmatrix} y_1 \\ y_2 \\ \vdots \\ y_\eta \end{pmatrix}$ implies the following explicit relations:

$$|Y\rangle = \langle X'|\beta\rangle + \epsilon \equiv (X')'\beta + \epsilon = X\beta + \epsilon, \quad i.e., \quad y_i = \sum_{j=1}^{\kappa} x_{i,j}\beta_j = X_i\beta + \epsilon_i, \forall 1 \le i \le \eta$$

$$|\epsilon\rangle = \begin{pmatrix} \epsilon_1 \\ \epsilon_2 \\ \vdots \\ \epsilon_\eta \end{pmatrix} \sim \mathcal{D}(\mu, \Sigma).$$

This model represents a linear function of the parameter vector (β), not the observed predictors X.

2 Univariate Random Variables (Processes)

We will present the core definitions first in the context of a single variable. Later, we will generalize these to multi-feature random vectors and multivariate processes.

To estimate the conditional expectation of Y given X, $f(X) = \mathbb{E}(Y|X)$, we need to estimate the parameter vector (effect size), β. This can be accomplished by least squares estimation of the model fit (fidelity term), or by optimizing an additive objective function involving fidelity and regularization terms:

$$L(\beta) = |||Y\rangle - \langle X'|\beta\rangle||^2 = \sum_{i=1}^{\eta}\left(y_i - \sum_{j=1}^{\kappa} x_{i,j}\beta_j\right)^2.$$

The Jacobian of the objective function $L(.)$ is defined by:

$$\langle J| = \left(\frac{\partial}{\partial\beta_1}, \frac{\partial}{\partial\beta_2}, \ldots, \frac{\partial}{\partial\beta_\kappa}\right).$$

For a symmetric matrix, A, and an objective function $L(v) = \langle v|\ Av\rangle \equiv v'Av$, the Jacobian of L is $\langle J|L\rangle = \frac{\partial L}{\partial\beta} = 2'A' = 2(Av)'$. Here $L(\beta) = |||Y\rangle - \langle X'|\beta\rangle||^2 = ||Y - X\beta||^2$, which is minimized when $0 = \langle J|L\rangle = \frac{\partial L}{\partial\beta} = -2X'(Y - X\beta) = -2\langle X|(Y - X\beta)\rangle$.

Therefore, $0 = \langle X|Y - X\beta\rangle$ implying $0 = \langle X|Y\rangle - \langle X|X\beta\rangle$ and $\langle X|Y\rangle = \langle X|X\beta\rangle$, i.e., $X'Y = X'X\beta$. The least squares solution is

$$\hat{\beta} = (X'X)^{-1}X'Y$$

i.e.,

$$\hat{\beta}^{OLS} = \langle X|X\rangle^{-1}\langle X|Y\rangle.$$

$\hat{\beta}^{OLS}$ represents the least squares estimates, or effects, for the linear model $|Y\rangle = \langle\beta|X\rangle + \epsilon$.

3 Conditional Probability

For a pair of events, $A, B \subseteq \Omega$, their probabilities of being observed are $\langle A|$ and $\langle B|$. To define the conditional probability in bra-ket notation, observed (given) evidence will be denoted by $|B\rangle$ and represent the probability of the evidence. *Conditional probability* is defined by:

$$\underbrace{\langle A|}_{\text{event}} \underbrace{|B\rangle}_{\text{evidence}} = \langle A|B\rangle \equiv P(A|B) = \frac{P(AB)}{P(B)}.$$

Note the following traditional properties:

$$\begin{array}{c} 1 = \langle A|B\rangle \Leftrightarrow \varnothing \subset B \subseteq A \\ 0 = \langle A|B\rangle \Leftrightarrow P(AB) = 0, \text{ and } B \cap A = \varnothing \Rightarrow 0 = \langle A|B\rangle = 0 \\ \langle A|B\rangle = \langle A| \equiv \langle A|\Omega\rangle \Leftrightarrow B \text{ and } A \text{ are mutually independent}, P(AB) = P(A)P(B) \\ \langle A|B\rangle = \langle A| \Leftrightarrow \langle B|A\rangle = \langle B| \\ \langle A \cup B| = \langle A| + \langle B| \Leftrightarrow B \text{ and } A \text{ are disjoint}, B \cap A = \varnothing \\ \langle A \cup B| = \langle A| + \langle B| - \langle A \cap B| \\ \langle A^c| = 1 - \langle A|, \text{where } A^c \equiv \Omega \backslash A \\ \langle A|B\rangle = \langle B|A\rangle \frac{\langle A|}{\langle B|}, \text{Bayesian rule} \\ \forall \omega \in \Omega, \text{elementary events}, \langle \omega_i|\omega_j\rangle = \delta_{i,j} = \begin{cases} 1, i = j \\ 0, i \neq j \end{cases} (\text{orthonormality}) \end{array}$$

The *unitary operator* is defined by:

$$\hat{I} \equiv \sum_i |\omega_i\rangle \langle\omega_i|,$$

which preserves the conditional probability for any pairs of (composite) events $A, B \subseteq \Omega$:

$$\langle A|\hat{I}|B\rangle = \sum_{\omega \in \Omega} \langle A|\omega\rangle \langle\omega|B\rangle =$$

$$= \sum_{\omega\in A} \langle \omega|B\rangle \underbrace{=}_{P(\omega|B)=\frac{P(\omega\cap B)}{P(B)}} \sum_{\omega\in A\cap B} \langle \omega|B\rangle \underbrace{=}_{\langle A\cup B| = \langle A| + \langle B| - \langle A\cap B|} \langle A|B\rangle .$$

The last equality is due to $\langle A\cup B| = \langle A| + \langle B| - \langle A\cap B|$, since

$$\langle A\cap B|B\rangle = \langle A|B\rangle + \langle B|B\rangle - \langle A\cup B|B\rangle = \langle A|B\rangle + 1 - 1 = \langle A|B\rangle .$$

Also, the trivial *event* and *evidence* for the entire sample space, Ω, are as expected:

$$|\Omega\rangle = \hat{I}|\Omega\rangle = \sum_i |\omega_i\rangle \underbrace{\langle\omega_i|\Omega\rangle}_{1} = \sum_i |\omega_i\rangle,$$

$$\langle\Omega| = \langle\Omega|\hat{I} = \sum_i \underbrace{\langle\Omega|\omega_i\rangle}_{1} \langle\omega_i| = \sum_i \langle\omega_i| .$$

These properties ensure the probability measure normalization requirement:

$$1 = \langle\Omega| = \sum_i \langle\Omega|\omega_i\rangle \langle\omega_i| = \sum_i \left(\langle\Omega|\omega_i\rangle \sum_j \langle\omega_j|\Omega\rangle \right) =$$

$$= \sum_{i,j} \langle\Omega|\omega_i\rangle\langle\omega_j|\Omega\rangle = \langle\Omega|\Omega\rangle = \left\langle\Omega|\hat{I}|\Omega\right\rangle = 1.$$

The duality of the classical physics bra and ket operators on Hilbert spaces simply means that they are Hermitian conjugates of one another. This duality is also true for any kevent, a subset of the sample space, $A \subseteq \Omega$.

$$|A\rangle = \hat{I}|A\rangle = \sum_i |\omega_i\rangle \langle\omega_i|A\rangle = \sum_i |\omega_i\rangle \frac{A|\omega_i\rangle\langle\omega_i|\Omega\rangle}{\langle A|\Omega\rangle} = \sum_{\omega\in A} \underbrace{|\omega\rangle}_{\substack{\text{base}\\ \text{events}}} \times \underbrace{\frac{\langle\omega|\Omega\rangle}{\langle A|\Omega\rangle}}_{\text{conditional}} .$$

Thus, $|A\rangle$ explicitly identifies what atomic outcomes are part of the composite event A as well as the conditional probability of each base-event under evidence A.

Plugging the entire sample space in the above, we get the expected identity:

$$\langle\Omega|A\rangle = \sum_{\omega\in A} \underbrace{\langle\Omega|\omega\rangle}_{1} \frac{\langle\omega|\Omega\rangle}{\langle A|\Omega\rangle} = \sum_{\omega\in A} \frac{\langle\omega|\Omega\rangle}{\langle A|\Omega\rangle} = \frac{1}{\langle A|\Omega\rangle} \sum_{\omega\in A} \langle\omega|\Omega\rangle = \frac{\langle A|\Omega\rangle}{\langle A|\Omega\rangle} = 1.$$

Similarly, we can expand the bra component for the composite event A:

$$\langle A| = \langle A|\hat{I} = \sum_i \langle A|\omega_i\rangle \langle\omega_i| = \sum_{\omega\in A} \underbrace{\langle A|\omega\rangle}_{1} \langle\omega| = \sum_{\omega\in A} \langle\omega| .$$

Thus, $|A\rangle$, identifies both its base-events as well as their conditional probabilities under the evidence A, whereas $\langle A|$ only explicates its atomic constituents (base events).

4 Evolution: Time-Varying Processes, Probability, and Data

Sometimes, the probability distribution function may be time-dependent (TD), e.g., Markov chain transition probabilities, $m(\omega, t)$. In these situations,

$$|A_t\rangle = \hat{I}|A_t\rangle = \sum_i |\omega_i\rangle \langle\omega_i|A_t\rangle = \sum_{\omega\in A} m(\omega, t)|\omega\rangle.$$

However, $\langle A|$ represents the set of all possible outcomes at all times:

$$\langle A_t| = \langle A_t|\hat{I} = \sum_i \langle A_t|\omega_i\rangle \langle\omega_i| = \sum_{\omega\in A} \underbrace{\langle A_t|\omega\rangle}_{1} \langle\omega| = \sum_{\omega\in A} \langle\omega| = \langle A|.$$

Hence, the *energy* of the composite event, A_t, is:

$$\langle A_t|A_t\rangle = \sum_{\omega\in A} m(\omega, t)\langle\omega|\omega\rangle = \sum_{\omega\in A} m(\omega,\ t).$$

The time-dependent sample space Ω_t represents a cross-section of the entire sample space Ω. Whereas Ω_t contains all possible outcomes observed at a fixed time $= t$, the whole sample space, Ω, contains all possible outcomes over the entire time range.

In particular, given the time-dependent process mass/distribution function, $m(\omega,\ t)$, $\langle\Omega|\Omega_t\rangle = \sum_{\omega\in\Omega} m(\omega,\ t) = 1$. We can see this by the transformation mapping ket-to-ket and bra-to-bra:

$$\text{Bra: } \langle\Omega| = \langle\Omega|\hat{I} = \sum_i \underbrace{\langle\Omega|\omega_i\rangle}_{1} \langle\omega_i| = \sum_i \langle\omega_i| = \left[\underbrace{1, 1, 1, \ldots, 1}_{|\Omega| = \eta}\right],$$

$$\text{Bra: } \langle\Omega_t| = \sum_{\omega\in\Omega} \left(\underbrace{m(\omega, t)}_{\substack{\text{time - dependent}\\ \text{mass}}} \langle\omega| \right) = \big(m(\omega_1, t), m(\omega_2, t), m(\omega_3, t), \ldots, m(\omega_\eta, t)\big),$$

$$\text{Ket: } |\Omega_t\rangle = I|\Omega_t\rangle = \sum_i |\omega_i\rangle\langle\omega_i|\Omega_t\rangle = \sum_{\omega\in\Omega} m(\omega, t)|\omega\rangle = \begin{pmatrix} m(\omega_1, t) \\ m(\omega_2, t) \\ \vdots \\ m(\omega_\eta, t) \end{pmatrix}.$$

This establishes the normalization over the entire sample space, Ω:

$$\langle\Omega|\Omega_t\rangle = \langle\Omega|\hat{I}|\Omega_t\rangle = \sum_{\omega\in\Omega} \langle\omega|m(\omega,\ t)|\omega\rangle = \sum_{\omega\in\Omega} m(\omega, t) = 1.$$

However, the time-dependent inner product is not necessarily normalized, and often may be less than 1:

$$\langle \Omega_t | \Omega_t \rangle = \langle \Omega_t | \hat{I} | \Omega_t \rangle = \sum_{\omega \in \Omega} \langle \omega | (m(\omega, t))^2 | \omega \rangle = \sum_{\omega \in \Omega} (m(\omega, t))^2 \leq 1.$$

We will come back to *time-varying stochastic processes* later.

Open Problem

Extend the concepts of time-based evolution, time-varying processes, and probability to the 2D kime manifold.

5 Preservation of Energy and Total Probability

The preservation of energy of an event B, i.e., law of total probability, is stated in terms of a partition, $\{B_i\}$ of the sample space, $\Omega = \cup B_i$:

$$\langle B| = \sum_i \langle B|B_i\rangle \langle B_i|.$$

Thus, we can expand the Bayesian formula:

$$\langle A|B\rangle = \langle B|A\rangle \frac{\langle A|}{\langle B|} = \langle B|A\rangle \frac{\langle A|}{\sum_i \langle B|B_i\rangle \langle B_i|}.$$

6 Expectation and Variance

Random variables on a sample space Ω represent the *observable* quantities. An instantiation of the random variable X corresponds to observing its value as a static number x. In effect, X can be considered as an operator, $\hat{X}$, acting on the bras and kets by:

$$\langle x|\hat{X} = \langle x|x \text{ and } \hat{X}|x\rangle = x|x\rangle.$$

Note that for each random variable, X, the bras and the kets of its possible values form a base of Ω. In other words, the $\langle x|$ and $|x\rangle$ represent the atomic outcomes of the sample space, in terms of the random variable X. This is because:

$$\langle x_1|x_2\rangle = \delta_{x_1,x_2} \text{ and } 1 = \langle\Omega| = \sum\nolimits_{x\in\Omega} \langle\Omega|x\rangle\langle x| = \sum\nolimits_{x\in\Omega} |x\rangle\,\langle x| = 1.$$

The (marginal) *expectation* of X is defined by:

$$\langle X\rangle = \langle\Omega|X|\Omega\rangle = \sum_{x\in\Omega}\langle\Omega|X|x\rangle\langle x|\Omega\rangle = \sum_{x\in\Omega}\langle\Omega|x|x\rangle\langle x|\Omega\rangle = \sum_{x\in\Omega}\underbrace{\langle\Omega|x\rangle}_{1}x\langle x|\Omega\rangle =$$

$$= \sum_{x\in\Omega} x\langle x|\Omega\rangle = \sum_{x\in\Omega} x \underbrace{m(x)}_{X \text{ distribution}} = \mathbb{E}(X) = \bar{X}.$$

Also, for each continuous function $f()$, the expectation of $f()$ is:

$$\langle f(X)\rangle = \mathbb{E}(f(x)) = \langle\Omega|f(X)|\Omega\rangle = \sum_{x\in\Omega} f(x)m(x).$$

The *variance* of X is defined as the expectation of the square deviance operator $((X - \langle X\rangle)^2)$:

$$\langle(X - \langle X\rangle)^2\rangle = \langle\Omega|(X - \langle X\rangle)^2|\Omega\rangle = \sum_{x\in\Omega}\langle\Omega|\underbrace{X^2 - 2X\langle X\rangle + \langle X\rangle^2}_{(X-\langle X\rangle)^2 \text{expanded}}|x\rangle\langle x|\Omega\rangle =$$

$$= \sum_{x\in\Omega}\langle\Omega|X^2|x\rangle\langle x|\Omega\rangle - 2\sum_{x\in\Omega}\langle\Omega|2X\langle X\rangle|x\rangle\langle x|\Omega\rangle + \langle X\rangle^2\sum_{x\in\Omega}\underbrace{\langle\Omega|x\rangle}_{1}\langle x|\Omega\rangle =$$

$$= \langle X^2\rangle - 2\langle X\rangle \underbrace{\sum_{x\in\Omega}\langle\Omega|X|x\rangle\langle x|\Omega\rangle}_{\langle X\rangle} + \langle X\rangle^2 = \langle X^2\rangle - 2\langle X\rangle^2 + \langle X\rangle^2 = \langle X^2\rangle - \langle X\rangle^2 =$$

$$= \sum_{x\in\Omega} x^2 \underbrace{m(x)}_{X \text{ distribution}} - \left(\sum_{x\in\Omega} x \underbrace{m(x)}_{X \text{ distribution}}\right)^2 = \mathbb{E}(X^2) - \mathbb{E}((X))^2 = V(X) = \sigma^2.$$

Using the above definition for conditional probability, we can define *conditional expectation* for the random variable X given any event $A \subseteq \Omega$:

$$\langle X|A\rangle = \sum_{x\in\Omega} x \underbrace{\langle X = x|A\rangle}_{\substack{\text{conditional} \\ \text{probability distribution}}}$$

Clearly, $\langle X|A\rangle = \langle\Omega|X|A\rangle$:

$$\langle X|A\rangle \equiv \sum_{x\in\Omega} x\langle X=x|A\rangle = \sum_{x\in\Omega} \underbrace{\langle \Omega|x\rangle}_{1} x\langle x|A\rangle =$$

$$\sum_{x\in\Omega} \langle \Omega|x|x\rangle\ \langle x|A\rangle = \sum_{x\in\Omega} \langle \Omega|X|x\rangle\ \langle x|A\rangle = \langle \Omega|X|A\rangle.$$

Suppose we have a random variable X and a finite partition of the sample space, $\Omega = \cup B_j$, where $B_i \cap B_j = \emptyset, \forall i \neq j$. The following equation connects (marginal) expectation and conditional expectation:

$$\underbrace{\langle X\rangle}_{\text{marginal expectation}} = \sum_{x\in\Omega} x\langle x|\Omega\rangle = \sum_{x\in\Omega} x \sum_{j} \langle x|B_j\rangle\langle B_j|\Omega\rangle = \sum_{j}\sum_{x} x\langle x|B_j\rangle\langle B_j|\Omega\rangle =$$

$$\sum_{j} \langle \Omega|X|B_j\rangle\langle B_j|\Omega\rangle = \sum_{j} \underbrace{\langle X|B_j\rangle}_{\text{conditional expectations}} \langle B_j|.$$

7 Multiple Features, Random Vectors, Independent Random Variables

Many real experiments and most observable processes involve vectors of random observables. We can use the same notation as in the previous section, where κ features, $\{X_i\}_{i=1}^{\kappa}$ are used to model, predict, explain, forecast, or quantify an independent outcome variable Y.

$$\langle X| = (X_1, X_2, \ldots, X_\kappa) = \begin{bmatrix} x_{1,1} & \cdots & x_{1,\kappa} \\ \vdots & \ddots & \vdots \\ \underbrace{x_{\eta,1}}_{X_1} & \cdots & \underbrace{x_{\eta,\kappa}}_{X_\kappa} \end{bmatrix} \text{ and } |Y\rangle = \begin{pmatrix} y_1 \\ y_2 \\ \vdots \\ y_\eta \end{pmatrix}.$$

For each feature, X_i, let's denote the atomic base-events r_i for the outcome space, i.e., $r_i \in \Omega_i$. When the features X_i are statistically independent, i.e., $\langle X_i|X_j\rangle = \langle X_i|$, $\forall i \neq j$, then:

$$|r_1, r_2, \ldots, r_\kappa\rangle = \prod_{i=1}^{\kappa} |r_i\rangle.$$

Recall that the features $\{X_i\}_{i=1}^{\kappa}$ are mutually independent if and only if for any κ-tuple in the state space, $(r_1, r_2, \ldots, r_\kappa) \in \Omega_1 \times \Omega_2 \times \ldots \times \Omega_\kappa = \Omega$, the joint probability function factors out as a product of marginal probabilities:

$$P(X_1=r_1,\ X_2=r_2,\ \ldots,\ X_\kappa=r_\kappa)=\prod_{i=1}^{\kappa}P(X_i=r_i).$$

In general, whether or not the features are independent,

$$P(X_1=r_1,\ X_2=r_2,\ \ldots,\ X_\kappa=r_\kappa)=$$

$$P(X_1=r_1|X_{-1})P\big(X_2=r_2|X_{-(1,2)}\big)\cdots P(X_{\kappa-1}=r_{\kappa-1}|X_\kappa)P(X_\kappa=r_\kappa).$$

In the very special case of mutual independence, the conditional probabilities equal their marginal counterparts, by independence, and therefore:

$$P(X_1=r_1,\ X_2=r_2,\ \ldots,\ X_\kappa=r_\kappa)=$$

$$P(X_1=r_1)P(X_2=r_2)\cdots P(X_{\kappa-1}=r_{\kappa-1})P(X_\kappa=r_\kappa)=\prod_{i=1}^{\kappa}P(X_i=r_i).$$

Let's formulate these properties for observables in terms of the bra-ket notation.

$$\left| \begin{array}{l} X_i\ |\ r_1,r_2,\ \ldots,\ r_\kappa\rangle=r_i\ |\ r_1,r_2,\ \ldots,\ r_\kappa\rangle \\ \langle r_1,r_2,\ \ldots,\ r_\kappa\ |\ X_i=\langle r_1,r_2,\ \ldots,\ r_\kappa\ |\ r_i \end{array} \right. .$$

Therefore, for all additive (linear) and multiplicative (power-product) combinations of the features we have the following two general rules:

$$\Bigg\langle \underbrace{\sum_i c_i f_i(X_i)}_{\text{linear combination}} \Bigg\rangle = \Big\langle \Omega \Big| \sum_i c_i f_i(X_i) \Big| \Omega \Big\rangle \underbrace{=}_{\substack{\text{linear}\\ \text{operators}}} \sum_i c_i f_i(\langle X_i\rangle)$$

$$\Bigg\langle \underbrace{\prod_i X_i^{p_i}}_{\text{power combination}} \Bigg\rangle = \Big\langle \Omega \Big| \prod_i X_i^{p_i} \Big| \Omega \Big\rangle \underbrace{=}_{\text{independence}} \prod_i \langle X_i\rangle^{p_i}$$

The above equations rely on the expectation property that $\langle X_i\rangle \equiv \langle \Omega|X_i|\Omega\rangle = \langle \Omega_i|X_i|\Omega_i\rangle$.

8 Time-Varying Stochastic Processes

A stochastic process $X(t),\ t\in T$, is a time-dependent observable and we can define the bra-ket operators as follows:

$$X(t)|X(t)=x_i\rangle \equiv X(t)|t,x_i\rangle = x_i|t,x_i\rangle.$$

As we indicated earlier, the time-dependent sample space Ω_t represents a snapshot (cross-section in time) of the entire sample space Ω. Whereas Ω_t contains all possible outcomes observed at time $=t$, the total sample space, Ω, contains all possible outcomes over the entire time range.

The following identities are clear by the definitions:

$$\Omega_t \subseteq \Omega,$$
$$\langle \Omega | \Omega_t \rangle = 1,$$
$$\Omega | t, x \rangle = 1,$$
$$\langle t, x | \Omega \equiv \langle x | \Omega_t.$$

For any time-dependent (TD) observable $X(t)$, the corresponding TD identity operator, $\hat{I}(t)$, probability mass/distribution function, $m(t, x)$, and expectation, $\mathbb{E}_t(X(t))$, can be expressed as:

$$\text{TD identity operator: } \hat{I}(t) = \sum_i |t, x_i\rangle \langle t, x_i|,$$

$$\text{TD mass: } m(t,\ x_i) = \langle t, x_i | \Omega = \langle x_i | \Omega_t,$$

$$\text{TD expectation: } \mathbb{E}_t(X(t)) = \langle \Omega | X(t) | \Omega \rangle = \sum_i \langle \Omega | X(t) | t, x_i \rangle \langle x_i | \Omega_t = \sum_i m(t, x_i)\ x_i.$$

The time parameter reflects the transition probability, or time-increment, which is defined in the time-incremental direction. Inserting the identity operator, $\hat{I}(t)$, selects the appropriate time. For instance, at time t, the observed measurement, $X(t)$, picks up its value from Ω_t, not Ω or Ω_{t+1}. That is:

$$X_t = X_t\ \hat{I}(t) = \hat{I}(t) X_t.$$

Therefore,

$$X(t)|\Omega\rangle = X_t\ \hat{I}(t)|\Omega\rangle = X_t \sum_i |t, x_i\rangle \langle t, x_i | \Omega \rangle = X_t \sum_i |t, x_i\rangle \langle x_i | \Omega_t \rangle =$$
$$X_t\ \hat{I}(t)|\Omega_t\rangle = X_t\ |\Omega_t\rangle.$$

And

$$\langle \Omega | X(t) | \Omega \rangle = \langle \Omega | X_t\ \hat{I}(t) | \Omega \rangle = \langle \Omega | X_t\ | \Omega_t \rangle.$$

This shift of time dependence from the observable is related to the shift from the *algebraic Heisenberg* uncertainty principle (where operators change in time while the basis of the Hilbert space remains fixed) to the *analytical Schrödinger* equation (where operators remain fixed while the Schrödinger equation changes with time).

9 Bracket Notation: Probabilistic Inference vs. Wavefunctions

Table 5.3 outlines the synergies between the notation and interpretation of probabilistic inference-based (left) vs. physics-based (right) Dirac bracket notation.

Table 5.3: Parallels between quantum mechanics and data inference concepts and notations.

Concepts	Inference-based	Physics-based
Space	Sample space Ω, associated with a random feature vector X	$\boldsymbol{H}$, Minkowski spacetime, Hilbert space
Bra	$\langle A\|$: an observable event in Ω $\langle \Omega\|$: state bra	$\langle \psi_A\|$: row vector in $\boldsymbol{H}$ $\langle \psi(\mathrm{t})\|$: state bra
Ket	$\|B\rangle$: an *evidence* set in Ω $\|\Omega_t\rangle$: state ket	$\|\psi_B\rangle$: column vector in $\boldsymbol{H}$ $\|\psi(\mathrm{t})\rangle = \langle\psi(\mathrm{t})\|^{\dagger}$: state ket (conjugate pairs)
Bracket	$\langle A\|B\rangle \equiv P(A\|B)$: inner product bracket (conditional probability)	$\langle\psi_A\|\psi_B\rangle = \langle\psi_A, \psi_B\rangle$: bracket (inner product)
Bracket transposition	Bayes formula $\langle A\|B\rangle = \langle B\|A\rangle \frac{\langle A\|}{\langle B\|}$	$\langle\psi_A\|\psi_B\rangle = \langle\psi_B\|\psi_A\rangle^*$
Bases	Mutually disjoint sets associated with the random variable X: $\omega_i \cap \omega_j = \delta_{i,j}\omega_i$ and $\Omega = \sum_i \omega_i$	Eigenvectors of a Hermitian Operator H : $\hat{H}\|\psi_i\rangle = E_i\|\psi_i\rangle$
Ortho-normality	$\langle\omega_i\|\omega_j\rangle = \delta_{i,j}$	$\left\langle\psi_i\|\psi_j\right\rangle = \delta_{i,j}$
Unitary operator	$\hat{I} \equiv \sum_i \|\omega_i\rangle\langle\omega_i\|$	$\hat{I} \equiv \sum_i \|\psi_i\rangle\langle\psi_i\|$
State normalization	$\langle\Omega\|\Omega_t\rangle = \sum_i mi(t) = 1$	$\langle\psi(t)\|\psi(t)\rangle = \sum_i \|c_i(t)\|^2$
Observable	$X\|\omega_i\rangle = x_i\|\omega_i\rangle$	$\hat{H}\|\psi_i\rangle = E_i\|\psi_i\rangle$
Expectation	$\langle \mathrm{X}\rangle = \langle\Omega\|X\|\Omega\rangle = \sum_{x\in\Omega} x\langle x\|\Omega\rangle = \sum_{x\in\Omega} x \underbrace{m(x)}_{X\ \mathrm{prob/mass}}$	$\langle\mathrm{H}\rangle = \left\langle\psi(t)\|\hat{H}\|\psi(t)\right\rangle = = \sum_i \|c_i(t)\|^2 E_i$
Special relations	$1 = \langle A\|B\rangle \Leftrightarrow \emptyset \subset B \subseteq A$ $0 = \langle A\|B\rangle \Leftrightarrow B \cap A = \emptyset$ $\langle A\|B\rangle = \langle A\| \equiv \langle A\|\Omega\rangle \Leftrightarrow$ B and A are mutually independent, $P(AB) = P(A)P(B)$ $\langle A\|B\rangle = \langle A\| \Leftrightarrow \langle B\|A\rangle = \langle B\|$ $\langle A \cup B\| = \langle A\| + B\| \Leftrightarrow$ B and A are disjoint, $B \cap A = \emptyset$ $\langle A \cup B\| = \langle A\| + B\| - \langle A \cap B\|$ $\langle A^c\| = 1 - \langle A\|$, where $A^c = \Omega \backslash A$ $\langle A\|B\rangle = \langle B\|A\rangle \frac{\langle A\|}{\langle B\|}$, Bayesian rule $\forall \omega \in \Omega$, elementary events, $\langle\omega_i\|\omega_j\rangle = \delta_{i,j} = \begin{cases} 1, & i=j \\ 0, & i \neq j \end{cases}$ (orthonormality)	

10 Derivation of the Transformed Metric and the Corresponding Cosmological Constant (Λ)

Below, we derive the transformed metric and the corresponding cosmological constant (Λ) for the 5D Ricci-flat spacekime equations extending the corresponding 4D Einstein spacetime equations in vacuum, see **Section 5.4** (*Uncertainty in 5D Spacekime*).

Assume unitary speed of light c and gravitational constant G, and denote with lower-case Greek letters $\alpha, \beta, \gamma, \mu \in \left\{ \underbrace{0}_{\text{time}}, \underbrace{1,2,3}_{\text{space}} \right\}$ and upper-case Latin letters $A, B \in \left\{ \underbrace{\overbrace{0}^{t}}_{\text{time}}, \underbrace{1,2,3}_{\text{space}}, \underbrace{\overbrace{4}^{l}}_{\text{time}} \right\}$ the spacetime coordinate references in 4D spacetime and 5D spacekime, respectively.

The 4D and 5D proper time intervals are defined in terms of the corresponding metric tensors:

$$ds^2 = g_{\alpha\beta}\ dx^\alpha dx^\beta = \sum_{\alpha=0}^{3}\sum_{\beta=0}^{3} g_{\alpha\beta}\ dx^\alpha dx^\beta,$$

$$dS^2 = g_{AB}\ dx^A dx^B = \sum_{A=0}^{4}\sum_{B=0}^{4} g_{AB}\ dx^A dx^B.$$

By isolating the $x^4 \equiv l$ part of the 5D interval and using the given implicit restrictions in the field equations, we can derive a special case of the Campbell-Wesson embedding theorem, which explicates how the fifth dimension constrains the 4D spacetime dynamics.

The derivations below are based on this definition of the rank-2 *Ricci tensor of the first kind* [306]:

$$\underbrace{R_{\mu\nu} \equiv R_{\nu\mu}}_{\text{index symmetry}} = \sum_{\lambda=0}^{3} \left(\underbrace{\frac{\partial \Gamma^\lambda_{\mu\lambda}}{\partial x^\upsilon} - \frac{\partial \Gamma^\lambda_{\mu\upsilon}}{\partial x^\lambda} + \sum_{\beta=0}^{3} \left(\Gamma^\beta_{\mu\lambda}\Gamma^\lambda_{\upsilon\beta} - \Gamma^\beta_{\mu\upsilon}\Gamma^\lambda_{\beta\lambda} \right)}_{R^{\underline{\lambda}}_{\mu\nu\underline{\lambda}}} \right).$$

which contracts the *third index* (last contravariant index) of the rank-4 Riemannian tensor of the second kind, $R^{\underline{\lambda}}_{\mu\nu\underline{\lambda}}$. By raising the first index of the Ricci tensor of the first kind, we obtain the *Ricci tensor of the second kind*, R^λ_μ, which naturally leads to defining the *Ricci scalar*, R, as a curvature invariant resulting from contracting the indices of the Ricci tensor of the second kind:

$$R_{\mu}^{\lambda} \equiv g^{\lambda\beta} R_{\beta\mu} = \sum_{\beta=0}^{3} (g^{\lambda\beta} R_{\beta\mu}), \quad R \equiv R_{\mu}^{\mu} = \sum_{\beta=0}^{3} \sum_{\mu=0}^{3} (g^{\mu\beta} R_{\beta\mu}).$$

The Riemannian tensor characterizes the properties of spaces and surfaces in general and in non-Euclidean geometries [307]:

$$R^{\gamma}_{\mu\nu\lambda} = \partial_{\nu}\Gamma^{\gamma}_{\mu\lambda} - \partial_{\lambda}\Gamma^{\gamma}_{\mu\nu} + \Gamma^{\delta}_{\mu\lambda}\Gamma^{\gamma}_{\delta\nu} - \Gamma^{\delta}_{\mu\nu}\Gamma^{\gamma}_{\delta\lambda},$$

$$\Gamma^{\mu}_{\beta\gamma} = \sum_{\alpha=0}^{3} \frac{1}{2} g^{\mu\alpha} \left(\frac{\partial g_{\alpha\beta}}{\partial x^{\gamma}} + \frac{\partial g_{\alpha\gamma}}{\partial x^{\beta}} - \frac{\partial g_{\beta\gamma}}{\partial x^{\alpha}} \right) \quad \text{(Christoffel symbols)}.$$

Note that by lowering the contravariant index, we can transform the Riemann tensor of the first kind to the Riemann tensor of the second kind and vice-versa (by raising the covariant index):

$$R_{\gamma\mu\nu\lambda} = g_{\gamma\delta} R^{\delta}_{\mu\nu\lambda},$$

$$R^{\delta}_{\mu\nu\lambda} = g^{\delta\gamma} R_{\gamma\mu\nu\lambda}.$$

By index symmetry and antisymmetry properties:

$$R_{\gamma\mu\nu\lambda} = R_{\nu\lambda\ \gamma\mu} = -R_{\mu\gamma\nu\lambda} = -R_{\gamma\mu\ \lambda\nu}.$$

In spacetime, the 4D Einstein field equations of general relativity are written as

$$G_{\alpha\beta} + \Lambda g_{\alpha\beta} = 8\ \pi\ T_{\alpha\beta},$$

where the Einstein tensor $G_{\alpha\beta} = R_{\alpha\beta} - R\ \frac{g_{\alpha\beta}}{2}$ is expressed in terms of the Ricci tensor, $R_{\alpha\beta} \equiv R^{4D}_{\alpha\beta}$. Also, for the specific metric signature, the 4D *Ricci scalar* is calculated in terms of the embedding $R = \pm \frac{12}{l^2} = 4\Lambda$, the cosmological constant $\Lambda = \pm \frac{3}{l^2}$, and $T_{\alpha\beta}$ is the energy-momentum tensor, which contains the material sources and in vacuum may be considered negligible [256, 261]. Note that $\Lambda = \pm \frac{3}{L^2}$ on the spacetime leaf hypersurface $l = L$, where the de Sitter solution of general relativity with $\Lambda < 0$ is topologically equivalent to a closed hypersphere of radius $L = \left(\frac{3}{|\Lambda|}\right)^{\frac{1}{2}}$ corresponding to 4D spacetime.

Hence, the field equations in Einstein space can be explicated as:

$$R_{\alpha\beta} - R\frac{g_{\alpha\beta}}{2} + \Lambda g_{\alpha\beta} = 0 \Rightarrow \{\text{Multiply by the inverse metric } g^{\alpha\beta}\} \Rightarrow$$

$$\underbrace{\overbrace{R_{\alpha\beta}g^{\alpha\beta}}^{\sum_\alpha \sum_\beta R_{\alpha\beta}g^{\alpha\beta} = R}}_{\text{scalar curvature, } R} - R.\frac{1}{2}\left(g_{\alpha\beta}g^{\alpha\beta}\right) + \Lambda \cdot \left(\underbrace{g_{\alpha\beta}g^{\alpha\beta}}_{\sum_\alpha \delta^\alpha_\alpha = 4} \right) = 0 \Rightarrow$$

$$-R + 4\Lambda = 0 \Rightarrow \{\text{Plug back}\} \Rightarrow$$

$$R_{\alpha\beta} - 2\Lambda g_{\alpha\beta} + \Lambda g_{\alpha\beta} = R_{\alpha\beta} - \Lambda g_{\alpha\beta} = 0, \quad \forall \alpha,\ \beta \in \{0,\ 1, 2, 3\}.$$

In the 5D spacekime Ricci-flat space, $R_{AB} \equiv R_{AB}^{5D} = 0$, $\forall A < B \in \{0,\ 1, 2, 3,\ 4\}$. The field equations provide a canonical description of the gravitational, electromagnetic and scalar interactions in classical physics.

Campbell's theorem suggests that the Ricci-flat space equations naturally reduce to the Einstein space equations, however, we need to use a specific 5D metric to explicate the 5D spacekime physics.

Without loss of generality, we use the constraints on the five available coordinate degrees of freedom to nullify the electromagnetic potentials, i.e., $g_{4\alpha} = 0$ and fix the scalar potential, i.e., $g_{44} = \epsilon = \pm 1$, relative to the metric signature $(+, - - -, \epsilon)$. By replacing l with t, these restrictions make the coordinate system analogous to the synchronous general relativity framework where l-lines are congruent to geodesic normals to the 4D spacetime and $x^4 \equiv l$.

The Ricci flat space 5D spacekime metric $dS^2 = ds^2 + \epsilon dl^2$, where the spacetime interval ds^2 depends on x^γ and $x^4 \equiv l$. It's useful to factorize the 4D component of the 5D metric by l^2 to synergize with previous reports and compare with other models of general relativity where the spatial 3D component is modulated by an analogous temporal factor of t^2.

Let's introduce a pair of cases using alternative variable transformations to explicate the spacekime interval (line element) $dS^2 = ds^2 + \epsilon dl^2$ when $x^4 \equiv l$ is either spacelike (case 1) or timelike (case 2). In either metric signature, the transformed interval is $dS^2 = \left(\frac{l}{L}\right)^2 ds^2 + \epsilon dl^2$.

Case 1: The fifth dimension ($x^4 \equiv l$) is spacelike, i.e., $g_{44} = \epsilon = -1$ and the metric signature is $(+, - - -, -)$. We can apply a *hyperbolic* coordinate transformation:

$$s \to l \sinh\left(\frac{s}{L}\right), l \to l \cosh\left(\frac{s}{L}\right),$$

$$ds \to \left(\sinh\left(\frac{s}{L}\right)\right) dl + l\left(\cosh\left(\frac{s}{L}\right)\right)\frac{ds}{L},$$

$$dl \to \left(\cosh\left(\frac{s}{L}\right)\right) dl + l\left(\sinh\left(\frac{s}{L}\right)\right)\frac{ds}{L},$$

where L is a constant length introduced for consistency of physical dimensions. For spacelike fifth dimension, $\epsilon = -1$ and

$$dS^2 = ds^2 - dl^2 =$$

$$\left(\left(\sinh\left(\frac{s}{L}\right)\right) dl + l\left(\cosh\left(\frac{s}{L}\right)\right)\frac{ds}{L}\right)^2 - \left(\left(\cosh\left(\frac{s}{L}\right)\right) dl + l\left(\sinh\left(\frac{s}{L}\right)\right)\frac{ds}{L}\right)^2 \underbrace{=}_{\cosh^2\varphi - \sinh^2\varphi = 1}$$

$$\left(\frac{l}{L}\right)^2 ds^2 - dl^2 = \left(\frac{l}{L}\right)^2 ds^2 + \epsilon dl^2.$$

Case 2: The fifth dimension ($x^4 \equiv l$) is timelike, i.e., $g_{44} = \epsilon = +1$ and the metric signature is $(+, - - -, +)$. We can apply a *polar* coordinate transformation:

$$s \to l \sin\left(\frac{S}{L}\right),\ l \to l \cos\left(\frac{S}{L}\right),$$

$$ds \to \left(\sin\left(\frac{S}{L}\right)\right) dl + l\left(\cos\left(\frac{S}{L}\right)\right)\frac{ds}{L},$$

$$dl \to \left(\cos\left(\frac{S}{L}\right)\right) dl + l\left(-\sin\left(\frac{S}{L}\right)\right)\frac{ds}{L}.$$

Then,

$$dS^2 = ds^2 + dl^2 =$$

$$\left(\left(\sin\left(\frac{S}{L}\right)\right) dl + l\left(\cos\left(\frac{S}{L}\right)\right)\frac{ds}{L}\right)^2 + \left(\left(\cos\left(\frac{S}{L}\right)\right) dl - l\left(\sin\left(\frac{S}{L}\right)\right)\frac{ds}{L}\right)^2 \underbrace{=}_{\cos^2\varphi + \sin^2\varphi = 1}$$

$$\left(\frac{l}{L}\right)^2 ds^2 + dl^2 = \left(\frac{l}{L}\right)^2 ds^2 + \epsilon dl^2.$$

Therefore, in either metric signature:

$$dS^2 = \left(\frac{l}{L}\right)^2 g_{\alpha\beta}(x^\gamma, l) dx^\alpha dx^\beta + \epsilon dl^2 \equiv \sum_{\alpha=0}^{3}\sum_{\beta=0}^{3}\left(\frac{l}{L}\right)^2 g_{\alpha\beta}(x^\gamma, l) dx^\alpha dx^\beta + \epsilon\ dl^2.$$

Before we formulate the 5D Ricci tensor, let's define the Christoffel symbols of the second kind, $\Gamma^\alpha_{\mu\beta}$, in terms of the metric $g_{\alpha\beta}(x^\gamma, l)$. These symbols represent the Levi-Civita connection coefficients [308] on the 5D Ricci flat manifold. The Christoffel scalar coefficients form a pseudo-tensor that allows computing distances and facilitates differentiation of tangent vector fields as functions on the manifold. Denote the operators $\partial_\mu \equiv \nabla_\mu, \forall 0 \le \mu \le 4,$ to represent the 4D covariant derivatives of a vector field defined in a neighborhood of a point P, where the derivative at P is computed along the direction of a specified manifold tangent vector. Then, the six 5D Christoffel symbols (left) and their 4D counterparts (right) [144, 309] are:

$$\overbrace{\begin{aligned}\Gamma_{AC}^{C} &= \frac{1}{2}\sum_{C=0}^{4}\sum_{L=0}^{4}\left(g^{CL}(\partial_A g_{CL} + \partial_C g_{LA} - \partial_L g_{AC})\right)\\ &\equiv \tfrac{1}{2}g^{CL}(\partial_A g_{CL} + \partial_C g_{LA} - \partial_L g_{AC})\\ \Gamma_{AB}^{C} &= \tfrac{1}{2}g^{CM}(\partial_A g_{BM} + \partial_B g_{MA} - \partial_M g_{AB})\\ \Gamma_{AC}^{K} &= \tfrac{1}{2}g^{KJ}(\partial_A g_{CJ} + \partial_C g_{JA} - \partial_J g_{AC})\\ \Gamma_{BK}^{C} &= \tfrac{1}{2}g^{CN}(\partial_B g_{KN} + \partial_K g_{NB} - \partial_N g_{BK})\\ \Gamma_{AB}^{K} &= \tfrac{1}{2}g^{KM}(\partial_A g_{BM} + \partial_B g_{MA} - \partial_M g_{AB})\\ \Gamma_{KC}^{C} &= \tfrac{1}{2}g^{CL}(\partial_K g_{CL} + \partial_C g_{LK} - \partial_L g_{KC})\end{aligned}}^{5D\,\text{spacekime}} \qquad \overbrace{\begin{aligned}\Gamma_{\mu\lambda}^{\lambda} &= \frac{1}{2}\sum_{\lambda=0}^{3}\sum_{k=0}^{3}\left(g^{\lambda k}(\partial_\mu g_{\lambda k} + \partial_\lambda g_{k\mu} - \partial_k g_{\mu\lambda})\right)\\ &\equiv \tfrac{1}{2}g^{\lambda k}(\partial_\mu g_{\lambda k} + \partial_\lambda g_{k\mu} - \partial_k g_{\mu\lambda})\\ \Gamma_{\mu\upsilon}^{\lambda} &= \tfrac{1}{2}g^{\lambda t}(\partial_\mu g_{\upsilon t} + \partial_\upsilon g_{t\mu} - \partial_t g_{\mu\upsilon})\\ \Gamma_{\mu\lambda}^{\beta} &= \tfrac{1}{2}g^{\beta l}(\partial_\mu g_{\lambda l} + \partial_\lambda g_{l\mu} - \partial_l g_{\mu\lambda})\\ \Gamma_{\upsilon\beta}^{\lambda} &= \tfrac{1}{2}g^{\lambda z}(\partial_\upsilon g_{\beta z} + \partial_\beta g_{z\upsilon} - \partial_z g_{\upsilon\beta})\\ \Gamma_{\mu\upsilon}^{\beta} &= \tfrac{1}{2}g^{\beta t}(\partial_\mu g_{\upsilon t} + \partial_\upsilon g_{t\mu} - \partial_t g_{\mu\upsilon})\\ \Gamma_{\beta\lambda}^{\lambda} &= \tfrac{1}{2}g^{\lambda k}(\partial_\beta g_{\lambda k} + \partial_\lambda g_{k\beta} - \partial_k g_{\beta\lambda})\end{aligned}}^{4D\,\text{spacetime}}$$

The 5D → 4D index correspondence maps include:

$A \to \{\mu,4\}, B \to \{\nu,4\}, C \to \{\lambda,4\}, J \to \{l,4\}, K \to \{\beta,4\}, L \to \{k,4\}, M \to \{t,4\}, N \to \{z,4\}.$

The notation convention uses uppercase Latin letters to denote 5-tuples $\left\{\underbrace{\overbrace{0}^{t}}_{\text{time}}, \underbrace{1,2,3}_{\text{space}}, \underbrace{\overbrace{4}^{t}}_{\substack{\text{timelike}\\ \text{spacelike}}}\right\}$, and correspondingly lowercase Greek/Latin letters α, β, γ, μ, ν to denote 4-tuples $\left\{\underbrace{0}_{\text{time}}, \underbrace{1,2,3}_{\text{space}}\right\}$

Our strategy is to first examine the Ricci tensor $R_{\mu\nu}^{5D}$ sequentially with the corresponding metric:

$$\left(\begin{array}{cc} \underbrace{\overbrace{R_{\mu\nu}^{5D} = R_{\mu\nu}^{4D}}^{\text{Part (3)}}}_{0\le\mu,\nu\le3} & \underbrace{\overbrace{R_{\mu4}^{5D}}^{\text{Part (2)}}}_{0\le\mu\le3} \\ \underbrace{\overbrace{R_{4\mu}^{5D}}^{\text{Part (2)}}}_{0\le\mu\le3} & \overbrace{R_{44}^{5D}}^{\text{Part (1)}} \end{array}\right), \quad \underbrace{\left(\begin{array}{cc} \underbrace{g_{\alpha\beta}(x^{\gamma}, l)}_{0\le\alpha,\beta\le3} & 0 \\ 0 & \epsilon \end{array}\right)}_{\substack{\text{corresponding}\\ \text{metric tensor}}}.$$

Next, we can explicate the field equations corresponding to the spacekime metric by expressing the 5D Ricci tensor (R_{AB}^{5D}):

$$R_{AB}^{5D} = \sum_{C=0}^{4}\left(\frac{\partial\Gamma_{AC}^{C}}{\partial x^{B}} - \frac{\partial\Gamma_{AB}^{C}}{\partial x^{C}} + \sum_{K=0}^{4}\left(\Gamma_{AC}^{K}\Gamma_{BK}^{C} - \Gamma_{AB}^{K}\Gamma_{KC}^{C}\right)\right) \equiv \frac{\partial\Gamma_{AC}^{C}}{\partial x^{B}} - \frac{\partial\Gamma_{AB}^{C}}{\partial x^{C}} + \Gamma_{AC}^{K}\Gamma_{BK}^{C} - \Gamma_{AB}^{K}\Gamma_{KC}^{C},$$

$$\overbrace{0 \le A \le B \le 4}^{\text{15 pairs}}.$$

We will fragment these 15 Ricci tensor elements into three complementary parts:
(1) R^{5D}_{44}, corresponding to one *wave equation* for a scalar field;
(2) $R^{5D}_{\mu 4}$, representing a set of $4 = \binom{4}{1}$ *conservation equations*; and
(3) $R^{5D}_{\mu\nu}$, another set of $10 = \binom{4}{1} + \binom{4}{2}$ *Einstein equations*:

$$R^{5D}_{\mu\nu} = R^{4D}_{\mu\nu} = \sum_{\lambda=0}^{4} \left(\frac{\partial \Gamma^{\lambda}_{\mu\lambda}}{\partial x^{\upsilon}} - \frac{\partial \Gamma^{\lambda}_{\mu\upsilon}}{\partial x^{\lambda}} + \sum_{\beta=0}^{4} \left(\Gamma^{\beta}_{\mu\lambda}\Gamma^{\lambda}_{\upsilon\beta} - \Gamma^{\beta}_{\mu\upsilon}\Gamma^{\lambda}_{\beta\lambda} \right) \right) \equiv \underbrace{\frac{\partial \Gamma^{\lambda}_{\mu\lambda}}{\partial x^{\upsilon}}}_{U} - \underbrace{\frac{\partial \Gamma^{\lambda}_{\mu\upsilon}}{\partial x^{\lambda}}}_{V} + \underbrace{\Gamma^{\beta}_{\mu\lambda}\Gamma^{\lambda}_{\upsilon\beta}}_{W} - \underbrace{\Gamma^{\beta}_{\mu\upsilon}\Gamma^{\lambda}_{\beta\lambda}}_{Z},$$

$$\overbrace{0 \le \nu \le \mu \le 3}^{\text{10 pairs}}.$$

To solve the 5D spacekime field equations in the Ricci flat space, $R^{5D}_{AB} = 0$, we can separate the general metric tensor using $g_{\alpha\beta}(x^{\gamma}, l) \equiv \chi(x^{\gamma}, l) \overset{*}{g}_{\alpha\beta}(x^{\gamma})$. Let's examine separately each of the three types of equations and their 15 corresponding Ricci flat tensor elements, R^{5D}_{AB}.

Part 1: Let's first examine R^{5D}_{44}, corresponding to a wave equation for a scalar field, where $A = B = 4$. By the separability condition, $g_{\alpha\beta}(x^{\gamma}, l) \equiv \chi(x^{\gamma}, l)\ \overset{*}{g}_{\alpha\beta}(x^{\gamma})$, $g_{k4} = 0$, $g_{4k} = 0, \forall 0 \le k \le 3$, $g_{44} = \epsilon$, and we have:

$$\Gamma^{C}_{AC} = \Gamma^{C}_{4C} = \left\{ \begin{array}{c} \Gamma^{4}_{A4} = 0 \\ C = \{\lambda, 4\} \end{array} \right\} = \Gamma^{\lambda}_{4\lambda} = \tfrac{1}{2} g^{\lambda k} (\partial_4 g_{\lambda k} + \partial_\lambda g_{k4} - \partial_k g_{4\lambda}) = \tfrac{1}{2} g^{\lambda k} \left(\tfrac{\partial}{\partial l} g_{\lambda k}\right) \Rightarrow$$

$$U = \frac{\partial \Gamma^{C}_{AC}}{\partial x^{B}} = \frac{\partial}{\partial l}\left(\frac{1}{2} g^{\lambda k} \left(\frac{\partial}{\partial l} g_{\lambda k} \right) \right), \quad 0 \le \lambda, k \le 3,$$

$$\Gamma^{4}_{A4} = \frac{1}{2} g^{4L} (\partial_A g_{4L} + \partial_4 g_{LA} - \partial_L g_{A4}) = \frac{1}{2} g^{44} \partial_4 g_{4A} = 0,$$

$$A, B = 4 \Rightarrow \Gamma^{C}_{AB} = \left\{ \Gamma^{4}_{44} = 0 \right\} = \Gamma^{\lambda}_{44} = \frac{1}{2} g^{\lambda t} (\partial_4 g_{4t} + \partial_4 g_{t4} - \partial_t g_{44}) = 0 \Rightarrow V = \frac{\partial \Gamma^{C}_{AB}}{\partial x^{C}} = 0.$$

Also,

$$\Gamma^{K}_{AC} = \Gamma^{K}_{4C} = \left\{ \begin{array}{c} \Gamma^{K}_{44} = 0 \\ \underbrace{\Gamma^{4}_{4C} = \Gamma^{4}_{C4}}_{\text{torsionless}} = 0 \end{array} \right\} = \Gamma^{\beta}_{4\lambda} = \frac{1}{2} g^{\beta l} (\partial_4 g_{\lambda l} + \partial_\lambda g_{l4} - \partial_l g_{4\lambda}) = \frac{1}{2} g^{\beta l} \partial_4 g_{\lambda l} = \frac{1}{2} g^{\beta l} \frac{\partial}{\partial l} g_{\lambda l},$$

$$\Gamma^{C}_{BK} = \Gamma^{C}_{4K} = \Gamma^{\lambda}_{4\beta} = \frac{1}{2} g^{\lambda z} (\partial_4 g_{\beta z} + \partial_\beta g_{z4} - \partial_z g_{4\beta}) = \frac{1}{2} g^{\lambda z} \left(\frac{\partial}{\partial l} g_{\beta z} \right),$$

$$W \equiv \Gamma^K_{AC}\Gamma^C_{BK} = \Gamma^\beta_{4\lambda}\Gamma^\lambda_{4\beta} = \left(\frac{1}{2}g^{\beta l}\frac{\partial}{\partial l}g_{\lambda l}\right)\left(\frac{1}{2}g^{\lambda z}\frac{\partial}{\partial l}g_{\beta z}\right),$$

$$A, B = 4 \Rightarrow \Gamma^K_{AB} = \Gamma^K_{44} = \Gamma^\beta_{44} = \frac{1}{2}g^{\beta t}(\partial_4 g_{4t} + \partial_4 g_{t4} - \partial_t g_{44}) = 0 \Rightarrow Z \equiv \Gamma^K_{AB}\Gamma^C_{KC} = 0.$$

Thus,

$$R^{5D}_{44} = U + W = \frac{\partial\Gamma^C_{AC}}{\partial x^B} + \Gamma^K_{AC}\Gamma^C_{BK} = \frac{\partial}{\partial l}\left(\frac{1}{2}g^{\lambda k}\left(\frac{\partial}{\partial l}g_{\lambda k}\right)\right) + \frac{1}{2}g^{\beta l}\frac{\partial}{\partial l}g_{\lambda l}\frac{1}{2}g^{\lambda z}\left(\frac{\partial}{\partial l}g_{\beta z}\right).$$

Using the metric separability, $g_{\alpha\beta}(x^\gamma, l) \equiv \chi(x^\gamma, l)\, \overset{*}{g}_{\alpha\beta}(x^\gamma)$, the warp factor $\frac{l^2}{L^2}$, and $\chi' = \frac{\partial}{\partial l}\chi(x^\gamma, l)$ we have:

$$R^{5D}_{44} = \overbrace{\frac{\partial}{\partial l}\left(\frac{1}{2}\frac{L^2}{l^2}\frac{1}{\chi}g^{(\lambda k)*}\left(\frac{\partial}{\partial l}\frac{l^2}{L^2}\chi\overset{*}{g}_{\lambda k}\right)\right)}^{U} + \overbrace{\frac{1}{2}\frac{L^2}{l^2}\frac{1}{\chi}g^{(\beta l)*}\left(\frac{\partial}{\partial l}\frac{l^2}{L^2}\chi\overset{*}{g}_{\lambda l}\right)\frac{1}{2}\frac{L^2}{l^2}\frac{1}{\chi}g^{(\lambda z)*}\left(\frac{\partial}{\partial l}\frac{l^2}{L^2}\chi\overset{*}{g}_{\beta z}\right)}^{W} =$$

$$\frac{4}{2}\frac{\partial}{\partial l}\left(\frac{1}{l^2}\frac{1}{\chi}\left(\frac{\partial}{\partial l}l^2\chi\right)\right) + \frac{4}{4}\frac{1}{l^4}\frac{1}{\chi^2}\left(\frac{\partial}{\partial l}l^2\chi\right)\left(\frac{\partial}{\partial l}l^2\chi\right) = 2\frac{\partial}{\partial l}\left(\frac{1}{l^2}\frac{1}{\chi}2l\chi + \frac{1}{\chi}\chi'\right) + \frac{1}{l^4}\frac{1}{\chi^2}(2l\chi + l^2\chi')^2 =$$

$$2\frac{\partial}{\partial l}\left(\frac{2}{l} + \frac{1}{\chi}\chi'\right) + \frac{1}{l^4}\frac{1}{\chi^2}(2l\chi + l^2\chi')^2 = -\frac{4}{l^2} + 2\frac{\partial}{\partial l}\left(\frac{\chi'}{\chi}\right) + \frac{4}{l^2} + 2\frac{\chi'}{l\chi} + \left(\frac{\chi'}{\chi}\right)^2.$$

Hence, this 5D Ricci flat space equation $R^{5D}_{44} = 0$ has a special kind of solution:

$$R^{5D}_{44} = 2\frac{\partial\left(\frac{\chi'}{\chi}\right)}{\partial l} + 4\frac{\chi'}{l\chi} + \left(\frac{\chi'}{\chi}\right)^2 = 0 \Rightarrow \underbrace{\chi(x^\gamma, l)}_{\substack{\text{quadratic}\\ \text{warpfactor}}} = \left(1 - \frac{\overbrace{l_o(x^\gamma)}^{\text{length}}}{l}\right) \times \underbrace{k(x^\gamma)}_{\substack{\text{function of}\\ \text{integration}}}.$$

The general solution to this second-order non-linear ordinary differential equation, $2\frac{\partial}{\partial l}\left(\frac{\chi'}{\chi}\right) + 4\frac{\chi'}{l\chi} + \left(\frac{\chi'}{\chi}\right)^2 = 0$, may be expressed in terms of a function of integration $k(x^\gamma) \equiv \frac{f_2(x^\gamma)}{f_1^2(x^\gamma)}$, where f_1 and f_2 are arbitrary functions independent of l, and the negative reciprocal function $l_o(x^\gamma) = -\frac{1}{f_1(x^\gamma)}$:

$$\chi \equiv \chi(x^\gamma,\ l) = f_2(x^\gamma) \times e^{-2\,(\log l - \log(f_1(x^\gamma)\, l + 1))} = f_2(x^\gamma)\left(\frac{1}{l^2}\right)(l\, f_1(x^\gamma) + 1)^2 =$$

$$f_2(x^\gamma)\left(f_1(x^\gamma) + \frac{1}{l}\right)^2 = \underbrace{\frac{f_2(x^\gamma)}{f_1^2(x^\gamma)}}_{k(x^\gamma)}\left(1 + \overbrace{\frac{1}{f_1(x^\gamma)}}^{-l_o(x^\gamma)}\frac{1}{l}\right)^2 = k(x^\gamma)\left(1 - \frac{l_o(x^\gamma)}{l}\right)^2.$$

Part 2: Next, we examine the solutions to the four conservation equations:

$$R^{5D}_{\mu 4}=0, \qquad \forall 0\le\mu\le 3, \qquad B=4.$$

Recall the difference between the partial derivative $\chi'=\frac{\partial}{\partial l}\chi(x^{\gamma},l)$ with respect to the univariate fifth timelike dimension, and the 4D covariant derivatives of different vector fields, e.g., $\nabla_\mu\chi$ and $\nabla_\mu g_{\lambda k}$, along the direction of a specified manifold tangent vector, x^μ. Then we have:

$$\frac{\partial}{\partial l}\left(\frac{\nabla_\mu\chi}{\chi}\right)=\frac{\partial}{\partial l}(\nabla_\mu\chi)\left(\frac{1}{\chi}\right)-\chi'\frac{1}{\chi^2}\nabla_\mu(\chi),$$

$$R^{5D}_{\mu B}=\sum_{C=0}^{4}\left(\frac{\partial\Gamma^C_{\mu C}}{\partial x^B}-\frac{\partial\Gamma^C_{\mu B}}{\partial x^C}+\sum_{K=0}^{4}\left(\Gamma^K_{\mu C}\Gamma^C_{BK}-\Gamma^K_{\mu B}\Gamma^C_{KC}\right)\right)\equiv\underbrace{\frac{\partial\Gamma^C_{\mu C}}{\partial x^B}}_{U}-\underbrace{\frac{\partial\Gamma^C_{\mu B}}{\partial x^C}}_{V}+\underbrace{\Gamma^K_{\mu C}\Gamma^C_{BK}}_{W}-\underbrace{\Gamma^K_{\mu B}\Gamma^C_{KC}}_{Z},$$

$$U=\frac{\partial\Gamma^C_{\mu C}}{\partial x^B}=\frac{\partial\Gamma^C_{\mu C}}{\partial x^4}=\frac{\partial\Gamma^C_{\mu C}}{\partial l}=\frac{\partial}{\partial l}\left(\frac{1}{2}g^{CL}\left(\partial_\mu g_{CL}+\underbrace{\partial_C g_{L\mu}-\partial_L g_{\mu C}}_{0}\right)\right)=\frac{\partial}{\partial l}\left(\frac{1}{2}g^{CL}(\partial_\mu g_{CL})\right)=$$

$$\frac{\partial}{\partial l}\left(\frac{1}{2}g^{\lambda k}\,\partial_\mu g_{\lambda k}\right)=\frac{1}{2}\frac{\partial}{\partial l}\sum_{\lambda=0}^{3}\sum_{k=0}^{3}\left(((\nabla_\mu\chi)\overset{*}{g}_{\lambda k}+(\nabla_\mu\overset{*}{g}_{\lambda k})\chi)\chi^{-1}g^{(\lambda k)*}\right)=$$

$$\frac{1}{2}\sum_{\lambda=0}^{3}\sum_{k=0}^{3}\left(\frac{\partial}{\partial l}\left((\nabla_\mu\chi)\chi^{-1}\overset{*}{g}_{\lambda k}g^{(\lambda k)*}\right)\right)+\frac{1}{2}\frac{\partial}{\partial l}\sum_{\lambda=0}^{3}\sum_{k=0}^{3}\underbrace{\overbrace{(\nabla_\mu\overset{*}{g}_{\lambda k})g^{(\lambda k)*}}^{0}}_{\text{independent of } l}=2\frac{\partial}{\partial l}\left(\frac{\nabla_\mu\chi'}{\chi}\right),$$

$$V=\frac{\partial\Gamma^C_{\mu B}}{\partial x^C}=\frac{\partial\Gamma^C_{\mu 4}}{\partial x^C}=\left\{\frac{\partial\Gamma^4_{\mu 4}}{\partial x^4}=0\right\}=\frac{\partial\Gamma^\lambda_{\mu 4}}{\partial x^\lambda}=\frac{\partial}{\partial x^\lambda}\left(\frac{1}{2}g^{\lambda t}\partial_4 g_{t\mu}\right)=\frac{\partial}{\partial x^\lambda}\left(\chi'\frac{1}{2\chi}\delta^\lambda_\mu+\underbrace{\overbrace{\frac{1}{l}}^{0}}_{\mu\neq 4}\delta^\lambda_\mu\right)=$$

$$\frac{\partial}{\partial x^\mu}\left(\chi'\frac{1}{2\chi}\right)=\frac{1}{2}\left(\nabla_\mu(\chi')\frac{1}{\chi}-\chi'\frac{1}{\chi^2}\nabla_\mu(\chi)\right)=\frac{\partial}{2\partial l}\left(\frac{\nabla_\mu\chi}{\chi}\right).$$

To verify that for $\upsilon=4$ we have $W-Z\equiv\Gamma^K_{\mu C}\Gamma^C_{BK}-\Gamma^K_{\mu B}\Gamma^C_{KC}=0$, we expand the products of Christoffel coefficients and utilize summation index contraction:

$$\Gamma^C_{BK}=\Gamma^C_{4K}=\frac{1}{2}g^{CN}\left(\partial_4 g_{KN}+\underbrace{\partial_K g_{NB}-\partial_N g_{BK}}_{0}\right)=\frac{1}{2}g^{CN}(\partial_4 g_{KN})\Rightarrow$$

$$W=\Gamma^K_{\mu C}\Gamma^C_{BK}=\frac{1}{2}g^{CN}(\partial_4 g_{KN})\frac{1}{2}g^{KJ}\left(\partial_\mu g_{CJ}+\underbrace{\overbrace{\partial_C g_{J\mu}-\partial_J g_{\mu C}}^{0}}_{C\text{ and }J\text{ index symmetry}}\right)=$$

$$\frac{1}{2}g^{CN}(\partial_4 g_{KN})\frac{1}{2}g^{KJ}(\partial_\mu g_{CJ})=\frac{1}{2}g^{\lambda z}(\partial_4 g_{\beta z})\frac{1}{2}g^{\beta l}(\partial_\mu g_{\lambda l}).$$

$$\Gamma^K_{\mu B} = \frac{1}{2} g^{KM}\left(\partial_\mu g_{4M} + \partial_4 g_{M\mu} - \partial_M g_{\mu 4}\right) = \frac{1}{2} g^{KM}\left(\partial_4 g_{M\mu}\right),$$

$$\Gamma^C_{KC} = \frac{1}{2} g^{CL}\left(\partial_K g_{CL} + \partial_C g_{LK} - \partial_L g_{KC}\right) = \frac{1}{2} g^{CL}\left(\partial_K g_{CL}\right) \Rightarrow$$

$$Z = \Gamma^K_{\mu B}\Gamma^C_{KC} = \frac{1}{2} g^{KM}\left(\partial_4 g_{M\mu}\right)\frac{1}{2} g^{CL}\left(\partial_K g_{CL}\right) = \left\{\begin{array}{c} g^{44} = g_{44} = \epsilon \\ g_{4\mu} = 0 \end{array}\right\} = \frac{1}{2} g^{\beta t}\left(\partial_4 g_{t\mu}\right)\frac{1}{2} g^{\lambda k}\partial_\beta g_{\lambda k}.$$

Using the metric separability, $g_{\alpha\beta}(x^\gamma, l) \equiv \underbrace{\chi(x^\gamma, l)}_{\text{quadratic warp factor}} \times \underbrace{\overset{*}{g}_{\alpha\beta}(x^\gamma)}_{\text{pure-canonical metric}}$, we show that $W - Z = 0$:

$$W - Z = \Gamma^K_{\mu C}\Gamma^C_{BK} - \Gamma^K_{\mu B}\Gamma^C_{KC} = \frac{1}{2} g^{\lambda z}\left(\partial_4 g_{\beta z}\right)\frac{1}{2} g^{\beta l}\left(\partial_\mu g_{\lambda l}\right) - \frac{1}{2} g^{\beta t}\left(\partial_4 g_{t\mu}\right)\frac{1}{2} g^{\lambda k}\partial_\beta g_{\lambda k} =$$

$$\frac{1}{4}\left(\underbrace{g^{\lambda z}\left(\partial_4 g_{\beta z}\right) g^{\beta l}\left(\partial_\mu g_{\lambda l}\right)}_{A} - \underbrace{g^{\beta t}\left(\partial_4 g_{t\mu}\right) g^{\lambda k}\partial_\beta g_{\lambda k}}_{B}\right).$$

By demonstrating that the terms A and B are identical we infer that $W - Z = 0$:

$$A = g^{\lambda z}\left(\partial_4 g_{\beta z}\right) g^{\beta l}\left(\partial_\mu g_{\lambda l}\right) = \frac{L^2}{l^2}\frac{1}{\chi} g^{(\lambda z)*}\left(\partial_4 \frac{l^2}{L^2}\chi\right)\overset{*}{g}_{\beta z}\frac{L^2}{l^2}\frac{1}{\chi} g^{(\beta l)*}\left(\partial_\mu \frac{l^2}{L^2}\chi \overset{*}{g}_{\lambda l}\right) =$$

$$\frac{1}{l^4}\frac{1}{\chi^2} g^{(\lambda l)*}\left(\partial_4 l^2\chi\right)\left(\partial_\mu l^2\chi \overset{*}{g}_{\lambda l}\right),$$

$$B = g^{\beta t}\left(\partial_4 g_{t\mu}\right) g^{\lambda k}\partial_\beta g_{\lambda k} = \frac{L^2}{l^2}\frac{1}{\chi} g^{(\beta t)*}\left(\partial_4 \frac{l^2}{L^2}\chi\right)\overset{*}{g}_{t\mu}\frac{L^2}{l^2}\frac{1}{\chi} g^{(\lambda k)*}\left(\partial_\beta \frac{l^2}{L^2}\chi \overset{*}{g}_{\lambda k}\right) =$$

$$\left\{\text{since } g^{(\beta t)*}\overset{*}{g}_{t\mu} = \delta^\beta_\mu\right\}$$

$$\frac{1}{l^4}\frac{1}{\chi^2} g^{(\lambda k)*}\left(\partial_4 l^2\chi\right)\delta^\beta_\mu\left(\partial_\beta l^2\chi \overset{*}{g}_{\lambda k}\right) = \frac{1}{l^4}\frac{1}{\chi^2} g^{(\lambda k)*}\left(\partial_4 l^2\chi\right)\left(\partial_\mu l^2\chi \overset{*}{g}_{\lambda k}\right) \equiv A.$$

Therefore, the set of four conservation laws represent second order differential equations:

$$0 = R^{5D}_{\mu 4} = U - V + \underbrace{W - Z}_{0} = 2\frac{\partial}{\partial l}\left(\frac{\nabla_\mu \chi}{\chi}\right) - \frac{1}{2}\frac{\partial}{\partial l}\left(\frac{\nabla_\mu \chi}{\chi}\right) = \frac{3}{2}\frac{\partial}{\partial l}\left(\frac{\nabla_\mu \chi}{\chi}\right) \Rightarrow \frac{\partial}{\partial l}\left(\frac{\nabla_\mu \chi}{\chi}\right) = 0.$$

The general solutions of $R^{5D}_{\mu 4} = 0$ are a special type, namely $l_o(x^\gamma) = \text{constant}$, of the solutions to the previous wave equation solution of $R^{5D}_{44} = 0$ (part 1):

$$\chi(x^\gamma, l) = \left(1 - \frac{l_o(x^\gamma)}{l}\right)^2 k(x^\gamma) = \left\{\begin{array}{c} k(x^\gamma),\ l_o(x^\gamma) \equiv 0,\ l \neq 0 \\ \left(1 - \frac{l_o}{l}\right)^2 k(x^\gamma),\ l = l_o(x^\gamma) \neq 0 \end{array}\right..$$

Part 3: Finally, using the above expressions, we can solve the ten Einstein equations:

$$0 = R_{\mu\nu}^{5D} = \sum_{\lambda=0}^{4}\left(\frac{\partial\Gamma_{\mu\lambda}^{\lambda}}{\partial x^{\nu}} - \frac{\partial\Gamma_{\mu\nu}^{\lambda}}{\partial x^{\lambda}} + \sum_{\beta=0}^{4}\left(\Gamma_{\mu\lambda}^{\beta}\Gamma_{\nu\beta}^{\lambda} - \Gamma_{\mu\nu}^{\beta}\Gamma_{\beta\lambda}^{\lambda}\right)\right) \equiv \underbrace{\frac{\partial\Gamma_{\mu\lambda}^{\lambda}}{\partial x^{\nu}}}_{U} - \underbrace{\frac{\partial\Gamma_{\mu\nu}^{\lambda}}{\partial x^{\lambda}}}_{V} + \underbrace{\Gamma_{\mu\lambda}^{\beta}\Gamma_{\nu\beta}^{\lambda}}_{W} - \underbrace{\Gamma_{\mu\nu}^{\beta}\Gamma_{\beta\lambda}^{\lambda}}_{Z},$$

$$0 \leq \nu \leq \mu \leq 3.$$

$$R_{\mu\nu}^{5D} = \sum_{C=0}^{4}\left(\frac{\partial\Gamma_{\mu C}^{C}}{\partial x^{\nu}} - \frac{\partial\Gamma_{\mu\nu}^{C}}{\partial x^{C}} + \sum_{K=0}^{4}\left(\Gamma_{\mu C}^{K}\Gamma_{\nu K}^{C} - \Gamma_{\mu\nu}^{K}\Gamma_{KC}^{C}\right)\right) = \left\{\begin{array}{l} C \to \{\lambda, 4\} \\ K \to \{\beta, 4\} \end{array}\right\} =$$

$$\sum_{\lambda=0}^{3}\left(\underbrace{\frac{\partial\Gamma_{\mu\lambda}^{\lambda}}{\partial x^{\nu}} - \frac{\partial\Gamma_{\mu\nu}^{\lambda}}{\partial x^{\lambda}} + \sum_{\beta=0}^{3}\left(\Gamma_{\mu\lambda}^{\beta}\Gamma_{\nu\beta}^{\lambda} - \Gamma_{\mu\nu}^{\beta}\Gamma_{\beta\lambda}^{\lambda}\right)}_{R_{\mu\nu}^{4D}} + \left(\Gamma_{\mu\lambda}^{4}\Gamma_{\nu4}^{\lambda} - \Gamma_{\mu\nu}^{4}\Gamma_{4\lambda}^{\lambda}\right)\right)$$

$$+\left(\frac{\partial\Gamma_{\mu4}^{4}}{\partial x^{\nu}} - \frac{\partial\Gamma_{\mu\nu}^{4}}{\partial x^{4}} + \sum_{\beta=0}^{4}\left(\Gamma_{\mu4}^{\beta}\Gamma_{\nu\beta}^{4} - \Gamma_{\mu\nu}^{\beta}\Gamma_{\beta4}^{4}\right)\right) \equiv$$

$$R_{\mu\nu}^{4D} + \underbrace{\frac{\partial\Gamma_{\mu4}^{4}}{\partial x^{\nu}}}_{V_1=0} - \underbrace{\frac{\partial\Gamma_{\mu\nu}^{4}}{\partial x^{4}}}_{V_2} + \underbrace{\Gamma_{\mu\lambda}^{4}\Gamma_{\nu4}^{\lambda}}_{V_3} + \underbrace{\Gamma_{\mu4}^{\beta}\Gamma_{\nu\beta}^{4}}_{V_4} + \underbrace{\Gamma_{\mu4}^{4}\Gamma_{\nu4}^{4}}_{V_5=0} - \underbrace{\Gamma_{\mu\nu}^{4}\Gamma_{4\lambda}^{\lambda}}_{V_6} - \underbrace{\Gamma_{\mu\nu}^{\beta}\Gamma_{\beta4}^{4}}_{V_7=0} - \underbrace{\Gamma_{\mu\nu}^{4}\Gamma_{44}^{4}}_{V_8=0}.$$

Note that $g^{4k} = 0, \forall 0 \leq k \leq 3$ and $\Gamma_{\mu4}^{4} = 0, \forall 0 \leq \mu \leq 3$, hence:

$$\Gamma_{\mu4}^{4} = \tfrac{1}{2}g^{4k}\left(\partial_{\mu}g_{4k} + \partial_{4}g_{k\mu} - \partial_{k}g_{\mu4}\right) = \tfrac{1}{2}g^{44}\partial_{4}g_{4\mu} = 0 \Rightarrow$$

$$V_1 = V_5 = V_7 = V_8 = 0.$$

Let's examine the remaining four terms starting with V_2:

$$V_2 = \frac{\partial\Gamma_{\mu\nu}^{4}}{\partial x^{4}} = \frac{\partial}{\partial l}\left(\frac{1}{2}g^{4t}\left(\partial_{\mu}g_{\nu t} + \partial_{\nu}g_{t\mu} - \partial_{t}g_{\mu\nu}\right)\right) = \frac{\partial}{\partial l}\left(\frac{1}{2}g^{44}\left(\partial_{\mu}g_{\nu4} + \partial_{\nu}g_{4\mu} - \partial_{4}g_{\mu\nu}\right)\right) =$$

$$\frac{\partial}{\partial l}\left(\frac{1}{2}\epsilon\left(-\partial_{4}g_{\mu\nu}\right)\right) = -\frac{1}{2}\epsilon\frac{\partial}{\partial l}\left(\frac{\partial}{\partial l}\left(\frac{l^2}{L^2}\chi\right)\right)\overset{*}{g}_{\mu\nu} = -\frac{1}{2}\epsilon\frac{\partial^2}{(\partial l)^2}\left(\frac{l^2}{L^2}\chi\right)\underbrace{\chi^{-1}g_{\mu\nu}}_{\overset{*}{g}_{\mu\nu}}.$$

Similar calculations simplify the term V_3:

$$\Gamma_{\mu\lambda}^{4} = \frac{1}{2}g^{4l}\left(\partial_{\mu}g_{\lambda l} + \partial_{\lambda}g_{l\mu} - \partial_{l}g_{\mu\lambda}\right) =$$

$$\frac{1}{2}g^{44}\left(\partial_{\mu}g_{\lambda4} + \partial_{\lambda}g_{4\mu} - \partial_{4}g_{\mu\lambda}\right) = -\frac{1}{2}\epsilon\partial_{4}g_{\mu\lambda},$$

$$\Gamma_{\nu4}^{\lambda} = \frac{1}{2}g^{\lambda z}\left(\partial_{\nu}g_{4z} + \partial_{4}g_{z\nu} - \partial_{z}g_{\nu4}\right) = \frac{1}{2}g^{\lambda z}\partial_{4}g_{z\nu},$$

$$V_3 = -\frac{1}{2}\epsilon\partial_4 g_{\mu\lambda}\frac{1}{2}g^{\lambda z}\partial_4 g_{zv} \underbrace{=}_{\substack{\text{since}\\ \overset{*}{g}_{\mu\lambda}g^{(\lambda z)^*}\overset{*}{g}_{zv}=\overset{*}{g}_{\mu v}}}$$

$$-\frac{1}{2}\epsilon\partial_4\frac{l^2}{L^2}\chi\overset{*}{g}_{\mu\lambda}\frac{1}{2}\frac{L^2}{l^2}\chi^{-1}g^{(\lambda z)^*}\partial_4\frac{l^2}{L^2}\chi\overset{*}{g}_{zv} = -\frac{1}{4}\frac{1}{L^2 l^2}\epsilon\left(\frac{\partial}{\partial l}(l^2\chi)\right)^2\chi^{-1}\underbrace{\overset{*}{g}_{\mu v}}_{\chi^{-1}g_{\mu v}} =$$

$$-\frac{1}{4}\frac{1}{L^2 l^2}\epsilon\left(\frac{\partial}{\partial l}(l^2\chi)\right)^2\chi^{-2}g_{\mu v}.$$

Next we evaluate is the term V_4:

$$\Gamma^{\beta}_{\mu 4} = \frac{1}{2}g^{\beta t}\left(\partial_\mu g_{4t} + \partial_4 g_{t\mu} - \partial_t g_{\mu 4}\right) = \frac{1}{2}g^{\beta t}\partial_4 g_{t\mu},$$

$$\Gamma^{4}_{v\beta} = \frac{1}{2}g^{4z}\left(\partial_v g_{\beta z} + \partial_\beta g_{zv} - \partial_z g_{v\beta}\right) = \frac{1}{2}g^{4z}\left(\partial_v g_{\beta 4} + \partial_\beta g_{4v} - \partial_4 g_{v\beta}\right) = -\frac{1}{2}\epsilon\partial_4 g_{v\beta},$$

$$V_4 = -\frac{1}{4}\epsilon g^{\beta t}\partial_4 g_{t\mu}\partial_4 g_{v\beta} = -\frac{1}{4}\frac{1}{l^2L^2}\epsilon\left(\frac{\partial}{\partial l}(l^2\chi)\right)^2\chi^{-1}g_{\mu v}.$$

The last term is V_6:

$$\Gamma^{4}_{\mu v} = \frac{1}{2}g^{4t}\left(\partial_\mu g_{vt} + \partial_v g_{t\mu} - \partial_t g_{\mu v}\right) = \frac{1}{2}\epsilon\left(\partial_\mu g_{v4} + \partial_v g_{4\mu} - \partial_4 g_{\mu v}\right) = -\frac{1}{2}\epsilon\partial_4 g_{\mu v},$$

$$\Gamma^{\lambda}_{4\lambda} = \frac{1}{2}g^{\lambda k}(\partial_4 g_{\lambda k} + \partial_\lambda g_{k4} - \partial_k g_{4\lambda}) = \frac{1}{2}g^{\lambda k}(\partial_4 g_{\lambda k}) \underbrace{=}_{\substack{\text{since}\\ 0\le\lambda\le 3}} 2\frac{1}{l^2}\left(\partial_4\left(l^2\chi\right)\right),$$

$$V_6 = \Gamma^{4}_{\mu v}\Gamma^{\lambda}_{4\lambda} = -\frac{1}{l^2L^2}\epsilon\left(\frac{\partial}{\partial l}(l^2\chi)\right)^2\chi^{-2}g_{\mu v}.$$

Plugging all eight terms back into the 10 Einstein's equations and simplifying yields:

$$0 = R^{5D}_{\mu v} = R^{4D}_{\mu v} + \frac{1}{2}\epsilon\frac{\partial^2}{(\partial l)^2}\left(\frac{l^2}{L^2}\chi\right)\underbrace{\chi^{-1}g_{\mu v}}_{\overset{*}{g}_{\mu v}} + \frac{1}{2}\frac{1}{l^2L^2}\epsilon\left(\frac{\partial}{\partial l}(l^2\chi)\right)^2\chi^{-2}g_{\mu v} =$$

$$R^{4D}_{\mu v} + \frac{1}{2}\epsilon\frac{1}{L^2}(2\chi + 4l\chi' + l^2\chi'')\underbrace{\chi^{-1}g_{\mu v}}_{\overset{*}{g}_{\mu v}} + \frac{1}{2}\frac{1}{l^2L^2}\underbrace{\left(4l^2\chi^2 + 4l^3\chi\chi' + l^4\left(\chi'\right)^2\right)}_{\left(2l\chi + l^2\chi'\right)^2}\chi^{-2}g_{\mu v} =$$

$$R^{4D}_{\mu v} + \underbrace{\epsilon\left(\frac{l^2}{2L^2}\frac{\chi''}{\chi} + \frac{4l}{L^2}\frac{\chi'}{\chi} + \frac{l^2}{2L^2}\left(\frac{\chi'}{\chi}\right)^2 + \frac{3}{L^2}\right)}_{S_{\mu v}}g_{\mu v}.$$

Before we finalize the solutions of these differential equations, let's examine the following three fractions:

$$\frac{\chi''}{\chi}=\frac{\frac{6l_0^2}{l^4}-\frac{4l_0}{l^3}}{\left(1-\frac{l_0}{l}\right)^2}\Rightarrow\frac{l^2}{2L^2}\frac{\chi''}{\chi}=\frac{\frac{3l_0^2}{L^2l^2}-\frac{2l_0}{L^2l}}{\left(1-\frac{l_0}{l}\right)^2},$$

$$\frac{4l}{L^2}\frac{\chi'}{\chi}=\frac{\frac{8l_0}{lL^2}-\frac{8l_0^2}{l^2L^2}}{\left(1-\frac{l_0}{l}\right)^2},$$

$$\frac{l^2}{2L^2}\left(\frac{\chi'}{\chi}\right)^2=\frac{l^2}{2L^2}\frac{4\frac{l_0^2}{l^4}}{\left(1-\frac{l_0}{l}\right)^2}=\frac{2\frac{l_0^2}{L^2l^2}}{\left(1-\frac{l_0}{l}\right)^2}.$$

These relations allow us to simplify the term $S_{\mu\nu}$:

$$S_{\mu\nu}=\epsilon\left(\frac{l^2}{2L^2}\frac{\chi''}{\chi}+\frac{4l}{L^2}\frac{\chi'}{\chi}+\frac{l^2}{2L^2}\left(\frac{\chi'}{\chi}\right)^2+\frac{3}{L^2}\right)g_{\mu\nu}=$$

$$\epsilon\frac{1}{\left(1-\frac{l_0}{l}\right)^2}\left(\frac{3l_0^2}{L^2l^2}-\frac{2l_0}{L^2l}+\frac{8l_0}{lL^2}-\frac{8l_0^2}{l^2L^2}+2\frac{l_0^2}{L^2l^2}+\frac{3}{L^2}\left(1-\frac{l_0}{l}\right)^2\right)g_{\mu\nu}=\underbrace{\epsilon\frac{1}{\left(1-\frac{l_0}{l}\right)^2}\frac{3}{L^2}}_{-\Lambda}g_{\mu\nu}.$$

Therefore, the solutions of the Einstein equations in terms of the 5D Ricci tensor $R_{\mu\nu}^{5D}=R_{\mu\nu}^{4D}+S_{\mu\nu}=0$ correspond directly to the 4D spacetime components of the Ricci tensor $\Lambda g_{\mu\nu}\equiv R_{\mu\nu}^{4D}=-S_{\mu\nu}=-\epsilon\frac{3}{L^2}\left(\frac{l}{l-l_0}\right)^2 g_{\mu\nu}$. Thus, $\Lambda=-\epsilon\frac{3}{L^2}\left(\frac{l}{l-l_0}\right)^2$, where $\epsilon=-1$ (spacelike) or $\epsilon=+1$ (timelike) fifth dimension.

Since the 4D Ricci tensor is invariant under a constant conformal transformation of the spacetime metric, $R_{\mu\nu}^{4D}$ remains unchanged and the field equations in Einstein space are indeed $R_{\alpha\beta}\equiv R_{\alpha\beta}^{4D}=\Lambda g_{\alpha\beta}$, $\forall\alpha,\beta\in\{0,1,2,3\}$ and in the 5D spacekime Ricci-flat space as $R_{AB}=R_{AB}^{5D}=0$, $\forall A<B\in\{0,1,2,3,4\}$.

For each 4D solution $\bar{g}_{\mu\nu}$ of these equations and the cosmological constant $\frac{3}{L^2}$, there is a corresponding 5D solution of the Ricci-flat field equations $R_{AB}=0$ with an interval

$$dS^2=\left(\frac{l-l_o}{L}\right)^2\bar{g}_{\alpha\beta}(x^\gamma)dx^\alpha dx^\beta-dl^2\equiv\left(\frac{l-l_o}{L}\right)^2\sum_{\alpha=0}^{3}\sum_{\beta=0}^{3}\left(\bar{g}_{\alpha\beta}(x^\gamma)dx^\alpha dx^\beta\right)-dl^2,$$

where the 4D spacetime metric $g_{\mu\nu}=\left(\frac{l-l_0}{L}\right)^2\bar{g}_{\mu\nu}(x^\gamma)$ has a cosmological constant $\Lambda=\frac{3}{L^2}\left(\frac{l}{l-l_0}\right)^2$.

This derivation of the 5D solutions of the Einstein equations using the Campbell-Wesson embedding theorem utilizes the following definitions of the 5D and 4D Ricci tensors:

$$R^{4D}_{\mu\nu} = \sum_{\lambda=0}^{3} \left(\frac{\partial \Gamma^{\lambda}_{\mu\lambda}}{\partial x^{\upsilon}} - \frac{\partial \Gamma^{\lambda}_{\mu\upsilon}}{\partial x^{\lambda}} + \sum_{\beta=0}^{3} \left(\Gamma^{\beta}_{\mu\lambda}\Gamma^{\lambda}_{\upsilon\beta} - \Gamma^{\beta}_{\mu\upsilon}\Gamma^{\lambda}_{\beta\lambda} \right) \right), \quad 0 \le \nu \le \mu \le 3,$$

$$R^{5D}_{AB} = \sum_{C=0}^{4} \left(\frac{\partial \Gamma^{C}_{AC}}{\partial x^{B}} - \frac{\partial \Gamma^{C}_{AB}}{\partial x^{C}} + \sum_{K=0}^{4} \left(\Gamma^{K}_{AC}\Gamma^{C}_{BK} - \Gamma^{K}_{AB}\Gamma^{C}_{KC} \right) \right), \quad 0 \le A \le B \le 4.$$

This convention is used by Wolfgang Rindler [306] and others to express the Ricci tensor by contracting the *third index* of Riemannian tensor $R^{\lambda}_{\mu\nu\underline{\lambda}}$. Other alternative formulations of the Ricci tensor [62, 147, 257, 309] include an extra factor of -1 and contract on the *second index* $R^{\lambda}_{\mu\underline{\lambda}\nu}$:

$$R^{4D}_{\mu\nu} = \sum_{\lambda=0}^{3} \left(-\frac{\partial \Gamma^{\lambda}_{\mu\lambda}}{\partial x^{\upsilon}} + \frac{\partial \Gamma^{\lambda}_{\mu\upsilon}}{\partial x^{\lambda}} - \sum_{\beta=0}^{3} \left(\Gamma^{\beta}_{\mu\lambda}\Gamma^{\lambda}_{\upsilon\beta} - \Gamma^{\beta}_{\mu\upsilon}\Gamma^{\lambda}_{\beta\lambda} \right) \right).$$

$$R^{5D}_{AB} = \sum_{C=0}^{4} \left(-\frac{\partial \Gamma^{C}_{AC}}{\partial x^{B}} + \frac{\partial \Gamma^{C}_{AB}}{\partial x^{C}} - \sum_{K=0}^{4} \left(\Gamma^{K}_{AC}\Gamma^{C}_{BK} - \Gamma^{K}_{AB}\Gamma^{C}_{KC} \right) \right).$$

In this scenario, the sign of $S_{\mu\nu}$ in **Part 3** derivation is flipped, $S_{\mu\nu} = \epsilon \frac{3}{L^2} \left(\frac{l}{l - l_o} \right)^2 g_{\mu\nu}$ and $\Lambda = \epsilon \frac{3}{L^2} \left(\frac{l}{l - l_o} \right)^2$, where $\epsilon = +1$ for timelike 5th dimension corresponds to $\Lambda > 0$, and conversely, $\epsilon = -1$ for spacelike 5th dimension results in $\Lambda < 0$.

Chapter 6
Applications

In this chapter, we will illustrate several examples of utilizing complex time, spacekime representation, and data science techniques described earlier for interpreting the content, analyzing the information, and graphically visualizing digital datasets. For brevity, we will only show a few instances of the many different types of time-to-kime transformations and alternative strategies to transform longitudinal data from spacetime to spacekime. In general, transforming longitudinal signals to kime functions may be accomplished by many different methods. A purely data-driven strategy relies on phase aggregation, a process that pools repeated measurements, possibly of other similar processes, to estimate the missing kime phases. Alternative computational-model based approaches rely on symmetric distribution priors, which employ theoretical distribution models to represent the probability densities of the phases. Analytic duality techniques represent a third class of methods based on analytic bijective mappings transforming functions of time into corresponding kime-functionals, e.g., using the Laplace transform.

Let us start by drawing parallels between classical statistical inference approaches, based on independent and identically distributed (IID) samples, and spacekime-driven analytics that rely on observing lightlike and kimelike kevents at a fixed longitudinal order and varying kime orientations.

There is a key difference between spacekime data analytics and spacetime data modeling and inference. This contrast is based on the fact that in spacetime, statistical results are obtained by aggregating repeated datasets (random samples) or measuring identical replicate cohorts under controlled equivalent conditions. In particular, population characteristics (like the mean and variance) can be estimated in spacetime by collecting IID samples and then *aggregating* the observations (via standard formulas like arithmetic average and sample variance) to approximate the population characteristics of interest. In spacekime, reliable inference may be obtained based on a single observation, if the perfect kime-phase distribution is known. Unfortunately, the phase angles are generally not observable, however, they can be estimated, inferred, or approximated.

Similar to classical spacetime inference where population characteristics are generally unknown, kime-phases may generally not be available to support reliable spacekime analytics. Recall the last example in **Chapter 1**, which involved spacekime estimation of the consumer sentiment index. In spacekime, the analogue of the spacetime statistical inference process corresponds to generating a sample of observable kevents for a fixed kevent longitudinal order (r_o) with randomly dispersed kime-phases (varying θ_j's). As the kevents' order is observed, even if their phases may not be, we can estimate the kime directions. Consider the analogy with Fourier transformation (FT) of observed spacetime signals into k-space where their

https://doi.org/10.1515/9783110697827-006

phases are synthetically generated. The *aggregation* of the approximate phases will parallel the formulas used for estimating population characteristics in spacetime. To clarify this point, let us consider the spacekime counterpart of arithmetic averaging used to estimate the population mean parameter in spacetime inference:

$$\underbrace{\mu}_{\substack{\text{population}\\ \text{parameter}}} \underbrace{\sim}_{\mu=\mathbb{E}(\bar{x})} \underbrace{\bar{x}}_{\substack{\text{point}\\ \text{estimate}}} = \underbrace{\frac{1}{n}\sum_{j=1}^{n} \underbrace{x_j}_{\text{IIDs}}}_{\substack{\text{arithmetic averaging}\\ \text{(aggregator)}}} .$$

The corresponding spacekime analogue will be the *kime geometric mean*, which effectively averages complex numbers in polar coordinates. As the polar radius is fixed, we are only estimating the center of the phase orientations, not the point of gravitational balance of the kime values provided by arithmetic averaging. Let us denote the kime components of the observed kevents as $k_j = r_o e^{i\theta_j}$. Of course, the kime phases (θ_j) are not known, however, we can synthetically estimate them by using the FT. We will employ these kime-phase estimates $(\mathcal{K}: k \to \hat{k})$ to obtain an aggregate estimate $(\hat{\theta})$ of the unobserved population kime phase (θ), which facilitate spacekime analytics using the kime reconstruction:

$$\hat{k} = r_o e^{i\hat{\theta}}.$$

One example of kime-estimation involves the application of an *aggregator operator* to the native kime-phases. A very natural instance of such an aggregator is based on the geometric mean of the kime-phases, $\{k_j\}_{j=1}^{n}$, which can be expressed as:

$$\hat{k} = \underbrace{\sqrt[n]{\prod_{j=1}^{n} k_j}}_{\substack{\text{geometric}\\ \text{mean}}} = \sqrt[n]{\prod_{j=1}^{n} r_o e^{i\theta_j}} = \sqrt[n]{r_o^n \prod_{j=1}^{n} e^{i\theta_j}} = r_o \sqrt[n]{e^{i\left(\sum_{j=1}^{n}\theta_j\right)}} = r_o e^{i\left(\underbrace{\frac{1}{n}\sum_{j=1}^{n}\theta_j}_{\substack{\text{arithmetic}\\ \text{mean}}}\right)} = r_o e^{i\hat{\theta}}.$$

Finally, the resulting spacekime analytics will be based on the spacetime-reconstructed data using the inverse Fourier transform (IFT) of the k-space corrected kime-phases. In other words, the ultimate model-based inference, model-free prediction, or cluster labeling will be conducted on the IFT-synthesized data $(\hat{\hat{f}})$:

$$\hat{\hat{f}} = IFT(\mathcal{K}(\hat{f})),$$

where the original data, f, is transformed into k-space, $\hat{f} = FT(f)$, and acted on by the kime-estimation operator, $\mathcal{K}$, which transforms the synthetically generated Fourier transform kime-phases, k, into aggregate kime approximations, $\hat{k}$. Note that during the spacekime analytics, the specific mathematical model or statistical inference underpinning the analysis may either utilize the entire spacekime complex data, or collapse the kime-phases and only use the modified kime magnitudes.

Figure 6.1 shows simulations demonstrating the three general scenarios when estimating the kime-phase using sample data. In each case, a sample of $n = 1,000$ kime-phases is randomly drawn from von Mises circular distribution, subject to different mean parameters controlling the centrality, and concentration parameters that govern the dispersion of the phases. The first scenario, ***black*** color, corresponds to kime-phases that are predominantly positive. In this situation, the kime-phase aggregator will yield a positive (non-trivial) phase estimate. The result of this, as expected, will manifest differences in the spacekime and spacetime data-analytical results. The *green* color shows kime estimates corresponding to a trivial phase (nil phase), which is a very special case where directly analyzing the data in spacetime would be equivalent to the corresponding spacekime analytics following the spacekime transformation and kime estimation. Finally, the *red* color corresponds to kime estimates that are predominantly

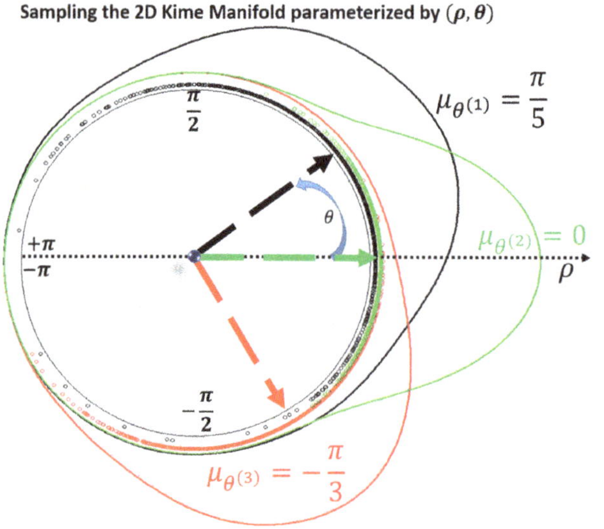

Figure 6.1: A simulation illustrating the (angular/circular) phase distributions of three alternative processes. The graph depicts (1) the raw kime-phases as points along S^1 presented as color-coded point scatters, and (2) the kernel-density estimations of the kime-phase distributions, presented as smooth circular density curves in the kime direction space. The three colors, *green*, ***black***, and *red*, correspond to three alternative *mean* kime-phases, 0, $\frac{\pi}{5}$, $-\frac{\pi}{3}$, respectively. Typically, alternative phase probability distribution models conform to specific sample characteristics, e.g., sample statistics, and directly relate to the unknown population kime-phases we are trying to estimate.

negative. Just like in the ***black*** sample simulation, these non-trivial (negative) kime-phase aggregators are expected to yield spacekime analytical results that are distinct from their spacetime counterparts.

Next, we will explore several specific examples illustrating the effects of spacekime transformation on the corresponding data-driven inference and predictive analytics.

6.1 Image Processing

By separability, the multidimensional FT (e.g., of functions defined on 4D Minkowski spacetime) is just a composition of several FTs, one for each of the dimensions. For instance, in 4D spacetime $(\mathbf{x},t)=(x,y,z,t)$, $n=4$, the FT of a (continuous) function $f(\mathbf{x},t):\mathbb{R}^4\to\mathbb{C}$ can be decomposed into four transforms, one for each of the four coordinate directions. Although the sign in the exponential term may be chosen either way, traditionally the sign convention represents a wave with *angular frequency* ω that propagates in the wavenumber direction $\mathbf{k}$ (*spatial frequency*):

$$(\text{Analysis})\ FT(f)=\hat{f}(\mathbf{k},\omega)=\frac{1}{(2\pi)^{\frac{n}{2}}}\int f(\mathbf{x},t)e^{i(\omega t-\mathbf{kx})}dtd^3\mathbf{x}=$$

$$\underset{(\text{separability})}{=}\frac{1}{(2\pi)^{\frac{n}{2}}}\int\left(\int f(\mathbf{x},t)e^{i(\omega t-\mathbf{kx})}d^3\mathbf{x}\right)dt,$$

$$(\text{Synthesis})\ IFT\left(\hat{f}\right)=\hat{\hat{f}}(\mathbf{x},t)=\frac{1}{(2\pi)^{\frac{n}{2}}}\int\hat{f}(\mathbf{k},\omega)e^{-i(\omega t-\mathbf{kx})}d\omega d^3\mathbf{k}.$$

The forward (analysis) and IFT (synthesis) formulas are separable, which allows the decomposition of the transformations of higher-dimensional functions $g(\mathbf{x}):\mathbb{R}^n\to\mathbb{C}$ into n 1D FTs. This allows us to decrease the number of computations. For instance, expressing the 2D FT $g(\mathbf{x}):\mathbb{R}^2\to\mathbb{C}$ in terms of a series of two 1D transforms.

The discrete Fourier transform (DFT) is the Fourier transform sampled on a discrete lattice. Therefore, it only includes a set of samples, which is large enough to fully describe the spatial domain image, rather than containing all frequencies of the original signal. The number of frequencies included corresponds to the number of samples (e.g., pixels, voxels) in the spatial domain of the signal. For 2D images, the signal representation in the spatial and Fourier domains are arrays of the same sizes. Suppose we have an $M\times N$ square image, $f(m,n):\{(m,n)\in M\times N\}\to\mathbb{R}$ is a discretely sampled and periodic image. Then, the 2D forward (analysis) and inverse (synthesis) DFTs are given by:

$$(\text{Analysis})\ DFT(f)=\hat{f}(k,l)=\frac{1}{\sqrt{MN}}\sum_{m=0}^{M-1}\sum_{n=0}^{N-1}f(m,n)e^{-i\,2\pi\left(\frac{k}{M}m+\frac{l}{N}n\right)},$$

$$\text{(Synthesis) } IDFT(\hat{f}) = \hat{\hat{f}}(m,n) = \frac{1}{\sqrt{MN}} \sum_{m=0}^{M-1} \sum_{n=0}^{N-1} \hat{f}(k,l) e^{i\, 2\pi\left(\frac{k}{N}m + \frac{l}{N}n\right)}.$$

The computational complexity of the discrete Fourier transform (DFT) of a 1D function is $O(N^2)$. The use of the fast Fourier transform (FFT), which assumes that the dimensions are powers of 2, i.e., $N = 2^d$, reduces that complexity to $O(N \log N)$.

Let us demonstrate the FT of 2D images,

$$\hat{f}(k,l) = \underbrace{Real(k,l)}_{Re(k,l)} + i\,\underbrace{Imaginary(k,l)}_{Im(k,l)}.$$

The *magnitudes* of the Fourier coefficients are computed by:

$$Magnitude = \sqrt{Real^2 + Imaginary^2}.$$

That is, the magnitude (amplitude) can be represented in terms of the *real* and *imaginary* components of the complex-valued FT.

The discrete FFT method, *fft()*, generates only *meaningful frequency* up to half the sampling frequency. The DFT values include both positive and negative frequencies. Although, as sampling a signal in discrete time intervals causes aliasing problems, *fft()* yields all frequencies up to the sampling frequency. For instance, sampling a *50 Hz* sine wave and *950 Hz* sine wave with *1,000 Hz* will generate identical results, as the FT cannot distinguish between the two frequencies. Hence, the sampling frequency must always be at least twice as high as the expected signal frequency. For each actual frequency in the signal, the FT will give two peaks (one at the "actual" frequency and one at sampling frequency minus "actual" frequency). This will make the second half of the magnitude vector a mirror image of the first half.

As long as the sampling frequency is at least twice as high as the expected signal frequency, all *meaningful information* will be contained in the first half of the magnitude vector. However, a peak in the low-frequency range might appear when high "noise" frequency is present in the signal (or image). At this point, the vector of extracted magnitudes is only *indexed* by the frequencies but has no associated frequencies. To calculate the corresponding frequencies, the FT simply takes (or generates) the index vector $(1, 2, 3, \ldots, \text{length}(\textit{magnitude vector}))$ and divides it by the length of the data block (in seconds).

In 1D, the phases would represent a vector of the same length as the magnitude vector with phase values of each frequency in $[-\pi, +\pi)$. Phase shifts correspond to translations in space (e.g., x-axis) for a given wave component that are measured in angles (radians). For instance, shifting this wave $f(x) = 0.5\sin(3wt) + 0.25\sin(10wt)$ by $\frac{\pi}{2}$ would produce the following Fourier series:

$$f(t) = 0.5\sin\left(3wt + \frac{\pi}{2}\right) + 0.25\sin\left(10wt + \frac{\pi}{2}\right).$$

In 2D, the Fourier transform (FT/IFT) for images is defined by:

$$\hat{f}(u,v) = F(u,v) = \int_{-\infty}^{\infty}\int_{-\infty}^{\infty} f(x,y)e^{-i\,2\pi(ux+vy)}dxdy,$$

$$f(x,y) = \hat{\hat{f}}(x,y) = \hat{F}(x,y) = \int_{-\infty}^{\infty}\int_{-\infty}^{\infty} F(u,v)e^{i\,2\pi(ux+vy)}dudv,$$

where u and v are the spatial frequencies, $F(u,v) = \underbrace{F_R(u,v)}_{\text{Real}} + i\underbrace{F_I(u,v)}_{\text{Imaginary}}$ is a complex number for each pair of arguments, the *magnitude* spectrum is

$$|F(u,v)| = \sqrt{F_R^2(u,v) + F_I^2(u,v)},$$

and the *phase angle* (direction) spectrum is

$$Phase = \arctan\left(\frac{F_I(u,v)}{F_R(u,v)}\right).$$

The Euler formula allows us to express a complex exponential:

$$e^{-i\,2\pi(ux+vy)} = \cos(2\pi(ux+vy)) - i\sin(2\pi(ux+vy))$$

using the real and imaginary (complex) sinusoidal terms in the 2D plane. The extrema of its real part, $\cos(2\pi(ux+vy))$, occur at $2\pi(ux+vy) = n\pi$. Using vector notation, the extrema are attained at:

$$2\pi(ux+vy) = 2\pi\,\langle U|X\rangle = n\pi,$$

where the extrema points $U = (u,v)^T$ and $X = (x,y)^T$ represent sets of equally spaced parallel lines with a normal U and wavelength $\frac{1}{\sqrt{u^2+v^2}}$.

Let us define the index *shifting* paradigm associated with the discrete FT, which is simply used for convenience and *better visualization*. It has no other relevance to the actual calculation of the FT and its inverse, IFT. When applying the forward or reverse generalized discrete FT, it is possible to shift the transform sampling in time and frequency domain by some real offset values, a, b. Symbolically,

$$\hat{f}(k) = \sum_{n=0}^{N-1} f(n)e^{-\frac{2\pi i}{N}(k+b)(n+a)}, \quad k = 0, \ldots, N-1.$$

As the TCIU package (https://github.com/SOCR/TCIU, accessed January 29, 2021) is developed using R, we will present some of the technical details using R syntax where indices start with 1, not 0, as in some other languages.

In the TCIU R package, the function *fftshift()* is useful for visualizing the FT in k-space with the zero-frequency component in the *middle* of the spectrum, rather than in the corner. Its inverse counterpart, *ifftshift()*, is needed to rearrange the indices appropriately after the IFT is employed, so that the image is correctly reconstructed in spacetime. The FT only computes half of the frequency spectrum corresponding to the non-negative frequencies (positive and zero if the *length (f)* is odd) in order to save computation time. To preserve the dimensions of the output, $\hat{f} = FT(f)$, the second half of the frequency spectrum (the complex conjugate of the first half) is just added at the end of this vector.

Earlier in **Chapter 3**, we showed the effects of the 2D FT on simple binary images (squares, circles), see **Figure 3.3**. Let us now examine the effects of the phases on the image synthesis in spacetime. **Figure 6.2** depicts the effects of synthesizing, or reconstructing, a pair of 2D alphabet images using alternative phase estimates. This experiment shows that both the kime-order (magnitudes) and kime-direction (phase angles) are critical for correctly interpreting, modeling, and analyzing 2D images. In row one of this figure, the second and third columns represent the image reconstructions of the Cyrillic (30-character Bulgarian) alphabet and the Latin (26-character English) alphabet, respectively. Different rows in this figure illustrate alternative kime-phase estimates used in the spacetime image synthesis.

In practical data analytics, solely using time, i.e., accounting only for kime-magnitude but ignoring kime-phase, corresponds to interpreting, visualizing, and analyzing datasets corresponding to the flat priors for kime phases, as shown in the last row in **Figure 6.2**. Clearly, a substantial component of the data energy may be wasted by discarding the kime-directions. Imagine performing statistical inference (e.g., optical character recognition) on a series of such images representing text from one or several languages. Certain parts of the images, e.g., letter "O," may be well preserved in the time-only reconstructions; however, the recognition of more complex characters would be inhibited by suppressing the kime-phase information.

6.2 fMRI Data

The next spacekime demonstration involves longitudinal functional magnetic resonance imaging (fMRI) data. We show spacekime data analytics using real 4D fMRI data with dimension sizes $\left(\underbrace{64}_{x} \times \underbrace{64}_{y} \times \underbrace{21}_{z} \times \underbrace{180}_{t}\right)$. For simplicity of the presentation, analysis, and visualization, we will focus on a longitudinal time series over a 2D spatial domain, rather than the entire 4D fMRI hypervolume. In other words, we'll (artificially) reduce the native 3D spatial domain ($\boldsymbol{x} = (x, y, z) \in \mathbb{R}^3$) to 2D ($\boldsymbol{x} = (x, y) \in \mathbb{R}^2 \subset \mathbb{R}^3$) by focusing only on one mid-axial (transverse) slice through the brain ($z = 11$). This reduction of 3D to 2D spatial dimension does not reduce the

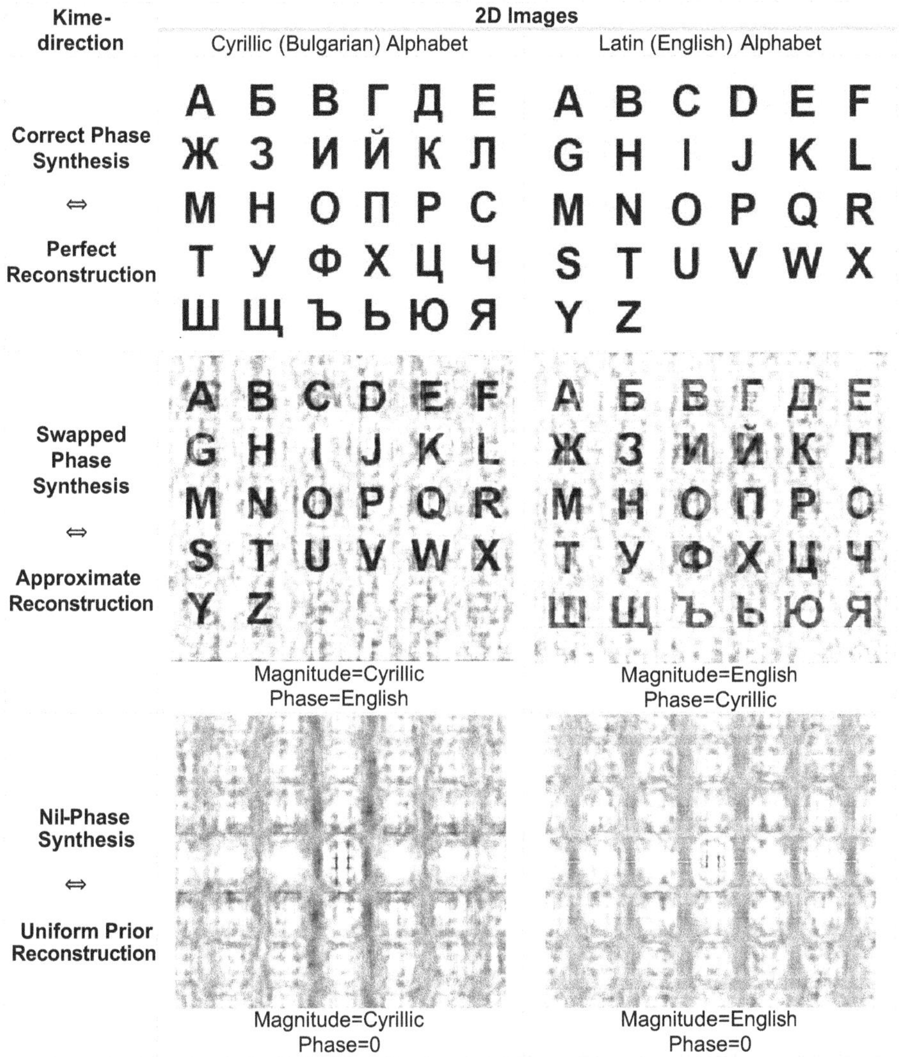

Figure 6.2: Effects of kime-direction estimation on the reconstruction of 2D images representing the 30-character Bulgarian (Cyrillic) and 26-character English (Latin) alphabets.

generality of the experiment and provides a more intuitive way to illustrate the effects of spacekime analytics under alternative phase-estimation strategies.

Figure 6.3 shows a superposition of three time points ($1 \leq t_1 < t_2 < t_3 \leq 180$) of the same 2D spatial domain, where the image intensities at location (x, y) are shown as vertical heights of the surface reconstructions. Brighter colors correspond to larger image intensities and higher surface elevations (altitudes). The surface at the middle time point (t_2) is intentionally shown opaque, whereas the first and last surfaces

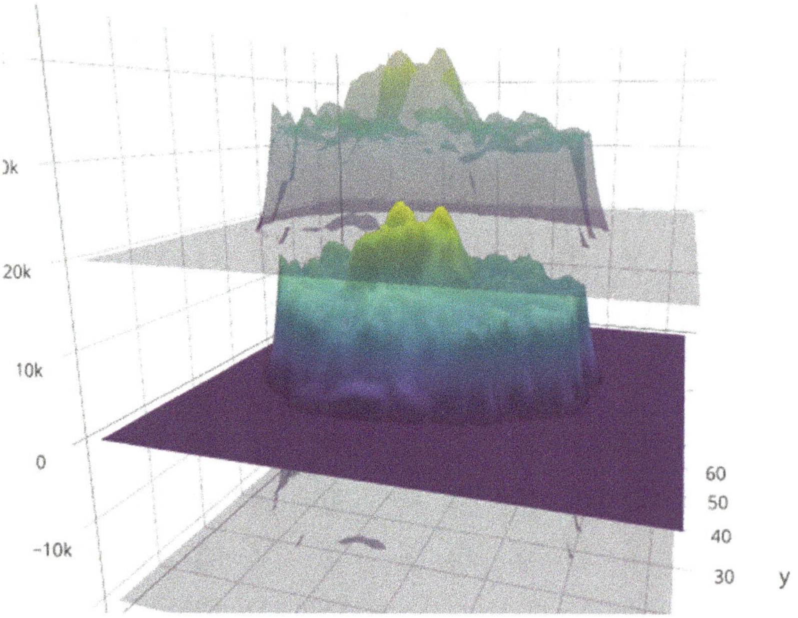

Figure 6.3: 3D rendering of three time cross-sections of the fMRI series over the reduced 2D spatial domain.

$(t_1 < t_3)$ are rendered semitransparent to see through the fMRI intensities at the middle time point.

Suppose the 2D fMRI time series is represented analytically by:

$$f(\mathbf{x}, t) = f(x, y, t) : \mathbb{R}^2 \times \mathbb{R}^+ \to \mathbb{R}$$

and computationally as a third-order tensor, or a 3D array, $f[x, y, t]$. Then, each of the following are also 3D arrays: the complex-valued $\hat{f} = FT(f)$, and the real-valued magnitude of the FT $\left(|\hat{f}| = \sqrt{\left(Re(\hat{f})\right)^2 + \left(Im(\hat{f})\right)^2} \right)$ and phase angle, $\theta = \arctan\left(\frac{Im(\hat{f})}{Re(\hat{f})}\right)$.

We will focus on the function $\hat{f} = \hat{f}\ [f_1, f_2, f_3]$ where the 3rd wavenumber corresponds to the *temporal frequency*. Specifically, we will consider the magnitude of its 3rd dimension as $time = |f_3|$ and we will pretend its phase is unknown, e.g., $\theta_3 = 0$. Thus, inverting the FT of the modified function $\tilde{\hat{f}}$, where $\theta_3 = 0$, will yield an estimate $\hat{\tilde{\hat{f}}}$ of *kime* for the original 2D fMRI time series.

As *time* is observable, the kime magnitude is measurable, however, the kime-*phase* angles are not. They can be either estimated from other similar data, provided by an oracle, modeled using some prior distribution, or fixed according to some experimental conditions. For simplicity, we'll consider two specific instances:

- *Case 1*: When the time-dimension phases are indeed the actual FT phases. Although in general, the phases are unknown, in our fMRI simulation example, they are actually computed from the original spacetime fMRI time series via the FT, and
- *Case 2*: When the time-dimension phases are provided by the investigator, e.g., trivial (*nil*) phases or phases derived from other similar datasets.

Figure 6.4 shows the extracted time-course of one voxel location $(x = 25,\ y = 25,\ z = 12)$ along with a pair of kernel-based smooth reconstructions of the fMRI series at that location. This example is interesting, as in fMRI data, the signal-to-noise ratio (SNR) is really low. Depending on the voxel spatial location and the specific functional stimulus applied at a particular time, less than 0.08% of the observed signal variability may actually be due to the underlying brain physiologic hemodynamic response (HDR) to the stimulus at the particular location. The remaining part of the signal is just random noise, Rician or Rayleigh noise mixed with some bio-stochasticity. **Figure 6.4** shows the raw signal and a couple of smoothed versions of the fMRI time course at one voxel location $(x = 25,\ y = 25,\ z = 12)$. Smoothing effectively corresponds to denoising, which in turn tends to increase the SNR and results in more clear visual patterns, reproducible models, and tractable analytics.

Figure 6.5 depicts the inverse FT reconstruction (synthesis) using the fMRI *real*-magnitude and *nil*-phase estimates. The low correlation $(\rho = 0.16)$ between the two time series (original fMRI series and nil phase reconstructed signal) indicates poor approximation of the original data by the synthesized time series.

Let us now try the oracle approach of kime-phase reconstruction. We can employ the phases of another neighboring voxel with highly correlated $(\rho = 0.79)$ intensities to reconstruct the fMRI time course at the location $(25,\ 25,\ 12)$, see **Figure 6.6**. Note that the correlation between the original and the synthesized signal is now much higher $(\rho = 0.88)$.

These experiments demonstrate that time series representation and analysis work well in spacekime. However, in various situations where one may or may not be able to observe or estimate the kime-direction (phase angle), the results may vary widely based on how reasonable the synthesis of information is without explicit knowledge of the phase measures. As an observable, the *time* (kime-order) is measurable; however, the *phase* angle (kime-direction) needs to be either estimated from other similar data, provided by an oracle, or fixed according to some experimental conditions or prior knowledge.

We can demonstrate the manifold foliation we discussed in **Chapter 5** in terms of fMRI kime-series. In 4D spacetime, classical time series are represented as real-valued functions defined over the positive real time domain $(\mathbb{R}^+)$. In the 5D spacekime manifold, such time series curves extend to *kime-series* that are represented geometrically as kimesurfaces. **Figure 6.7** illustrates an example of one such kimesurface extension of a standard time series anchored at one fixed spatial location. On the https://www.spacekime.org/ website, there are 3D animations that provide

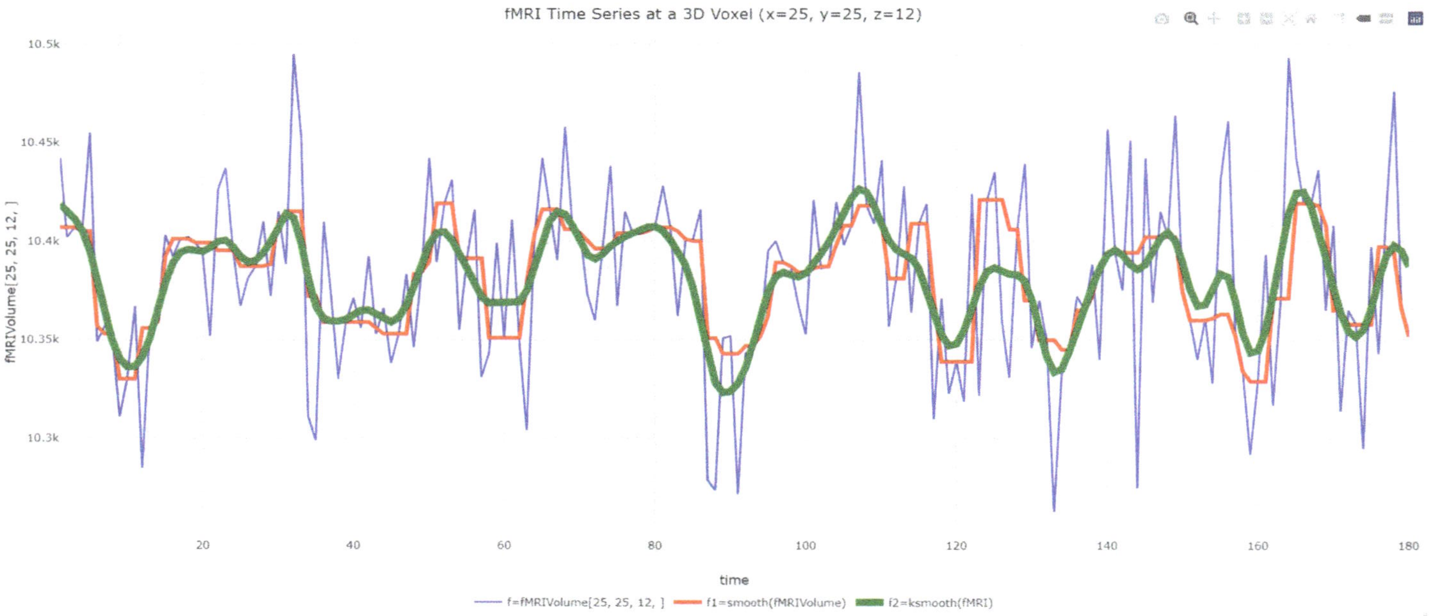

Figure 6.4: The raw and smoothed versions of the time course of the fMRI signal at just one specific voxel location, ($x = 25$, $y = 25$, $z = 12$).

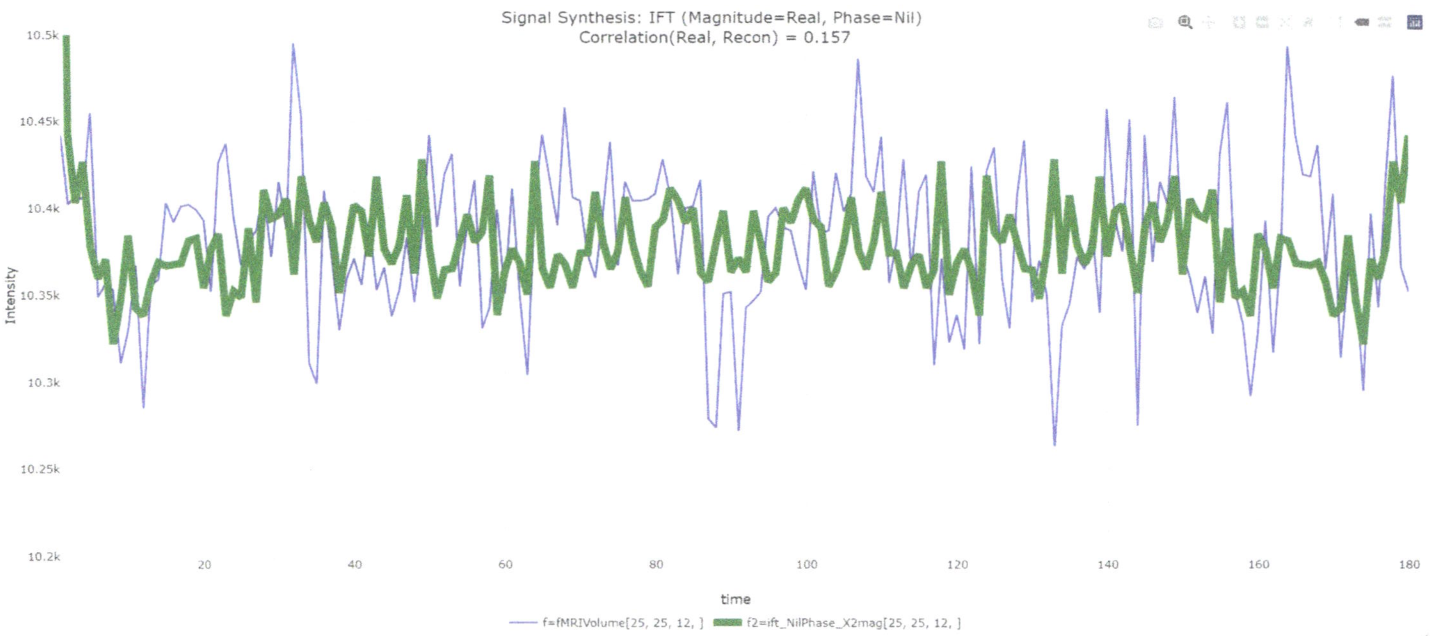

Figure 6.5: Nil-phase-based synthesis of the fMRI signal does not correlate well with the original time series. This suggests that ignoring the kime-phase significantly distorts the fMRI time course.

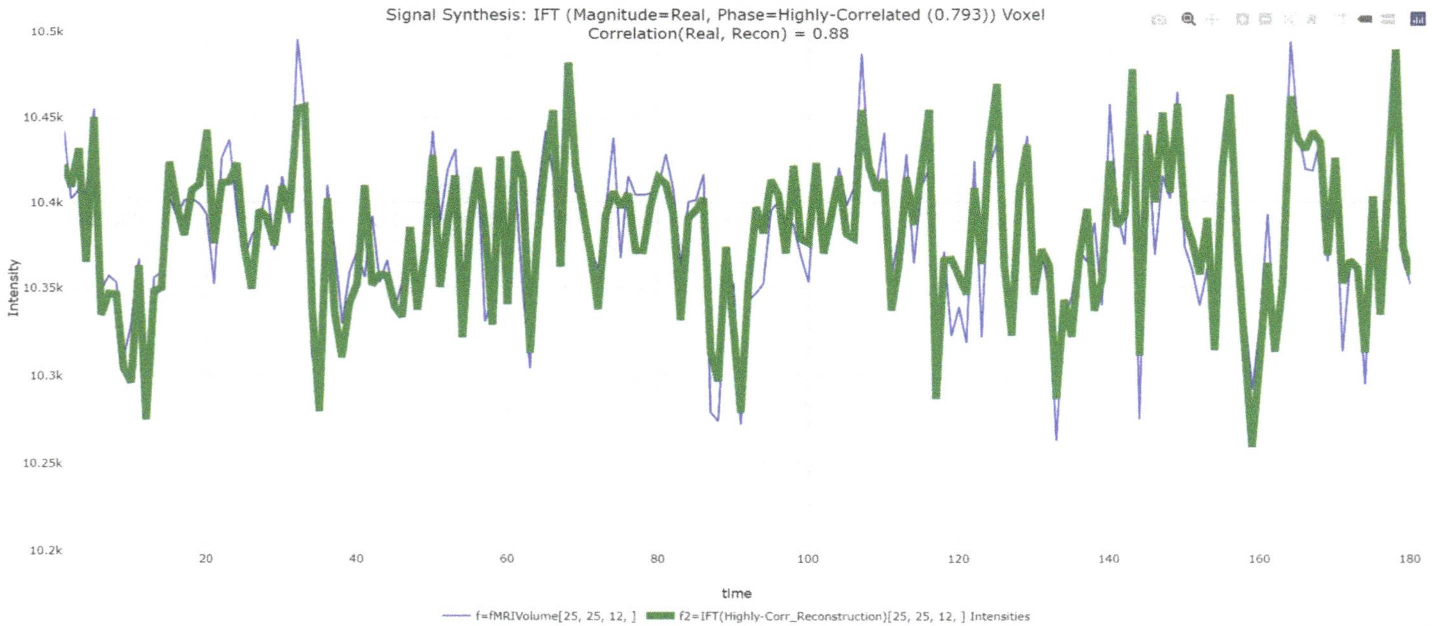

Figure 6.6: Signal synthesis via the inverse FT (magnitude = real, phase = derived from a neighboring highly correlated voxel). The resulting high-correlation between the original and reconstructed signals suggests that one may expect similar data analytics applied to both, real and reconstructed signals to yield similar results. Contrast this to the results reported in Figure 6.5, which used nil phase in the signal reconstruction.

additional visual cues to the relation between kime-series (kimesurfaces) and their lower-dimensional kime-foliations time series (time curves). At any given kime, i.e., specifying the kime-magnitude (t) and the kime-phase (φ), the value of the kimesurface (height) represents the intensity of the kimesurface, which is coded in rainbow color on **Figure 6.7**. A parametric kime grid is superimposed on the surface of the kime-series to show the kime-magnitude (time) and kime-direction (phase). Various kime-phase aggregating operators that can be employed to transform traditional time series curves to kimesurfaces. The latter can then be modeled, interpreted, and used to forecast the kime-longitudinal behavior (kime-course) of the fMRI signal using advanced spacekime analytical techniques. For a given kime-series surface, all of its corresponding time series curves are embedded in concrete kime-foliations, i.e., all time series at $(x, y, z) \in \mathbb{R}^3$ are projections of the kimesurface at the same spatial location onto leaves of the kime-foliation planes. Having an appropriate kime-phase model (or estimate) allows us to interpret and analyze a single kimesurface instead of analyzing a large number of (repeated) time series representing the same process (e.g., fMRI signal at specific voxel for a known stimulus under controlled, event-related, experimental conditions).

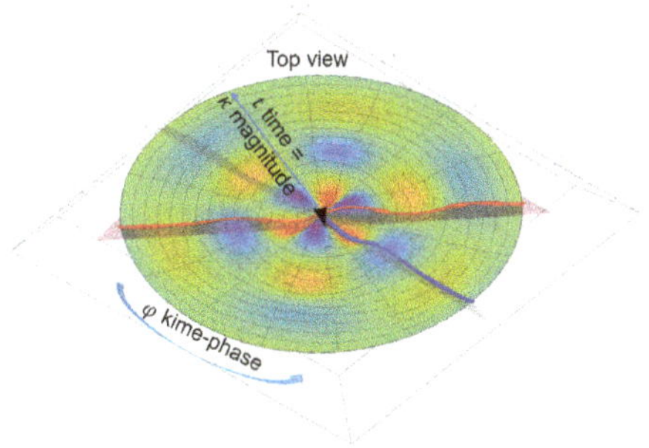

Top view of an fMRI kime-surface.

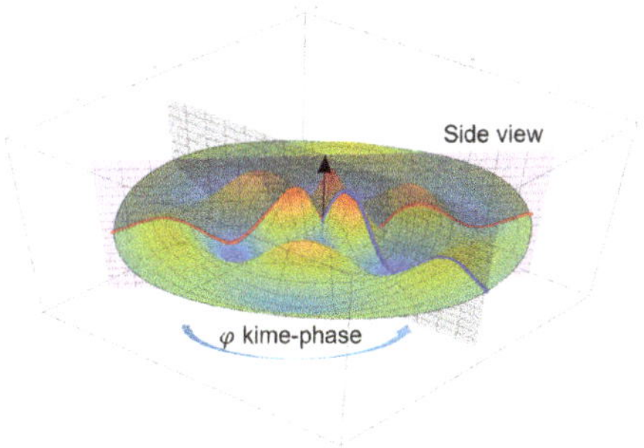

Side view of an fMRI kime-surfacealong with the corresponding classical fMRI time-series as kime-foliation projections.

Figure 6.7: fMRI *kime-series* at a single spatial voxel location. The rainbow color represents the fMRI kime intensities. Examples of 1D time series are shown as projections (red and blue curves) of the kimesurface onto the lower-dimensional kime-foliation leaves.

Recall that in **Chapter 4**, we discussed tensor-based linear modeling as a generalization of classical linear modeling (**Chapter 4, Section 4.6**). There we also presented alternative strategies for representing longitudinal time series as kimesurfaces and showed differences between the *ON* (stimulus) and OFF (rest) conditions of this event-related fMRI study (**Chapter 4, Section 4.7**). For instance, in **Figure 4.4**, we showed reconstructions of the real spacetime fMRI time series as kimesurfaces. This spacekime representation allows us to statistically analyze the kimesurface differences between different experimental conditions (e.g., on–off stimuli in this fMRI study).

Let us apply spacekime tensor-based linear modeling, see **Chapter 4, Section 4.6,** to expand this fMRI data analytics example and derive 3D voxel-based and region-of-interest based statistical maps associated with this finger-tapping somatosensory motor task. There are many alternative spacekime analytics approaches that can be employed to determine the brain regions activated during different experimental conditions. Below, we will demonstrate one specific three-tier analytical strategy. The fMRI data can first be spatially co-registered or aligned with the brain atlas using a number of linear and non-linear spatial normalization techniques [311–314]. Then, the preprocessing continues by using the Laboratory of Neuro Imaging (LONI) probabilistic brain atlas [315] to partition the entire brain anatomical space into 56 regions of interest (ROIs). For each voxel location in the brain, the probabilistic atlas assigns likelihoods of the voxel to be inside of any of the 56 ROIs. This effectively tessellates the spatial domain (3D lattice) into complementary regions. Each voxel location in the domain is tagged with a label representing the most likely ROI it is part of along with a corresponding probability value. Our preprocessing protocol relies on using an affine registration to spatially normalize all time points of 3D fMRI volumes into the probabilistic brain atlas space and superimpose the 56 brain ROI labels onto the functional signal at each time point. In general, this can be done for both real- and complex-valued fMRI data.

The *first tier* of analysis aims to identify any of the ROIs that may be potentially activated by the rest (off) versus activation (on) stimulus tasks. This process pinpoints only the most important brain regions that can subsequently be interrogated voxel-by-voxel to further localize the statistical significance maps reflecting the brain-location association with the event-related (finger-tapping) fMRI task. Although many alternative models can be used for this tier-one analysis, in this experiment, we used a measure called temporal contrast-to-noise ratio (CNR) [316], which represents the quotient of a contrast $\left(\mu_{i,ON-OFF}\right)$ and a corresponding noise estimate $\left(\sigma_{i,ON-OFF}\right)$ over time. The contrast numerator represents the average signal-change or task-related variability. The noise denominator captures the non-task-related variability over time. For a fixed ROI, the CNR statistic is computed by:

$$Y_{i,ON-OFF} = Y_{i,ON} - Y_{i,OFF},\quad \underbrace{\mu_{i,ON-OFF}}_{\text{contrast}} = \frac{1}{n}\sum_{\text{time}}\left(Y_{i,ON-OFF}\right),$$

$$\underbrace{\sigma_{i,ON-OFF}}_{\text{noise}} = \left(\frac{1}{n-1}\sum_{\text{time}}\left(Y_{i,ON-OFF} - \mu_{i,ON-OFF}\right)^2\right)^{\frac{1}{2}},\quad CNR_i = \frac{\mu_{i,ON-OFF}}{\sigma_{i,ON-OFF}},$$

where $Y_{i,ON}$ and $Y_{i,OFF}$ are the fMRI intensities over the ROI voxels i corresponding to the ON and OFF stimulus conditions, respectively, and n is the number of ON–OFF pairs in a time series.

By the central limit theorem (CLT), parametric statistical tests can be used to quantify the ROI-wide stimulus-related brain activation. In our case, we will use a t-test to assess the statistical significance in each of the 56 ROIs independently. The test statistics can be computed using the observed CNR vector $(CNR_1, CNR_2, \cdots, CNR_N)^T$, where N is the number of voxels in the ROI. To control the type 1 error rate, we corrected for multiple testing by Bonferroni adjusting the resulting ROI p-values ($\alpha = 0.000892$).

The *second tier* in the analysis employs tensor-on-tensor linear regression [230] only over the ROIs identified in the first step. The resulting statistical maps will isolate specific voxel locations within the "active regions" that are most significantly associated with the stimuli. Let us explicate the corresponding dimensions of all tensor components and the complete tensor linear model. Using the probabilistic brain brain atlas, the tensor linear regression models are estimated either globally, jointly on all 56 ROIs, or separately on individual ROIs and then aggregating the results into one ROI-conditional model. Once we estimate the effect-size tensor elements, $B = (\beta)$, we can compute the corresponding p-values of the effects (e.g., on–off stimulus) by either parametric t-tests or by running non-parametric tests (e.g., Wilcoxon test). Adjustments for multiple testing across all voxels within each ROI may be necessary to control the false positive rate.

As shown in **Chapter 4**, tensor-based linear modeling requires a structured representation of space and time encapsulating each (irregularly shaped) ROI with the smallest bounding box. This facilitates the tensor arithmetic by matricizing the fMRI signal where the tensor elements are equal to the actual fMRI values inside each ROI or zero outside the ROI boundary. We will denote the general dimensions of the smallest bounding box for each ROI by $a \times b \times c$. Then, for a given ROI, we can implement a tensor linear model using the metricized fMRI tensors:

$$Y = \underbrace{\langle X, B \rangle}_{\text{tensor product}} + E\,.$$

The dimensions of the tensor Y are $160 \times a \times b \times c$, where the tensor elements represent the response variable $Y[t, x, y, z]$, i.e., fMRI intensity. For fMRI magnitude (real-valued signal), the design tensor X dimensions are:

$$\underbrace{160}_{\text{time}} \times \underbrace{4}_{\text{effects}} \times \underbrace{1}_{\mathbb{R}}\,.$$

The X tensor dimensions are $160 \times 4 \times 1$ and include elements corresponding to the on–off stimulus, i.e., finger-tapping task indicator $\begin{Bmatrix} +1(On) \\ -1(Off) \end{Bmatrix}$, adjusting for the HDR function to create the expected BOLD signal x_1. In addition, the design tensor includes polynomial drift terms of order one (linear, x_2) and two (quadratic, x_3):

$$X\left[\underbrace{t}_{160}, \underbrace{(intercept = 1, x_1 = BOLD_{signal}, x_2 = linear_{trend}, x_3 = quadratic_{trend})}_{4}, \underbrace{1}_{1}\right].$$

By fitting the tensor linear model, we can estimate the corresponding effect-size coefficient tensor $\hat{B}$, whose dimensions are $4 \times 1 \times a \times b \times c$, make tensor-model-driven predictions using $\hat{Y} = \langle X, \hat{B}\rangle$ of dimensions $160 \times a \times b \times c$, and the compute the residual tensor $\hat{E} = Y - \underbrace{\langle X, \hat{B}\rangle}_{\hat{Y}}$, whose dimensions are also $160 \times a \times b \times c$.

Again, parametric or non-parametric statistical tests can be used to make inference using the estimated BOLD signal effects $\hat{B}[x_1, 1, a, b, c]$ at spatial location $v = (a, b, c)$ within each ROI. These effects correspond to p-values, $p[a, b, c]$, quantifying the significance of the linear relation. In this case, we ran t-tests on the estimated coefficients $\hat{B}$ in each ROI and our results demonstrate significant effects (smaller corresponding p-values) in the motor area, which is expected by the finger-tapping task. For each ROI, we calculate the p-values based on the estimated tensor coefficients of the effect of $BOLD_{signal}$ and combine these using the spatial information into a global effect-size tensor across all brain spatial locations. In this finger-tapping experimental, the smallest p-values localize the strongest paradigm-specific brain activation brain areas.

The last *third tier* of the analysis involves a post-hoc process. This is necessary for avoiding false positive findings and for exposing only the major brain activation sites associated with the finger-tapping stimulus. Again, there are many alternative approaches to accomplish these post-hoc processes. We will reduce the number of potential false positive voxel locations by correcting for multiple testing using false discovery rate [317, 318]. Then, will apply a spatial clustering filter to remove smaller in size clusters of voxels, which may reflect sporadic effects or low SNR ratio. The effect of these post-processing adjustments on the final statistical maps is twofold. It tempers erratic noise and uncovers only a few major brain areas that have the strongest association with the underlying event related stimulus task.

The strength of the *statistical significance* is determined in terms of quantifying the effect-size tensor, i.e., computing the p-values corresponding to the on–off stimulus. The corresponding *practical importance* of the effect can be visualized in terms of the size and the overall intensity of the resulting brain activation blocks within the specific ROIs. **Figure 6.8** depicts 2D projections and 3D renderings visualizing the results of each of the three sequential steps in the complete tensor-based statistical analysis protocol.

The results on **Figure 6.8** show a couple of important points regarding the three-tier tensor-based analytical protocol. First is the dichotomy between ROI-based and voxel-by-voxel statistical inference. The former relies on aggregate statistics covering all voxels within each ROI. Voxel-by-voxel inference analyzes the fMRI data time course independently for each voxel location. Second, there is a progressive reduction of the number of activation regions, complemented by an emphasis on larger

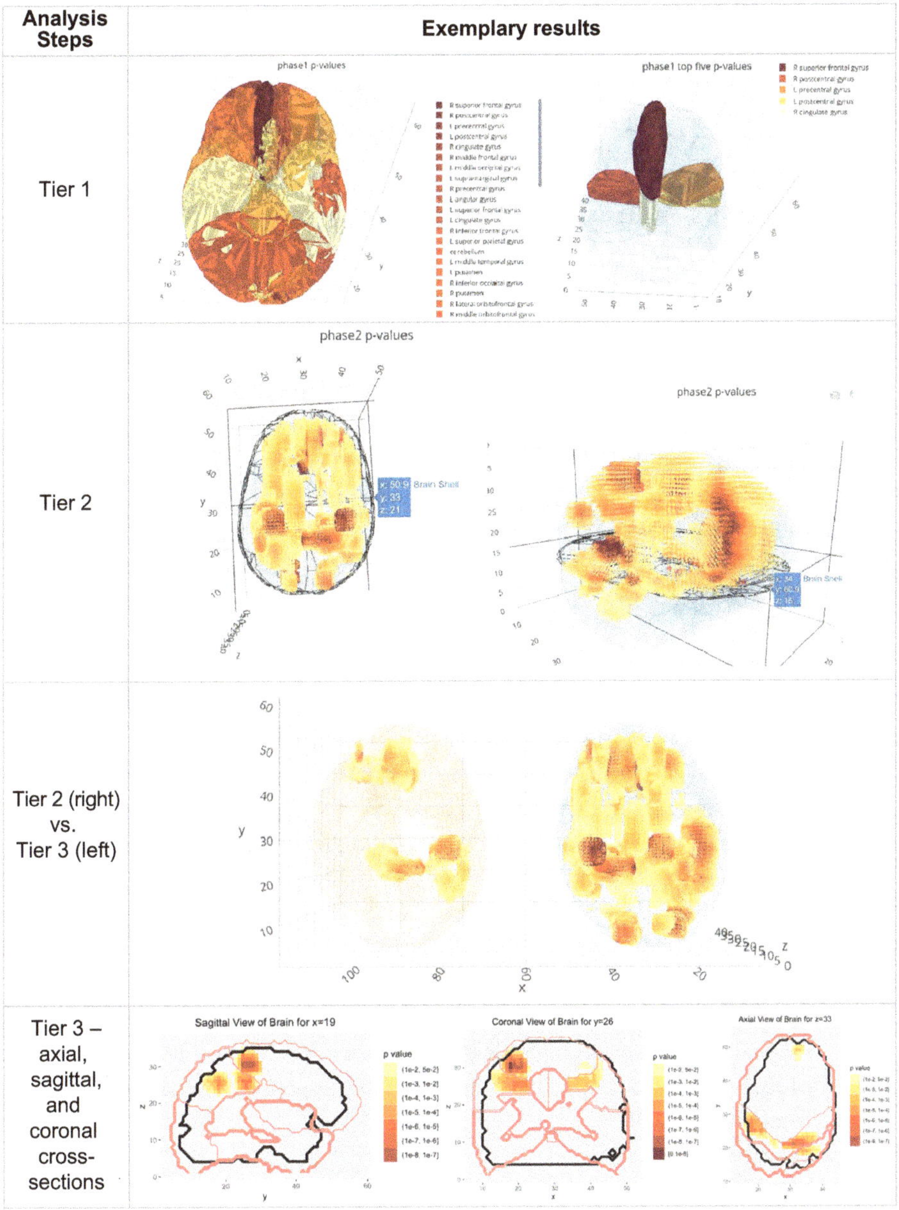

Figure 6.8: Results of the three-tier statistical analysis protocol that starts with registering the fMRI data into a canonical brain atlas space, parcellating the signal into 56 ROIs, determining the regional significance of each ROI, identifying voxel-based statistics using tensor linear models, and post-hoc processing to control the false positive error rate and ensure that large to moderate clusters of significant voxels are identified within the important ROIs.

activation clusters of significant voxels. In this finger-tapping event-related stimulus task, we see strong evidence of observing activation in the somatosensory cortex (motor area), which provides validation of the entire tensor-based statistical analysis of fMRI brain activation.

6.3 Exogenous Feature Time Series Analysis

The next application uses the UCI ML Air Quality Dataset to demonstrate the effect of kime-direction on the analysis of the longitudinal data (https://archive.ics.uci.edu/ml/datasets/Air+quality, accessed January 29, 2021). This *Air Quality dataset* consists of 9,358 hourly averaged responses from an array of 5 sensors embedded in an Air Quality Chemical Multisensor Device. These measurements were obtained in a significantly polluted area during a one-year period (March 2004 to February 2005). The features include concentrations for carbon monoxide (CO), non-methane hydrocarbons (a variety of chemically different organic compounds, e.g., benzene, ethanol, formaldehyde, cyclohexane, acetone), benzene, total nitrogen oxides (NOx), and nitrogen dioxide (NO_2).

The UCI ML Air Quality Dataset features include:
- Date (DD/MM/YYYY).
- Time (HH.MM.SS).
- True hourly averaged concentration CO in mg/m^3 (reference analyzer).
- PT08.S1 (tin oxide) hourly averaged sensor response (nominally CO targeted).
- True hourly averaged overall non-metanic hydrocarbons concentration in $\mu g/m^3$ (reference analyzer).
- True hourly averaged benzene concentration in $\mu g/m^3$ (reference analyzer).
- PT08.S2 (Titania) hourly averaged sensor response (nominally NMHC targeted).
- True hourly averaged NOx concentration in ppb (reference analyzer).
- PT08.S3 (tungsten oxide) hourly averaged sensor response (nominally NOx targeted).
- True hourly averaged NO2 concentration in $\mu g/m^3$ (reference analyzer).
- PT08.S4 (tungsten oxide) hourly averaged sensor response (nominally NO2 targeted).
- PT08.S5 (indium oxide) hourly averaged sensor response (nominally O3 targeted).
- Temperature in °C.
- Relative humidity (%).
- AH absolute humidity.

We will start by exploring the harmonics (frequency–space decomposition) of the time course of carbon monoxide (CO) concentration, **Figure 6.9**. This figure shows the superimpositions of the first 3 harmonics (top row), the mixture of the first 12 harmonics (middle row), and the breakdown of the first 14 harmonics (bottom row) of the CO time series.

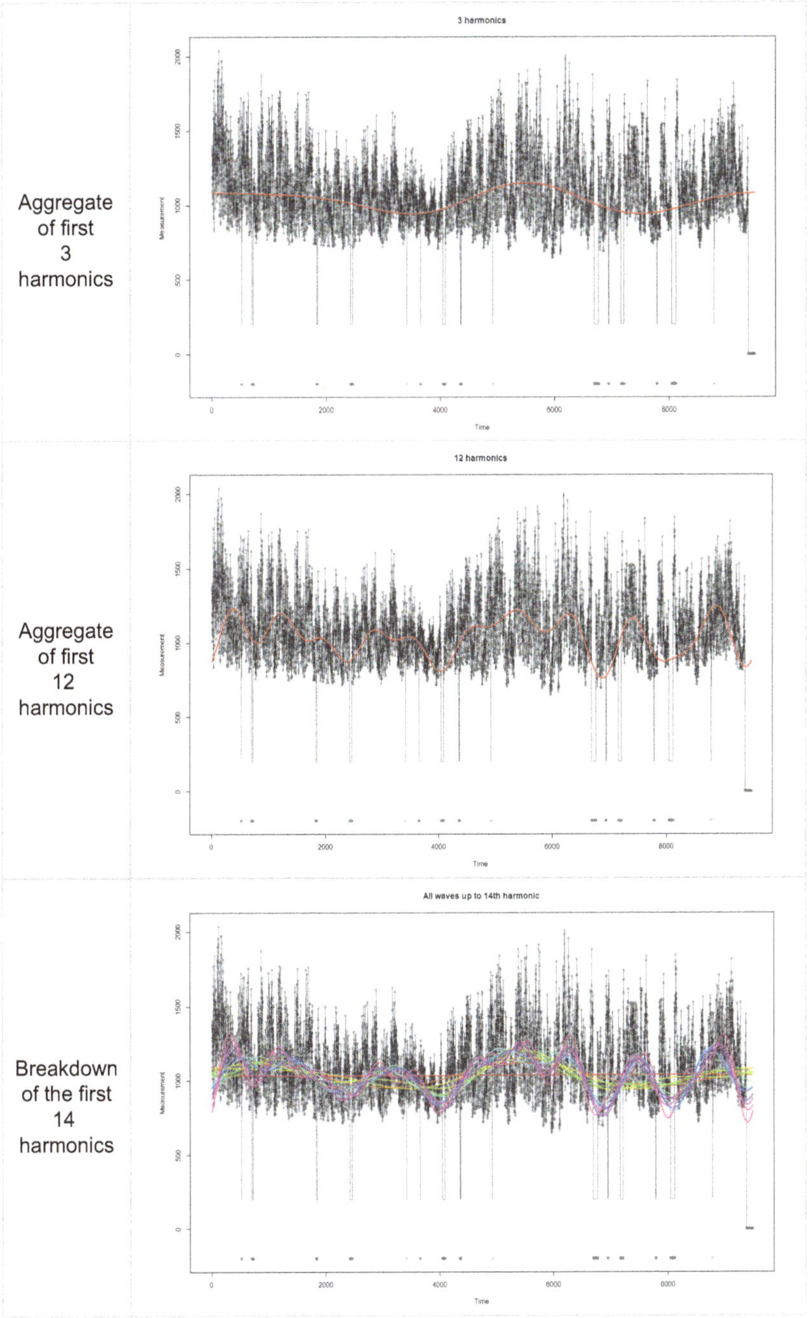

Figure 6.9: Harmonics of CO concentration. The vertical axes in each panel represent CO concentration and the horizontal axes represent time. The top, middle, and bottom rows show the aggregate of the first 3 and 12 harmonics, and the breakdown of the individual harmonics of order 1 through 14.

As the FT and IFT are linear functionals, addition, averaging, and multiplication by constants are preserved by the forward and IFT. Therefore, if we have a number of phase estimates in k-space, we can aggregate these (e.g., by averaging them) and use the resulting *ensemble phase* to synthesize the data in spacekime. If the composite phases are indeed representative of the process kime orientation, then the reconstructed spacekime inference is expected to be valid even if we use a single sample. This way, spacekime inference provides a dual analogy to the CLT [206]. The CLT guarantees the convergence of sample averages to their corresponding population mean counterparts; the spacekime inference ensures that having a large number of phases of the process yields stable, reliable, and reproducible inference. However, in spacekime, ensembling of the known phases may not be as trivial as taking their arithmetic average.

In this air pollution case study, we will deeply explore this *analytic duality* by performing a traditional exogenous ARIMAX modeling [319] of CO concentration (*outcome*: PT08.S1.CO.) based on several covariates, e.g., the following *predictors* NMHC.GT, C6H6.GT, PT08.S2.NMHC, NOx.GT, PT08.S3.NOx, NO2.GT, PT08.S4.NO2, PT08.S5.O3, T, RH, and AH.

The data are interpreted as nine epochs of the same process. We will estimate three complementary ARIMAX models using only epoch 1 data. Each model will utilize alternative phase estimation strategy – true, nil, or average phase angles. **Table 6.1** includes the results of the model fitting under the three experimental conditions with the corresponding estimates of model quality. Note that the model Akaike Information Criterion [320] (AIC) measure of the nil-phase reconstruction is smaller than the model computed using the average-phase synthesis; suggesting the nil-phase model may be better. At first glance, this observation may be counterintuitive. However, deeper investigation uncovers that the lower AIC of the nil-phase signal reconstruction simply suggests that this model captures more of the *nil-phase signal energy*. In other words, better AIC does not necessarily indicate that the nil-phase estimated ARIMA (2,0,1) model represents the *right energy* (information content) of the real CO concentration level. Indeed, it is clear that overall, the parameters obtained using the average-phase ARIMA (2,0,3) model are much closer to their real-signal counterparts. For instance, the estimates of the exogenous feature effects (xreg's) obtained via the average-phase ARIMA(2,0,3) model are closer to their true-model counterparts computed via the ARIMA(1,1,4) model. This suggests that any subsequent data analytic inference based on different phase-estimates will affect the final interpretation of the resulting model-based forecasting of CO concentration.

Figure 6.10 provides a complementary graphical evidence of the impact of kime-direction estimation on the time series forecasting. Notice the improvement of prediction accuracy of the ARIMA model based on the average-phase reconstruction, relative to the no phase-angle (nil) model. These results clearly show improvements of the data analytics using better estimates of the kime-phases. In addition to the visual appeal of phase averaging, we see quantitative evidence (improved correlations between models

Table 6.1: Comparison of the ARIMAX models derived on three different signal reconstructions based on alternative kime-phase estimates.

	Phase estimates		
	Nil	**Average**	**True = original**
Model estimate	ARIMA(2,0,1)	ARIMA(2,0,3)	ARIMA(1,1,4)
AIC	13,179	14,183	10,581
ar1	1.11406562	0.329482302	0.2765312
ar2	−0.14565048	0.238363531	.
ma1	−0.78919188	0.267291585	−0.88913497
ma2	.	−0.006079386	0.12679494
ma3	.	0.15726556	0.03043726
ma4	.	.	−0.17655728
intercept	503.3455144	742.800113	.
xreg1	−0.40283891	0.58379483	0.08035744
xreg2	0.13656613	0.280936931	6.14947902
xreg3	−0.51457636	−0.649722755	0.09859223
xreg4	1.09611981	1.239910298	0.01634736
xreg5	1.21946209	−0.026110332	−0.04816591
xreg6	1.30628469	1.081777956	−0.01104142
xreg7	1.20868397	0.254018471	0.1832854
xreg8	1.14905809	0.306524131	0.17648482
xreg9	−0.48233756	−0.405204908	6.53739782
xreg10	0.03145281	0.351063312	1.79388326
xreg11	−0.46395772	−0.457689796	−12.06965578

ARIMAX (*p*,*d*,*q*)
p = order (# of time lags) of the autoregressive (AR) part
d = differencing (# of past values subtractions)
q = order of moving average (MA) part
"." denotes trivial effect estimates
Highlighted rows indicate better parameter estimates derived using the average-phase signal reconstruction, ARIMA(2,0,3) model, relative to the baseline nil-phase estimated model, ARIMA(2,0,1).

based on true and average phase estimates) of better forecasting when the kime directions can be well estimated.

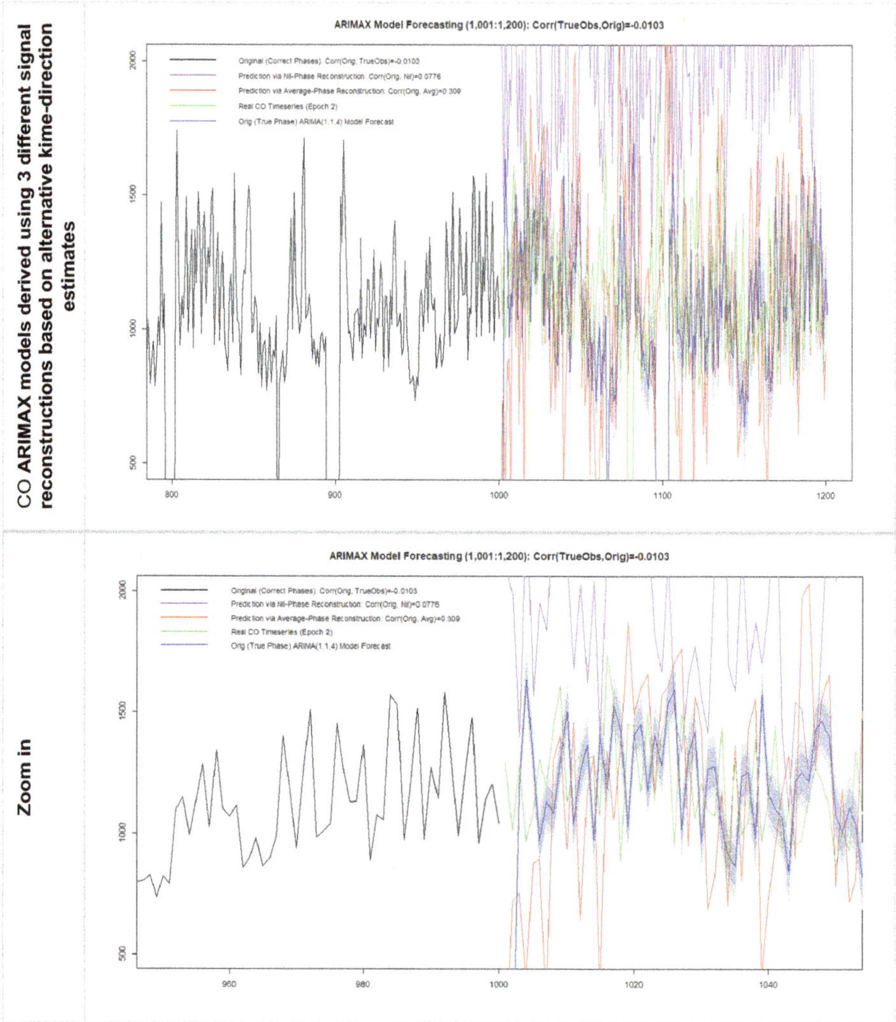

Figure 6.10: Forward in time prediction of the CO concentration using alternative phase estimates.

An alternative data analytic approach involves using the FT applied to the complete 2D data matrix (rows = time, columns = features), inverting it back in spacetime, and investigating the effect of the time series analysis *with* and *without* using the correct phases. Knowing the true phases or ability to better estimate them would be expected to yield better analytical results (e.g., lower bias, reduced dispersion).

We will demonstrate this approach by examining the effects of the phase angles on the inference obtained via regularized linear modeling using least absolute shrinkage and selection operator (LASSO) penalty [10, 238]. Again, we will focus on modeling the CO concentration:

$$CO.GT \sim PT08.S1.CO + NMHC.GT + C6H6.GT + PT08.S2.NMHC + NOx.GT + PT08.S3.NOx \\ + NO2.GT + PT08.S4.NO2 + PT08.S5.O3 + T + RH + AH.$$

Figure 6.11 shows some of the graphical outputs for the nil-phase and true-phase LASSO models. Note the substantial model improvement using the correctly reconstructed signal based on the correct kime-angle directions. In addition to attaining lower mean square error, fewer regression coefficients are needed to model the data and their magnitudes are reduced.

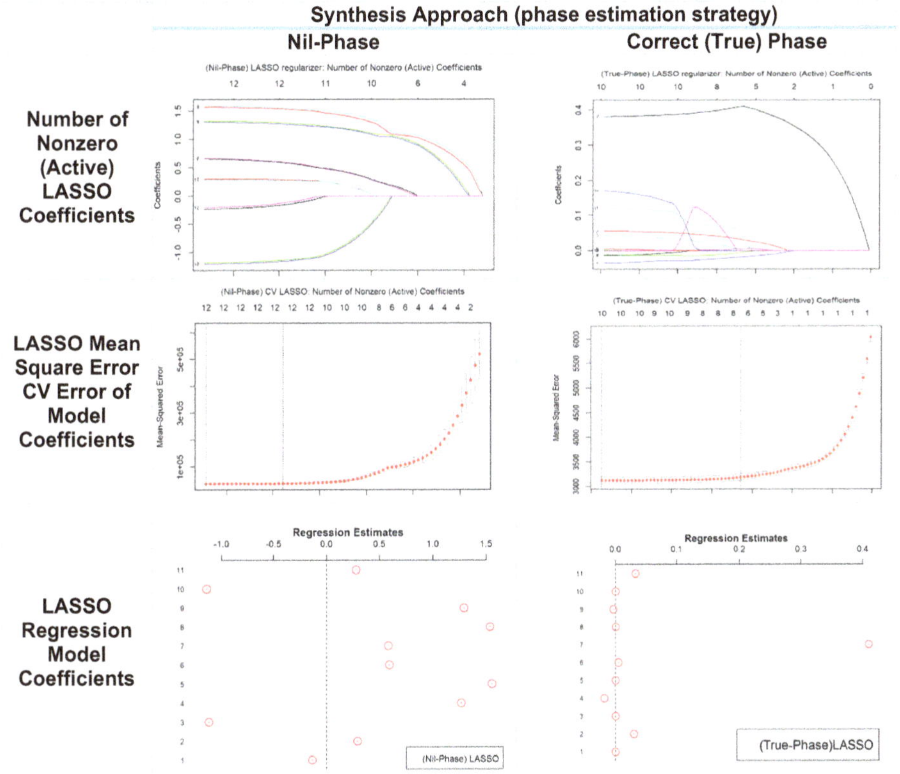

Figure 6.11: Results of alternative phase estimation strategies on the resulting data analytic inference derived using regularized linear modeling of CO concentration with LASSO penalty.

These experiments clearly illustrate the benefits of accurate phase estimation on the final CO concentration analytic inference.

6.4 Structured Big Data Analytics Case Study

Next, we will look at another interesting example of a large structured tabular dataset. The goal remains the same – examine the effects of indexing complex data only using kime-order (time) and comparing the data representations as well as the subsequent data analytics. In this case study, we will use the UK Biobank (UKBB) data archive (https://www.ukbiobank.ac.uk, accessed January 29, 2021) [321].

A previous investigation [57] based on 7,614 variables including clinical information, phenotypic features, and neuroimaging data of 9,914 UKBB subjects, reported the 20 most salient derived imaging biomarkers associated with mental health conditions (e.g., depression, anxiety). By jointly representing and modeling the significant clinical and demographic variables along with the derived salient neuroimaging features, the researchers predicted the presence and progression of depression and other mental health disorders in the cohort of UKBB participating volunteers. We will explore the effects of kime-direction on the findings based on the same data and similar analytical methods. To streamline the demonstration, enable efficient calculations, and facilitate direct interpretation, we will transform the data into a tight computable object of dimensions $9,914 \times 107$ (participants $\times$ features) that can be processed, interpreted and visualized more intuitively. An interactive demonstration of this tensor data showing linear and non-linear dimensionality reduction is available online (https://socr.umich.edu/HTML5/SOCR_TensorBoard_UKBB, accessed January 29, 2021).

The UKBB archive contains incomplete records. We employed multiple imputation by chained equations [322] to obtain complete instances of the data and avoid issues with missing observations. Other preprocessing steps were necessary prior to the main data analytics. As the derived neuroimaging data elements include morphometry measures of vastly different units, e.g., linear, quadratic, and fractal dimensions [323], we normalized these data elements to make the neuroimaging features unitless. The UKBB archive was split into 11 time epochs where the modeling and inference is based on epoch 1, whereas the other epochs were used to estimate the kime-directions. Again, many alternative kime-direction aggregator functions can be employed. In this demonstration, we used phase averaging aggregation to estimate the unknown phase-angles.

To assess the quality of the analytical inference, we will focus on predicting a specific clinical phenotype – *Ever depressed for a whole week 1*, "X4598.2.0." A number of model-based and model-free methods can be used for supervised and unsupervised, retrodictive and predictive strategies for binary, categorical, or continuous regression, clustering and classification. We will demonstrate a decision-tree classification approach for forecasting the binary clinical depression phenotype.

Figures 6.12, **6.13**, and **6.14** show the results of the data analytics for the three alternative kime-phase reconstruction schemes corresponding to the true phases, multi-epoch phase average, and trivial (nil) phases. Note the graphical representation (topology, depth structure, and complexity) of the corresponding decision trees, as well as the quantitative measures capturing the classification accuracy, reliability, and predictive power of each of the corresponding models.

One would expect to see the pruning-based observed simplification of the decision-tree graphs to be tightly coupled with a corresponding decrease of the forecasting reliability (e.g., quantitative metrics such as the Kappa measure). Comparing the results across the three alternative kime-phase reconstruction strategies, we also see the effects of the a priori knowledge about the kime-direction on the final statistical inference.

Finally, **Figure 6.15** shows a summary comparing all analytical results from the three complementary Big Data spacekime reconstruction approaches. The horizontal and vertical axes in this figure show the feature indexing (x-axis) and the feature average across all 9,914 participants (y-axis). Note the substantial discrepancy between the original data (green) and the reconstructions using no-prior nil-phase kime estimates (red). There is better agreement between the true (original, green) signal and the spacekime reconstruction using phase-aggregation (in this case using phase averaging across epochs, blue). This provides strong indirect evidence of the importance of correctly estimating the kime-phases.

In this study, we have the benefit of hindsight, as a prior study [57] had already identified the most salient derived neuroimaging and clinical biomarkers ($k = 107$). The earlier findings include physician-derived clinical outcomes and the computed phenotypes using unsupervised machine learning methods. Therefore, this application shows a simplified demonstration only using the previously selected salient features. In general, a feature-selection preprocessing step may be necessary, or alternatively, different specific data analytic processes may also jointly conduct inference and feature selection. Our purpose in this instance was to examine the effects of the kime-phase estimation on the scientific inference, rather than complete de novo predictive data analytics on the original high-dimensional UKBB data archive.

6.5 Spacekime Analytics of Financial Market and Economics Data

It is known that there are a lot of social and economic phenomena which cannot be predicted long-term with enough accuracy; whether empirically or theoretically. Examples of such processes include forecasting stock markets, economic cycles, inflation dynamics, unemployment, etc. Various mathematical models treating essential aspects of these phenomena have been developed but there are no deterministic theories that allow timely, accurate, and reliable predictions. Unfortunately, contemporary economic theories cannot go beyond detailed explanations of specific past or present states of various observed micro- and macro-economic systems.

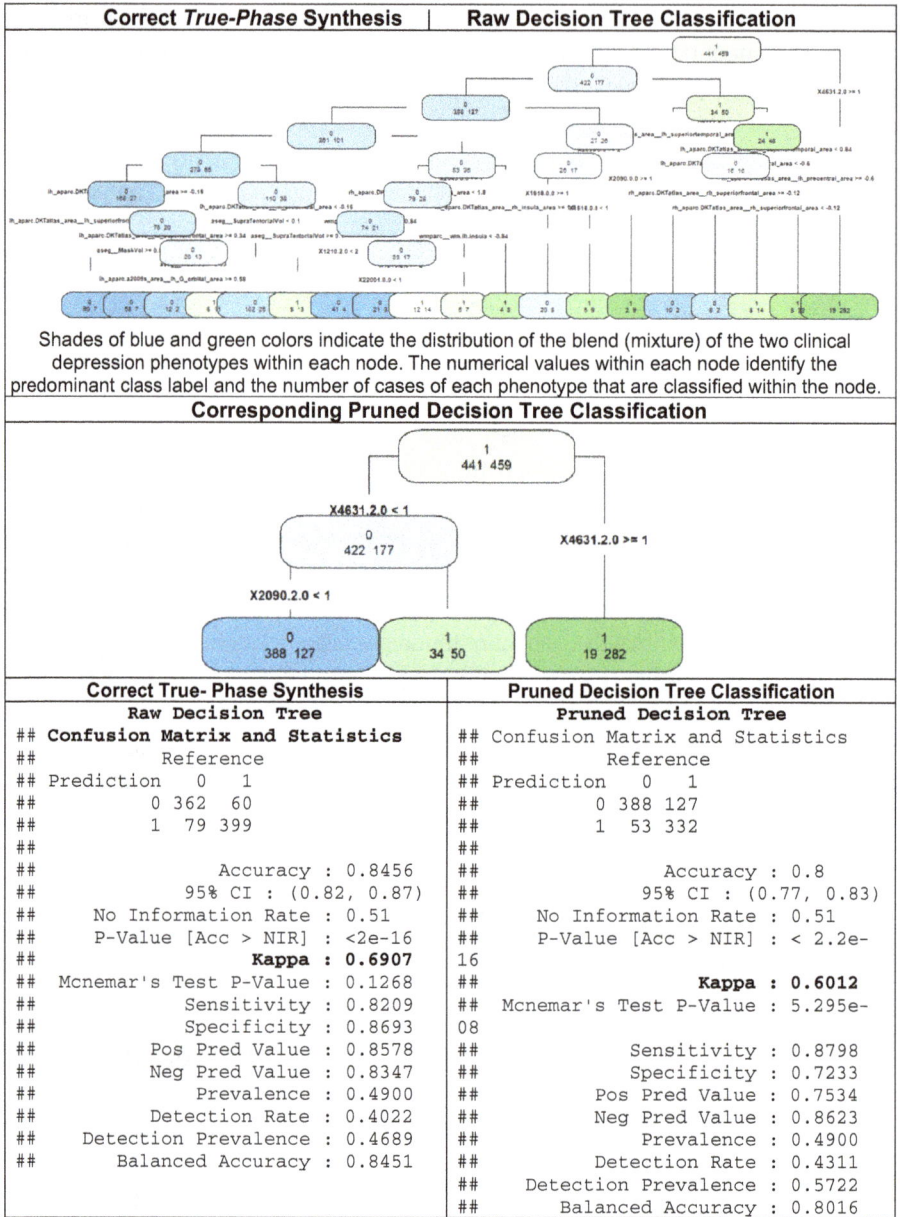

| Correct *True-Phase* Synthesis | Raw Decision Tree Classification |

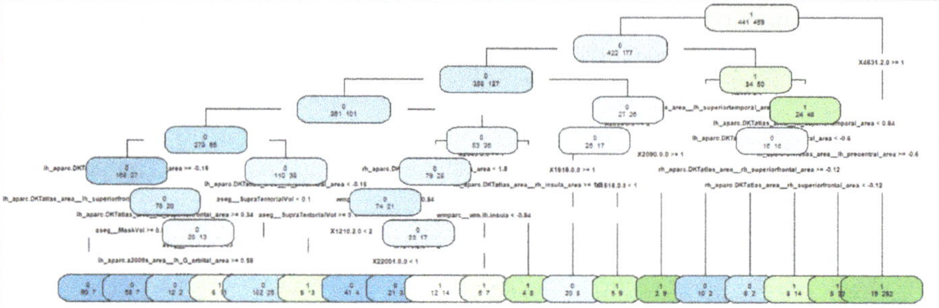

Shades of blue and green colors indicate the distribution of the blend (mixture) of the two clinical depression phenotypes within each node. The numerical values within each node identify the predominant class label and the number of cases of each phenotype that are classified within the node.

Corresponding Pruned Decision Tree Classification

Correct True- Phase Synthesis Raw Decision Tree	Pruned Decision Tree Classification Pruned Decision Tree
## Confusion Matrix and Statistics	## Confusion Matrix and Statistics
## Reference	## Reference
## Prediction 0 1	## Prediction 0 1
## 0 362 60	## 0 388 127
## 1 79 399	## 1 53 332
##	##
## Accuracy : 0.8456	## Accuracy : 0.8
## 95% CI : (0.82, 0.87)	## 95% CI : (0.77, 0.83)
## No Information Rate : 0.51	## No Information Rate : 0.51
## P-Value [Acc > NIR] : <2e-16	## P-Value [Acc > NIR] : < 2.2e-16
## **Kappa : 0.6907**	## **Kappa : 0.6012**
## Mcnemar's Test P-Value : 0.1268	## Mcnemar's Test P-Value : 5.295e-08
## Sensitivity : 0.8209	## Sensitivity : 0.8798
## Specificity : 0.8693	## Specificity : 0.7233
## Pos Pred Value : 0.8578	## Pos Pred Value : 0.7534
## Neg Pred Value : 0.8347	## Neg Pred Value : 0.8623
## Prevalence : 0.4900	## Prevalence : 0.4900
## Detection Rate : 0.4022	## Detection Rate : 0.4311
## Detection Prevalence : 0.4689	## Detection Prevalence : 0.5722
## Balanced Accuracy : 0.8451	## Balanced Accuracy : 0.8016

Figure 6.12: (*True kime-phase* reconstruction): Results of using decision-tree classification on epoch 1 data to predict depression using the correct (true) kime-phase reconstruction.

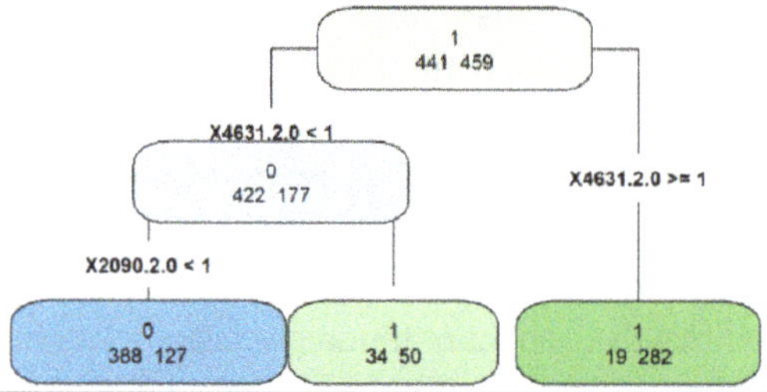

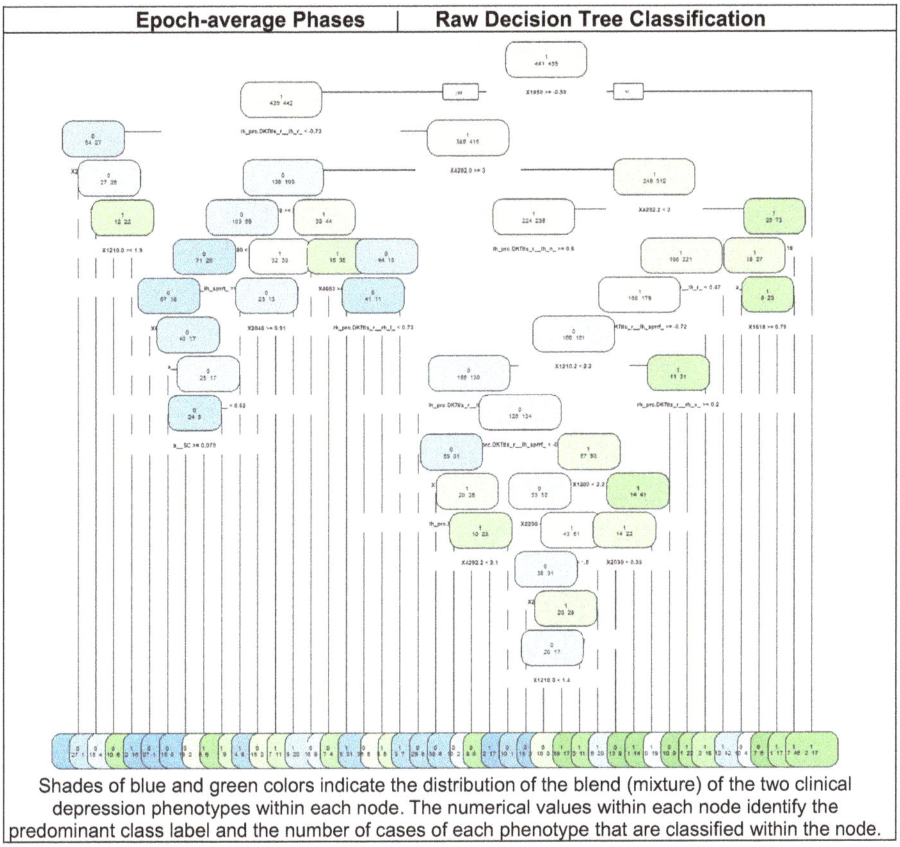
Epoch-average Phases | Raw Decision Tree Classification
Shades of blue and green colors indicate the distribution of the blend (mixture) of the two clinical depression phenotypes within each node. The numerical values within each node identify the predominant class label and the number of cases of each phenotype that are classified within the node.

Figure 6.13: (*Epoch-average* phases): Results of using decision-tree (binary) classification on epoch 1 data to predict depression using synthesis based on an aggregate (average) kime-phases of the 11 epochs.

Murray Gell-Mann [324] defines the concept of "effective complexity," with regard to a given rational being that observes it and builds a schema, as the "length of the compressed description of the regularities of this entity, identified in the schema". The effective complexity of Maxwell's equations is equal to zero insofar as the length of the scheme used to represent them is practically trivial. The same applies to other physical equations. As Murray Gell-Mann [324] remarks, Maxwell's equations describe in just a few lines the behavior of electromagnetism in the entire universe. From this point of view, they provide an amazingly powerful, canonical, and analytical modeling scheme.

Unlike physical phenomena, in general, economic phenomena do not obey certain symmetries. As Focardi and Fabozzi [325] noted, the field of economics does not study immutable laws of nature. Rather this discipline examines human artefacts that are subject to emotion and change due to human decision-making. Therefore, economic

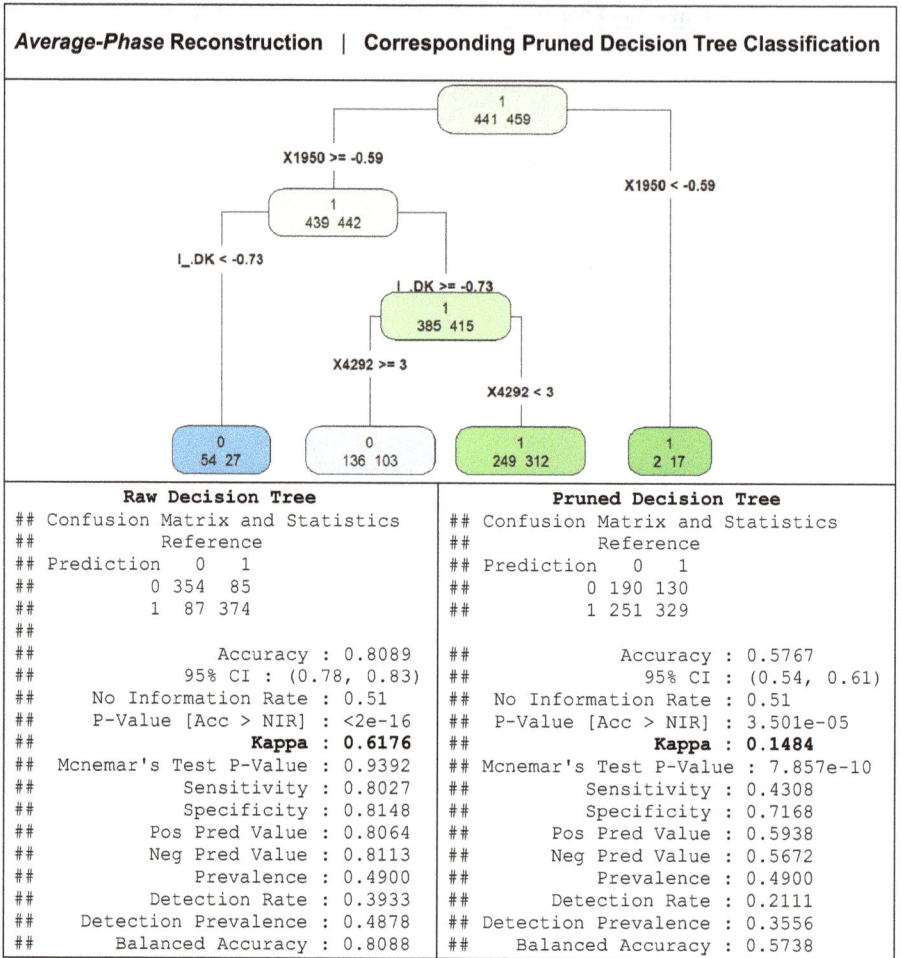

***Average-Phase* Reconstruction | Corresponding Pruned Decision Tree Classification**

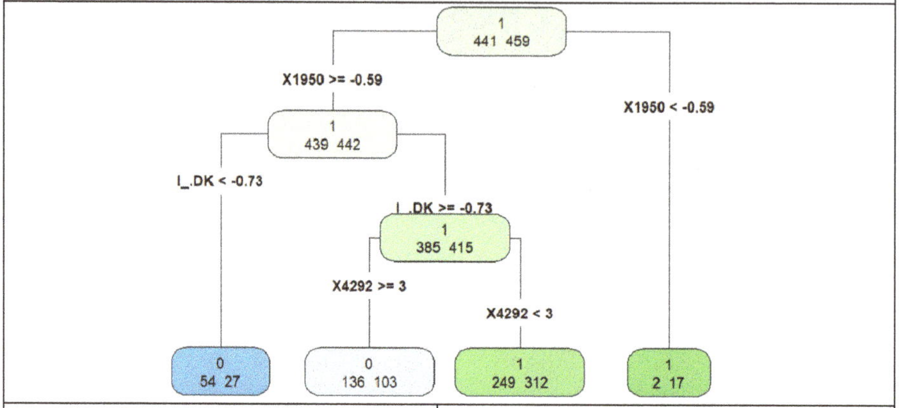

Raw Decision Tree

```
## Confusion Matrix and Statistics
##           Reference
## Prediction   0   1
##          0 354  85
##          1  87 374
##
##                Accuracy : 0.8089
##                  95% CI : (0.78, 0.83)
##     No Information Rate : 0.51
##     P-Value [Acc > NIR] : <2e-16
##                   Kappa : 0.6176
##  Mcnemar's Test P-Value : 0.9392
##             Sensitivity : 0.8027
##             Specificity : 0.8148
##          Pos Pred Value : 0.8064
##          Neg Pred Value : 0.8113
##              Prevalence : 0.4900
##          Detection Rate : 0.3933
##    Detection Prevalence : 0.4878
##       Balanced Accuracy : 0.8088
```

Pruned Decision Tree

```
## Confusion Matrix and Statistics
##           Reference
## Prediction   0   1
##          0 190 130
##          1 251 329
##
##                Accuracy : 0.5767
##                  95% CI : (0.54, 0.61)
##     No Information Rate : 0.51
##     P-Value [Acc > NIR] : 3.501e-05
##                   Kappa : 0.1484
##  Mcnemar's Test P-Value : 7.857e-10
##             Sensitivity : 0.4308
##             Specificity : 0.7168
##          Pos Pred Value : 0.5938
##          Neg Pred Value : 0.5672
##              Prevalence : 0.4900
##          Detection Rate : 0.2111
##    Detection Prevalence : 0.3556
##       Balanced Accuracy : 0.5738
```

Figure 6.13 (continued)

models based on hard metrics and static laws can only be moderately accurate. For this reason, in economic theory, there are no concise schemes describing the entire diversity of phenomena as it is possible in physics. According to Gell-Mann, the effective complexity of economic systems may be greater than the apparent effective complexity of various physical systems. In addition, economic systems tend to be *self-reflecting*, i.e., the knowledge accumulated on the system perturbs the very system itself. To some extent, this is true in quantum physics, as well, where the act of making an observation affects the particle system itself, see the earlier discussions of Heisenberg's uncertainty principle. However, unlike in physics, in economics, there is no repeatability of experiments, and experimentation for certain phenomena may be impractical or even impossible. This fact creates quite a few limitations, both on the direction of

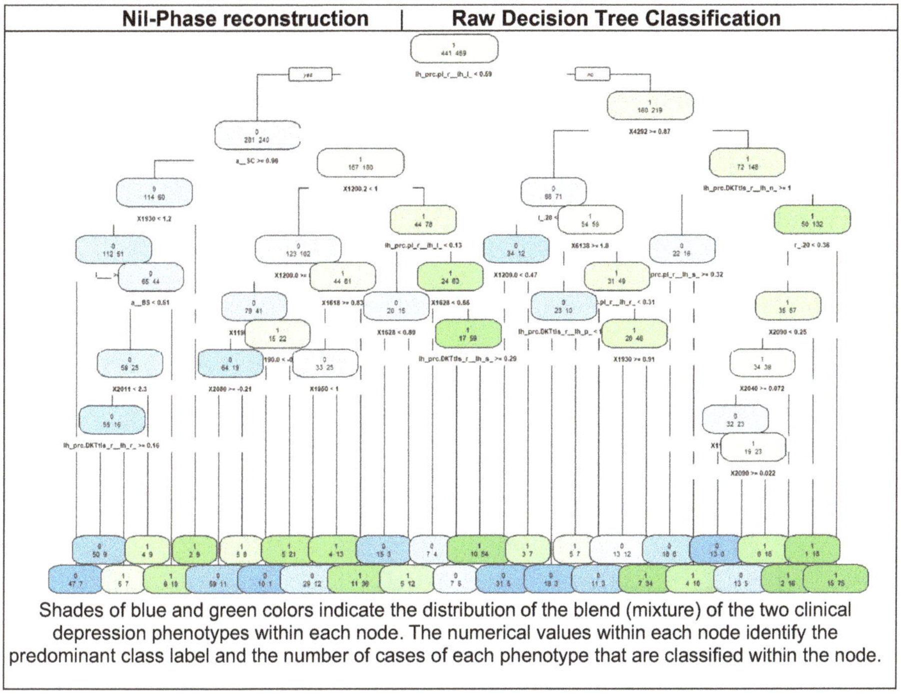
Nil-Phase reconstruction | Raw Decision Tree Classification
Shades of blue and green colors indicate the distribution of the blend (mixture) of the two clinical depression phenotypes within each node. The numerical values within each node identify the predominant class label and the number of cases of each phenotype that are classified within the node.

Figure 6.14: (***Nil-phase* forecasting**): Results of using decision-tree classification on epoch 1 data to predict depression using synthesis based on nil kime-phase estimates.

empirical research and on the development of theoretical foundations. Aside from economics, in most modern scientific disciplines, the development of elaborate theoretical models is based on much broader *understandings of factual experimental evidence* and deeper representation of the underlying theoretical principles [326].

Frequently, economic systems represent elaborate processes with a large number of unknowns and multiplex multivariate relationships. Most often, it is hard to isolate and investigate meaningfully smaller simplified subsystems that lead to comprehensive generalizations. The entire complex system, along with all its interactions, has to be investigated as a holistic economic challenge. The intuitive idea that in economics everything interacts with everything else is not entirely groundless. Across time, different schools of economics have regarded alternative factors as major, minor, or irrelevant, leading to inconsistent and sometimes contradictory conclusions. Some of the economic models purport to express the way in which the entire economic system functions by means of a few highly generalized (aggregate) variables. However, such oversimplified models are generally of little value as substitutes for deeper theoretical explorations.

A great number of the principles derived in economics may not be universal or generally applicable in many situations. Some assumptions underlying various economic theories are intrinsically qualitative and appear to be either too vague or not

Nil-Phase Reconstruction	Corresponding Pruned Decision Tree Classification

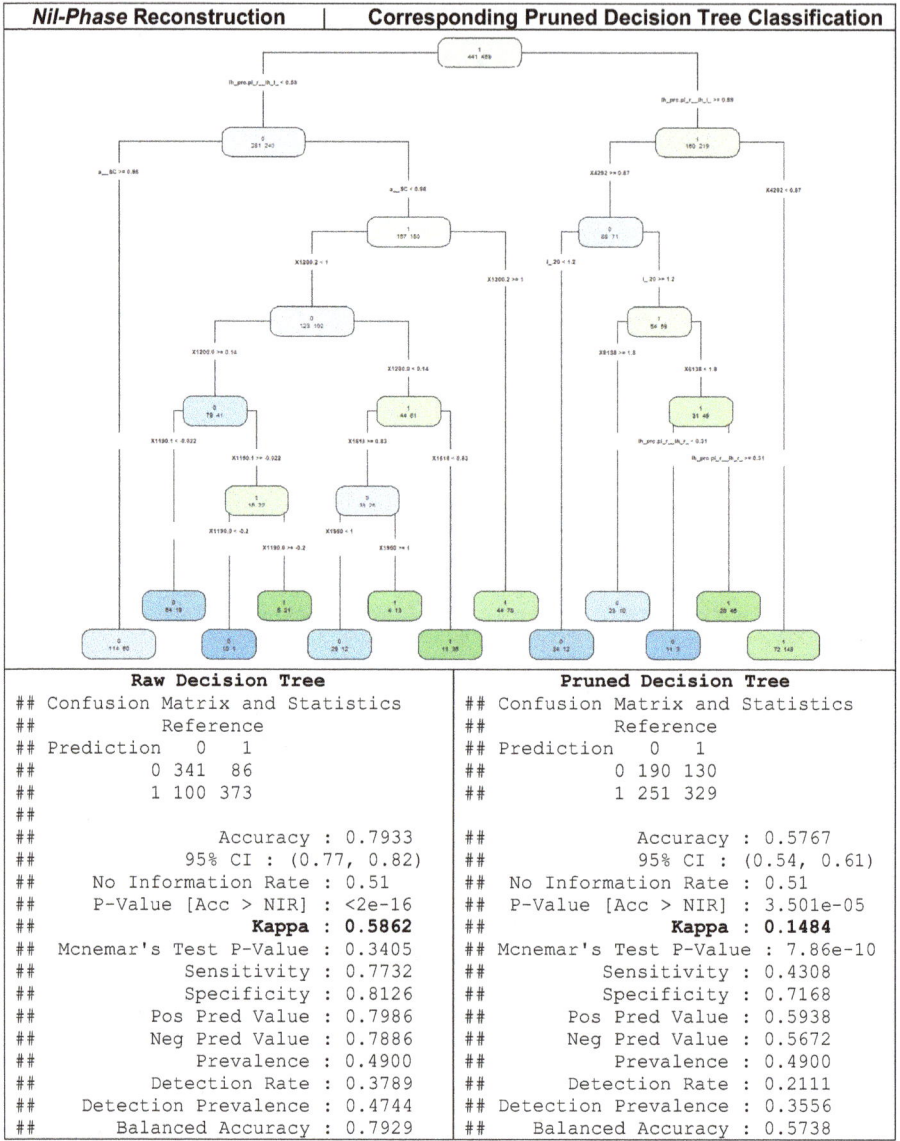

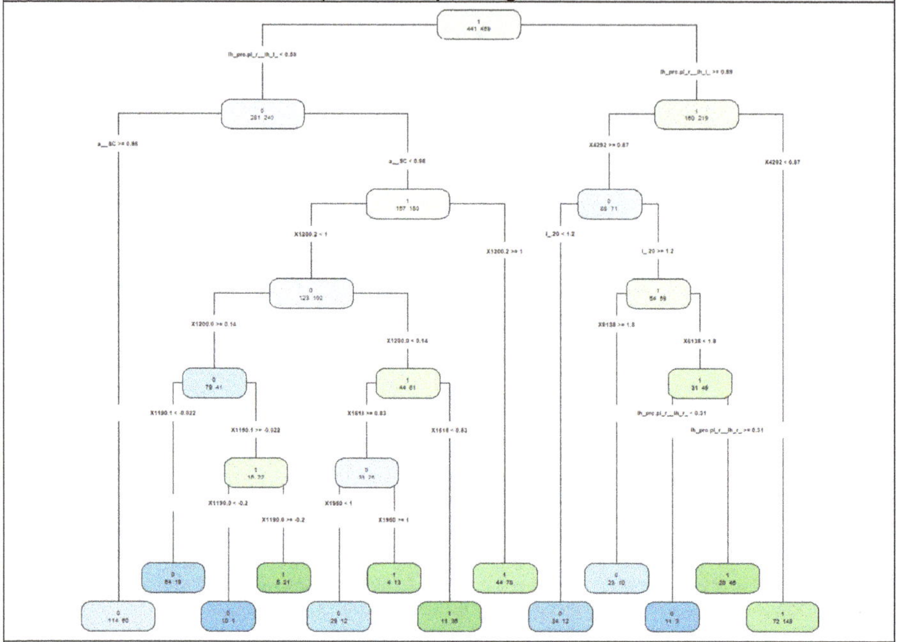

Raw Decision Tree

```
## Confusion Matrix and Statistics
##            Reference
## Prediction   0   1
##          0 341  86
##          1 100 373
##
##                Accuracy : 0.7933
##                  95% CI : (0.77, 0.82)
##     No Information Rate : 0.51
##     P-Value [Acc > NIR] : <2e-16
##                   Kappa : 0.5862
##  Mcnemar's Test P-Value : 0.3405
##             Sensitivity : 0.7732
##             Specificity : 0.8126
##          Pos Pred Value : 0.7986
##          Neg Pred Value : 0.7886
##              Prevalence : 0.4900
##          Detection Rate : 0.3789
##    Detection Prevalence : 0.4744
##       Balanced Accuracy : 0.7929
```

Pruned Decision Tree

```
## Confusion Matrix and Statistics
##            Reference
## Prediction   0   1
##          0 190 130
##          1 251 329
##
##                Accuracy : 0.5767
##                  95% CI : (0.54, 0.61)
##     No Information Rate : 0.51
##     P-Value [Acc > NIR] : 3.501e-05
##                   Kappa : 0.1484
##  Mcnemar's Test P-Value : 7.86e-10
##             Sensitivity : 0.4308
##             Specificity : 0.7168
##          Pos Pred Value : 0.5938
##          Neg Pred Value : 0.5672
##              Prevalence : 0.4900
##          Detection Rate : 0.2111
##    Detection Prevalence : 0.3556
##       Balanced Accuracy : 0.5738
```

Figure 6.14 (continued)

generalizable. What are the consequences of this situation in economics? Unlike in physics and mathematics, in economics an axiomatic system has not yet been formulated. Most of the mathematical and econometric models used in this field are not sufficiently effective, make substantial assumptions, or add little to our knowledge of the underlying mechanics of the economic phenomena. Such models typically involve

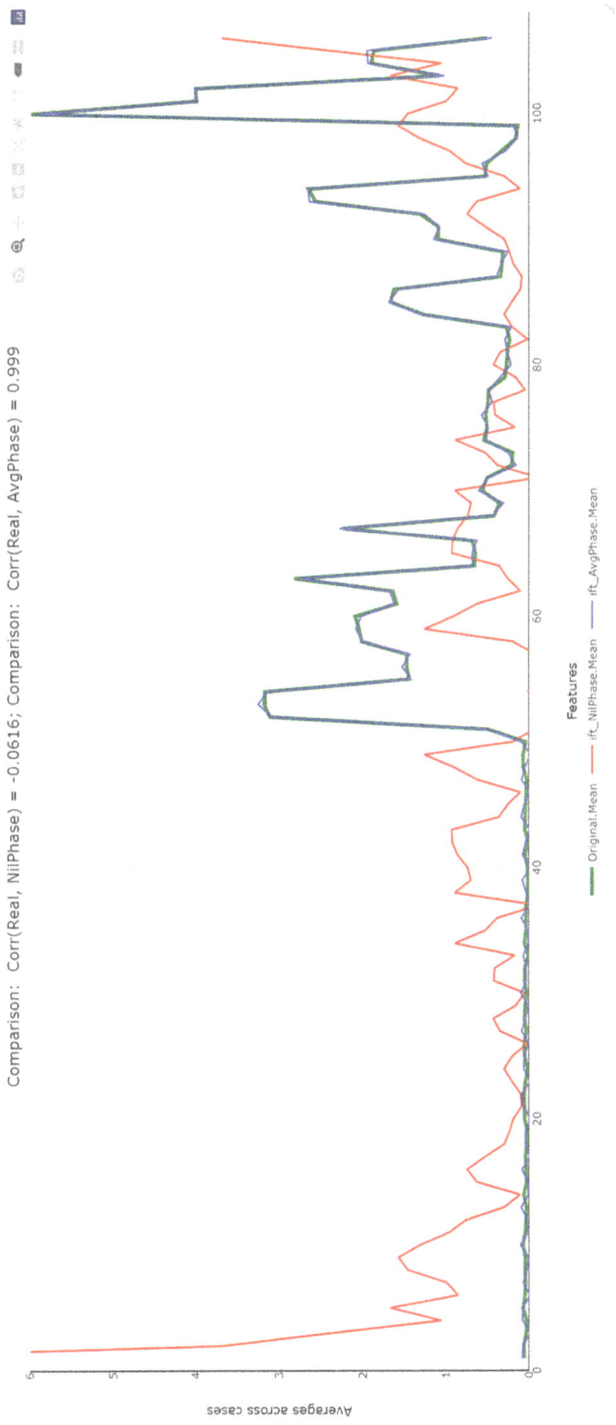

Figure 6.15: Comparison of the overall feature averages across cases for the three complementary data analytic strategies. The close similarity of the original UKBB data (green) to the average aggregate kime-phase estimates (blue) and the relative dispersion of the nil-phase reconstructed data (red) suggest that correct representation or estimation of kime direction plays an important role in the derived scientific inference.

pure theoretical probability and statistics laws, rather than relying on fundamental economic principles.

Mathematical models cannot always be expected to generate reliable predictions for future events under all possible conditions. They mostly reflect a priori explicit quantitative relationships encoded in the models. The predictive capabilities of the economic theories are restricted to abstract characteristics or to the most general schemes of the spontaneously changing system. The positive contributions of mathematical methods to the determination of the specific quantitative properties of various economic systems may be quite limited.

As Leontief [326] remarks, indirect statistical conclusions, however methodologically refined, may not be sufficient to study the quantitative relationships underlying a modern economy. Direct empirical research appears to be the most reliable approach to profound understanding of (1) the functional characteristics of modern economy, (2) the quantitative descriptions of the structural properties of the economic system, and (3) the development and application of new suitable mathematical structures.

Below, we will work with a large economic dataset and discuss examples of the challenges, algorithms, processes, and tools necessary to manage, aggregate, and interpret such data. In data science, time discordance frequently manifests as sampling incongruency, heterogeneous scales, and intricate dependencies. Specifically, we will apply the concept of 2D complex-time (kime) to an economic case-study to illustrate how the kime-order (time) and kime-direction (phase) affect the advanced predictive analytics and the resulting scientific inference.

This case study examines the financial and economic market conditions of the core 31 countries part of the European Union (EU). The 2000–2017 data includes quarterly measures for a large number of indicators, for each country separately, as well as for the entire EU block. The data were retrieved in 2018 from the Luxembourg-based Statistical Office of the European Union, EuroStat (https://ec.europa.eu/eurostat/home, accessed January 29, 2021), and then preprocessed, harmonized, and aggregated locally. The archive includes multivariate and longitudinal features (economic indicators) that will be used in supervised classification, prediction, and unsupervised clustering.

6.5.1 Longitudinal Modeling

Let us start by applying (*exogenous features) AR integrated moving average* (ARIMAX) models to predict specific univariate outcomes. **Figure 6.16** shows the trajectories of a number of econometrics for the EU countries across time in a common 3D canonical space (country, by feature, by time). There were a number of complications associated with this data archive including incompleteness, different sampling rates, longitudinal dependencies, highly correlated time-courses of various

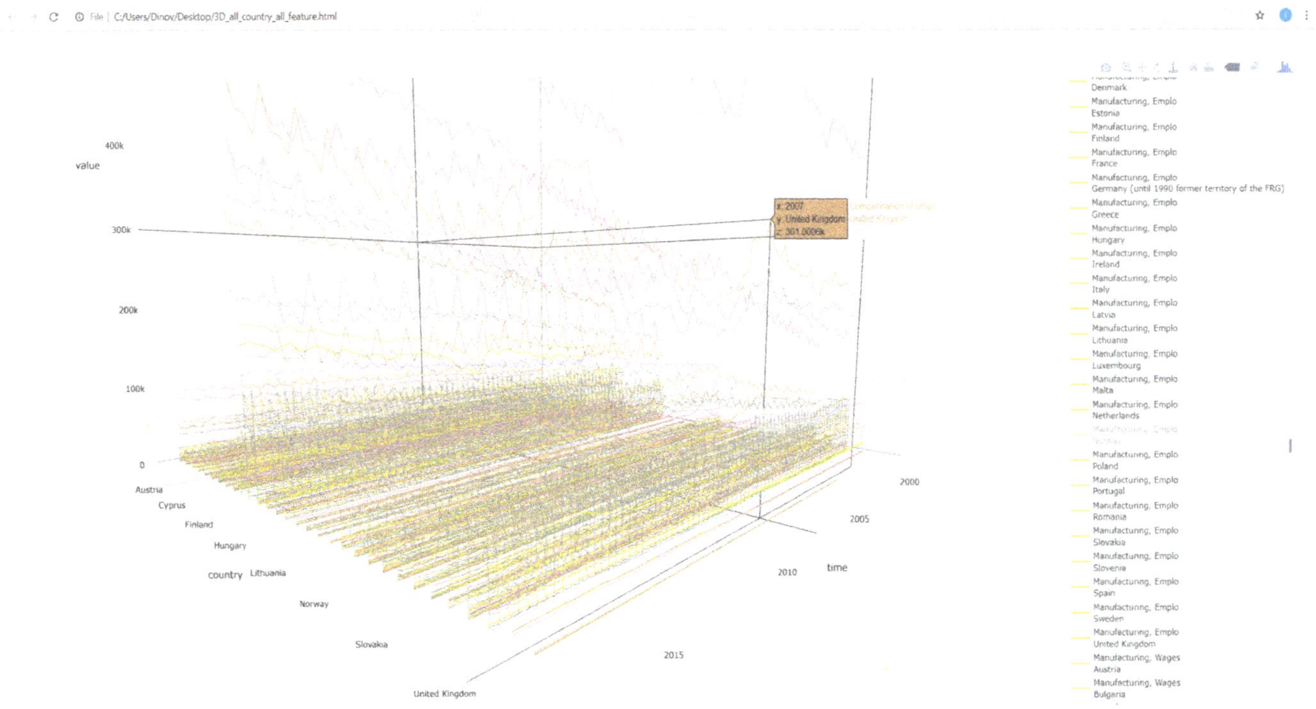

Figure 6.16: An interactive 3D visualization of the EU econometric data is available on the Spacekime TCIU website (https://tciu.predictive.space, accessed January 29, 2021). The scene shows the trajectories of a number of econometric features (vertical z-axis) across a 2D plane indexed by $x = time$ (quarterly measurements for the period 2000–2017) for $y = EU\ country$ (alphabetically ordered). Ignore the scale of the vertical z-axis. This plot does not attempt to contrast between different econometric indicators; this is not meaningful as each indicator is relative to a different unit of measurement. The plot aims to illustrate the relative motifs between countries, the longitudinal nature of the data, the compounding properties of the time courses, and the data heterogeneities.

econometric indicators, etc. This 3D scene is also available as an interactive 3D visualization scheme on the textbook website (https://tciu.predictive.space, accessed January 29, 2021). Additionally, the R code generating this plot is provided with the supplementary materials.

A typical spacetime analytics involving ARIMAX modeling of this longitudinal econometric data is shown in **Figure 6.17**. In this case, we chose to model the outcome variable gross domestic product (GDP) at market prices, which represents a major indicator for a nation's economic situation. The GDP reflects the total value of all goods and services produced, less the value of goods and services used for intermediate consumption in their production. Expressing GDP in purchasing power standards (PPS) eliminates differences in price levels between the 31 EU countries. GDP-PPS is calculated on a per head basis to allow direct comparison of economies that may be significantly different in their absolute sizes. As GDP-PPS represents current prices, in euro per capita, its volume facilitates cross-country comparisons rather than for temporal longitudinal tracking within a country. GDP-PPS eliminates the differences in price levels between countries allowing meaningful volume comparisons of gross GDP between the EU countries. Per country GDP-PPS is expressed relative to the overall European Union ($EU31\ GDP = 100$). This means that countries with GDP-PPS indices above or below 100 have per head GDP over or under the EU average, respectively.

Figure 6.17 includes several alternative strategies to forecast Belgium GDP using all available data over the model training range (2000–2014) and alternative predictors (*Xreg*) over the validation (testing phase). The Belgium GDP Training Data (2000–2014) is shown in **black**; the GDP model-fit ($ARIMAX(4,0,2)$) forecasting (2015–2017) using Belgium's own economic indicators (*Xreg*) is shown in **blue and green**; the Belgium GDP (2015–2017) predictions using the same ARIMAX model along with the 131 prospective covariates for a different EU country (Bulgaria) is shown in **purple**; a modified Belgium GDP prediction using an *offset* of the Belgium trained model with prospective Bulgarian covariates (2015–2017) is shown in **orange**; and the actually reported Belgium GDP-PPS is in **red**.

The resulting $ARIMAX(4,0,2)$ GDP longitudinal model clearly indicates some expected findings. For instance, both unemployment and labor costs are inversely proportional to the GDP-PPS. In other words, increasing employment (decreasing unemployment) and improving productivity (lowering labor costs) drive up GDP. The rank order of the top 10 indicators driving up the (relative) GDP-PPS along with their effect sizes are listed below:

1) Unemployment, females, from 15 to 64 years, from 18 to 23 months: effect = −1.52,
2) ar2, effect = −1.182,
3) Labor cost other than wages and salaries, effect = −0.84,
4) Unemployment, females, from 15 to 64 years, from 12 to 17 months, effect = −0.70,

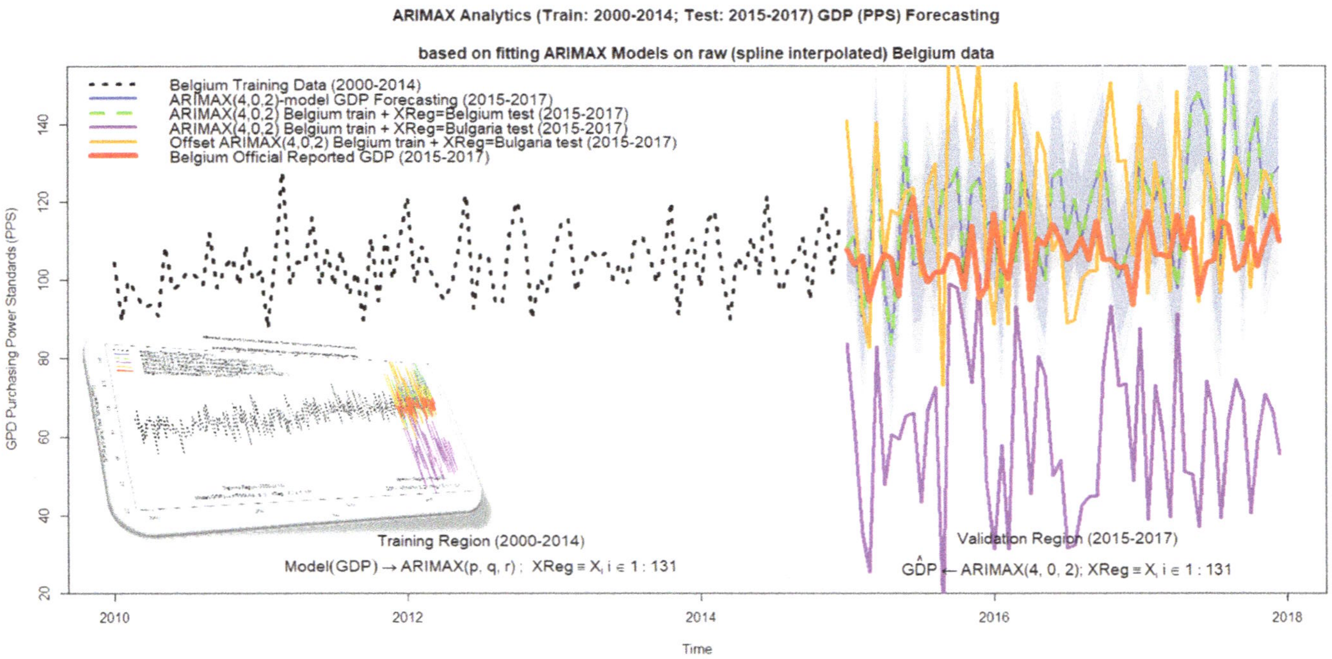

Figure 6.17: Spacetime longitudinal modeling of the Belgium gross domestic product in purchasing power standards (GDP-PPS). The ARIMAX model used 15 years of data on Belgium GDP and 131 other economic factors (2000–2014, training data), and produced future projections over a 3-year period (2015–2017, testing range). For clarity, we only zoom in on the range 2010–2017 to illustrate the similarities and differences between alternative spacetime GDP models and the actual reported Belgium GDP values over the validation timeframe (2015–2017). The insert image shows the entire GDP time-course covering the full training range (2000–2014) and the testing range (2015–2017).

5) ar4, effect = −0.64,
6) sar2, effect = −0.63,
7) Unemployment, females, from 15 to 64 years, from 3 to 5 months, effect = −0.53,
8) Labor cost for LCI (compensation of employees + taxes − subsidies), effect = −0.29,
9) Unemployment, females, from 15 to 64 years, 48 months or over, effect = −0.28,
10) Unemployment, males, from 15 to 64 years, from 3 to 5 months, effect = −0.26.

The training-range estimated full ARIMAX model, $ARIMAX(4, 0, 2, 0, 20, 0, 0)$, includes the following seven parameters: (non-seasonal) AR and MA, (seasonal) AR and MA, the period, and the number of non-seasonal and seasonal differences.

Let us now examine the alternative spacekime analytics of this EU economics forecasting problem. As we did earlier in **Chapter 1**, we can verify the three characteristics of the kime-phases for the Belgium economic indicators: (1) the angular phase distributions for all features are different, (2) all phases are in the range $[-\pi, +\pi)$, and (3) the distributions are zero mean and symmetric, **Figure 6.18**, which we saw the theoretical evidence for in **Chapter 5**.

The results of the spacekime analytics forecasting Belgium GDP-PPS using ARIMAX longitudinal models are shown on **Figure 6.19**. These results show the prospective prediction of Belgium's GDP, relative to the average EU indicator, by fitting the exogenous variables ARIMAX models on spacekime transformed data. In addition to the exact spacetime model, we saw earlier, ARIMA(4,0,2), and the subsequently officially reported Belgium GDP (**red**), there are three alternative ARIMA models derived based on different kime-phase aggregators – nil-phase reconstruction **ARIMA (2,0,1)**, swap-phase reconstruction where the phases of covariate features are randomly swapped **ARIMA(0,0,2)**, and random phase reconstructions where for each feature, we randomly draw phases from the corresponding feature phase distributions ARIMA(4,0,2). And **Table 6.2** shows some of the quantitative model comparison metrics.

One interesting finding is that for all models, the strong negative correlations between relative GDP growth and covariates like unemployment and labor costs are always clearly identifiable.

In this experiment, we used correlation to compare the observed prospective longitudinal GDP-PPS for Belgium against four alternative time series reconstructions (classical spacetime time series modeling, nil-phase, swapped phase, and random phase estimation strategies). These correlations represent the associations between ARIMA model predictions and the observed Belgium GDP over the testing data range. Note that the resulting ARIMA models vary between the four alternative strategies. While all model forecasts of the time series over the follow-up 24-month period are in the right range, the actual correlations between the predicted and the observed GDP-PPS values are rather weak, $|\rho(pred, obs)| \le 0.15$. The key covariate features, and their effect-sizes, also vary across the four different models. This may be explained by

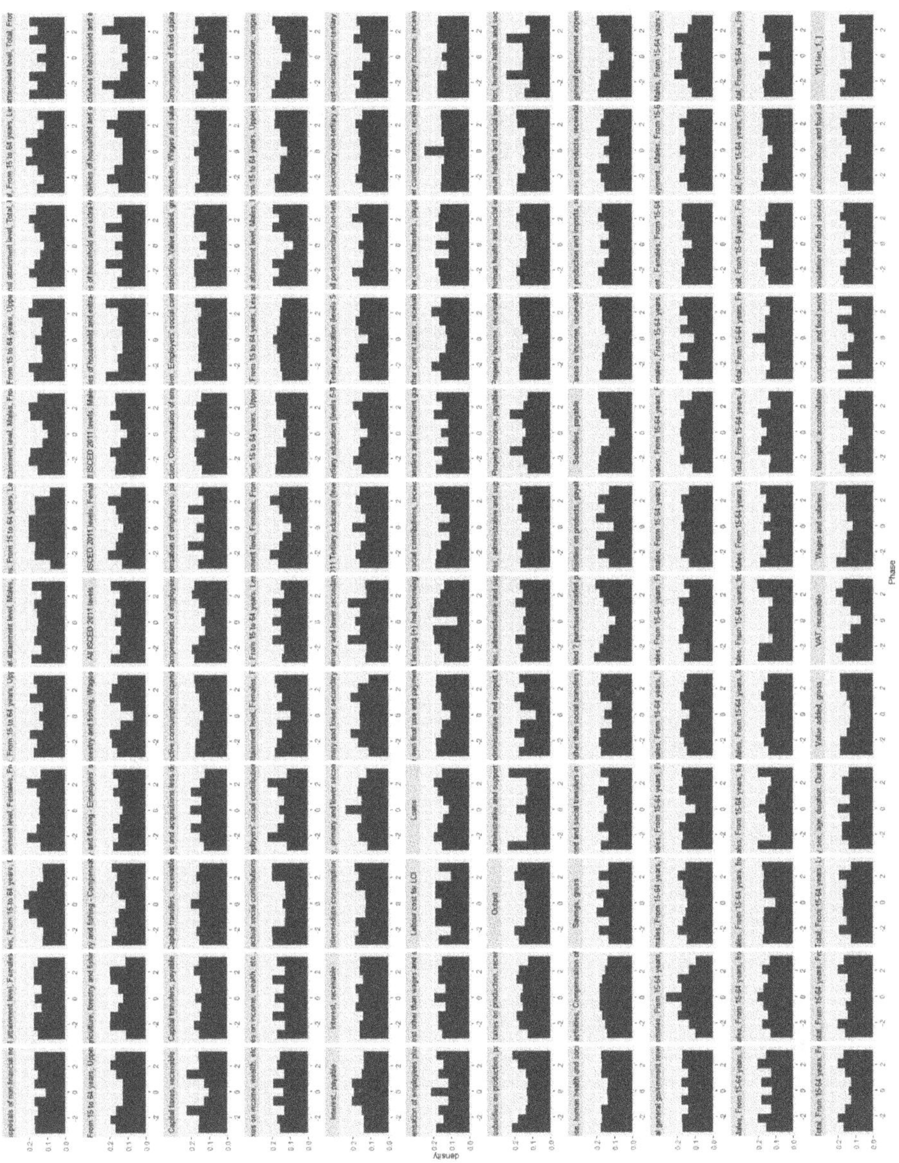

Figure 6.18: Kime phase distributions for the Belgium economic indicators exhibit the following three properties – varying angular phase distributions across features, phases are in the range $[-\pi:+\pi)$, and zero-mean and symmetric distributions.

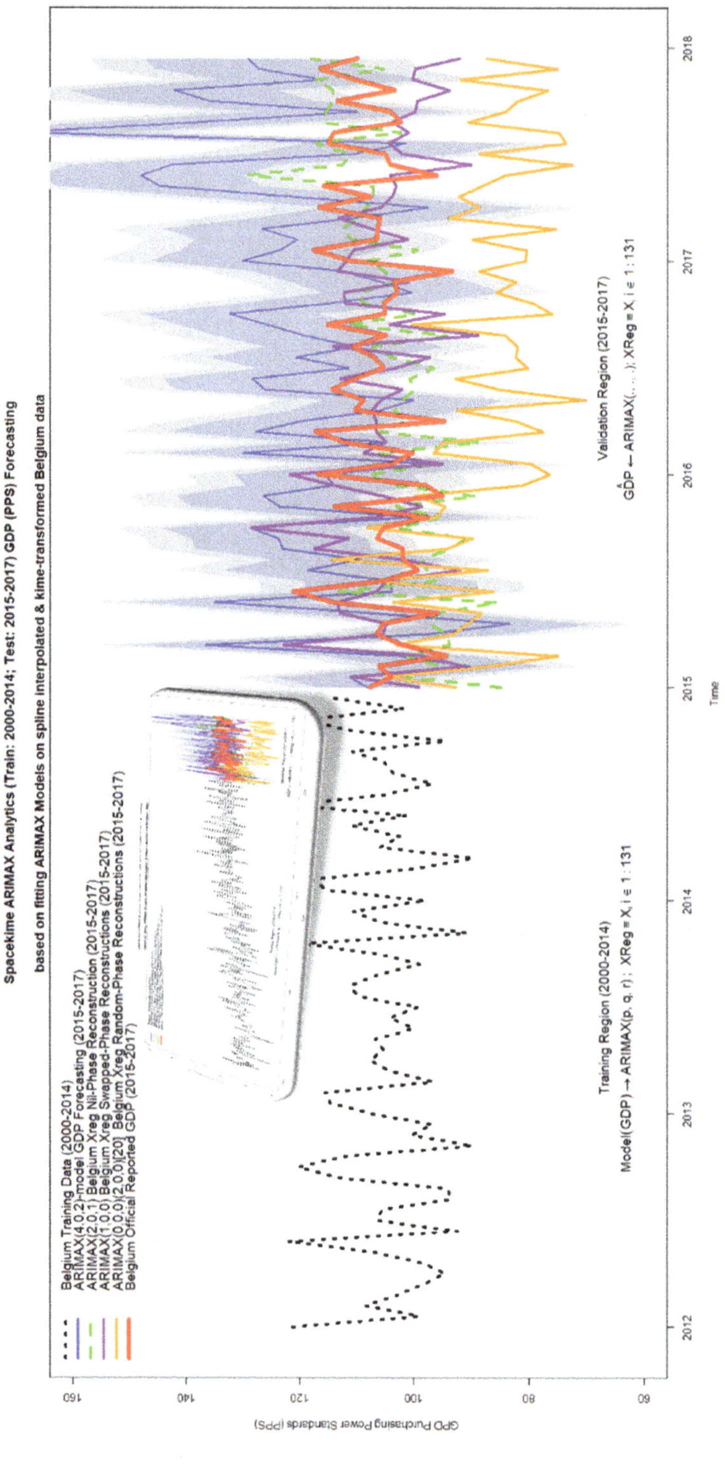

Figure 6.19: Spacekime analytics forecasting Belgium GDP-PPS using ARIMAX longitudinal models fit on spacekime transformed data.

Table 6.2: Summary results for the main spacekime ARIMAX models of Belgium longitudinal GDP-PPS.

Model	**Top predictors of GDP-PPS**	**Model assessment metrics**	**Prospective correlations** $\rho(\boldsymbol{pred}, \boldsymbol{obs})$
Spacetime reconstruction: ARIMA (4,0,0)(2,0,0)[20]	Unemployment, females, 15–64 years, effect = −1.52 ar2, effect = −1.18 Labor cost, effect = −0.84 ar4, effect = −0.64 sar2, effect = −0.63 Unemployment, males, 15–64 years, effect = −0.26	Prediction interval: [2015 – 2017] $\mu = 118$ $\sigma^2 = 46$ $LogLike = -919$ $AIC = 2116$ $AICc = 2360$ $BIC = 2631$	$\rho = 0.11$
Nil-phase reconstruction: ARIMA (2,0,1)(2,0,0)[20]	Labor cost, effect = −1.60 ar1, effect = −1.42 ma1, effect = −0.99 ar2, effect = −0.86 Unemployment, females, 15–64 years, effect = −0.71 Unemployment, total, 15–64, effect = −0.58 Unemployment, males, 15–64 years, effect = −0.50 sar2, effect = −0.35	Prediction interval: [2015 – 2017] $\mu = 104$ $\sigma^2 = 0.86$ $LogLike = -320$ $AIC = 914$ $AICc = 1148$ $BIC = 1422$	$\rho = 0.14$

(continued)

Table 6.2 (continued)

Model	**Top predictors of GDP-PPS**	**Model assessment metrics**	**Prospective correlations** $\rho(\boldsymbol{pred}, \boldsymbol{obs})$
Swapped-phase reconstruction: ARIMA (1,0,0)(2,0,0)[20]	ar1, effect = −0.38 Unemployment, females, 15–64 years, effect = −0.31 Labor cost, effect = −0.31 sar1, effect = −0.22 Unemployment, males, 15–64 years, effect = −0.11 Unemployment, total, effect = −0.09	Prediction interval: [2015 – 2017] $\mu = 105$ $\sigma^2 = 70$ $LogLike = -976$ $AIC = 2224$ $AICc = 2453$ $BIC = 2728$	$\rho = 0.0$
Random-phase reconstructions: ARIMA (0,0,0)(2,0,0)[20]	Labor cost, effect = −0.72 Unemployment, females, 15–64 years, effect = −0.55 Unemployment, males, 15–64 years, effect = −0.25 Unemployment, total, From 15–64 years, effect = −0.19 Agriculture, forestry and fishing – Employers' social contributions, effect = −0.12	Prediction interval: [2015 – 2017] $\mu = 88$ $\sigma^2 = 72.17$ $LogLike = -988$ $AIC = 2244$ $AICc = 2463$ $BIC = 2740$	$\rho = -0.15$

the heterogeneity of the effects on the response, as well as low signal-to-noise ratio of Belgium's temporal GDP-PPS data.

6.5.2 Regularized Linear Modeling

We can also attempt generalized linear modeling (GLM) using LASSO regularization [10, 238]. To simplify the situation, we will transform the longitudinal data features into a cross-sectional data object by fitting ARIMA models for each indicator variable separately. The resulting computable data object represents a tensor with 31 rows, one for each of the 31 EU countries, and 379 features. The features include 378 derived ARIMA model parameters representing a signature vector of size 9 for each of the 42 economic indicators that are commonly observed for all 31 countries. The last (379th) feature is an overall (OA) country ranking [327] (see https://wiki.socr.umich.edu/index.php/SOCR_Data_2008_World_CountriesRankings, accessed January 29, 2021). For each pair (*country*, *economic indicator*), the signature vector of length 9 encoding the ARIMA-model longitudinal characteristics contains:
(1) average time series value (retrospective),
(2) average ARIMA 3-year forecast (prospective),
(3) non-seasonal AR,
(4) non-seasonal MA,
(5) seasonal AR,
(6) seasonal MA,
(7) period,
(8) non-seasonal difference,
(9) seasonal differences.

Of course, this signature vector of capturing the longitudinal characteristics of each economic indicator for each country is not unique, and there are many other alternative mechanisms to model the data. The observed 42 common EU country indicators include:
(1) *"Active population by sex, age and educational attainment level, Females, From 15 to 64 years, All ISCED 2011 levels"*
(2) *"Active population by sex, age and educational attainment level, Females, From 15 to 64 years, Less than primary, primary and lower secondary education (levels 0–2)"*
(3) *"Active population by sex, age and educational attainment level, Females, From 15 to 64 years, Tertiary education (levels 5–8)"*
(4) *"Active population by sex, age and educational attainment level, Females, From 15 to 64 years, Upper secondary and post-secondary non-tertiary education (levels 3 and 4)"*

(5) *"Active population by sex, age and educational attainment level, Males, From 15 to 64 years, All ISCED 2011 levels"*
(6) *"Active population by sex, age and educational attainment level, Males, From 15 to 64 years, Less than primary, primary and lower secondary education (levels 0–2)"*
(7) *"Active population by sex, age and educational attainment level, Males, From 15 to 64 years, Tertiary education (levels 5–8)"*
(8) *"Active population by sex, age and educational attainment level, Males, From 15 to 64 years, Upper secondary and post-secondary non-tertiary education (levels 3 and 4)"*
(9) *"Active population by sex, age and educational attainment level, Total, From 15 to 64 years, All ISCED 2011 levels"*
(10) *"Active population by sex, age and educational attainment level, Total, From 15 to 64 years, Less than primary, primary and lower secondary education (levels 0–2)"*
(11) *"Active population by sex, age and educational attainment level, Total, From 15 to 64 years, Tertiary education (levels 5–8)"*
(12) *"Active population by sex, age and educational attainment level, Total, From 15 to 64 years, Upper secondary and post-secondary non-tertiary education (levels 3 and 4)"*
(13) *"All ISCED 2011 levels"*
(14) *"All ISCED 2011 levels, Females"*
(15) *"All ISCED 2011 levels, Males"*
(16) *"Capital transfers, payable"*
(17) *"Capital transfers, receivable"*
(18) *"Compensation of employees, payable"*
(19) *"Current taxes on income, wealth, etc., receivable"*
(20) *"Employment by sex, age and educational attainment level, Females, From 15 to 64 years, All ISCED 2011 levels"*
(21) *"Employment by sex, age and educational attainment level, Females, From 15 to 64 years, Less than primary, primary and lower secondary education (levels 0–2)"*
(22) *"Other current transfers, payable"*
(23) *"Other current transfers, receivable"*
(24) *"Property income, payable"*
(25) *"Property income, receivable"*
(26) *"Savings, gross"*
(27) *"Subsidies, payable"*
(28) *"Taxes on production and imports, receivable"*
(29) *"Total general government expenditure"*
(30) *"Total general government revenue"*
(31) *"Unemployment, Females, From 15–64 years, Total"*
(32) *"Unemployment, Males, From 15–64 years"*

(33) *"Unemployment, Males, From 15–64 years, from 1 to 2 months"*
(34) *"Unemployment, Males, From 15–64 years, from 3 to 5 months"*
(35) *"Unemployment, Males, From 15–64 years, from 6 to 11 months"*
(36) *"Unemployment, Total, From 15–64 years, From 1 to 2 months"*
(37) *"Unemployment, Total, From 15–64 years, From 12 to 17 months"*
(38) *"Unemployment, Total, From 15–64 years, From 3 to 5 months"*
(39) *"Unemployment, Total, From 15–64 years, From 6 to 11 months"*
(40) *"Unemployment, Total, From 15–64 years, Less than 1 month"*
(41) *"Unemployment by sex, age, duration. Duration NA not started"*
(42) *"VAT, receivable"*.

We will fit regularized linear models using LASSO penalty and use tenfold internal statistical cross-validation. In the first (supervised) model-based prediction approach, we will forecast the outcome *overall country ranking (OA)* based on the 378 features, 42(indicators) * 9(ARIMA signature vector). In the second (unsupervised) model-free analytical strategy, we will compute derived cluster labels that will be explicated in terms of the specific country economic indicators.

Using only limited data (378 time series derivatives), **Figure 6.20** summarizes the classical spacetime inference and the various spacekime analytics employing alternative phase-aggregators. Notice the similarities and differences between the classical spacetime inference (GLM with LASSO regularization) and the spacekime analytics (nil-phase and swapped-phase kime direction aggregators, again using similar GLM with LASSO regularization models). All three models capture the OA country ranking trend and clearly discriminate between the top-30 rank and other (non-top-30) countries. However, for approximately the same model complexity, the spacekime GLM model applied to the IFT-synthesized data, which was derived via swapped-phase reconstruction, achieved the highest correlation between observed and predicted OA country ranking (correlation = 0.86). This can be compared to the spacetime GLM model (correlation = 0.84) and the nil-phase reconstructed data (correlation = 0.64). The binary top-30 and non-top-30 labels are derived from a worldwide country ranking. These binary labels are not used in any of the quantitative analytics. They are just used to label and visually discriminate between the "best countries" (top-30, world ranking) and the rest of the European countries.

In this example, to forecast the OA country ranking, we utilized relatively weak information (9-parameter signature vectors derived from fitting autoregressive models for the 42 economic indicators common for all 31 EU countries). Would this change if we strengthened the data energy (increase the given information signal) by adding country meta-data augmenting the available derived longitudinal model signature vectors for the 42 economic indicators? We would like to explore if the spacetime and spacekime analytics change significantly if we increase the energy, i.e., information content, in the dataset.

Data analytics based only on the 378 time-series derivatives

Analytical Strategy	Model quantification & Prediction Results
	Classical linear modeling (*spacetime* analytics)

GLM with LASSO regularization

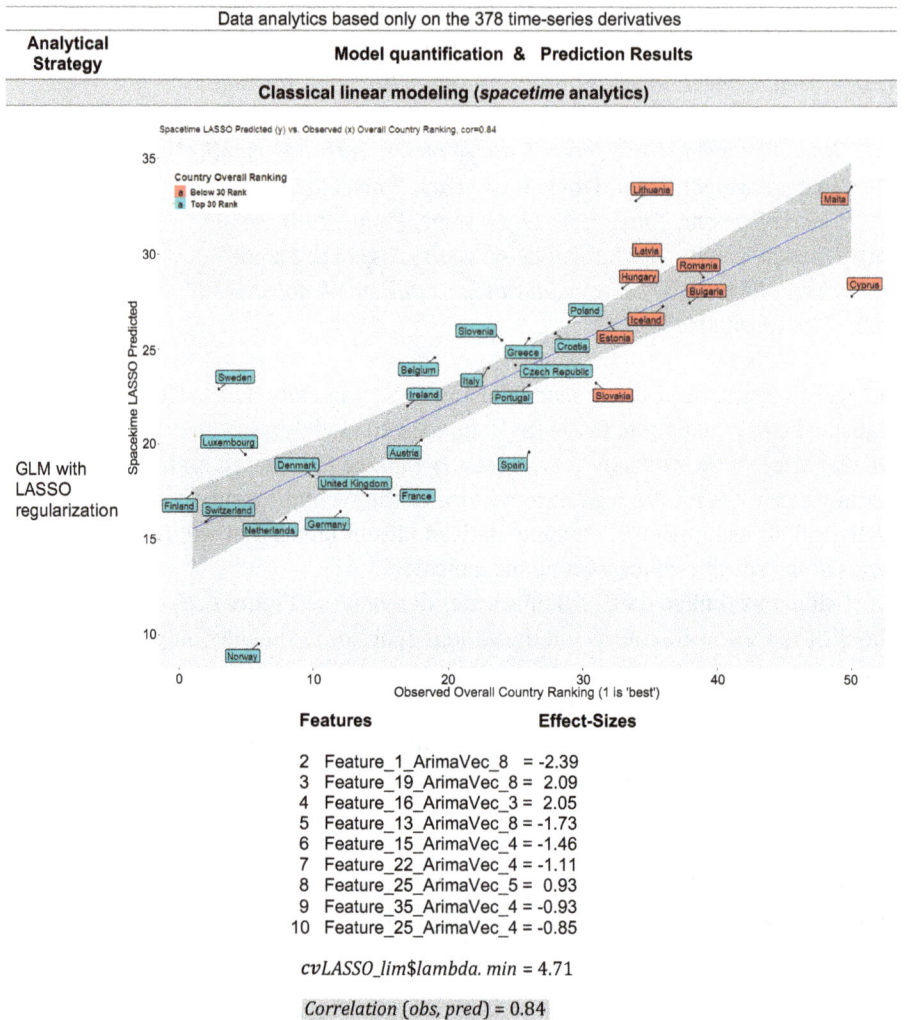

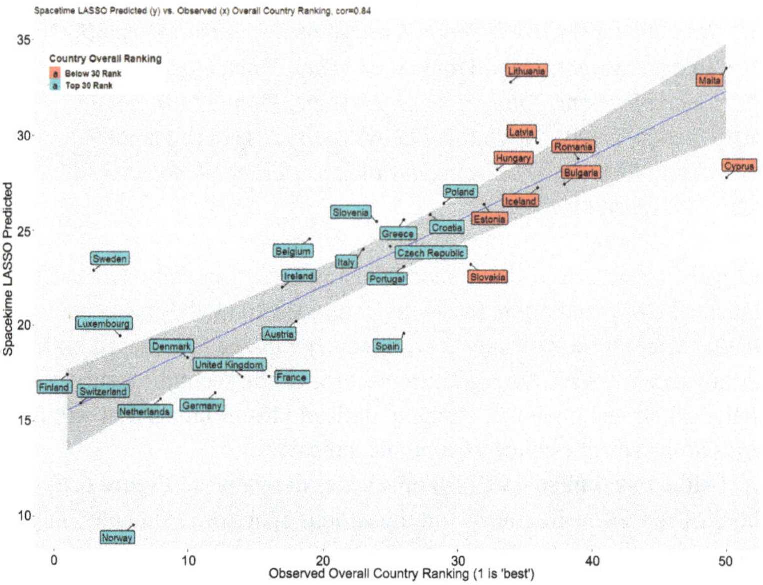

Features	**Effect-Sizes**
2	Feature_1_ArimaVec_8 = -2.39
3	Feature_19_ArimaVec_8 = 2.09
4	Feature_16_ArimaVec_3 = 2.05
5	Feature_13_ArimaVec_8 = -1.73
6	Feature_15_ArimaVec_4 = -1.46
7	Feature_22_ArimaVec_4 = -1.11
8	Feature_25_ArimaVec_5 = 0.93
9	Feature_35_ArimaVec_4 = -0.93
10	Feature_25_ArimaVec_4 = -0.85

cvLASSO_lim$lambda. min = 4.71

Correlation (obs, pred) = 0.84

Figure 6.20: Prediction of the overall country ranking using traditional regularized linear modeling and comparing the results to spacekime-transformed analytics utilizing alternative phase aggregators. The graphs illustrate the association between the country observed Overall ranking (horizontal axis) and its forecasted ranking (vertical axis). The countries are split in two dichotomous classes – top-30 ranking (best overall scores) and others. The results of this model-based inference are best validated by comparing the correlations between *observed* and *predicted* overall country ranking and via the categorical color coding of the countries as top-30 rank and other.

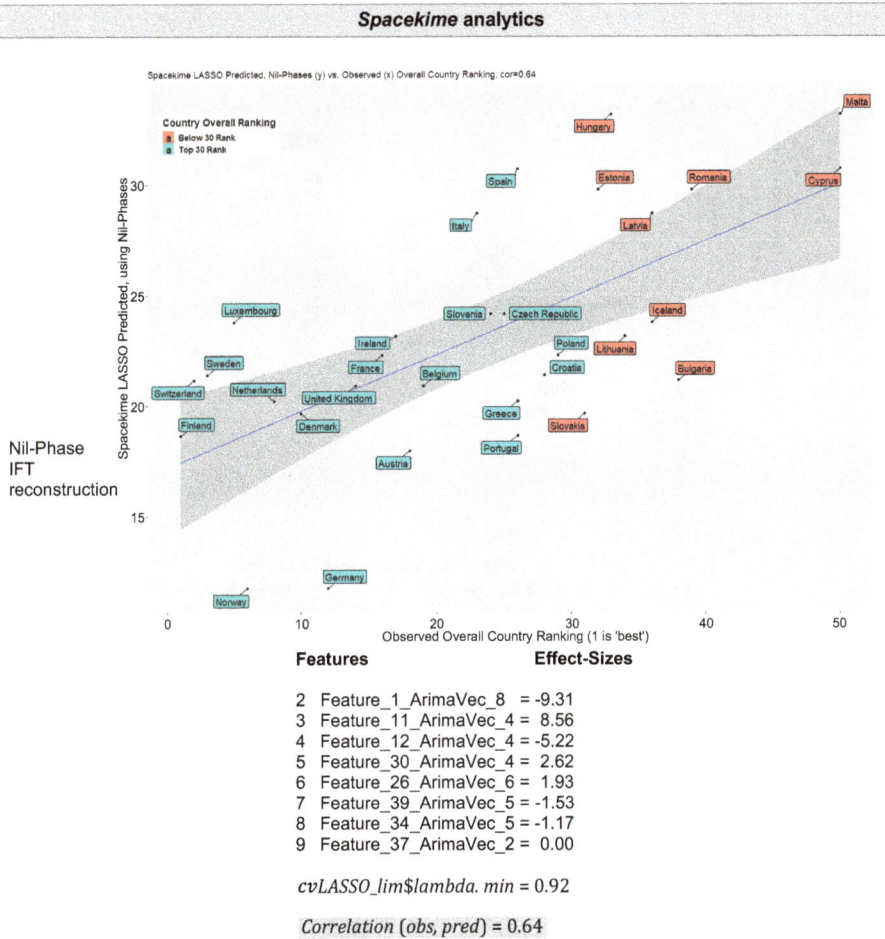

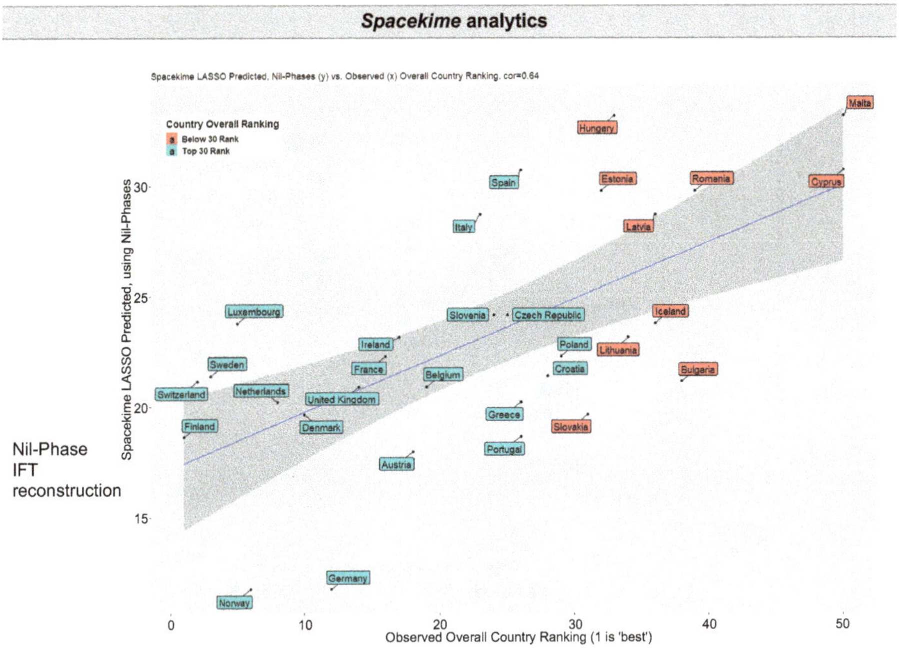

Features	Effect-Sizes
2 Feature_1_ArimaVec_8	= -9.31
3 Feature_11_ArimaVec_4	= 8.56
4 Feature_12_ArimaVec_4	= -5.22
5 Feature_30_ArimaVec_4	= 2.62
6 Feature_26_ArimaVec_6	= 1.93
7 Feature_39_ArimaVec_5	= -1.53
8 Feature_34_ArimaVec_5	= -1.17
9 Feature_37_ArimaVec_2	= 0.00

cvLASSO_lim$lambda. min = 0.92

Correlation (obs, pred) = 0.64

Figure 6.20 (continued)

Figure 6.21 shows the results of applying the same three analytical strategies to the expanded dataset consisting of 386 predictors and the single outcome (OA country ranking). To be more specific, we enhanced the data by augmenting the initial 378 ARIMA-derived features with eight additional country meta-data elements:

(1) *Income group: low: GNI per capita < $3,946; middle: $3,946 < GNI per capita < $12,195; high: GNI per capita > $12,196*

(2) *Population group: small: population < 20 million; medium: 20 million < population < 50 million; large: population > 50 million*

(3) *Economic dynamism: Index of Productive growth in dollars (based on GDP/capita at PPP, average of GDP/capita growth rate over last 10 years, GDP/capita*

Swapped-Phase IFT reconstruction

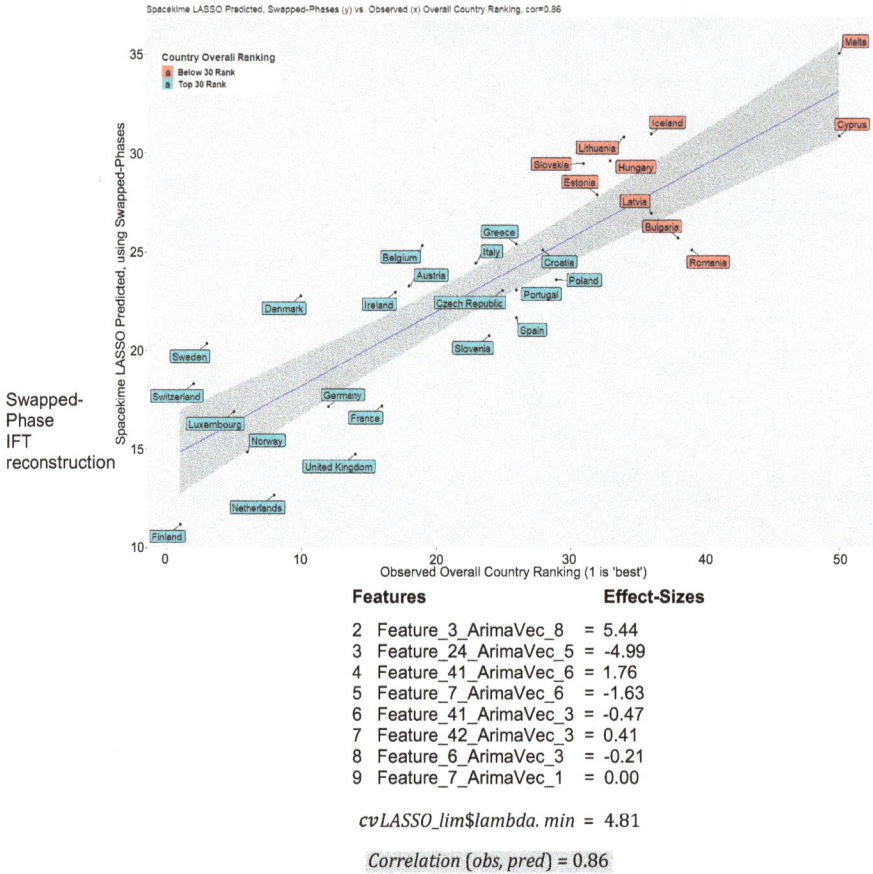

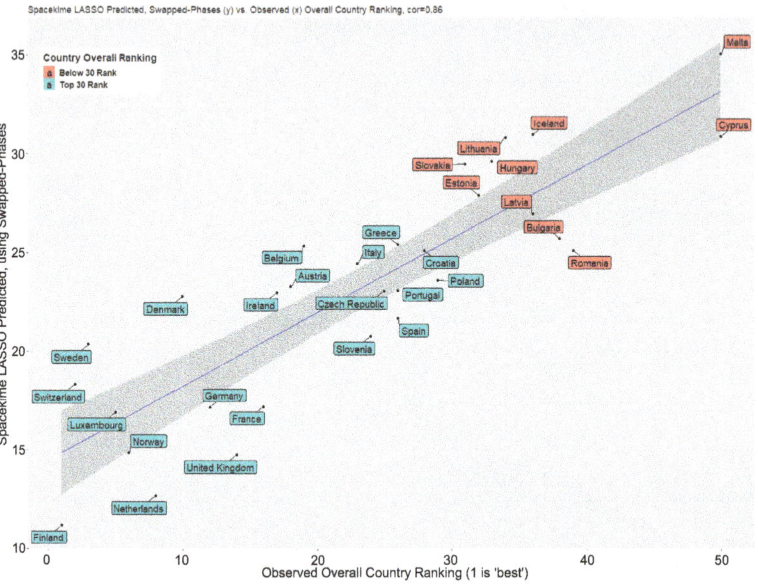

	Features		Effect-Sizes
2	Feature_3_ArimaVec_8	=	5.44
3	Feature_24_ArimaVec_5	=	-4.99
4	Feature_41_ArimaVec_6	=	1.76
5	Feature_7_ArimaVec_6	=	-1.63
6	Feature_41_ArimaVec_3	=	-0.47
7	Feature_42_ArimaVec_3	=	0.41
8	Feature_6_ArimaVec_3	=	-0.21
9	Feature_7_ArimaVec_1	=	0.00

$cvLASSO\_lim\$lambda.min = 4.81$

$Correlation\ (obs, pred) = 0.86$

Figure 6.20 (continued)

growth rate over next 10 years, economic dynamism: manufacturing percent of GDP, services percent of GDP percent (100 = best, 0 = worst).

(4) *Education/literacy rate (percent of population able to read and write at a specified age)*

(5) *Health index: the average number of years a person lives in full health, taking into account years lived in less than full health*

(6) *QOL: quality of life: population percent living on < $2/day*

(7) *Political environment: freedom house rating of political participation (qualitative assessment of voter participation/turn-out for national elections, citizens' engagement with politics)*

(8) *Religiosity of the country as a percent (%) of the population.*

Clearly, when using these merged and harmonized data consisting of 386 predicting covariates, one would expect to obtain much better forecasting results. Note the marked tightening of the paired observed and predicted OA country rankings around the forecasting linear model line (see points on the graph, **Figure 6.21**). This is true for all three models, but the improvement is most noticeable for the spacetime GLM analytics. This experiment suggests that for strong-signals, adding spacekime manipulations may not be necessary, or beneficial. However, for weaker signals, as in the previous example, **Figure 6.20**, spacekime transformation may improve the model quality and enhance the resulting inference. Note that in this supervised model-based machine learning strategy, the performance of all three methods is enhanced by adding additional information to strengthen the SNR ratio in the data. However, the increases in the observed-to-predicted country ranking correlations are not uniform. The three alternative strategies – spacetime, spacekime-nil-phase, and spacekime-swapped-phase reconstructions – exhibit different rates of improvement, in terms of the increased correlation coefficient between the predicted and actual OA country ranking for each of the three strategies, respectively:

$$\overbrace{\rho = 0.84 \underbrace{\qquad\longrightarrow\qquad}_{\substack{\text{increase due to adding}\\ \text{8 meta - data elements}}} \rho' = 0.98}^{\text{spacetime GLM}}$$

$$\overbrace{\rho = 0.64 \underbrace{\;\longrightarrow\;}_{\text{increase}} \rho' = 0.65}^{\text{spacekime - nil - phase GLM}} .$$

$$\overbrace{\rho = 0.86 \underbrace{\;\longrightarrow\;}_{\text{increase}} \rho' = 0.9}^{\text{spacekime - swapped - phase GLM}}$$

The experimental results shown in **Figures 6.20** and **6.21** suggest that spacekime analytics may be practically useful and sometimes enhance the quality of the forecasting results for weak signal data. For high-energy data, phase-aggregation may be less impactful; however, this is just one experiment utilizing a pair of oversimplified zeroth-order phase aggregators. Deeper studies of the theoretical properties of various kime-phase aggregators need to be conducted to determine the expected performance of spacekime analytics, in general.

Figure 6.22 provides a comprehensive graphical summary of the performance of the different methods for supervised prediction of the EU country overall ranking. Note the reported correlations between the predicted overall country ranking and the actual reported ranking:

- predLASSO_spacetime LASSO Predicted (386): $cor(predLASSO, Y) = 0.98$,
- predLASSO_lim LASSO Predicted (378): $cor(predLASSO\_lim, Y) = 0.84$,
- predLASSO_nil (spacekime) LASSO Predicted: $cor(predLASSO\_kime, Y) = 0.66$,
- predLASSO_swapped (spacekime) LASSO Predicted: $cor(predLASSO\_kime\_swapped, Y) = 0.90$.

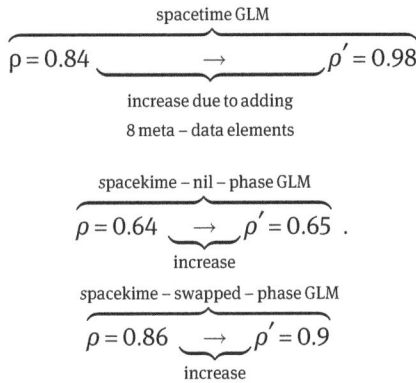

Augmented data analytics based on 386 features (378 time-series derivatives and 8 meta-data features)

Analytical Strategy	**Model quantification & Prediction Results**
	Classical linear modeling (*spacetime* analytics)

GLM with LASSO regularization

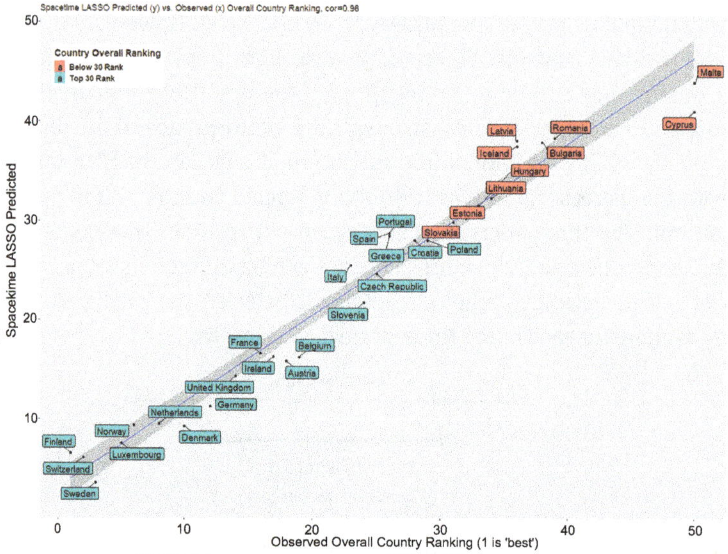

Features		**Effect-Sizes**
2	Feature_1_ArimaVec_8	= 2.75
3	Feature_9_ArimaVec_4	= 0.27
4	Feature_9_ArimaVec_8	= 1.09
5	Feature_20_ArimaVec_8	= 1.69
6	Feature_25_ArimaVec_5	= 0.51
7	IncomeGroup	= 1.18
8	ED	= 0.75
9	QOL	= 0.21
10	PE	= 0.51

$cvLASSO\$lambda.\ min = 1.1$

$Correlation\ (obs, pred) = 0.98$

Figure 6.21: Prediction of the overall country ranking based on the augmented dataset. Again, we employ the same traditional regularized linear modeling to predict country ranking and compare the results to their spacekime-transformed analytics counterparts. The graphs illustrate the association between the country observed Overall ranking (horizontal axis) and its forecasted ranking (vertical axis). The countries are split into dichotomous classes – top-30 ranking (best overall country scores) and others. This model-based inference directly relates to the correlation between observed and predicted overall country ranking. The color coding of the countries shows the top-30 rank versus other status and provides additional context of the modeling and results.

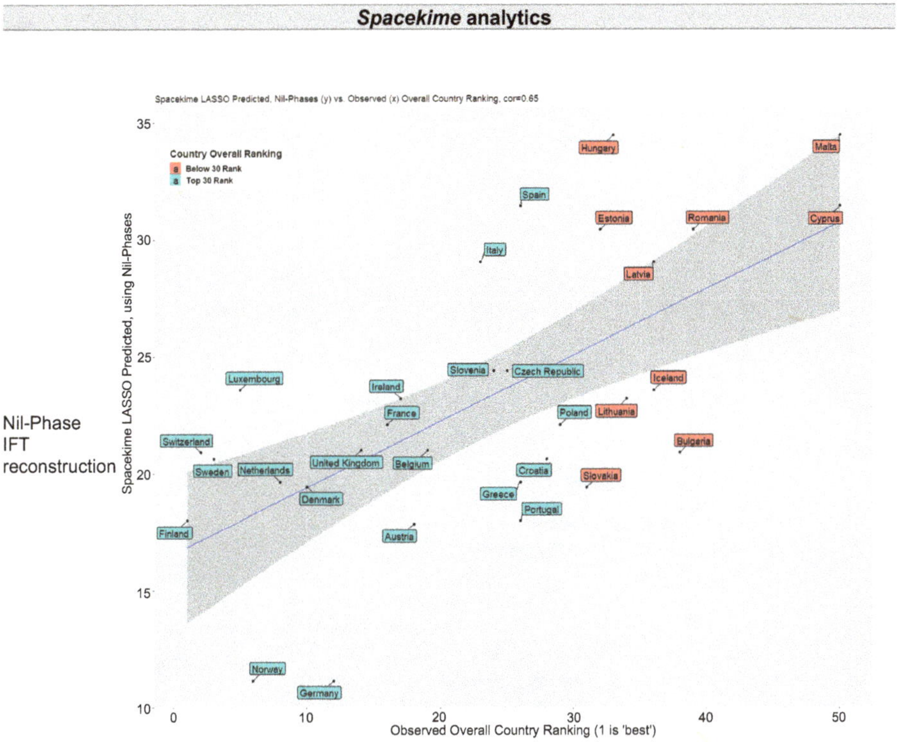

Features **Effect-Sizes**

2 Feature_12_ArimaVec_8 = -10.38
3 Feature_11_ArimaVec_4 = 8.45
4 Feature_12_ArimaVec_4 = -5.38
5 Feature_30_ArimaVec_4 = 3.30
6 Feature_39_ArimaVec_5 = -2.09
7 Feature_26_ArimaVec_6 = 2.07
8 Feature_34_ArimaVec_5 = -0.97
9 Feature_6_ArimaVec_6 = -0.50

cvLASSO_kim$lambda. min = 0.78

Correlation (obs, pred) = 0.65

Figure 6.21 (continued)

6.5.3 Laplace Transform of GDP Time Series to Kimesurfaces

We can examine some country-specific differences in the gross domestic product at market price. **Figure 6.23** illustrates the Laplace-transformed market-price GDP time series represented as kimesurfaces for 30 EU countries. Note that the common upward trajectory of GDP growth through the EU zone manifests as global kimesurface shape similarities, which exhibit fine local detail differences unique to each country. Topological data analysis, manifold distance metrics, or other methods

Swapped-Phase IFT reconstruction

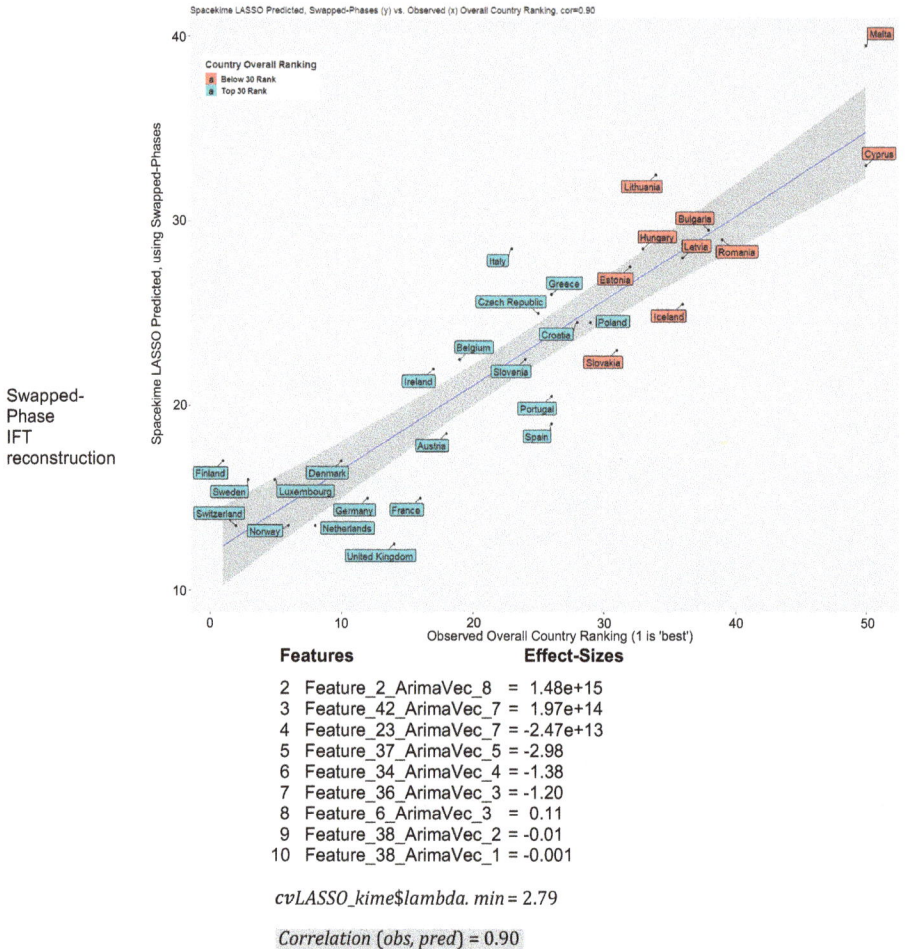

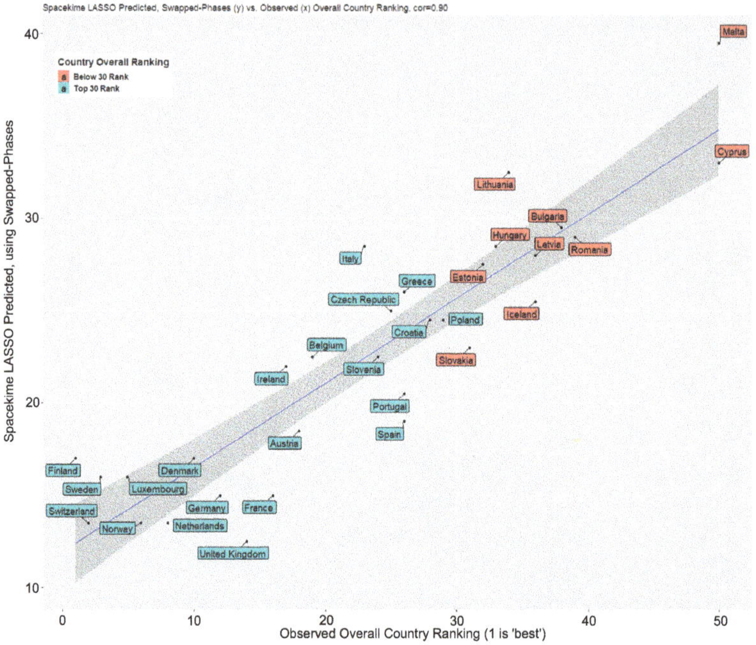

	Features	Effect-Sizes
2	Feature_2_ArimaVec_8	= 1.48e+15
3	Feature_42_ArimaVec_7	= 1.97e+14
4	Feature_23_ArimaVec_7	= -2.47e+13
5	Feature_37_ArimaVec_5	= -2.98
6	Feature_34_ArimaVec_4	= -1.38
7	Feature_36_ArimaVec_3	= -1.20
8	Feature_6_ArimaVec_3	= 0.11
9	Feature_38_ArimaVec_2	= -0.01
10	Feature_38_ArimaVec_1	= -0.001

$cvLASSO\_kime\$lambda.\ min = 2.79$

$Correlation\ (obs, pred) = 0.90$

Figure 6.21 (continued)

may be applied to the corresponding kimesurfaces to quantify their statistical variability and derive inference about multiple feature associations and between-country differences. Clearly, a more holistic multivariate analysis will utilize the kimesurface representations of all available economic metrics. For simplicity, in this example, we illustrate the derivation of just one feature, GDP, however, all economic markers can be similarly encoded in spacekime.

Figure 6.24 depicts the spacetime inverse Laplace transform reconstructions of the GDP time series from the corresponding kimesurfaces for a set of four EU countries. The close relations between the original and the reconstructed time series indicate that the kimesurface representations encode at least as much information as the original time series. In general, the geometry and topology of manifolds, such as kimesurfaces,

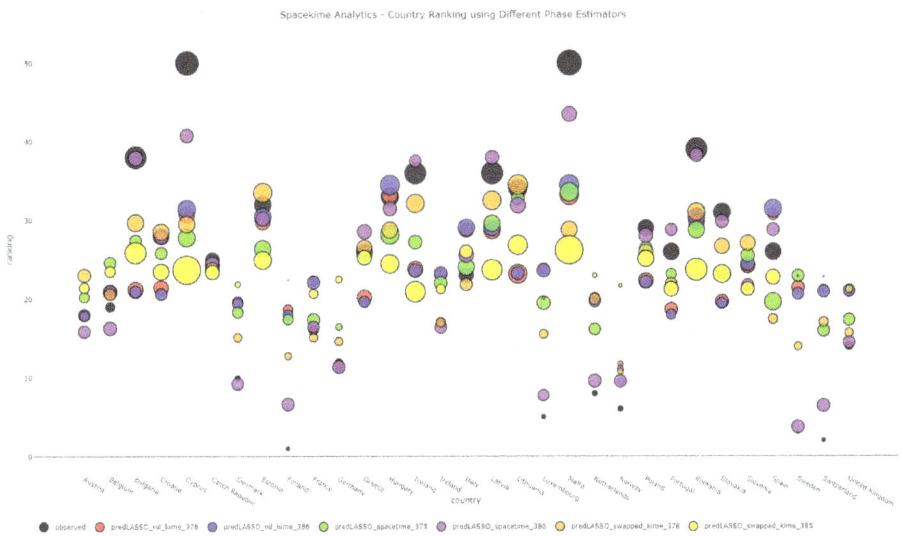

Figure 6.22: Summary of the performance of several alternative analytical strategies to forecast the overall country ranking based on the ARIMA signature vectors for the 42 common EU economic indicators and the additional meta-data elements. The horizontal axis shows the alphabetical order of the EU countries, and the vertical axis depicts the *observed* and *predicted* overall country ranking using LASSO-based spacetime and spacekime analytics with alternative kime-phase estimators. The size and color of the bubbles reflect the country-ranking prediction according to the specific analytical strategy. Black color is used for the actual country-ranking.

are much richer than the collective sum of their corresponding lower dimensional foliation leaves, such as time series. Hence, it is reasonable to expect that in certain situations, compared to classical spacetime inference, spacekime analytics may yield more reliable results, increased precision, reduced bias, or improved prediction forecasts.

6.5.4 Dimensionality Reduction

Next, we will explore some 2D and 3D linear (e.g., principal component analysis, PCA) and non-linear (e.g., t-distributed stochastic neighbor embedding, t-SNE) projections of the spacetime and spacekime-transformed data. Using swapped-phase estimation, the spacekime dimensionality reduction yields very stable and reproducible simplifications of the high-dimensional data (386 features). **Figure 6.25** and **6.26** show the spacetime and spacekime t-SNE manifold projections, respectively.

The last two figures illustrate that machine learning and dimensionality reduction methods can be employed to analyze spacekime transformed data and obtain clustering or classification results that correspond with specific phenotypic

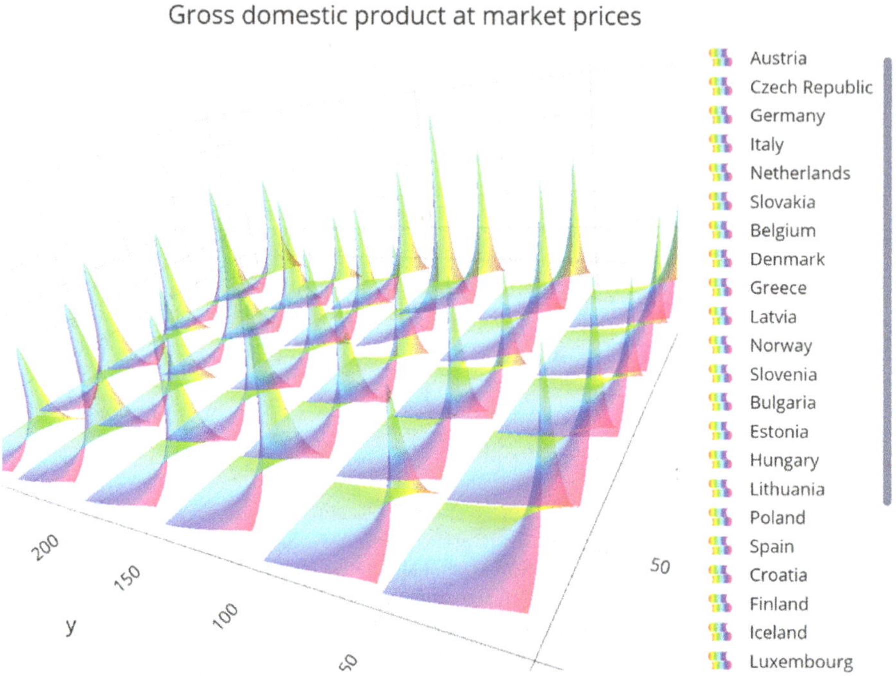

Figure 6.23: This 5 × 6 grid shows the GDP kimesurfaces corresponding to the GDP time series of 30 EU countries. The kimesurfaces' shapes and intensities (heights) reflect the value of the kimesurface magnitude and their colors correspond to the canonical kimesurface phases. Country-specific economic differences are reflected in the shape variations of the corresponding kimesurfaces.

population cohorts, in this case, **top-30** versus **not-top-30** overall country ranking. These experiments suggest that there is strong signal captured in the spacekime transformed data that can be exploited further by quantitative artificial intelligence methods.

The readers are encouraged to download the raw spacetime data, the spacekime transformed data, and the binary **top-30** / **not-top-30** country labels from the book website (https://SpaceKime.org, accessed January 29, 2021) and to try these hands-on visualization and analytical methods.

6.5.5 Unsupervised Clustering

In this section, we will examine the variable importance by feature selection and identify derived computed country phenotypes (clusters) based on the available information. **Figure 6.27** shows traditional spacetime analytics using the complete data (left) and the reduced feature set (right). These results illustrate the accuracy

Figure 6.24: Spacetime reconstructions of the GDP time series of Belgium, Bulgaria, France, and the Netherlands using their corresponding spacekime kimesurface representations. For each country, the reconstructed time series is superimposed on the original GDP time-course.

of supervised binary classification (top-30 country vs. not-top-30 country). We employ supervised methods like decision-tree bagging, random forest, decision-tree adaboost, GLM, and SVM with gamma cost function. In addition, we tested an unsupervised hierarchical clustering method that naturally groups countries with similar phenotypes and separates others that have district traits.

Figure 6.28 shows the *spacekime* analytics based on swapped kime-phase reconstructions that correspond to the analogous spacetime results depicted in **Figure 6.27**. Clearly the accuracy of the spacekime analytics to predict the binary country ranking (top-30 vs. not-top-30) has slightly decreased. The same level of forecasting accuracy improvement can be expected by assuming the data is initially acquired in the k-space, instead of spacetime, with no knowledge of the phases (trivial, entangled, or swapped phases), and we compare the analytics in the Fourier domain obtained with or without the correct kime-phases.

All demonstrations shown in this chapter illustrate that estimating the kime phases by Fourier domain estimation, by phase-modeling, or via phase aggregation or ensembling, may improve the derived inference and enhance the predictive data analytic process. This approach assumes that either we have a very large number of samples that

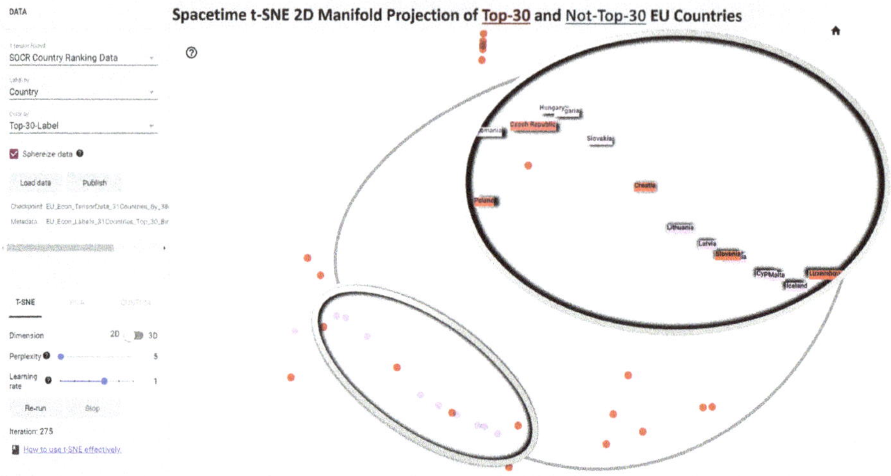

Figure 6.25: 2D t-SNE projection of the *raw 386-dimensional spacetime data* showing the clustering of the **not-top-30** countries in the middle mixed with some of the **top-30** countries like Czech Republic, Croatia, and Slovenia. The curved arcs connect the corresponding point-based projection representations and their magnified counterparts where country names replace the points corresponding to each country. This magnification is added to illustrate the compact central clustering of most of the **not-top-30** EU countries in the sample of 31.

effectively span the range of kime-directions, an a priori phase model is used, or the kime-phase space sampling is uniform. If these assumptions are violated, other phase aggregation, or phase ensembling, methods (e.g., weighted mean, non-parametric measures of centrality, or Bayesian strategies) may need to be utilized to ensure the reliability of the final analytical inference.

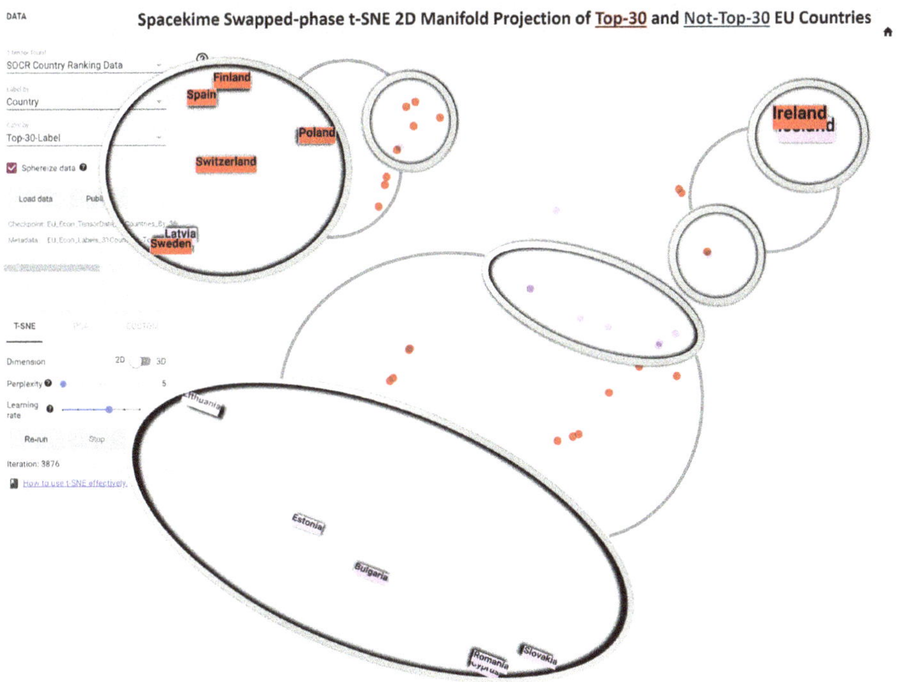

Figure 6.26: 2D t-SNE projection of the *spacekime-transformed 386-dimensional data* showing the clustering of the not-top-30 countries in the middle with two notable exceptions – Latvia, which is coupled with Sweden (on the top-left), and Ireland, which is paired with Iceland (on the top-right). Curved arcs connect the corresponding point-based projection representations and their magnified counterparts where country-names replace the points corresponding to each country. These magnifications are added to illustrate the compact central clustering of most of the not-top-30 EU countries in the sample of 31.

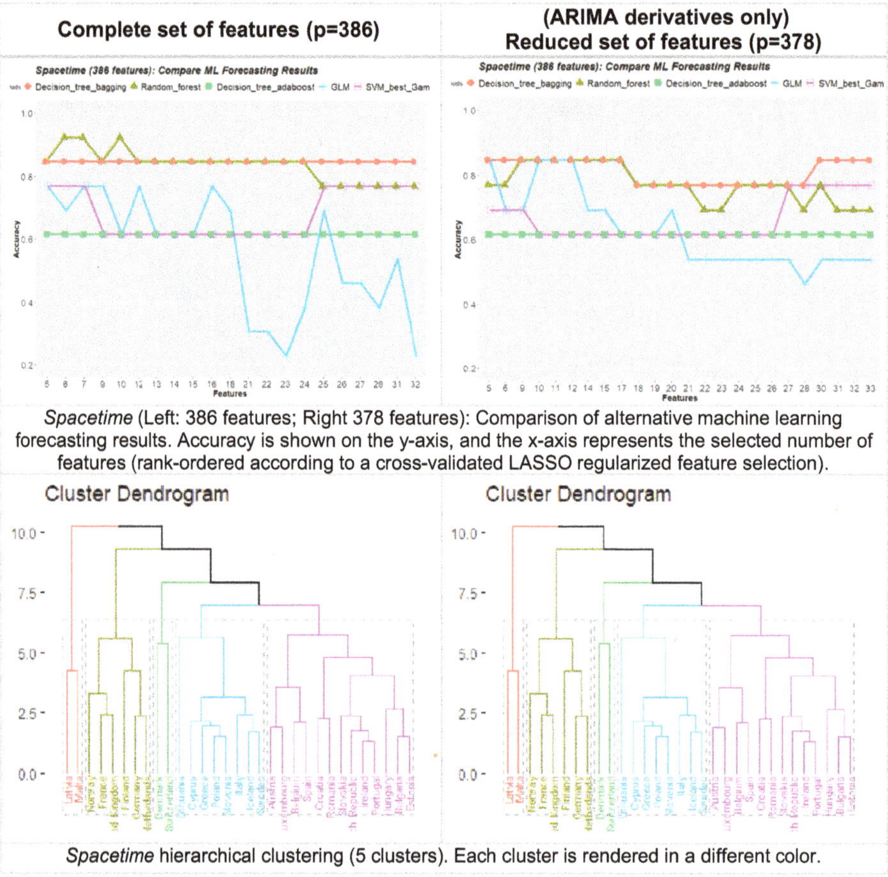
Complete set of features (p=386)
(ARIMA derivatives only)
Reduced set of features (p=378)
Spacetime (386 features): Compare ML Forecasting Results
Decision_tree_bagging
Random_forest
Decision_tree_adaboost
GLM
SVM_best_Gam
Accuracy
Features
Spacetime (Left: 386 features; Right 378 features): Comparison of alternative machine learning forecasting results. Accuracy is shown on the y-axis, and the x-axis represents the selected number of features (rank-ordered according to a cross-validated LASSO regularized feature selection).
Cluster Dendrogram
Cluster Dendrogram
Spacetime hierarchical clustering (5 clusters). Each cluster is rendered in a different color.

Figure 6.27: *Spacetime* analytics using the complete and reduced feature sets illustrating the accuracy of (1) supervised classification using decision-tree bagging, random forest, decision-tree adaboost, GLM, and SVM gamma cost function, and (2) unsupervised hierarchical clustering.

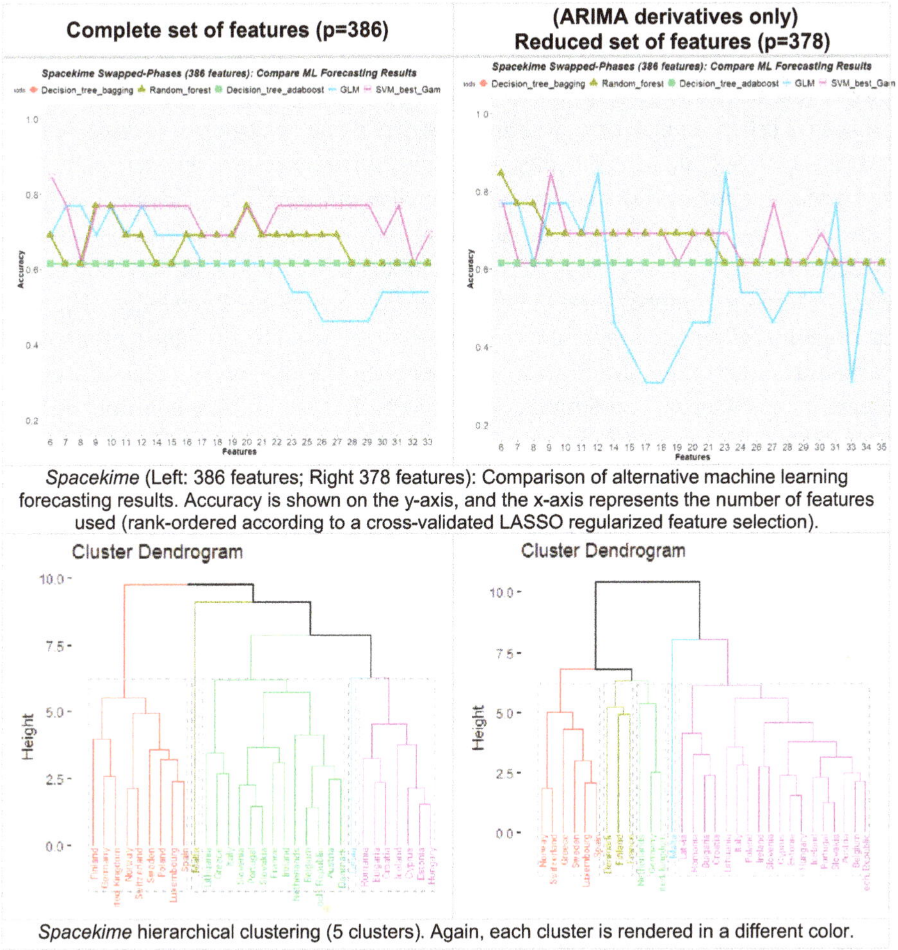

Spacekime (Left: 386 features; Right 378 features): Comparison of alternative machine learning forecasting results. Accuracy is shown on the y-axis, and the x-axis represents the number of features used (rank-ordered according to a cross-validated LASSO regularized feature selection).

Spacekime hierarchical clustering (5 clusters). Again, each cluster is rendered in a different color.

Figure 6.28: *Spacekime* analytics using the complete (left) and reduced feature sets (right) illustrating the accuracy of (1) supervised classification using decision-tree bagging, random forest, decision-tree adaboost, GLM, and SVM gamma cost function, and (2) unsupervised hierarchical clustering.

7 Summary

Observational data comes in many sizes, forms, shapes, and dimensions. A non-exhaustive list of common data types includes digital and analogue waveforms, (hyper) images, voice, text, optical and video recordings, surveys and experiments, instrument and device outputs, and spatial and longitudinal information. The field of data science provides a new pillar of scientific discovery that is fundamentally focused on representation and management of vast amounts of multiplex information along with its mining, analysis, and compression into actionable knowledge. The first component, *information representation*, reflects the need to efficiently preprocess, harmonize, aggregate, structurize, and represent the observed compound measurements as cohesive computable data objects. In general, any solutions to this monumental task are neither unique, nor universal, canonical, lossless, or necessarily optimal. However, under certain conditions, various approaches may yield tractable solutions for the information representation task. Often, ad hoc protocols need to be designed to fit each specific case study. The second aspect of data science, *AI analytics and knowledge extraction*, could involve a number of alternative strategies including supervised learning, regression, classification, unsupervised clustering, dimensionality reduction, retrospective or prospective forecasting, and general artificial intelligence decision-making.

No data science analytic strategy or computational technique could ever be both universal and optimal. A method's *universality* and *optimality* refer respectively to the technique's applicability to any observational dataset and its analytic utility for extraction of maximum valuable knowledge. At the extremes of the *universality-optimality* balance scale are trivial-analytical strategies that may not be particularly useful, for instance, assigning a fixed outcome of a desired result-type for all possible input states. Such methods are certainly universally applicable for all data states, however, their utility will be minimal, i.e., these methods would rarely yield optimal decision-making results. On the other extreme, a simple linear model represents a perfect (optimal) solution for a very narrow scope of problems satisfying all appropriate (parametric) assumptions, e.g., presence of an underlying linear association, joint multivariate normality, lack of multi-collinearity or auto-correlation, and homoscedasticity [328]. The most powerful data science techniques attempt to expand the breadth of applicability simultaneously to increasing the utility of the findings. Such balanced approaches scale up the scope of the applications and maximize the knowledge gain of the corresponding inferential results.

In this book, we focus on a specific data representation expanding the longitudinal dimension of time and lifting the classical 4D universal spacetime to a 5D spacekime manifold. The rationale for this extension is multifold. The spacekime representation allows us to resolve some of the problems of time, generalize the mathematical equations describing the natural laws of physics, and show that the corresponding

https://doi.org/10.1515/9783110697827-007

analytical expressions (e.g., equation solutions) agree with their standard 4D counterparts. In addition, we demonstrate that spacekime data representation, reconstruction, modeling, and analytics could potentially expose supplemental information that may enhance traditional spacetime observation-based scientific inference, improve data-driven predictions, and refine evidence-based decision-making processes.

The authors attempted to make this monograph as self-contained as possible, however, some readers may find that various sections require additional mathematical, statistical, and computational background. As spacekime analytics blends techniques from multiple domains, e.g., physics, mathematics, statistics, data science, computing, and artificial intelligence, it may be occasionally necessary to reference the inline citations and explore outside resources. The book content is organized using one specific linear order where chapters, and sections within chapters, transition from motivation to applications. This organization starts with foundations of mathematical physics and progressively builds the definition of complex time (kime), extensions of classical laws of physics, transformations of longitudinal time-series into complex-valued kimesurface, spacekime analytics, inferential uncertainty in spacekime, and finally demonstrates some applications.

As data science is tremendously transdisciplinary, there is no unique, linear, or optimal strategy to cover all concepts in the book. Depending on their background, expertise, or interests, instructors, readers, and learners may opt to cover the material in a different order utilizing the extensive indexing, glossary, and citation references. Some students and trainees reported that starting with the applications in the last chapter may provide motivation, additional justification, and contextualization for subsequently covering the previous more technical chapters. Open-problems and supplementary appendices included throughout may facilitate deeper investigations and community contributions.

From a data science perspective, some of the most interesting spacekime analytic ideas reflected in this treatise relate to four broad topics. The first two cover a Bayesian formulation of spacekime inference and the duality between classical large-size random sampling and small-size spacekime data acquisition using kime-phase prior distributions. The other two ideas address the intriguing conceptualization of data scientific uncertainty and the tensor-based linear modeling of complex-valued kimesurfaces. The spacekime representation of data offers significant research opportunities. Examples of these include design and validation of novel statistical models and computational algorithms using kimesurfaces for risk estimation, probabilistic modeling, projection forecasting, parametric and non-parametric inference, and supervised and unsupervised artificial intelligence.

The broader spacekime domain of mathematical-physics reaches far beyond data science. It bridges between deep theoretical principles, experimental science, astrophysics, and philosophy. In the early twentieth century, to account for the joint universal curvature and gravitational ripples, physicists extended the Newtonian notion of an absolute and static Euclidean space. This extension coupled with a unidirectional past-to-present-to-future time dimension led to formulation of the canonical Minkowski spacetime and development of the theory of relativity. Yet, it's still unsettling that something as humanly-intuitive and predictable as time can also feel so perplexing. For instance, the unidirectional arrow of time is not explicit in the fundamental equations describing observed physical phenomena. The psychological concept of "present" remains elusive as real observations are not really instantaneous. Human interpretations of "now" actually reflect the past, not "present", as the brain takes over 80 milliseconds to interpret the ambient environment. Lastly, deeper investigations into the neuropsychological interpretation of time are bound to be subjective, change with age, and possibly vary according to the different time directions (kime-phases).

The observable universe is full of repetitive notions, stochastic processes, and quantifiable events. Examples of these include orbital rotations of interstellar bodies, radioactive decay distributions, and measurable particle properties, e.g., energy, momentum, spin, and position. These naturally lead to the human interpretation of the unidirectional arrow of time as a longitudinal progression of event ordering. However, time is not intrinsic in some of the fundamental equations describing mathematically the laws of physics. Another important driver of the enigmatic positive-direction of time is related to the global cooling and accelerated expansion of the universe. This ties directly to the second law of thermodynamics, which dictates that *closed* systems naturally evolve from order to disorder, e.g., think of the natural process of diffusion. In more specific terms, the universe is transitioning from an extremely low entropy (highly ordered state) to a more natural high entropy (extremely chaotic and disordered state). In the middle of these two extremes there exists a small window where various life-forms and cautious-thought are possible. However, these very special states of matter, and mind, can thrive in particular periods where the entropy is tolerant of life's reorganization of matter. It's natural to think of life as feeding on negative entropy. Life tends to organize matter, artificially lowering (locally and temporarily) the entropy of its ambient system. The current understanding of the role of dark matter as a relentless force, invariant of spatial distances and time intervals, is to pull space apart and in the process, increase its entropy. However, it's also possible that dark matter may eventually turn into a contracting force. For instance, if spacekime is a curved and compact manifold, the initial diffusion of matter and dispersion of energy eventually may lead to a corresponding compactification of spacekime, e.g., via Big Pop, and possibly repeated recurrences of cyclical universal expansions (bangs) and contractions (pops).

In the classical Newtonian description of the universe, it is inconceivable that two different alternative future outcomes are simultaneously observed under the same initial conditions. In 1814, this deterministic view of classical physics led the polymath Pierre Simon Laplace to hypothesize the existence of a *divine calculator* (demon) that yields perfect estimates of particle position and velocity, which effectively reduces the universal past, present, and future to a static longitudinal model.

However, contemporary experiments and the development of the modern theory of quantum mechanics shattered this deterministic model of the universe and suggested that there may be many alternative (probabilistically defined) futures for each fixed present state of a system. For example, radioactivity is defined probabilistically in terms of exponential decay and half-life, not as binary outcomes such as present or not-present states. Similarly, quantum mechanical descriptions of the position, momenta, spin, and energy of particles dictate that prior to the act of making an observation, all states of the system exist synchronously and simultaneously, albeit with potentially different loading likelihoods.

In 1957, Hugh Everett postulated that all the possible futures are actualized. In other words, all possible states are observed, perhaps in different kime directions, and hence a single observer cannot really detect and acknowledge simultaneously all states as they take place in alternate universes [329]. This became the theoretical foundation of the "many worlds" (multiverse) interpretation of quantum mechanics. From a statistical perspective, we can only observe discrete random samples or detect finite instances of a stochastic process. It's not possible to instantaneously observe the entire probability distribution or measure the entire wavefunction representing the complete state-space. By definition, samples and observations are intrinsically discrete measurements, limited simulations, finite experimental results, or countable interactions arising from known models, unknown distributions, or partially understood state-spaces.

The purpose of this book is to lay down some of the spacekime mathematical foundations that lead to designing novel analytical techniques. While many challenges remain to be worked out and numerous open-mathematical problems are yet to be solved, the early evidence suggests that computational, probabilistic, or analytical extensions of spacetime to spacekime provide mechanisms to advance data science and evidence-based scientific inference. Applications of spacekime analytical methods include parameter estimation, model-based statistical inference, and model-free artificial intelligence. For instance, spacekime representations may provide mechanisms for replacing classical random sampling (IID statistical drawing strategies) with alternative data acquisition schemes. Spacekime measurement programs may involve acquiring only a few observations that can be paired with a series of approximate kime-directions (phase estimates) to enable reliable spacekime reconstructions giving rise to robust scientific inference. We also presented a Bayesian formulation of spacekime analytics that facilitates the calculation of posterior predictive probability distributions given specific kime-phase priors. Some of the examples in the book illustrate the synergies between classical model-based statistical inference and model-free machine learning forecasting and their corresponding spacekime counterparts derived by using appropriate phase aggregators, probability priors, or Laplace transformations.

When explaining difficulties in understanding the world and the theory of quantum mechanics, Richard Feynman reflected "*you know how it always is, every new idea, it takes a generation or two until it becomes obvious that there's no real problem*" [87]. In that sense, it may take some time and significant community effort to complete the spacekime representation, determine the optimal kime-phase aggregators, and develop the most accurate and highly reliable spacekime analytical techniques. This monograph contains a number of open problems and conjectures that may need to be investigated further by the entire scientific community. The ultimate impact of the spacekime representation of the universe has yet to be fully understood. In support of *open-science*, the authors are maintaining a supplementary website (https://SpaceKime.org, accessed January 23, 2021) that contains additional materials, software code, case studies, and protocols used to analyze the presented data and generate all figures, tables, and findings reported in the book.

Illustrations of model-based and model-free spacekime analytic techniques applied to economic forecasting, identification of functional brain activation, and high-dimensional biomedical population census phenotyping are shown throughout different sections and chapters. Specific case study examples include unsupervised clustering using the Michigan Consumer Sentiment Index (MCSI), model-based inference using functional magnetic resonance imaging (fMRI) data, and model-free inference using the UK Biobank data archive.

The steady increase of the volume and complexity of observed and recorded digital information drives the urgent need to develop novel data analytical strategies. Spacekime analytics represents one new data-analytic approach, which provides a mechanism to understand compound phenomena that are observed as multiplex longitudinal processes and computationally tracked by multivariate proxy measures. Some of the materials in this book may resonate with philosophers, futurists, astrophysicists, space industry technicians, biomedical researchers, health practitioners, and the general public. However, the primary audience may include transdisciplinary researchers, academic scholars, graduate students, postdoctoral fellows, artificial intelligence and machine learning engineers, biostatisticians, and data analysts.

References

[1] Krebs, R.E., Krebs, C.A. (2003). *Groundbreaking Scientific Experiments, Inventions, and Discoveries of the Ancient World*. Greenwood Publishing Group.

[2] McClellan iii, J.E., Dorn, H. (2015). *Science and Technology in World History: An Introduction*. JHU Press.

[3] Blair, A., et al. (2021). *Information: A Historical Companion*. Princeton University Press.

[4] Berggren, J.L., et al. (1990). *Innovation and Tradition in Sharaf al-Dīn al-Ṭūsī's Muʿādalāt*. Journal of the American Oriental Society, **110**(2): 304–309.

[5] Kryder, M. (2005). *Kryder's law*. Scientific American, 32–33.

[6] Dinov, I.D. (2016). *Volume and value of big healthcare data*. Journal of Medical Statistics and Informatics, **4**(1): 1–7.

[7] Dinov, I.D., Petrosyan, P., Liu, Z., Eggert, P., Zamanyan, A., Torri, F., Macciardi, F., Hobel, S., Moon, S.W., Sung, Y.H., Toga, A.W. (2014). *The perfect neuroimaging-genetics-computation storm: Collision of petabytes of data, millions of hardware devices and thousands of software tools*. Brain Imaging and Behavior, **8**(2): 311–322.

[8] Mollick, E. (2006). *Establishing Moore's law*. Annals of the History of Computing, IEEE, **28**(3): 62–75.

[9] Schwanholz, J., Graham, T., Stoll, P.-T. (2018). *Managing Democracy in the Digital Age*. Springer.

[10] Dinov, I. (2018). *Data Science and Predictive Analytics: Biomedical and Health Applications Using R. Computer Science*. Springer International Publishing. 800.

[11] Hey, T., Tansley, S., Tolle, K.M. (2009). *The Fourth Paradigm: Data-intensive Scientific Discovery*. Vol. 1. Microsoft research Redmond, WA.

[12] Hilgevoord, J., Uffink, J. (1990). A New View on the Uncertainty Principle. In: Miller A.I. (eds.), *Sixty-Two Years of Uncertainty. NATO ASI Series (Series B: Physics)*, vol 226. Springer, Boston, MA.

[13] Müller-Merbach, H. (2006). *Heraclitus: Philosophy of change, a challenge for knowledge management?* Knowledge Management Research & Practice, **4**(2): 170–171.

[14] Carlsson, G. (2009). *Topology and data*. Bulletin of the American Mathematical Society, **46**(2): 255–308.

[15] Marino, S., et al. (2018). *Controlled Feature Selection and Compressive Big Data Analytics: Applications to Biomedical and Health Studies*. PLoS Bioinformatics, **13**(8): e0202674.

[16] Kuang, L., et al. (2014). *A tensor-based approach for big data representation and dimensionality reduction*. IEEE Transactions on Emerging Topics in Computing, **2**(3): 280–291.

[17] Zhou, Z.-H., Wu, J., Tang, W. (2002). *Ensembling neural networks: Many could be better than all*. Artificial Intelligence, **137**(1–2): 239–263.

[18] Najafabadi, M.M., et al. (2015). *Deep learning applications and challenges in big data analytics*. Journal of Big Data, **2**(1): 1.

[19] Condie, T., Mineiro, P., Polyzotis, N., Weimer, M. *(2013). Machine learning on Big Data*. In *2013 IEEE 29th International Conference on Data Engineering (ICDE)*, pp. 1242–1244.

[20] Mamourian, A.C. (2010). *Practical MR Physics*. Oxford University Press.

[21] Atzori, L., Iera, A., Morabito, G. (2010). *The internet of things: A survey*. Computer Networks, **54**(15): 2787–2805.

[22] Urbach, N., Ahlemann, F. (2018). *IT Management in the Digital Age: A Roadmap for the IT Department of the Future*. Springer.

[23] Gödel, K. (1931). *Über formal unentscheidbare Sätze der Principia Mathematica und verwandter Systeme I*. Monatshefte Für Mathematik Und Physik, **38**(1): 173–198.

https://doi.org/10.1515/9783110697827-008

[24] Deng, X., et al. (2018). *Support high-order tensor data description for outlier detection in high-dimensional big sensor data*. Future Generation Computer Systems, **81**: 177–187.
[25] Chien, J.-T., Bao, Y.-T. (2018). *Tensor-factorized neural networks*. IEEE Transactions on Neural Networks and Learning Systems, **29**(5): 1998–2011.
[26] Maaten, L. v.d., Hinton, G. (2008). Visualizing Data Using t-SNE. Journal of Machine Learning Research, **9**(Nov): 2579–2605.
[27] Roweis, S.T., Saul, L.K. (2000). *Nonlinear Dimensionality Reduction by Locally Linear Embedding*. Science, **290**(5500): 2323–2326.
[28] McInnes, L., Healy, J., Melville, J. (2018). *Umap: Uniform manifold approximation and projection for dimension reduction*. arXiv preprint arXiv:1802.03426.
[29] Brahim, A.B., Limam, M. (2018). *Ensemble feature selection for high dimensional data: A new method and a comparative study*. Advances in Data Analysis and Classification, **12**(4): 937–952.
[30] Kogan, J. (2007). *Introduction to Clustering Large and High-dimensional Data*. Cambridge University Press.
[31] Abpeykar, S., Ghatee, M., Zare, H. (2019). *Ensemble decision forest of RBF networks via hybrid feature clustering approach for high-dimensional data classification*. Computational Statistics & Data Analysis, **131**: 12–36.
[32] Bertozzi, A.L., et al. (2018). *Uncertainty Quantification in Graph-Based Classification of High Dimensional Data*. SIAM/ASA Journal on Uncertainty Quantification, **6**(2): 568–595.
[33] Williams, C.K.I., Rasmussen, C.E. (1996). Gaussian processes for regression. In: Touretzky, D. S., Mozer, M. C., Hasselmo, M. E. (eds.), Advances in Neural Information Processing Systems 8. MIT
[34] Donoho, D.L., Grimes, C. (2003). *Hessian eigenmaps: Locally linear embedding techniques for high-dimensional data*. Proceedings of the National Academy of Sciences, **100**(10): 5591–5596.
[35] Li, B., Li, Y.-R., Zhang, X.-L. (2018). *A survey on Laplacian eigenmaps based manifold learning methods*. Neurocomputing, accessed January 29, 2021.
[36] Smith, K., Vul, E. (2014). Looking forwards and backwards: Similarities and differences in prediction and retrodiction. *Proceedings of the Annual Meeting of the Cognitive Science Society*, 36. Retrieved from https://escholarship.org/uc/item/3vq840dp
[37] Cleveland, W.S. (2001). *Data Science: An Action Plan for Expanding the Technical Areas of the Field of Statistics*. International Statistical Review, **69**(1): 21–26.
[38] Donoho, D. (2017). *50 Years of Data Science*, Journal of Computational and Graphical Statistics, 26(4):745–766.
[39] Voytek, B. (2017). *Social Media, Open Science, and Data Science Are Inextricably Linked*. Neuron, **96**(6): 1219–1222.
[40] Dunn, M.C., Bourne, P.E. (2017). *Building the biomedical data science workforce*. PLoS Biology, **15**(7): e2003082.
[41] Dinov, I.D. (2019). *Quant data science meets dexterous artistry*. International Journal of Data Science and Analytics, **7**(2): 81–86.
[42] Sternberg, R.J., Sternberg, R.J. (1982). *Handbook of Human Intelligence*. CUP Archive.
[43] Hall, E.C. (1996). *Journey to the Moon: The History of the Apollo Guidance Computer*. Aiaa.
[44] Yager, R.R. (1997). *Fuzzy logics and artificial intelligence*. Fuzzy Sets and Systems, **90**(2): 193–198.
[45] Shortliffe, E.H., et al. (1973). *An Artificial Intelligence program to advise physicians regarding antimicrobial therapy*. Computers and Biomedical Research, **6**(6): 544–560.
[46] Kasparov, G. (2017). *Deep Thinking: Where Machine Intelligence Ends and Human Creativity Begins*. PublicAffairs.

[47] Callaway, E. (2020). *'It will change everything': DeepMind's AI makes gigantic leap in solving protein structures*. Nature, **588**: 203–204.
[48] Kaplan, A., Haenlein, M. (2019). *Siri, Siri, in my hand: Who's the fairest in the land? On the interpretations, illustrations, and implications of artificial intelligence*. Business Horizons, **62**(1): 15–25.
[49] Farahmand, A. M., Shademan, A., Jagersand, M., Szepesvari, C. (2009). Model-based and model-free reinforcement learning for visual servoing. In *2009 IEEE International Conference on Robotics and Automation*, pp. 2917–2924.
[50] Geffner, H. (2018). *Model-free, model-based, and general intelligence*. arXiv preprint arXiv:1806.02308.
[51] Gao, C., et al. (2018). *Model-based and Model-free Machine Learning Techniques for Diagnostic Prediction and Classification of Clinical Outcomes in Parkinson's Disease*. Scientific Reports, **8**(1): 7129.
[52] Lehnert, L., Littman, M.L. (2019). *Successor features support model-based and model-free reinforcement learning*. arXiv preprint arXiv:1901.11437.
[53] Rawat, W., Wang, Z. (2017). *Deep convolutional neural networks for image classification: A comprehensive review*. Neural Computation, **29**(9): 2352–2449.
[54] Alom, M.Z., et al. (2018). *The history began from alexnet: A comprehensive survey on deep learning approaches*. arXiv preprint arXiv:1803.01164.
[55] Guo, Y., et al. (2016). *Deep learning for visual understanding: A review*. Neurocomputing, **187**: 27–48.
[56] Jack, C., et al. (2008). *The Alzheimer's disease neuroimaging initiative (ADNI): MRI methods*. Journal of Magnetic Resonance Imaging, **27**(4): 685–691.
[57] Zhou, Y., et al. (2019). *Predictive Big Data Analytics using the UK Biobank Data*. Scientific Reports, **9**(1): 6012.
[58] Dekker, J., et al. (2017). *The 4D nucleome project*. Nature, **549**: 219.
[59] Zheng, G., et al. (2018). *Hypothesis: Caco-2 cell rotational 3D mechanogenomic turing patterns have clinical implications to colon crypts*. Journal of Cellular and Molecular Medicine, **22**(12): 6380–6385.
[60] Kalinin, A.A., et al. (2018). *3D Shape Modeling for Cell Nuclear Morphological Analysis and Classification*. Scientific Reports, **8**(1): 13658.
[61] Carroll, S.M. (2010). *From Eternity to Here: The Quest for the Ultimate Theory of Time*. Penguin.
[62] Bars, I., Terning, J., Nekoogar, F. (2010). *Extra Dimensions in Space and Time*. Springer.
[63] Wesson, P., Overduin, J. (2018). *Principles of Space-Time-Matter Cosmology, Particles and Waves in Five Dimensions. Principles of Space-Time-Matter: Cosmology, Particles and Waves in Five Dimensions*. World Scientific, 276. ISBN: 9813235799, 9789813235793. https://books.google.com/books?hl=en&lr=&id=i9WEDwAAQBAJ
[64] Köhn, C. (2017). *The Planck Length and the Constancy of the Speed of Light in Five Dimensional Spacetime Parametrized with Two Time Coordinates*. Journal of High Energy Physics, Gravitation and Cosmology, **3**: 635–650.
[65] Mazzola, G. (2012). *The Topos of Music: Geometric Logic of Concepts, Theory, and Performance*. Birkhäuser.
[66] Mannone, M., Mazzola, G. (2015). *Hypergestures in complex time: Creative performance between symbolic and physical reality*. In *International Conference on Mathematics and Computation in Music*. Springer.
[67] Gilboa, G., Sochen, N., Zeevi, Y.Y. (2004). *Image enhancement and denoising by complex diffusion processes*. IEEE Transactions on Pattern Analysis and Machine Intelligence, **26**(8): 1020–1036.

[68] Araújo, R.d.A. (2012). *A robust automatic phase-adjustment method for financial forecasting*. Knowledge-Based Systems, **27**: 245–261.
[69] Schwartz, C., Zemach, C. (1966). *Theory and calculation of scattering with the Bethe-Salpeter equation*. Physical Review, **141**(4): 1454.
[70] Wesson, P.S. (1999). *Space-time-matter: Modern Kaluza-Klein Theory*. World Scientific.
[71] Taubes, C.H. (1987). *Gauge theory on asymptotically periodic {4}-manifolds*. Journal of Differential Geometry, **25**(3): 363–430.
[72] O'shea, D. (2008). *The Poincaré Conjecture: In Search of the Shape of the Universe*. Bloomsbury Publishing USA.
[73] Kirby, R.C., et al. (1977). *Foundational Essays on Topological Manifolds, Smoothings, and Triangulations*. Princeton University Press.
[74] Stallings, J. (1962). *The piecewise-linear structure of Euclidean space*. In *Mathematical Proceedings of the Cambridge Philosophical Society*. Cambridge University Press.
[75] Scorpan, A. (2005). *The Wild World of 4-manifolds*. American Mathematical Soc.
[76] Curtin, R.T. (2019). *Consumer Expectations: Micro Foundations and Macro Impact*. Cambridge University Press.
[77] Anggraeni, W., Aristiani, L. Using Google Trend data in forecasting number of dengue fever cases with ARIMAX method case study: Surabaya, Indonesia. In *2016 International Conference on Information & Communication Technology and Systems (ICTS)*, pp. 114–118.
[78] Jordan, T.F., Shaji, A., Sudarshan, E. (2006). *Mapping the Schrödinger picture of open quantum dynamics*. Physical Review A, **73**(1): 012106.
[79] Brody, D.C., Hughston, L.P. (2001). *Geometric quantum mechanics*. Journal of Geometry and Physics, **38**(1): 19–53.
[80] Goodrich, R.K. (1970). *A Riesz representation theorem*. Proceedings of the American Mathematical Society, **24**(3): 629–636.
[81] Wigner, E. (2012). *Group Theory: And Its Application to the Quantum Mechanics of Atomic Spectra*. Vol. 5. Elsevier.
[82] P. Busch: The Time–Energy Uncertainty Relation, Lecture Notes in Physics, 734, 73–105 (2008).
[83] Summers, S.J. (1990). *On the Independence of Local Algebras in Quantum Field Theory*. Reviews in Mathematical Physics, **02**(02): 201–247.
[84] Florig, M., Summers, S.J. (1997). *On the statistical independence of algebras of observables*. Journal of Mathematical Physics, **38**(3): 1318–1328.
[85] Bell, J.S. (1964). *On the Einstein Podolsky Rosen paradox*. Physics Physique Fizika, **1**(3): 195.
[86] Bunce, L., Hamhalter, J. (2004). *C*-independence, product states and commutation*. In *Annales Henri Poincaré*. Springer.
[87] Feynman, R.P. (1982). *Simulating physics with computers*. International Journal of Theoretical Physics, **21**(6): 467–488.
[88] Rodriguez, J.A., et al. (2015). *Structure of the toxic core of α-synuclein from invisible crystals*. Nature, **525**(7570): 486.
[89] Chiang, L.-Y. (2001). *The importance of Fourier phases for the morphology of gravitational clustering*. Monthly Notices of the Royal Astronomical Society, **325**(1): 405–411.
[90] Poincaré, H. (1906). *On the dynamics of the electron, Rend.* Circolo Mat. Palermo, **21**: 129–176.
[91] Poincaré, H. (2007). *On the dynamics of the electron (Excerpts)*, In: Janssen M., Norton J.D., Renn J., Sauer T., Stachel J. (eds) *The Genesis of General Relativity*. Springer: 1179–1198.
[92] Minkowski, H. (1910). *Die Grundgleichungen für die elektromagnetischen Vorgänge in bewegten Körpern*. Mathematische Annalen, **68**(4): 472–525.

[93] Theodor, K. (1921). *Zum Unitätsproblem in der Physik*. Sitzungsber. Preuss. Akad. Wiss. Berlin.(Math. Phys.), **1921**: 966–972.
[94] Klein, O. (1926). *Quantentheorie und fünfdimensionale Relativitätstheorie*. Zeitschrift Für Physik, **37**(12): 895–906.
[95] McLaughlin, D.W. (1972). *Complex time, contour independent path integrals, and barrier penetration*. Journal of Mathematical Physics, **13**(8): 1099–1108.
[96] Argyris, J., Ciubotariu, C. (1997). *On El Naschie's complex time and gravitation*. Chaos, Solitons & Fractals, **8**(5): 743–751.
[97] Czajko, J. (2000). *On conjugate complex time – I: Complex time implies existence of tangential potential that can cause some equipotential effects of gravity*. Chaos, Solitons & Fractals, **11**(13): 1983–1992.
[98] Czajko, J. (2000). *On conjugate complex time – II: Equipotential effect of gravity retrodicts differential and predicts apparent anomalous rotation of the sun*. Chaos, Solitons & Fractals, **11**(13): 2001–2016.
[99] Haddad, M., Ghaffari-Miab, M., Faraji-Dana, R. (2010). *Transient analysis of thin-wire structures above a multilayer medium using complex-time Green's functions*. IET Microwaves, Antennas & Propagation, **4**(11): 1937–1947.
[100] Ponce de Leon, J. (2002). *Equations of Motion in Kaluza-Klein Gravity Reexamined*. Gravitation and Cosmology, **8**: 272–284.
[101] Velev, M.V. (2012). *Relativistic mechanics in multiple time dimensions*. Physics Essays, **25**(3).
[102] Dobbs, H.A.C. (1958). *Multidimensional Time*. The British Journal for the Philosophy of Science, **9**(35): 225–227.
[103] Bunge, M. (1958). *On Multi-dimensional Time*. The British Journal for the Philosophy of Science, **IX**(33): 39–39.
[104] Chari, C. (1957). *A Note on Multi-Dimensional Time*. The British Journal for the Philosophy of Science, **8**(30): 155–158.
[105] Hartle, J.B., Hawking, S.W. (1983). *Wave function of the universe*. Physical Review D, **28**(12): 2960.
[106] Bars, I. (2007) *The Standard Model as a 2T-physics Theory*. In *AIP Conference Proceedings*. AIP.
[107] Hawking, S. (1996). *The Illustrated a Brief History of Time*. Bantam.
[108] Deutsch, D. (2002). *The structure of the multiverse*. Proceedings of the Royal Society of London A: Mathematical, Physical and Engineering Sciences, **458**(2028): 2911–2923.
[109] Page, D.N., Wootters, W.K. (1983). *Evolution without evolution: Dynamics described by stationary observables*. Physical Review D, **27**(12): 2885.
[110] Alonso-Serrano, A., et al. (2013). *Interacting universes and the cosmological constant*. Physics Letters B, **719**(1–3): 200–205.
[111] Hagura, N., et al. (2012). *Ready steady slow: Action preparation slows the subjective passage of time*. Proceedings of Royal Soceity B, rspb20121339.
[112] Bars, I. (2005). *Twistors and 2T-physics*. In *AIP Conference Proceedings*. AIP.
[113] Dorling, J. (1970). *The dimensionality of time*. American Journal of Physics, **38**(4): 539–540.
[114] Foster, J.G., Müller, B. (2010). *Physics with two time dimensions*. arXiv preprint arXiv:1001.2485.
[115] Tegmark, M. (1997). *On the dimensionality of spacetime*. Classical and Quantum Gravity, **14**(4): L69.
[116] Carr, B., Ellis, G. (2008). *Universe or multiverse?* Astronomy & Geophysics, **49**(2): 2.29–2.33.
[117] Jech, T.J. (2008). *The Axiom of Choice*. Courier Corporation.
[118] Bars, I. (2006). *Standard model of particles and forces in the framework of two-time physics*. Physical Review D, **74**(8): 0085019.

[119] Adler, M., van Moerbeke, P. (1989). *The complex geometry of the Kowalewski-Painlevé analysis*. Inventiones Mathematicae, **97**(1): 3–51.

[120] Gromak, V.I., Laine, I., Shimomura, S. (2008). *Painlevé Differential Equations in the Complex Plane*. Vol. 28. Walter de Gruyter.

[121] Weinberg, S. (2008). *Cosmology*. Oxford university press.

[122] Gerchberg, R., Saxton, W. (1973). *Comment onA method for the solution of the phase problem in electron microscopy'*. Journal of Physics D: Applied Physics, **6**(5): L31.

[123] Fienup, J.R. (1982). *Phase retrieval algorithms: A comparison*. Applied Optics, **21**(15): 2758–2769.

[124] Elser, V. (2003). *Phase retrieval by iterated projections*. Josa A, **20**(1): 40–55.

[125] Netrapalli, P., Jain, P., Sanghavi, S. (2013). *Phase retrieval using alternating minimization*. In *Advances in Neural Information Processing Systems*.

[126] Zhang, K. (1993). *SQUASH–combining constraints for macromolecular phase refinement and extension*. Acta Crystallographica Section D: Biological Crystallography, **49**(1): 213–222.

[127] Blackledget, J.M. (2006). *Chapter 4 – The Fourier Transform*, In Woodhead Publishing Series in Electronic and Optical Materials, Digital Signal Processing (Second Edition), J.M. Blackledget, Editor. Woodhead Publishing. 75–113.

[128] Van der Laan, M.J., Polley, E.C., Hubbard, A.E. (2007). *Super learner*. Statistical Applications in Genetics and Molecular Biology, **6**(1).

[129] Marino, S., et al. (2020). *Compressive Big Data Analytics: An Ensemble Meta-Algorithm for High-dimensional Multisource Datasets*. PLoS, **15**(8): e0228520.

[130] Pizer, S.M., et al. (1987). *Adaptive histogram equalization and its variations*. Computer Vision, Graphics, and Image Processing, **39**(3): 355–368.

[131] Sorkin, R.D. (1994). *Quantum mechanics as quantum measure theory*. Modern Physics Letters A, **9**(33): 3119–3127.

[132] Schrödinger, E. (1926). *An undulatory theory of the mechanics of atoms and molecules*. Physical Review, **28**(6): 1049.

[133] Glimm, J., Jaffe, A. (2012). *Quantum Physics: A Functional Integral Point of View*. Springer Science & Business Media.

[134] Sakurai, J., Napolitano, J. (2017). *Modern Quantum Mechanics*. Cambridge University Press.

[135] Feit, M., Fleck Jr, J., Steiger, A. (1982). *Solution of the Schrödinger equation by a spectral method*. Journal of Computational Physics, **47**(3): 412–433.

[136] Nelson, R.A. (1987). *Generalized Lorentz transformation for an accelerated, rotating frame of reference*. Journal of Mathematical Physics, **28**(10): 2379–2383.

[137] Vay, J.-L., et al. (2011). *Effects of hyperbolic rotation in Minkowski space on the modeling of plasma accelerators in a Lorentz boosted frame*. Physics of Plasmas, **18**(3): 030701.

[138] Valentini, A. (1997). *On Galilean and Lorentz invariance in pilot-wave dynamics*. Physics Letters A, **228**(4–5): 215–222.

[139] Ortega, R., et al. (2013). *Passivity-based Control of Euler-Lagrange Systems: Mechanical, Electrical and Electromechanical Applications*. Springer Science & Business Media.

[140] Tulczyjew, W. (1980). *The Euler-Lagrange resolution*, In *Differential Geometrical Methods in Mathematical Physics*. Springer, 22–48.

[141] Norbury, J.W. (1998). *From Newton's laws to the Wheeler-DeWitt equation*. European Journal of Physics, **19**(2): 143.

[142] DeWitt, B.S. (1967). *Quantum theory of gravity. I. The canonical theory*. Physical Review, **160**(5): 1113.

[143] Chen, X. (2005). *Three dimensional time theory: To unify the principles of basic quantum physics and relativity*. arXiv preprint quant-ph/0510010.

[144] Korn, G.A., Korn, T.M. (2000). *Mathematical Handbook for Scientists and Engineers: Definitions, Theorems, and Formulas for Reference and Review.* Courier Corporation.

[145] Wesson, P.S. (2006). *Five-dimensional Physics: Classical and Quantum Consequences of Kaluza-Klein Cosmology.* World Scientific.

[146] Cardone, F., Mignani, R. (2007). *Deformed Spacetime: Geometrizing Interactions in Four and Five Dimensions. Volume 157 of Fundamental Theories of Physics. Vol. 157.* Springer Science & Business Media.

[147] De Leon, J.P. (2001). *Equations of Motion in Kaluza-Klein Gravity Revisited.* arXiv preprint gr-qc/0104008.

[148] Wesson, P., Ponce de Leon, J. (1995). *The equation of motion in Kaluza-Klein cosmology and its implications for astrophysics.* Astronomy and Astrophysics, **294**: 1–7.

[149] Seahra, S. (2003). *Physics in Higher-Dimensional Manifolds.* UWSpace. http://hdl.handle.net/10012/1276

[150] Linge, S., Langtangen, H.P. (2017). *Wave equations,* In *Finite Difference Computing with PDEs.* Springer, 93–205.

[151] Courant, R., Hilbert, D. (1962). *Methods of Mathematical Physics, Volume II.* John Wiley & Sons, New York.

[152] Craig, W., Weinstein, S. (2009). *On determinism and well-posedness in multiple time dimensions.* Proceedings of the Royal Society A: Mathematical, Physical and Engineering Sciences, **465**(2110): 3023–3046.

[153] Katz, V.J. (1979). *The history of Stokes' theorem.* Mathematics Magazine, **52**(3): 146–156.

[154] Rudin, W. (2006). *Real and Complex Analysis.* Tata McGraw-hill education.

[155] Asgeirsson, L. (1948). *Uber Mittelwertgleichungen, Die Mehreren Partiellen Differentialgleichungen 2. Ordnung Zugeordnet Sind (German). Studies and Essays.* Interscience, New York.

[156] Hörmander, L. (2001). *Asgeirsson's mean value theorem and related identities.* Journal of Functional Analysis, **184**(2): 377–401.

[157] Mackinnon, L. (1978). *A nondispersive de Broglie wave packet.* Foundations of Physics, **8**(3): 157–176.

[158] Wirtinger, W. (1927). *Zur formalen theorie der funktionen von mehr komplexen veränderlichen.* Mathematische Annalen, **97**(1): 357–375.

[159] Candes, E.J., Li, X., Soltanolkotabi, M. (2015). *Phase retrieval via Wirtinger flow: Theory and algorithms.* IEEE Transactions on Information Theory, **61**(4): 1985–2007.

[160] Moore, E.H. (1900). *A simple proof of the fundamental Cauchy-Goursat theorem.* Transactions of the American Mathematical Society, **1**(4): 499–506.

[161] Stapp, H.P. (1972). *The copenhagen interpretation.* American Journal of Physics, **40**(8): 1098–1116.

[162] Abellán, C., et al. (2018). *Challenging local realism with human choices.* Nature, **557**(7704): 212–216.

[163] Lombardi, O., et al. (2017). *What Is Quantum Information?* Cambridge University Press.

[164] Cabello, A. (2017). *Interpretations of quantum theory: A map of madness.* What is quantum information, 138–144.

[165] Dinov, I., Christou, N., Gould, R. (2009). *Law of Large Numbers: The Theory, Applications and Technology-based Education.* Journal of Statistical Education, **17**(1): 1–15.

[166] Uhlig, H. (1996). *A Law of Large Numbers for Large Economies.* Economic Theory, **8**: 41–50.

[167] De Faria, E., De Melo, W. (2010). *Mathematical Aspects of Quantum Field Theory.* Vol. 127. Cambridge University Press.

[168] Kallenberg, O. (2006). *Foundations of Modern Probability.* Springer Science & Business Media.

[169] Velev, M.V. (2012). *Relativistic mechanics in multiple time dimensions*. Physics Essays, **25**(3): 403.

[170] Gueorguiev, V. (2019). *Reparametrization-Invariance and Some of the Key Properties of Physical Systems*. arXiv preprint arXiv:1903.02483.

[171] Foster, J.G., Müller, B. (2010). *Physics with two time dimensions*.

[172] Dainton, B. (2016). *Time and Space*. Routledge.

[173] Mach, E. (1915). *The Science of Mechanics: A Critical and Historical Account of Its Development. Supplement*. Open court publishing Company.

[174] Fahnestock, J. (1998). *Accommodating science: The rhetorical life of scientific facts*. Written Communication, **15**(3): 330–350.

[175] Krippendorff, K. (2009). *The Content Analysis Reader*. Sage.

[176] Dantzig, J.A.S.a.D.v. (1932). *Generelle Feldtheorie*. Zeitschrift Fur Physik, **78**(9–10): 639–667.

[177] Pauli, W. (1933). *Uber die Formulierung der Naturgesetze mit funf homogenen Koordinaten. Teil II: Die Diracschen Gleichungen fur die Materiewellen*. Annalen Der Physik, **410**(4): 337–372.

[178] Dirac, P.A.M. (1935). *The electron wave equation in de-Sitter space*. Annals of Mathematics, **36**(3): 657–669.

[179] Lubanski, J. (1942). *Sur la theorie des particules elementaires de spin quelconque. I*. Physica, **9**(3): 10–324.

[180] Lubanski, J. (1942). *Sur la theorie des particules elementaires de spin quelconque. II*. Physica, **9**(3): 325–338.

[181] Bhabha, H.J. (1945). *Relativistic wave equations for the elementary particles*. Reviews of Modern Physics, **17**(2–3): 200–216.

[182] Lee, F.G.a.T. (1963). *Spin 1/2 wave equation in De-Sitter Space*. Proceedings of the National Academy of Sciences of the United States of America, **49**: 179–186.

[183] Kocinski, J. (1999). *A five-dimensional form of the Dirac equation*. Journal of Physics A, **32**: 4257–4277.

[184] Zhang, G.M., Wu, Y.L., Wang, X.Z., Sun, J.L. (2000). *Dirac field of Kaluza-Klein theory in Weitzenbock space*. International Journal of Theoretical Physics, **39**(8): 2051–2062.

[185] Lodhi, N.R.a.M.A.K. (2007). *Simple five-dimensional wave equation for a Dirac particle*. Journal of Mathematical Physics, **48**: 022303.

[186] Dong, S.H. (2011). *Wave Equations in Higher Dimensions*. New York: Springer.

[187] Breban, R. (2018). *The Four Dimensional Dirac Equation in Five Dimensions*. Annalen Der Physik, 1800042.

[188] Arcodía, M.R., Bellini, M. (2019). *Particle-antiparticle duality from an extra timelike dimension*. The European Physical Journal C, **79**(9): 796.

[189] Dirac, P.A.M. (1935). *The electron wave equation in de-Sitter space*. Annals of Mathematics, 657–669.

[190] Bars, I. (2001). *Survey of two-time physics*. Classical and Quantum Gravity, **18**(16): 3113.

[191] Bars, I., Deliduman, C. (2001). *High spin gauge fields and two-time physics*. Physical Review D, **64**(4): 045004.

[192] Bars, I., Kounnas, C. (1997). *String and particle with two times*. Physical Review D, **56**(6): 3664.

[193] Redington, N., Lodhi, M. (2007). *Simple five-dimensional wave equation for a Dirac particle*. Journal of Mathematical Physics, **48**(2): 022303.

[194] Hestenes, D. (1966). *Space-time Algebra*. Vol. 1. Springer.

[195] Morimoto, T., Furusaki, A. (2013). *Topological classification with additional symmetries from Clifford algebras*. Physical Review B, **88**(12): 125129.

[196] Abłamowicz, R. (1998). *Spinor representations of Clifford algebras: A symbolic approach*. Computer Physics Communications, **115**(2–3): 510–535.

[197] Ziino, G. (1996). *Fermion Quantum Field Theory with a New Gauge Symmetry, Related to a Pseudoscalar-Conserved-Charge Variety*. International Journal of Modern Physics A, **11**(12): 2081–2109.

[198] Recami, E., Ziino, G. (1976). *About new space-time symmetries in relativity and quantum mechanics*. Il Nuovo Cimento A (1971–1996), **33**(2): 205–215.

[199] Nan, F.Y., Nowak, R.D. (1999). *Generalized likelihood ratio detection for fMRI using complex data*. IEEE Transactions on Medical Imaging, **18**(4): 320–329.

[200] Calhoun, V.D., et al. (2002). *Independent component analysis of fMRI data in the complex domain*. Magnetic Resonance in Medicine: An Official Journal of the International Society for Magnetic Resonance in Medicine, **48**(1): 180–192.

[201] Rowe, D.B., Logan, B.R. (2004). *A complex way to compute fMRI activation*. NeuroImage, **23**(3): 1078–1092.

[202] Adrian, D.W., Maitra, R., Rowe, D.B. (2018). *Complex-valued time series modeling for improved activation detection in fMRI studies*. The Annals of Applied Statistics, **12**(3): 1451–1478.

[203] McKeown, M.J., Hu, Y.-j., Wang, Z.J. (2006). *ICA Denoising for Event-Related fMRI Studies*. In *2005 IEEE Engineering in Medicine and Biology 27th Annual Conference*, pp. 157–161.

[204] Dinov, I., Siegrist, K., Pearl, D.K., Kalinin, A., Christou, N. (2015). *Probability Distributome: A web computational infrastructure for exploring the properties, interrelations, and applications of probability distributions*. Computational Statistics, **594**: 1–19.

[205] Sijbers, J., den Dekker, A.J. (2005). *Generalized likelihood ratio tests for complex fMRI data: A simulation study*. IEEE Transactions on Medical Imaging, **24**(5): 604–611.

[206] Dinov, I., Christou, N., Sanchez, J. (2008). *Central Limit Theorem: New SOCR Applet and Demonstration Activity*. Journal of Statistical Education, **16**(2): 1–12.

[207] Ver Hoef, J.M. (2012). *Who invented the delta method?* The American Statistician, **66**(2): 124–127.

[208] Efron, B., Hinkley, D.V. (1978). *Assessing the accuracy of the maximum likelihood estimator: Observed versus expected Fisher information*. Biometrika, **65**(3): 457–483.

[209] Myung, J.I., Navarro, D.J. (2005). *Information Matrix*. Encyclopedia of Statistics in Behavioral Science.

[210] Lafaye de Micheaux, P., Liquet, B. (2009). *Understanding convergence concepts: A visual-minded and graphical simulation-based approach*. The American Statistician, **63**(2): 173–178.

[211] Gilat, D. (1972). *Convergence in distribution, convergence in probability and almost sure convergence of discrete martingales*. The Annals of Mathematical Statistics, **43**(4): 1374–1379.

[212] Allen, M.P. (2007). *Understanding Regression Analysis*. Springer US.

[213] Graham, A. (2018). *Kronecker Products and Matrix Calculus with Applications*. Courier Dover Publications.

[214] DeYoe, E.A., et al. (1994). *Functional magnetic resonance imaging (FMRI) of the human brain*. Journal of Neuroscience Methods, **54**(2): 171–187.

[215] Arfken, G.B., Weber, H.J., Harris, F.E. (2013). *Chapter 14 – Bessel functions*, In *Mathematical Methods for Physicists (Seventh Edition)*, G.B. Arfken, H.J. Weber, and F.E. Harris, Editors. Academic Press: Boston, 643–713.

[216] Wilks, S.S. (1938). *The Large-Sample Distribution of the Likelihood Ratio for Testing Composite Hypotheses*. Annals of Mathematics and Statistics, **9**(1): 60–62.

[217] Cormen, T.H., et al. (2009). *Introduction to Algorithms*. MIT press.

[218] Winitzki, S. (2003). *Uniform approximations for transcendental functions*. in *International Conference on Computational Science and Its Applications*. Springer.

[219] Abramowitz, M., Stegun, I.A. (1972). *Modified Bessel functions I and K*. Handbook of mathematical functions with formulas, graphs, and mathematical tables, 9th printing, 374–377.

[220] Bender, C.M., Orszag, S.A. (2013). *Advanced Mathematical Methods for Scientists and Engineers I: Asymptotic Methods and Perturbation Theory*. Springer Science & Business Media.

[221] Laplace, P.S. (1986). *Memoir on the Probability of the Causes of Events*. Statistical Science. **1**(3): 364–378.

[222] Fog, A. (2008). *Calculation methods for Wallenius' noncentral hypergeometric distribution*. Communications in Statistics – Simulation and Computation®, **37**(2): 258–273.

[223] Arfken, G. (1985). *Hankel functions11.4 in mathematical methods for physicists*, in *Mathematical Methods for Physicists*. Academic Press: Orlando, FL, 604–610.

[224] Amemiya, T. (1985). *Advanced Econometrics*. Harvard university press.

[225] Nyquist, H. (1991). *Restricted Estimation of Generalized Linear Models*. Journal of the Royal Statistical Society. Series C (Applied Statistics), **40**(1): 133–141.

[226] Gourieroux, C., Holly, A., Monfort, A. (1982). *Likelihood ratio test, Wald test, and Kuhn-Tucker test in linear models with inequality constraints on the regression parameters*. Econometrica: Journal of the Econometric Society, 63–80.

[227] Stuart, A., Ord, J.K., Arnold, S.F. (2004). *Kendall's Advanced Theory of Statistics: Classical Inference and the Linear Model. Volume 2A*. John Wiley.

[228] Andersen, A., Kirsch, J. (1996). *Analysis of noise in phase contrast MR imaging*. Medical Physics, **23**(6): 857–869.

[229] Miranda, M.F., et al. (2018). *TPRM: Tensor partition regression models with applications in imaging biomarker detection*. The Annals of Applied Statistics, **12**(3): 1422.

[230] Lock, E.F. (2018). *Tensor-on-tensor regression*. Journal of Computational and Graphical Statistics: A Joint Publication of American Statistical Association, Institute of Mathematical Statistics, Interface Foundation of North America, **27**(3): 638–647.

[231] Kolda, T.G., Bader, B.W. (2009). *Tensor decompositions and applications*. SIAM Review, **51**(3): 455–500.

[232] Henderson, H.V., Searle, S.R. (1981). *The vec-permutation matrix, the vec operator and Kronecker products: A review*. Linear and Multilinear Algebra, **9**(4): 271–288.

[233] Ryan, R.A. (2013). *Introduction to Tensor Products of Banach Spaces*. Springer Science & Business Media.

[234] Huang, G.B., et al. (2008). *Labeled Faces in the Wild: A Database for Studying Face Recognition in Unconstrained Environments*.

[235] Kawulok, M., Celebi, E., Smolka, B. (2016). *Advances in Face Detection and Facial Image Analysis*. Springer.

[236] Bader, B.W., Kolda, T.G. (2006). *Algorithm 862: MATLAB tensor classes for fast algorithm prototyping*. ACM Transactions on Mathematical Software (TOMS), **32**(4): 635–653.

[237] Savas, B., Eldén, L. (2013). *Krylov-type methods for tensor computations I*. Linear Algebra and Its Applications, **438**(2): 891–918.

[238] Tibshirani, R. (1996). *Regression shrinkage and selection via the lasso*. Journal of the Royal Statistical Society: Series B (Methodological), **58**(1): 267–288.

[239] Zhou, H., Li, L., Zhu, H. (2013). *Tensor regression with applications in neuroimaging data analysis*. Journal of the American Statistical Association, **108**(502): 540–552.

[240] Guo, W., Kotsia, I., Patras, I. (2011). *Tensor learning for regression*. IEEE Transactions on Image Processing, **21**(2): 816–827.

[241] Mukherjee, A., Zhu, J. (2011). *Reduced rank ridge regression and its kernel extensions*. Statistical Analysis and Data Mining: The ASA Data Science Journal, **4**(6): 612–622.

[242] Harris, J. (2013). *Algebraic Geometry: A First Course*. Vol. 133. Springer Science & Business Media.
[243] Sidiropoulos, N.D., Bro, R. (2000). *On the uniqueness of multilinear decomposition of N-way arrays*. Journal of Chemometrics: A Journal of the Chemometrics Society, **14**(3): 229–239.
[244] Hitchcock, F.L. (1927). *The expression of a tensor or a polyadic as a sum of products*. Journal of Mathematics and Physics, **6**(1–4): 164–189.
[245] Chen, B., et al. (2012). *Maximum block improvement and polynomial optimization*. SIAM Journal on Optimization, **22**(1): 87–107.
[246] Peltier, S.J., et al. (2009). *Support vector machine classification of complex fMRI data*. In *2009 Annual International Conference of the IEEE Engineering in Medicine and Biology Society*, pp. 5381–5384.
[247] Peltier, S., et al. (2013). *Multivariate Classification of Complex and Multi-echo fMRI Data*. in *2013 International Workshop on Pattern Recognition in Neuroimaging*. 2013, pp. 229–232.
[248] Tabelow, K., Polzehl, J. (2010). *Statistical Parametric Maps for Functional MRI Experiments in R: The Package Fmri*. WIAS.
[249] Eloyan, A., et al. (2014). *Analytic Programming with fMRI Data: A Quick-Start Guide for Statisticians Using R*. PLOS One, **9**(2): e89470.
[250] Spiegel, M.R. (1965). *Laplace Transforms*. McGraw-Hill, New York.
[251] Dyke, P.P. (2014). *An Introduction to Laplace Transforms and Fourier Series*. Springer.
[252] Agarwal, R.P., Perera, K., Pinelas, S. (2011). *Cauchy-Goursat Theorem, in an Introduction to Complex Analysis*. Springer, 96–101.
[253] Ahlberg, J.H., Nilson, E.N., Walsh, J.L. (2016). *The Theory of Splines and Their Applications: Mathematics in Science and Engineering: A Series of Monographs and Textbooks, Vol. 38*. Elsevier.
[254] Schölkopf, B., Herbrich, R., Smola, A.J. (2001). *A generalized representer theorem*. In *International conference on computational learning theory*. Springer.
[255] Ralaivola, L., et al. (2005). *Graph kernels for chemical informatics*. Neural Networks, **18**(8): 1093–1110.
[256] De Leon, J.P. (2009). *Effective spacetime from multidimensional gravity*. Gravitation and Cosmology, **15**(4): 345.
[257] Romero, C., Tavakol, R., Zalaletdinov, R. (1996). *The embedding of general relativity in five dimensions*. General Relativity and Gravitation, **28**(3): 365–376.
[258] Seahra, S.S., Wesson, P.S. (2003). *Application of the Campbell–Magaard theorem to higher-dimensional physics*. Classical and Quantum Gravity, **20**(7): 1321.
[259] de Leon, J.P. (2006). *Extra symmetry in the field equations in 5D with spatial spherical symmetry*. Classical and Quantum Gravity, **23**(9): 3043.
[260] Wesson, P.S. (2004). *Space-time uncertainty from higher-dimensional determinism*. General Relativity and Gravitation, **36**(2): 451–457.
[261] Weyl, H. (1922). *Space--time--matter*. Dutton.
[262] Wesson, P.S. (2010). *The embedding of general relativity in five-dimensional canonical space: A short history and a review of recent physical progress*. arXiv preprint arXiv:1011.0214.
[263] Goddard, A. (1977). *Foliations of space-times by spacelike hypersurfaces of constant mean curvature*. Communications in Mathematical Physics, **54**(3): 279–282.
[264] Mashhoon, B., Wesson, P. (2007). *An embedding for general relativity and its implications for new physics*. General Relativity and Gravitation, **39**(9): 1403–1412.
[265] Campbell, J.E. (1926). *A Course of Differential Geometry*. Oxford Claredon. xv + 261.
[266] Lidsey, J.E., Tavakol, R., Romero, C. (1997). *Campbell's Embedding Theorem*. Modern Physics Letters A, **12**(31): 2319–2323.

[267] Dahia, F., Romero, C. (2002). *The embedding of the space–time in five dimensions: An extension of the Campbell–Magaard theorem*. Journal of Mathematical Physics, **43**(11): 5804–5814.
[268] Lidsey, J.E., et al. (1997). *On applications of Campbell's embedding theorem*. Classical and Quantum Gravity, **14**(4): 865.
[269] De Broglie, L. (1968). *La réinterpretation de la mécanique ondulatoire*. Physics Bulletin, **19**(5): 133.
[270] Bohm, D. (1952). *A Suggested Interpretation of the Quantum Theory in Terms of "Hidden" Variables. I.* Physical Review, **85**(2): 166–179.
[271] Holland, P.R. (1995). *The Quantum Theory of Motion: An Account of the De Broglie-Bohm Causal Interpretation of Quantum Mechanics*. Cambridge university press.
[272] Bohm, D., Hiley, B. (1988). *Nonlocality and the Einstein-Podolsky-Rosen experiment as understood through the quantum-potential approach*, in *Quantum Mechanics versus Local Realism*. Springer, 235–256.
[273] Noshad, M., Choi, J., Sun, Y., Hero, A., Dinov, I.D. (2021). A data value metric for quantifying information content and utility, Journal of Big Data, **8**(82).
[274] Meng, X.-L. (2018). *Statistical paradises and paradoxes in big data (I): Law of large populations, big data paradox, and the 2016 US presidential election*. The Annals of Applied Statistics, **12**(2): 685–726.
[275] Neuenschwander, D.E. (2014). *Resource letter NTUC-1: Noether's theorem in the undergraduate curriculum*. American Journal of Physics, **82**(3): 183–188.
[276] Shang, C., et al. (2014). *Data-driven soft sensor development based on deep learning technique*. Journal of Process Control, **24**(3): 223–233.
[277] Smith, S.M., Nichols, T.E. (2018). *Statistical challenges in "Big Data" human neuroimaging*. Neuron, **97**(2): 263–268.
[278] Putnam, C.R. (1967). *Commutators of Bounded Operators, in Commutation Properties of Hilbert Space Operators and Related Topics*. Springer Berlin Heidelberg: Berlin, Heidelberg, 1–14.
[279] Lenth, R.V. (2001). *Some practical guidelines for effective sample size determination*. The American Statistician, **55**(3): 187–193.
[280] Biegler, L., et al. (2011). *Large-scale Inverse Problems and Quantification of Uncertainty*. Vol. 712. Wiley Online Library.
[281] Levy, J.K. (2001). *Computer support for environmental multiple criteria decision analysis under uncertainty*, In *Systems Design Engineering*, University of Waterloo Waterloo, Ontario, Canada
[282] Shannon, C.E. (1956). *The bandwagon*. IRE Transactions on Information Theory, **2**(1): 3.
[283] Shannon, C.E. (1949). *Communication in the presence of noise*. Proceedings of the IRE, **37**(1): 10–21.
[284] Naber, G.L. (2012). *The Geometry of Minkowski Spacetime: An Introduction to the Mathematics of the Special Theory of Relativity*. Vol. 92. Springer Science & Business Media.
[285] Beckner, W. (1975). *Inequalities in Fourier Analysis*. Annals of Mathematics, **102**(1): 159–182.
[286] Bialynicki-Birula, I. (2006). *Formulation of the uncertainty relations in terms of the R\'enyi entropies*. Physical Review A, **74**(5): 052101.
[287] Lehmann, E.L., Casella, G. (2006). *Theory of Point Estimation*. Springer Science & Business Media.
[288] Cramér, H. (1999). *Mathematical Methods of Statistics*. Vol. 43. Princeton university press.
[289] Lukacs, E. (1955). *A characterization of the gamma distribution*. The Annals of Mathematical Statistics, **26**(2): 319–324.

[290] Slepian, D. (1954). *Estimation of signal parameters in the presence of noise*. Transactions of the IRE Professional Group on Information Theory, **3**(3): 68–89.
[291] Rife, D., Boorstyn, R. (1974). *Single tone parameter estimation from discrete-time observations*. IEEE Transactions on Information Theory, **20**(5): 591–598.
[291] Yao, S., Wu, Q., Fang, S. (2019). *An Improved Fine-Resolution Method for Frequency Estimation of Real-Valued Single-Tone Using Three DFT Samples*. IEEE Access, **7**: 117063–117074.
[293] Campobello, G., Segreto, A., Donato, N. (2020). *A new frequency estimation algorithm for IIoT applications and low-cost instrumentation*. In *2020 IEEE International Instrumentation and Measurement Technology Conference (I2MTC)*.
[294] Rao, C.R. (1992). *Information and the accuracy attainable in the estimation of statistical parameters*, In *Breakthroughs in Statistics*, Springer, 235–247.
[295] Kagan, A. (2001). *Another look at the Cramér-Rao inequality*. The American Statistician, **55**(3): 211–212.
[296] Baum, L.E., Katz, M. (1965). *Convergence rates in the law of large numbers*. Transactions of the American Mathematical Society, **120**(1): 108–123.
[297] Tretter, S. (1985). *Estimating the frequency of a noisy sinusoid by linear regression (Corresp.)*. IEEE Transactions on Information Theory, **31**(6): 832–835.
[298] Coles, P.J., et al. (2017). *Entropic uncertainty relations and their applications*. Reviews of Modern Physics, **89**(1): 015002.
[299] Hörmander, L. (2015). *The Analysis of Linear Partial Differential Operators I: Distribution Theory and Fourier Analysis*. Springer.
[300] Strichartz, R.S. (2003). *A Guide to Distribution Theory and Fourier Transforms*. World Scientific Publishing Company.
[301] Sandryhaila, A., Moura, J.M. (2013). *Discrete signal processing on graphs: Graph fourier transform*. In *IEEE International Conference on Acoustics, Speech and Signal Processing*.
[302] Tsitsvero, M., Barbarossa, S., Lorenzo, P.D. (2016). *Signals on Graphs: Uncertainty Principle and Sampling*. IEEE Transactions on Signal Processing, **64**(18): 4845–4860.
[303] Wilkinson, M.D., et al. (2016). *The FAIR Guiding Principles for scientific data management and stewardship*. Scientific Data, **3**.
[304] Dwork, C. (2009). *The differential privacy frontier*. In *Theory of Cryptography Conference*. Springer.
[305] Gentry, C. (2009). *A Fully Homomorphic Encryption Scheme*. Stanford University.
[306] Papalexakis, E.E., Faloutsos, C. (2016). *Unsupervised Tensor Mining for Big Data Practitioners*. Big Data, **4**(3): 179–191.
[307] Rindler, W. (2006). *Relativity: Special, General, and Cosmological*. Oxford University Press on Demand.
[308] Synge, J.L., Schild, A. (1978). *Tensor Calculus*. Vol. 5. Courier Corporation.
[309] Chatterjee, U., Chatterjee, N. (2010). *Vector & Tensor Analysis*. Academic Publishers.
[310] Wesson, P.S., Ponce de Leon, J. (1992). *Kaluza-Klein equations, Einstein's equations, and an effective energy-momentum tensor*. Journal of Mathematical Physics, **33**(11): 3883–3887.
[311] Lederman, C., et al. (2015). *A Unified Variational Volume Registration Method Based on Automatically Learned Brain Structures*. Journal of Mathematical Imaging and Vision, 1–20.
[312] Pantazis, D., et al. (2010). *Comparison of landmark-based and automatic methods for cortical surface registration*. Neuroimage, **49**(3): 2479–2493.
[313] Leung, K., et al. (2008). IRMA: An Image Registration Meta-algorithm. In: Ludäscher B., Mamoulis N. (eds) Scientific and Statistical Database Management. SSDBM 2008. Lecture Notes in Computer Science, vol 5069. Springer-Verlag, Berlin, Heidelberg.

[314] Dinov, I., et al. (2002). *Quantitative comparison and analysis of brain image registration using frequency-adaptive wavelet shrinkage*. IEEE Transactions on Information Technology in Biomedicine, **6**(1): 73–85.

[315] Shattuck, D.W., et al. (2008). *Construction of a 3D probabilistic atlas of human cortical structures*. NeuroImage, **39**(3): 1064–1080.

[316] Geissler, A., et al. (2007). *Contrast- to- noise ratio (CNR) as a quality parameter in fMRI*. Journal of Magnetic Resonance Imaging: An Official Journal of the International Society for Magnetic Resonance in Medicine, **25**(6): 1263–1270.

[317] Barber, R.F., Candès, E.J. (2015). *Controlling the false discovery rate via knockoffs*. The Annals of Statistics, **43**(5): 2055–2085.

[318] Benjamini, Y., Hochberg, Y. (1995). *Controlling the False Discovery Rate: A Practical and Powerful Approach to Multiple Testing*. Royal Statistical Society, **57**(1): 289–300.

[319] Weiss, A.A. (1984). *Systematic sampling and temporal aggregation in time series models*. Journal of Econometrics, **26**(3): 271–281.

[320] Akaike, H. (1976). *Canonical correlation analysis of time series and the use of an information criterion*, In *Mathematics in Science and Engineering*. Elsevier, 27–96.

[321] Sudlow, C., et al. (2015). *UK biobank: An open access resource for identifying the causes of a wide range of complex diseases of middle and old age*. PLoS Medicine, **12**(3): e1001779.

[322] Buuren, S., Groothuis-Oudshoorn, K. (2011). *MICE: Multivariate imputation by chained equations in R*. Journal of Statistical Software, **45**(3).

[323] Dinov, I., et al. (2010). *Neuroimaging Study Designs, Computational Analyses and Data Provenance Using the LONI Pipeline*. PLoS ONE, **5**(9): e13070.

[324] M., G.-M. (1994). *The Quark and the Jaguar*. New York: W.H. Freeman.

[325] Focardi, S.a.F., F. (2010). *The Reasonable Effectiveness of Mathematics in Economics*. American Economist, **49**(1): 3–15.

[326] Leontief, W. (1966). *Essays in Economics: Mathematics in Economics*. Oxford: Oxford University Press.

[327] Foroohar, R. (2010). *The best countries in the world*. Newsweek, **156**(9): 30–38.

[328] Berry, W.D. (1993). *Understanding Regression Assumptions*. Vol. 92. Sage Publications.

[329] Everett III, H. (1956). *The many-worlds interpretation of quantum mechanics. the theory of the universal wavefunction*. Thesis, Princeton University.

Index

https://doi.org/10.1515/9783110697827-009